Approximation, Optimization and Computing

Theory and Applications

International Association for
Mathematics and Computers in Simulation

edited by

A.G. LAW and C.L. WANG
University of Regina
Saskatchewan, Canada

1990

NORTH-HOLLAND
AMSTERDAM • NEW YORK • OXFORD • TOKYO

ELSEVIER SCIENCE PUBLISHERS B.V.
Sara Burgerhartstraat 25
P.O. Box 211, 1000 AE Amsterdam, The Netherlands

Distributors for the United States and Canada:

ELSEVIER SCIENCE PUBLISHING COMPANY, INC.
655 Avenue of the Americas
New York, N.Y. 10010, U.S.A.

Library of Congress Cataloging-in-Publication Data

Approximation, optimization, and computing : theory and applications / International Association for Mathematics and Computers in Simulation ; edited by A.G. Law and C.L. Wang.
p. cm.
Includes bibliographical references.
ISBN 0-444-88693-1
1. Approximation theory--Data processing. 2. Mathematical optimization--Data processing. I. Law, Alan G. (Alan Greenwell), 1936- . II. Wang, C. L. (Chung-lie) III. International Association for Mathematics and Computers in Simulation.
QA221.A64 1990
511'.4--dc20 90-6893
CIP

ISBN: 0 444 88693 1

PRINTED IN THE NETHERLANDS

Approximation, Optimization and Computing

Theory and Applications

INTERNATIONAL ASSOCIATION FOR MATHEMATICS AND COMPUTERS IN SIMULATION
Association Internationale pour les Mathématiques et Calculateurs en Simulation

FOREWORD

This volume consists of 101 papers in the general areas of approximation, optimization and computing, and applications. Under the sponsorship of IMACS, it represents a collaborative venture, initiated in 1986, between the Dalian University of Technology and the University of Regina. A primary goal of the joint program and publication committee was to provide an opportunity for research papers reflecting emerging directions within theory or applications. The papers accepted for publication represent the varied and substantial efforts of 157 authors towards this goal.

The four invited papers are followed by 57 contributions in approximation theory. The two groupings for optimization (22) and computing (8) precede the collection of applications in several areas. Some papers could belong to more than one of the groupings and the final arrangement reflects the editors' decision.

A. G. Law and C.L. Wang
December, 1989

PROGRAM AND

PUBLICATION COMMITTEE

CHAIRMAN: A.G. Law (Canada)

CO-CHAIRMAN: L.C. Hsu (China)

MEMBERS:

Dalian University of Technology

G. D. Cheng
R. H. Wang

University of Regina

D. M. Secoy
C. L. Wang

LIST OF CONTENTS

IV. APPLICATIONS

ACKNOWLEDGEMENTS

The editors wish to thank members of the Program and Publication Committee for carrying out the demanding and diverse roles which led to preparation of this volume. The University of Regina and the Dalian University of Technology deserve special acknowledgement for continued support, as does IMACS for ongoing interest and encouragement. Our final thank you goes to Colette Grundy and Karen Hitchcock for their considerable efforts in advertising, correspondence and typing.

INVITED PAPERS

Approximation, Optimization and Computing:
Theory and Applications, A.G. Law and C.L. Wang (eds.)
Elsevier Science Publishers B.V. (North-Holland)

MULTIPLE NUMERICAL INTEGRATION FORMULA USING REGULAR LATTICES

Sin HITOTUMATU

Research Institute for Mathematical Sciences*)
Kyoto University
Kitashirakawa, Sakyo-ku, Kyoto, Japan 606

Abstract

As is known, the trapezoidal rule with equidistant nodes has the highest accuracy when the influences at the end points may be neglected. Under the similar situations, we propose multiple numerical integration formulas over the vertices of regular dense lattices, such as equilateral triangular lattice on the plane, face-central cubic or body-central cubic lattices in R^3, or the D_4-lattice in R^4. We also give a method of error analysis and some numerical examples.

1. INTRODUCTION

Prof. M. Mori [7] proved that the trapezoidal rule with equidistant nodes has the highest accuracy among all the numerical quadrature formulas, when the influences at the end points may be neglected. Such situation occurs when the integrand decays very rapidly or the integration of a periodic function along a whole period. We would like to generalize it for multiple integrals.

In the Euclidean space R^n of higher dimension, the nodes of the simple cubic lattice SC_n are quite sparse. In fact, the centre of a hypercubic cell is apart $\sqrt{n}/2$ times to its side-length from the surrounding vertices, which is longer than the side when $n \geq 5$. Therefore, we must select more dense regular lattices for multiple integration. Although the precise densest lattices in R^n are seldomly known, we have several dense lattices not difficult to construct. For example, we have the *equilateral triangular lattice* L_2 in R^2, the *face-central cubic lattice* FC_3 or the *body-central cubic lattice* BC_3 in R^3, and in R^4 a very dense lattice called *D_4-lattice* consisting of the vertices of tesselation by the regular 24 cells (e.g. Coxeter [1]). Also there are *E_8-lattice* in R^8 and the *Leech lattice* in R^{24}, though they are less useful for practical applications.

In the present paper, we first give our general formulas (§2) and discuss the construction of such regular lattices in normalized forms (§3). We then give a method of error analysis (§4). We shall mainly concern with the L_2-lattice on the plane, but the similar procedure holds for other lattices in higher dimensional spaces at least approximately. Finally in §5, we show some numerical examples.

The idea of using regular dense lattices for a multiple integral is not new. Such methods has been suggested in various literatures, such as [2], [3] and [5]. The author is particularly grateful to Prof. M. Sugihara whose lecture ([8]) at a meeting in RIMS has suggested the present algorithm. This paper is essentially a revised version of the note [4].

2. GENERAL FORMULAS

Let $\Omega = \Omega(L)$ be the set of vertices in a given lattice L in R^n. The *fundamental region* A_P of a vertex P is the set of points X in R^n such that the distance $\overline{PX}$ does not exceed the distance $\overline{QX}$ for any other vertices $Q \neq P$ in Ω. We say that the lattice L is *regular* if the volume of the fundamental region $V(A_P)$ does not depend upon the selection of the vertex P. We denote the constant volume by V.

Now, in order to compute approximately a multiple integral $\int_D f(x)dx$, we first cover the integration domain D by a regular lattice L,

MOS classification 65D30;
key words: multiple numerical integration, trapezoidal rule, regular lattice, equilateral triangular lattice, face-central cubic lattice, body-central cubic lattice, D_4-lattice.

*) In RIMS until 1989 March. After 1989 April his address will be: Dept. of Informatics, Tokyo Electro-Mechanical College at Hatoyama Campus, Saitama Pref., Japan 350-03.

and take the summation over all lattice points Ω in D; say

$$(1)\quad T = V \cdot \sum_{x\in\Omega\cap D} f(x).$$

We mainly concern with a rapidly decreasing function f(x), where the integral near the boundary ∂D may be almost negligible. If necessary, we can modify the integral near the boundary by the method in §4. Hence, we may assume that the integration domain D itself is a union of the fundamental regions A_P, $P\in\Omega(L)$.

Our formula is actually a generalization of the *midpoint rule* to higher dimensional spaces; i.e., the integral over a fundamental region A_P is replaced by the product of the representative value of the integrand at its centre P and the volume V of A_P.

As we mainly concern with rapidly decreasing functions, it may be more convenient to arrange the set of vertices along the spherical layers around the origin. We denote by N the set of the square of the normilized distance of a vertex from the origin, except 0. For abbreviation, we shall call N simply the *set of distances*, although the actual distance is $\sqrt{c}$, $c\in N$.

Then our formula (1) reads

$$(2)\quad T = V\,[f(0) + \sum_{c\in N} (\sum_{\|x\|^2=c} f(x))].$$

For very regular lattices such as L_2, D_4 or E_8, the number of the vertices over a spherical layer $\|x\|^2 = c \neq 0$ is always a multiple of the number m of the nearest neighbouring verices in the lattice. This property does not hold for BC_3 and FC_3.

3. CONSTRUCTION OF THE REGULAR LATTICES

3.1. Equilateral triangular lattice on a plane

On the Euclidean plane R^2, we construct the equilateral triangular lattice L_2 as follows. We take one of the vertices to be the origin of the coordinates, and select one side issuing from it to be the x-axis. We denote by h the side-length of each equilateral triangle. Then each lattice point x of L_2 is uniquely represented by

$$X = (ae+bW)h$$

where e is the unit vector to the positive x-axis, $W = (1/2, \sqrt{3}/2)$, and a,b are integers. The number of the neighboring points is evidently 6. The fundamental region A is a regular hexagon of the side-length $h/\sqrt{3}$ with the centre at the lattice point. The volume V is $\sqrt{3}h^2/2$.

The set of distances N consists of the integers

$$c = a^2 + ab + b^2.$$

It is well-known that a number $c\in N$ as above is characterized by the properties that it is of the form $c = \ell^2\cdot q$, where q is either 1 or square-free whose prime factors are 3 or primes $p \equiv 1 \pmod 6$. The first part of N is 1,3,4,7,9,12,13,16,19,21,25,27,28,31,36,37,39, 43,48,49,52,57,...

Further it is known that the primitive solutions of integers for $c^2=a^2+ab+b^2$ is given just by

$$a = (2m+n)n,\ b = m^2-n^2,\ c = m^2+mn+n^2$$

where m and n are relatively prime positive integers (m>n) whose difference is not divisible by 3.

Assume that the integrand f(x,y) is represented in the form

$$\sum_{i=1}^{m} \varphi_i(r)\cdot g_i(\theta)$$

in polar coordinates (r,θ), where $\varphi_i(r)$ is rapidly decreasing with respect to r, and $g_i(\theta)$ varies rather slowly in θ. Then the following summation procedure may be useful.

$$(3)\quad T = \frac{\sqrt{3}}{2} h^2\{f(0,0) + \sum_{i=1}^{m} (\sum_{c\in N} \varphi_i(\sqrt{c})$$

$$\times \sum_{k=0}^{5} [g_i(\frac{k\pi}{3}+\theta_0)+g_i(\frac{k\pi}{3}-\theta_0)])\},$$

where $c = a^2+ab+b^2\in N$; a,b are integers satisfying $0\leq b\leq a$, $(a,b) \neq (0,0)$, and $\theta_0 = \arccos[(2a+b)/2r]$. The last term in (3) should be replaced by

$$\sum_{k=0}^{5} g_i(\frac{k\pi}{3}+\theta_0)$$

when $b = 0$ $(\theta_0 = 0)$ or $a = b$ $(\theta_0 = \pi/6)$. Note that the pair (a,b) is not unique for a given c, and the summation must be computed with respect to all indices (a,b).

3.2. Face-central and Body-central lattices in R^3

The vertices of a face-central cubic lattice FC_3 in R^3 are given, after suitable normalization, by the points (ah,bh,ch) where a,b and c are integers whose sum $a+b+c$ is even. The fundamental region is the rhombic dodecahedron with 12 congruent rhombic faces whose volume V is $2h^3$. The set of distances N coincides with the set of all even numbers.

Similarly, the body-central cubic lattice BC_3 in R^3 are given by the points (ah,bh,ch), where a,b and c are integers which are all even or odd simultaneously. The fundamental region is the truncated octahedron with 14 faces (6 squares and 8 regular hexagons) whose volume is $4h^3$. The set of distances N covers all multiples of 4 and the numbers of the form $8p+3$.

3.3. D_4-lattice and higher dimensional cases

The D_4-lattice in R^4 is actually the *body-central cubic lattice* BC_4 in R^4. In a normalized form, the vertex is given by (ah,bh,ch,dh), where a,b,c and d are integers which are all even or odd simultaneously. The fundamental region is a smaller regular 24-cells whose volume is $8h^4$. The set of distances N coincides with the set of all multiples of 4. The number of vertices over a spherical layer $\|x\|^2 = 4\ell>0$ is always a multiple of 24.

It is known (Hurwitz [6] and Coxeter [1], Chap. 5) that the D_4-lattice diminished by half to the above one is given by the totality of *quaternion-integers*, i.e., the quaternions of the form $a+bi+cj+dk$ where a,b,c and d are all integers or odd multiples of 1/2 simultaneously. In this representation, the set N coincides with the set of all positive integers.

Similarly the E_8-lattice in R^8 is represented by $(a_1h,a_2h,\dots,a_8h)$ with integers $a_1,\dots,a_8$ satisfying the conditions that the sum $a_1+\dots+a_8$ is a multiple of 4 and all components are even or odd simultaneously. The volume of the fundamental region is $256h^8 = (2h)^8$. The set of distances N is the set of all multiples of 8. The number of the neighboring vertices is 240. Remark that the very dense property of the vertices does not contradict that they all lie over particular hyperplanes of the form $x_1+\dots+x_8 = 4\ell$.

4. ERROR ANALYSIS

4.1. Fundamental formulas

We assume that the integrand $f(x_1,\dots,x_n)$ is *analytic*, or at least a sufficiently smooth function. We expand it into Taylor series and integrate it term-by-term in the fundamental region A. We remark that the region around the origin for the above lattices have the property that it is symmetric with respect to the variables $x_1,x_2,\dots,x_n$, except L_2. Further it is symmetric with respect to every coordinate hyperplanes except E_8.

For simplicity, we first assume that the vertex P is the origin. By the symmetric properties, all terms containing odd powers vanish except for the case E_8. Even for E_8, since we see that the fundamental region A of E_8 is symmetric with respect to every linear subspace of codimension 2 $(x_i=x_j=0,\ i\neq j)$, the first few odd terms also vanish. Hence we may write the first part as

$$(4)\quad \int_A f(x)dx - V\cdot f(0) = \frac{h^{2+n}}{2!}\sum_{i=1}^{n}\alpha_i^{(1)}\frac{\partial^2 f}{\partial x_i^2}(0)$$

$$+\frac{h^{4+n}}{4!}\Big[\sum_{i=1}^{n}\alpha_i^{(2)}\frac{\partial^4 f}{\partial x_i^4}(0)+6\sum_{i<j}\beta_{ij}^{(2)}\frac{\partial^4 f}{\partial x_i^2\partial x_j^2}(0)\Big]$$

$$+\frac{h^{6+n}}{6!}\Big[\sum_{i=1}^{n}\alpha_i^{(3)}\frac{\partial^6 f}{\partial x_i^6}(0) + 15\sum_{i,j}\beta_{ij}^{(3)}\frac{\partial^6 f}{\partial x_i^4 x_j^2}(0)$$

$$+ 90\sum_{i<j<k}\gamma_{ijk}^{(3)}\frac{\partial^6 f}{\partial x_i^2\partial x_j^2\partial x_k^2}(0)\Big] + \cdots$$

where the coefficients are constants given by the integrals of the corresponding terms x_i^2, x_i^4, $x_i^2x_j^2,\dots$ in the fundamental region A normalized as $h = 1$. By the above symmetricity, and by the fact that

$$\alpha_1^{(1)} = \alpha_2^{(1)} = 5\sqrt{3}/144$$

for L_2, the first term is always written in in the form

$$(5)\quad V\cdot c_1h^2\Delta f(0),\ \Delta = \text{Laplacean}.$$

Note that V is the integral of 1 in A and proportional to h^n. The values of the coefficients c_1 are given in Table 1 below. For L_2, it is 5/144.

If the fundamental domain A is actually a hyperbowl around the centre, we always have the relations

(6-a) $\alpha_i^{(2)} = 3\beta_{ij}^{(2)}$

and

(6-b) $\alpha_i^{(3)} = 5\beta_{ij}^{(3)} = 15\gamma_{ijk}^{(3)}$,

so that the second and the third terms in the right hand side of (4) are written as

(7-a) $V\cdot c_2 h^4 \Delta^2 f(0)$

and

(7-b) $V\cdot c_3 h^6 \Delta^3 f(0)$

respectively.

4.2. Pseudo-spheric lattices

For the fundamental domains of the above lattices, the terms in (4) after the second one are not necessarily written precisely in similar forms as (7-a) and (7-b). Fortunately, for the lattices L_2 and D_4, we have the relation (6-a), and then the second term is arranged just in the form as (7-a). This means that the fundamental domains of L_2 and D_4 are quite close to a bowl, and we shall call such domains to be *pseudo-spheric*. In other cases, the relation (6-a) does not hold. But we may write the second term in a form (7-a), at least in some approximational sense, taking the coefficient c_2 as a suitable average of the true coefficients. We give such values also in Table 2, where * means that the given value c_2 is not an approximation, but a correct value.

Even in the cases of L_2 and D_4, the relation (6-b) does not hold, i.e., we cannot write the third term precisely in the form as (7-b). However, through the above cases, the third term is approximated fairly well by the from (7-b), if we take suitable weighted average of the coefficients for c_3. We give such values also in Table 2.

Therefore, we may assume, at least approximately, that the error E_0 is expressed in the form

$$(8)\quad E_0 = \int_A f(x)dx - V\cdot f(0)$$

$$= c_1 h^2 V\Delta f(0) + c_2 h^4 V\Delta^2 f(0)$$

$$+ c_3 h^6 V\Delta^3 f(0) + \cdots$$

4.3. Total error

Up to now, we have considered only the fundamental domain around the origin. But similar formulas are valid for the integrals in other fundamental regions A_P, if we replace the origin 0 by the point P. Summing up these formulas we have

$$(9)\quad E = \int_D f(x) - T = c_1 h^2 T_1 + c_2 h^4 T_2 + c_3 h^6 T_3 + \cdots$$

where T_k is the value of our integration formula (1) whose integrand is replaced by $\Delta^k f$ instead of f. Therefore, T_k is an approximation of the integral I_k of the function $\Delta^k f$. The errors are also given by similar formulas as (9). Replacing T_k by the integral I_k of $\Delta^k f$, we have

Table 1. Constants for lattices

No.		V	$\alpha^{(1)}$	$\alpha^{(2)}$	$\beta^{(2)}$	$\alpha_1^{(3)}$	$\alpha_2^{(3)}$	$\beta_{2,1}^{(3)}$	$\beta_{1,2}^{(3)}$	$\gamma^{(3)}$
0	L_1	h	$\frac{1}{12}h^3$	$\frac{1}{80}h^5$	—	$\frac{1}{486}h^7$		—	—	—
1	SC_1	h^2	$\frac{1}{12}h^4$	$\frac{1}{80}h^6$	$\frac{1}{144}h^6$	$\frac{1}{486}h^8$		$\frac{1}{960}h^8$		—
2	L_2	$\frac{\sqrt{3}}{2}h^2$	$\frac{5\sqrt{3}}{144}h^4$	$\frac{7\sqrt{3}}{1440}h^6$	$\frac{7\sqrt{3}}{4320}h^6$	$81s$	$85s$	$\frac{93}{5}s$	$\frac{73}{5}s$	$(s=\frac{\sqrt{3}h^8}{96768})$
3	FC_3	$2h^3$	$\frac{1}{4}h^5$	$\frac{3}{40}h^7$	$\frac{7}{360}h^7$	$\frac{85}{2688}h^9$		$\frac{31}{8064}h^9$		$\frac{37}{24192}h^9$
4	BC_3	$4h^3$	$\frac{19}{24}h^5$	$\frac{89}{240}h^7$	$\frac{41}{420}h^7$	$\frac{1237}{5376}h^9$		$\frac{151}{4480}h^9$		$\frac{4223}{1451520}h^9$
5	D_4	$8h^4$	$\frac{26}{15}h^6$	$\frac{6}{7}h^8$	$\frac{2}{7}h^8$	$\frac{173}{315}h^{10}$		$\frac{16}{135}h^{10}$		$\frac{19}{630}h^{10}$

Table 2. Coefficients for error terms

No.	$c_1=b_1$	c_2	c_3	b_2	b_3
0	1/24	1/1920*	1/322560	-7/5760	13/241920
1	1/12	1/1440	1/193536	-1/160	151/322560
2	5/144	7/17280*	83/34836480	-83/103680	421/26127360
3	1/16	1/720	1/61440	-29/11520	1/11520
4	19/192	1/288	1/15360	-233/36864	1213/11796480
5	13/120	1/224*	1/10368	-733/100800	5393/12096000

* in c_2 means that it is exact value of the second term $c_2 \cdot \Delta^2 f(0)$. Other values are approximate ones.

$$(10) \quad E = b_1 h^2 I_1 + b_2 h^4 I_2 + b_3 h^6 I_3 + \cdots,$$

where $b_1 = c_1$, $b_2 = c_2 - c_1^2$,

$$b_3 = c_3 - 2c_1c_2 + c_1^3.$$

The values of b_k are given in Table 2.

Using Stokes' theorem, the integral $I_1 = \int_D \Delta f(x)dx$ is replaced by the surface integral along the boundary ∂D. Under the notations of differential forms, we have

$$(11) \quad I_1 = \sum_{k=1}^{n} (-1)^k \frac{\partial f}{\partial x_k} dx_1 \wedge \cdots \wedge dx_{k-1} \wedge dx_{k+1} \wedge \cdots \wedge dx_n.$$

If the function f is precisely a function $\varphi(r)$ of the readius r only, we have $\partial f/\partial x_k = \varphi'(r) \cdot x_k/r$. Inserting this to (11), we note that

$$(-1)^k x_k dx_1 \wedge \cdots \wedge dx_{k-1} \wedge dx_{k+1} \wedge \cdots \wedge dx_n$$

gives the volume of the domain $V(D)$ for each k. If the integration domain D is close to a sphere of radius R around the origin, we have then approximately

$$(12) \quad I_1 = [\varphi'(R)/R]nV(D).$$

Similar expressions are valid for I_2, I_3 replacing f by Δf or $\Delta^2 f$, respectively. When f strongly depends upon r and changes rather slowly over the hyperspherical surfaces, the above approximation may be fairly good, and usually the correction by the first term $b_1 h^2 I_1$ alone improves the approximation considerably.

If the integrand f decreases rapidly in r, we may select h such that the first two terms mutually cancel out. Then the remaining third term will give an estimation of error E. This is just the case for $f = \exp(-\|x\|^2/a)$ as is shown in the next paragraph.

5. NUMERICAL EXAMPLES

First we see the effect of boundary modification for a function not rapidly decreasing.

<u>Exp. 1.</u> Let us consider the integral $\iint_D y^2 dxdy$ on the unit circle $D = \{|x^2+y^2| \leq 1\}$. The true value is $\pi/4 = 0.785398\ldots$ Taking $h = 1/\sqrt{50}$, there are totally 187 vertices in D. The sum is

$$S = 2 \times 1608 \times (3/200), \quad V = \sqrt{3}/100$$

and then the approximating value $T=S\cdot V$ is about $0.55704\ldots$ The first modification term I_1 is $\int_{\partial D} 2y\, dx = 2 \times$ (area of D). It may be better to take the area of the union of fundamental region $187 \times \sqrt{3}/100$, rather than to take the area of the original circle D itself. Then the boundary modification is

$$\frac{2\times5}{144} \cdot \frac{\sqrt{3}\times187}{100} = 0.22493\ldots,$$

and the modified value is $0.78197\ldots$, which is much closer to the true value than the first approximation T.

Table 3

No.	Computed Value	Theoretical Value	a	c	h
1	0.78555529587	0.78539816398=$\pi/4$	1	4	1
1	0.785398171805	c→2c*	2	8	
2	0.604604393372	0.604599788078=$\pi/\sqrt{27}$	1	6	1
2	0.604599788090	c→2c*	2	12	
3	1.85610933222	1.85610933228=$\pi\sqrt{\pi}/3$	4	12	$\sqrt{2}$
4	1.39208204382	1.39208199921=$\pi\sqrt{\pi}/4$	4	8	$\sqrt{2}$
5	0.822467086066	0.822467033424=$\pi^2/12$	4	24	2
6	0.405871459469	0.405871212642=$\pi^4/240$	4	240	$2\sqrt{2}$
6	0.405871212642	c→16c*	8	3840	

*) c → 2c means that we have normalized such as to replace c by 2c for the comparison.

Exp. 2. We next consider a very favarable function, say

$$f(x) = (1/c)\exp(-\|x\|^2/a),$$

a and c being constants.

Instead of changing the side-length h, we have changed the constatn a and normalization parameter c. We have included the valume V into the factor c, so that we show the values of the sum itself without multiplying V. A few results are shown in Table 3, where No.6 corresponds to E_8-lattice in R^8. The underlined part in the table means the essentially correct digits.

REFERENCES

[]] H.S.M. Coxeter, Regular Complex Polytopes, Cambridge Univ. Press, 1974

[2] R.J. Davis and P. Rabinowitz, Methods of Numerical Integration, Academic Press, Second Ed., 1984.

[3] W. Gautschi, Multidimensional Euler enumeration formulas and numerical cubuture, Numerical Integration (G. Hammerlin ed.) p.77-88, Birkhäuser, 1982.

[4] S. Hitotumatu, Multiple integration using regular lattices (in Japanese), Res. Inst. Math. Sci. Lecture Note Series 553 (1985), p.159-165.

[5] Loo-Keng Hua and Yuan Wang, Applications of number theory to numerical analysis, Springer, 1981

[6] A. Hurwitz, Vorlesungen über Zahlentheorie der Quaternionen, Springer, 1919.

[7] M. Mori, On the superiority of the trapezoidal rule for the integration of periodic analytic functions, Mem. Num. Math., 1 (1974), p.11-20.

[8] M. Sugihara, Lectures at RIMS on Numerical Analysis, 1984 June.

Approximation, Optimization and Computing:
Theory and Applications, A.G. Law and C.L. Wang (eds.)
Elsevier Science Publishers B.V. (North-Holland)

ASYMPTOTIC EXPANSIONS OF MULTIPLE INTEGRALS OF RAPIDLY OSCILLATING FUNCTIONS

L.C. Hsu(Lizhi Xu)

Inst. of Math., Dalian Univ. of Technology, Dalian 116024, China

ABSTRACT: Here presented is a kind of general asymptotic expansion involving a remainder estimate for a class of multiple integrals of rapidly oscillating functions.

1. INTRODUCTION

We are concerned with an extensive class of strongly oscillatory n-tiple integrals of the form

$$I[F;(\lambda)] = \int_{[0,1]^n} F[(x),(\lambda x)]d(x)$$

$$= \int_0^1 \cdots \int_0^1 F(x_1,\cdots,x_n;\lambda_1 x_1,\cdots,\lambda_n x_n)dx_1\cdots dx_n \tag{1}$$

where $F[(x),(y)] \equiv F(x_1\cdots,x_n;y_1,\cdots,y_n)$ is a continuous periodic function of period 1 in each y_i ($i = 1,2,\cdots,n$), and λ_i are positive integer parameters so that $F[(x),(\lambda x)]$ will be rapidly oscillating when λ_i become large numbers. It is always assumed that the following condition

$$F_{(x)}^{(k)}[(x),(y)] \equiv \partial^{|(k)|}F/\partial x_1^{k_1}\cdots\partial x_n^{k_n} \in C(D) \tag{2}$$

is satisfied for all sets $(k) \equiv (k_1,\cdots,k_n)$ of integers $k_i \geq 0$ with $|(k)| = k_1+\cdots+k_n \leq r+1$, $D \equiv [0,1]^n \times [0,\infty)^n$ and $C(D)$ denoting the class of continuous functions on D. Also, the L-norm of $F_{(x)}^{(k)}$ on $[0,1]^n \times [0,1]^n$ will be utilized, viz.

$$||F_{(x)}^{(k)}||_1 = \int_{[0,1]^{2n}} |F_{(x)}^{(k)}[(x),(y)]|d(x)d(y)$$

Throughout $B_j(t)$ and $\overline{B}_j(t)$ will denote the Bernoulli polynomial of j-th degree and the corresponding periodic function of period 1 with $\overline{B}_j(t) = B_j(t)$ $(0 \leq t < 1)$, respectively. Also, we will use partial defferential operators with respect to variables x_i, and take defferences between the derivatives at $x_i = 0$ and $x_i = 1$. Thus, for instance

$$[(\frac{\partial}{\partial x_2})^s]_0^1\,[(\frac{\partial}{\partial x_1})^r]_0^1\, f(x_1,x_2,\cdots)$$

$$= [\frac{\partial^s}{\partial x_2^s}(\frac{\partial^r f}{\partial x_1^r})_{x_1=0}^{x_1=1}]_{x_2=0}^{x_2=1}$$

where $[f(x_1,\cdots)]_{x_1=0}^{x_1=1} = f(1,\cdots) - f(0,\cdots)$, etc. In particular, differentiation of negative order just means integration, for example

$$[(\frac{\partial}{\partial x_2})^{-1}]_0^1\,[(\frac{\partial}{\partial x_1})^{-1}]_0^1\, f(x_1,x_2,\cdots)$$

$$= \int_0^1 dx_2 \int_0^1 f(x_1,x_2,\cdots)dx_1$$

What we are doing to show is that $I[F;(\lambda)]$ can be expanded in terms of negative powers of $\lambda_i (i = 1,\cdots,n)$ with an estimate for the remainder.

2. STATEMENT OF RESULT

Let us introduce a useful operator $\Delta^{(k)}\ [(y)]$ with the product of n Bernoulli polynomials as a weight function of the following form

$$\Delta^{(k)}[(y)] = \prod_{i=1}^n B_{k_i}(y_i) \prod_{i=1}^n [(\frac{\partial}{\partial x_i})^{k_i-1}]_0^1 \tag{3}$$

where $k_i \geq 0$. For brevity we denote $(k)! = k_1!\cdots k_n!$ and $(\lambda)^{-(k)} = \lambda_1^{-k_1}\cdots\lambda_n^{-k_n}$. Then our main result is contained in the following

THEOREM Let λ_i be positive integer parameters that are allowed to tend to $+\infty$ of the same order, viz. there are positive constants α and β such that $\alpha < \lambda_i/\lambda_j < \beta$ for all $i \neq j$. Let $\lambda = \min\{\lambda_1,\cdots,\lambda_n\}$. Then we have the expansion formula

$$I[F;(\lambda)] = \sum_{0\leq|(k)|\leq r} \frac{(\lambda)^{-(k)}}{(k)!} \int_{[0,1]^n} \Delta^{(k)}[(y)]F[(x),(y)]d(y)$$

$$+\rho_r[(\lambda)] \tag{4}$$

where the summation extends over all the nonnegative integers k_i such that $|(k)| = k_1+\cdots+k_n \leq r$, and the remainder satisfies the inequality

$$|\rho_r[(\lambda)]| \le 2(1/\lambda)^{r+1} \sum_{|(k)|=r+1} \frac{1}{(k)!} ||F^{(k)}_{(x)}||_1 \cdot \prod_{i=1}^{n} ||\overline{B}_{k_i}||_\infty \tag{5}$$

the sum of the right-side of (5) being taken over all $k_i \ge 0$ such that $|(k)| = r+1$; and $||\overline{B}_j||_\infty = max|\overline{B}_j(t)|$ so that $||\overline{B}_0||_\infty = 1$ and $||\overline{B}_1||_\infty = 1/2$.

Notice that the principal term contained in the right-side of (4) is the term with $|(k)| = 0$. viz. $k_1 = \cdots = k_n = 0$. This term is just the $2n$-tiple integral of $F[(x),(y)]$ taken over $[0,1]^n \times [0,1]^n$.

As may be observed, the expansion formula (4) reduces to that of [3] for the particular case $F[(x),(y)] = g[(x)]\ w[(y)]$. However, the explicit form of the remainder $\rho_r[(\lambda)]$ is too complicated to be given here.

3. SKETCH OF PROOF

We have to make use of an early result which was proved in the author's paper [2] in 1963. Let us restate it as a lemma (cf. loc. cit. p. 84-85).

LEMMA Let $G(x,y)$ have continuous partial derivatives with respect to x of orders up to $r+1$, and be periodic of period 1 in y. Then for any given positive integer parameter λ we have the expansion

$$\int_0^1 G(x,\lambda x)dx = \sum_{j=0}^{r+1} \frac{\lambda^{-j}}{j!} \int_0^1 B_j(y)[\frac{\partial^{j-1}G}{\partial x^{j-1}}]_{x=0}^{x=1} dy + \delta_r(\lambda) \tag{6}$$

were $G \equiv G(x,y)$ and the remainder is given by

$$\delta_r(\lambda) = -\frac{\lambda^{-r-1}}{(r+1)!} \int_0^1 dx \int_{\lambda x}^{\lambda x+1} B_{r+1}(y-\lambda x)[\frac{\partial^{r+1}G}{\partial x^{r+1}}]dy$$

$$= -\frac{1}{(r+1)!}(\frac{1}{\lambda})^{r+1} \int_0^1 dy \int_0^1 \overline{B}_{r+1}(y-\lambda x) G_x^{(r+1)}(x,y)dx \tag{7}$$

COROLLARY Let (6) be rewritten in the form

$$\int_0^1 G(x,\lambda x)dx = \sum_{j=0}^{r} \frac{\lambda^{-j}}{j!} \int_0^1 B_j(y)[(\frac{\partial}{\partial x})^{j-1}]_0^1$$

$$\times G(x,y)dy + \rho_r(\lambda) \tag{8}$$

Then the remainder term $\rho_r(\lambda)$ satisfies the inequality

$$|\rho_r(\lambda)| \le \frac{2}{(r+1)!}(\frac{1}{\lambda})^{r+1} ||G_x^{(r+1)}||_1 \cdot ||\overline{B}_{r+1}||_\infty \tag{9}$$

Note that the main result of the joint paper by Banerjee, Lardy and Lutoborski [1] in 1987 is in fact implied by the above lemma as a special case with $G(x,y) = f(x)\overline{w}(y)$ (cf.[2], p.84-85).

For proof of our theorem, it will be convenient to express the operator defined by (3) in the form

$$\Delta^{(k)}[(y)] = \prod_{i=1}^{n} [B_{k_i}(y_i)(\frac{\partial}{\partial x_i})^{k_i-1}]_{x_i=0}^{x_i=1}$$

$$\equiv \prod_{i=1}^{n} \Delta^{(k_i)}(y_i) \tag{10}$$

As is seen, (8) shows that the oscillatory integral can be expanded in terms of negative powers of λ using the operator $\Delta^{(j)}(y)$. Thus in accordance with (6) or (8) one may apply the operators $\Delta^{(k_1)}(y_1)$. $(0 \le k_1 \le r+1)$ to $I[F;\ (\lambda)]$, obtaining

$$I[F;(\lambda)] = \int_{[0,1]^{n-1}} (d(x)/dx_1)$$

$$\times \int_0^1 F(x_1,\cdots;\lambda_1 x_1,\cdots)dx_1$$

$$= \sum_{k_1=0}^{r} \frac{\lambda_1^{-k_1}}{k_1!} \int_{[0,1]^n} \Delta^{(k_1)}(y_1)F[(x),y_1,\lambda_2 x_2,\cdots]dy_1$$

$$\times[\frac{d(x)}{dx_1}] + \tilde{\rho}_r(\lambda_1) \tag{11}$$

Here it is easy to get an estimate of the remainder by means of (9), namely

$$|\tilde{\rho}_r(\lambda_1)| \le \frac{2}{(r+1)!}(\frac{1}{\lambda_1})^{r+1}$$

$$\times ||F^{(r+1,\,0,\cdots,0)}_{(x_1,\cdots,x_n)}||_1 \cdot ||\overline{B}_{r+1}||_\infty \tag{12}$$

Clearly the integrals appearing on the rightside of (11) can still be expanded in terms of negative powers of λ_2 using the operator $\Delta^{(k_2)}(y_2)$ with $0 \le k_1 + k_2 \le r+1$ so that all the remainders attached can be estimated totally with the upper bound

$$2 \sum_{k_1+k_2=r+1} \frac{\lambda_1^{-k_1}\lambda_2^{-k_2}}{k_1!k_2!} ||F^{(k_1,k_2,0,\cdots,0)}_{(x_1,x_2,\cdots,x_n)}||_1 \cdot ||\overline{B}_{k_1}||_\infty \cdot ||\overline{B}_{k_2}||_\infty$$

where the summation is taken over all $k_1, k_2 \ge 0$ such that $k_1 + k_2 = r+1$.

Proceeding in the similar manner as above, one may find that successive applications of the operators $\Delta^{(k_i)}$ (y_i) will lead to the final result of the form (4) with an estimate (upper bound) of $\rho_r[(\lambda)]$ given by (5) in which $\lambda = \min\{\lambda_1,\cdots,\lambda_n\}$.

4. APPROXIMATE COMPUTATION

For the computation of $I[F;\ (\lambda)]$ involving large parameters λ_i, some approximation formulas can be at once derived from (4), dropping the remainder term and substituting explicit expressions of Bernoulli polynomials into $\Delta^{(k)}[(y)]$. It is obvious that the formulas so obtained consist of only ordinary integrals which can be computed with ordinary numerical integration methods.

REFERENCES

[1] Banerjee, U., Lardy, L.J. and Lutoborski, A., Asymptotic expansions of integrals of certain rapidly oscillating functions, *Mathematics of Computation*, 49, No. 179 (1987), 243-249.

[2] Hsu, L.C., Approximate integration of rapidly oscillating functions and of periodic functions, *Proc. Cambridge Phil. Soc.*, 59 (1963), 81-88.

[3] Iwaniec, T. and Lutoborski, A., Asymptotic expansions of multiple integrals of rapidly oscillating functions, *Mathematics of Computation*, 50, No. 181 (1988), 215-228.

Approximation, Optimization and Computing:
Theory and Applications, A.G. Law and C.L. Wang (eds.)
Elsevier Science Publishers B.V. (North-Holland)

OPTIMIZATION THEORY FOR INFINITE DIMENSIONAL SET FUNCTIONS*

Hang–Chin LAI†

Institute of Mathematics
National Tsing Hua University
Hsinchu, Taiwan 30042

The Farkas theorem is extended to the cases of nonlinear convex set functions, and several necessary optimality conditions of programming problems with set functions are established under the generalized Farkas theorems. Moreau–Rockafellar type theorems and the Kuhn-Tucker/Fritz John types of optimality conditions are presented.

1. INTRODUCTION

Let (X,Γ,μ) be a finite atomless positive measure space, Y and Z be topological real vector spaces which are ordered by the pointed convex cones C and D respectively. In this paper, we shall present some properties for convex set functions $F : \mathscr{S} \longrightarrow Y$ and $G : \mathscr{S} \longrightarrow Z$, where $\mathscr{S}$ is a subset of Γ which are measurable subsets of X. We investigate the optimization theory for the following programming problem with set functions which take values in infinite dimensional vector spaces:

(P) minimize $F(S)$
subject to $S \in \mathscr{S} \subset \Gamma$,
$G(S) \leq_D \theta$ or $G(S) \in (-D)$,
S is the zero vector.

Optimization problems containing set functions arise naturally in many situations dealing with optimal constrained selection of measurable subsets. Specifically, problems of this type have been encountered in, for example, regional design (Corley and Roberts [9–10]), optimal plasma confinement (Wang [29]), etc. In most cases, the theory and methods which have been proposed for treating such problems are of ad hoc nature, and address only special classes of problems. Recently, Morris [26] defined the notions of convexity, differentiability for set functions, and established optimality conditions and Lagrangian duality relations for a general nonlinear programming problem with set functions. The ideas and results advanced by Morris have been utilized and further extended by many authors, who obtained many interesting results, in [2, 3, 4, 5, 6], [13], [18, 19], [21, 22, 23], [27], etc. In most of these works, the range spaces of set functions are taken in $\mathbb{R}$ or $\mathbb{R}^n$. The optimization theory with respect to abstract cones in infinite dimensional topological vector spaces, or minimizing/maximizing cone convex objective functions is investigated in [1], [7, 8], [14, 15], [17], [23, 24], [31, 32] (see also [28], [30]), etc.

In Section 2, we gather, for convenience of reference, some basic definitions and concepts of the convex subfamily of measurable subsets of X; convex set functions; order structure; weak*–lower/upper semicontinuous (w^*–l.s.c./w^*–u.s.c.) for set functions. In Section 3, we extend the Farkas theorem to convex set functions, and establish some necessary optimality conditions for set functions. In Section 4, we present the Moreau-Rockafellar type theorems for the subdifferentiability of convex set functions. Finally, we establish the Kuhn-Tucker type optimality conditions for set functions.

2. PRELIMINARIES

Throughout this paper let Y and Z be locally convex Hausdorff (topological real) vector spaces, C and D be closed convex pointed cones which determine the partial

---

*The work was partially supported by NSC, Taipei, R.O.C.

†This paper was written while the author was a Visiting Professor at the University of Iowa for the academic year 1987–1988.

orders of Y and Z respectively. We further assume that Y and Z are ordered complete vector lattices and that the cones C and D are *normal* (see Zowe [31, 32]).

We adopt the following notations (see [23, 24]). For $x, y, \in Y$,

$$x \leq_C y \Longleftrightarrow y-x \in C;$$

$$x <_C y \Longleftrightarrow y-x \in C\backslash\{\theta\}, \quad \theta \text{ is zero vector;}$$

$$x <<_C y \Longleftrightarrow y-x \in \overset{\circ}{C}, \text{ the set of all interior points of } C.$$

For a set $A \subset Y$,

$y_0 \in \min A$, a *minimum* of A, if and only if there is no $y \in A$ such that $y <_C y_0$;
$y_0 \in \text{w}\min A$, a *weak minimum* of A, if and only if there is no $y \in A$ such that $y <<_C y_0$.

C^* is the dual cone of C, that is,

$$C^* = \{y^* \in Y^* : \langle y^*, y\rangle \geq 0 \text{ for all } y \in C\}$$

where Y^* is the topological dual of Y. And $B^+(Z,Y)$, the set of all positive linear operators from Z to Y, is defined by

$$B^+(Z,Y) = \{W \in B(Z,Y) : W(D) \subset C\}$$

where $B(Z,Y)$ is the set of all continuous linear operators from Z to Y. Denote ∞ as the element larger than all vectors in Y and Z;

$$Y_\infty = Y \cup \{\infty\}\ ; \quad Z_\infty = Z \cup \{\infty\}.$$

Now let (X,Γ,μ) be an atomless finite measure space with $L^1 = L^1(X,\Gamma,\mu)$ separable. For any $S \in \Gamma$, there corresponds a characteristic function $I_S \in L^\infty = L^\infty(X,\Gamma,\mu) \subset L^1$ since $\mu(X) < \infty$. The integral $\int_S h d\mu$ for $h \in L^1$ will denote the dual pair $\langle h, I_S\rangle$. By the separability of L^1, for any $(R,S,\lambda) \in \mathscr{S} \times \mathscr{S} \times [0,1]$, there exist sequences, $\{R_n\}$ and $\{S_n\}$ in Γ, such that

$$I_{R_n} \xrightarrow{w^*} \lambda I_{R\backslash S} \quad \text{and} \quad I_{S_n} \xrightarrow{w^*} (1-\lambda)I_{S\backslash R} \tag{2.1}$$

where w^* stands for the convergence in weak*–topology. In [26, Proposition 3.2], Morris showed that (1) implies

$$I_{R_n \cup S_n \cup (R\cap S)} \xrightarrow{w^*} \lambda I_R + (1-\lambda)I_S. \tag{2.2}$$

A subfamily $\mathscr{S}$ of measurable subsets in Γ is said to be *convex* if, for any $(R,S,\lambda) \in \mathscr{S} \times \mathscr{S} \times [0,1]$, any sequences $\{R_n\}$ and $\{S_n\}$ in Γ having the property (2.1), there exist subsequences, $\{R_{nk}\}$ of $\{R_n\}$ and $\{S_{nk}\}$ of $\{S_n\}$, such that

$$R_{nk} \cup S_{nk} \cup (R\cap S) \in \mathscr{S} \text{ for all } k.$$

This is equivalent to saying that $\mathscr{S}$ is *convex* if for any $(R,S,\lambda) \in \mathscr{S} \times \mathscr{S} \times [0,1]$, there exist sequences $\{R_n\}$ and $\{S_n\}$, with

$$R_n \cup S_n \cup (R\cap S) \in \mathscr{S} \text{ for all } n$$

satisfying (2.1) and (2.2). For convenience, we call such sequences

$$M_n = R_n \cup S_n \cup (R\cap S)$$

a *Morris sequence* (it is something different from the Morris sequence in [2]) associated with (R,S,λ).

<u>Definition 1</u>. A set function $F : \mathscr{S} \longrightarrow Y$ is said to be *convex* on the convex subfamily $\mathscr{S}$ if for any $(R,\mathscr{S},\lambda)$ in $\mathscr{S} \times \mathscr{S} \times [0,1]$,

$$\limsup_{n\to\infty} F(M_n) \leq_C \lambda F(R)+(1-\lambda)F(S)$$

for any Morris sequence $\{M_n\}$ in $\mathscr{S}$ associated with (R,S,λ).

Denote $I_\Gamma = \{I_S : S \in \Gamma\}$. Then the σ–field Γ is identified as a subset of L^∞. For a convex set function $F : \mathscr{S} \longrightarrow Y$, we admit $F(S) = F(R)$ if $I_R = I_S$ μ–a.e. Whence F can be regarded as a Y–*functional* on the subset $I_{\mathscr{S}} = \{I_S : S \in \mathscr{S}\}$ of L^∞. Let $\overline{\Gamma} = \{f \in L^\infty : 0 \leq f \leq 1\}$ be the weak*–closure of Γ (*i.e.*, I_Γ) in L^∞ (see [4, Corollary 3.6]). Let $A \subset Y \times \Gamma$ and $\overline{A}$ be the w^*–closure of A in $Y \times L^\infty$. Let $\mathscr{N}(f)$ be the family of all w^*–neighborhoods of f in $\overline{\Gamma}$.

<u>Definition 2</u>. Let $F : \Gamma \longrightarrow Y_\infty$ with $\text{dom}\, F = \{S \in \Gamma \mid F(S) \in Y\} = \mathscr{S}$. We say that

(i) F is w^*–*lower (upper) semicontinuous* at $S \in \mathscr{S}$ if

$$F(S) \leq_C \liminf_{n\to\infty} F(S_n)$$

$$(F(S) \geq_C \limsup_{n\to\infty} F(S_n)).$$

(ii) F is w^*–*continuous* at $S \in \mathscr{S}$ if

$$F(S) = \lim_{n\to\infty} F(S_n)$$

for any sequence $\{S_n\}$ in $\mathscr{S}$ with

$$I_{S_n} \xrightarrow{w^*} I_S.$$

Any cone convex set function F is w^*–u.s.c. (see Lai/Lin [22, Lemma 3.3 and Corollary 3.4]), and hence any cone convex w^*–l.s.c. set function is w^*–continuous.

Definition 3. For a set function $F : \mathscr{S} \longrightarrow Y$, the w^*–lower (w^*–*upper*) *semicontinuous hull* $\overline{F}$ ($\hat{F}$) of F is defined by

$$\overline{F}(f) = \sup_{V\in\mathscr{N}(f)} \inf_{S\in V\cap\mathscr{S}} F(S) \quad \text{for } f \in \overline{\mathscr{S}}$$

$$(\hat{F}(f) = \inf_{V\in\mathscr{N}(f)} \sup_{S\in V\cap\mathscr{S}} F(S) \quad \text{for } f \in \overline{\mathscr{S}})$$

where $\overline{\mathscr{S}}$ is the weak*–closure of $I_{\mathscr{S}}$ in L^∞.

By Definitions 2 and 3, it follows that, for $F : \mathscr{S} \longrightarrow Y$,

(i) $\overline{F}|_{\mathscr{S}} \leq_C F \leq_C \hat{F}|_{\mathscr{S}}$;

where $\cdot|_{\mathscr{S}}$ means the restriction function on $\mathscr{S}$.

(ii) If F is w^*–l.s.c (resp. w^*–u.s.c.), then

$$\overline{F}|_{\mathscr{S}} = F \quad (\text{resp. } \hat{F}|_{\mathscr{S}} = F);$$

(iii) If F is w^*–continuous on $\mathscr{S}$, then $\overline{F} = \hat{F}$ on $\overline{\mathscr{S}}$, and so $\overline{F}$ is a unique w^*–continuous extension of F;

(iv) If F is convex on a convex subfamily $\mathscr{S}$, then $\overline{\mathscr{S}}$ is convex in L^∞, and $\overline{F}$ is convex on $\overline{\mathscr{S}}$ (see [4, Corollary 3.10]).

For C–convex set function $F : \mathscr{S} \longrightarrow Y$, denote

$$[F,\mathscr{S}] = \{(y,S) \in Y \times \Gamma \mid s \in \mathscr{S}, F(S) \leq_C y\}.$$

Then, from Proposition 3.9 and Corollary 3.10 of [4], we get

$$\overline{[F, \mathscr{S}]} = [\overline{F}, \overline{\mathscr{S}}]$$

provided $F : \mathscr{S} \longrightarrow Y$ is a cone convex set function.

3. GENERALIZED FARKAS THEOREMS AND OPTIMALITY CONDITIONS

Fan [11, Theorem 4] interpreted the Farkas theorem [12, pp. 5–7] as follows:

> Let f(x) be a linear functional on a vector space E. In order that the inequality $f(x) \geq \alpha$ be a consequence of a consistant linear system
>
> $$g_i(x) \geq \beta_i, \quad i = 1, 2, \cdots, p,$$
>
> it is necessary and sufficient that there exist p nonnegative numbers λ_i, $1 \leq i \leq p$ such that

$$f = \sum_{i=1}^{p} \lambda_i g_i \quad \text{and} \quad \alpha \leq \sum_{i=1}^{p} \lambda_i \beta_i. \tag{3.1}$$

This theorem is equivalent to saying that the linear system of inequalities,

$$\begin{aligned} &f(x) - \alpha < 0 \\ &\beta_i - g_i(x) \leq 0, \quad i = 1,2,\cdots,p, \end{aligned} \tag{3.2}$$

has no solution if and only if there exist p nonnegative numbers λ_i, $1 \leq i \leq p$, such that

$$(f(x)-\alpha) + \sum_{i=1}^{p} \lambda_i(\beta_i - g_i(x)) \geq 0. \tag{3.3}$$

We shall extend the Farkas theorem to convex set functions as in the following theorems. We assume throughout this section that the cones C and D in Y and Z have nonempty interior points.

Theorem 3.1. Suppose that $F : \mathscr{S} \to Y_\infty$ and $G : \mathscr{S} \to Z_\infty$ are cone convex, not always infinite, set functions on the convex subfamily $\mathscr{S}$ of Γ. Then the nonlinear system

$$\begin{aligned} F(S) &<_C \theta \\ G(S) &<_D \theta \end{aligned} \tag{3.4}$$

has no solution in $\mathscr{S}$ if and only if there exists nonzero (y^*, z^*) in $C^* \times D^*$ such that

$$\langle y^*, F(S)\rangle + \langle z^*, G(S)\rangle \geq 0 \quad \text{for all } S \in \mathscr{S}. \tag{3.5}$$

Proof. The "if" part is obvious, we only prove the "only if" part. Let

$$A = \{(y,z) \in Y \times Z \mid F(S) <_C y \text{ and } G(S) <_D z \text{ for some } s \in \mathscr{S}\}.$$

Since (3.4) has no solution, (θ, θ) is not in A.

Claim: A is convex. Let (y,z) and $(\tilde{y}, \tilde{z})$ be in A. Then there exist $R,S \in \mathscr{S}$ such that

$$F(R) <_C y, \quad G(R) <_D z$$

and

$$F(S) <_C \tilde{y}, \quad G(S) <_D \tilde{z}.$$

Since F and G are convex, there exists a Morris sequence $\{M_n\}$ in $\mathscr{S}$ corresponding to $(R,S,\lambda) \in \mathscr{S} \times \mathscr{S} \times [0,1]$ such that

$$\limsup_{n\to\infty} F(M_n) \leq_C \lambda F(R)+(1-\lambda)F(S) <_C \lambda y+(1-\lambda)\tilde{y}$$

$$\limsup_{n\to\infty} G(M_n) \leq_D \lambda G(R)+(1-\lambda)G(S) <_D \lambda z+(1-\lambda)\tilde{z},$$

thus

$$F(M_n) <_C \lambda y+(1-\lambda)\tilde{y}$$

$$G(M_n) <_D \lambda z+(1-\lambda)\tilde{z} \qquad \text{for all } n.$$

It follows that

$$\lambda(y,z)+(1-\lambda)(\tilde{y}, \tilde{z}) = (\lambda y+(1-\lambda)\tilde{y}, \lambda z+(1-\lambda)\tilde{z}) \in A.$$

Hence A is convex.

Since C and D have nonempty interiors, A possesses the interior point. Hence, as (θ, θ) is not in A, applying the separation theorem, we have a nonzero (y^*, z^*) in $Y^* \times Z^*$ such that

$$\langle y^*,y\rangle + \langle z^*,z\rangle \geq 0 \quad \text{for all } (y,z) \in A.$$

Since F and G are not always infinite, there is a $S \in \mathscr{S}$ such that

$$F(S) <_C F(S) + c = y$$

$$G(S) <_D G(S) + d = z$$

for any $c \neq \theta$ in C and $d \neq \theta$ in D. It follows that

$$\langle y^*,F(S)\rangle+\langle z^*,G(S)\rangle+\langle y^*,c\rangle+\langle z^*,d\rangle \geq 0. \qquad (3.6)$$

Claim: $(y^*,z^*) \in C^* \times D^*$. For if there is a $\tilde{c} \in C$ such that $\langle y^*,\tilde{c}\rangle < 0$, then by closeness of C, $n\bar{c} \in C$ for any positive integer n, and (3.6) becomes

$$0 \leq \langle y^*,F(S)\rangle+\langle z^*,G(S)\rangle+n\langle y^*,\tilde{c}\rangle+\langle z^*,d\rangle < 0$$

for sufficiently large n, this is a contradiction. So $y^* \in C^*$. Similarly $z^* \in D^*$. Finally, by continuity of y^* and z^*, letting $c \longrightarrow \theta$ and $d \longrightarrow \theta$, we obtain (3.5). Q.E.D.

Theorem 3.2. As in the assumptions of Theorem 3.1, we further assume that the dual cone C^* has nonempty interior, and $G(\hat{S}) <<_D \theta$ for some $\hat{S} \in \mathscr{S}$ (this condition is called the Slater condition). Then the system

$$\begin{aligned} F(S) &<_C \theta \\ G(S) &\leq_D \theta \end{aligned} \qquad (3.7)$$

has no solution in $\mathscr{S}$ if and only if there exists $W_0 \in B^+(Z,Y)$ such that

$$F(S) + W_0(G(S)) <_C \theta \qquad (3.8)$$

is not true for any $S \in \mathscr{S}$.

Proof. The "if" part is obvious, we only prove the "only if" part. If (3.7) has no solution in $\mathscr{S}$, the system (3.5) has no solution, and, by Theorem 3.1, there exists a nonzero (y^*, z^*) in $C^* \times D^*$ such that (3.5) holds.

Claim: $y^* \neq \theta$. For otherwise, $z^* \neq \theta$ and $\langle z^*, z\rangle > 0$ for any nonzero z in $\overset{o}{D}$. Since $G(\hat{S}) <<_D \theta$ for some $\hat{S} \in \mathscr{S}$,

$$0 > \langle z^*, G(\hat{S})\rangle = \langle \theta, F(\hat{S})+\langle z^*, G(\hat{S})\rangle \geq 0$$

is a contradiction. Hence $y^* \neq \theta$.

As C^* has interior point, we can choose y^* in $(C^*)^0$ and a nonzero $y_0 \in C$ such that $\langle y^*, y_0\rangle = 1$. Define $W_0 : Z \longrightarrow Y$ by

$$W_0(z) = \langle z^*, z\rangle y_0.$$

Then $W_0 \in B^+(Z,Y)$ and

$$\langle y^*,F(S)+W_0(G(S))\rangle = \langle y^*,F(S)\rangle+\langle z^*,G(S)\rangle \geq 0$$

for all $S \in \mathscr{S}$. This shows that $F(S)+W_0(G(S)) \notin (-C)\backslash\{\theta\}$ for all $S \in \mathscr{S}$. Q.E.D.

Note that if the system (3.7) has no solution, then the system

$$F(S) <<_C \theta$$
$$G(S) \leq_D \theta \qquad (3.9)$$

has no solution. But the converse is not valid. However, we can apply the same argument used in the proof of Theorem 3.1 (change $<_C$ to $<<_C$ if necessary) to obtain the following theorem.

__Theorem 3.3__. Under the assumption of Theorem 3.1, we further assume $G(\hat{S}) <<_D \theta$ for some $S \in \mathscr{S}$. Then the system (3.9) has no solution if and only if there exists $W_0 \in B^+(Z,Y)$ such that

$$F(S) + W_0(G(S)) <<_C \theta \qquad (3.10)$$

does not hold for any $S \in \mathscr{S}$.

We will employ the generalized Farkas theorems to establish the necessary optimality conditions of problem (P). Recall that $S_0 \in \mathscr{S}$ is a *minimal* (or *weakly minimal*) point of (P) if there does not exist $S \in \mathscr{S}$ such that

$$F(S) <_C F(S_0) \quad (F(S) <<_C F(S_0)).$$

We denote the Lagrangian by $L(S,W) = F(S)+W(G(S))$, $W \in B(Z,Y)$. We say that (S_0, W_0) is a *saddle* (*weakly saddle*) *point* of $L(S, W)$ if $W_0 \in B^+(Z,Y)$, $S_0 \in \mathscr{S}$; there does not exist $W \in B^+(Z, Y)$ such that

$$L(S_0, W_0) <_C L(S_0, W) \quad (L(S_0, W_0) <<_C L(S_0, W_0))$$

and there is no $S \in \mathscr{S}$ such that

$$L(S, W_0) <_C L(S_0, W_0) \quad (L(S, W_0) <<_C L(S_0, W_0)).$$

__Theorem 3.4__. As in the assumptions of Theorem 3.2, if S_0 is a minimal point of (P), then there exists $W_0 \in B^+(Z, Y)$ such that $W_0(G(S_0)) = \theta$, and (S_0, W_0) is saddle point of the Lagrangian $L(S, W)$.

__Proof__. If $S_0 \in \mathscr{S}$ is a minimal point of (P), then there does not exist $S \in \mathscr{S}$ such that

$$F(S) <_C F(S_0) \quad \text{and} \quad G(S) \leq_D \theta.$$

That is, the system

$$F(S) - F(S_0) <_C \theta$$
$$G(S) \leq_D \theta$$

has no solution. By Theorem 3.2, there exists $W_0 \in B^+(Z, Y)$ such that

$$F(S) - F(S_0) + W_0(G(S)) <_C \theta \qquad (3.11)$$

is not true for any $S \in \mathscr{S}$. Since $W_0 \in B^+(Z, Y)$ and $G(S_0) \leq_D \theta$, we have

$$W_0(G(S_0)) \leq_C \theta. \qquad (3.12)$$

Setting $S = S_0$ in (3.11), we have $W_0(G(S_0)) <_C \theta$ is not true. It follows from (3.12) that only

$$W_0(G(S_0)) = \theta. \qquad (3.13)$$

From (3.11) and (3.13), we see that

$$F(S) - F(S_0) + W_0(G(S)) - W_0(G(S_0)) <_C \theta$$

is not true for any $S \in \mathscr{S}$, that is, there does not exist $S \in \mathscr{S}$ such that

$$L(S, W_0) <_C L(S_0, W_0).$$

On the other hand, if there is a $W \in B^+(Z, Y)$ such that $L(S_0, W_0) <_C L(S_0, W)$, then

$$\theta = W_0(G(S_0)) <_C W(G(S_0))$$

is a contradiction since $G(S_0) \leq_D \theta$ and $W \in B^+(Z, Y)$. Whence there does not exist $W \in B^+(Z, Y)$ such that

$$L(S_0, W_0) <_C L(S_0, W).$$

Hence (S_0, W_0) is a saddle point of (P). Q.E.D.

Note that a minimal point S_0 of (P) is also a weakly minimal point of (P), but the converse is not necessarily true. However, if S_0 is a weakly minimal point of (P), we have the following result.

__Theorem 3.5__. As in the assumptions of Theorem 3.4, except the condition that $(C^*) \neq \emptyset$, if S_0 is a weakly minimal point of (P), then there exists $W_0 \in B^+(Z, Y)$ such that (S_0, W_0) is a weak saddle point of the Lagrangian $L(S, W)$.

__Proof__. If S_0 is a weakly minimal point of (P), then the system

$$F(S) - F(S_0) <<_C \theta$$
$$G(S) \leq_D \theta$$

has no solution in $\mathscr{S}$. By Theorem 3.3, there exists $W_0 \in B^+(Z, Y)$ such that

$$F(S) - F(S_0) + W_0(G(S)) \notin (-C)^0. \quad (3.14)$$

Since $W_0 \in B^+(Z, Y)$ and $G(S_0) \leq_D \theta$, $W_0(G(S_0)) \leq_C \theta$. Setting $S = S_0$ in (3.14), we have $W_0(G(S_0)) \notin (-C)^0$. It follows that

$$W_0(G(S_0)) \in \partial(-C), \text{ the boundary of } -C. \quad (3.15)$$

From (3.14) and $W_0(G(S_0)) \notin (-C)^0$, we obtain

$$F(S) - F(S_0) + W_0(G(S)) - W_0(G(S_0)) \notin (-C)^0$$

for all $S \in \mathscr{S}$. Hence there does not exist $S \in \mathscr{S}$ such that

$$L(S,W_0) <<_C L(S_0, W_0).$$

On the other hand, if there is a $W \in B^+(Z, Y)$ such that $L(S_0, W_0) <<_C L(S_0, W)$, *i.e.*, $W_0(G(S_0)) - W(G(S_0)) \in (-\overset{\circ}{C})$. Then from (3.15),

$$W(G(S_0)) \in \overset{\circ}{C} + W_0(G(S_0)) \subset \overset{\circ}{C} + \partial(-C).$$

Evidently, $W(G(S_0)) \in (-C)$, so that

$$W(G(S_0)) \in \{\overset{\circ}{C}+\partial(-C)\} \cap (-C)$$

is a contradiction if we have shown $\{\overset{\circ}{C}+\partial(-C)\} \cap -C) = \emptyset$. In fact, if $y_1 \in \overset{\circ}{C}$, $y_2 \in \partial(-C)$ and $y_1+y_2 \in (-C)$, then

$$y_2 \in -C-y_1 \subset -C-\overset{\circ}{C} \subset (-\overset{\circ}{C})$$

since C is a convex pointed cone, and $C+\overset{\circ}{C} \subset \overset{\circ}{C}$. Thus $y_2 \in (-C)^0 \cap \partial(-C)$. This is impossible since $(-C)^0 \cap \partial(-C) = \emptyset$. Therefore there does not exist $W \in B^+(Z, Y)$ such that

$$L(S_0, W_0) <<_C L(S_0, W),$$

and so (S_0, W_0) is a weakly saddle point. Q.E.D.

If $Y = \mathbb{R}$, then weakly minimal point coincides with minimal point, and Theorems 3.4 and 3.5 are reduced to the following corollary.

<u>Corollary 3.6</u>. In Theorem 3.5, if $Y = \mathbb{R}$, then S_0 is a minimal point of (P) if and only if there exists $z^* \in D^*$ such that $\langle z_0^*, G(S_0)\rangle = 0$, and (S_0,z_0^*) is a saddle point of the Lagrangian $L(S,z^*)$.

If $Y = Z = \mathbb{R}$, Corollary 3.6 is reduced to Theorem 3.2 of [2].

The following theorem may be regarded as a special case of Theorem 3.4 for scalarly Lagrangian.

<u>Theorem 3.7</u>. As in the assumptions of Theorem 3.4, we drop the condition $(C^*)^0 \neq \emptyset$ and use $G(\hat{S}) <_D \theta$ instead of $G(\hat{S}) <<_D \theta$ for some $\hat{S} \in \mathscr{S}$. Then when S_0 is a minimal point of (P), there exists $(y_0^*,z_0^*) \in C^* \times D^*$ such that $\langle z_0^*, G(S_0)\rangle = 0$ and (S_0,z_0^*) is a saddle point of the Lagrangian:

$$\tilde{L}(S,z^*) = \langle y_0^*, F(S)\rangle + \langle z^*, G(S)\rangle$$

for all $S \in \mathscr{S}$ and $z^* \in D^*$.

<u>Proof</u>. If S_0 is a minimal point of (P), then the system

$$F(S) - F(S_0) <_C \theta$$
$$G(S) \leq_D \theta$$

has no solution. Thus the proof follows Theorem 3.1 and the fact $\langle z_0^*, G(S_0)\rangle \leq 0$. Q.E.D.

4. MOREAU–ROCKAFELLAR TYPE THEOREM AND OPTIMALITY CONDITIONS

For a linear operator $T \in B(L^\infty_{w^*}, Y)$, we mean that

$$g_n \xrightarrow{w^*} g \text{ in } L^\infty \Longrightarrow T(g_n) \longrightarrow T(g) \text{ in } Y.$$

<u>Definition 4</u>. A C–convex set function $F : \Gamma \longrightarrow Y_\infty$ is *subdifferentiable* (resp. *weak subdifferentiable*) at $S_0 \in \Gamma$ if there exists $T \in B(L^\infty_{w^*}, Y)$, namely, *subgradient* (resp. *weak subgradient*) of F at S_0, such that

$$F(S) \nless_C F(S_0) + T(I_S - I_{S_0}) \text{ for all } S \in \Gamma.$$

(resp. $F(S_0)-T(I_{S_0}) \in$ w–min$\{F(S)-T(I_S) \mid S \in \Gamma\}$). The set of all subgradients (resp. weak subgradients) of F at S_0 is denoted by $\partial F(S_0)$ (resp. $\partial_w F(S_0)$), namely, the *subdifferential* (resp. *weak subdifferential*) of F at S_0.

Evidently, if $T \in \partial_w F(S_0)$ (resp. $\partial F(S_0)$) then there does not exist $S \in \Gamma$ such that

$$F(S) <<_C F(S_0) + T(I_S - I_{S_0})$$

$$\text{(resp. } F(S) <_C F(S_0) + T(I_S - I_{S_0})).$$

Remark.

(i) $F(S_0) \in$ w–min$\{F(S) : S \in \Gamma\}$ if and only if $0 \in \partial_w F(S_0)$.

(ii) $\partial F(S_0) \subset \partial_w F(S_0)$.

(iii) If $Y = \mathbb{R}$, then $\partial F(S_0) = \partial_w F(S_0)$ for $S_0 \in$ dom F.

(iv) If dom $F = \mathscr{S} \subset \Gamma$ and $S_0 \notin \mathscr{S}$, then

$F(S_0) = \infty$ and $\partial_w F(S_0) = B(L^\infty_{w*}, Y)$

$= \partial F(S_0)$. Indeed, any $T \in B(L^\infty_{w*}, Y)$,

$\infty = F(S_0) \nless_C F(S) + T(I_S - I_{S_0})$ for all $S \in \Gamma$,

that is, there does not exist $S \in \Gamma$ such that

$$F(S_0) - F(S) + T(I_S - I_{S_0}) \in \mathring{C}.$$

We have the following generalized Moreau–Rockafellar theorem.

Theorem 4.1 (Lai/Lin [23, Theorem 3.1]). Let F_1, F_2 be C–convex set functions on $\mathscr{S} = \text{dom } F_1 = \text{dom } F_2$, and F_1 be w^*–continuous on $\mathscr{S}$. Suppose that $\mathscr{S}$ is convex and $\overline{\mathscr{S}}$ has a relative interior point. Then

$$\partial_w(F_1+F_2)(S) \subset \partial_w F_1(S) + \partial_w F_2(S) \quad \text{for all } s \in \Gamma.$$

The reverse inclusion is not true for vector valued set functions (see Example 3.1 of Lai and Lin [23]). However if the subdifferential is used instead of the weak subdifferential and both F_1 and F_2 are w^*–continuous, then we have

(i) $\partial(F_1+F_2)(S) = \partial F_1(S) + \partial(F_2(S)$ for all $S \in \mathscr{S}$,

(ii) $\partial(\overline{F}_1+\overline{F}_2)(f) = \partial\overline{F}_1(f) + \partial(\overline{F}_2(f)$ for all $f \in L^\infty$

where $\overline{F}_i : L^\infty \longrightarrow Y_\infty$, $i = 1,2$ are the w^*–l.s.c. hull of F_i and dom $\overline{F}_i = \overline{\mathscr{S}}$, $i = 1,2$. This result can be proved similarly to Lai and Lin [21, Theorem 9]. Theorem 4.1 can be easily extended to a more general case as follows:

Corollary 4.2. If $F_i : \Gamma \longrightarrow Y_\infty$, $i = 1,2,\cdots,n$ are C-convex set functions on dom $F_i = \mathscr{S}$, $i = 1,2,\cdots,n$ and $\mathscr{S}$ is a convex subfamily of Γ with relative interior $\text{ri}\overline{\mathscr{S}} \neq \emptyset$; if all set functions $F_1,\cdots,F_n$, except possibly one, are w^*–continuous on $\mathscr{S}$, then

$$\partial_w(F_1+\cdots+F_n)(S) \subset \partial_w F_1(S)+\cdots+\partial_w F_n(S)$$

for all $S \in \mathscr{S}$.

Definition 5. Let $W \in B^+(Z, Y)$ and $G : \mathscr{S} \longrightarrow Z$ be a D–convex set function. We say that G is *weakly regular subdifferentiable* at $S_0 \in \mathscr{S}$ if

$$\partial_w(W \circ G)(S_0) = W \circ \partial_w G(S_0) \quad \text{for } S_0 \in \mathscr{S}.$$

If $Y = \mathbb{R}$, then G is referred to *regular subdifferentiable* at $S_0 \in \mathscr{S}$ if

$$\partial(z^* \circ G)(S_0) = z^* \circ \partial G(S_0)$$

where $z^* \in D^* = B^+(Z, \mathbb{R})$.

Applying the Moreau–Rockafellar type theorem, we obtain a theorem of Kuhn–Tucker type condition for existence of weak–minimal point to problem (P) as follows:

Theorem 4.3 (Lai and Lin [23, Theorem 4.1]). Let $F : \mathscr{S} \longrightarrow Y_\infty$ be C–convex, $G : \mathscr{S} \longrightarrow Z_\infty$ D–convex, where $\mathscr{S} = \text{dom } F = \text{dom } G$ is a convex subfamily of Γ. Assume that one of F and G is w^*–continuous on $\mathscr{S}$ and $(\text{ri } \mathscr{S}) \neq \emptyset$. Furthermore, assume that there exists $\hat{S} \in \mathscr{S}$ with $G(\hat{S}) <<_D \theta$; $S_0 \in$ w–min(P) and G is weakly regular subdifferentiable at S_0. Then there exist $W_0 \in B^+(Z, Y)$ such that

$$W_0(G(S_0)) = \theta$$

and

$$0 \in \partial_w F(S_0) + W_0 \circ \partial_w G(S_0) + N_w(S_0)$$

where

$$N_w(S_0) = \{T \in B(L^\infty_{w*}, Y) : \text{there is no } S \in \mathscr{S} \text{ such that } \theta <<_C T(I_S - I_{S_0})\}$$

is the weakly normal cone at S_0.

If we drop the constraint qualification in Theorem 4.1, using the same argument in the proof of Theorem 4.1, we obtain the Fritz John type condition (see Theorem 3 of Lai and Ho [20]) as follows:

Theorem 4.4. Under the assumptions, except the constrained qualification, of Theorem 4.1, we further assume that F and G are regularly differentiable at S_0. Then there exists nonzero (y_0^*, z_0^*) in $C^* \times D^*$ such that $\langle z_0^*, G(S_0)\rangle = 0$ and

$$0 \in y_0^* \circ \partial F(S_0) + z_0^* \circ \partial G(S_0) + N(S_0)$$

where $N(S_0) = \{f \in L^1 \mid \langle f, I_S - I_{S_0}\rangle \leq 0 \text{ for all } S \in \mathscr{S}\}$.

If $Y = \mathbb{R}$, Theorem 4.3 reduces to the following corollary.

Corollary 4.5. In Thorem 4.3, let $Y = \mathbb{R}$. Then $S_0 \in \min(P)$ if and only if there exists $z_0^* \in D^*$ such that $\langle z_0^*, G(\Omega_0)\rangle = 0$ and

$$0 \in \partial F(S_0) + z_0^* \circ \partial G(S_0) + N(S_0)$$

where $N(S_0) = \{f \in L^1 : \langle f, I_S - I_{S_0}\rangle \leq 0 \text{ for all } S \in \mathscr{S}\}$.

Remark.

(i) If $Y = Z = \mathbb{R}$, then Corollary 4.5 is reduced to Theorem 11 of Lai and Lin [21].

(ii) If $Y = \mathbb{R}$ in Theorem 4.4, it reduces a similar result of [21, Theorem 12], but here we need not have Slater's condition.

REFERENCES

[1] Borwein, J., Proper efficient points for maximization with respect to cones, *SIAM J. Control Optim.* 15(1977) 57–63.

[2] Chou, J.H., Hsia, W.S. and Lee, T.Y., On multiple objective programming problems with set functions, *J. Math. Anal. Appl.* 105(1985) 383-394.

[3] Chou, J.H., Hsia, W.S. and Lee, T.Y., Second order optimality conditions for mathematical programming with set functions, *J. Austral. Math. Soc.* (Ser.B) 26(1985) 284–292.

[4] Chou, J.H., Hsia, W.S. and Lee, T.Y., Epigraphs of convex set functions, *J. Math. Anal. Appl.* 118 (1986) 247–254.

[5] Chou, J.H., Hsia, W.S. and Lee, T.Y., Convex programming with set function, *Rocky Mountain J. Math.* 17(1987), 535–543.

[6] Corley, H.W., Optimization theory for n–set functions, *J. Math. Anal. Appl.* 127(1987) 193–205.

[7] Corley, H.W., Duality theory for maximization with respect to cones, *J. Math. Anal. Appl.* 84 (1981) 560–568.

[8] Corley, H. W., An existence result for maximization with respect to cones, *J. Optim. Theory Appl.* 31(1980) 277–281.

[9] Corley, H.W. and Roberts, S.D., A partitioning problem with applications in regional design, *Operation Research* 20(1972) 1010–1019.

[10] Corley, H.W. and Roberts, S.D., Duality relationships for a partitioning problem, *SIAM J. Appl. Math.* 23(1972) 490–494.

[11] Fan, K., On systems of linear inequalities, in: Kuhn, H.W. and Tucker, A.W. (eds.), Linear Inequalities and Related System (Ann. Math. Studies 38, Princeton Univ., N.J., 1956) pp. 99–156.

[12] Farkas, J., Uber die theorie der einfachen ungleichungen, *J. Reine Angew. Math.* 124(1902) 1–27.

[13] Hsia, W.S. and Lee, T.Y., Proper D–solutions of multiobjective programming problems with set functions, *J. Optim. Theory Appl.* 53(1987) 247–258.

[14] Jahn, J., A characterization of properly minimal elements of a set, *SIAM J. Control Optim.* 23 (1985) 649–656.

[15] Jahn, J., Existence theorem in vector optimization, *J. Optim. Theory Appl.* 50(1986) 397–406.

[16] Koshi, S., Lai, H.C. and Komuro, N., Convex programming on spaces of measurable functions, *Hokkaido Math. J.* 14(1985) 75–84.

[17] Lai, H.C., Conjugate operators and subdifferentials for convex operators in ordered vector spaces, *Soochow J. Math.* 9(1983) 127–135.

[18] Lai, H.C. and Yang, S.S., Saddle point and duality in the optimization theory of convex set functions, *J. Austral. Math. Soc.* (Ser.B) 24(1982) 130–137.

[19] Lai, H.C., Yang, S.S. and Hwang, G.R., Duality in mathematical programming of set functions—On Fenchel duality theorem, *J. Math. Anal. Appl.* 95(1983) 223–234.

[20] Lai, H.C. and Ho, C.P., Duality theorem of nondifferentiable convex multiobjective programming, *J. Optim. Theory Appl.* 50(1986) 407–420.

[21] Lai, H.C. and Lin, L.J., Moreau–Rockafellar type theorem for convex set functions. *J. Math. Anal. Appl.* 132(1988), 558-571.

[22] Lai, H.C. and Lin, L.J., The Fenchel–Moreau theorem for set functions, *Proc. Amer. Math. Soc.* 103(1988), 85-90.

[23] Lai, H.C. and Lin, L.J., Optimality for set functions with values in ordered vector spaces, *J. Optim. Theory Appl.*, 63(1989). No. 3.

[24] Lai, H.C. and Yang, L.S., Strong duality for infinite–dimensional vector–valued programming problems, *J. Optim. Theory Appl.*, 62(1989) 447-464.

[25] Luenberger, D.G., Optimization by Vector Space Methods (Wiley, New York, 1969).

[26] Morris, R.J.I., Optimal constrained selection of a measurable subset, *J. Math. Anal. Appl.* 70(1979) 546–562.

[27] Tanaka, K. and Maruyama, Y., The multiobjective optimization problem of set functions, *J. Inform. Optim. Sci.* 5(1984) 293–306.

[28] Tanino, T. and Sawaragi, Y., Conjugate maps and dualilty in multiobjective optimization, *J. Optim. Theory Appl.* 31(1980) 473–499.

[29] Wang, P.K.C., On a class of optimization problems involving domain variations, Lecture Notes in Control and Information Sciences, No. 2 (Springer–Verlag, Berlin, 1977).

[30] Yu, P.L., Cone convexity, cone extreme points, and nondominated solution in decision problem with multiobjective, *J. Optim. Theory Appl.* 14 (1974) 319–377.

[31] Zowe, J., A duality theorem for a convex programming problem in order complete vector lattices, *J. Math. Anal. Appl.* 50(1975) 273–287.

[32] Zowe, J., The saddle point theorem of Kuhn and Tucker in order vector spaces, *J. Math. Anal. Appl.* 57(1977) 47–55.

Approximation, Optimization and Computing:
Theory and Applications, A.G. Law and C.L. Wang (eds.)
Elsevier Science Publishers B.V. (North-Holland)

ITERATIVE COMPUTATION IN DYNAMIC OPTIMIZATION

E. S. Lee

Department of Industrial Engineering, Kansas State University,
Manhattan, KS 66506

The advantages and disadvantages of six prominent iterative or approximation techniques are compared based on both numerical results and their basic characteristics. The effectiveness of these techniques to overcome the computational difficulties in dynamic optimization are examined based on the boundary value difficulty for the classical approach and the dimensionality difficulty for the dynamic programming approach. Various numerical results are presented to illustrate the techniques.

1. INTRODUCTION

The two classes of techniques for attacking dynamic optimization problems are the classical calculus of variations or the maximum principle and the more recent dynamic programming. Both classes have their computational difficulties. The classical approach has boundary value difficulty and dynamic programming has dimensionality difficulty. Both of these difficulties will be considered and the numerical techniques for overcoming these difficulties will be compared.

Iterative computational techniques, also known as successive approximation, are well-known numerical tools. Some of the more prominent iterative techniques are the Newton's method and the gradient technique. With the development of computers, these techniques have further developed into functional techniques which can be used to solve functional equations. Furthermore, new and more powerful iterative techniques are also developed. Some of these techniques are very useful for solving dynamic optimization problems numerically.

In this paper, six iterative or approximation techniques will be considered. These six numerical techniques are the functional gradient technique, the second variational technique, the conjugate gradient technique, quasilinearization, invariant imbedding and stochastic approximation. Numerical examples will be used to illustrate the different approaches. These techniques will be compared based on both the numerical results and the basic characteristics of the technique.

2. THE DYNAMIC OPTIMIZATION PROBLEM

To introduce nomenclature, we shall review briefly the essentials of the problem and the functional iterative techniques. The dynamic optimization problem can be stated as: find the control functions $u(t)$ such that the functions $x(t)$ defined by the equations

$$\frac{dx_i}{dt} = f_i(x(t), u(t), t), \quad 0 \le t \le T, \quad i=1,2,\dots,n, \tag{1}$$

with initial conditions

$$x(0) = X^0. \tag{2}$$

Maximize the function

$$J = g(x(T), T) \tag{3}$$

where $x(t)$ is an n-dimensional state vector, $u(t)$ is a one-dimensional control variable and f is an n-dimensional function. This is the problem of Mayer in calculus of variations [1,2].

3. THE ITERATIVE APPROXIMATION TECHNIQUES

3.1 Functional Gradient Technique

To solve the problem defined by Equations (1)-(3) by the functional gradient technique, let us define

$S(x)$ = The final value of the objective function using a nominal control sequence $u^0(t)$ (4)

The function S has the following semigroup property.

$$S(x(t)) = S(x(t+\Delta t)) \tag{5}$$

Using this semigroup property, the recurrence relation for computing the gradient can be written as

$$\left.\frac{\partial s}{\partial u}\right|_t = \sum_{i=1}^{n} \left.\frac{\partial s}{\partial x_i}\right|_{t+\Delta t} \left.\frac{\partial f_i}{\partial u}\right|_t \Delta t \tag{6}$$

where

$$\left.\frac{\partial s}{\partial x_j}\right|_t = \left.\frac{\partial s}{\partial x_j}\right|_{t+\Delta t} + \sum_{i=1}^{n+1} \left.\frac{\partial s}{\partial x_i}\right|_{t+\Delta t} \left.\frac{\partial f_i}{\partial x_j}\right|_t \Delta t,$$

$j=1,2,\ldots,n$ (7)

The final conditions for Equation (7) are

$$\left.\frac{\partial s}{\partial x_j}\right|_T = \left.\frac{\partial J}{\partial x_j}\right|_T - \left.\left(\frac{dJ}{dt}\Big/\frac{\partial\psi}{dt}\right)\right|_T \left.\frac{\partial\psi}{\partial x_j}\right|_T,$$

$j=1,2,\ldots,n$ (8)

where ψ is an additional condition which must be satisfied. For the problem defined above, this condition simplifies to

$$\psi = t - T. \tag{9}$$

The computational procedure can be summarized as

1. assume a nominal control sequence $u^o(t), 0\le t\le T$.
2. integrate Equation (1.)
3. Using Equation (8), solve Equation (7) by backward recursion
4. Using Equation (6), compute the gradient at every grid point
5. Obtain an improved control sequence by

$$u^{k+1}(t) = u^k(t) + \Delta J \left.\left(\frac{\partial s}{\partial u}\right)\right|_t \Big/ \sum_{t=0}^{T} \left.\left(\frac{\partial s}{\partial u}\right)^2\right|_t \tag{10}$$

6. repeat steps (2) through (5) until the desired accuracy is obtained

where ΔJ is the step size or the required improvement. For a detailed discussion of the functional gradient technique, the reader can refer to [3].

3.2 Direct Second Variational Method

The Hamiltonian function for the problem defined by Equations (1)-(3) can be defined as

$$H(x,\lambda,u,t) = \sum_{i=1}^{n} \lambda_i f_i(x,u,t). \tag{11}$$

The n-dimensional adjoint vector λ is governed by

$$\frac{d\lambda_i}{dt} = -\frac{\partial H}{\partial x_i}, \quad i=1,2,\ldots,n \tag{12}$$

with the final conditions

$$\lambda_i(T) = \frac{\partial J}{\partial x_i(T)}, \quad i=1,2,\ldots,n \tag{13}$$

If (x^*,u^*,λ^*) represent the optimum vector, then the maximum principle states that

$$H_u(x^*,u^*,\lambda^*) = 0 \tag{14}$$

Notice that Equations (1) and (12) constitute a boundary value problem which is frequently unstable and difficult to solve numerically. Various numerical techniques have been devised to solve this problem. The direct second variational technique and quasilinearization (or generalized Newton-Raphson technique) are devised to overcome this difficulty.

Instead of using the optimality condition, Equation (14), the direct second variational technique generates a sequence of controls $u^k(t)$ which approaches the optimum with quadratic convergence rate.

To obtain the desired sequence, the Taylor series with terms of second and higher order neglected is used to expand the Hamiltonian function. After introducing a relaxation factor η, the desired equation becomes

$$u^{k+1} = u^k - \eta^k (H^k_{uu})^{-1}(H^k_u + H^k_{ux}\delta x^k + H^k_{u\lambda}\delta\lambda^k) \tag{15}$$

which can be represented as

$$u^{k+1} = \ell(x^{k+1},x^k,\lambda^{k+1},\lambda^k,u^k,t) \tag{16}$$

where η^k is a relaxation factor whose value is between 0 and 1.

Substituting Equation (16) into the original system, we have

$$\frac{dx_i^{k+1}}{dt} = f_i(x^{k+1},\ell,t), \quad i=1,2,\ldots,n \tag{17}$$

$$\frac{d\lambda_i^{k+1}}{dt} = -\frac{\partial H(x^{k+1},\ell,\lambda^{k+1},t)}{\partial x_i}, \quad i=1,2,\ldots,n \tag{18}$$

The computational procedure can be summarized as

1. assume or obtain an initial sequence
2. integrate Equations (1) and (12)
3. compute and store H_u, H_{un}, H_{xx} and $H_{u\lambda}$ at all grid points
4. integrate Equations (17) and (18)
5. obtain improved control using Equation (15)
6. if the desired accuracy is obtained, stop. Otherwise to to step 2

For a detailed discussion on the direct second variational method, the reader can refer to [4].

3.3 Conjugate Gradient Method

We shall not go into any derivation of the conjugate gradient method. The reader can refer to the literature [5]. We shall only summarize the computational steps as follows:

1. assume a nominal control sequence $u^0(t)$
2. integrate Equations (1) and (12)
3. calculate the gradient direct g from the equation

$$g(t) = \frac{\partial H}{\partial u} \tag{19}$$

4. define $S_K = - g_K$ for minimization problem or $S_K = g_k$ for maximization problem
5. find the best step size α^* so that $\alpha^* = \alpha$ minimizes $J(u+\alpha S)$
6. set $u_{K+1} = u_K + \alpha^* S_K$
7. repeat steps (2) and (3)
8. compute $\beta_K = (g_{k+1}, g_{k+1})/(g_k, g_k)$ where $(g_j, g_j) = \int_0^T g_j(t) g_j(t)dt$
9. compute the conjugate gradient direction $S_{K+1} = - g_{K+1} + \beta_K S_K$ for minimization and $S_{K+1} = - g_{k+1} + \beta_K S_K$ for maximization
10. repeat steps (2) through (9) until the desired accuracy is obtained

3.4 Quasilinearization

Quasilinearization is essentially a generalized functional Newton-Raphson method. In addition to the basic iterative approximation characteristics, quasilinearization also linearizes the original nonlinear problem iteratively [2]. This technique can also be used effectively to solve the Mayer's problem represented by Equations (1)-(3) and (11)-(13). However, to illustrate the versatility of the approach, we shall use quasilinearization to overcome the dimensionality problem in dynamic programming.

To illustrate the approach, consider the following linear difference equations

$$x(n) = A(n)x(n-1)+h(u(n),n), \quad n=1,2,...,N \tag{20}$$

with initial conditions

$$x_i(0) = x_i^0, \quad i=1,2,..,m$$

We wish to minimize the following objective function.

$$H(x_1(N), x_2(N), ..., x_M(N)) \tag{21}$$

subject to the constraints of Equation (20) and

$$u_{j,min}(n) \le u_j(n) \le u_{j,max}(n) \tag{22}$$

with $j=1,2,...,g$ and $M < m$.

If this problem is solved by dynamic programming and if m is large, dimensionality difficulty would appear. However, since Equation (20) is linear, the closed form solution of Equation (20) can be obtained. Using this closed form solution instead of Equation (20) directly, the dimensionality of the problem can be reduced from m to M. The closed form solution of Equation (20) can be represented by

$$x(n) = \left[\prod_{s=0}^{n-1} A(n-s)\right] x(0) + \left[\prod_{\ell=0}^{n-1} A(n-\ell)\right] \sum_{j=s}^{n-1} \left[\left\{\prod_{j=s}^{n-1} A(n-j)\right\}^{-1} h(n-s)\right]$$

At the final stage N, the above equation becomes

$$x(N) = c + \sum_{s=0}^{N-1} K(N-s)\, h(N-s) \tag{23}$$

with

$$c = \prod_{s=0}^{N-1} A(N-s)\, x(0)$$

$$K(N-s) = \prod_{\ell=0}^{N-1} A(N-\ell) \left[\prod_{j=s}^{N-1} A(N-j)\right]^{-1}$$

Substituting Equation (23) into Equation (21), we have

$$H\left[c_1 + \sum_{s=0}^{N-1} \left[\sum_{j=1}^{m} k_{1j}(n-s) h_j(N-s)\right], ...,\; c_M + \sum_{s=0}^{N-1} \left[\sum_{j=1}^{m} k_{Mj}(n-s) h_j(N-s)\right]\right] \tag{24}$$

The problem now becomes the minimization of Equation (24) subject to the constraints of Equation (22) with the initial conditions $c_1, c_2, ..., c_M$.

Since the minimum value of H depends only on the values of $c_1, c_2, ..., c_M$ and the number of stages N, let us define

$$g_N(c_1, c_2, ..., c_M) = \min_{\{u(n)\}} H \tag{25}$$

where $\{u(n)\}$ denotes the sequence $u(1), u(2), ..., u(N)$. Applying the principle of optimality, we have

$$g_N(c_1, c_2, .., c_M) = \min_{\{u(n)\}} \left\{ g_{N-1}(c_1 + \right.$$

$$\sum_{j=1}^{m} k_{1j}(1)h_j(1),\ldots,C_M + \sum_{j=1}^{m} k_{Mj}(1)h_j(1))\Big],$$

$$N=2,3,\ldots,N \qquad (26)$$

For a one-stage process, we have

$$g_1(c_1,c_2,\ldots,c_M) = \min_{u(1)} \left\{ H\left[c_1 + \sum_{j=1}^{m} k_{ij}(1)h_j(1),\ldots,C_M + \sum_{j=1}^{m} k_{Mj}(1)h_j(1)\right]\right\} \qquad (27)$$

Now the problem is reduced to the computation and storage of functions of M variables. If M is small, the above scheme can be used to overcome the dimensionality difficulty of the problem. If M is large, further reduction in dimensionality to one can be achieved by introducing an additional variable as:

$$x_{m+1}(n) = H(x_1(n),x_2(n),\ldots,x_M(n)) \qquad (28)$$

Instead of Equation (21), the problem now becomes the minimization of x_{m+1}. If the transformation equations such as Equation (20) are nonlinear, quasilinearization can be used to linearize the nonlinear equation iteratively [2]. Thus, the dimensionality difficulty can be overcome by the combined use of dynamic programming and quasilinearization.

3.5 Invariant Imbedding

Invariant imbedding can also be used to overcome the two-point boundary value problem in the classical approach. In fact, dynamic programming is essentially invariant imbedding with the addition of a maximum or minimum operation.

Only the essential equations will be introduced. For a detailed treatment, the reader is referred to the literature [2]. For simplicity, instead of using Equations (1) and (12), let us consider the following scalar equations

$$\frac{dx}{dt} = f(x,y,t),\ x(0) = C$$
$$\frac{dy}{dt} = g(x,y,t),\ y(T) = 0 \qquad (29)$$

with $0 \le t \le T$. Let us define

$$\gamma(c,a) = \begin{cases}\text{the missing initial condition for} \\ \text{Equation (29) where the process} \\ \text{begins at } t=a \text{ with } x(a)=c\end{cases}$$

Thus $\quad y(a) = r(c,a) \qquad (30)$

Using the Taylor series and after some manipulation, we have

$$f(c,r(c,a),a)\,\frac{\partial r(c,a)}{\partial c} + \frac{\partial r(c,a)}{\partial a} = g(c,r(c,a),a) \qquad (31)$$

with initial condition

$$r(c,t_f) = 0 \qquad (32)$$

Thus, the two-point boundary value represented by Equation (29) reduced to an initial value problem represented by Equations (31) and (32) where r represents the missing initial condition.

Notice that Equation (31) is essentially the functional dynamic programming equation without maximization or minimization. Thus, Equation (31) also has the dimensionality difficulty if the numbers of state variables are large. However, since there is no maximization or minimization, this dimensionality difficulty is much easier to overcome.

3.6 Stochastic Approximation

Stochastic approximation can also be used to solve deterministic two-point boundary value problems. This approach was originally developed by Robbins and Monro [6] for solving stochastic problems. Recently, Sugiyama and Lee [7,8] have shown that this approach is also a very powerful tool for solving deterministic problems. In fact, it is better than the Newton's method because of its generality and no differentiation is required. Only a short summary will be given here. For a detailed discussion, the reader is referred to the literature [6-8].

To solve the two-point boundary value problem, consider the problem

$$y'' = g(y',y,t) \qquad (33)$$

with boundary conditions

$$y(0) = c,\ y(1) = b \qquad (34)$$

the missing initial condition can be obtained by the iterative algorithm [7,8]

$$u_{n+1} = u_n + a_n[b-y(1|u_n)] \qquad (35)$$

where u_n is the nth approximation of the missing initial condition $y'(0)$, or

$$y'(0) = \lim_{n\to\infty} u_n$$

if the process converges. Notice that $y(1|u_n)$ is a conditional expression and is equal to the value of y at x=1 provided $y'(0)=u_n$. Thus, if the correct value of $y'(0)$ is used, then $y(1|u_n)$ is equal to the given final condition.

Notice that no differentials are needed in the above equations. Another advantage is that convergence can be proved for a general unknown function.

4. NUMERICAL EXAMPLES

To illustrate the advantages and disadvantages of the various approximation and iteration techniques, some simple numerical examples will be solved.

4.1 Inventory and Advertising Scheduling

Let the production at time t be

$$p(t) = a+bt \tag{36}$$

If $Q(t)$ represents the number of customers at time t and c_q represents the number of items bought by each customer, the rate of change of inventory is

$$\frac{dI(t)}{dt} = p(t) - c_q\, Q(t) = f_1 \tag{37}$$

Using the diffusion equation to represent advertising and the transformation of information, the change of informed customer, $Q(t)$, with time can be represented by

$$\frac{dQ(t)}{dt} = (c+u(t))Q(t)(1- \frac{Q(t)}{L}) = f_2 \tag{38}$$

where L represents the total number of persons, c represents the natural contact coefficient and u is the increase of this contact coefficient due to advertising. We wish to maximize the net profit

$$\phi = \int_0^T [Fc_q\, Q(t)-C_I(c_i-I(t))^2-c_u u^2 Q(t)]dt \tag{39}$$

where F is the revenue from the sale of one unit of product, C_I is the inventory carrying cost, c_c the desired inventory. Introducing an additional variable $x(t)$ to represent the cumulation of profit until time t, the transformation equation for x can be written as

$$\frac{dx}{dt} = Fc_q\, Q(t) - C_I(c_c-I(t))^2-c_u u^2 Q(t) = f_3 \tag{40}$$

with initial condition $x(0)=0$. The problem now becomes the maximization of $x(T)$ subject to the constraints of Equations (36)-(38) and (40) with u as the control variable.

4.2 Production and Advertising Scheduling

Instead of a fixed production rate, let us assume that production can be controlled. Let us assume that raw material is fed into two chemical reactors in series from which the desired product B is obtained. The raw material is a mixture of A,B, and C. The reactions in the reactors are $A \rightarrow B \rightarrow C$ and can be represented by

$$V_1 \frac{dx_1}{dt} = q(x_0-x_1) - V_1 k_{a1} x_1 \tag{41}$$

$$V_1 \frac{dy_1}{dt} = q(y_0-y_1) - V_1 k_{b1} y_1 + V_1 k_{a1} x_1 \tag{42}$$

$$V_2 \frac{dx_2}{dt} = q(x_1-x_2) - V_2 k_{a2} x_2 \tag{43}$$

$$V_2 \frac{dy_2}{dt} = q(y_1-y_2) - V_2 k_{b2} y_2 + V_2 K_{a2} x_2 \tag{44}$$

where x and y are concentrations of A and B, respectively V represents the volume of the reactor. Subscripts 1 and 2 represent reactors 1 and 2, respectively, q is the flow rate. The reaction rate constants, k, have the usual Arrhenices expression and hence are functions of the temperatures T_1 and T_2 in reactors 1 and 2, respectively. Let I represent the inventory of the desired product B, we have

$$\frac{dI}{dt} = q\, y_2 - c_q Q(t) \tag{45}$$

with initial condition $I(0) = I^0$. The equation to be maximized is

$$J = \int_0^T [c_1 c_q Q(t) + c_2\, Q\, x_2 + c_3\, q(1-x_2-y_2) - c_I(c_c - I(t)^2 - c_A\, u^2(t)\, Q(A) - c_T(cT_{1m}-T_1)^2 + (T_1-T_2)^2]\, dt \tag{46}$$

where c_1, c_2 and c_3 are the revenues from sales of a unit of product B, A, and C, respectively. Introducing an additional variable $x(t)$, we have

$$\frac{dx}{dt} = c_1 c_q Q(t) + c_2 q\, x_2 + c_3\, q(1-x_2-y_2) - c_I(I_m-I(t)^2 - c_A(a(t)-Q(t))^2 - c_T[T_{1m}-T_1)^2 + (T_1-T_2)^2] \tag{47}$$

with the initial condition $x(0)=0$. The problem now reduces to find the temperature profiles in reactors 1 and 2 and the advertisement rate in order that the profit $x(T)$ is maximized subject to the constraints of Equations (38), (41)-(45) and (47). This problem has 7 state variables x_1, x_2, y_1, y_2, I, Q, and x and 3 control variables T_1, T_2 and $u(t)$.

4.3 Inventory and Advertising Scheduling by Functional Gradient Techniques.

Recurrence equations similar to Equations (5)-(10) can be obtained for the system represented by Equations (36)-(40). For example, the recurrence relation for the gradient of the objective function with respect to the control for the present system becomes

$$\left.\frac{\partial s}{\partial u}\right|_t = \left[\left.\frac{\partial s}{\partial Q}\right|_{t+\Delta t}(Q - \frac{Q^2}{L}) - \left.\frac{\partial s}{\partial x}\right|_{t+\Delta t}(2uQ(u)\right]\Delta 0t$$

Using these recurrence relations and the performance equations given by Equations (36)-(40), this problem is solved with the following numerical constants.

$$\begin{array}{ll} C_c = 50 & b = 100 \\ C_I = 0.15 & c = 2 \\ C_u = 1.5 & F = 10.0 \\ L = 150 & C_q = 1.0 \\ a = 70 & \Delta t = 0.01 \\ T = 1 & \end{array} \tag{48}$$

The value of ΔJ used were $\Delta J=40$ for the first 17 iterations and $\Delta J=0.05$ for the remaining iterations. The initial conditions used were: $I(0)=Q(0)=20$, $x(0)=0$.

With an initial guess of the control variable, advertising, $u(t)=2.5$ for all values of t, the control variable converged to the optimal profile in 80 iterations. The profit function had a value of 25.0 with the assumed control profile and converged to the optimal of 584.2 after 83 iterations. As expected, the rate of convergence is very fast during the initial stages of iteration and becomes very slow during the final stages.

4.4 Inventory and Advertising Scheduling by Second Variational Technique

Introducing a three-dimensional adjoint vector λ and formulating the Hamiltonian function for Equations (37), (38) and (40), we have

$$H = \sum_{i=1}^{3} \lambda_i f_i \tag{49}$$

The differential equations governing the adjoint vector are

$$\frac{d\lambda_1}{dt} = - 2\lambda_3 C_I(C_c - I(t) \tag{50}$$

$$\frac{d\lambda_2}{dt} = C_q\lambda_1 - \lambda_2[c+u(t)][1- \frac{2Q(t)}{L}] - \lambda_3[FC_q-C_u{}^2(t)] \tag{51}$$

$$\frac{d\lambda_3}{dt} = 0 \tag{52}$$

with final conditions:

$$\lambda_1(T) = 0$$
$$\lambda_2(T) = 0$$
$$\lambda_3(T) = 1$$

Differentiating Equation (49), the partial derivatives of the Hamiltonian can be obtained as

$$\frac{\partial H}{\partial u} = \lambda_2[Q(t) - \frac{Q^2(t)}{L}] - 2\lambda_3 C_u u(t)Q(t) \tag{53}$$

$$\frac{\partial 2_H}{\partial u^2} = -2\ \lambda_3\ C_u Q(t) \tag{54}$$

$$H_{ux} = \begin{bmatrix} 0 \\ \lambda_2[1- \frac{2Q(t)}{L}]\ -2\lambda_3 C_u u(t) \\ 0 \end{bmatrix} \tag{55}$$

$$H_{u\lambda} = \begin{bmatrix} 0 \\ Q(t)\ \ [1- \frac{Q(t)}{L}] \\ -2C_u u(t)Q(t) \end{bmatrix} \tag{56}$$

Using Equations (53)-(56), improved controls can be computed from Equation (15).

Using the same numerical constants as that listed in Equation (48), the above problem was solved. The initial control variable profile used was obtained from the 15th iteration of the functional gradient results. Runge-Kutta integration scheme with step size 0.01 was used to integrate the equations. Seven additional iterations were needed to obtain the optimal.

4.5 Inventory and Advertising Scheduling by the Conjugate Gradient Technique.

For the conjugate gradient approach, one additional equation needed is the gradient direction of the Hamiltonian which is defined by Equation (19). Obviously, this direction is the same as Equation (53). Using the procedure discussed earlier, this problem is solved. The initial nominal control profile assumed was $u(t)=2.5$ for all t. The numerical values used are listed in Equation (48). The problem converges very fast even though a very rough initial nominal control profile was used. Only seven iterations are needed to obtain the optimal. The optimal profit obtained was 584.18.

4.6 Production and Advertising Scheduling by Functional Gradient Technique.

The recurrence equations for this problem are very long, we shall not list them here. These equations can be obtained in the same way as that discussed before. These procedures are essentially partial differentiation of Equations (38), (41)-(45), and (47). The improved control can be obtained from

$$\theta_{j,new}(t) = \theta_{j,old}(t) + \frac{\Delta\phi_j \left.\frac{\partial S}{\partial \theta_j}\right|_t}{\left\| \sum_{t=0}^{T} \left(\frac{\partial S}{\partial \theta_j}\right)^2 \right\|_2}, \quad j=1,2,3 \tag{57}$$

where $\Delta\phi_j$ is the desired improvement in the objective function due to the jth control. The numerical values used are

R	2.0	N	100.0
q	60.0	C_1	5.0
V_1	12.0	C_2	0.0
G_a	0.535×10^{11}	C_3	0.0
E_a	18000.0	C_I	1.0
V_2	12.0	C_A	0.01
G_b	0.461×10^{18}	C_T	0.0005
E_b	30000.0	x_0	0.53
C_q	1.0	y_0	0.43
C	1.0	T_{1m}	340.0^0K
		I_m	20.0

The initial conditions for the seven state variables are

$$x_1(0) = x_2(0) = 0.53$$

$$y_1(0) = y_2(0) = 0.43$$

$$I(0) = 8.0, \quad Q(0) = 1.0, \quad x(0) = 0$$

The initial assumed profiles for the three control variables are: $T_1(t) = T_2(t) = 345^0$K, $u(t) = 3.0$ for all $0 \le t \le 1$. A maximum profit of 66.05 was obtained after 260 iterations. The convergence rates of T_1 and T_2 are shown in Figures 1 and 2, respectively.

4.7 Production and Advertising Scheduling by the Second Variational Technique

Define the Hamiltonian function as

$$H(x,\lambda,t) = \sum_{i=1}^{7} \lambda_i f_i \tag{59}$$

where f_i, i=1,2,...,7 corresponds to the right-hand side of Equations (38), (41)-(45) and (47). The seven differential equationsgoverning the seven adjoint variables, λ, can be obtained easily. The first and second order partial derivatives of the Hamiltonian also can be obtained by differentiation. These equations are very long and will not be listed here. Using the adjoint equations and the partial derivatives of the Hamiltonian, the improved control vector can be obtained from

$$\begin{bmatrix} T_1 \\ T_2 \\ u \end{bmatrix}^{k+1} = \begin{bmatrix} T_1 \\ T_2 \\ u \end{bmatrix}^{k} - \begin{bmatrix} \eta_1 \\ \eta_2 \\ \eta_3 \end{bmatrix} [H^k_{TT}]^{-1} [H^k_T + H^k_{TX} (x_{k+1} - x_k) + H^k_{T\lambda}(\lambda_{k+1} - \lambda_k)] \tag{60}$$

where η is a three-dimensional vector which represents the step size of the three controls. Attempt was made to solve this problem by using the numerical values listed in Equation (58) and by using the following initial guessed nominal control profiles: $T_1(t) = T_2(t) = 345$, $u(t) = 2.5$, $0 \le t \le 1$. Depending on the values chosen for the step size η, the problem either diverges or converges extremely slowly. The value used for the first two step sizes is $\eta_1 = \eta_2 = 0.5$. The problem diverges if η_3 is above 0.003. With $\eta_3 = 0.003$, the problem was allowed to iterate for 730 iterations. The control profile fluctuates widely during the during the first fifty iterations.

4.8 Production and Advertising Scheduling by the Conjugate Gradient Technique

Using the procedure discussed earlier and the numerical values listed in Equation (58), this problem was solved by the conjugate gradient technique. The initially guessed nominal control variables are $T_1(t) = 345^0$k, $u(t)=3.0$ for $0 \le t \le 1$. The problem converged to optimum in 34 iterations. Notice the very approximate nature of the initial nominal controls. The optimal profit obtained was 66.022.

In order to obtain the optimal step sizes in step 5 of the procedure, a three-dimensional random search procedure was used. In performing this random search, the correct range of

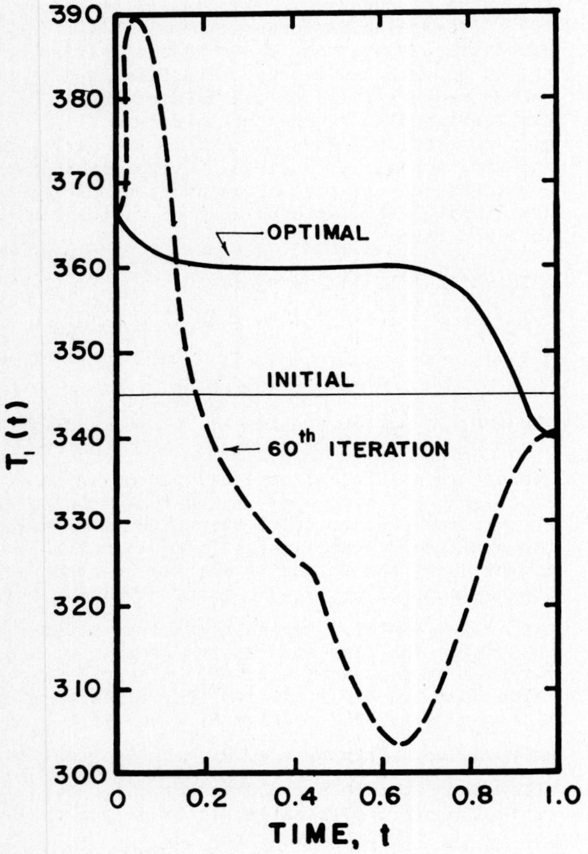

Figure 1. Convergence rate of T_1 by functional gradient technique

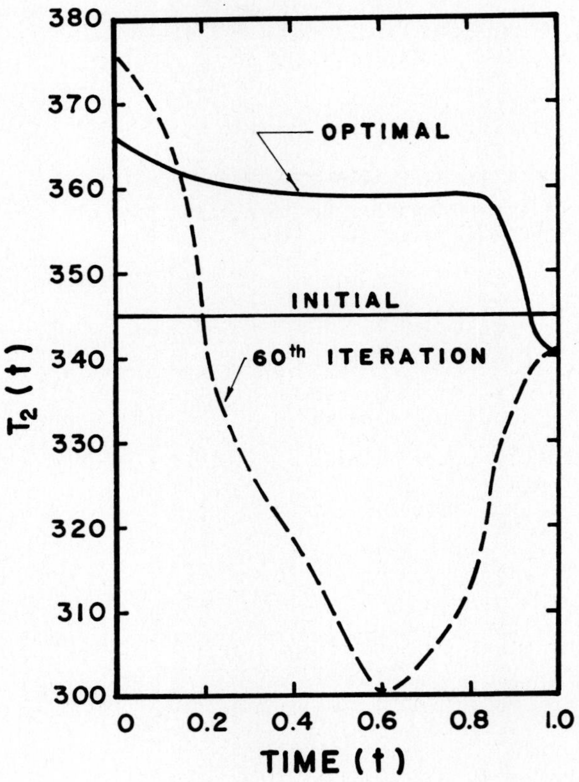

Figure 2. Convergence rate of T_2 by functional gradient technique

the random number used is very important. Another important factor is the number of random search for each iteration. If a large number of random search is used, the optimal step size obtained would be accurate but at the expense of more computational effort. Thirty random searches per iterations were used for the first 28 iterations and 100 were used for the remaining iterations. The actual convergence rates for T_1, T_2 and advertising are shown in Figures 3,4, and 5 respectively.

4.9 Inventory and Advertising scheduling by Quasilinearization and Dynamic Programming

Equations (37), (38) and (40) can be discretized into difference equations

$$I(n) - I(n+1) + [p(n) - C_q Q(n)]\Delta t \qquad (61)$$

$$Q(n) = Q(n+1) + Q(n+1)\ [C+u(n)][1-\frac{Q(n)}{L}\Delta t \qquad (62)$$

$$x(n) = x(n+1) + [FC_q Q(n) - C_I(C_c - I(n))^2$$

$$-C_u u(n)^2 Q(n)]\Delta t \qquad (63)$$

with $N\Delta t = T$. In the above equations, we have numbered the stage backward. This is a three-dimensional problem which can be reduced to a one-dimensional problem as discussed earlier. For this problem, the function to be maximized becomes

$$H(x(N) = x(N) = C_3 + \sum_{S=0}^{N-1} [\sum_{j=1}^{3} k_{3j}$$

$$(N-S)hj(N-S)] \qquad (64)$$

Since, the maximum of Equation (64) depends only on C_3, a one-dimensional functional equation in dynamic programming can be obtained. This problem was solved as a one dimensional problem using the combined procedure. The numerical values used are the same as that listed in Equation (48). In addition, the following inequality constraint is assumed for the control variable: $u(n) \le 6$, $n=1,2,\ldots,N$. The initially guessed values for the control for step (1) are $u(n)=2$, $n=1,2,\ldots,N$.

This problem was solved for N=5, 10, 15 and 20. The convergence rate for N=5 is shown in Table 1. The convergence rates for N=10,15, and 20 are about the same as that shown in Table 1. Notice that fairly accurate results were obtained in only two iterations in spite of the very approximate nature of the initially guessed control profile. The maximum value of ϕ obtained are 620.5, 667.0 and 691.0 for N=10,15, and 20, respectively.

4.10 Stochastic Approximation and Boundary Value Problems

To illustrate the approach, consider the boundary value problem

$$y'' + (1-2/t^2)y=0 \qquad (65)$$

with boundary conditions

$$y(1) = 0.5061 \text{ and } y(1.5) = 1.0 \qquad (66)$$

Equation (35) becomes

$$u_{n+1} = u_n + a_n\{1-y(1.5|u_n)\}. \qquad (67)$$

Table 1

Convergence Rate with N = 5

Iter.	u(5)	u(4)	u(3)	u(2)	u(1)	ϕ
0	2.0	2.0	2.0	2.0	2.0	323.8
1	6.0	6.0	0.6	0	0	498.9
2	6.0	6.0	0.2	0	0	500.6
3	6.0	6.0	0.8	0	0	497.9
4	6.0	6.0	0.6	0	0	498.9
5	6.0	6.0	0.6	0	0	498.9
6	6.0	6.0	0.11	0	0	501.0
7	6.0	6.0	0.0	0	0	501.3
8	6.0	6.0	0.0	0	0	501.3

With $a_n = 1.0$ for all n, this problem is solved. The convergence rate is listed in Table 2.

Table 2

n	u_n	$y(1.5\|u_n)$
1	1.0	1.0452
2	0.954	1.0224
3	0.932	1.0110
4	0.921	1.0054
5	0.916	1.0027
6	0.913	1.0013
10	0.911	1.0001
11	0.911	1.0000

The fourth order Runge-Kutta technique is used to integrate Equation (65).

5. *COMPARISONS AND DISCUSSIONS*

The six numerical techniques discussed can be classified into two categories: the iterative approach and the invariant imbedding approach. The invariant imbedding approach contains no iteration and it wolves the problem once for all. The accuracy of the solution depends only on the step size. The other five approaches are all iterative techniques.

The two most important properties of an iterative technique are (1) Whether the procedure converges or not and (2) the convergence rate. For most practical problems, the first question

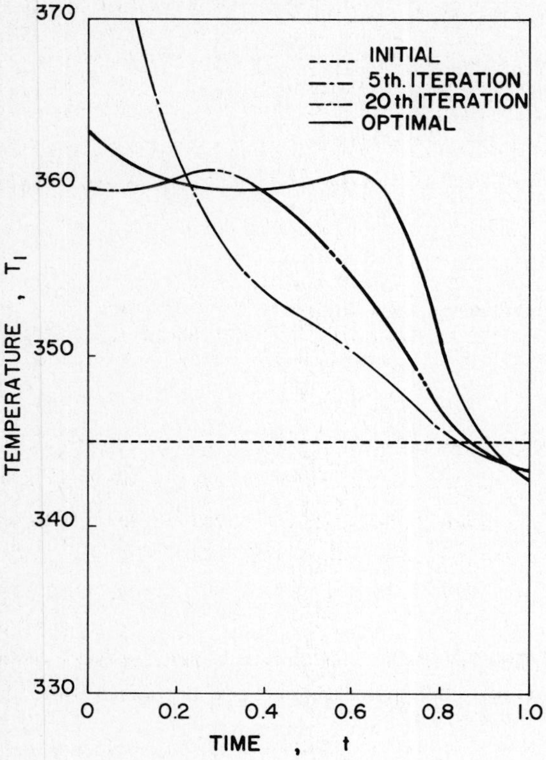

Figure 3. Convergence rate of T_1 by conjugate gradient technique

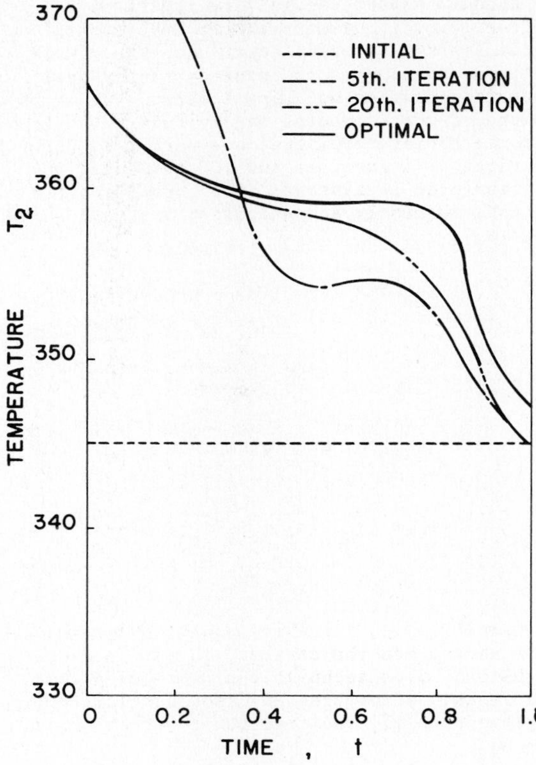

Figure 4. Convergence rate of T_2 by conjugate gradient technique

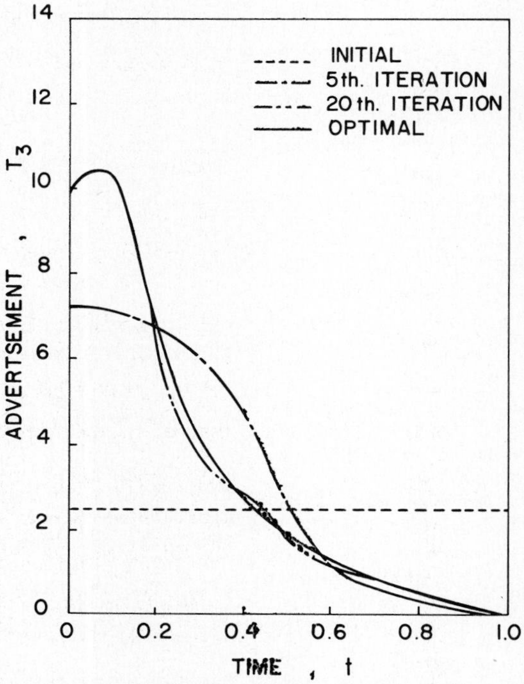

Figure 5. Convergence rate of advertisement by conjugate gradient technique.

cannot be answered easily. It is true that some techniques are more stable and thus converges for most problems while other techniques are less stable and frequently diverge. Since convergence can seldom be guaranteed, the convergence rate becomes an important consideration in choosing an iterative technique. In fact, if the rate of convergence or divergence is fast, the first question can be answered easily and quickly by numerical means.

Techniques such as the functional gradient technique are known to have first order convergence properties. The second variation, conjugate gradient and quasilinearization have the second order or quadratic convergence property. But, sometimes the quadratic convergence property cannot be obtained due to approximations. By quadratic convergence, we mean that the error of the (n+1)st iteration is proportional to the square of the error of the nth iteration. For first order convergence, the error of the (n+1)st iteration is only proportional to the error of the nth iteration. Thus, second order convergence is a very desirable property.

Table 3 compares the convergence rates of some of the iterative techniques based on the numerical results. As expected, the first order convergence technique, functional gradient

Table 3. Number of Iterations to Reach Optimum

Technique	Inventory and Advertising Problem	Production and Advertising Problem
Functional gradient	83	260
Second variation	7	unstable
Conjugate gradient	7	34
Quasilinearization	4	--

converges the slowest and quasilinearization which is essentially a generalized Newton's method, converges the quickest. Second variational techniques appear to be unstable for large problems. These conclusions are in general agreement with past experiences.

Another important factor controlling the desirability of a technique is the ease of use. From this standpoint, the stochastic approximation is most desirable among the iterative techniques. Notice that both the functional gradient technique and quasilinearization require the first order derivatives which can be very tedious to obtain for large problems. This is especially true since there exist no accurate numerical techniques to compute the first derivatives. This problem is even more severe for the second variational technique which requires both the first and the second order derivatives and thus this technique is the most tedious to use. Although the conjugate gradient technique needs the conjugate gradient, it does not cause much more tedious work. This is because of the fact that the conjugate gradient direction is generated iteratively. However, for most of the versions of the conjugate technique, the first order derivative is needed. Thus, stochastic approximation is the simplest and easiest to use and it does not require any derivatives. Furthermore, the function concerned even can be an unknown function as long as experiments or observations can be obtained from this unknown function.

The noniterative invariant imbedding technique is certainly very easy to use and to program. It is a simple recurrence relation. The accuracy of the results can be increased by simply increasing the number of steps or decreasing the step size. Since there is no iteration involved, there does not exist the uncertainty of convergence or divergence. With the increasing availability of super computers, the dimensionality difficulty of invariant imbedding can also be overcome. This is especially true because invariant imbedding does not have the maximization or minimization operation as in dynamic programming.

REFERENCES

[1] Bliss, G. A., Lectures on the Calculus of Variations, (University of Chicago Press, 1946).

[2] Lee, E. S., Quasilinearization and Invariant Imbedding, (Academic Press, 1968).

[3] Lee, E. S., Optimization by a Gradient Technique, in: IEC Fundamentals, 3, (1964) p. 4.

[4] Padmanabhan, L. and S. G. Bankoff, A Direct Second Variational Method, in: Int. J. Control, 9, (1969) p. 167.

[5] Lasdon, L. S., S. K. Mitter and A. D. Warren, The Conjugate Gradient Method for Optimal Control, in: IEEE Trans. Automatic Control, AC-12, (1967) p. 132.

[6] Robbins, H. and S. Monro, A Stochastic Approximation Method, in: Ann. Math. Statist., 22, (1951) p. 400.

[7] Sugiyama, H. and E. S. Lee, Solving Deterministic Problems by Stochastic Approximation, in: J. Math. Analysis Appl., 138, (1989) p. 569.

[8] Sugiyama, H. and E. S. Lee, Stochastic Approximation--A Powerful Method for Solving Deterministic Numerical Problems, in: Computers Math. with Appls., 15, (1988) p. 963.

I
APPROXIMATION
Theory and Techniques

Approximation, Optimization and Computing:
Theory and Applications, A.G. Law and C.L. Wang (eds.)
Elsevier Science Publishers B.V. (North-Holland)

Some Pál Type Interpolation Problems

M.R. Alhlaghi and A. Sharma

1. Introduction. Let $\{x_k\}_{k=1}^n$ denote the zeros of the polynomial $\pi_n(x) = (1-x^2)P'_{n-1}(x)$, arranged in increasing order where $P_n(x)$ is the Legendre polynomial of degree n with normalization $P_n(1) = 1$. If $\{\xi_k\}_{k=1}^{n-1}$ are the zeros of $\pi'_n(x)$, we can formulate the following closely related interpolation problems:

Problem A. Does there exist a unique polynomial $P(x)$ of degree $\leq 2n-2$ such that

$$(1.1)\qquad \begin{aligned} &P(x_k) = a_k, \quad k = 1, \ldots, n; \\ &P^{(\nu)}(\xi_k) = b_k, \quad k = 1, \ldots, n-1 \end{aligned}$$

where $\{a_k\}_{k=1}^n$ and $\{b_k\}_{k=1}^n$ are arbitrary given real or complex numbers and ν $(1 \leq \nu \leq n-1)$ is a given integer? If $P(x)$ is unique what are the fundamental polynomials of interpolation?

When $\nu \geq n$, the problem is not regular (i.e., there does not exist a unique $P(x)$) because then the Polyá condition (see [4]) is not satisfied. When $\nu = 1$, it is clear that the above problem is not regular for any $n \geq 2$. The unique existence of $P(x)$ in Problem A can be proved if we require $P(x)$ to be of degree $2n-1$ and add to (1.1) the requirement that $P(a) = 0$ where $a \neq x_k$ $(k = 1, \ldots, n)$ is a given real number. This was done by Pál [5] and later the convergence problem was studied by Eneduanya [3].

A problem analogous to Problem A can be obtained by interchanging x_k with ξ_k in (1.1). More precisely, we have

Problem B. Does there exist a unique polynomial $P(x)$ of degree $\leq 2n-2$ such that

$$(1.2)\qquad \begin{aligned} &P(\xi_k) = a_k, \quad k = 1, \ldots, n-1 \\ &P^{(\nu)}(x_k) = b_k, \quad k = 1, \ldots, n, \end{aligned}$$

for arbitrary given sequences $\{a_k\}$ and $\{b_k\}$ and a given integer ν $(1 \leq \nu \leq n-2)$? If the problem is regular, find the fundamental polynomials.

Here Polyá condition requires that $\nu \leq n-2$. Recently L. Szili [7] showed that Problem B is regular for $\nu = 1$ only when n is even.

To put the problems in a historical perspective, we observe that J. Balazs and P. Turán [2] initiated the problem of (0,2) interpolation on the zeros of $\pi_n(x)$ in 1957 and proved that it is regular only when n is even. To distinguish this problem from Problems A and B, we use the notation (0;2) where the semicolon indicates that values are prescribed on one set and second derivatives on another set of nodes.

The object of this note is to discuss the regularity of the problems A and B when $\nu = 2$. It will be shown that when $\nu = 2$, both problems are regular irrespective of the parity of n. This contrasts with the situation when $\nu = 1$ discussed by Pál [5] and Szili [7]. In Section 2 we state the preliminaries and the main results. In Section 3, we sketch the proofs of the two theorems. Sections 4 and 5 are devoted to the fundamental polynomials and how to find them.

2. Preliminaries and Main Results. We recall (see [6]) that $\pi_n(x)$ satisfies the differential equation

$$(2.1)\qquad (1-x^2)y'' + n(n-1)y = 0,$$

with $\pi'_n(1) = -n(n-1) = (-1)^{n+1}\pi'_n(-1)$. We can verify that

$$(2.2)\qquad \int_{-1}^{1}(1-x^2)P'_{k'-1}(t)P'_{j-1}(t)dt = \frac{2k(k-1)}{2k-1}\delta_{jk},$$

and

$$(2.3)\qquad \int_{-1}^{1}(1-t)(P'_{n-1}(t) - P'_{n-1}(-1))P'_{k-1}(t)dt = (n-k)(n+k-1)(-1)^{n+k-1}.$$

We formulate the following lemma which is easy to prove:

Lemma 1. *If $g(x)$ is a polynomial of degree $\leq m$ and if $L_g(x)$ denotes the linear function interpolating $g(x)$ at ± 1, then the only polynomial solution of the differential equation*

$$(2.4) \qquad (1-x^2)y'' - n(n-1)y = g(x)$$

is given by

(2.5)

$$y = -\frac{1}{n(n-1)}L_g(x) + \int_{-1}^{1}(f(t) - L_g(t))K(x,t)dt$$

where

$$(2.6) \qquad K(x,t) = -\sum_{k=2}^{m}\frac{(2k-1)\pi_k(x)P'_{k-1}(t)}{2k(k-1)\lambda_{n,k}}.$$

In particular if $g(x) = cx + D + \sum_{j=2}^{m} A_j\pi_j(x)$, *then*

$$(2.7) \qquad y = -\frac{cx+D}{n(n-1)} - \sum_{j=2}^{m}\frac{A_j}{\lambda_{n,j}}\pi_j(x)$$

where $\lambda_{n,j} = j(j-1) + n(n-1)$.

We shall prove

THEOREM 1. *For* $\nu = 2$, *Problem A is regular when* $\{x_k\}_1^n$ *are the zeros of* $\pi_n(n)$ *and* $\{\xi_k\}_1^{n-1}$ *are the zeros of* $\pi'_n(x)$.

THEOREM 2. *For* $\nu = 2$, *Problem B is regular when the nodes* $\{x_k\}_1^n$, *and* $\{\xi_k\}_1^{n-1}$ *are as in Theorem 1.*

For the proof of Theorem 2, we shall need

LEMMA 2. *If* $P(x) \in \pi_{2n-2}$ *and if*

$$(2.8) \qquad \begin{cases} P(\xi_k) = 0, \ k = 1, \ldots, n-1; \\ P(-1) = P(1) = 0 \\ P''(x_k) = 0, \ k = 2, \ldots, n-1. \end{cases}$$

then $P(x)$ *is identically zero.*

PROOF: From the first set of conditions in (2.5) we may set $P(x) = \pi'_n(x)q(x)$ where $q(x) \in \pi_{n-1}$. Since $P(-1) = P(1) = 0$, it follows that $q(-1) = q(1) = 0$. From the last set of $n-2$ conditions in (2.8), we can see easily (since $q(-1) = q(1) = 0$) that

$$(1-x_k^2)q''(x_k) - n(n-1)q(x_k) = 0, \ k = 1, \ldots, n-1.$$

Since $q(x) \in \pi_{n-1}$, we obtain

$$(1-x^2)q''(x) - n(n-1)q(x) = 0$$

whence we get $q(x) \equiv 0$ by Lemma 1. □

We shall now determine the polynomial $V_1(x)$ of degree $2n-2$ which satisfies the following conditions:

$$(2.9) \qquad \begin{cases} V_1(\xi_k) = 0, \quad k = 1, \ldots, n-1 \\ V_1''(x_k) = 0, \quad k = 2, \ldots, n-1 \\ V_1(-1) = 1, \ V_1(1) = 0. \end{cases}$$

LEMMA 3. *The polynomial* $V_1(x)$ *satisfying (2.9) is given by*

$$(2.10) \ V_1(x) = \frac{\pi'_n(x)}{2\pi'_n(-1)}\Big[1 - x - \sum_{k=2}^{n-1}(-1)^k d_k\pi_k(x)\Big]$$

where $d_k = \frac{(n-k)(n+k-1)(2k-1)}{k(k-1)\lambda_{n,k}}$.

PROOF: In view of (2.9), we set

$$V_1(x) = \frac{(1-x)\pi'_n(x)}{2\pi'_n(-1)} + \pi'_n(x)q_1(x),$$

$$(2.11) \qquad q_1(x) \in \pi_{n-1}$$

where $q_1(1) = q_1(-1) = 0$. From the second set of conditions on $V_1(x)$, after some simplification, we see that $q_1(x)$ must satisfy the following differential equation:

$$(1-x^2)q_1''(x) - n(n-1)q_1(x) = -\frac{(1-x)(P'_{n-1}(x) - P'_{n-1}(-1))}{n(n-1)}.$$

From (2.2) and (2.3), it is easy to see that

$$(1-x)(P'_{n-1}(x) - P'_{n-1}(-1)) = \sum_{j=2}^{n-1}A_j\pi_j(x),$$

where $A_k = \frac{(2k-1)(n-k)(n+k-1)(-1)^{n+k}}{2k(k-1)}$. The result now follows from Lemma 1 and (2.11). □

Because the zeros of $\pi_n(x)$ and also those of $\pi'_n(x)$ are symmetric about the origin, the polynomial $V_n(x)$ of degree $2n-2$ given by $V_n(x) = V_1(-x)$ will satisfy all the conditions of (2.9) except the last two which become $V_n(1) = 1, \ V_n(-1) = 0$.

3. Proof of Theorem 1. We shall show that the only polynomial $P(x)$ of degree $\leq 2n-2$ satisfying the homogeneous interpolation problem A is identically zero. Clearly then we may set $P(x) = \pi_n(x)q(x)$ where $q(x) \in \pi_{n-2}$. Since we require $P''(\xi_k) = 0, \ k = 1, \ldots, n-1$, it follows after using (2.1) that

$$(1-\xi_k^2)q''(\xi_k) - n(n-1)q(\xi_k) = 0, \quad k = 1, \ldots, n-1.$$

Since $q(x) \in \pi_{n-2}$, we get

$$(1-x^2)q''(x) - n(n-1)q(x) \equiv 0$$

so that from Lemma 1, $q(x) \equiv 0$, which completes the proof. □

Proof of Theorem 2. We shall show that if $P(x)$ of degree $\le 2n-2$ satisfies $P(\xi_k) = 0$, $k = 1, \ldots, n-1$ and $P''(x_k) = 0$, $k = 1, \ldots, n$, then $P(x)$ is identically zero. By Lemma 3, we set

$$P(x) = AV_1(x) + BV_n(x).$$

Then $P(\xi_k) = 0$, $k = 1, \ldots, n-1$ and $P''(x_k) = 0$, $k = 2, \ldots, n-1$. From $P''(-1) = P''(1) = 0$, we obtain

$$\text{(3.1)} \qquad \begin{cases} AV_1''(-1) + BV_n''(-1) = 0 \\ A_1V_1''(1) + BV_n''(1) = 0. \end{cases}$$

If Δ is the determinant of this system of equations, then since $V_n(x) = V_1(-x)$, we get

$$\Delta = \big(V_1''(-1) - V_1''(1)\big)\big(V_1''(-1) + V_1''(1)\big).$$

From (2.10), after some elementary calculations we obtain

$$V_1''(-1) = \frac{n(n-1)}{4}\Big[\frac{n^2-n+2}{2} + \sum_{k=2}^{n-1} \alpha_{nk}\Big]$$

$$V_1''(1) = \frac{(-1)^n n(n-1)}{4}\Big[2 - \sum_{k=2}^{n-1} (-1)^k \alpha_{nk}\Big]$$

where

$$\alpha_{nk} := \frac{(2k-1)(n-k)(n+k-1)}{n(n-1)\lambda_{nk}}(\lambda_{nk} + n(n-1)).$$

From this we see easily that $V_1''(-1) \pm V_1''(1)$ are both > 0. Then $\Delta > 0$ implies that $A = B = 0$ which completes the proof. □

4. Fundamental Polynomials for Problem A.

We shall denote the fundamental polynomials by $\{L_{k,0}(x)\}_{k=1}^{n}$ and $\{L_{k,2}(x)\}_{k=1}^{n-1}$. Then every polynomial $Q(x) \in \pi_{2n-2}$ has the unique representation

$$Q(x) = \sum_{k=1}^{n} Q(x_k)L_{k,0}(x) + \sum_{k=1}^{n-1} Q''(\xi_k)L_{k,2}(x).$$

(i) Polynomials $L_{k,2}(x)$, $(k = 1, \ldots, n-1)$.

Since $L_{k,2}(x) \in \pi_{2n-2}$ and since $L_{k,2}(x_j) = 0$, $j = 1, \ldots, n$ we have $L_{k,2}(x) = \pi_n(x)q_k(x)$, where $q_k(x) \in \pi_{n-2}$. From $L''_{k,2}(\xi_j) = \delta_{kj}$, $j = 1, \ldots, n-1$, we obtain the conditions

$$\text{(4.1)} \quad (1-\xi_j^2)q_k''(\xi_j) - n(n-1)q_k(\xi_j) = \frac{(1-\xi_j^2)}{\pi_n(\xi_j)}\delta_{kj}.$$

From (4.1) it follows that the polynomial $q_k(x)$ satisfies the equation

$$\text{(4.2)} \quad (1-x^2)q_k''(x) - n(n-1)q_k(x) = \frac{\ell_k^*(x)}{P'_{n-1}(\xi_k)}$$

where

$$\ell_k^*(x) = \frac{P_{n-1}(x)}{(x-\xi_k)P'_{n-1}(\xi_k)}.$$

If we set

$$\ell_k^*(x) = c_0 + c_1x + \sum_{k=2}^{n-2} c_j\pi_j(x),$$

where $c_0 = \frac{1}{2}(\ell_k^*(1) + \ell_k^*(-1))$, $c_1 = \dfrac{\ell_k^*(1) - \ell_k^*(-1)}{2}$, and c_j are determined on using (2.2). Then $q_k(x)$ is given by (2.7) so that

$$\text{(4.3)} \qquad L_{k,2}(x) = -\frac{\pi_n(x)}{P'_{n-1}(\xi_k)} \sum_{j=0}^{n-2} \frac{c_j\pi_j(x)}{\lambda_{n,j}},$$

where $\pi_0(x) = 1, \pi_1(x) = x$.

(ii) Polynomials $L_{k,0}(x), (k = 1, \ldots, n)$.

Since $L_{k,0}(x) \in \pi_{2n-2}$ and satisfies the conditions $L_{k,0}(x_j) = \delta_{kj}$, $j = 1, \ldots, n$ and $L''_{k,0}(\xi_j) = 0$, $j = 1, \ldots, n-1$, we set

$$\text{(4.4)} \quad L_{k,0}(x) = \ell_k(x) + \pi_n(x)q_k(x), \; q_k(x) \in \pi_{n-2}$$

where

$$\ell_k(x) = \frac{\pi_n(x)}{(x-x_k)\pi_n'(x_k)}.$$

Following the method in (i), we can derive the equations

$$\text{(4.5)} \quad \begin{aligned} &(1-\xi_j^2)q_k''(\xi_j) - n(n-1)q_k(\xi_j) \\ &\qquad = -\frac{(1-\xi_j^2)}{\pi_n(\xi_j)}\ell_k''(\xi_j) \\ &\qquad (j = 1, \ldots, n-1). \end{aligned}$$

For $2 \le k \le n-1$, we can write the right side of (4.5) as below:

$$\frac{1}{\pi_n'(x_k)}\Big[\frac{n^2-n+2}{\xi_j - x_k} + \frac{4x_k}{(\xi_j-x_k)^2} - \frac{2(1-x_k^2)}{(\xi_j-x_k)^3}\Big].$$

If we set

(4.6) $$Q_{\nu,k}(x) := \frac{P_{n-1}(x_k) - P_{n-1}(x)}{P_{n-1}(x_k)(x - x_k)^\nu}, \quad \nu = 1,2$$

and

$$Q_{3,k}(x) := \frac{Q_{1,k}(x)}{(x-x_k)^2} - \frac{n(n-1)}{2(1-x_k^2)} \cdot \frac{P_{n-1}(x)}{P_{n-1}(x_k)(x-x_k)},$$

then it can be verified that each of the polynomials $Q_{\nu,k}(x), (\nu = 1,2,3)$ is of degree $\leq n-2$ and that

$$Q_{\nu,k}(\xi_j) = \frac{1}{(\xi_j - x_k)^\nu}, \quad \nu = 1,2,3.$$

It follows that if $2 \leq k \leq n-1$, then $q_k(x)$ satisfies the differential equation

$$(1-x^2)q_k''(x) - n(n-1)q_k(x) = g_k(x)$$

where

$$g_k(x) = \frac{1}{\pi_n'(x_k)}\big[(n^2 - n + 2)Q_{1,k}(x) + 4x_k Q_{2,k}(x) - 2(1-x_k^2)Q_{3,k}(x)\big].$$

It is easy to see from the symmetry of the nodes about the origin that $L_{n,0}(x) = L_{1,0}(-x)$. These polynomials can be determined in a similar way with suitable modification of $Q_{2,1}(x)$. More precisely when $k = 1$,

$$Q_{2,1}(x) = [(-1)^n(P_{n-1}(x) - P_{n-1}(-1)) + (1+x)P_{n-1}'(-1)P_{n-1}(x)]/(1+x)^2$$

and

$$g_1(x) = \frac{1}{\pi_n'(-1)}\Big[(n^2 - n + 2)Q_{1,1}(x) - 4Q_{2,1}(x)\Big].$$

The value of $q_k(x)$ can now be obtained from Lemma 1. We spare the reader the gruesome details.

5. Fundamental Polynomials (Problem B).

In this case we use the device used in the proof of Theorem 2. This method has also been used successfully in another case earlier [1]. More precisely we first determine the fundamental polynomials of a modified form of Problem B. We shall denote the fundamental polynomials of the modified problem by $\{U_{k,0}(x)\}_{k=1}^{n-1}$, $\{U_{k,2}(x)\}_{k=2}^{n-1}$, $V_1(x)$ and $V_n(x)$. The polynomial $V_1(x)$ is given by Lemma 3 and $V_n(x) = V_1(-x)$. Thus every polynomial $Q(x) \in \pi_{2n-2}$ can be uniquely represented thus:

$$Q(x) = \sum_{k=1}^{n-1} Q(\xi_k)U_{k,0}(x) + \sum_{k=1}^{n} Q''(x_k)U_{k,2}(x) + Q(-1)V_1(x) + Q(1)V_n(x).$$

The explicit forms of $U_{k,0}(x)$, $U_{k,2}(x)$ can be obtained by the method used in Section 4. We can use these to find the basic polynomials of the Problem B which we denote by $\{\tilde{L}_{k,0}(x)\}_{k=1}^{n-1}$ and $\{\tilde{L}_{k,2}(x)\}_{k=1}^{n}$. It is easy to verify (as in [1]) that $\tilde{L}_{k,j}(x) =$

$$\frac{1}{\Delta}\begin{vmatrix} U_{k,j}(n) & V_1(x) & V_n(x) \\ U_{k,j}''(-1) & V_1''(-1) & V_n''(-1) \\ U_{k,j}''(1) & V_1''(1) & V_n''(1) \end{vmatrix}, \quad j = 0,2.$$

where $\Delta > 0$ is the determinant of the system (3.1).

References

[1] Akhlaghi, M.R., Chak, A.M. and Sharma, A., $(0,3)$ *Interpolation on the zeros of* $\pi_n(x)$, Rocky Mountain J. Math. (to appear).

[2] Balazs, J. and Turán, P., Notes on Interpolation III. Acta Math. Acad. Sci. Hungar. **8** (1957), 201–215.

[3] Eneduanya, S.A., *On the convergence of interpolation polynomials*, Analysis Math. **11** (1985), 13–22.

[4] Lorentz, G.G., Jetter, K. and Riemenschneider, S., Birkhoff Interpolation, in "Encyclopaedia of Math.", Addison-Wesley, 1983.

[5] Pál, I.G., *A new modification of the Hermite-Fejer interpolation*, Analysis Math. **1** (1975), 197–205.

[6] Szegö, G., Orthogonal polynomials, Amer. Math. Soc. Coll. Publ. (New York, 1959).

[7] Szili, L., *An interpolation process on the roots of the integrated Legendre polynomials*, Analysis Math. **9** (1983), 235–245.

Department of Mathematics
University of Alberta
Edmonton, Alberta, Canada T6G 2G1

Approximation, Optimization and Computing:
Theory and Applications, A.G. Law and C.L. Wang (eds.)
Elsevier Science Publishers B.V. (North-Holland)

ON A METHOD OF APPROXIMATION BY MEANS OF SPLINE FUNCTIONS (+)

Laura Gori N.Amati (*) **Elisabetta Santi** (**)

SUMMARY. We discuss the problem of approximating a function f on the interval [0,1] by a spline function of degree m, with n (variable) knots and defects k_i+1 $i=1(1)n$, matching as many of the initial moments of f as possible.
We can also impose additional constraints on the derivatives of the approximation at one end point of [0,1].
We show that, if the approximations exist, they can be represented in terms of Stancu-Lobatto and Stancu-Radau quadrature rules. We prove pointwise convergence as $n\to\infty$ for fixed $m\geq 2$.

1. INTRODUCTION

Recently some Authors [1,2,3,7,8] considered the problem of finding a spline with variable knots which reproduces, on an interval (finite or infinite), as many moments of a given function f as possible.

The practical motivation for this problem is discussed in [2] where the Author considers the approximation of spherically symmetric distributions in R^d, $d\geq 1$, by linear combinations of Heaviside step functions or Dirac delta functions.
Later, some moment preserving spline approximations on $[0,\infty)$ were obtained, by means of polynomial splines in [3] and polynomial splines with defect in [8]. In both cases, under suitable assumptions on f, it has been shown that the problem has an unique solution if and only if certain Gaussian quadrature rules exist corresponding to a moment functional depending on f. Pointwise convergence of this approximation process depends on a convergence property of the Gauss or Turán quadrature rules.
More recently the analogous problem was treated in [1] on a finite interval.
Finally, Micchelli [7] pointed out that a problem, somewhat more general than this one, can be conveniently related to questions concerning existence of suitable monosplines.
In this paper we discuss the problem of approximating a function f on the finite interval [0,1] by a spline function of degree $m\geq 2$, with n knots x_i, each of defects $k_i+1\leq m-1$, $i=1(1)n$. We consider two different aspects of the problem and we show that both can be related to Gauss-Stancu quadrature formulae with two or one fixed nodes at the extremes of [0,1]. The connection between pointwise convergence of approximating spline and the convergence property of the Gauss-Stancu quadrature rule is proved in Section 5.

2. SOME PRELIMINARY STATEMENTS

Let $I\equiv[0,1]$, and let

$$(2.1)\quad \Delta = \{x_i\}_{i=1}^{n}$$

with

$$(2.2)\quad 0 = x_0 < x_1 < x_2 < \dots < x_n < x_{n+1} = 1$$

be a partition of I into n+1 subinter-

(+) Work supported by Ministero della Pubblica Istruzione of Italy
(*) Department of "Metodi e Modelli per le Scienze Applicate", University of Rome "La Sapienza", Via Antonio Scarpa 10, 00161 Roma, ITALY
(**) Department of "Energetica", University of L'Aquila, Località Monteluco, 67040 Roio Poggio (L'Aquila), ITALY

vals

$$(2.3)\quad \begin{cases} I_i=[x_i,x_{i+1}) & i=0(1)n-1, \\ I_n=[x_n,x_{n+1}] . \end{cases}$$

Let $m\geq 2$ be a positive integer and let

$$(2.4)\quad M=(\sigma_1,\ \sigma_2,\ \dots,\ \sigma_n)$$

be a vector of integers, such that

$$(2.5)\quad 0\leq 2\sigma_i=k_i\leq m-1, \qquad i=1(1)n .$$

A spline function of degree $m\geq 2$, with n knots x_i $i=1(1)n$ in the interior of [0,1] and defects k_i+1, can be written in the form

$$(2.6)\quad S_{nm}(x) = p_m(x) + \sum_{i=1}^{n}\sum_{j=0}^{k_i} a_{ji}(x_i-x)_+^{m-j}, \quad 0\leq x\leq 1$$

where a_{ji} are real numbers, p_m is a polynomial of degree $\leq m$ and

$$(2.7)\quad \begin{cases} (x_i-x)_+ := \max\limits_{x\in[0,1]} ((x_i-x),0) & \\ & i=1(1)n, \\ (x_i-x)_+^j = [(x_i-x)_+]^j, & \end{cases}$$

We consider two related problems:

A - Determine S_{nm} in (2.6) such that

$$(2.8)\quad \int_0^1 x^j\, S_{nm}(x)\,dx = \int_0^1 x^j\, f(x)\,dx\ , \qquad j=0(1)N+m$$

where

$$(2.9)\quad N = 2\,(\sum_{i=1}^{n} \sigma_i + n) .$$

B - Determine S_{nm} in (2.6) such that

$$(2.10)\quad \begin{cases} S_{nm}^{(k)}(1) = f^{(k)}(1), & k=0(1)m \\ \int_0^1 x^j S_{nm}(x)dx = \int_0^1 x^j f(x)dx, & \\ & j=0(1)N-1 . \end{cases}$$

The former problem requires only the existence and knowledge of the involved moments of f; the latter requires some additional regularity of f, because we must assume that f has m known derivatives at t=1.

Both problems will be solved in two ways: first, in terms of moment functionals; then, with some additional hypothesis on f, in terms of Gauss-Stancu quadratures connected with σ-orthogonal polynomials, that are those polynomials satisfying the conditions (3.6).

3. SOLUTION OF PROBLEM A

3.1. Solution by moment functionals

Let

$$(3.1)\quad \mu_j = \frac{(m+j+1)!}{m!j!}\int_0^1 x^j\ f(x)\ dx \qquad j=0(1)N+m,$$

where the moments of f on the right are assumed to exist.

We define a linear functional L on the set of polynomials of the form $x^{m+1}p(x)$, $p\in\mathbb{P}_{n+m}$, by

$$(3.2)\quad L(x^{m+1}x^j) = \mu_j \qquad j=0(1)N+m\ ,$$

and the inner product for any couple of polynomials q,t with $q\cdot t\in\mathbb{P}_{N-1}$ by

$$(3.3)\quad (q,t) = L(x^{m+1}(1-x)^{m+1}qt) .$$

Let $\sigma=(\sigma_1,\sigma_2,..,\sigma_h,..)$ be a sequence of natural numbers σ_i that satisfy the conditions

$$(3.4)\quad 0\leq 2\sigma_i=k_i\leq m-1 \qquad \forall\ i\in N/\{0\} .$$

Let us fix a natural number $n\geq 1$ and the first n terms of the above defined sequence σ_i, i=1(1)n. Then, given the monic polynomial

$$(3.5)\quad \pi_n(x) = \prod_{j=1}^{n} (x-y_{jn})$$

of degree n with $0<y_{in}<1$, i=1(1)n, we say it is σ-orthogonal with respect to the inner product (3.3), if it satisfies

$$(3.6)\quad (\prod_{j=1}^{n} (x-y_{jn})^{k_j+1},q)=0 \qquad q\in\mathbb{P}_{n-1}.$$

Let us introduce the linear functional L_0 so defined:

$$(3.7)\quad L_0(g) = \sum_{k=0}^{m} b_k g^{(m-k)}(1) + \sum_{i=1}^{n}\sum_{h=0}^{k_i} B_{hi} g^{(h)}(x_i)$$

where

$$(3.8)\quad b_k = (-1)^k/m!\; p_m^{(k)}(1) \qquad k=0(1)m$$

$$(3.9)\quad B_{hi} = \frac{h!}{(m-h)!} a_{hi} \quad i=1(1)n,\; h=0(1)k_i .$$

We have the following :

Theorem 1: There exists a unique spline function on [0,1]

$$(3.10)\quad S_{nm}(x) = p_m(x) + \sum_{i=1}^{n}\sum_{h=0}^{k_i} a_{hi}(x_i - x)_+^{m-h}$$

$$(3.11)\quad 0<x_i<1 \qquad x_i \neq x_r \quad \text{for} \quad i\neq r$$

satisfying moment equations (2.8) if and only if the σ-orthogonal polynomial $\pi_n = \pi_n(\cdot;L)$ that satisfies (3.6), exists uniquely and has n distinct real zeros y_{in} i=1(1)n all belonging to (0,1). The knots of the spline are precisely these zeros, while the coefficients b_k and B_{hi} in (3.7) are determined uniquely solving the linear system

$$(3.12)\quad L_0(x^{m+1}p) = L(x^{m+1}p) \qquad p \in \mathbb{P}_{N+m-n} .$$

<u>Proof :</u> Substituting (3.10) in (2.8) and observing that $0<x_i<1$, we have

$$(3.13)\quad \int_0^1 x^j f(x)\,dx = \int_0^1 x^j p_m(x)\,dx + \sum_{i=1}^{n}\sum_{h=0}^{k_i} a_{hi}\int_0^{x_i} x^j (x_i - x)^{m-h} dx =$$

$$= I_1 + I_2 , \qquad j=0(1)N+m.$$

With some calculations we can write:

$$(3.14)\quad I_1 = \frac{m!j!}{(m+j+1)!}\sum_{k=0}^{m} b_k \left(x^{m+1+j}\right)^{(m-k)}_{x=1}$$

where b_k is defined in (3.8), and

$$(3.15)\quad I_2 = j!\sum_{i=1}^{n}\sum_{h=0}^{k_i} \frac{a_{hi}(m-h)!}{(j+m-h+1)!} x_i^{j+m-h+1} .$$

Inserting (3.14) and (3.15) in (3.13), and dividing through by $j!m!/(m+j+1)!$ we have

$$(3.16)\quad L_0(x^{m+1}x^j) = \mu_j \qquad j=0(1)N+m$$

where μ_j is defined by (3.1) and L_0 by (3.7).
Using (3.2), (3.16) and the linearity of L and L_0, we conclude that equations (2.8) and

$$(3.17)\quad L_0(x^{m+1}p) = L(x^{m+1}p) \qquad p\in \mathbb{P}_{N+m}$$

are equivalent.
The condition (3.17) is a condition for the knots: in fact, if we consider the polynomial

$$(3.18)\quad \pi_n(x) = \prod_{i=1}^{n} (x-x_i)$$

that have the zeros coincident with the knots of the spline, we have for any $q\in \mathbb{P}_{N-1}$

$$(3.19)\quad \left(\prod_{i=1}^{n} (x-x_i)^{k_i+1}, q\right) =$$

$$= L\left(x^{m+1}(1-x)^{m+1}\prod_{i=1}^{n} (x-x_i)^{k_i+1} q\right) =$$

$$= L_0\left(x^{m+1}(1-x)^{m+1}\prod_{i=1}^{n} (x-x_i)^{k_i+1} q\right) = 0.$$

Thus the knots x_i i=1(1)n must be the zeros of the σ-orthogonal polynomial $\pi_n(\cdot;L)$ of (3.6).
If $x_i = y_{in}$, the system (3.12) is non singular. Thus, the existence of σ-orthogonal polynomial $\pi_n(\cdot;L)$ together with (3.12) imply (3.17). ■

3.2. Solution by Gauss-Stancu quadrature

Theorem 2: Assume $f \in AC^m[0,1]$. There exists a unique spline function (3.10) satisfying problem A, if and only if there exists uniquely the σ-orthogonal polynomial $\pi(\cdot;\bar{L})$ relative to the inner product

$$(3.20)\quad (p,q) = \bar{L}(x^{m+1}(1-x)^{m+1}pq) \qquad pq \in \mathbb{P}_{N-1}$$

where

(3.21) $\bar{L}(g)=\sum_{k=0}^{m} \Phi_k g^{(m-k)}(1)+\int_0^1 g(x)\,d\lambda_m(x)$

(3.22) $\Phi_k = \frac{(-1)^k f^{(k)}(1)}{m!} \qquad k=0(1)m$

and

(3.23) $d\lambda_m(x) = \frac{(-1)^{m+1}}{m!} f^{(m+1)}(x)\,dx$.

The zeros y_{in} $i=1(1)n$ of π_n must be distinct and belonging to $(0,1)$. These zeros are the knots of the spline, the constants B_{hi} and b_k in (3.7) are obtained uniquely solving the linear system

(3.24) $L_0(x^{m+1}p) = \bar{L}(x^{m+1}p) \qquad p\in \mathbb{P}_{N+m-n}$

where L_0 was defined by (3.7).

<u>Proof :</u> We point out that, by (3.21), the σ-orthogonal conditions with respect to $\bar{L}$ are the same than those ones with respect to $\tilde{L}$ where

(3.25) $\tilde{L}(g) = \int_0^1 g(x)\,d\alpha_m(x)$

and

(3.26) $d\alpha_m(x) = x^{m+1}(1-x)^{m+1}\,d\lambda_m(x)$.

Integrating m+1 times by parts all the integrals in (3.13) we have

(3.27) $L_0(x^{m+1}x^j)=\bar{L}(x^{m+1}x^j) \qquad j=0(1)N+m$.

The claim is reached in the same way as in Theorem 1. ■

By (3.24) the result of Theorem 2 can be interpreted in terms of the generalized Stancu-Lobatto quadrature formula

(3.28) $\int_0^1 g(x)\,d\lambda_m(x) =$

$= \sum_{k=0}^{m} [C_k g^{(k)}(0) + W_k g^{(k)}(1)] +$

$+\sum_{i=1}^{n} \sum_{k=0}^{k_i} E_{ki} g^{(k)}(y_{in}) + R_{nm}(g;d\lambda_m)$

having precision degree

(3.29) $\vartheta = N+2m+1$.

Several expressions of the coefficients of (3.28) are given in [5].
It is well known that this formula is related with the formula

(3.30) $\int_0^1 g(x)\,d\alpha_m(x) =$

$= \sum_{i=1}^{n} \sum_{k=0}^{k_i} F_{ki} g^{(k)}(y_{in}) +$

$+ R_{nm}(g;d\alpha_m)$

where

(3.31) $R_n(g;d\alpha_m) = 0 \qquad \forall g\in \mathbb{P}_{N-1}$.

The knots y_{in} in (3.28) and (3.30) are the same; they coincide with the zeros of polynomial σ-orthogonal respect to the inner product

(3.32) $(p,q) = \int_0^1 x^{m+1}(1-x)^{m+1} pq\,d\lambda_m(x)=$

$= \int_0^1 pq\,d\alpha_m(x)$.

The coefficients F_{ki} in (3.30) are expressible in terms of coefficients E_{ki} in (3.28) by [5]

(3.33) $F_{ri}= \sum_{k=r}^{k_i} \binom{k}{r} E_{ki} \left(x^{m+1}(1-x)^{m+1}\right)^{(k-r)}_{x=y_{in}}$

$r=0(1)k_i$.

We can obtain the coefficients C_k and W_k in (3.28) by solving the linear system

(3.34) $R_{n,m}(p;d\lambda_m) = 0 \qquad p\in \mathbb{P}_{2m+1}$.

4. SOLUTION OF PROBLEM B

Following the same line of reasoning we may state Theorems 3, 4 which solve Problem B.

Theorem 3: There exists a unique spline function on [0,1]

(4.1) $S^*_{nm}(x)= p^*_m(x)+ \sum_{i=1}^{n} \sum_{j=0}^{k_i} a^*_{ji}(x^*_i-x)^j_+$

satisfying the conditions

$$(4.2)\quad \int_0^1 x^j s^*_{nm}(x)\,dx = \int_0^1 x^j f(x)\,dx, \qquad j=0(1)N-1$$

and

$$(4.3)\quad s^{*(j)}_{nm}(1) = f^{(j)}(1), \qquad j=0(1)m$$

if and only if there exists uniquely the σ-orthogonal polynomial $\pi^*_n(\cdot;L^*)$ with

$$(4.4)\quad L^*(x^{m+1}x^j) = \mu^*_j\ , \qquad j=0(1)N-1\ .$$

where

$$(4.5)\quad \mu^*_j = \frac{(m+j+1)!}{m!j!}\int_0^1 x^j\Big[f(x) + - \sum_{k=0}^{m} \frac{f^{(k)}(1)}{k!}(x-1)^k\Big]dx, \quad j=0(1)N-1.$$

$\pi^*_n(\cdot;L^*)$ has n distinct zeros y^*_{in}, $i=1(1)n$, belonging to $(0,1)$. The knots x^*_i are exactly these zeros; the polynomial p^*_m is given by

$$(4.6)\quad p^*_m(x) = \sum_{k=0}^{m} f^{(k)}(1)/k!\ (x-1)^k;$$

the coefficient a^*_{ji} are the solution of linear system

$$(4.7)\quad L^*_0(x^{m+1}p) = L^*(x^{m+1}p) \qquad p\in \mathbb{P}_{N-n-1}$$

where

$$(4.8)\quad L^*_0(g) = \sum_{i=1}^{n}\sum_{j=0}^{k_i} B^*_{ji}g^{(j)}(x^*_i)$$

and

$$(4.9)\quad B^*_{ji} = \frac{(m-j)!}{m!}a^*_{ji}, \quad i=1(1)n,\ j=0(1)k_i\ .$$

Theorem 4: Assume $f\in AC^m[0,1]$. There exists a unique spline function (4.1) on [0,1] satisfying (4.2) and (4.3), if and only if the σ-orthogonal polynomial $\pi^*_n(\cdot;L^*)$ exists uniquely, has n distinct zeros $y^*_{i,n}$ $i=1(1)n$ belonging to $(0,1)$.

These zeros are the knots of the spline; the polynomials p^*_m is given by (4.6) and the coefficients a^*_{ji} are the solution of

$$(4.10)\quad L^*_0(x^{m+1}p) = \bar{L}^*(x^{m+1}p) \qquad p\in \mathbb{P}_{N-n-1}$$

where

$$(4.11)\quad \bar{L}^*(g) = \int_0^1 g(x)\,d\lambda^*_m(x)$$

and

$$(4.12)\quad d\lambda^*_m(x) = (-1)^{m+1}/m!\ f^{m+1}(x)\,dx\ .$$

In this case the quadrature rule which occurs instead of (3.28) is the Stancu-Radau formula

$$(4.13)\quad \int_0^1 g(x)\,d\lambda^*_m(x) = \sum_{k=0}^{m} A^*_k g^{(k)}(0) + \sum_{i=1}^{m}\sum_{k=0}^{2\sigma_i} E^*_{ki}g^{(k)}(y^*_{in}) + R^*_{nm}(g;d\lambda^*_m)$$

where

$$(4.14)\quad R^*_{nm}(g;d\lambda^*_m) = 0 \quad \text{if} \quad g\in \mathbb{P}_{N+m},$$

as easily may be seen.

<u>Remarks.</u> A sufficient condition for existence and uniqueness of (3.28) and (4.13) is $(-1)^{m+1}f^{(m+1)}(x)>0$, $x\in I$.

5. CONVERGENCE OF APPROXIMATION

Theorem 5: Assume that $d\lambda_m$ is a positive measure on the interval [0,1], then the approximations S_{nm} and S^*_{nm} constructed in sections 3, 4 converge pointwise to f in (0,1), as $n\to\infty$ for fixed $m\geq 2$.

<u>Proof :</u> In the hypothesis of Theorems 2,4 we can prove that for $0<x<1$ we have for Problem A

$$(5.1)\quad f(x) - S_{nm}(x) = R_{nm}(\rho_x;d\lambda_m)$$

and for Problem B

$$(5.2)\quad f(x) - S^*_{nm}(x) = R^*_{nm}(\rho_x;d\lambda^*_m)$$

where $R_{nm}(\rho_x;d\lambda_m)$ and $R^*_{nm}(\rho_x;d\lambda^*_m)$ are respectively the remainder terms of Stancu-Lobatto and Stancu-Radau quadrature formulae given in (3.28) and (4.13).
We define

$$(5.3)\quad \rho_x(t) = (t-x)^m_+ \qquad 0\le t\le 1 .$$

We prove only (5.1); the proof of (5.2) is similar.
By Taylor's Theorem we can write:

$$(5.4)\quad f(x) = \sum_{k=0}^{m} \frac{f^{(k)}(1)}{k!}(x-1)^k + \frac{(-1)^{m+1}}{m!}\int_x^1 (t-x)^m f^{(m+1)}(t)\,dt = \sum_{k=0}^{m} \frac{f^{(k)}(1)}{k!}(x-1)^k + \frac{(-1)^{m+1}}{m!}\int_x^1 \rho_x(t) f^{(m+1)}(t)\,dt .$$

By Theorem 2 we have

$$(5.5)\quad S_{nm}(x) = \sum_{k=0}^{m} \frac{p_m^{(k)}(1)}{k!}(x-1)^k + \sum_{i=1}^{n}\sum_{j=0}^{k_i} \frac{m!}{j!} E_{ji}(y_{in}-x)^j_+$$

and thus

$$(5.5)\quad f(x)-S_{nm}(x) = \int_x^1 \rho_x(t)\,d\lambda_m(t) + \sum_{k=0}^{m} \frac{[f^{(k)}(1)-p^{(k)}(1)]}{k!}(x-1)^k + - \sum_{i=1}^{n}\sum_{j=0}^{k_i} \frac{m!}{j!} E_{ji}(y_{in}-x)^j_+ .$$

We have [see (3.8), (3.22)]

$$(5.6)\quad f(x)-S_{nm}(x) = \int_x^1 \rho_x(t)\,d\lambda_m(t) + - \sum_{k=0}^{m} \frac{(-1)^k}{k!} m!\,(b_k-\Phi_k)(x-1)^k + - \sum_{i=1}^{n}\sum_{j=0}^{k_i} \frac{m!}{j!} E_{ji}(y_{in}-x)^j_+ .$$

By (5.3) we can rewrite (5.6) in the form

$$(5.7)\quad f(x)-S_{nm}(x) = \int_0^1 \rho_x(t)\,d\lambda_m(t) + - \sum_{k=0}^{m} (b_k-\Phi_k)\,(\rho_x^{(m-k)}(t))_{t=1} + - \sum_{i=1}^{n}\sum_{j=0}^{k_i} E_{ji}\,(\rho_x^{(j)}(t))_{t=y_{in}} = R_{nm}(\rho_x;d\lambda_m) .$$

By means of a result given in [4] for the case $\sigma_i = s$ $i=1(1)n$, we have pointwise convergence to f of the sequence $\{S_{nm}\}$, as

$$\lim_{n\to\infty} R_{nm}(\rho_x;d\lambda_m) = 0 .$$

This property holds also when the σ_i $i=1(1)n$ are different, [6]. So the claim follows. ■

REFERENCES

[1] Frontini,M.,Gautschi,W.,Milovanovic, G.V., Moment preserving spline approximation on finite intervals, Num.Math. 50 (1987) 503-518

[2] Gautschi,W., Discrete approximations to spherically symmetric distributions, Num.Math. 4 (1984) 53-60

[3] Gautschi,W.,Milovanovic,G.W., Spline approximations to spherically symmetric distributions, Num.Math. 49 (1986) 111-121

[4] Gori,L.,Santi,E., On the convergence of quasi-Gaussian functionals, J. of Appr. Theory, in print

[5] Gori,L., Santi,E., Monospline collegata ai funzionali quasi-Gaussiani, Calcolo, in print

[6] Gori,L., Santi,E., A remark on the convergence of Gauss-Stancu quadrature rules, preprint

[7] Micchelli,C., Monosplines and moment preserving spline approximation, Int. Series of Num.Math. 85 (1988) 131-139

[8] Milovanovic, G.V., Kovacevic, M.A., Moment preserving spline approximation and Turán quadratures, in ICNM 88 "Numerical Mathematics, Singapore 1988"

Approximation, Optimization and Computing:
Theory and Applications, A.G. Law and C.L. Wang (eds.)
Elsevier Science Publishers B.V. (North-Holland)

WEAK CONVERGENCE AND THE PROKHOROV RADIUS FOR PROBABILITY MEASURES

George A. Anastassiou

Department of Mathematical Sciences, Memphis State University,
Memphis, TN 38152 U.S.A.

The *Prokhorov radius* for a set of probability measures satisfying basic moment conditions is introduced, through the Prokhorov distance of these measures from the Dirac measure at a fixed point of the real line. This is calculated precisely by the use of standard tools from the Kemperman geometric moment theory. The above radius gives the exact rate of weak convergence of these measures to the Dirac measure.

1. INTRODUCTION

Here we consider probability measures μ on $\mathbb{R}$ such that both $\int|t|d\mu$, $\int t^2 d\mu < \infty$. We consider

$$M(\varepsilon_1,\varepsilon_2) = \{\mu: |\int t^j d\mu - \alpha^j| \le \varepsilon_j,\ j = 1,2\} \tag{1}$$

where α is a given point in $\mathbb{R}$, also $0 < \varepsilon_j < 1$, $j = 1,2$ and $0 < \varepsilon_2 + 2|\alpha|\varepsilon_1 < 1$. We would like to measure the "size" of $M(\varepsilon_1,\varepsilon_2)$ to be given by a simple formula involving only ε_1, ε_2, α.

Since weak convergence of probability measures is of central importance and their standard weak topology is well described by the Prokhorov distance π, it is natural to define the *Prokhorov radius* for $M(\varepsilon_1,\varepsilon_2)$ as

$$D = \sup_{\mu\in M(\varepsilon_1,\varepsilon_2)} \pi(\mu,\delta_\alpha) \tag{2}$$

where δ_α is the Dirac measure at α. Using a geometric moment theoretical method due to Kemperman [2] we are able to calculate the exact value of D.

It will be helpful to mention

Definition 1 (see [3]). Let U be a Polish space with a metric d and C is the set of all nonempty closed subsets of U. Let $A \in C$, then for $\varepsilon > 0$

$$A^\varepsilon = \{x : d(x,A) < \varepsilon\} .$$

Consider μ,ν probability measures on U. Prokhorov (1956) introduced his famous metric

$$\pi(\mu,\nu) = \inf\{\varepsilon > 0: \mu(A) \le \nu(A^\varepsilon) + \varepsilon, \nu(A) \le \mu(A^\varepsilon) + \varepsilon,\ \forall A \in C\}.$$

When μ is a probability measure on $\mathbb{R}$, then

$$\pi(\mu,\delta_\alpha) = \inf\{r > 0: \mu([\alpha - r, \alpha + r]) \ge 1 - r\},\quad \alpha \in \mathbb{R}. \tag{3}$$

Remark 1. One can restate that

$$D = \sup_{\mu\in M(\varepsilon_1,\varepsilon_2)} (\inf\{r > 0: \mu([\alpha - r, \alpha + r]) \ge 1 - r\}). \tag{4}$$

Thus

$$r \ge D \text{ iff } \mu([\alpha - r, \alpha + r]) \ge 1 - r, \text{ all } \mu \in M(\varepsilon_1,\varepsilon_2) \tag{5}$$

iff

$$\lambda_r = \inf_{\mu\in M(\varepsilon_1,\varepsilon_2)} \mu([\alpha - r, \alpha + r]) \ge 1 - r.$$

Obviously $0 < D \le 1$, therefore we are interested only for $r \in (0,1]$. One can easily see that

$$D = \min\{r \in (0,1]: \inf_{\mu\in M(\varepsilon_1,\varepsilon_2)} \mu([\alpha - r, \alpha + r]) \ge 1 - r\}. \tag{6}$$

Remark 2. When $\varepsilon_1,\varepsilon_2 \to 0$, then $\int t d\mu \to \alpha$ and $\int t^2 d\mu \to \alpha^2$. Thus $\int(t - \alpha)^2 d\mu \to 0$, implying that $\int|t - \alpha|d\mu \to 0$. Hence for arbitrarily small $\varepsilon > 0$ we have

$$\varepsilon\mu(\{t: |t - \alpha| > \varepsilon\}) \le \int_{\{t:|t-\alpha|>\varepsilon\}} |t - \alpha| d\mu \le \int |t - \alpha| d\mu \le \varepsilon^2.$$

That is, $\mu(\{t: |t - \alpha| > \varepsilon\}) \le \varepsilon$. Therefore $\pi(\mu,\delta_\alpha) \le \varepsilon$ and $D = D(\varepsilon_1,\varepsilon_2, \alpha) \le \varepsilon$. Clearly $\pi(\mu,\delta_\alpha) \to 0$ and $D \to 0$ as $\varepsilon_1,\varepsilon_2 \to 0$, giving us that $\mu \to \delta_\alpha$ weakly. The knowledge of D gives the rate of weak convergence of μ to δ_α.

2. RESULTS

Our main result has as follows:

Theorem 1.

1) If $\alpha = 0$ we get $D = \varepsilon_2^{1/3}$.
2) If $|\alpha| \ge 1$ we get $D = (\varepsilon_2 + 2|\alpha|\varepsilon_1)^{1/3}$.

3) For $0 < |\alpha| < 1$ and sufficiently small ε_1, ε_2 we obtain again

$$D = (\varepsilon_2 + 2|\alpha|\varepsilon_1)^{1/3}.$$

To prove Theorem 1 among others we need

Theorem 2. Let μ be probability measures on $\mathbb{R}$ such that $\int t d\mu = y_1$, $\int t^2 d\mu = y_2$. Let $\alpha \in \mathbb{R}$, $0 < r \leq 1$. Set

$$L = L_{[\alpha-r,\alpha+r]}(y_1,y_2) = \inf_{\mu}(\mu([\alpha-r, \alpha+r])).$$

Then we find

1) $L = 0$, if $y_2 + \alpha^2 - 2\alpha y_1 > r^2$;
2) $L = 1$, if $y_2 = y_1^2$ with $y_1 \in [\alpha-r, \alpha+r]$;
3) Let $y_2 \neq y_1^2$ and $r|y_1 - \alpha| \leq y_2 + \alpha^2 - 2\alpha y_1 \leq r^2$, then
$$L = 1 - \frac{[y_2 + \alpha^2 - 2\alpha y_1]}{r^2}$$
4) Let $y_2 \neq y_1^2$ and $y_2 + \alpha^2 - 2\alpha y_1 \leq r(y_1 - \alpha)$, then
$$L = \frac{((\alpha + r) - y_1)^2}{(\alpha + r)^2 - 2(\alpha + r)y_1 + y_2};$$
5) Let $y_2 \neq y_1^2$ and $y_2 + \alpha^2 - 2\alpha y_1 \leq r(\alpha - y_1)$, then
$$L = (y_1 - (\alpha-r))^2/((\alpha-r)^2 - 2(\alpha-r)y_1 + y_2).$$

Comment 1. $L = 0$ and $L = 1$ mean nothing towards the calculation of D. Because of their length we intend to present proofs elsewhere.

REFERENCES

1. G. Anastassiou, "The Levy radius of a set of probability measures satisfying basic moment conditions involving $\{t,t^2\}$", *Constructive Approx. Journal* (1987) 3: 257-263.
2. J. H. B. Kemperman, "The general moment problem, a geometric approach", *The Annals of Mathematical Statistics* (1968), Vol. 39, No. 1, 93-122.
3. S. T. Rachev and R. M. Shortt, "Classification problem for probability metrics", Proceedings of the Conference in Honor of Dorothy Maharam Stone, University of Rochester, Sept. 1987, A.M.S. Contemporary Mathematics, to appear.

Approximation, Optimization and Computing:
Theory and Applications, A.G. Law and C.L. Wang (eds.)
Elsevier Science Publishers B.V. (North-Holland)

THE ESTIMATION OF N-WIDTHS OF CERTAIN PERIODIC FUNCTION CLASSES

Han-lin Chen & Chun Li

Institute of Mathematics, Academia Sinica, Beijing, P.R.China

In this paper, we give the strong asymptotic estimates of the values $d_n(\widetilde{N}^r_{p,\tau}; L_p)$, $d^n(\widetilde{N}^r_{p,\tau}; L_p)$, $\delta_n(\widetilde{N}^r_{p,\tau}; L_p)$ and $b_n(\widetilde{N}^r_{p,\tau}; L_p)$ as n tends to infinity for $1 < p < \infty$, where d_n, d^n, δ_n and b_n denote the Kolmogorov, Gelfand, linear and Bernstein n-widths, respectively, $\widetilde{N}^r_{p,\tau}$ is a class of periodic functions $\{f\}$ for which $\tau\|f\|_p + \|\mathcal{L}_r f\|_p \leq 1$ and $\mathcal{L}_r$ is the differential operator $D^\sigma \prod_{j=1}^l (D^2 - t_j^2)$; $\sigma + 2l = r$, $\{t_j\}_1^l$ real numbers and $\tau \geq 0$. We also find the optimal upper bound of $\delta_{2n}(\widetilde{\mathcal{M}}^r_p; L_q)$ for $1 < q \leq p < \infty$ and the optimal lower bound of $b_{2n-1}(\widetilde{\mathcal{M}}^r_p; L_q)$ for $1 < p \leq q < \infty$, where $\widetilde{\mathcal{M}}^r_p := \{f : f^{(r-1)}$ abs. cont. on $[0,1]$, $f^{(i)}(0) = f^{(i)}(1)$, $i = 0, \cdots, r-1$ and $\|\mathcal{L}_r f\|_p \leq 1\}$, where $\mathcal{L}_r = D_r \cdots D_1$ and $D_j(\cdot) = \alpha_{j-1}(t) D\{\beta_{j-1}(t)\cdot\}$, $\alpha_{j-1}(t)$, $\beta_{j-1}(t)$ are some positive smooth functions.

1. INTRODUCTION

The Kolmogorov n-width of a set $\mathcal{M}$ in a normed linear space X is defined as

$$d_n(\mathcal{M}; X) = \inf_{X_n} \sup_{f \in \mathcal{M}} \inf_{g \in X_n} \|f - g\|_X$$

where the last infimum is taken over all n-dimensional linear subspace $X_n \subset X$. The Gelfand n-width of $\mathcal{M}$ is defined as

$$d^n(\mathcal{M}; X) = \inf_{\Gamma^n} \sup_{f \in \mathcal{M} \cap \Gamma^n} \|f\|_X$$

where the infimum is taken over all linear subspaces of codimension n. The linear n-width of $\mathcal{M}$ in X is defined by

$$\delta_n(\mathcal{M}; X) = \inf_{P_n} \sup_{f \in \mathcal{M}} \|f - P_n(f)\|_X$$

where the P_n ranges over the set of continuous linear operators of X into X of rank at most n. The Bernstein n-width of $\mathcal{M}$ in X is given by

$$b_n(\mathcal{M}; X) = \sup_{X_{n+1}} \sup\{\lambda | \lambda S(X_{n+1}) \subseteq \mathcal{M}\}$$

where $S(X_{n+1})$ is the unit ball in X_{n+1} with center at origin and X_{n+1} ranges over all subspaces of dimension at least $n+1$.

The differential operator

$$\mathcal{L}_j = D_j \cdots D_1 \quad (j = 1, \cdots, r) \tag{1.0}$$

is defined by $D_j(\cdot) = \alpha_{j-1}(t) D\{\beta_{j-1}(t)\cdot\}$, where $\alpha_{j-1}(t)$, $\beta_{j-1}(t)$ are positive smooth functions satisfying

$$\alpha_j(t+h) = C_j \alpha_j(t), \quad \beta_j(t+h) = C_j^{-1} \beta_j(t) \tag{1.1}$$

for all $t \in R^1$ and $j = 0, \cdots, r$ and $h = \frac{1}{2n}$, n positive integer.

It is worthwhile to mention that the differential operator $\mathcal{L}_j$ extends those with constant coefficients and real roots. Indeed, by setting $\alpha_j(t) = e^{r_j t} = \beta_j^{-1}(t)$, r_j real, we have $\mathcal{L}_j(\cdot) = \prod_{k=0}^{j-1} (D - r_k)(\cdot) = \prod_{k=0}^{j-1} e^{r_k t} D(e^{-r_k t}\cdot)$, it is admissible since $\alpha_k(t) = \beta_k^{-1}(t) = e^{r_k t}, \alpha_k, \beta_k$ satisfy formulas (1.1). If $\alpha_{j-1}(t)$, $\beta_{j-1}(t)$ are positive, with period h, then $D_j(\cdot) = \alpha_{j-1}(t) D(\beta_{j-1}(t)\cdot)$ is also an admissible operator, etc.

Now We define some function classes:

$$\begin{aligned} \widetilde{\mathcal{M}}^r_p := \; & \{f \mid f^{(r-1)} \text{abs. cont. on} [0,1], \|\mathcal{L}_r f\|_p \leq 1, \\ & f^{(i)}(0) = f^{(i)}(1),\ i = 0, \cdots, r-1\} \end{aligned} \tag{1.2}$$

$$\widetilde{B}^r_p := \{f \mid f \in \widetilde{\mathcal{M}}^r_p,\ \mathcal{L}_r = D^r\} \tag{1.3}$$

$$\begin{aligned} \widetilde{N}^r_p := \; & \{f \mid f \in \widetilde{\mathcal{M}}^r_p,\ \mathcal{L}_r = D^\sigma \prod_{j=1}^l (D^2 - t_j^2), \\ & r = \sigma + 2l,\ \sigma = 0 \text{ or } 1,\ t_j \text{ reals}\} \end{aligned} \tag{1.4}$$

$$\widetilde{N}^r_{p,\tau} := \{f \mid f \in \widetilde{N}^r_p,\ \tau\|f\|_p + \|\mathcal{L}_r f\|_p \leq 1\} \tag{1.5}$$

where $\tau \geq 0$ is fixed.

For some special values of p and q, the n-width $d_n(\widetilde{B}^r_p; L_q)$ were studied in [1—14, 17]. We do not know whether the four n-widths of $\widetilde{B}^r_p$ in L_p are equal or not, but for the subset $\widetilde{B}^{r,1}_p := \{f \mid f \in \widetilde{B}^r_p,\ f(x_1) = 0\}$, where x_1 is any fixed point on $[0,1]$, we proved that [17]

$$d_n(\widetilde{B}^{r,1}_p; L_p) = d^n(\widetilde{B}^{r,1}_p; L_p) = \delta_n(\widetilde{B}^{r,1}_p; L_p) = b_n(\widetilde{B}^{r,1}_p; L_p),$$

where n is odd and $1 < p < \infty$.

The exact determination of the values of n-widths which abound in the literature is generally impossible to obtain. Much effort has nonetheless been devoted toward this goal, as well as to the related task of determining the asymptotic behavaor of these values as $n \to \infty$. This latter question is of importance not only in approximation theory, but also in the study of ideals of operators between Banach spaces [8], [13].

Any function in $\widetilde{\mathcal{M}}_p^r$ may be represented by the following formula [18]:

$$\left.\begin{array}{ll} f(t) & = \int_0^1 G_{r-1}(t,\tau)\mathcal{L}_r f(\tau)d\tau \\ & \text{if} \quad C_j \neq 1, \quad j=0,1,\cdots,r-1; \quad \text{or} \\ f(t) & = c + \int_0^1 H_{r-1}(t,\tau)\mathcal{L}_r f(\tau)d\tau \\ & \text{if some of } \{C_j\} \text{ are equal to } 1, \end{array}\right\} \tag{1.6}$$

where $\{C_j\}$ are constants appeared in (1.1), G_{r-1} and H_{r-1} are Green's function with respect to the periodic boundary conditions and differential operator $\mathcal{L}_r$ defined as (1.0), (1.1).

We say that the function $G_{r-1}(t,\tau)$ $(H_{r-1}(t,\tau))$ satisfies Property A if

$$\begin{array}{rcl} G_{r-1}(-t,-\tau) & = & G_{r-1}(t,\tau) \\ (H_{r-1}(-t,-\tau) & = & \varepsilon H_{r-1}(t,\tau)), \end{array} \tag{1.7}$$

where $\varepsilon \in \{+1,-1\}$ is fixed. As a special case, the Green's function $G(x)$ for the differential operator defined in (1.4) is

$$G(x) = \frac{1}{2\pi}\sum_{k\in Z}\frac{e^{2\pi ikx}}{P_r(2\pi ki)}, \tag{1.8}$$

where $P_r(x) = \prod_{j=0}^{r-1}(x-2\pi t_j)$, $P_r(0) \neq 0$, $\{t_j\}_0^{r-1}$ the points on R^1 which are symmetrically spaced about the origin, as for the case $P_r(0) = 0$, one may refer to (2.5).

In section 2, we shall consider an extremal problem and explore some properties of the extremal function. In section 3, the optimal upper bound of linear n-width is obtained. In section 4, we estimate the optimal lower bound of the value of Bernstein n-width. In section 5, we obtain the strong asymptotic value of $\xi_{n+s}(\tilde{N}_{p,r}^r, L_p)$ as n tends to infinity, where ξ_k denotes any of the four widths: d_k, d^k, δ_k and b_k, s is a fixed integer.

For ease of exposition, we shall consider the simple case $\mathcal{L}_r = D^\sigma \prod_{j=1}^{l}(D^2 - t_j^2)$ in the proof of the lemmas presented in section 2, 3 and 4, but the consequences are still true for the more general case as will be pointed out in the sequel.

2. AN EXTREMAL PROBLEM

Let E_n and F_n be the classes of functions and $h = \frac{1}{2n}$.

$$E_n := \{\varphi \mid \varphi(x+h) = -\varphi(x),\ \varphi(-x) = \varphi(x) \text{ a.e. } x \in R^1\},$$

$$F_n := \{\varphi \mid \varphi(x+h) = -\varphi(x),\ \varphi(-x) = -\varphi(x) \text{ a.e. } x \in R^1\},$$

$$\lambda_n := \sup\{\|f\|_q \mid f \in \tilde{N}_p^r \cap E_n\}, \tag{2.0}$$

$$\mu_n := \sup\{\|f\|_q \mid f \in \tilde{N}_p^r \cap F_n\}. \tag{2.1}$$

It is easy to verify $\lambda_n = \mu_n$. By weak* compactness theorem, we assert there is a function φ_n such that $\|\varphi_n\|_p = 1$, $\varphi_n \in F_n$ and it solves (2.1):

$$\lambda_n = \|G * \varphi_n\|_q, \tag{2.2}$$

where $(G*\varphi_n)(x) = \int_0^1 G(x-y)\varphi(y)dy$ and φ_n satisfies the following equation

$$H_n(x) = \int_0^1 G(y-x)\Phi_n(y)dy, \tag{2.3}$$

where $H_n(x) = |\varphi_n(x)|^{p-1}\text{sgn}(\varphi_n(x))$, $\Phi_n(y) = \lambda_n^{-q}|(G*\varphi_n)(y)|^{q-1}\text{sgn}[(G*\varphi_n)(y)]$. We have the following important lemma.

Lemma 2.1. The extremal function φ_n for the extremal problem (2.1) has the following sign pattern

$$\text{sgn}\varphi_n(x) = \varepsilon\text{sgn}[\sin 2n\pi x], \quad \varepsilon \in \{-1,1\}. \tag{2.4}$$

Corollary 2.2. $H_n(x)$ has only simple zeros on [0,1).

Remark. In this section we study the extremal problem (2.0) for the set $\tilde{N}_p^r$ which corresponds to the differential operator defined in (1.4). If we consider the more general case $\widetilde{M}_p^r$ which relates to the operator (1.0), and in addition, the corresponding Green's function satisfies Property A, we also have the similar conclusions as stated in Lemma 2.1 and Corollary 2.2. We emphasize that all the conclusions and consequences in section 3 and section 4 are valid for the general case as stated above.

3. ESTIMATION OF LINEAR N-WIDTH FROM ABOVE

From the lemma in previous section we can prove the following

Lemma 3.1. Let $r \geq 2$ and $1 < q \leq p < \infty$, then

$$\delta_{2n}(\tilde{N}_p^r; L_q) \leq \lambda_n. \tag{3.0}$$

Set

$$E(\tilde{N}_p^r, S_{2n,r-1})_q := \sup_{f\in\tilde{N}_p^r}\inf_{g\in S_{2n,r-1}}\|f-g\|_q.$$

We have

Proposition 3.2. Assume $r \geq 2,\ 1 < q \leq p < \infty$. Then

$E(\tilde{N}_p^r, S_{2n,r-1})_q = \sup\{\|G * \varphi\|_q \mid \varphi \in F_n,\ \|\varphi\|_p \leq 1\} = \lambda_n.$

Remark. If $q > p$, we have $E(\tilde{N}_p^r, S_{2n,r-1})_q \geq \lambda_n$.

4. ESTIMATION OF BERNSTEIN N-WIDTH FROM BELOW

Set

$$f_j(x) = \begin{cases} |\varphi_n(x)|, & jh \leq x < (j+1)h, \\ 0, & \text{otherwise}. \end{cases}$$
$$g_j(x) = (G * f_j)(x), \quad j = 0, \cdots, 2n-1,$$
$$g_{2n}(x) = 1.$$

Lemma 4.1. $r \geq 2$, r integer, and $1 < p \leq q < \infty$, then

$$\begin{aligned} \min \ & \{\|\textstyle\sum_{j=0}^{2n} a_j g_j\|_q \big/ \|\sum_{j=0}^{2n-1} a_j f_j\|_p \ \big|\ a_j \in R^1, \\ & j = 0, \cdots, 2n, \ \textstyle\sum_{j=0}^{2n-1} a_j = 0\} \ = \ \lambda_n \end{aligned} \tag{4.0}$$

Lemma 4.2. $r \geq 2,\ r$ integer, $1 < p \leq q < \infty$, then

$$b_{2n-1}(\tilde{N}_p^r;\ L_q) \geq \lambda_n.$$

The proof of Lemma 4.1 is analogous to that in [15, 16], as for the periodic case, one may refer to [17]. Lemmas 4.2 follows from the previous lemma immediately.

Corollary 4.3. Assume that $1 < p < \infty$, and $r \geq 2$, ξ_n denotes any of the four n-widths: b_n, d_n, d^n and δ_n. Then

$$\xi_{2n}(\tilde{N}_p^r; L_p) \leq \lambda_n \leq \xi_{2n-1}(\tilde{N}_p^r; L_p). \tag{4.1}$$

5.THE STRONG ASYMPTOTIC ESTIMATES FOR N-WIDTHS

At first, we consider the special case $\mathcal{L}_r(D) = D^r$ and the extremal problem:

$$\omega_n := \sup\{\|f\|_q \mid f \in \tilde{B}_p^r \cap F_n\} \tag{5.0}$$

where $\tilde{B}_p^r$, F_n are defined in section 1 and section 2, respectively. We prove the following important

Lemma 5.1. $\omega_n = \omega / n^r, n = 1, 2, \cdots$.
where $\omega = \omega_1 = \sup\{\|f\|_q \mid f \in \tilde{B}_p^r \cap F_1\}$.

Proof. Assume $f \in \tilde{B}_p^r \cap F_n$, set $g(x) = n^r f(x/n)$, then we have $g \in F_1$. On the other hand, if $g \in F_1$, then $f \in \tilde{B}_p^r \cap F_n$. It can be shown $\|f\|_q = \frac{1}{n^r}\|g\|_q$. Thus $\omega_n = \omega_1/n^r$. ∎

From (4.1), we have

$$\omega_{n+1} \leq d_{2n+1}(\tilde{B}_p^r; L_p) \leq d_{2n}(\tilde{B}_p^r; L_p) \leq \omega_n. \tag{5.1}$$

From (5.1) and Lemma 5.1, we have

Theorem 1. Let ξ_n denote any of the four widths : b_n, d_n, d^n and δ_n. Then

$$\lim_{n\to\infty} n^r \xi_n(\tilde{B}_p^r; L_p) = 2^r \omega,$$

where $\omega = \omega_1$ is a constant appeared in Lemma 5.1.

Let $A = D^r + \sum_{k=0}^{r-1} a_k(t) D^k$, $a_k \in C^k$, be a differential operator. We define some classes of functions as follows

$$\begin{aligned} W_p(A) &:= \{f \mid \|Af\|_p \leq 1\}, \\ \widetilde{W}_p(A) &:= \{f \mid f \in W_p(A), \\ & \qquad f^{(i)}(0) = f^{(i)}(1),\ i = 0, \cdots, r-1\}, \\ \widetilde{W}_{p,o}(A) &:= \{f \mid f \in \tilde{W}_p(A), \\ & \qquad f^{(i)}(0) = 0,\ i = 0, \cdots, r-1\}. \end{aligned}$$

We prove the following

Lemma 5.2. Let ω be the constant in Lemma 5.1. Then

$$\begin{aligned} &\lim_{n\to\infty} n^r d_{2n}(\widetilde{W}_p(A); L_p) = \omega, \\ &\lim_{n\to\infty} n^r d_{2n-1}(\widetilde{W}_p(A); L_p) = \omega. \end{aligned} \tag{5.2}$$

Proof. It can be shown[6] $d_n(B_p^r; L_q) \sim n^{-\alpha}$, and

$$\lim_{n\to\infty} \frac{d_n(W_p(A); L_q)}{d_n(B_p^r; L_q)} = 1,$$

where $B_p^r := \{f \mid f^{(r-1)}\text{abs. cont. on } [0,1],\ \|f^{(r)}\|_p \leq 1\}$, and for all p, q $(1 \leq p,\ q \leq \infty)$ we have

$$\begin{aligned} & d_n(W_p(A); L_q) \geq d_n(\widetilde{W}_p(A); L_q) \\ \geq \ & d_n(\widetilde{W}_{p,o}(A); L_q) \geq d_{n+2r}(W_p(A); L_q). \end{aligned}$$

The last inequality follows from the duality between d_n and d^n and the definition of Gelfand n-width. We then easily obtain (5.2). ∎

Now we consider the of class functions $\tilde{N}_p^r$ defined in (1.4). Since $\mathcal{L}_r(D) = D^\sigma \prod_{j=1}^{l}(D^2 - t_j^2)$ is a special case of differential operator $A = D^r + \sum_{k=0}^{r-1} a_k(t) D^k$, from (4.1) and Lemma 5.2, we obtain the following

Theorem 2. Let ξ_n denote any of the four widths d_n, d^n, b_n and δ_n. Then

$$\lim_{n\to\infty} n^r \xi_n(\tilde{N}_p^r; L_p) = 2^r \omega.$$

We extent Theorem 2 to more general function class $\widetilde{N}^r_{p,\tau}$ (see (1.5)).

Theorem 3. Let ξ_n denote any of the four widths d_n, d^n, b_n and δ_n. Then

$$\lim_{n\to\infty} n^r \xi_{n+s}(\widetilde{N}^r_{p,\tau}; L_p) = 2^r \omega \tag{5.3}$$

where s is any fixed integer.

Proof. We estimate the values of two n-widths.
i) $\delta_{2n}(\widetilde{N}^r_{p,\tau}; L_p) \le \delta_{2n}(\widetilde{N}^r_p; L_p) \le \lambda_n$.
ii)

$$\begin{aligned} & b_{2n-1}(\widetilde{N}^r_{p,\tau}; L_p) \\ = & \sup_{X_{2n}} \inf_{f\in \widetilde{N}^r_{p,\tau}\cap X_{2n}} \frac{\|f\|_p}{\tau\|f\|_p + \|\mathcal{L}_r f\|_p} \\ = & \sup_{X_{2n}} \frac{\inf_{f\in\widetilde{N}^r_p\cap X_{2n}} \|f\|_p/\|\mathcal{L}_r f\|_p}{1+\tau \inf_{f\in\widetilde{N}^r_p\cap X_{2n}} \|f\|_p/\|\mathcal{L}_r f\|_p} \\ \ge & \frac{\lambda_n}{1+\tau\lambda_n} \quad \text{(see Lemma 4.2).} \end{aligned}$$

From i) and ii) we have $\dfrac{\lambda_{n+1}}{1+\tau\lambda_{n+1}} \le \xi_{2n+1} \le \xi_{2n} \le \lambda_n$,

where ξ_n denotes any of the four n-widths d_n, d^n, b_n and δ_n. Thus (5.3) follows immediately. ∎

REFERENCES

[1] Pinkus, A., n-Widths in Approximation Theory, Berlin Heidelberg, New York, Spring-Verlag, (1985).

[2] Ligun, A.A., Diameters of certain calsses of differentiable periodic functions, Mat. Zametki 27(1980), 61–75.

[3] Makovoz, Yu.I., Diameters of certain function classes in the space L, Vesci Akad, Navuk BSSR Ser. Fiz-Mat. Navuk 4(1969), 19–28.

[4] ——, On a method for estimation from below of diameters of sets in Banach spaces, Mat. Sb. (N.S.) 87(1972), 136–142.

[5] ——, Diameters of Sobolev classes and splines deviating least from zero, Mat. Zametki 26(1979), 805–812.

[6] ——, On n-widths of certain functional classes defined by linear differential operators. Proceeding Amer. Math. Soc., Vol. 89, No. 1 (1983), 109–112.

[7] Melkman, A.A., Micchelli, C.A., Spline spaces are optimal for L^2 n-width, Illinois J. Math. 22(1978), 541–564.

[8] Pietsch, A., s-Numbers of operators in Banach spaces, Studia Math. 51 (1974), 201–223.

[9] Pinkus, A., On n-widths of periodic functions, J. Analyse Math. 35(1979), 209–235.

[10] Subbotin, Yu. N., Diameters of class W^rL in $L(0, 2\pi)$ and spline function approximation, Mat. Zametki, 7 (1970), 43–52.

[11] Taikov, L.V., On approximating some classes of periodic functions in mean, Proc. Steklov Inst. Meth., 88(1967), 65–74.

[12] Tichomirov, V.M., Diameters of sets in function spaces and the theory of best approximations, Uspehi Mat. Nauk., 15(1960), 81–120.

[13] Triebel, H., Interpolationseigenschaffen von entropic- und duschmesseridealen kompakter operatoren, Sstudia Math. 34 (1970), 89–107.

[14] Zensybaev, A.A., On the best quadrature formulas on the class W^rL_p, Dokl. Akad. Nauk SSSR 227(1976), 277–279

[15] Chun Li, n-widths of Ω^r_p in L^p, Approx. Theory & Its Appl., Vol. 5, No. 1 (1989), 47–62.

[16] Pinkus, A., n-Widths of Sobolev Spaces, Constr. Approx., Vol. 1, No. 1 (1985), 15–62.

[17] H.L.Chen, On n-width of periodic functions, in print.

[18] H.L.Chen, On integral representation of certain classes of periodic functions, in print.

Approximation, Optimization and Computing:
Theory and Applications, A.G. Law and C.L. Wang (eds.)
Elsevier Science Publishers B.V. (North-Holland)

AN ERROR APPROXIMATION

Yi-Ling Chiang

Computer and Information Science
New Jersey Institute of Technology
Newark, New Jersey

ABSTRACT

An error approximation of semi-discretized scheme in the solution of a linear partial differential equation initial-valued problem has been established. In the deviation, a modified von Neumann solution is introduced and the problem solution is assumed to be both bounded and band-limited. This approximation relates the error in the numerical solution of a problem to both equations and conditions of that problem. Hence, it can be considered as a problem-oriented error bound.

1. INTRODUCTION

In this paper, an error approximation of a semi-discretized scheme in the solution of a linear partial differential equation initial-valued problem is established. This error approximation relates the error to both equations and conditions of a problem, hence, it can serve as a problem-oriented error bound, [6]. We considered the errors in the numerical solution of a linear partial differential equation initial-valued problem in one space dimension, [10].

$$\frac{\partial \bar{U}(x,t)}{\partial t} = \tilde{X} \cdot \bar{U}(x,t) \tag{1.1}$$

$\bar{U}(x,o) = \bar{U}_0(x)$ is given,

where for a positive integer M, the differential operator $\tilde{X}\cdot$ is:

$$\tilde{X} \cdot = \sum_{i=0}^{M} C_{1i}(x) C_{2i}(t) \frac{\partial^i}{\partial x^i}, \tag{1.2}$$

and the initial condition is either continuous or piecewise continuous. If in (1.2), for all i, $C_{1i}(x)$ is a constant and $C_{2i}(t) = C_i g(t)$ for a constant C_i and a function g(t), then (1.1) is satisfied in the frequency domain, ([9], [11]) by the von Neumann solution, i.e.,

$$\bar{U}(x,t) = \hat{a}_w(t) e^{iwx}. \tag{1.3}$$

However, for more general cases, solution (1.3) does not satisfy (1.1), and other forms of solution are necessary.

Consider the following, that

$$\bar{U}(x,t) = \hat{a}_w(t) \phi_w(x) e^{iwx} \tag{1.4}$$

where $\phi_w(x)$ is arbitrary. If $\phi_w(x)$ exists to satisfy

$$\frac{d \log a_w(t)}{dt} = \frac{\hat{X} \cdot [\phi_w(x) e^{iwx}]}{\phi_w(x) e^{iwx}} \tag{1.5}$$

then (1.4) is a solution of (1,1) and is called the modified von Neumann solution. The modified von Neumann solution will be described in detail, and an example is given to show the existence of a nontrivial $\phi_w(x)$ as the coefficient functions, $C_{1i}(x)$ in (1.2) are not constants for all i. The error approximation is then established with the assumption that the problem solution is both bounded and band-limited ([1], [2]).

2. MODIFIED VON NEUMANN SOLUTION

Let $\bar{U}(x,t)$ be the exact solution of problem (1.1), and $\{x_n\}$ a set of equally-spaced mesh point, Δx apart. Let $\tilde{A}\cdot$ be a finite difference operator defined on $\{x_n\}$ to represent the differential operator $\tilde{X}\cdot$ of (1.2) and function $\bar{U}_h(x_n,t)$ satisfy:

$$\frac{d\ \bar{U}_h(x_n,t)}{dt} = \tilde{A} \cdot \bar{U}_h(x_n,t)$$
$$\bar{U}_h(x_n,0) = \bar{U}_o(x_n). \tag{2.1}$$

Then $\bar{U}_h(x_n,t)$ is said to be a numerical solution of problem (1.1) with respect to scheme $\tilde{A}\cdot$.

To find a solution of (1.1) in general, let us consider $\bar{U}(x,t)$ of (1.4), which can be a solution of (1.1) if function $\phi_w(x)$ satisfied (1.5). For a given problem $\phi_w(x)$ may not always exist. If $\phi_w(x)$ does exist to satisfy (1.5), then (1.4) is called a modified von Neumann solution of (1.1). The existence of $\phi_w(x)$ indicates that both sides of (1.5) are independent of the space variable x, and a separation of variable in the problem solution of (1.1) is possible. In general, $\phi_w(x)$ may exist for only a single equation (1.1), but most unlikely for a system of equations. the existence of $\phi_w(x)$ does not imply the uniqueness of $\phi_w(x)$, but does imply a possible space transformation such that von Neumann solution is applicable in the transformed spaces.

$$y = x + \frac{1}{iw}\log \phi_w(x) \tag{2.2}$$

In general, if $\phi_w(x)$ exists, then (1.1) has a solution for all w, [4], and

$$\bar{U}(x,t) = \int_{-\infty}^{\infty} \hat{a}_w(0)\phi_w(x)e^{iwx + \int_0^t \hat{X}(w,\xi)d\xi}dw, \tag{2.3}$$

where

$$\hat{X}(w,t) = \frac{\tilde{X}\cdot\ [\phi_w(x)e^{iwx}]}{\phi_w(x)e^{iwx}} \tag{2.4}$$

and ξ is used as a dummy variable throughout the entire paper. In (2.3), function $\hat{a}_w(0)$ is the Fourier transform of the initial condition $\bar{U}_o(x)$ in the transformed y-space. In (2.4), the existence of $\phi_w(x)$ shows that function $\hat{X}(w,t)$ is independent of the space variable x.

To show the validity of the introduced modified von Neumann solution, an example of a nontrivial $\phi_w(x)$ existing in the solution of (1.1) is given. In the example, the differential operator $\tilde{X}\cdot$ in (1.2) has the following form:

$$\tilde{X}\cdot = C_1(x)C_2(t)\frac{\partial}{\partial x}\cdot \tag{2.5}$$

In (2.5), if the reciprocal of function $C_1(x)$ is integrable, then $\phi_w(x)$ exists and satisfies,

$$\log \phi_w(x) = \int (iw - \frac{\lambda_w}{C_1(x)})\, dx + c \tag{2.6}$$

where c is an arbitrary constant and λ_w a parameter. Substituting (2.6) into (2.4), we have:

$$\hat{X}(w,t) = -\lambda_w \cdot C_2(t) \tag{2.7}$$

and substituting both (2.6) and (2.7) into (2.3), we have the solution of (1.1),

$$\bar{U}(x,t) = \int_{-\infty}^{\infty} e^{\lambda_w[\int \frac{dx}{C_1(x)} - \int_0^t C_2(\xi)d\xi + c]}dw. \tag{2.8}$$

Choose λ_w =iw. Then at t = 0, the factor e^{iwc} represents the Fourier transform of the initial condition $\bar{U}_o(x)$ in the transformed space,

$$y = \int \frac{dx}{C_1(x)}. \tag{2.9}$$

Let $\hat{a}_w(0) = e^{iwc}$. Then solution (2.8) can be written as:

$$\bar{U}(x,t) = \int_{-\infty}^{\infty} \hat{a}_w(0) e^{iw[\int \frac{dx}{C_1(x)} - \int_0^t C_2(\xi)d\xi]} dw. \qquad (2.10)$$

To find an expression for the numerical solution $\bar{U}_h(x_n,t)$ of (2.1) raises a little difficulty. We considered the following: Let $\{y_n\}$ be a set of equally-spaced points, h apart in the transformed y-space and the difference operator $\tilde{A}\cdot$ be defined on $\{y_n\}$. Let $y = g(x)$ and the inverse function of $g(x)$ exist. Then $x = g^{-1}(y)$ and for an n, the distance Δx_n between two consecutive points x_n and x_{n+1} can be measured as:

$$\Delta x_n = g^{-1}(y_{n+1}) - g^{-1}(y_n). \qquad (2.11)$$

By the law of mean, we have [5]

$$\Delta x_n = \left.\frac{dg^{-1}(y)}{dy}\right|_{y=\xi_n} \cdot h, \qquad (2.12)$$

where $y_n \le \xi_n \le y_{n+1}$. As given in (2.12), Δx_n may not be a constant. However, the numerical approximation can be written as:

$$\bar{U}_h(x_n,t) = \int_{-\infty}^{\infty} \hat{a}_w(0)\phi_w(x_n) e^{iwx_n + \int_0^t \hat{A}(w,\xi)d\xi} dw \qquad (2.13)$$

where

$$\hat{A}(w,t) = \frac{\tilde{A}\cdot [\phi_w(x_n)e^{iwx_n}]}{\phi_w(x_n)e^{iwx_n}} \qquad (2.14)$$

The numerical approximation $\bar{U}_h(x_n,t)$ in (2.13) is defined on a set of unequally-spaced mesh points $\{x_n\}$ in the x-space. However, function $\hat{a}_w(0)$ in both (2.10) and (2.13) and the difference operator $\tilde{A}\cdot$ of (2.1) are all still defined in the transformed y-space. since h is a constant, function $\hat{A}(w,t)$ in (2.14) is independent of the space variable x.

Let the solution of (1.1) be:

$$\bar{U}(x,t) = e^{\psi[Z_w(x,t),w]} \qquad (2.15)$$

where function $\psi[\cdot,\cdot]$ is arbitrary, [8]. It is well known that if $Z_w(x,t)$ satisfied (1.1) and ψ-function is not a constant with respect to $Z_w(x,t)$, then solution (2.15) satisfies also (1.1). For a frequency w, problem (1.1) has a solution,

$$Z_w(x,t) = \phi_w(x)e^{iwx} + \int_0^t \hat{X}(w,\xi)d\xi \qquad (2.16)$$

Hence, in general, for all w, the solution of (1.1) can be given as:

$$\bar{U}(x,t) = \int_{-\infty}^{\infty} e^{\psi[\phi_w(x)e^{iwx} + \int_0^t \hat{X}(w,\xi)d\xi, w]} dw. \qquad (2.17)$$

Choose ψ such that the Fourier transform of $\bar{U}_0(x)$ is

$$\hat{a}_w(0) = e^{\psi[\phi_w(x)e^{iwx},w] - iwx} \qquad (2.18)$$

Then the restriction that $\hat{a}_w(0)$ has to be defined in the transformed y-space is removed.

To show the validity of the technique, in (2.5) the coefficient functions are assumed to be $C_1(x) = 1 + (\frac{x}{a})^2$ and $C_2(t) = C_0$. Choose $\lambda_w = i$ in both (2.6) and (2.7). Then a $\phi_w(x)$ exists to be:

$$\phi_w(x) = e^{i[\tan^{-1}(\frac{x}{a}) - wx]}. \qquad (2.19)$$

This gives:

$$\hat{X}(w,t) = -\frac{iC_0}{a} \qquad (2.20)$$

and

$$Z_w(x,t) = e^{i[\tan^{-1}(\frac{x}{a}) - \frac{C_0 t}{a}]} \qquad (2.21)$$

Choose ψ-function:

$$\psi[Z_w(x,t),w] = \alpha(w) + aw\,\frac{Z_w(x,t) - 1/Z_w(x,t)}{Z_w(x,t) + 1/Z_w(x,t)} \qquad (2.22)$$

Then the solution of (1.1) is:

$$\bar{U}(x,t) = \int_{-\infty}^{\infty} e^{\alpha(w) + iwa\,\tan[\tan^{-1}(\frac{x}{a}) - \frac{C_o t}{a}]} dw\,. \tag{2.23}$$

Let $\hat{a}_w(0) = e^{\alpha(w)}$. Then $\hat{a}_w(0)$ represents the Fourier transform of $\bar{U}_o(x)$ in the x-space and

$$\bar{U}(x,t) = \int_{-\infty}^{\infty} \hat{a}_w(0) e^{iwa \cdot \tan[\tan^{-1}(\frac{x}{a}) - \frac{C_o t}{a}]} dw\,. \tag{2.24}$$

Correspondingly, the numerical solution is:

$$\bar{U}_h(x_n,t) \sim \int_{-\infty}^{\infty} e^{\psi[\phi_w(x_n) e^{iwx_n} + \int_0^t \hat{A}(w,\xi)d\xi, w]} dw\,. \tag{2.25}$$

3. ERROR APPROXIMATION

The errors are defined at the mesh points only. For simplicity, we consider first the case that $\phi_w(x) \equiv 1$ and then extend the results to the case that $\phi_w(x) \not\equiv 1$. In the former case, let $\{x_n\}$ be a set of equally-spaced mesh points, Δx apart, then $x = x_n$, the truncation and accumulated errors, $E_T(x_n,t)$ and $e(x_n,t)$, respectively can be defined as:

$$E_T(x_n,t) = \tilde{X} \cdot \bar{U}(x_n,t) - \tilde{A} \cdot \bar{U}(x_n,t) \tag{3.1}$$

and

$$e(x_n,t) = \bar{U}(x_n,t) - \bar{U}_h(x_n,t). \tag{3.2}$$

An analysis on sets of discrete values, $\{E_T(x_n,t)\}$ and $\{e(x_n,t)\}$ is not easy, and other approaches are necessary.

Consider the following functions [7]:

$$\bar{U}^*(x,t) = \sum_{n=-\infty}^{\infty} \bar{U}_h(x_n,t) \frac{\sin \frac{\pi}{\Delta x}(x-x_n)}{\frac{\pi}{\Delta x}(x-x_n)}, \tag{3.3}$$

$$e^*(x,t) = \sum_{n=-\infty}^{\infty} e(x_n,t) \frac{\sin \frac{\pi}{\Delta x}(x-x_n)}{\frac{\pi}{\Delta x}(x-x_n)}, \tag{3.4}$$

and

$$E_T^*(x,t) = \sum_{n=-\infty}^{\infty} E_T(x_n,t) \frac{\sin \frac{\pi}{\Delta x}(x-x_n)}{\frac{\pi}{\Delta x}(x-x_n)}. \tag{3.5}$$

If stable schemes are employed in the solutions of the problems, then for all n, at $x = x_n$, the functional values, $\bar{U}_h(x_n,t)$, $e(x_n,t)$ and $E_T(x_n,t)$ are bounded. This implies that the infinite series of (3.3) and (3.5) are all uniformly convergence, hence, the functions, $\bar{U}^*(x,t)$, $e^*(x,t)$ and $E_T^*(x,t)$ are continuous. Furthermore, at any mesh point, $x = x_n$,

$$\bar{U}^*(x_n,t) = \bar{U}_h(x_n,t), \tag{3.6}$$

$$e^*(x_n,t) = e(x_n,t), \tag{3.7}$$

and

$$E_T^*(x_n,t) = E_T(x_n,t)\,. \tag{3.8}$$

Hence, the functions $\bar{U}^*(x,t)$, $e^*(x,t)$ and $E_T^*(x,t)$, respectively can be considered as the numerical approximation, the accumulated and the truncated errors everywhere.

As defined in (3.3) to (3.5), the functions have also the property that their Fourier transforms vanish if the frequency w satisfied the condition $|w| \geq \frac{\pi}{\Delta x}$, ([1], [2], [3]), where Δx is the mesh size. This says that the functions are band-limited. On the other hand, if the problem solution $\bar{U}(x,t)$ is band-limited, then by Whittaker's sampling theory [12], $\bar{U}(x,t)$ can be written as:

$$\bar{U}(x,t) = \sum_{n=-\infty}^{\infty} \bar{U}(x_n,t) \frac{\sin \frac{\pi}{\Delta x}(x-x_n)}{\frac{\pi}{\Delta x}(x-x_n)}. \tag{3.9}$$

Substituting (3.2) into (3.4), and (3.1) into (3.5), we have:

$$e^*(x,t) = \sum_{n=-\infty}^{\infty} [\bar{U}(x_n,t) - \bar{U}_h(x_n,t)] \frac{\sin \frac{\pi}{\Delta x}(x-x_n)}{\frac{\pi}{\Delta x}(x-x_n)}. \tag{3.10}$$

and

$$E_T^*(x,t) =$$

$$\sum_{n=-\infty}^{\infty} [\tilde{X} \cdot \bar{U}(x_n,t) - \tilde{A} \cdot \bar{U}(x_n,t)] \frac{\sin \frac{\pi}{\Delta x}(x-x_n)}{\frac{\pi}{\Delta x}(x-x_n)}. \tag{3.11}$$

Substituting (2.9) and (2.1) into (3.10) and (3.11), we obtain:

$$e^*(x,t) \sim \sum_{n=-\infty}^{\infty} \left[\int_{-\infty}^{\infty} \hat{a}_w(0) \left\{ e^{iwx_n + \int_0^t \hat{X}(w,\xi)d\xi} - e^{iwx_n + \int_0^t \hat{A}(w,\xi)d\xi} \right\} dw \right] * \frac{\sin \frac{\pi}{\Delta x}(x-x_n)}{\frac{\pi}{\Delta x}(x-x_n)}, \tag{3.12}$$

and

$$E_T^*(x.t) \sim$$

$$\sum_{n=-\infty}^{\infty} [\tilde{X} \cdot - \tilde{A} \cdot] \left[\int_{-\infty}^{\infty} \hat{a}_w(0) e^{iwx_n + \int_0^t \hat{X}(w,\xi)d\xi} dw \right] * \frac{\sin \frac{\pi}{\Delta x}(x-x_n)}{\frac{\pi}{\Delta x}(x-x_n)}. \tag{3.13}$$

Both (3.12) and (3.13) can be simplified further by using the following relationships:

$$\frac{\sin \frac{\pi}{\Delta x}(x-x_n)}{\frac{\pi}{\Delta x}(x-x_n)} = \frac{\Delta x}{2\pi} \int_{-\frac{\pi}{\Delta x}}^{\frac{\pi}{\Delta x}} e^{iw(x-x_n)} dw, \tag{3.14}$$

$$\frac{\Delta x}{2\pi} \sum_{n=-\infty}^{\infty} e^{-i(w-w')x_n} = \delta(w-w'), \tag{3.15}$$

and

$$\int_{-\infty}^{\infty} f(w)\delta(w-w')dw = f(w') \tag{3.16}$$

for any continuous f(w).

In (3.12), the integrand is a continuous function of w and the infinite series is also convergent uniformly, hence, the integration and the summation can be interchanged [4]. Furthermore, the functions, $\hat{X}(w,t)$ and $\hat{A}(w,t)$ are independent of the space variable x, hence, the relationships, (3.14) to (3.16) can be applied to both (3.12) and (3.13). This gives:

$$e^*(x,t) \sim$$

$$\int_{-\frac{\pi}{\Delta x}}^{\frac{\pi}{\Delta x}} \hat{a}_w(0) e^{iwx} \left\{ e^{\int_0^t \hat{X}(w,\xi)d\xi} - e^{\int_0^t \hat{A}(w,\xi)d\xi} \right\} dw \tag{3.17}$$

By using similar reasoning to simplify (3.12) and the additional relationships,

$$\tilde{X} \cdot e^{iw'x} = \hat{X}(w',t)e^{iw'x} \tag{3.18}$$

and

$$\tilde{A} \cdot e^{iw'x_n} = \hat{A}(w',t)e^{iw'x_n}, \tag{3.19}$$

we also have:

$$E_T^*(x,t) \sim$$

$$\int_{-\frac{\pi}{\Delta x}}^{\frac{\pi}{\Delta x}} \hat{a}_w(0)[\hat{X}(w,t) - \hat{A}(w,t)] e^{iwx + \int_0^t \hat{X}(w,\xi)d\xi} dw \tag{3.20}$$

In (3.17) and (3.20), both functions, $\hat{a}_w(0)$ and $\hat{A}(w,t)$ are defined on the mesh points $\{x_n\}$.

In case that $\phi_w(x)$ exists to be nontrivial, then the space transform y of (2.2) is possible. Let $\{y_n\}$ be a set of equally spaced mesh points defined in the transformed y-space.

Then in the y-space, corresponding to (3.17) and (3.20), we have:

$$e^*(y,t) \sim \int_{-\frac{\pi}{h}}^{\frac{\pi}{h}} \hat{a}_w(0)e^{iwy}\left\{e^{\int_0^t \hat{X}(w,\xi)d\xi} - e^{\int_0^t \hat{A}(w,\xi)d\xi}\right\}dw \tag{3.21}$$

and

$$E_T^*(y,t) \sim \int_{-\frac{\pi}{h}}^{\frac{\pi}{h}} \hat{a}_w(0)e^{iwy}\left\{\hat{X}(w,t) - \hat{A}(w,t)\right\}e^{\int_0^t \hat{X}(w,\xi)d\xi} dw, \tag{3.22}$$

where the functions $\hat{a}_w(0)$ and $\hat{A}(w,t)$ are both defined in the transformed y-space. The inverse transformation gives:

$$e^{iwy} = \phi_w(x)e^{iwx}. \tag{3.23}$$

Substituting (3.23) into (3.21) and (3.22), we have:

$$e^*(x,t) \sim \int_{-\frac{\pi}{h}}^{\frac{\pi}{h}} \hat{a}_w(0)\phi_w(x)e^{iwx}\left\{e^{\int_0^t \hat{X}(w,\xi)d\xi} - e^{\int_0^t \hat{A}(w,\xi)d\xi}\right\}dw \tag{3.24}$$

and

$$E_T^*(x,t) \sim \int_{-\frac{\pi}{h}}^{\frac{\pi}{h}} \hat{a}_w(0)\phi_w(x)e^{iwx}\left\{\hat{X}(w,t) - \hat{A}(w,t)\right\}e^{\int_0^t \hat{A}(w,\xi)d\xi} dw \tag{3.25}$$

where both $\hat{a}_w(0)$ and $\hat{A}(w,t)$ are still defined in the transformed y-space. However, as described in the previous section, this restriction that $\hat{a}_w(0)$ and $\hat{A}(w,t)$ are defined in the y-space can be removed.

REFERENCES

1) Cambell, L.L., A Comparison of the Sampling Theorem of Kramer and Whittaker, Jour. of SIAM. Appl. Math. Vol. 12, No. 1, 1964, pp. 117-130.

2) Cambell, L.L., Sampling Theorem for the Fourier Transform of a Distribution with Bounded Support, pp. 353-372, SIAM J. Appl. Math. Vol. 16, No. 3, 1968, pp. 626-636.

3) Chiang, Y.F., Truncation and Accumulated Errors in Wave Propagation Problems, Journal of Computational Physics, No. 79, 1988, 353-372.

4) Courant, R., Differential and Integral Calculus, Vol. I d II, McGraw Hill, N.Y., 1936.

5) Hammaing, R.W., Numerical Methods for Scientists and Engineers, McGraw Hill, N.Y., 1962.

6) Majda, A., and Osher S., Propagation of Errors into Regions of Smoothness for Accurate Difference Approximations to Hyperbolic Equations, Comm. on Pure and Applied Math. Vol. XXX 1977, pp. 671-705.

7) Papoulis, H., The Fourier Integral and its Application, McGraw Hill, N.Y., 1962.

8) Richtmyer, R.D., and Morton, K.W., Difference Methods for Initial Value Problems, Interscience Publishers, N.Y., 1969.

9) Vichnevetsky, R., and Bowles, J. B., Fourier Analysis of Numerical Approximations of Hyperbolic Equations, SIAM, Phil. 1982.

10) Vichnevetsky, R., Computer Methods for Partial Differential Equations, Prentice Hall, Englewood Cliffs, N.J., 1981.

11) von Neumann, J., and Richtmeyer, R.D., A Method for the Numerical Calculation of Hyperdynamic Shocks, J. Appl. Physics, 21, 1950, pp. 232-237.

12) Whittaker, J.M., The Fourier Theory of the Cardinal Function, Proc. Roy. Soc. Edin., 1928, pp. 169-176.

Approximation, Optimization and Computing:
Theory and Applications, A.G. Law and C.L. Wang (eds.)
Elsevier Science Publishers B.V. (North-Holland)

UNIFORM APPROXIMATION OF CHRISTOFFEL NUMBERS FOR LAGUERRE WEIGHT*

Isabella Cravero and Luigi Gatteschi

Dipartimento di Matematica
Universitá di Torino
Via Carlo Alberto 10
I - 10123 Torino , Italy

Abstract. By using some uniform asymptotic expansions of Laguerre polynomials $L_n^{(\alpha)}(x)$ and their zeros, we obtain two estimates for the weights of the Gauss-Laguerre quadrature rule. Such estimates hold, as $n \to \infty$, for $k = 1, 2, \ldots, [qn]$ with fixed q , $0 < q < 1$, and for $k = [pn], [pn]+1, \ldots, n$ with fixed $p, 0 < q < 1$, respectively.

1. INTRODUCTION

Many problems in numerical integration theory and more generally in approximation theory require estimates of the weights associated with the Laguerre weight

$$w(x) = x^\alpha e^{-x} \quad , \qquad \alpha > -1 \quad ,$$

that is of the so-called Christoffel numbers $w_{n,k}^{(\alpha)}$ defined by

$$(1.1) \qquad w_{n,k}^{(\alpha)} = \frac{\Gamma(n+\alpha+1)}{\Gamma(n+1)} \lambda_{n,k}^{-1} \left\{ L_n^{(\alpha)\prime}(\lambda_{n,k}) \right\}^{-2},$$

for $k = 1, 2, \ldots, n$ and where $\lambda_{n,k} \equiv \lambda_{n,k}^{(\alpha)}$ is the k-th zero, in increasing order, of the Laguerre polynomial $L_n^{(\alpha)}(x)$. It is well-known that the zeros $\lambda_{n,k}$ lie in the oscillatory region $0 < x < \nu$, where

$$(1.2) \qquad \nu = 4n + 2\alpha + 2$$

is the turning point of the differential equation satisfied by $L_n^{(\alpha)}(x)$. Throughout this paper we shall assume that $\alpha > -1$ and ν will be defined by (1.2) .

For the weights $w_{n,k}^{(\alpha)}$ the following asymptotic representations hold (see Szegö [5], p.354) .

Let $\epsilon \le \lambda_{n,k} \le \omega$, then

$$(1.3) \qquad w_{n,k}^{(\alpha)} \cong \pi e^{-\lambda_{n,k}} \lambda_{n,k}^{\alpha+1/2} n^{-1/2}, \qquad n \to \infty,$$

where ϵ and ω are fixed positive numbers.

AMS(MOS)subject classifications. Primary 41A60, secondary 33A65.

*This work was supported by the Consiglio Nazionale delle Ricerche of Italy and by the Ministero della Pubblica Istruzione of Italy.

Let k be fixed and $n \to \infty$, then

$$(1.4) \qquad w_{n,k}^{(\alpha)} \cong (j_{\alpha,k}/2)^{2\alpha} \left\{ J'_\alpha(j_{\alpha,k}) \right\}^{-2} n^{-\alpha-1} \quad ,$$

where $j_{\alpha,k}$ is the k-th positive zero of the Bessel function $J_\alpha(x)$.

We notice that the two formulas (1.3) and (1.4) are not sufficient to represent $w_{n,k}^{(\alpha)}$, as $n \to \infty$, for all values of k. In this paper we shall obtain two asymptotic formulas for $w_{n,k}^{(\alpha)}$ which are of uniform type, in the sense that they hold for $k = 1, 2, \ldots, [qn]$ and for $k = [pn], [pn]+1, \ldots, n$ respectively, with q and p fixed numbers in (0,1). So, by assuming $q \ge p$ we can get estimates of $w_{n,k}^{(\alpha)}$ for all values of k.

As a particular case of our results we shall obtain in Section 5 two representations for the Christoffel numbers $w_{n,k}$ associated to the Hermite weight $w(x) = \exp(-x^2)$. For this case Whitney [6] has proved that,if $h_{n,k}$ is the k-th positive zero of the Hermite polynomials $H_n(x)$, then for the corresponding weight $w_{n,k}$ we have

$$(1.5) \qquad w_{n,k} \cong \frac{\pi}{\sqrt{2n}} e^{h_{n,k}^2}, \; n \to \infty,$$

with a relative error estimate

$$\frac{h_{n,i}^2}{2n - \delta} \qquad \text{if} \quad h_{n,i}^2 < 2(1+\delta),$$

being δ a fixed positive number.

2. SOME ASYMPTOTIC RESULTS

In this section we recall two well-known Erdélyi's [2] of the Laguerre polynomials $L_n^{(\alpha)}(x)$. For the notations we refer to a recent paper of Frenzen and Wong [3] on two general asymptotic expansions of $L_n^{(\alpha)}(x)$.

In the sequel the variable x will be assumed to satisfy the condition $0 < x < \nu$, where ν is defined by (1.2).

Theorem 2.1. Let

$$(2.1) \qquad A(t) = \frac{1}{2} \left\{ \arcsin \sqrt{t} + \sqrt{t(1-t)} \right\}, 0 \le t < 1.$$

Then

$$(2.2) \qquad 2^\alpha e^{-\nu t/2} L_n^{(\alpha)}(\nu t) = \{A(t)\}^{-\alpha} \cdot \alpha_0(t) J_\alpha\{\nu A(t)\} + \varepsilon(\nu, \alpha, t),$$

where

$$(2.3) \qquad \alpha_0(t) = \{A(t)\}^{\alpha+1/2} t^{-(2\alpha+1)/4} (1-t)^{-1/4}.$$

For the remainder $\varepsilon(\nu,\alpha,t)$ we have, as $n \to \infty$ uniformly with respect to t on $0 \le t \le b < 1, b$ being a fixed constant on $(0,1)$,

$$(2.4)\qquad |\,\varepsilon(\nu,\alpha,t)\,| = t^{1/2}\left|\frac{\overline{J}_{\alpha+1}\{\nu A(t)\}}{\{A(t)\}^{\alpha}}\right| O\left(\nu^{-1}\right),$$

where

$$\overline{J}_{\alpha+1}(u) = \begin{cases} J_{\alpha+1}(u), & \text{if} \quad 0 \le u \le \delta, \\ \left\{J^2_{\alpha+1}(u) + Y^2_{\alpha+1}(u)\right\}^{1/2}, & \text{if} \quad u > \delta. \end{cases}$$

Here δ is chosen such that $J_{\alpha+1}(u) \neq 0$ when $0 < u \le \delta$ and $\delta > 0$.

Theorem 2.2. Let

$$(2.5)\qquad B(t) = i\Big(\frac{3}{4}\{\arccos\sqrt{t} - \sqrt{t(1-t)}\}\Big)^{1/3},$$

with $0 < t \le 1$. Then, as $n \to \infty$,

$$(2.6)\qquad e^{-\nu t/2} L_n^{(\alpha)}(\nu t) = (-1)^n 2^{-\alpha+1/2}\beta_0(t)\cdot \cdot\, \nu^{-1/3}\mathrm{Ai}\left\{\nu^{2/3}B^2(t)\right\} + O\left(\nu^{-5/3}\right),$$

where

$$(2.7)\qquad \beta_0(t) = t^{-(2\alpha+1)/4}(1-t)^{-1/4}\,|\,B(t)\,|^{1/2}$$

and $\mathrm{Ai}(z)$ is the Airy function defined as in Abramowitz and Stegun [1] handbook. Here the bound for the error term holds uniformly for $0 < a \le t \le 1$ with a fixed. Correspondingly, by using the results obtained by one of us [4] for the zeros $\lambda_{n,k}$, we can derive the following one-term approximations.

Theorem 2.3. Let $\tau_{n,k}$ be the root of the equation

$$(2.8)\qquad \arcsin\sqrt{t} + \sqrt{t(1-t)} = \frac{2j_{\alpha,k}}{\nu}, 0 \le t < 1 \quad .$$

Then we have, as $n \to \infty$,

$$(2.9)\qquad \lambda_{n,k} = \nu\tau_{n,k} + \tau_{n,k}O\left(\nu^{-1}\right),$$

uniformly for $k = 1, 2, \ldots, [qn]$, where q is a fixed number in the interval $(0,1)$.

Theorem 2.4. Let a_j be the j-th zero of the Airy function $\mathrm{Ai}(x)$ and let $\tau_{n,k}$ be the root of the equation

$$(2.10)\qquad \arccos\sqrt{t} - \sqrt{t(1-t)} = \frac{4}{3\nu}\left(-a_{n-k+1}\right)^{3/2} \quad ,$$

with $0 < t \le 1$. Then, as $n \to \infty$,

$$(2.11)\qquad \lambda_{n,k} = \nu\tau_{n,k} + O\left(\nu^{-1}\right) \quad ,$$

for all values of $k = [pn], [pn]+1, \ldots, n$, where p is a fixed number in $(0,1)$.

3. THE FIRST ASYMPTOTIC FORMULA

We refer to the reciprocal of Christoffel numbers defined by (1.1), that is, by applying the derivation formula

$$\frac{d}{dx}L_n^{(\alpha)}(x) = -L_{n-1}^{(\alpha+1)}(x),$$

to the numbers

$$(3.1)\qquad \frac{1}{w^{(\alpha)}_{n,k}} = \frac{\Gamma(n+1)}{\Gamma(n+\alpha+1)}\lambda_{n,k}\left\{L_{n-1}^{(\alpha+1)}(\lambda_{n,k})\right\}^2.$$

Taking into account the asymptotic representation (2.2), it is easily seen that we must evaluate the function $J_{\alpha+1}(z)$ for

$$z = \overline{\nu}A\left(\lambda_{n,k}/\overline{\nu}\right),$$

where $A(t)$ is defined by (2.1) and

$$\overline{\nu} = 4n + 2\alpha = \nu - 2.$$

The Taylor series for $A(t)$ gives

$$A\Big(\frac{\lambda_{n,k}}{\overline{\nu}}\Big) = A\Big(\frac{\lambda_{n,k}}{\nu} + \frac{2\lambda_{n,k}}{\nu(\nu-2)}\Big) = = A\Big(\frac{\lambda_{n,k}}{\nu}\Big) + \frac{\lambda_{n,k}}{\nu(\nu-2)}\Big(\frac{\nu}{\lambda_{n,k}} - 1\Big)^{1/2} + \ldots,$$

that is

$$(3.2)\quad A\Big(\frac{\lambda_{n,k}}{\overline{\nu}}\Big) = A\Big(\frac{\lambda_{n,k}}{\nu}\Big) + + \frac{1}{\nu-2}\Big(\frac{\lambda_{n,k}}{\nu}\Big(1 - \frac{\lambda_{n,k}}{\nu}\Big)\Big)^{1/2} + O\left(\nu^{-2}\right),$$

as $n \to \infty$ uniformly for $k = 1, 2, \ldots, [qn]$, where q is fixed on $(0,1)$. Since [4, (2.8)]

$$A\Big(\frac{\lambda_{n,k}}{\nu}\Big) = \frac{j_{\alpha,k}}{\nu} + O\left(\nu^{-2}\right),$$

by using (2.9) we obtain from (3.2), as $n \to \infty$,

$$(3.3)\qquad z_{n,k} = \overline{\nu}A\Big(\frac{\lambda_{n,k}}{\overline{\nu}}\Big) = j_{\alpha,k} - \frac{2j_{\alpha,k}}{\nu} + + \left\{\tau_{n,k}\left(1 - \tau_{n,k}\right)\right\}^{1/2} + O\left(\nu^{-1}\right),$$

for $k = 1, 2, \ldots, [qn]$.
The recurrence relation

$$J_{\alpha+2}(u) = \frac{2(\alpha+1)}{u}J_{\alpha+1}(u) - J_{\alpha}(u),$$

applied to the Bessel function in the right-hand member of (2,4) , shows that

$$\left|\varepsilon(\overline{\nu}, \alpha+1, \frac{\lambda_{n,k}}{\overline{\nu}})\right| = \left|\frac{J_{\alpha+1}\left(z_{n,k}\right)}{\{A(\frac{\lambda_{n,k}}{\overline{\nu}})\}^{\alpha+1}}\right| O\left(\nu^{-1}\right),$$

where $z_{n,k}$ is given by (3.3). Hence, taking into account of the behavior of the function $\alpha_0(t)$ on $0 \le t \le b < 1$, the asymptotic representation of $L_{n-1}^{(\alpha+1)}(\lambda_{n,k})$, for $n \to \infty$ and $k = 1, 2, \ldots, [qn]$, can be put in the form

$$2^{\alpha+1} e^{-\lambda_{n,k}/2} L_{n-1}^{(\alpha+1)}(\lambda_{n,k}) = z_{n,k}^{-\alpha-1} \overline{\nu}^{\alpha+1} \cdot$$
$$\cdot \overline{\alpha}_0\left(\frac{\lambda_{n,k}}{\overline{\nu}}\right) J_{\alpha+1}(z_{n,k}) \left\{1 + O\left(\nu^{-1}\right)\right\},$$

where

$$\overline{\alpha}_0(u) = \{A(u)\}^{\alpha+3/2} u^{-(2\alpha+3)/4} (1-u)^{-1/4}.$$

Now we observe that

$$\frac{\lambda_{n,k}}{\overline{\nu}} = \tau_{n,k}\left\{1 + O\left(\nu^{-1}\right)\right\}$$

and that

$$\frac{\Gamma(n+1)}{\Gamma(n+\alpha+1)} = 4^\alpha \nu^{-\alpha}\left\{1 + O\left(\nu^{-1}\right)\right\}.$$

After a great deal of computation, substitution into (3.1) yields the main result of this section.

Theorem 3.1. Let $\tau_{n,k}$ be the root of the equation

$$\text{(3.4)} \qquad \arcsin\sqrt{t} + \sqrt{t(1-t)} = \frac{2 j_{\alpha,k}}{\nu}, 0 \le t < 1 \quad ,$$

and let

$$\text{(3.5)} \qquad j_{\alpha,k}^* = j_{\alpha,k} - \frac{2 j_{\alpha,k}}{\nu} + \left\{\tau_{n,k}\left(1 - \tau_{n,k}\right)\right\}^{1/2}.$$

Then, as $n \to \infty$,

$$\text{(3.6)} \left\{\exp\left(\lambda_{n,k}\right) w_{n,k}^{(\alpha)}\right\}^{-1} = \frac{1}{4} \nu^{-\alpha} \tau_{n,k}^{-\alpha-1/2} \cdot$$
$$\cdot \left(1 - \tau_{n,k}\right)^{-1/2} j_{\alpha,k}^* J_{\alpha+1}^2\left(j_{\alpha,k}^*\right)\left\{1 + O\left(\nu^{-1}\right)\right\},$$

uniformly for $k = 1, 2, \ldots, [qn]$, where q is fixed on $(0, 1)$.

If k is fixed as $n \to \infty$, then (3.4) and (3.5) give

$$\tau_{n,k} = \frac{j_{\alpha,k}^2}{\nu^2}\{1 + O(\nu^{-1})\},$$
$$j_{\alpha,k}^* = j_{\alpha,k}\{1 + O(\nu^{-1})\},$$

respectively, and (3.6) reduces to (1.4).

4. THE SECOND ASYMPTOTIC FORMULA

In this section we shall evaluate the numbers $L_{n-1}^{(\alpha+1)}(\lambda_{n,k})$ which occur in (3.1) by means of the Theorems 2.2 and 2.4. We give only the essential steps of the procedure. We first evaluate the function $B(t)$ defined by (2.5) at the point $t = \lambda_{n,k}/\overline{\nu}$, where $\overline{\nu}$ has the same meaning as in the previous section. In terms of the solution of the equation (2.10), we find

$$B^2\left(\frac{\lambda_{n,k}}{\overline{\nu}}\right) = a_{n-k+1}\nu^{-2/3} +$$
$$+ \frac{\{\tau_{n,k}(1-\tau_{n,k})\}^{1/2}}{(-a_{n-k+1})^{1/2}} \nu^{-2/3} + O(\nu^{-2}).$$

Hence, taking also into account that the zeros of the Airy function $\mathrm{Ai}(z)$ are negative,

$$\text{(4.1)} \; \overline{\nu}^{2/3} B^2\left(\frac{\lambda_{n,k}}{\overline{\nu}}\right) = \left(1 - \frac{4}{3\nu}\right) \cdot$$
$$\cdot \left\{a_{n-k+1} + \left(\frac{\tau_{n,k}(\tau_{n,k}-1)}{a_{n-k+1}}\right)^{1/2}\right\} + O(\nu^{-4/3}),$$

and

$$\text{(4.2)} \qquad \left|B\left(\frac{\lambda_{n,k}}{\overline{\nu}}\right)\right| = \frac{1}{\nu^{1/3}}\{-a_{n-k+1} -$$
$$- \left(\frac{\tau_{n,k}(\tau_{n,k}-1)}{a_{n-k+1}}\right)^{1/2}\} \cdot \{1 + O(\nu^{-4/3})\}.$$

Next, changing α into $\alpha + 1$ in the function $\beta_0(t)$ defined by (2.7), we set

$$\overline{\beta}_0(t) = t^{-(2\alpha+3)/4}(1-t)^{-1/4} \mid B(t) \mid^{1/2}$$

Then, observing that from (2.11)

$$\frac{\lambda_{n,k}}{\overline{\nu}} = \tau_{n,k}\left\{1 + O(\nu^{-1})\right\},$$

we obtain

$$\text{(4.3)} \qquad \overline{\beta}_0\left(\frac{\lambda_{n,k}}{\overline{\nu}}\right) = \tau_{n,k}^{-(2\alpha+3)/4}\left(1 - \tau_{n,k}\right)^{-1/4} \cdot$$
$$\cdot \left|B\left(\frac{\lambda_{n,k}}{\overline{\nu}}\right)\right|^{1/2}\left\{1 + O(\nu^{-1})\right\}.$$

Substitution into (3.1) gives the final result :

Theorem 4.1. Let $\tau_{n,k}$ be the root of equation

$$\text{(4.4)} \qquad \arccos\sqrt{t} - \sqrt{t(1-t)} = \frac{4}{3\nu}\left(-a_{n-k+1}\right)^{3/2},$$

with $0 < t \le 1$, and let

$$\text{(4.5)} \qquad a_{n-k+1}^* = a_{n-k+1} + \frac{\sqrt{\tau_{n,k}\left(1 - \tau_{n,k}\right)}}{\sqrt{-a_{n-k+1}}}.$$

Then, as $n \to \infty$,

$$\text{(4.6)} \qquad \left\{\exp\left(\lambda_{n,k}\right) w_{n,k}^{(\alpha)}\right\}^{-1} = \frac{1}{2}\nu^{-\alpha}\tau_{n,k}^{-\alpha-1/2} \cdot$$
$$\cdot (1 - \tau_{n,k})^{-1/2}(-a_{n-k+1}^*)^{1/2} \cdot$$
$$\cdot \mathrm{Ai}^2\left\{\left(1 - \frac{4}{3\nu}\right) a_{n-k+1}^*\right\}\left\{1 + O(\nu^{-1})\right\},$$

for $k = [pn], [pn] + 1, \ldots n$ where p is a fixed number in the interval (0,1).

We notice that, if k increases with n, and more precisely is such that the difference $n-k$ remains fixed as n increases, from (4.4) we obtain

$$1-\tau_{n,k}=\frac{-2^{2/3}a_{n-k+1}}{\nu^{2/3}}\Big(1+\frac{2^{2/3}a_{n-k+1}}{5\nu^{2/3}}\Big)\cdot \cdot\{1+O(\nu^{-4/3})\},$$

$$\tau_{n,k}=\Big(1+\frac{2^{2/3}a_{n-k+1}}{\nu^{2/3}}\Big)\{1+O(\nu^{-4/3})\}.$$

Hence, for $n\to\infty$

$$a^*_{n-k+1}=\{a_{n-k+1}+\frac{2^{1/3}}{\nu^{1/3}}\Big(1+\frac{3\cdot 2^{2/3}a_{n-k+1}}{5\nu^{2/3}}\Big)\}\cdot \cdot\{1+O(\nu^{-4/3})\},$$

and

$$\Big(1-\frac{4}{3\nu}\Big)a^*_{n-k+1}=a_{n-k+1}+\frac{2^{1/3}}{\nu^{1/3}}- -\frac{2a_{n-k+1}}{15\nu}+O(\nu^{-4/3}),$$

being $n-k$ fixed. Then, observing that

$$\mathrm{Ai}''(a_{n-k+1})=-a_{n-k+1}\,\mathrm{Ai}(a_{n-k+1})=0,$$

and

$$\mathrm{Ai}'''(a_{n-k+1})=-a_{n-k+1}\,\mathrm{Ai}'(a_{n-k+1}),$$

we have

$$\mathrm{Ai}\{\Big(1-\frac{4}{3\nu}\Big)a^*_{n-k+1}\}=\frac{2^{1/3}}{\nu^{1/3}}\Big(1-\frac{7a_{n-k+1}}{15\cdot 2^{1/3}\nu^{2/3}}\Big)\cdot \cdot\mathrm{Ai}'(a_{n-k+1})\{1+O(\nu^{-1})\},$$

and the following statement holds.

Corollary 4.1. Let $n-k$ fixed. Then, as $n\to\infty$,

$$\text{(4.7)}\quad \{\exp(\lambda_{n,k})\,w^{(\alpha)}_{n,k}\}^{-1}=2^{-2/3}\nu^{-\alpha-1/3}\cdot \cdot\{1+\frac{2^{-2/3}}{\nu^{1/3}a_{n-k+1}}-\Big(\alpha+\frac{16}{15}\Big)\frac{2^{2/3}a_{n-k+1}}{\nu^{2/3}}\}\cdot \cdot\mathrm{Ai}'^2(a_{n-k+1})\{1+O(\nu^{-1})\}.$$

5. THE HERMITE WEIGHT.

It is well-known that Hermite polynomials $H_n(x)$, which are orthogonal on $(-\infty,+\infty)$ with respect to the weight

$$w(x)=e^{-x^2/2},$$

can be reduced to Laguerre polynomials with $\alpha=\pm 1/2$ (see Szegö [5], p.106). More precisely, for the positive zeros $h_{n,1}, h_{n,2},\dots h_{n,[n/2]}$ in increasing order we have

$$h_{2m,k}=\{\lambda^{(-1/2)}_{m,k}\}^{1/2},$$
$$h_{2m+1,k}=\{\lambda^{(1/2)}_{m,k}\}^{1/2},$$

with $k=1,2,\dots m$ .The corresponding Christoffel numbers $w_{n,k}$ are given by (see Szegö [5], p.353)

$$w_{n,k}=\pi^{1/2}2^{n+1}n!\{H'_n(h_{n,k}\}^{-2},$$

and using Szegö's [5,p.106] relationships and (3.1) it is easily seen that

$$\text{(5.1)}\qquad w_{2m,k}=\frac{1}{2}w^{(-1/2)}_{m,k},$$

$$\text{(5.2)}\qquad w_{2m+1,k}=\frac{1}{2\lambda^{(1/2)}_{m,k}}w^{(1/2)}_{m,k}.$$

The representations of $w_{n,k}$ can be put in a particularly simple form, which involves only elementary functions, when we apply Theorem 3.1. Indeed, in this case, we have

$$J_{-1/2}(x)=\Big(\frac{2}{\pi x}\Big)^{1/2}\cos x,$$
$$J_{1/2}(x)=\Big(\frac{2}{\pi x}\Big)^{1/2}\sin x.$$

References

[1] - M.Abramowitz and I.A.Stegun,*Handbook of Mathematical Functions*, Applied Mathematics Series,55, National Bureau of Standards, Washington,DC, 1964.
[2] - A.Erdélyi and C.A.Swanson,*Asymptotic forms of Whittaker's confluent hypergeometric functions*,Mem. Amer. Math. Soc. No. 25, Providence, RI,1957.
[3] - C.L.Frenzen and R.Wong,*Uniform asymptotic expansions of Laguerre polynomials*,SIAM J.Math. Anal. 19 (1988), 1232-1248.
[4] - L.Gatteschi,*Uniform approximations for the zeros of Laguerre polynomials*,Numerical Mathematics, Proc.Int.Conf.Singapore 1988, Internat. Ser. Numer. Math. 86(1988), 137-148.
[5] - G.Szegö,*Orthogonal Polynomials*,Colloquium Publications, Vol.23,4th ed., American Mathematical Society,Providence,RI,1975.
[6] - E.L.Whitney,*Estimates of weights in Gauss-type quadrature*,Math. Comp., 19(1965),277-286.

Approximation, Optimization and Computing:
Theory and Applications, A.G. Law and C.L. Wang (eds.)
Elsevier Science Publishers B.V. (North-Holland)

SOME CONVERGENCE ESTIMATES FOR THE EXTENDED INTERPOLATION

Giuliana CRISCUOLO
Istituto per Applicazioni della Matematica - CNR
Via P.Castellino 111
80131 Napoli Italy

Giuseppe MASTROIANNI
Istituto di Matematica - Università della Basilicata
Via N.Sauro 85
85100 Potenza Italy

Donatella OCCORSIO
Istituto per le Applicazioni del Calcolo "M.Picone" - CNR
V.le del Policlinico 137
00161 Roma Italy

Abstract. The authors consider as knots of interpolation the zeros of orthogonal polynomials corresponding to a weight w and construct formulas of extended interpolation adding the zeros of orthogonal polynomials with respect to a weight u related to w. Some pointwise estimates for the extended interpolating polynomials are also given.

1. EXTENDED INTERPOLATION FORMULAS

Let w be a weight function and $\{p_m(w)\}$ be the corresponding system of orthonormal polynomials defined by

$p_m(w;x)=\gamma_m(w)x^m$+lower degree terms, $\gamma_m(w)>0$

and

$\int_{-1}^{1} p_m(w;x)p_n(w;x)w(x)dx=\delta_{m,n}$.

For a given bounded function f the correspon-ding Lagrange interpolating polynomial on the zeros $x_{m,i}(w)$ of $p_m(w)$ is denoted by $L_m(w;f)$.

Moreover, let u be another weight function (not necessarily $u \neq w$), $\{p_n(u)\}$ be the corresponding system of orthonormal polynomials, $x_{n,i}(u)$ be the zeros of $p_n(u)$ and let $\{n=n(m),\ m\in N\}$ be a sequence of integers such that $\lim_{m\to\infty} n(m)=\infty$.

Thus, we define the "Extended Interpolation Polynomial" (E.I.P.) $L_{m+n}(w,u;f)$ as the Lagrange polynomial of degree $m+n-1$ which interpolates the function f at the points $x_{m,i}(w)$, $i=1,\dots,m$ and $x_{n,i}(u)$, $i=1,\dots,n$.

Since

$L_{m+n}(w,u;f)=p_m(w)L_n(u;fp_m^{-1}(w))+p_n(u)L_m(w;fp_n^{-1}(u))$

the E.I.P. makes sense when the polynomials $p_m(w)$ and $p_n(u)$ have not common zeros for every fixed $m\in N$.

In the case $u=w$, we can make the choice $n(m)=m+1$; indeed the polynomials $p_m(w)$ and $p_{m+1}(w)$ have not common zeros for every fixed $m\in N$; then, for every weight w we obtain the interpolation formula (cf.[2])

$$L_{2m+1}(w,w;f;x)=\frac{\gamma_m(w)}{\gamma_{m+1}(w)}p_m(w;x)p_{m+1}(w;x)\cdot$$
$$\cdot\left\{\sum_{i=1}^{m+1}\frac{\lambda_{m+1,i}(w)}{x-x_{m+1,i}(w)}f(x_{m+1,i}(w))-\right.$$
$$\left.-\sum_{i=1}^{m}\frac{\lambda_{m,i}(w)}{x-x_{m,i}(w)}f(x_{m,i}(w))\right\},$$

where $\lambda_{m,i}(w)$, $i=1,\dots,m$, are the Christoffel constants defined by $\lambda_{m,i}(w)=\lambda_m(w;x_{m,i}(w))$, being $\lambda_m(w;x) = [\sum_{i=0}^{m-1} p_i^2(w;x)]^{-1}$ the m-th Christoffel function.

Furthermore, in the case $u \neq w$ the following theorem holds.

Theorem 1. Let w be a weight function such that

$$\lim_{m\to\infty} 2^{-m}\gamma_m(w) = \frac{1}{\sqrt{\pi}} \exp\{-\frac{1}{2\pi}\int_{-1}^{1} \frac{\log w(t)}{\sqrt{1-t^2}}dt\}.$$

If $w_1(x)=(1-x)w(x)$, $w_2(x)=(1+x)w(x)$ then the polynomials $p_m(w_1)$ and $p_m(w_2)$ have not common zeros for every fixed $m \in N$.

Moreover, if $\bar{w}(x)=(1-x^2)w(x)$ then the polynomials $p_{m+1}(w)$ and $p_m(\bar{w})$ have not common zeros for every fixed $m \in N$.

The previous theorem, the proof of which will be published elsewhere, allow us to construct E.I.P. on an odd or even number of knots. In addition, since the knots are zeros of orthogonal polynomials it is possible to add the points ± 1 or, following a procedure used in [1], to add p points $z_{m,j} \in (-1,-1+cm^{-2})\cup(1-cm^{-2},1)$, $j=1,2,...,p$, with c a suitable constant.

It is interesting to consider also extended interpolation formulas starting by a Hermite interpolating polynomial. In particular, we consider the Hermite interpolating polynomial $H_m^{(r,s)}(w;f)$ defined as follows:

if r and s are positive integers then $H_m^{(r,s)}(w;f)$ is the unique polynomial of degree $m+r+s-1$ which satisfies

$H_m^{(r,s)}(w;f;x_{m,i}(w)) = f(x_{m,i}(w))$, $1\le i\le m$,

$H_m^{(r,s)}(w;f;1)^{(j)} = f^{(j)}(1)$, $0\le j\le r-1$,

$H_m^{(r,s)}(w;f;-1)^{(j)} = f^{(j)}(-1)$, $0\le j\le s-1$.

We observe that the interpolating procedure used by Balázs, Kilgore and Vértesi in [1] becomes the Hermite procedure before defined when the add points $z_{m,j} \to \pm 1$, $j=1,2,...,p$.

Now, assuming that $w(x) = \psi(x)(1-x)^\alpha(1+x)^\beta$, $\alpha,\beta > -1$ and $\psi > 0$ is continuous and the modulus of continuity ω of ψ satisfies $\int_0^1 \omega(\psi;t)t^{-1}dt < \infty$, we give two easy examples of extended interpolation formulas starting by the Hermite polynomial $H_m^{(r,s)}(w;f)$:

$$H_{2m+1}^{(1,1)}(w,\bar{w};f;x)=p_{m+1}(w;x)p_m(\bar{w};x)\cdot \tag{1}$$

$$\cdot\{[\gamma_m(\bar{w})/\gamma_{m+1}(w)+\gamma_{m+1}(w)/\gamma_m(\bar{w})]^{-1}(1-x^2)\cdot$$

$$\cdot[\sum_{i=1}^{m+1}\frac{\lambda_{m+1,i}(w)}{x-x_{m+1,i}(w)} f(x_{m+1,i}(w)) -$$

$$-\sum_{i=1}^{m}\frac{\lambda_{m,i}(\bar{w})}{(x-x_{m,i}(\bar{w}))(1-x_{m,i}^2(\bar{w}))} f(x_{m,i}(\bar{w}))]+$$

$$+\frac{1+x}{2p_m(\bar{w};1)p_{m+1}(w;1)}f(-1)+$$

$$+\frac{1-x}{2p_m(\bar{w};-1)p_{m+1}(w;-1)}f(1)\},$$

where $\bar{w}(x)=(1-x^2)w(x)$;

$$H_{2m+1}^{(s,s)}(w,w;f;x)=p_{m+1}(w;x)p_m(w;x)\cdot \tag{2}$$

$$\cdot\{\gamma_m(w)/\gamma_{m+1}(w)(1-x^2)^s\cdot$$

$$\cdot[\sum_{i=1}^{m+1}\frac{\lambda_{m+1,i}(w)}{(x-x_{m+1,i}(w))(1-x_{m+1,i}^2(w))^s}f(x_{m+1,i}(w))-$$

$$-\sum_{i=1}^{m}\frac{\lambda_{m,i}(w)}{(x-x_{m,i}(w))(1-x_{m,i}^2(w))^s}f(x_{m,i}(w))]+$$

$$+\sum_{j=0}^{s-1}A_j(x)f^{(j)}(1)+\sum_{j=0}^{s-1}B_j(x)f^{(j)}(-1)\},$$

where A_j and B_j are the polynomials of degree $2s-1$ such that $A_j^{(i)}(1)=\delta_{i,j}$, $A_j^{(i)}(-1)=0$, $B_j^{(i)}(-1)=\delta_{i,j}$, $B_j^{(i)}(1)=0$, $i=0,1,...,s-1$, $j=0,1,...,s-1$.

2. SOME ESTIMATES OF THE ERROR

The main aim of this paper is to give some

pointwise estimates of the remainder terms corresponding to the formulas (1) and (2).

Theorem 2. Let $w(x)=(1-x)^{\alpha}(1+x)^{\beta}$, $\alpha,\beta>-1$. For any function $f\in C^{s}[-1,1]$, $s\geq 0$, we have

$$|f(x)-H_{2m+1}^{(1,1)}(w,\bar{w};f;x)|\leq \text{const}\,[\log m + (\sqrt{1-x}+m^{-1})^{-2\alpha} + (\sqrt{1+x}+m^{-1})^{-2\beta}\,\omega_s(f;m^{-1}),\quad |x|\leq 1,$$

where $\omega_s(f;\delta)=\sup_{0\leq h\leq\delta}||\Delta_h f||_{[-1,1-sh]}$, $\delta>0$, is the s-th modulus of continuity and const is independent of f and m.

The next theorem gives a pointwise evaluation of the simultaneous approximation by the interpolatory formula (2).

Theorem 3. Let $w(x)=(1-x)^{\alpha}(1+x)^{\beta}$, $\alpha=\beta>-1$. For any function $f\in C^{s}[-1,1]$, $s\geq 1$, if $-1+s/2\leq\alpha=\beta\leq s/2$ then

$$|(f(x)-H_{2m+1}^{(s,s)}(w,w;f;x))^{(k)}|\leq \tag{3}$$

$$\leq \text{const}\,\left[\frac{\sqrt{1-x^2}}{m}\right]^{s-1-k}\frac{\log m}{m}\,\omega(f^{(s)};m^{-1}),\quad |x|\leq 1,\quad k=0,1,\ldots,s-1,$$

where const is independent of f and m.

For the sake of brevity, we will give elsewhere the proofs of the previous theorems. In particular, the inequalities (3) follow by some results of Gopengauz [3].

Finally, the hypothesis $\alpha=\beta$ of Theorem 3 is not essential; indeed, the same result holds also in the case of weights belonging to a class more large than the class of Gegenbauer weights.

REFERENCES

[1] Balázs, K. and Kilgore, T. and Vértesi, P., An Interpolatory Version of Timan's Theorem on Simultaneous Approximation, Manuscript.

[2] Criscuolo, G. and Mastroianni, G. and Occorsio, D., Convergence of Extended Lagrange Interpolation, Submitted.

[3] Gopengauz, I., A Theorem of A.F.Timan on the Approximation of Functions by Polynomials on a Finite Segment, Mat. Zametki 1 (1967), pp.163-172 (in Russian); Math. Notes 1 (1967), pp.110-116 (English translation).

Approximation, Optimization and Computing:
Theory and Applications, A.G. Law and C.L. Wang (eds.)
Elsevier Science Publishers B.V. (North-Holland)
© IMACS, 1990

MODIFIED SUMMATION OPERATORS AND THEIR APPLICATIONS

Jizheng Di

Inst. of Math., Dalian Univ. of Technology, Dalian 116024, China

ABSTRACT: In this paper, the author constructs some new operators based on a kind of summation operators, and studies their applications in approximation theory.

1. OPERATORS WITH VARIABLE KNOTS

Let Ω be a Banach space and $\mathbf{B}$ be a $\sigma-$algebra of Borel subsets of Ω induced by the norm on Ω. For $A \in \mathbf{B}, C(A)$ denotes the set of bounded real-valued functions defined on Ω, which are continuous in the following sense: For every $\varepsilon > 0$, there is a $\delta > 0$, when $\|s - t\| < \delta$, $s \in A$, $t \in \Omega$, we have $|f(s) - f(t)| < \varepsilon$.

LEMMA 1. Suppose that $\varphi_{k,n}(x) \geq 0$ $(x \in \Omega)$, $x_{k,n} \in \Omega$, $k = 0, 1, \cdots, n = 1, 2, \cdots$. If for any $\delta > 0$

$$\lim_{n\to\infty} \sum_{\|x_{k,n}-x\|<\delta} \varphi_{k,n}(x) = 1$$

$$\lim_{n\to\infty} \sum_{\|x_{k,n}-x\|\geq\delta} \varphi_{k,n}(x) = 0$$

hold uniformly on $x \in A$, then for any $f \in C(A)$

$$\lim_{n\to\infty} \sum_{k=0}^{\infty} \varphi_{k,n}(x)f(x_{k,n}) = f(x)$$

is valid uniformly on $x \in A$.

LEMMA 2. Under the conditions of Lemma 1, if there are $x'_{k,n} \in \Omega$, $k = 0, 1, \cdots, n = 1, 2, \cdots$, such that $\|x'_{k,n} - x_{k,n}\| \to 0$ $(n\to \infty)$ uniformly on $k = 0, 1, \cdots$, then for any $f \in C(A)$

$$\lim_{n\to\infty} \sum_{k=0}^{\infty} \varphi_{k,n}(x)f(x'_{k,n}) = f(x)$$

holds uniformly on $x \in A$.

In the following, we give some results related to the Bernstein operator.

COROLLARY. Let the following conditions be fulfiled:

(i) $\alpha_i(n)/\beta_i(n) = O(1)$ $(\beta_i(n) \neq 0)$, $\beta_i(n)/n \to 0$, $n \to \infty, i = 1, 2, \cdots, q$; $\beta_j(n) \to 0$, $\alpha_j(n)/n \to 0$,

$n \to \infty$, $j = q + 1, \cdots, r$.

(ii) $\lambda_i(n) \to \lambda_i, n \to \infty$, $i = 1, 2, \cdots, r$, $\sum_{i=1}^{r} \lambda_i(n) \equiv 1$.

Let

$$B_{i,n}(f,x) = \sum_{k=0}^{n} \binom{n}{k} f((k + \alpha_i(n))/(n + \beta_i(n)))$$

$$\times x^k(1 - x)^{n-k},$$

$$B_n^{\star}(f,x) = \sum_{i=1}^{r} \lambda_i(n)B_{i,n}(f,x).$$

Then for $f \in C([0,1])$,

$$\lim_{n\to\infty} B_n^{\star}(f,x) = f(x)$$

holds uniformly on $x \in [0,1]$.

For $q = r = 1$ and $\alpha_1(n) = 0$, the operator $B_n^{\star}(f,x)$ becomes Shikkema's (cf.[1]).

From this corollary we can obtain an affirmative answer to the question: For any positive integer m, whether there exists a sequence of linear operators $\{L_n(f,x)\}$ that converges to $f(x)$ $(n \to \infty)$ uniformly on $x \in [0,1]$ for $f \in C([0,1])$ and $L_n(P,x) \equiv P(x)$ $(x \in [0,1], n \geq m)$ for every polynomial $P(x)$ of degree $\leq m$?

THEOREM 1. For $m \geq 2$, there exist $\lambda_i(n)$, $n = 1, 2, \cdots, \sum_{i=1}^{p} \lambda_i(n) \equiv 1$, and constants $c_i, d_i, i = 1, 2, \cdots, p$, $p = (m + 1)(m + 2)/2$, such that

$$\lim_{n\to\infty} \overline{B}_n(f,x) = f(x)$$

holds uniformly on $x \in [0,1]$ for any $f \in C([0,1])$, and $\overline{B}_n(P,x) \equiv P(x)(x \in [0,1], n \geq m)$ for every polynomial $P(x)$ of degree $\leq m$, where

$$\overline{B}_n(f,x) = \sum_{i=1}^{p} \lambda_i(n) \sum_{k=0}^{n} \binom{n}{k}$$

$$\times x^k(1-x)^{n-k} f\left(\frac{k+c_i n^{(m-1)/m}}{n+d_i n^{(m-1)/m}}\right).$$

NOTE: $c_i, d_i, i = 1,2,\cdots,p$, can be any real numbers but the zero points of the polynomial

$$\begin{vmatrix} \cdots & \cdots & \cdots \\ d_1^k & \cdots & d_p^k \\ d_1^k c_1 & \cdots & d_p^k c_p \\ \cdots & \cdots & \cdots \\ d_1^k c_1^{m-k} & \cdots & d_p^k c_p^{m-k} \\ \cdots & \cdots & \cdots \end{vmatrix}, \quad k = 0,1,\cdots,m.$$

$\lambda_i(n)$ can be obtained by the choice of $c_i, d_i, i = 1,2, \cdots, p$.

2. COMPOUND SUMMATION OPERATORS

THEOREM 2. Under the conditions of Lemma 1, let $L_{k,n}(f), k = 0,1,\cdots, n = 1,2,\cdots$, be positive linear functionals of f, and $L_{k,n}(1) \to 1$, $L_{k,n}(\|*-x_{k,n}\|^2) \to 0$ $(n \to \infty)$ uniformly on $k = 0,1,\cdots$. Denote

$$L_n^*(f,x) = \sum_{k=0}^{\infty} \varphi_{k,n}(x) L_{k,n}(f).$$

Then for any $f \in C(A)$,

$$\lim_{n\to\infty} L_n^*(f,x) = f(x)$$

holds uniformly on $x \in A$.

From Theorem 2 we obtain the main results of [2],[3].

Theorem 3. Under the conditions of Theorem 2, let

$$a_n(x) = \sum_{k=0}^{\infty} \varphi_{k,n}(x),$$

$$\alpha_n^2(x) = \sum_{k=0}^{\infty} \varphi_{k,n}(x)\|x - x_{k,n}\|^2,$$

$$b_{k,n} = L_{k,n}(\|*-x_{k,n}\|^2),$$

$$\beta_n^2 = \sup_{0\le k<\infty} \beta_{k,n}^2, k = 0,1,2,\cdots, n = 1,2,\cdots, x \in A.$$

Then we have

$$|L_n^*(f,x) - f(x)| \le |f(x)||L_n^*(1,x) - 1|$$

$$+(b_n a_n(x) + 2\delta^{-2}(\beta_n^2 a_n(x) + b_n \alpha_n^2(x)))\omega(f,\delta),$$

where

$$\omega(f,\delta) = \sup_{\substack{s\in A, t\in\Omega \\ \|s-t\|\le\delta}} |f(s) - f(t)|.$$

3. ERROR SUMMATION OPERATORS

Under the conditions of Lemma 1, if $f(x_{k,n}), k = 0,1,\cdots$, $n = 1,2,\cdots$, are not known exactly, we have

THEOREM 4. Let $L_n(f,x) = \sum_{k=0}^{\infty} \varphi_{k,n}(x) f(x_{k,n})$, where $\varphi_{k,n}(x)$, $x_{k,n}$ with the same meaning as in Lemma 1, $\overline{L}_n(f,x) = \sum_{k=0}^{\infty} \varphi_{k,n}(x)(f(x_{k,n}) + \alpha_{k,n}(f))$, $n = 1,2,\cdots$, with $\lim_{n\to\infty} \alpha_{k,n}(f) = 0$ uniformly on $k = 0,1,\cdots$. Then for $f \in C(A)$,

$$\lim_{n\to\infty} \overline{L}_n(f,x) = f(x)$$

holds uniformly on $x \in A$. Further more, we have

$$|\overline{L}_n(f,x) - f(x)| \le |f(x)||L_n(1,x) - 1|$$

$$+(L_n(1,x) + \delta^{-2}\alpha_n^2(x))\omega(f,\delta)$$

$$+ \sup_{0\le k<\infty} |\alpha_{k,n}(f)| L_n(1,x),$$

where

$$\alpha_n^2(x) = \sum_{k=0}^{\infty} \varphi_{k,n}(x)\|x - x_{k,n}\|^2.$$

NOTE: $\overline{L}_n(f,x)$ may be not a linear operator.

REFERENCES

[1] Shikkema, P. C., Über die schurerschen linearen positiven operatoren. II., *Proc. Kon. Ned. Akad. Wetenschappen, Amsterdam, Auch in Indagationes Math.*, 78 (1975), 243-253.

[2] Hsu, L.C., Yang J. X. and Chen W. Z., On a class of summattion integral operators, *Approximation Theorey, Proc. 4th Nat. Conf. Approx. Theory, Dalian, China.* 91-92.

[3] Di, J. Z., On sum-integral operators and their applications in the spaces of vector-valued functions, *J. Dalian Univ. Tech.* 1989, 29 (4), 505-512.

Approximation, Optimization and Computing:
Theory and Applications, A.G. Law and C.L. Wang (eds.)
Elsevier Science Publishers B.V. (North-Holland)

EXISTENCE OF NON-LINEAR UNIFORM APPROXIMATIONS UNDER HERMITE (LAGRANGE) TYPE INTERPOLATORY CONSTRAINTS

Charles B. Dunham

Department of Computer Science
Middlesex College
The University of Western Ontario
London, Ontario, Canada N6A 5B7

ABSTRACT

Best uniform (ordinary) rational and exponential polynomial approximations can fail to exist under Hermite (Lagrange) type interpolatory constraints. This carries over to mean approximation. This phenomenon is examined by considering approximation by the more general dense compact families. Finally, sufficient conditions for existence are given.

EXAMPLES OF NON-EXISTENCE

Let $\beta > 0$ and consider Chebyshev approximation of f on $[0, \beta]$ by ordinary rational functions R under the Hermite-type constraints [1:2]

(*) $R^{(i)}(A, 0) = f^{(i)}(0) \quad i = 0, \ldots, n$.

This particular constraint is of interest in approximation of the negative exponential as applied to the solution of differential equations, see for example Lau [6] and Trickett [8]. The author [4] has studied numerical approximation under (*). Lawson and Lau [15] also study numerical approximation under (*) and make mistaken claims of existence.

We show that best approximations may not exist for any $n \geq 0$ with

$R(A, x) = a_0/(1 + a_1x + \ldots + a_{n+1} x^{n+1})$.

THEOREM: Let $f(0) > 0$, $f'(0) = \ldots = f^{(n)}(0) = 0$, and f attain its norm on $[0, \beta]$ only with negative values. Then no best approximation exists to f under the constraint (*).

Proof: As approximations are of one sign, ones satisfying (*) must be > 0. Hence for A satisfying (*)

$|| f - R(A,.)|| > || f - 0 || = || f ||$.

Define $R(A^k, x) = f(0)/(1 + kx^{n+1})$. By Taylor expansion of $1/(1 + y)$ it is seen by letting $y = kx^{n+1}$ that the Taylor series of $R(A^k, x)$ has constant term f(0) and all terms in x of order 1 to n have coefficient zero, hence (*) is satisfied by A^k. There exists $\delta > 0$ such that f is closer to f(0) than to zero on $[0, \delta]$. As $R(A^k,.) > 0$ and $R(A^k,.)$ is monotone decreasing on $[0, \beta]$, $f - R(A^k,.)$ is smaller than f on $[0, \delta]$. On $[\delta, \beta]$, $R(A^k,.)$ converges uniformly to zero. Thus $|| f - R(A^k,.) || \to || f ||$.

The same proof applies to approximation by $R^0_m[0, \beta]$, $m > n + 1$.

We can also let $\beta = \infty$ or approximate on a subset of $[0, \beta]$.

Non-existence in the case of Lagrange-type interpolation was shown by Loeb, who gave a single (non-differentiable) example. His result was generalized by the author [5]. D. Schmidt [7] has further results on non-existence with Lagrange-type interpolation.

Next consider mean (L_p) approximation for $0 < p < \infty$ [9:10, Problem 13]. The theorem extends if we choose f to be in addition such that 0 is closer to f than any positive function, which can be achieved by making f large and negative over most of the interval. As $R(A^k,.) \to 0$ pointwise on $(0, \beta]$, $|| f - R(A^k,.)||_p \to || f ||_p$.

Non-existence can be generalized to approximation by powered rationals [13]

$R(A, x) = a_0/(1 + a_1 x + \dots + a_{n+1} x^{n+1})^r$

r a fixed natural number: we let $R(A^k, x) = f(0)/(1 + kx^{n+1})^r$.

Non-existence also extends to transformed rationals [12]

$R(A, x) = [a_0/(1 + a_1 x + \dots + a_{n+1} x^{n+1})]^r$, r an odd natural number, by identical arguments, if we let $f(0) = 1$.

If instead we have the exponential-polynomial [14]

$R(A, x) = a_0 \exp(1 + a_1 x + \dots + a_{n+1} x^{n+1})$

the theorem likewise holds with

$R(A^k, x) = f(0) \exp(1 - kx^{n+1})$

and Taylor expansion of exp $(1 + y)$. It extends to mean approximation similarly.

DENSE COMPACTNESS

Let X be a compact space. Let $\|\ \|$ be the Chebyshev (sup) norm on C(X). Let $\mathcal{G}$ be a non-empty set of continuous functions on X. Let Z be a finite subset of X. The problem of approximation with (Lagrange-type) interpolation on Z [1] is, given $f \in C(X)$, to find an element G^* for which $e(G) = \|f-G\|$ attains its infimum $\rho(f)$ under the interpolatory constraints

(*) $G(z) = f(z) \qquad z \in Z$

Such an element G^* is called best to f with (Lagrange-type) interpolation on Z.

A useful sufficient condition for existence in ordinary Chebyshev approximation is dense-compactness [16, 17, 18].

DEFINITION: $\mathcal{G}$ is dense-compact if for any bounded sequence $\{G_k\} \in \mathcal{G}$, there is a subsequence $\{G_{k(j)}\}$ converging pointwise to $G_0 \in \mathcal{G}$ on a dense subset Y of X.

Examples are given in the three references above and [14]. Ordinary rationals and exponential polynomials (both on intervals) are included.

DEFINITION: G_0 is a normal element of $\mathcal{G}$ if $\{G_k\} \to G_0$ uniformly.

THEOREM: Let $\mathcal{G}$ be dense-compact. Let G_k satisfy (*) and $e(G_k)$ be a decreasing sequence with limit $\rho(f)$. There exists $G_0 \in \mathcal{G}$ such that a subsequence $\{G_{k(j)}\} \to G_0$ pointwise on a dense subset Y of X and $\|f-G_0\| \leq \rho(f)$. If any such G_0 is normal, it is best.

Proof: $\{e(G_k)\}$ being bounded implies $\{G_k\}$ is a bounded sequence. Apply the definition of dense compactness. For $y \in Y$, $\limsup |f(y) - G_{k(j)}(y)| \leq \rho(f)$, hence $|f(y) - G_0(y)| \leq \rho(f)$. As Y is dense in X and $f - G_0$ is continuous, $\|f-G_0\| \leq \rho(f)$ follows. If G_0 is normal, $0 = f(z) - G_{k(j)}(z) \to f(z) - G_0(z)$, which must therefore, be zero. Hence G_0 satisfies (*).

As we saw in the first section, non-normal G_0 (specifically, zero) of the theorem may not satisfy (*) and hence may not be best. A partial converse follows shortly.

REMARK: It would suffice for existence in the previous theorem for $\{G_k\} \to G_0$ pointwise on X. However, the author's extensive experience with practical non-linear families suggests that if convergence is not uniform, it is not pointwise either.

Extension of this section to equality in some derivatives is straightforward.

A partial converse to the previous theorem follows:

THEOREM: Let X be a subset of the real line. Let $\mathcal{G}$ have (Rice's) property Z of degree n on an interval I containing X. Let a finite subset Z be given with a bounded sequence $\{G_k\}$ such that

(i) $\{G_k\}$ takes fixed values Ψ on Z

(ii) $\{G_k\} \to H \in G$ uniformly on a closed neighbourhood F of a set E of 2n+1 points distinct from Z, and

(iii) $H \not\equiv \Psi$ on Z.

Then there exists $f \in C(X)$ such that $e(G_k) \to \rho(f)$ but f possesses no best approximation under (*).

Proof: First require that $f = \psi$ on Z. Next require that f - H alternates 2n times on E with amplitude greater than 3 max $\{||f - H||_Z, ||H|| + \sup_k ||G_k||\}$. Extend f continuously to X such that

$$(1) \quad |(f-H)x| \leq ||f - H||_E \qquad x \in F$$
$$|(f-H)(x)| \leq ||f - H||_E/2 \qquad x \notin F$$

Suppose $||f - G|| \leq ||f - H||$ for G not H. By arguments of the strong lemma of de la Vallee-Poussin, G - H has n zeros on I, contradiction. By (1) $||f-G_k|| = ||f - G_k||_F \to ||f - H||$.

A SUFFICIENT CONDITION FOR EXISTENCE

Let X be a compact subset of the real line and $C^{(\ell)}(X)$ be the space of ℓ - times differentiable functions on X with norm

$$||h|| = \sup\{|h(x)| : x \in X\}.$$

Let $\mathcal{G}$ be a non-empty subset of $C^{(\ell)}(X)$. Let I be a set of pairs $\{x_i, j\}$ where $x_i \in X$ and j is an integer in $[0, \ell]$. The problem of approximation with Hermite-Birkhoff interpolation [1, 53] is given $f \in C^{(\ell)}(X)$, to find a $G \in \mathcal{G}$ to minimize $||f-G||$ subject to the constraints

$$(*) \quad G^{(j)}(x_i) = f^{(j)}(x_i) \qquad \{x_i, j\} \in I$$

Such a G is called a best (Hermite-Birkhoff) approximation to f with respect to I.

DEFINITION: A non-empty family $\mathcal{G}$ of continuous functions is ℓ-fold boundedly compact if every bounded sequence $\{G_k\}$ from $\mathcal{G}$ has a subsequence $\{G_{k(j)}\}$ such that $\{G_{k(j)}\} \to G_0 \in \mathcal{G}$ uniformly in all derivatives of order up to ℓ.

THEOREM: Let $\mathcal{G}$ be ℓ-fold boundedly compact. If there is a G in $\mathcal{G}$ satisfying (*) then a best approximation exists to f.

Proof: Let G_k satisfy (*) and $||f-G_k|| \to \rho(f) = \inf\{||f-G|| : G \text{ satisfying } (*)\}$. By arguments of [16], $\{G_k\}$ is bounded. Select $\{G_{k(j)}\} \to G_0$ then by the uniform convergence hypothesis, G_0 satisfies (*). By arguments of [16], G_0 is best.

A special case of Hermite-Birkhoff interpolation is Lagrange interpolation with $\ell = 0$. In this case X need not be a subset of the real line.

We can weaken the sufficient condition to require only pointwise convergence of $\{G_k^{(j)}\}$ to $G_0^{(j)}$ at x_i for every (x_i, j) in I, but this does not appear to widen the class of actual examples given next.

Unisolvent families are 0-fold boundedly compact by the uniform convergence theorem of Tornheim [21, 72-73]. Whether they are ℓ - fold boundedly compact for $\ell > 0$ is open.

(Finite dimensional) linear families are ℓ - fold boundedly compact for any $\ell > 0$: Some transformations of linear families are ℓ-fold boundedly compact for any $\ell > 0$: see Lemma 3 of [19]. Let P and Q be finite dimensional linear families, then $\{p/q : p \in P, q \in Q, \mu \leq q \leq \nu\}$ is ℓ - fold boundedly compact for any $\ell > 0$ providing $\mu > 0$. Weaker restrictions can be placed on restricted rationals: see [20, 113].

REFERENCES

[1] B.L. Chalmers and G.D. Taylor, Uniform approximation with constraints, Jber. Deutsch Math.-Verein 81, 1979, 49-86.

[2] A.L. Perrie, Uniform rational approximation with osculatory interpolation, J. Comp. Sys. Sci. 4 (1970), 509-522.

[3] A.L. Perrie, A note on rational L_p approximation, J. Approximation Theory 23 (1978), 199-200.

[4] C.B. Dunham, Rational Chebyshev approximation with Hermite interpolation at zero, Utilitas Math. 23 (1983), 241-244.

[5] C.B. Dunham, Difficulties in rational Chebyshev approximation, in "Constructive Theory of Functions", proceedings Varna conference, Bulgarian Academy of Science, Sofia, 1984, 319-327.

[6] T.C. Lau, "A class of approximations to the exponential function for the Numerical solution of stiff differential equations", Ph.D. thesis, Computer Science, University of Waterloo, 1974.

[7] D. Schmidt, Another Case of Non-existence in Rational Chebyshev approximation with interpolatory constraints, J. Approximation Theory, accepted.

[8] S.R. Trickett, "Rational approximations to the exponential function for the numerical solution of the heat conduction problem", Master's thesis, Applied Mathematics, University of Waterloo, 1984.

[9] C.B. Dunham, Mean approximation by transformed and constrained rational functions, J. Approximation Theory 10 (1974), 93-100.

[10] C.B. Dunham, Problems in Best Approximation, Technical Report No. 62, Department of Computer Science, The University of Western Ontario.

[11] J.D. Lawson and D.A. Swayne, High order near best uniform approximations to the solution of heat conduction problems, in S.H. Lavington (ed.), Information Processing 80, North Holland, 1980, 741-746.

[12] C.B. Dunham, Transformed rational Chebyshev approximation, Numer. Math. 10 (1967), 147-152.

[13] C.B. Dunham, Existence of Chebyshev approximations by transformations of powered rationals, J. Gilewicz, M. Pindor, W. Siemaszko (eds), in Rational Approximation and its applications in Mathematics and Physics, Proceedings of Lancut 1985, Lecture Notes in Mathematics, 1237, Springer, 1987, 63-67.

[14] C.B. Dunham, Chebyshev approximation by an exponential-polynomial. Appendix Linearization of best approximation, Congressus Numerantium 61 (1988) (proc. Manitoba Conf. on Numer. Math. and Computing), 91-106.

[15] J.D. Lawson and T.C. Lau, Order constrained Chebyshev rational approximation, Utilitas Math. 34 (1988), 45-52.

[16] C.B. Dunham, Existence and continuity of the Chebyshev operator, SIAM Rev. 10 (1968), 444-446.

[17] C.B. Dunham, Esistence and continuity of the Chebyshev operator, II, in S.P. Singh and J.H. Burry (eds.), Non-linear Analysis and Applications, Dekker, New York, 1982, 403-412.

[18] C.Z. Zhu and C.B. Dunham, Existence and continuity of Chebyshev approximations with restraints in S.P. Singh (ed.), Approximation Theory and Applications, Pitman, Boston, 1985, 1-9.

[19] C.B. Dunham, Transformed linear Chebyshev approximation, Aequationes Math. 12 (1975), 6-11.

[20] C.B. Dunham, Chebyshev approximation by restricted rationals, Approx. Theory Appl. 1 (1985), 111-118.

[21] J. Rice, The Approximation of Functions, Volume 1, Addison-Wesley, Reading, Mass., 1964.

Approximation, Optimization and Computing:
Theory and Applications, A.G. Law and C.L. Wang (eds.)
Elsevier Science Publishers B.V. (North-Holland)

PERTURBATION THEOREM FOR CONSTRAINED MEAN APPROXIMATION

C.B. DUNHAM

The University of Western Ontario
London, Ontario, Canada N6A 5B7

ZHU Changzhong*

Shanghai University of Science and Technology
China

Let Q be a quadrature rule or integral on closed interval I. Let τ be a continuous mapping of the real line into the non-negative real line. Let F be an approximating function with parameter $A=(a_1,...,a_n)$ taken from a parameter space P, a subset of n-space, such that $F(A,.) \varepsilon C(I)$, and K be a subset of P. The Q,τ,K,f problem is, given $f \varepsilon C(I)$ to choose a parameter A^* to minimize $e(A)=Q(\tau(f-F(A,.)))$ over $A \varepsilon K$. In this note we study the dependence of A^* on Q,τ,K,f.

Let $\| \|_c$ be a norm on n-space.

DEFINITION [3]: F satisfies the strong Young's condition if (1) $\|A^k\|_c \to \infty$ implies that for some subsequence $\{A^{k(j)}\}$ we have a closed neighbourhood U such that $\inf\{|F(A^{k(j)}, x)|:x \varepsilon U\} \to \infty$; (2) $\{A^k\} \to A$ implies $F(A^k,.)\to F(A,.)$ uniformly on I.

Examples of families satisfying the strong Young's condition can be found in 3.

Let $\{K_0,K_1,K_2,...\}$ satisfy the following conditions:

(H_1) $A^k \varepsilon K_k$, $\{A^k\} \to A^0$ imply $A^0 \varepsilon K_0$,

and

(H_2) For $A^0 \varepsilon K_0$ there exist $A^k \varepsilon K_k$ with $\{A^k\} \to A^0$.

THEOREM: Let $Q_k(h_k) \to \infty$ if $\inf\{h_k(x): x \varepsilon V\} \to \infty$ for $h_k \geqq 0$, V a non-degenerate interval, and $Q_k(h_k) \to Q_0(h)$ if $h_k,h \varepsilon C(I)$, $h_k \to h$ uniformly on I. Let $\tau_k(t_k) \to \infty$ as $|t_k| \to \infty$, and $\tau_k(h_k) \to \tau_0(h)$ uniformly on I if $h_k,h \varepsilon C(I)$, $h_k \to h$ uniformly on I. Let F satisfy the strong Young's condition, $\{f_k\} \to f_0$ uniformly on I, and $Q_0(\tau_0(f-F(B^0,.))) < \infty$ with $B^0 \varepsilon K_0$. Let A^k be best parameter in the Q_k,τ_k,K_k,f_k problem. Then $\{A^k\}$ has an accumulation point and any such accumulation point is a best parameter for Q_0,τ_0,K_0,f_0.

Proof of THEOREM: Suppose that $\{\|A^k\|_c\}$ is unbounded, then by taking a subsequence if necessary we can assume that $\|A^k\|_c \to \infty$. By the strong Young's condition, for some subsequence of $\{A^k\}$, without loss of generality, we can assume that it is $\{A^k\}$ itself, we have a closed neighbourhood U of I such that

$$\inf\{|F(A^k,x)|:x \varepsilon U\} \to \infty.$$

Thus, for any $x \varepsilon U$, $|F(A^k,x)| \to \infty$, $|f_k(x)-F(A^k,x)| \to \infty$ and $\tau_k(f_k(x)-F(A^k,x)) \to \infty$. Hence,

$$\inf\{\tau_k(f_k(x)-F(A^k,x)):x \varepsilon U\} \to \infty,$$

furthermore

$$Q_k(\tau_k(f_k-F(A^k,.))) \to \infty.$$

By condition (H_2), there exist $B^k \varepsilon K_k$ with $\{B^k\} \to B^0$, hence, by assumptions of the THEOREM, we have

$$Q_k(\tau_k(f_k-F(B^k,.))) \to Q_0(\tau_0(f_0-F(B^0,.))) < \infty.$$

Thus, for k sufficiently large,

$$Q_k(\tau_k(f_k-F(B^k,.))) < Q_k(\tau_k(f_k-F(A^k,.))),$$

a contradiction.

As $\{A^k\}$ is bounded, it has an accumulation point, say A^0. By taking a subsequence if necessary, we can assume $A^k \to A^0 \varepsilon K_0$. If A^0 is not best parameter for Q_0,τ_0,K_0,f_0, then there is $C^0 \varepsilon K_0$ such that

$$Q_0(\tau_0(f_0-F(C^0,.))) < Q_0(\tau_0(f_0-F(A^0,.))).$$

But

$$Q_k(\tau_k(f_k-F(A^k,.))) \to Q_0(\tau_0(f_0-F(A^0,.))).$$

---

* presenter

Hence, for k sufficiently large,

$$Q_k(\tau_k(f_k-F(A^k,.))) > Q_0(\tau_0(f_0-F(C^0,.))).$$

For C^0, there exist $C^k \in K_k$ with $\{C^k\} \to C^0$, we have

$$Q_k(\tau_k(f_k-F(C^k,.))) \to Q_0(\tau_0(f_0-F(C^0))).$$

Thus, for k sufficiently large,

$$Q_k(\tau_k(f_k-F(C^k,.))) < Q_k(\tau_k(f_k-F(A^k,.))),$$

again a contradiction. The proof is complete.

COROLLARY: If the best approximation for Q_0,τ_0,K_0,f_0, say $F(A^*,.)$, is unique, then $F(A^k,.) \to F(A^*,.)$ uniformly on I even if A^* is not unique.

In case when I is a finite set of real line, we only need to assume that F satisfies Young's condition (i.e. there is a subset Y of I such that $\|A^k\|_c \to \infty$ implies $\max\{|F(A^k,x)|: x \in Y\} \to \infty$) and that $Q_k(h_k) \to \infty$ if $\max\{h_k(x): x \in Y\} \to \infty$ for $h_k \geqq 0$, Y a finite set, and keep all other assumptions in above THEOREM unchanged, a similar THEOREM as above can be given.

REMARK If we let $Q_0(g) = \int g$ for $g \in C(I)$, fix p in $[1,\infty)$ and take "norm" $N_k(g) = [Q_k(|g|^p)]^{1/p}$ (setting $\tau(t) = |t|^p$), then $N_k(g)$ tends to the L_p norm of g on I.

Now let us consider two special cases. First consider linear approximation with varying Lagrange-type interpolatory constraints. Let $g_1,\ldots,g_n \in C(I)$ satisfy the Haar condition on I, and $F(A,x) = a_1g_1(x)+\ldots+a_ng_n(x)$. Let $x_1^k,\ldots,x_j^k$ be all distinct, and $\{x_i^k\} \to x_i^0$, $i = 1,\ldots,j$, and

$$K_k = \{A:\ F(A,x_i^k) = f(x_i^k),\ i=1,\ldots,j\},$$

where j is a fixed number between 1 and n. Now we verify condition (H_1,H_2). Let $A^k \in K_k$ and $\{A^k\} \to A^0$. By uniform convergence of $F(A^k,.)$ to $F(A^0,.)$ we must have $F(A^0,x_i^0) = f(x_i^0)$, $i=1,\ldots,j$, i.e. $A^0 \in K_0$, establishing (H_1). Let Y be a set of n-j points distinct from $\{x_1^0,\ldots,x_j^0\}$. Let $A^0 \in K_0$ be given and choose A^k to satisfy

$$F(A^k,x_i^k) = f(x_i^k),\ i=1,\ldots,j,$$
$$F(A^k,y) = F(A^0,y),\ y \in Y.$$

By standard arguments, e.g. that of Rice [1, 24-25], and by the Haar assumption, we can get $\{A^k\} \to A^0$, establishing (H_2).

In cases with coalescing of nodes or a non-Haar approximant, the conclusions of the THEOREM may be not true.

EXAMPLE 1. Approximate $f(x) = x^2$ by linear combinations of $\{1,x\}$. We require interpolation on $\{0,1/k\}$: the only interpolant is k^3x and no limit exists as $k \to \infty$. But any multiple of x interpolates f at zero.

EXAMPLE 2. Approximate $f(x) = x^2$ by multiples of x. We require interpolation on $\{1/k\}$: apply the rest of the above example.

Next consider changing restricted range (linear) approximation. Let $g_k,h_k,g_0,h_0 \in C(I)$, $g_k \to g_0$, $h_k \to h_0$ uniformly on I. Let

$$K_k = \{A:\ g_k(x) \leqq F(A,x) \leqq h_k(x),\ x \in I\}$$
$$(k=0,1,2,\ldots).$$

Providing K_0 is nonempty, (H_1) is satisfied: suppose $A^k \in K_k$, $A^k \to A^0$, if $A^0 \notin K_0$, without loss of generality we can assume that $F(A^0,\bar{x}) < g_0(\bar{x})-\varepsilon$, $\bar{x} \in I$, $\varepsilon > 0$, then for all k sufficiently large $F(A^k,\bar{x}) < g_k(\bar{x})-\varepsilon/2$, a contradiction. In order to ensure condition (H_2) will be satisfied, we need to make the additional assumption [2,196].

ASSUMPTION: Given $B \in K_0$ and $\delta > 0$, there is B^δ such that $\|B-B^\delta\|_C < \delta$ and

$$g_0(x) < F(B^\delta,x) < h_0(x),\ x \in I.$$

The additional assumption above is necessary.

EXAMPLE 3. Let $I = [0,1]$, choose

$$h_k(x) = 0, \qquad 0 \leqq x \leqq 1/k,$$
$$= x-1/k, \qquad x>1/k,$$
$$g_k(x) = -h_k(x).$$

It is easy to see that the additional assumption is not satisfied. Consider approximation by first degree polynomials, {0} is the only approximant in the restricted range g_k, h_k. But the approximants in the restricted range of g_0, h_0 are $\{ax: |a| \leq 1\}$.

REFERENCES

1. J. Rice, "The Approximation of Functions", Vol.1, Addison-Wesley, Reading, Mass., 1964.

2. C.B. Dunham, Nearby linear Chebyshev approximation under Constraints, J. of Approx. Theory, 42(1984), 193-199.

3. C.B. Dunham, The strong Young condition and restricted powered rationals, Approx. Theory & its Appl.,4(1988), 103-107.

Approximation, Optimization and Computing:
Theory and Applications, A.G. Law and C.L. Wang (eds.)
Elsevier Science Publishers B.V. (North-Holland)

The Local and Far Field Approximate Solutions of the Heat Conduction Equation with Cylindrical Symmetry

Chao-Kang Feng, Faculty of Aeronautical Engineering
Tamkang University, Tamsui, Taipei, Taiwan, R.O.C.

A systematic approach for obtaining the local and far field approximate solutions of problems involving unsteady heat conduction in infinite internally bounded cylindrical solids by using the similarity and perturbation methods is exhibited.
The local approximate solution is given to the case of the cylinder surface temperature U(t) varying with a power of time ($U=t^m$) and the far field approximate solution is obtained to the case of constant heat flux over cylinder surface.

I. Introduction

Solutions of the unsteady heat conduction equation in a region bounded internally by a cylinder are important in physical and engineering applications. In many physical situations and numerical computations, one is interested in the form of the temperature distribution for small values of time (local solution) or large values of time (far field solution).

In this paper we shall present a systematic approach for obtaining the local and far field solutions for various boundary conditions to the second order approximation.

II. Local Approximate Solution

The unsteady heat equation in a medium possessing constant diffusivity α and cylindrical symmetry may be written as [1]

$$u_t = \alpha\left(u_{rr} + \frac{1}{r}u_r\right) \tag{1}$$

I.C. $\quad u(r,0)=0 \qquad (2)$

B.C. $\quad u(a,t)=U(t) \qquad (3)$

$\quad u(\infty,t)=0 \qquad (4)$

where u is the temperature, r is the radial coordinate, t is the time, and a is the radius of the cylinder.

The similarity transformation for u in (1) is [2]

$$u/U(t) = f(s)\ , \quad s=r/2\sqrt{\tau} \tag{5}$$

with $U(t)=t^m$ and $\tau=\alpha t$. Unfortunately, the boundary condition is not fitted because of the appearance of the value of r=a in (3). But a deeper look shows that the similarity and perturbation methods can be used to get valuable information about asymptotic behavior in our problem [3] and [4].

In the early time, the heat from the surface r=a has not diffused very far ($r\to a$) and the curvature of the heat surface should not matter. Then we shall be able to determine the local approximate solution for small values of τ.

If $u(a,t)=U_0=\text{const.}$ in (3) and as $\tau\to 0$, $\delta(\tau)\to 0$, then the local asymptotic expansion of u has the following form [5]

$$u/U_0 = V_1(r,\eta)+\delta(\tau)V_2(r,\eta) \tag{6}$$

$$V_1(r,\eta)=r^n f_1(\eta) \tag{7}$$

$$V_2(r,\eta)=r^n f_2(\eta)+r^p g_2(\eta) \tag{8}$$

$$\eta = (r-a)/2\sqrt{\tau} \tag{9}$$

where n, p and $\delta(\tau)$ are unknowns.

Sub. (6) into (1), we obtain

$$(f_1''+2\eta f_1')+\delta(\tau)\Big[\Big(f_2''+2\eta f_2'-4\cdot\frac{\tau\delta'}{\delta}\cdot f_2\Big)$$

$$+r^{p-n}\Big[\Big(g_2''+2\eta g_2'-4\cdot\frac{\tau\delta'}{\delta}g_2+\frac{2\sqrt{\tau}}{\delta}(2n+1)r^{-p+n-1}f_1'\Big]$$

$$+\cdots = 0 \tag{10}$$

From (10), we find p=n-1 and $\tau\delta'/\delta = 1/2$ if $\delta(\tau)=\sqrt{\tau}$ and equating like powers of τ, we obtain a sequence of equations and solutions for boundary condition $u(\infty,t)=0$:

O(1) $\qquad f_1''+2\eta f_1' = 0 \qquad (11)$
$\qquad f_1 = c_1 \operatorname{erfc}\eta$

$O(\sqrt{\tau})$ $\qquad f_2''+2\eta f_2'-2f_2 = 0 \qquad (12)$
$\qquad f_2 = c_2 \operatorname{ierfc}\eta$

$$g_2''+2\eta g_2'-2g_2 = -2(2n+1)f_1' \tag{13}$$

$$g_2 = c_3 \operatorname{ierfc}\eta - \frac{c_1(2n+1)}{\sqrt{\pi}}\cdot e^{-\eta^2}$$

where $\operatorname{erfc}\eta = \dfrac{2}{\sqrt{\pi}}\displaystyle\int_\eta^\infty e^{-x^2}dx$ and

$$i^q\operatorname{erfc}\eta = \int_\eta^\infty i^{q-1}\operatorname{erfc}x\,dx\ , \qquad q=1,2,\cdots \tag{14}$$

Then (6) becomes

$$\frac{u}{U_0} = c_1 r^n \operatorname{erfc}\eta+\sqrt{\tau}\cdot\Big[\Big(c_2+\frac{c_3}{r}\Big)r^n \operatorname{ierfc}\eta\Big]$$

$$-\sqrt{\tau}\cdot\frac{c_1(2n+1)}{\sqrt{\pi}}\cdot r^{n-1}e^{-\eta^2}+O(\tau) \tag{15}$$

From b.c. (3), $u(a,t)=U_0$, (15) gives n=-1/2, $c_1=\sqrt{a}$ and $ac_2+c_3=0$ where c_2 and c_3 will be obtained from higher order approximations.

In general, if b.c. $u(a,t)=U(t)=t^m$, as $\tau\to 0$, the local asymptotic expansion for u in (1) is found as

$$u = t^{m}\cdot[r^{-1/2}f_1+\sqrt{\tau}\cdot r^{-1/2}(f_2+r^{-1}g_2)$$
$$+\tau\cdot r^{-1/2}\cdot(f_3+r^{-1}g_3+r^{-2}h_3)+O(\tau^{3/2})] \qquad (16)$$

Upon equating like powers of τ, we obtain a sequence of equations and solutions for $u(\infty,t)=0$:

$O(1)$ $\quad f_1''+2\eta f_1'-2(2m)f_1=0 \qquad (17)$
$\quad f_1 = c_1 i^{2m}\operatorname{erfc}\eta$

$O(\sqrt{\tau})$ $\quad f_2''+2\eta f_2'-2(2m+1)f_2=0 \qquad (18)$
$\quad f_2 = c_2 i^{2m+1}\operatorname{erfc}\eta$

$\quad g_2''+2\eta g_2'-2(2m+1)g_2=0 \qquad (19)$
$\quad g_2 = c_3 i^{2m+1}\operatorname{erfc}\eta$

$O(\tau)$ $\quad f_3''+2\eta f_3'-2(2m+2)f_3=0 \qquad (20)$
$\quad f_3 = c_4 i^{2m+2}\operatorname{erfc}\eta$

$\quad g_3''+2\eta g_3'-2(2m+2)g_3=0 \qquad (21)$
$\quad g_3 = c_5 i^{2m+2}\operatorname{erfc}\eta$

$\quad h_3''+2\eta h_3'-2(2m+2)h_3=(4g_2'-f_1) \qquad (22)$

$$h_3=c_6 i^{2m+2}+\frac{(c_1+4c_3)}{4}\cdot i^{2m}\operatorname{erfc}\eta$$

Finally, as $\tau\to 0$, the local approximate solutions of u and u_r to the second order are

$$u = t^{m}\Big[\frac{c_1}{r^{1/2}}\cdot i^{2m}\operatorname{erfc}\eta+\sqrt{\tau}\Big(\frac{c_2}{r^{1/2}}+\frac{c_3}{r^{3/2}}\Big)i^{2m+1}\operatorname{erfc}\eta$$
$$+\tau\cdot\frac{(c_1+4c_3)}{4r^{5/2}}i^{2m}\operatorname{erfc}\eta+\tau\Big(\frac{c_4}{r^{1/2}}+\frac{c_5}{r^{3/2}}+\frac{c_6}{r^{5/2}}\Big)\cdot$$
$$i^{2m+2}\operatorname{erfc}\eta+O(\tau^{3/2})\Big] \qquad (23)$$

$$u_r=\frac{-t^{m-1/2}}{2\sqrt{\alpha}}\Big\{\frac{c_1}{r^{1/2}}i^{2m-1}\operatorname{erfc}\eta+\sqrt{\tau}\Big[\frac{c_2}{r^{1/2}}+\frac{(c_1+c_3)}{r^{3/2}}\Big]i^{2m}\cdot$$
$$\operatorname{erfc}\eta+\frac{\tau}{4}\cdot\frac{(c_1+4c_3)}{r^{5/2}}i^{2m-1}\operatorname{erfc}\eta+\tau\Big[\frac{c_4}{r^{1/2}}$$
$$+\frac{(c_2+c_5)}{r^{3/2}}+\frac{(3c_3+c_6)}{r^{5/2}}\Big]i^{2m+1}\operatorname{erfc}\eta+O(\tau^{3/2})\Big\} \qquad (24)$$

Applications :

(i) Constant Surface Temperature $\quad u(a,t)=U_0$

Sub. $u(a,t)=U_0$ into (23), we obtain $m=0$, $c_1=U_0\sqrt{a}$, $c_2=U_0/4\sqrt{a}$ and $c_3=-U_0\sqrt{a}/4$, the local approximate solution to the second order is

$$\frac{u}{U_0}=\Big(\frac{a}{r}\Big)^{1/2}\cdot\operatorname{erfc}\eta+\sqrt{\tau}\frac{(r-a)}{4a^{1/2}r^{3/2}}\operatorname{ierfc}\eta+O(\tau) \qquad (25)$$

From (24), the local variation of the surface heat flux is

$$Q=-Ku_r|_{r=a}=\frac{KU_0}{a}\Big[\frac{1}{(\pi T)^{1/2}}+\frac{1}{2}-\frac{1}{4}\Big(\frac{T}{\pi}\Big)^{1/2}+O(T)\Big] \qquad (26)$$

where K is the thermal conductivity and $T=\tau/a^2$.

(ii) Constant Heat Flux $\quad -Ku_r|_{r=a}=Q_0$

Sub. $-Ku_r|_{r=a}=Q_0$ into (24), we obtain $m=1/2$, $c_1=2\sqrt{a\alpha}Q_0/K$, $c_2=-3\sqrt{\alpha}Q_0/2\sqrt{a}K$ and $c_3=-\sqrt{a\alpha}\,Q_0/2K$ and from (23), local solution to the second order is

$$u=\frac{2Q_0}{K}\Big(\frac{a}{r}\Big)^{1/2}\sqrt{\tau}\Big[\operatorname{ierfc}\eta-\sqrt{\tau}\frac{(3r+a)}{4ar}i^2\operatorname{erfc}\eta+O(\tau)\Big] \qquad (27)$$

and from (27), the local variation of the surface temperature is

$$\frac{Ku_s}{aQ_0}\Big|_{r=a}=\frac{2}{\sqrt{\pi}}\sqrt{T}-\frac{T}{2}+O(T^{3/2}) \qquad (28)$$

III. Far Field Approximate Solution

We now study the far field approximate solution for constant heat flux, then (1) becomes

$$u_T=\Big(u_{\rho\rho}+\frac{1}{\rho}u_{\rho}\Big) \qquad (29)$$

I.C. $\quad u(\rho,0)=0 \qquad (30)$

B.C. $\quad u_{\rho}(1,T)=-aQ_0/K \qquad (31)$

$\quad u(\infty,T)=0 \qquad (32)$

where $\rho=r/a$ and $T=\tau/a^2$.

For large values of time, the temperature profile has spread very far ($r \gg a$), i.e. $\eta=(r-a)/2\sqrt{\tau}\to r/2\sqrt{\tau}=\rho/2\sqrt{T}$ and the infinite long cylinder looks like a line source relatively.

Now we integrate of (29) over the region $a < r < \infty$ to obtain

$$\frac{d}{dt}\int_a^\infty rudr = \text{Constant} \qquad (33)$$

By using the linear group of transformations [6]

$$u' = \lambda^{\beta}u$$
$$r' = \lambda r$$
$$t' = \lambda^{2}t \qquad (34)$$
$$a' = \lambda a$$
$$Q_0' = \lambda^{\beta-1}Q_0$$

Then (33) becomes

$$\frac{d}{dt}\int_a^\infty rudr=\frac{1}{\lambda^{\beta}}\cdot\frac{d}{dt'}\int_{a'/\lambda}^{\infty}r'u'dr' \qquad (35)$$

For far field $r \gg a$, (35) should be little affected if we extend the lower limit to zero and we have a invariant problem. Then (35) requires $\beta=0$ and the far field asymptotic similarity solution of (29) with b.c.'s (30) to (32) is

$$u(\rho,T)\to F_1(\xi) \qquad (36)$$

with the similarity variable $\xi = \rho^2/4T$. In general, we shall be able to determine the far field approximate solution for large values of T.

As $T\to\infty$, $\Omega(T)\to 0$, $\sigma(T)\to 0$ and $\sigma/\Omega \to 0$, then the far field asymptotic expansion for u in (29) has the following form

$$u(\rho,T) = F_1(\xi)+\Omega(T)F_2(\xi)+\sigma(T)F_3(\xi)+\cdots \qquad (37)$$

$$\xi = \rho^2/4T \qquad (38)$$

As $T\to\infty$, from b.c. (31), $u_\rho \to \left.\dfrac{\partial F_1(\xi)}{\partial\rho}\right|_{\rho=1} \to$ const., it implies $F_1(\xi)$ has logarithmic behavior i.e. $F_1(\xi) \to \ln(4T/\rho^2)$. Then it suggests that $\sigma(T)<\Omega(T)<\ln T$ and as $T\to\infty$, $\mu(T)\to 0$, we assume that

$$\Omega(T) = \mu(T)\ln T \qquad (39)$$

Sub. (37) and (39) into (29), we obtain

$$[\xi F_1''+(1+\xi)F_1']+\mu(T)\ln T[\xi F_2''+(1+\xi)F_2'-T(\mu'/\mu)F_2]$$
$$+\sigma(T)[\xi F_3''+(1+\xi)F_3'-T(\sigma'/\sigma)F_3-(\mu/\sigma)F_2]+\cdots=0 \qquad (40)$$

From (40), we find the distinguished case $\sigma = \mu = 1/T$ and $\Omega=\ln T/T$ and the far field asymptotic expansion for u in (29) is found as

$$u=F_1+\frac{\ln T}{T}\cdot F_2+\frac{1}{T}\cdot F_3+\frac{\ln^2 T}{T^2}\cdot F_4+\frac{\ln T}{T^2}\cdot F_5+\frac{1}{T^2}\cdot F_6+\cdots \qquad (41)$$

Upon equating like powers of T, we obtain

$$\xi F_1''+(1+\xi)F_1'= 0 \qquad (42)$$

$$\xi F_2''+(1+\xi)F_2'+ F_2= 0 \qquad (43)$$

$$\xi F_3''+(1+\xi)F_3'+ F_3= F_2 \qquad (44)$$

$$\xi F_4''+(1+\xi)F_4'+ 2F_4= 0 \qquad (45)$$

$$\xi F_5''+(1+\xi)F_5'+ 2F_5= 2F_4 \qquad (46)$$

$$\xi F_6''+(1+\xi)F_6'+ 2F_6= F_5 \qquad (47)$$

From b.c.'s (31) and (32), we find

$$F_1 = C_1E_1(-\xi) \qquad (48)$$

$$F_2 = C_2e^{-\xi} \qquad (49)$$

$$F_3 = C_3e^{-\xi}-C_2e^{-\xi}\ln\sigma\xi \qquad (50)$$

where $$E_1(-\xi) = -\int_\xi^\infty \frac{e^{-x}}{x}\cdot dx \qquad (51)$$

$$C_1 = -aQ_0/2K \qquad (52)$$

$$C_2 = aQ_0/8K \qquad (53)$$

$$C_3 = aQ_0[1+\ln(4/\sigma)]/8K \qquad (54)$$

$$\ln c = \text{Euler's const.} = 0.5772\cdots \qquad (55)$$

Finally, from (41), we obtain the far field solution for the constant heat flux problem to the second order approximation

$$\frac{Ku}{aQ_0} = \frac{-1}{2}E_1(-\xi)+\frac{1}{4T}e^{-\xi}\left(\ln\frac{1}{c\xi}+\frac{1}{2}-\ln\frac{1}{\rho}\right)+\cdots \qquad (56)$$

The first order solution in (56) is the classical similarity solution of the line source with constant heat flux [1] and the second order solution was also obtained in terms of functions related to the derivatives of the reciprocal Gamma function by the operational method [7].

As $T\to\infty$, we have [8]

$$-E_1(-\xi)= \ln\frac{1}{c\xi}+\xi-\frac{1}{4}\cdot\xi^2+O(\xi^3) \qquad (57)$$

Sub. (57) into (56), we obtain

$$\frac{Ku}{aQ_0} = \frac{1}{2}\left[\ln\frac{1}{c\xi}+\frac{1}{2T}\ln\frac{1}{c\xi}+\frac{1}{4T}\left(1+\rho^2-2\ln\frac{1}{\rho}\right)\right]+O\left(\frac{\ln^2 T}{T^2}\right) \qquad (58)$$

and from (58), the far field variation of surface temperature to the second order approximation is

$$\left.\frac{Ku_s}{aQ_0}\right|_{\rho=1} = \frac{1}{2}\left[\ln\frac{4T}{c}+\frac{1}{2T}\cdot\left(\ln\frac{4T}{c}+1\right)\right] \qquad (59)$$

The complete solution of (29) with constant heat flux over cylinder surface in terms of infinite integral over various combinations of Bessel functions was given by Carslaw and Jaeger [1] and from [1], the complete surface temperature variation is

$$\left.\frac{Ku_s}{aQ_0}\right|_{\rho=1} = \frac{-2}{\pi}\int_0^\infty (1-e^{-Tv^2})\left\{\frac{J_0(v)Y_1(v)-Y_0(v)J_1(v)}{J_1^2(v)+Y_1^2(v)}\right\}\frac{dv}{v^2} \qquad (60)$$

The surface temperature variation of the second-order local solution (28) and far field solution (59) are compared with the numerical results of the complete solution (60) as shown in Fig. (1)

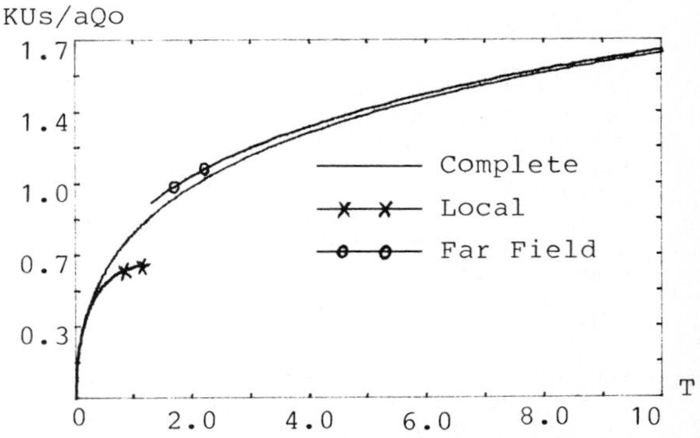

Fig. (1) KUs/aQo v.s. T

IV. CONCLUSIONS

A similarity representation is obtainable for a boundary value problem provided the governing differential equations and the associated boundary conditions are invariant under a group of transformations [9] and [10]. However, if any of the equations or boundary conditions is not invariant under a group, then the problem becomes nonsimilar [11].

In general, similarity olutions are often limiting or asymptotic solutions to given prob lem. In other words, similarity solutions typically appear as local or far field solutions near some singular points [12]. By using the perturbation techniques, attempts have been made in this paper to generate local and far field approximate solutions for our nonsimilar problems from known similarity solutions.

REFERENCES

[1] Carslaw, H.S. and Jaeger, J.C., Conduction of Heat in Solids (Oxford University Press, New York, 1959)

[2] Feng, C.K., A Study of Similarity Transformation for the Generalized Stokes-Rayleigh Problem, the Tran. of Aeronautical and Astronautical Society of R.O.C. (1972), PP. 33-40.

[3] Feng, C.K., Local Similarity Solution of the Tricomi Equation in the Elliptic Coordinates, Proceedings of International Symposium on Engineering Sciences and Mechanics National Cheng-Kung Univ., Taiwan, R.O.C., (1981) PP. 592-602.

[4] Feng,C.K., Second Order Transonic Solution Obtained from the Exact Solution of the Prandtl-Meyer Flow, The 30th Israel Annual Conference on Aviation and Astronautics, TelAviv, Israel, (1989) PP. 62-67.

[5] Kevorkian, J. and Cole, J.D., Perturbation Methods in Applied Mathematics (Springer-Verlag, New York, 1980)

[6] Hill, J.M., Solution of Differential Equations by Means of One-Parameter Groups (Pitman Advanced Publishing Program, London, 1982)

[7] Ritchie, R.H. and Sakakura, A.Y., Aymptotic Expansions of Solutions of the Heat Conduction in Internally Bounded Cylindrical Geometry, J. of Applied Physics, Vol. 27, (1956) PP. 1453-1459.

[8] Abramowitz, M. and Segun, I.A., Handbook of Mathematical Functions (Dover publications Inc., New York, 1968)

[9] Bluman, G. and Cole, J.D., Similarity Method for Differential Equations,(Springer -Verlag, New York, 1974)

[10] Feng,C.K., The General Similarity Solution of the Transonic Equation in the Hodograph Plane (Tricomi Equation), The 21th Israel Annual Conference on Aviation and Astronautics, Haifa, Israel, (1979) PP. 75-81.

[11] Seshadri, R. and Na,T.Y., Group Invariance in Engineering Boundary Value Problems (Springer-Verlag, New York, 1985)

[12] Feng,C.K., The General Similarity Solution of Laplace's Equation with Applications to Boundary-Value Problems, Proceedings of 1987 EQUADIFF Conference, Xanthi, Greece, (1987) PP. 237-248.

Approximation, Optimization and Computing:
Theory and Applications, A.G. Law and C.L. Wang (eds.)
Elsevier Science Publishers B.V. (North-Holland)

POINTWISE AND NODAL ERROR ESTIMATION IN LINEAR BOUNDARY VALUE PROBLEMS SOLVED BY THE STANDARD PARALLEL SHOOTING*

G. GHERI
Istituto di Matematiche Applicate, Facoltà di Ingegneria, Università di Pisa
Via Bonanno 25B, I-56126 Pisa (Italy)

This paper is concerned with an error estimation technique when the parallel shooting method is used in solving a linear boundary value problem for a system of first order equations. Formulas for the error at all mesh-points and in matrix form involving the shooting nodes only are given.

1. INTRODUCTION

We consider the boundary value problem

(1) $y'(x)=K(x)y(x)+f(x),\ 0<x<1,$

(2) $Ay(0)+By(1)=a,$

where $A, B, K(x) \in R^{m\times m}$, $y(x), a, f(x) \in R^m$ and $K(x), f(x) \in C^p[0,1]$. The problem has a unique solution $y(x) \in C^{p+1}[0,1]$ (Keller [1]) if and only if $Q=A+BW(1)$ is nonsingular being $W(x) \in R^{m\times m}$ the fundamental solution matrix satisfying $W'(x)=K(x)W(x)$, $W(0)=I$. Let

$$G(x,\tau)=W(x)Q^{-1}AW^{-1}(\tau),\ x\geq\tau,$$

$$G(x,\tau)=-W(x)Q^{-1}BW(1)W^{-1}(\tau),\ x\leq\tau,$$

be the Green's function for the problem (1)-(2). For somehow chosen nodes, $0=t_0<t_1<\ldots<t_M=1$, we solve the $(m+1)M$ initial value problems (IVP)

(3) $W_j'(x)=K(x)W_j(x),\ W_j(t_{j-1})=I,\ t_{j-1}\leq x\leq t_j,$

(4) $u_j'(x)=K(x)u_j(x)+f(x),\ u_j(t_{j-1})=0,\ t_{j-1}\leq x\leq t_j,$

$j=1,\ldots,M$. Then the solution of (1)-(2) is given by $y(x)\equiv y_j(x)=W_j(x)s_j+u_j(x)$ and satisfies the IVP

(5) $y_j'(x)=K(x)y_j(x)+f(x),\ y_j(t_{j-1})=s_j,\ t_{j-1}\leq x\leq t_j,$

$j=1,\ldots,M$. The m-vectors $s_1,\ldots,s_M$ are determined by setting $s^T=(s_1^T,\ldots,s_{M+1}^T)$ with $s_{M+1}=y_M(t_M)$,

$$R=\begin{pmatrix} A & & & B \\ -W_1(t_1) & I & & \\ & \cdot & \cdot & \\ & & \cdot & \cdot \\ & & -W_M(t_M) & I \end{pmatrix},\quad b=\begin{pmatrix} a \\ u_1(t_1) \\ \cdot \\ \cdot \\ u_M(t_M) \end{pmatrix}$$

and solving the linear system

(6) $Rs=b$

which expresses the continuity of $y(x)$ at internal nodes and the boundary condition $Ay_1(t_0)+By_M(t_M)=a$. Formally one has (Lentini, Osborne and Russell [2]) $(R^{-1})_{i1}=W_1(t_{i-1})Q^{-1}$, $i=1,\ldots,M+1$, and $(R^{-1})_{ij}=G(t_{i-1},t_{j-1})$, $i=1,\ldots,M+1$, $j=2,\ldots,M+1$.

2. ERROR ESTIMATION

We use a stable, accurate of order p, initial value method to compute an approximate solution of (3)-(4) in each interval $[t_{j-1},t_j]$ with mesh-points $x_n=t_{j-1}+nh$, $n=0,\ldots,N_j$, $N_j=(t_j-t_{j-1})/h$, $j=1,\ldots,M$, obtaining $W_{jn}=W_j(x_n)+O(h^p)$ and $u_{jn}=u_j(x_n)+O(h^p)$. With these values in R and b we denote $\bar{s}^T=(\bar{s}_1^T,\ldots,\bar{s}_{M+1}^T)$ the numerical solution

*Work supported in part by the italian Ministero della Pubblica Istruzione.

of (6) and assume $\bar{y}_{jn}=W_{jn}\bar{s}_j+u_{jn}\cong y_j(x_n)$.

THEOREM 1. *Setting* $\bar{y}_j(x)=W_j(x)\bar{s}_j+u_j(x)$, *the discretization error* $r_j(x_n)=\bar{y}_{jn}-\bar{y}_j(x_n)$ *arising in the numerical solution of the* (m+1)M *IVP (3)-(4) can be computed as the discretization error in the numerical solution of the* M *problems*

(7) $\bar{y}'_j(x)=K(x)\bar{y}_j(x)+f(x)$, $\bar{y}_j(t_{j-1})=\bar{s}_j$, $t_{j-1}\leq x\leq t_j$, $j=1,\ldots,M$.

PROOF. See Marzulli and Gheri [3] where the proposition is proved for the ordinary shooting.

The global discretization error $\bar{y}_{jn}-y_j(x_n)$ can be written as $e_j(x_n)=r_j(x_n)+\varepsilon_j(x_n)$ with $\varepsilon_j(x_n)=\bar{y}_j(x_n)-y_j(x_n)$, $n=0,\ldots,N_j$, $j=1,\ldots,M$. Using a suitable procedure (see, for example, Skeel [4]) and taking into account the theorem 1, one estimates $r_j(x_n)$: then $\varepsilon_j(x_n)$ can be appraised with an accuracy of the same order. In fact, setting $\sigma^T=(\sigma_1^T,\ldots,\sigma_{M+1}^T)$, $r^T=(r_0^T,\ldots,r_M^T)$ with $\sigma_j=\bar{s}_j-s_j$, $j=1,\ldots,M+1$, and $r_0=0$, $r_j=r_j(t_j)$, $j=1,\ldots,M$ the following theorem holds.

THEOREM 2. *For the global discretization error we have*

$$e_j(x_n)=r_j(x_n)+W_j(x_n)\sum_{i=1}^{M}G(t_{j-1},t_j)r_j,$$

$n=0,\ldots,N_j$, $j=1,\ldots,M$, *so that* $r_j(x_n)=O(h^p)$ *implies* $e_j(x_n)=O(h^p)$.

PROOF. Subtracting (5) from (7) we obtain $\varepsilon_j(x)=W_j(x)\sigma_j$ and thus $e_j(x_n)=r_j(x_n)+W_j(x_n)\sigma_j$. The system (6) can be expressed as

$$(8)\quad \begin{cases} As_1+Bs_{M+1}=a, \\ s_{j+1}=y_j(t_j),\ j=1,\ldots,M, \end{cases}$$

while using the computed values $\bar{y}_{j,N_j}$ it results

$$(9)\quad \begin{cases} A\bar{s}_1+B\bar{s}_{M+1}=a, \\ \bar{s}_{j+1}=\bar{y}_{j,N_j},\ j=1,\ldots,M. \end{cases}$$

By subtraction between (8) and (9) and taking into account the actual form of $e_j(t_j)$, we have, in compact form, $R\sigma=r$: the proof is achieved recalling the form of R^{-1} given above.

REMARK. Setting $e^T=(e_0^T,\ldots,e_M^T)$ with $e_0=0$, $e_j=e_j(t_j)$, $j=1,\ldots,M$, at the shooting nodes we can write

$$e=(I+VR^{-1})r,$$

where $V_{ij}=0$ if $i\neq j+1$ and $V_{j+1,j}=W_j(t_j)$, $j=1,\ldots,M$.

In practice, instead of R and V, we know the matrices $\bar{R}$ and $\bar{V}$, whose elements are the computed values W_{j,N_j} of $W_j(t_j)$. Therefore, setting $((\bar{R})^{-1})_{ij}=G_{i-1,j-1}$, $i=1,\ldots,M+1$, $j=2,\ldots,M+1$, according to the theorem 2, we obtain the following estimates

$$e_{jn}=r_j(x_n)+W_{jn}\sum_{i=1}^{M}G_{j-1,i}r_i,\quad n=0,\ldots,N_j, j=1,\ldots,M,$$

$$\bar{e}=(I+\bar{V}(\bar{R})^{-1})r,$$

which we call *pointwise* and *nodal* respectively.

COROLLARY. $\| e_j(x_n)-e_{jn}\| = O(h^{2p})$, $\|e-\bar{e}\|=O(h^{2p})$.

PROOF. Observe that $\| V-\bar{V}\| = O(h^p)$ and $\| R-\bar{R}\| = O(h^p)$: thus, by Banach lemma, $\| R^{-1}-(\bar{R})^{-1}\| = O(h^p)$, the proof is obtained.

EXAMPLE. We have considered the BVP (1)-(2) with

$$K(x)=\begin{pmatrix} 2c^2+24s^2 & -\pi-22cs \\ \pi-22cs & 2s^2+24c^2 \end{pmatrix},$$

$f(x)=0$, $y(0)+y(1)=(1,1)^T$, where $c=\cos(\pi x)$ and $s=\sin(\pi x)$.

We have employed the Euler's method to approximate the solutions of the IVP (3)-(4) and to compute the discretization error $r_j(x_n)$, according to the theorem 1. In fact, the discretization error satisfies (Prothero [5]) the IVP

$$r'_j(x)=K(x)r_j(x)+h\psi(x,\bar{y}_j(x)),\quad r_j(t_{j-1})=0,$$

where $\psi(x,y(x))$ is the principal error function. Defining $q_{jni}=(e_{jn})_i/(e_j(x_n))_i$, $i=1,2$, $n=0,\ldots,N_j$, $j=1,\ldots,M$, we have quoted in the following table the best value (the nearest to the unity) of q_{jni}, say q_b, and the worst one (the farthest from the unity), say q_w, corresponding to four values of the steplength.

h	0.1	0.02	0.01	0.002
q_b	0.95	0.98	1.00	1.00
q_w	0.88	0.90	0.95	0.97

REFERENCES

[1] Keller, H.B., Numerical Solution of Two Point Boundary Value Problems (SIAM Regional Conference Series in Applied Mathematics 24, Philadelphia, 1976).

[2] Lentini, M., Osborne, M.R. and Russell, R.D., The close relationships between methods for solving two-point boundary value problems, SIAM J. Numer. Anal., 22 (1985) pp. 280-309.

[3] Marzulli, P. and Gheri, G., Estimation of the global discretization error in shooting methods for linear boundary value problems, Journal of Computational and Applied Math., 27 (1989), in print.

[4] Skeel, R.D., Thirteen ways to estimate global error, Numer. Math., 48 (1986) pp. 1-20.

[5] Prothero, A., Estimating the accuracy of numerical solutions to ordinary differential equations, in: Gladwell, J. and Sayer, D.K. (eds.), Computational Technique for Ordinary Differential Equations (Academic Press, London, 1980) pp. 103-128.

Approximation, Optimization and Computing:
Theory and Applications, A.G. Law and C.L. Wang (eds.)
Elsevier Science Publishers B.V. (North-Holland)

ON APPROXIMATION OF CLASSES OF FUNCTIONS FOR MODIFIED MEYER-KÖNIG-ZILLER OPERATORS

SHUNSHENG GUO

Department of Mathematics, Hebei Normal University, Shijiazhuang, Hebei, China

In this note using some results of the probability theory, we discuss approximation of classes of functions for modified Meyer-König-Zeller operators L_n. The upper and lower estimates of $E(C_\omega(M), L_n)$ and $E(C_\omega(M, M_1), L_n)$ are given.

1. INTRODUCTION

Chen [1] introduced modified Meyer-König-Zeller operators

$$L_n(f,x)=\sum_{k=0}^{\infty} p_{nk}(x)C_{nk}^{-1}\int_0^1 p_{nk}(t)f(t)dt \qquad (1.1)$$

where $p_{nk}(x)=\binom{n+k-1}{k}x^k(1-x)^n$,

$$C_{nk}=\frac{n}{(n+k)(n+k+1)}.$$

In this note, we consider approximation of classes of functions for L_n. If A is a class of functions, we define the error in approximating A by L_n to be (see [2])

$$E(A,L_n)=\sup_{f\in A}\| f-L_n(f)\|.$$

Let ω be a modulus of continuity. We define (see[2])

$$C_\omega(M)=:\{f \mid f\in C[0,1],\ \omega(f,t)\leqslant M\omega(t)\},$$

$$C_\omega^1(M,M_1)=:\{f \mid f\in C^1[0,1],\ \| f'\| \leqslant M_1,\ \omega(f',t)\leqslant M\omega(t)\}.$$

We shall show the following theorem:

Theorem. For n sufficiently large, we have

$$\frac{M}{6}\omega\left(\frac{1}{8\sqrt{n}}\right)\leqslant E(C_\omega(M),\ L_n)\leqslant 2\omega\left(\frac{1}{\sqrt{n}}\right), \quad (1.2)$$

$$\frac{\bar{M}}{48\sqrt{n}}\omega\left(\frac{1}{8\sqrt{n}}\right)\leqslant E(C_\omega^1(M,M_1),\ L_n)\leqslant \frac{M_1}{n}+\frac{2M}{\sqrt{n}}\omega\left(\frac{1}{\sqrt{n}}\right), \qquad (1.3)$$

where $\bar{M}=\min\{\frac{M_1}{\omega(1/2)},\ M\}$.

2. LEMMAS

Lemma 1.[3] If $\{\xi_k\}$ $(k\geqslant 1)$ are independent random variables with same distribution functions and $0<D\xi_1<\infty$, $\beta_3=E(\xi_1-E\xi_1)^3<\infty$ then

$$\max_y \left| P\left(\frac{1}{b_1\sqrt{n}}\sum_{k=1}^{n}(\xi_k-a_1)\leqslant y\right)-\frac{1}{\sqrt{2\pi}}\int_{-\infty}^{y} e^{-t^2/2}dt\right| < \frac{\beta_3}{\sqrt{n}\ b_1^3} \qquad (2.1)$$

where $a_1=E\xi_1$ (expectation of ξ_1),

$b_1^2=D\xi_1=E(\xi_1-E\xi_1)^2$

(variance of ξ_1).

Lemma 2.[3] If $\{\xi_i\}$ $(i=1,2,\cdots)$ are independent random variables with same geometric distribution

$$P(\xi_1=k)=x^k(1-x),\quad 0<x<1,\quad i=1,2,\cdots$$

then

$$E\xi_1=\frac{x}{1-x},\qquad D\xi_1=\frac{x}{(1-x)^2}$$

and $\eta_n=\sum_{i=1}^{n}\xi_i$ is a random variable with distribution

$$p(\eta_n=k)=\binom{n+k-1}{k}x^k(1-x)^n. \qquad (2.2)$$

Lemma 3. For every $j\geqslant 0$, we have

$$\sum_{k=0}^{\infty}p_{n+1\ k}(x)=C_{nj}^{-1}\int_x^1 p_{nj}(t)dt. \qquad (2.3)$$

Proof. (2.3) can easily be proved by differentiating both its left-hand and right-hand sides.

Lemma 4. For n sufficiently large and every $j\geqslant 0$, we have

$$C_{nj}^{-1}\int_{|t-\frac{1}{2}|\leqslant\frac{1}{8\sqrt{n}}} p_{nj}(t)dt\leqslant\frac{5}{6}. \qquad (2.4)$$

Proof. By Lemma 3, one has

$$C_{nj}^{-1}\int_{|t-\frac{1}{2}|\leqslant\frac{1}{8\sqrt{n}}} p_{nj}(t)dt$$

$$= C_{nj}^{-1}[\int_{\frac{1}{2}-\frac{1}{8\sqrt{n}}}^{1} - \int_{\frac{1}{2}+\frac{1}{8\sqrt{n}}}^{1}]p_{nj}(t)dt$$

$$=\sum_{k=0}^{j}[p_{n+1\ k}(\frac{1}{2}-\frac{1}{8\sqrt{n}})-p_{n+1\ k}(\frac{1}{2}+\frac{1}{8\sqrt{n}})]. \tag{2.5}$$

Using Lemma 2, for $x\in(0,1)$ we have

$$\sum_{k=0}^{j}p_{n+1\ k}(x)=P(\eta_{n+1}\leqslant j)$$

$$=P(\frac{\eta_{n+1}-(n+1)x/(1-x)}{\sqrt{n+1}\ x/(1-x)}\leqslant\frac{j-(n+1)x/(1-x)}{\sqrt{n+1}\ x/(1-x)}).$$

Noting Lemma 1 and $b_1=\sqrt{x/(1-x)^2}$, $\beta_3\leqslant 16/(1-x)^3$ (see [3]), we have

$$|\sum_{k=0}^{j}p_{n+1\ k}(x)-\frac{1}{\sqrt{2\pi}}\int_{-\infty}^{\frac{j-(n+1)x/(1-x)}{\sqrt{n+1}\ x/(1-x)}}e^{-t^2/2}dt|$$

$$<\frac{\beta_3}{\sqrt{n+1}\ b_1^3}\leqslant\frac{16}{\sqrt{n+1}\ x^{3/2}}. \tag{2.6}$$

Write $G(x)=\frac{j-(n+1)x/(1-x)}{\sqrt{n+1}\ x/(1-x)}$ and

$$\alpha=G(\frac{1}{2}+\frac{1}{8\sqrt{n}}),\quad \beta=G(\frac{1}{2}-\frac{1}{8\sqrt{n}}).$$

From (2.5) and (2.6), we can show that

$$|C_{nj}^{-1}\int_{|t-\frac{1}{2}|\leqslant\frac{1}{8\sqrt{n}}}p_{nj}(t)dt|$$

$$\leqslant\frac{16}{\sqrt{n+1}(\frac{1}{2}-\frac{1}{8\sqrt{n}})^{3/2}}+$$

$$\frac{16}{\sqrt{n+1}(\frac{1}{2}+\frac{1}{8\sqrt{n}})^{3/2}}+$$

$$\frac{1}{\sqrt{2\pi}}\int_{\alpha}^{\beta}e^{-t^2/2}\,dt. \tag{2.7}$$

If $j\leqslant\frac{5(n+1)}{3}$, then

$$\beta-\alpha=\frac{j}{\sqrt{n(n+1)}\ [1-(16n)^{-1}]}$$

$$<\frac{5}{3}\sqrt{(n+1)/n}\ \frac{1}{1-(16n)^{-1}}$$

So for n sufficiently large, one has

$$\frac{1}{\sqrt{2\pi}}\int_{\alpha}^{\beta}e^{-t^2/2}\,dt\leqslant\frac{1}{\sqrt{2\pi}}(\beta-\alpha)\leqslant$$

$$\frac{1}{\sqrt{2\pi}}\ \frac{5}{\sqrt{6}}<\frac{5}{6}. \tag{2.8}$$

If $j\geqslant\frac{5(n+1)}{3}$, then $\alpha\geqslant 0$ and

$$\frac{1}{\sqrt{2\pi}}\int_{\alpha}^{\beta}e^{-t^2/2}dt\leqslant\frac{1}{\sqrt{2\pi}}\int_0^{\infty}e^{-t^2/2}dt=\frac{1}{2}. \tag{2.9}$$

Hence (2.4) can be proved from (2.7)-(2.9).

3. PROOF OF THE THEOREM

Since (see [1])

$$L_n(t,x)=x+\frac{(1-x)(1-3x)}{n}+0(n^{-2}),$$

$$L_n((t-x)^2,x)=\frac{2x(1-x)^2}{n}+0(n^{-2}),$$

by (2.3.2), (2.3.3) in [2], we can obtain the right-hand ineqalities of (1.2) and (1.3).

To establish the lower estimates in (1.2) and (1.3), we take

$$f_1(t)=M\omega(|t-\tfrac{1}{2}|),$$

$$f_2(t)=\bar{M}\int_{\frac{1}{2}}^{t}\mathrm{sign}(u-\tfrac{1}{2})\,\omega(|u-\tfrac{1}{2}|)du,$$

where $\bar{M}=\min\{\frac{M_1}{\omega(1/2)},M\}$. Obviously, $f_1\in C_\omega(M)$, $f_2\in C_\omega^1(M,M_1)$.

Using Lemma 4, we have

$$E(C_\omega(M),L_n)\geqslant M\ L_n(\omega(|t-\tfrac{1}{2}|),\tfrac{1}{2})$$

$$=M\sum_{j=0}^{\infty}p_{nj}(\tfrac{1}{2})C_{nj}^{-1}\int_0^1\omega(|t-\tfrac{1}{2}|)p_{nj}(t)dt$$

$$\geqslant M\sum_{j=0}^{\infty}p_{nj}(\tfrac{1}{2})C_{nj}^{-1}\int_{|t-\frac{1}{2}|\geqslant\frac{1}{8\sqrt{n}}}\omega(|t-\tfrac{1}{2}|)p_{nj}(t)dt$$

$$\geqslant M\omega(\frac{1}{8\sqrt{n}})\sum_{j=0}^{\infty}p_{nj}(\tfrac{1}{2})(1-C_{nj}^{-1}\int_{|t-\frac{1}{2}|\leqslant\frac{1}{8\sqrt{n}}}p_{nj}(t)dt)$$

$$\geqslant\frac{M}{6}\omega(\frac{1}{8\sqrt{n}})$$

and

$$E(C^1_\omega(M,M_1),L_n)\geqslant L_n(f_2(t),\frac{1}{2})$$

$$=\bar{M}\sum_{j=0}^{\infty}p_{nj}(\frac{1}{2})C_{nj}^{-1}\int_0^1[$$

$$\int_{\frac{1}{2}}^{t}\operatorname{sign}(u-\frac{1}{2})\,\omega(|u-\frac{1}{2}|)du]p_{nj}(t)dt$$

$$\geqslant\bar{M}\sum_{j=0}^{\infty}p_{nj}(\frac{1}{2})C_{nj}^{-1}\int_{|t-\frac{1}{2}|\geqslant\frac{1}{8\sqrt{n}}}[$$

$$\int_{\frac{1}{2}}^{t}\operatorname{sign}(u-\frac{1}{2})\,\omega(|u-\frac{1}{2}|)du]p_{nj}(t)dt$$

$$\geqslant\bar{M}\frac{1}{8\sqrt{n}}\,\omega(\frac{1}{8\sqrt{n}})\sum_{j=0}^{\infty}$$

$$p_{nj}(\frac{1}{2})C_{nj}^{-1}\int_{|t-\frac{1}{2}|\geqslant\frac{1}{8\sqrt{n}}}p_{nj}(t)dt$$

$$\geqslant\frac{\bar{M}}{48\sqrt{n}}\,\omega(\frac{1}{8\sqrt{n}}).$$

Thus, we complete the proof.

REFERENCES

[1] W. Chen, Approx. Th. & its Appl. 2:3, Sep. 1986, p7-18.

[2] R. DeVore, The Approximation of Continuous Functions by Positive Linear Operators, Lecture Notes of Math. 293, 1972.

[3] S. Guo, Approx. Th. & its Appl. 4:2, June, 1988, p9-18.

Approximation, Optimization and Computing:
Theory and Applications, A.G. Law and C.L. Wang (eds.)
Elsevier Science Publishers B.V. (North-Holland)

APPROXIMATION BY COMPOSITE POLYNOMIALS

Kazuo HATANO

Aichi Institute of Technology, Department of Electronic Engineering
1247 Yachigusa, Yakusa, Toyota, Japan・〒470-03 Phone <0565>48-8121 ext.231

The errors of truncated Fourier series are carefully analyzed. It is shown that the errors are composed from two terms. The first term is the major part of the errors in magnitude which is composed from the sum of algebraic and trigonometric polynomials, and can be calculated if requested. The other is the term which converges quite rapidly and small in magnitude. Based on their analysis, we propose a new approximation method for general functions by the sum of algebraic and trigonometric polynomials which we call composite polynomials.

1. INTRODUCTION

Since the advent of the FFT(Fast Fourier Transform) algorithm([1]), the importance of Fourier series is greatly increased. In general, sufficiently smooth periodic functions can be approximated well by truncated Fourier series. That is, finite truncation of Fourier series associated with such functions causes no difficulty. In contrast, periodic functions with discontinuous points cannot be approximated well by truncated Fourier series. Truncation of Fourier series implies the Gibb's phenomenon, that is, the violation occurs near discontinuous points. We analyzed this phenomenon carefully, which lead to the fact that the violation near discontinuous points is mainly caused from the sum of algebraic and trigonometric polynomials. Based on this fact, we construct a new approximation method for general functions.

In this article, we propose a method which we call composite polynomials approximation. Next we state computational scheme of composite polynomials. Some numerical examples are shown.

2. ERRORS OF TRUNCATED FOURIER SERIES

We first derive the errors of truncated Fourier series for a class of sufficiently smooth functions. Next, we estimate an error bound of truncated Fourier series for a class of sufficiently smooth periodic functions.

Let $W[0,2\pi]$ denote all the functions which satisfies $f(x)\in C^{2m}[0,2\pi]$ and $\|f^{(2m+1)}(x)\|_\infty<\infty$, where $\|f\|_\infty$ means the supremum of $f(x)$ on the interval $[0,2\pi]$.

Moreover, let $P[0,2\pi]$ denote all the periodic functions of period 2π which satisfies $f(x)\in C^{2m-1}(-\infty,\infty)$, $f^{(2m)}(x)\in C(0,2\pi)$ and $\|f^{(2m+1)}(x)\|_\infty<\infty$.

It is well known that $f(x)\in W[0,2\pi]$ can be expanded into Fourier series, that is,

$$f(x)=\frac{1}{2}a_0(f)+\sum_{j=1}^{\infty}\{a_j(f)\cos jx+b_j(f)\sin jx\}, \tag{1}$$

where

$$a_j(f)=\frac{1}{\pi}\int_0^{2\pi}f(x)\cos jx\cdot dx,$$
$$b_j(f)=\frac{1}{\pi}\int_0^{2\pi}f(x)\sin jx\cdot dx. \tag{2}$$

Repeated application of integration by parts to eq.(2) gives

$$a_j(f)=\sum_{i=1}^{m}\frac{(-1)^{i-1}}{j^{2i}}\omega_{2i-1}(f)+\frac{(-1)^{m+1}}{\pi j^{2m+1}}\int_0^{2\pi}f^{(2m+1)}(t)\times\sin jt\cdot dt,$$
$$b_j(f)=\sum_{i=0}^{m-1}\frac{(-1)^{i-1}}{j^{2i+1}}\omega_{2i}(f)+\frac{(-1)^{m+1}}{\pi j^{2m+1}}\int_0^{2\pi}f^{(2m+1)}(t)\times(1-\cos jt)\cdot dt. \tag{3}$$

where

$$\omega_i(f)=\{f^{(i)}(2\pi)-f^{(i)}(0)\}/\pi. \tag{4}$$

(Lyness calls eq.(3) "Asymptotic expansion of Fourier coefficients" in [3].)

Let $T_n(f;x)$ be the truncated Fourier series associated with the function $f(x)$, that is

$$T_n(f;x)=\frac{1}{2}a_0(f)+\sum_{j=1}^{n-1}\{a_j(f)\cos jx+b_j(f)\sin jx\}, \tag{5}$$

then the error of truncated Fourier series becomes

$$f(x)-T_n(f;x)=\sum_{j=n}^{\infty}\{a_j(f)\cos jx+b_j(f)\sin jx\}. \tag{6}$$

Applying eq.(3) to eq.(6), the error of truncated Fourier series can be rewritten to

$$f(x)-T_n(f;x)=\sum_{i=1}^{m}\omega_{2i-1}(f)\cdot\sum_{j=n}^{\infty}\frac{(-1)^{i-1}}{j^{2i}}\cos jx$$
$$+\sum_{i=0}^{m-1}\omega_{2i}(f)\cdot\sum_{j=n}^{\infty}\frac{(-1)^{i-1}}{j^{2i+1}}\sin jx$$

$$+\frac{1}{\pi}\int_0^{2\pi} f^{(2m+1)}(t)\cdot\sum_{j=n}^{\infty}\frac{(-1)^{m-1}}{j^{2m+1}}\{\sin jx-\sin j(x-t)\}dt. \tag{7}$$

We now introduce a set of functions, $\tilde{q}_i(x;n)$, as follows.

$$\tilde{q}_{2i}(x;n)=\sum_{j=n}^{\infty}\frac{(-1)^{i-1}}{j^{2i}}\cos jx,$$
$$\tilde{q}_{2i+1}(x;n)=\sum_{j=n}^{\infty}\frac{(-1)^{i-1}}{j^{2i+1}}\sin jx. \tag{8}$$

Then, the error of truncated Fourier series becomes to a compact form as

$$f(x)-T_n(f;x)=\sum_{i=1}^{2m}\omega_{i-1}(f)\cdot\tilde{q}_i(x;n)$$
$$+\frac{1}{\pi}\int_0^{2\pi} f^{(2m+1)}(t)\{\tilde{q}_{2m+1}(x;n)-\tilde{q}_{2m+1}(x-t;n)\}dt. \tag{9}$$

It is readily seen from eq.(9) that the first term $\omega_0(f)\cdot\tilde{q}_1(x;n)$ leads to the oscillation which is called "Gibb's phenomenon".

For $f(x)\in P[0,2\pi]$, the error of truncated Fourier series is given by

$$f(x)-T_n(f;x)$$
$$=\frac{1}{\pi}\int_0^{2\pi} f^{(2m+1)}(t)\{\tilde{q}_{2m+1}(x;n)-\tilde{q}_{2m+1}(x-t;n)\}dt. \tag{10}$$

This is the same as eq.(9) except for the absence of the first term in the right hand side of eq.(9).

Applying Hoelder's inequality to eq.(10), we obtain the error bound of truncated Fourier series associated with $f(x)\in P[0,2\pi]$.

$$|f(x)-T_n(f;x)|\leqq\frac{2}{\pi}\sqrt{\pi^2+4}\cdot\zeta(2m+1;n)\|f^{(2m+1)}\|_\infty, \tag{11}$$

where $\zeta(k;n)$ is the generalized Riemann Zeta function defined by([3])

$$\zeta(k;n)=\sum_{j=n}^{\infty}\frac{1}{j^k}. \tag{12}$$

For $k\geqq 2$, following inequality is easily derived.

$$\zeta(k;n)\leqq\frac{1}{(k-1)(n-1)^{k-1}} \tag{13}$$

This means that $O((n-1)^{-2m})$ is the convergence rate of the errors of truncated Fourier series associated with the function $f(x)\in P[0,2\pi]$.

3. COMPOSITE POLYNOMIALS

Now, we define composite polynomials. Let $D\equiv d/dx$. If $h(x)$ satisfies the differential equation

$$D^{2m+1}\prod_{j=1}^{n-1}(D^2+j^2)h(x)\equiv 0 \quad \text{on } (0,2\pi), \tag{14}$$

subject to the condition $h(0)=h(2\pi)=\{h(0+)+h(2\pi-)\}/2$, then we call $h(x)$ composite polynomials of degree $(2m,n)$. All composite polynomials are the sum of ordinary algebraic polynomials of degree $2m$ and trigonometric polynomials of degree $n-1$.

Bernoulli polynomial of degree i, $B_i(x)$ defined by

$$\frac{te^{xt}}{e^t-1}=\sum_{i=0}^{\infty}B_i(x)\frac{t^i}{i!} \quad : |t|<2\pi \tag{15}$$

satisfies the identity([3])

$$B_i'(x)=i\cdot B_{i-1}(x),$$
$$B_i(1-x)=(-1)^iB_i(x) \quad :i\geqq 1, \tag{16}$$
$$B_{2i+1}(0)=B_{2i+1}(1)=0.$$

Using $\{B_i(x)\}$, let's define $p_i(x)$ as follows.

$$p_0(x)=1/2 : x\in(0,2\pi),$$
$$p_1(x)=\begin{cases}(x-\pi)/2 & : x\in(0,2\pi)\\ 0 & : x=0,2\pi,\end{cases} \tag{17}$$
$$p_i(x)=\frac{(2\pi)^i}{2\cdot i!}B_i\left(\frac{x}{2\pi}\right) : i\geqq 2.$$

It is easily derived that

$$p_i'(x)=p_{i-1}(x),$$
$$p_i(2\pi-x)=(-1)^ip_i(x) : i\geqq 1, \tag{18}$$
$$p_{2i+1}(0)=p_{2i+1}(2\pi)=0.$$

and

$$p_i^{(l)}(0+)=p_i^{(l)}(2\pi-) : l=0,1,\ldots,i-2,i,\ldots,$$
$$-p_i^{(i-1)}(0+)=p_i^{(i-1)}(2\pi-)=\pi/2. \tag{19}$$

By $p_i(x)$, we can represent $h(x)$ as

$$h(x)=\frac{1}{2}a_0+\sum_{j=1}^{n-1}(\tilde{a}_j\cos jx+\tilde{b}_j\sin jx)+\sum_{i=1}^{2m}\tilde{c}_ip_i(x), \tag{20}$$

which implies that arbitrary composite polynomials can be represented in the form (20).

From eq.(19)

$$h^{(l-1)}(2\pi-)-h^{(l-1)}(0+)=\pi\tilde{c}_l \tag{21}$$

can easily be derived.

Next, we derive other representation of composite polynomials. From eq.(19), polynomial of degree i, $p_i(x)$ can be expanded into Fourier series([3])

$$p_{2i}(x)=\sum_{j=1}^{\infty}\frac{(-1)^{i-1}}{j^{2i}}\cos jx,$$
$$p_{2i+1}(x)=\sum_{j=1}^{\infty}\frac{(-1)^{i-1}}{j^{2i+1}}\sin jx. \tag{22}$$

which is slightly different from eq.(8). Applying eq.(22) to eq.(20), we can obtain

$$h(x)=\frac{1}{2}a_0+\sum_{j=1}^{n-1}(a_j\cos jx+b_j\sin jx)+\sum_{i=1}^{2m}\tilde{c}_i\tilde{q}_i(x). \tag{23}$$

where

$$\tilde{a}_j = a_j + \sum_{i=1}^{m} \frac{(-1)^{i-1}}{j^{2i}} \cdot c_{2i},$$
$$\tilde{b}_j = b_j + \sum_{i=0}^{m-1} \frac{(-1)^{i-1}}{j^{2i+1}} \cdot c_{2i+1}. \tag{24}$$

Concerning $\tilde{q}_i(x;n)$, the following orthogonality holds.

$$\int_0^{2\pi} \frac{1}{2} \cdot \tilde{q}_i(x;n)dx = \int_0^{2\pi} \tilde{q}_i(x;n) \cdot \cos jx dx$$
$$= \int_0^{2\pi} \tilde{q}_i(x;n) \cdot \sin jx dx = 0$$
$$: i \geqq 1, \ 1 \leqq j \leqq n-1, \tag{25}$$

$$\int_0^{2\pi} \tilde{q}_{2i}(x;n) \cdot \tilde{q}_{2j+1}(x;n)dx = 0 \quad : i \geqq 1, \ j \geqq 0. \tag{26}$$

When we examine the least squares approximation by composite polynomials, these orthogonality relationships are quite useful([7]).

4. APPROXIMATION BY COMPOSITE POLYNOMIALS

Eqs.(9),(10),(11) suggest a new approximation for $f(x) \in W[0,2\pi]$. If we approximate $f(x) \in W[0,2\pi]$ by

$$H_n^{2m}(f;x) = T_n(f;x) + \sum_{i=1}^{2m} \omega_{i-1}(f) \cdot \tilde{q}_i(x;n) \tag{27}$$

,that is, the sum of truncated Fourier series plus the major part of it's error, we can expect a good approximation. In this case, convergence rate of the error is $O((n-1)^{-2m})$.

We can write composite polynomials approximation for $f(x) \in W[0,2\pi]$ as follows.

$$H_n^{2m}(f;x) = \frac{1}{2}a_0(f) + \sum_{j=1}^{n-1} \{a_j(f)\cos jx + b_j(f)\sin jx\}$$
$$+ \sum_{i=1}^{2m} \omega_{i-1}(f) \cdot \tilde{q}_i(x;n) \tag{28}$$

In this equation, $a_j(f): j=0,1,\ldots,n-1$, $b_j(f): j=1,2,\ldots,n-1$ are given by eq.(2). These quantities are just Fourier coefficients of $f(x)$. The coefficients $\omega_{i-1}(f): i=1,2,\ldots,2m$ are given by eq.(4).

Here we can also represent composite polynomials approximation as follows.

$$H_n^{2m}(f;x) = \frac{1}{2}a_0(f) + \sum_{j=1}^{n-1} \{\tilde{a}_j(f)\cos jx + \tilde{b}_j(f)\sin jx\}$$
$$+ \sum_{i=1}^{2m} \omega_{i-1}(f) \cdot p_i(x) \tag{29}$$

where

$$\tilde{a}_j(f) = \frac{(-1)^{m+1}}{\pi j^{2m+1}} \int_0^{2\pi} f^{(2m+1)}(t) \sin jt \cdot dt,$$
$$\tilde{b}_j(f) = \frac{(-1)^{m+1}}{\pi j^{2m+1}} \int_0^{2\pi} f^{(2m+1)}(t)(1-\cos jt) \cdot dt. \tag{30}$$

The parameters $\tilde{a}_j(f)$, $\tilde{b}_j(f)$ are same as eq.(3) except for the absence of the first term in the right hand side of eq.(3).

It is easily derived that

$$|\tilde{a}_j(f)| \leqq \frac{4}{\pi j^{2m+1}} \cdot \|f^{(2m+1)}\|_\infty,$$
$$|\tilde{b}_j(f)| \leqq \frac{2}{j^{2m+1}} \cdot \|f^{(2m+1)}\|_\infty, \tag{31}$$

which mean that these quantities are decreasing of order $O(j^{-2m-1})$ as j increases. For small j's, the magnitude of $\tilde{a}_j(f)$ and $\tilde{b}_j(f)$ are often very large. Hence careless application of eq.(29) may cause serious cancellation.

5. COMPUTATION OF COMPOSITE POLYNOMIALS

Although it seems quite easy to compute composite polynomials by the use of eq.(29), but, as noted above, it is not the case.

In this section, we propose another method for calculating the composite polynomials. Instead of eq.(29), we use eq.(28). Let

$$h(x) = \frac{1}{2}a_0 + \sum_{j=1}^{n-1}(a_j \cos jx + b_j \sin jx) + \sum_{i=1}^{2m} c_i q_i(x;n), \tag{32}$$

where

$$q_i(x;n) = n^i \cdot \tilde{q}_i(x;n). \tag{33}$$

We compute h(x) only on the equi-spaced points on $[0,2\pi]$. Let

$$\bar{x}_r = \frac{2\pi r}{N} \quad : r=0,1,\ldots,N \quad (N>2n). \tag{34}$$

Now, from eq.(33) and eq.(8)

$$q_{2i}(\bar{x}_r;n) = n^{2i} \sum_{j=n}^{\infty} \frac{(-1)^{i-1}}{j^{2i}} \cos\frac{2\pi jr}{N}. \tag{35}$$

Let j's in eq.(35) divide into four categories.

j	:	j=n,n+1,...,N/2-1
kN	:	k=1,2,...
kN+j,kN-j	:	j=1,2,...,N/2-1 ; k=1,2,...
kN+N/2	:	k=0,1,... (36)

We compute partial sum for each group, and using periodicity and symmetry of trigonometric functions, we can get the following expression.

$$q_{2i}(\bar{x}_r;n) = (-1)^{i-1}\left(\frac{n}{N}\right)^{2i}\left\{\sum_{s=0}^{n-1(2)} \delta_{2i}\left(\frac{s}{N}\right)\cos s\bar{x}_r + \sum_{s=n}^{N/2(3)} \tau_{2i}\left(\frac{s}{N}\right)\cos s\bar{x}_r\right\}. \tag{37}$$

Similarly,

$$q_{2i+1}(\bar{x}_r;n) = (-1)^{i-1}\left(\frac{n}{N}\right)^{2i+1}\left\{\sum_{s=0}^{n-1} \delta_{2i+1}\left(\frac{s}{N}\right)\sin s\bar{x}_r + \sum_{s=n}^{N/2} \tau_{2i+1}\left(\frac{s}{N}\right)\sin s\bar{x}_r\right\}. \tag{38}$$

In eq.(37) and eq.(38),

$$\bar{\delta}_i(x)=\sum_{k=1}^{\infty}\{\frac{1}{(k+x)^i}+\frac{(-1)^i}{(k-x)^i}\}, \tag{39}$$

$$\bar{\tau}_i(x)=\bar{\delta}_i(x)+1/x^i, \tag{40}$$

and the conventional summations are adopted as

$$\sum_{s=0}^{n-1(2)} A(s)=\frac{1}{2}A(0)+\sum_{s=1}^{n-1}A(s),$$

$$\sum_{s=n}^{N/2(3)} A(s)=\sum_{s=n}^{N/2-1}A(s)+\frac{1}{2}A(\frac{N}{2}). \tag{41}$$

It is easily verified that

$$q_1(0;n)=q_1(2\pi;n)=0, \tag{42}$$

and

$$-q_1(0+;n)=q_1(2\pi-;n)=\pi n/2. \tag{43}$$

Combining these equations, we can get following.

$$\bar{h}(\bar{x}_r)=\frac{1}{2}\bar{u}_0+\sum_{j=1}^{N/2-1}\{\bar{u}_j\cos j\bar{x}_r+\bar{v}_j\sin j\bar{x}_r\}+\frac{1}{2}\bar{u}_{N/2}\cos\frac{N}{2}\bar{x}_r. \tag{44}$$

$$\bar{u}_j=\begin{cases} a_j+\sum_{i=1}^{m}c_{2i}(-1)^{i-1}(\frac{n}{N})^{2i}\bar{\delta}_{2i}(\frac{j}{N}) & :0\leqq j\leqq n-1,\\ \sum_{i=1}^{m}c_{2i}(-1)^{i-1}(\frac{n}{N})^{2i}\bar{\tau}_{2i}(\frac{j}{N}) & :n\leqq j\leqq N/2, \end{cases} \tag{45}$$

$$\bar{v}_j=\begin{cases} b_j+\sum_{i=0}^{m-1}c_{2i+1}(-1)^{i-1}(\frac{n}{N})^{2i+1}\bar{\delta}_{2i+1}(\frac{j}{N}) & :0\leqq j\leqq n-1,\\ \sum_{i=0}^{m-1}c_{2i+1}(-1)^{i-1}(\frac{n}{N})^{2i+1}\bar{\tau}_{2i+1}(\frac{j}{N}) & :n\leqq j\leqq N/2. \end{cases} \tag{46}$$

Applying inverse FFT of real numbers to eq.(44), we can get the values of composite polynomial at the points, $\bar{x}_r$:r=0,1,...,N, simultaneously.

Eq.(45) and (46) have one more practical significance.

Eq.(44) means that $\bar{u}_j:0\leqq j\leqq N/2$, $\bar{v}_j:1\leqq j\leqq N/2-1$ are discrete Fourier coefficients of h(x). Eq.(32) means that $a_j:0\leqq j\leqq n-1$, $b_j:1\leqq j\leqq n-1$ are Fourier coefficients of h(x) and

$$c_i=\{h^{(i-1)}(2\pi-)-h^{(i-1)}(0+)\}/\pi n^i. \tag{47}$$

For h(x), a high order approximation of f(x), eq.(45) and (46) give the relationship of continuous and discrete Fourier coeffficients of f(x). So, through correction of discrete Fourier coefficients by eq.(45) and (46), we can expect a good approximation of Fourier coefficients([7]).

6. NUMERICAL EXAMPLE

We compute the supremum of the errors of composite polynomials approximation, $\|f(x)-H_n^{2m}(f;x)\|_\infty$, for many cases. In this example, $f(x)=Y_0(2+1.5x):x\in[0,2\pi]$, where $Y_0(x)$ is the second kind Bessel function of degree zero. In this computation, we used only double precision arithmetic of i8086 and i8087 (about 16 significant digits). Following table shows the results.

2m \ n	8	16	32	64	128
0	0.365	0.365	0.365	0.365	0.365
2	3.9E-3	9.1E-4	2.2E-4	3.2E-5	7.0E-6
4	4.9E-5	2.5E-6	1.3E-7	7.1E-9	3.2E-10
6	4.7E-6	6.6E-8	8.9E-10	1.3E-11	1.4E-13
8	8.9E-7	3.2E-9	1.1E-11	4.4E-14	8.4E-15
10	4.9E-7	4.4E-10	4.0E-13	3.5E-15	8.4E-15
12	4.0E-7	9.1E-11	2.0E-14	3.4E-15	4.3E-15

7. CONCLUDING REMARKS

We have been discussed the errors of truncated Fourier series associated with the function which have only one discontinuous point. For the function which have finitely many discontinous points, almost same analysis can be made. In this case, the errors of truncated Fourier series are the sum of piecewise polynomials and trigonometric polynomials([4],[5]). Using these results, we can analyze the errors of trigonometric interpolation([6]). Error analysis of discrete Fourier coefficients yields an excellent scheme for calculating Fourier coefficients precisely([7]).

ACKNOWLEDGMENTS

I should like to thank Prof. Ichiuzo Ninomiya, Tatuo Torii, Taketomo Mitui for their support and help.

REFERENCES

[1] Colley, J.W. and Tukey, J.W.: An Algorithm for the Machine Calculation of Complex Fourier Series, Math. Comp., Vol.19, pp.297-301(1965).
[2] Abramowitz, M. and Stegun, I.(eds.): Handbook of Mathematical Functions, 10-th ed., Dover Publication Inc., New York (1972) 1046.
[3] Lyness, J.N.: The Calculation of Fourier Coefficients, SIAM J. Numer. Anal., Vol.4, No.2, pp.301-315(1967).
[4] Hatano, K., Hatano, Y., Ninomiya, I.: Approximation by Composite Polynomials(in Japanese), Trans. IPS Japan, Vol.23, No.6, pp.617-624(1982).
[5] Hatano, K., Hatano, Y., Ninomiya, I.: On calculating with Composite Polynomials(in Japanese), Trans. IPS Japan, Vol.23, No.6, pp.625-633(1982).
[6] Hatano, K.: Error Analysis of Trigonometric Interpolation(in Japanese), Trans. IPS Japan, Vol.30, No.2, pp.150-158(1989).
[7] Hatano, K.: The Least-Squares Approximation by Composite Polynomials(to appear in Japanese), Trans. IPS Japan, Vol.30, No.6, (1989).

Approximation, Optimization and Computing:
Theory and Applications, A.G. Law and C.L. Wang (eds.)
Elsevier Science Publishers B.V. (North-Holland)

CONSTRAINED APPROXIMATIONS

V. H. Hristov, K. G. Ivanov

Institute of Mathematics, Bulgarian Academy of Sciences, 1113 Sofia, Bulgaria

Using the notion of constrained K-functional we show that the problem for characterizing the order of constrained approximation can be separated to three independent steps - investigation of the non-constrained case, obtaining a Jackson type estimate for the constrained approximation and characterization of the constrained K-functional. Oneseded approximations are studed in more details.

1. NON-CONSTRAINED APPROXIMATION

First we briefly examine the case of best approximation for two reasons - to have it as a model for treating the constrained case and, later on, to make use of some of the results.

1.1. Direct and converse statements

Let (X_0, X_1) be a couple of normed spaces with (semi-)norms $\|\cdot\|_0$ and $\|\cdot\|_1$ respectively, $X_1 \subset X_0$. The Peetre *K-functional* for this couple is given by

$$K(f,t;X_0,X_1) := \inf\{\|f-g\|_0 + t\|g\|_1 : g \in X_1\}$$

for any $f \in X_0$ and $t>0$. Let $\{G_n\}_1^\infty$ be a family of subsets of X_1. For this family we assume: i) $0 \in G_1$ and $\|g\|_1 = 0$ $\forall g \in G_1$, ii) $G_n \subset G_{n+1}$, iii) $G_n = -G_n$, iv) the closure of $\{G_n\}_1^\infty$ in X_0 is X_0. The best approximation of $f \in X_0$ by the elements of G_n is given by

$$E_n(f)_0 := \inf\{\|f-g\|_0 : g \in G_n\}.$$

We suppose that this family satisfies *inequalities of Jackson and Bernstein type*, that is there are positive constants c_1, c_2 $(c_j \ge 1)$ and α such that for any n we have

$$(1) \qquad E_n(f)_0 \le c_1 n^{-\alpha}\|f\|_1 \qquad \forall f \in X_1;$$

$$(2) \qquad \|g_1-g_2\|_1 \le c_2 n^{\alpha}\|g_1-g_2\|_0 \qquad \forall g_1,g_2 \in G_n.$$

It is well known that inequalities (1) and (2) imply for any $f \in X_0$ the direct and converse estimates

$$(3) \qquad E_n(f)_0 \le c_1 K(f,n^{-\alpha};X_0,X_1),$$

$$(4) \quad K(f,n^{-\alpha};X_0,X_1) \le c(\alpha) n^{-\alpha} \sum_{k=1}^{n} k^{\alpha-1} E_k(f)_0$$

respectively. We also impose on the approximating sets the additional condition:

$$(5) \quad \inf\{\|f-g\|_0 + n^{-\alpha}\|g\|_1 : g \in G_n\} \le c_3 K(f,n^{-\alpha};X_0,X_1) \qquad \forall f \in X_0$$

for some positive $c_3 \ge 1$.

Now we state some conditions equivalent to (5) and describe approximating processes for which (1), (2) and (5) are fulfilled. By c we denote different positive numbers depending only on the parameters following in parentheses.

1.2. Equivalent conditions

It is natural to ask: Is condition (5) independent of Jackson and Bernstein inequalities (1), (2) or it is implied by them? We do not know the answer but as Theorem 1 shows (5) will follow if (2) is replaced by a somewhat stronger condition.

Denote by $P_n(f)$ an element of *best approximation* from G_n to f (assuming it exists) and by $Q_n(f)$ an element in G_n approximating f with the order of the K-functional, i.e.

$$\|f-P_n(f)\|_0 = E_n(f)_0, \quad \|f-Q_n(f)\|_0 \le AK(f,n^{-\alpha};X_0,X_1),$$

where A is a fixed positive constant. In view of (3), for every element g of *near-best approximation* from G_n, that is $\|f-g\|_0 \le A_1 E_n(f)_0$, we have $g = Q_n(f)$.

<u>Theorem 1.</u> *Let (1) and (2) hold. Then the following are equivalent:*

a) *(5) holds for* G_n;

b) $\|P_n(f)\|_1 \le cn^{\alpha}K(f,n^{-\alpha};X_0,X_1) \quad \forall f \in X_0$;

c) $\|Q_n(f)\|_1 \le cn^{\alpha}K(f,n^{-\alpha};X_0,X_1) \quad \forall f \in X_0$;

d) $\|g\|_1 \le cn^{\alpha}K(g,n^{-\alpha};X_0,X_1) \quad \forall g \in G_n$,

where constants c in b), c) and d) depend only on c_1, c_2, c_3 *and A (provided a) holds).*

Statements of the type of b) and c) in which

information for the structural properties of the function (behaviour of the K-functional or moduli of smoothness) implies knowledge for the growth of some seminorms of the polynomials of best approximation are often called Zamansky type theorems.

Theorem 2. *Let (1) and (2) hold. Assume that there is a subspace X_2 of X_0 with seminorm $\|\cdot\|_2$ such that $\|g\|_2=0$ for every $g\in G_1$ and family $\{G_n\}$ satisfies Jackson and Bernstein inequalities with parameter $\beta>\alpha$ with respect to the couple (X_0,X_2), i.e.*

$$E_n(f)_0 \le c_1' n^{-\beta}\|f\|_2 \qquad \forall f\in X_2;$$

$$\|g_1-g_2\|_2 \le c_2' n^{\beta}\|g_1-g_2\|_0 \qquad \forall g_1,g_2\in G_n.$$

Then (5) is equivalent to the condition:

$$\inf\{\|f-g\|_0+n^{-\alpha}\|g\|_1+n^{-\beta}\|g\|_2 : g\in X_1\cap X_2\}$$
$$\le c_4\, K(f,n^{-\alpha};X_0,X_1) \qquad \forall f\in X_0.$$

1.3. Approximation by trigonometric polynomials

Let r be natural, $1\le p\le\infty$, $X_0=L_p$ - the space of all 2π-periodic functions integrable in p-th power with the norm $\|f\|_p = \left\{\int_0^{2\pi}|f(x)|^p dx\right\}^{1/p}$ (with the usual change to sup norm for $p=\infty$) and $X_1=W_p^r$ - the space of all 2π-periodic functions with r-th derivative in L_p equipped with the seminorm $\|f^{(r)}\|_p$. The equivalence of the K-functional with the modulus of smoothness

$$\omega_r(f,t)_p = \sup\{\|\Delta_h^r f(\cdot)\|_p : |h|\le t\},$$
$$\Delta_h^r f(x) = \Sigma_{i=0}^r (-1)^{r-i}\binom{r}{i} f(x+ih),$$

is well known in this case:

(6) $c(r)\omega_r(f,t)_p \le K(f,t^r;L_p,W_p^r) \le c(r)\omega_r(f,t)_p$ $\forall f\in L_p[0,2\pi)$. Let G_n be the space of all trigonometric polynomials of degree n-1. It is well known (see e.g. [12, p.275, 230]) that in this case (1) and (2) hold with $\alpha=r$:

$$E_n(f)_p \le c(r)n^{-r}\|f^{(r)}\|_p \qquad \forall f\in W_p^r,$$
$$\|g^{(r)}\|_p \le n^r\|g\|_p \qquad \forall g\in G_n.$$

Condition (5) is also fulfilled and we have the equivalent to it inequalities

$$\|P_n(f)^{(r)}\|_p \le c(r)n^r\omega_r(f,n^{-1})_p \qquad \forall f\in L_p,$$
$$\|g^{(r)}\|_p \le c(r)n^r\omega_r(g,n^{-1})_p \qquad \forall g\in G_n.$$

As far as we know these two inequalities go back to Stechkin [11] ($p=\infty$).

All of the above statements hold (after the necessary modifications) in the case of multivariate trigonometric approximation too.

1.4. Approximation by algebraic polynomials

Let r be natural, $\varphi(x)=\sqrt{1-x^2}$, $1\le p\le\infty$, $X_0=L_p$ - the space of all functions integrable in p-th power with the norm $\|f\|_p = \left\{\int_{-1}^1 |f(x)|^p dx\right\}^{1/p}$ and let $X_1=W_p^r(\varphi)$ - the space of all functions f defined in $[-1,1]$ such that $\varphi^r f^{(r)}\in L_p$ with the seminorm $\|\varphi^r f^{(r)}\|_p$. A characterization of

$$K(f,t^r;L_p,W_p^r(\varphi))$$
$$:=\inf\{\|f-g\|_p+t^r\|\varphi^r g^{(r)}\|_p : g\in W_p^r(\varphi)\}$$

(similar to (6)) by two different type moduli of smoothness is given in [2, p.11] and [7].

Let G_n be the space of all algebraic polynomials of degree n-1. The Jackson type inequality (1) for this case with $\alpha=r$

$$E_n(f)_p \le c(r)n^{-r}\|\varphi^r f^{(r)}\|_p \qquad \forall f\in W_p^r(\varphi).$$

has been proved in [2, p.79] or [6]. What we call a Bernstein type inequality is now exactly the original Bernstein inequality (when $p=\infty$ and $r=1$, for the general case see e.g. [2, p.92])

$$\|\varphi^r g^{(r)}\|_p \le c(r)n^r\|g\|_p \qquad \forall g\in G_n.$$

A Zamansky type inequality

$$\|\varphi^r P_n(f)^{(r)}\|_p \le c(r)n^r K(f,n^{-r};L_p,W_p^r(\varphi))$$

is provided by Theorem 7.3.1 in [2]. Therefore (5) is valid in this case too.

2. APPROXIMATION WITH CONSTRAINTS

In the general setting of Section 1 assume that with every $f\in X_0^*$ $(X_0^*\subset X_0)$ we associate a subset $R(f)$ of X_0. $R(f)$ can be thought as the restriction which we impose on the functions approximating f. Various type of constraints are given in Section 3. The only assumptions (necessary for the direct result - Theorem 3) made for the constraint are

(7) $R(g)\subset R(f)$ $\forall g\in R(f)$; $X_1\cap R(f)\subset X_0^*$ $\forall f\in X_0^*$.

The constrained K-functional of $f\in X_0^*$ is

$$K^*(f,t;X_0,X_1) := \inf\{\|f-g\|_0+t\|g\|_1 : g\in X_1\cap R(f)\}$$

and the constrained best approximation of f by elements of G_n is given by

$$E_n^*(f)_0 := \inf\{\|f-g\|_0 : g\in G_n\cap R(f)\}.$$

The corresponding Jackson type inequality is

(8) $E_n^*(f)_0 \le c_1'' n^{-\alpha}\|f\|_1$ $\forall f\in X_1\cap X_0^*$.

Theorem 3. *If (7) and (8) are fulfilled then*

$$E_n^*(f)_0 \le c_1'' K^*(f,n^{-\alpha};X_0,X_1) \qquad \forall f\in X_0^*.$$

Theorem 4. *If (2) and (5) are fulfilled then for $f\in X_0^*$ we have*

$$K^*(f,n^{-\alpha};X_0,X_1)\le(1+c_2)E_n^*(f)_0+c_2c_3K(f,n^{-\alpha};X_0,X_1)$$

Theorem 4 looks as a "strong" type inverse result. Indeed, if for some f and n we have the constrained K-functional is several times bigger than the non-constrained one, e.g. $K^*(f,n^{-\alpha}) \ge 2c_2c_3K(f,n^{-\alpha})$ then Theorem 4 implies

$$K^*(f,n^{-\alpha};X_0,X_1) \le 2(1+c_2)E_n^*(f)_0.$$

But in general this is not so. Then from Theorem 4 and (4) we get

$$K^*(f,n^{-\alpha};X_0,X_1) \le c(\alpha)n^{-\alpha}\sum_{k=1}^{n} k^{\alpha-1}E_k^*(f)_0.$$

Moreover (5) implies a similar inequality and a Zamansky type inequality in the constrained case. Denote by $Q_n^*(f)\in G_n\cap R(f)$ an element in G_n approximating f with the order of the K^*-functional, i.e. $\|f-Q_n^*(f)\|_0\le AK^*(f,n^{-\alpha};X_0,X_1)$, where A is a fixed positive constant.

Theorem 5. *Let (1), (2), (5), (7) and (8) hold. Then*

$$\inf\{\|f-g\|_0+n^{-\alpha}\|g\|_1:g\in G_n\cap R(f)\}\le cK^*(f,n^{-\alpha};X_0,X_1);$$

$$n^{-\alpha}\|Q_n^*(f)\|_1 \le cK^*(f,n^{-\alpha};X_0,X_1).$$

Theorems 3 and 4 separate the investigation of the order of convergence of the constrained approximations into three practically independent parts:

a) obtaining of Jackson type results (8) for the constrained approximation;

b) investigating the non-constrained case - proving of (1), (2) and (5) with the same parameter α as in (8);

c) characterization of the constrained K-functional via proper moduli of functions.

The reason we require c) is the easier computation of moduli compared to the evaluation of K-functionals. If this program is fulfilled then we have a characterization of the constrained approximation. If we have a converse theorem better than (4) (see the next section) then, in view of Theorem 4, the corresponding inverse result is inherited by the constrained approximations.

3. ONESIDED AND SHAPE-PRESERVING APPROXIMATIONS

First we apply the scheme from the previous section in the following three cases of constraints for X_0 - a space of real-valued functions defined on a set Ω:

I. Approximation from below. X_0^* contains these functions from X_0 which are bounded from below,

$$R(f) = \{g\in X_0:g(x)\le f(x)\ \forall x\in\Omega\} = \{f\}-C_+,$$

where C_+ denote the cone of all non-negative functions, i.e. $C_+ = \{g\in X_0 : g(x)\ge 0\ \ \forall x\in\Omega\}$. The best onesided approximation from below with elements of G_n is given by

$$E_n^-(f)_0 := \inf\{\|f-g\|_0 : g\in G_n,\ g\le f\}.$$

II. Approximation from above. Similarly X_0^* contains these functions from X_0 which are bounded from above and

$$R(f) = \{g\in X_0:g(x)\ge f(x)\ \forall x\in\Omega\} = \{f\}+C_+.$$

The best onesided approximation from above with elements of G_n is given by

$$E_n^+(f)_0 := \inf\{\|f-g\|_0 : g\in G_n,\ g\ge f\}.$$

III. Onesided approximation. One approximates a function f simultaneously from above and from below. This case is a combination of I and II. The best onesided approximation with elements of G_n is given by

$$\tilde{E}_n(f)_0 := E_n^-(f)_0+E_n^+(f)_0.$$

In these cases the constraint is a cone with a vertex at f.

3.1. Trigonometric polynomials

Let, as in Section 1.3, G_n denote the set of all trigonometric polynomials of degree n-1, $X_0= L_p[0,2\pi)$ and $X_1=W_p^r[0,2\pi)$ ($1\le p\le\infty$, natural r). Condition (7) is satisfied because every function from X_1 is bounded. The constrained approximations corresponding to the three cases are denoted by $E_n^-(f)_p$, $E_n^+(f)_p$ and $\tilde{E}_n(f)_p$ respectively. The corresponding K-functionals are

$$K^-(f,t;L_p,W_p^r) := \inf\{\|f-g\|_p+t\|g^{(r)}\|_p:g\in W_p^r,g\le f\},$$

$$K^+(f,t;L_p,W_p^r) := \inf\{\|f-g\|_p+t\|g^{(r)}\|_p:g\in W_p^r,g\ge f\},$$

$$\tilde{K}(f,t;L_p,W_p^r) := K^-(f,t;L_p,W_p^r)+K^+(f,t;L_p,W_p^r).$$

As far as we know the onesided K-functional is introduced by Popov [8]. A characterization of the $\tilde{K}$ similar to (6) is provided by the average moduli of smoothness (see e.g. [9]) In a forthcoming paper [5] we give a characterization of the onesided K-functionals from below and from above via moduli of smoothness for r=1 and r=2. The problem for characterizing these functionals for $r\ge3$ is open.

For the Jackson type inequality

$$\tilde{E}_n(f)_p \le c(r)n^{-r}\|f^{(r)}\|_p, \qquad \forall f\in W_p^r$$

see e.g. [9, p.166]. This inequality implies similar inequalities for the onesided approximations from below and from above. Hence all statements of Section 2 hold for the onesided approximations (from below and from above) by trigonometric polynomials.

Using the inverse result of M.F.Timan [13] ($q=\min\{2,p\}$, $1<p<\infty$)

$$\omega_r(f,t)_p \le c(r,p)n^{-r}\left\{\sum_{k=1}^{n} k^{qr-1}E_k(f)_p^q\right\}^{1/q}$$

we get

$$K^*(f,n^{-r};L_p,W_p^r) \le c(r,p)n^{-r}\left\{\sum_{k=1}^{n} k^{qr-1}E_k^*(f)_p^q\right\}^{1/q},$$

where * stands for $^-$, $^+$ or $^\sim$.

Multivariate onesided approximation by trigonometric polynomials can be studied in a similar way. The necessary parts for the scheme can be found in [3].

3.2. Algebraic polynomials

Let G_n denote the set of all algebraic polynomials of degree n-1. Keeping the notations of Section 1.4 ($\varphi(x)=\sqrt{1-x^2}$, $X_0=L_p[-1,1]$, $X_1=W_p^r(\varphi)$) denote the onesided approximations by $\tilde{E}_n(f)_p$. A characterization of $\tilde{K}(f,n^{-r},;L_p,W_p^r(\varphi))$ is given in [4]. For the K-functionals from below and from above see [5].

The Jackson type inequality

$$\tilde{E}_n(f)_p \le c(r)\left\{n^{-r}\|\varphi^r f^{(r)}\|_p+n^{-2r}\|f^{(r)}\|_p\right\}, \quad \forall f\in W_p^r$$

is proved in [10], [4]. It implies similar inequalities for the onesided approximations from below and from above and we have the results of Section 2 in this case.

Some other examples of constraints fitting in the scheme of Section 2 are the shape-preserving approximations. Let X_0 be a set of functions defined on an interval Ω of the real line.

IV. Monotone and comonotone approximation.

Let X_0^* consists of all increasing (decreasing) functions from X_0. Then for $f\in X_0^*$ we set $R(f)=X_0^*$. In these cases we have monotone approximation.

Let $\Omega=\cup[a_i,a_{i+1}]$, where $a_i<a_{i+1}$. Let X_0^* consists of all functions $f\in X_0$ which are increasing or decreasing in $[a_i,a_{i+1}]$ if i is even or odd and $R(f)=X_0^*$ for every $f\in X_0^*$. In this case we have comonotone approximation.

V. Convex approximation.

As in IV X_0^* and $R(f)$ consist of all functions of the same convexity as f.

In the above cases condition (7) is obviously satisfied. The results on Jackson type inequalities and characterization of the shape-preserving K-functionals are less complete than in the case of onesided approximation. As far as we know the monotone K-functional is introduced by DeVore [1].

REFERENCES

[1] DeVore, R.A., Degree of approximation, in: Approximation theory II (Academic Press, New York, 1976) pp. 117-162.

[2] Ditzian, Z. and Totik, V., Moduli of Smoothness (Springer, New York, 1987).

[3] Hristov, V.H. and Ivanov, K.G., Operators for onesided approximation of functions, in: Constructive theory of functions'87, (Izd. BAS, Sofia, 1988) pp. 222-232.

[4] Hristov, V.H. and Ivanov, K.G., Operators for onesided approximation by algebraic polynomials in $L_p([-1,1]^d)$, Mathematica Balkanica (new series) **2** (1988) 374-390.

[5] Hristov, V.H. and Ivanov, K.G., Characterization of the best onesided approximation from below and from above, in print.

[6] Ivanov, K.G., A constructive characteristic of the best algebraic approximation in $L_p[-1,1]$ $(1\le p\le\infty)$, in: Constructive function theory'81 (Izd. BAS, Sofia, 1983) pp. 357-367.

[7] Ivanov, K.G., A characterization of weighted Peetre K-functionals, J. Appox. Theory **56** (1989) 185-211.

[8] Popov, V.A., Average moduli and their function spaces, in: Constructive function theory'81, (Izd. BAS, Sofia, 1983) pp. 482-487.

[9] Sendov, Bl. and Popov, V.A., The averaged moduli of smoothness (J.Wiley & sons, Chichester, 1988)

[10] Stojanova, M.P., The best onesided algebraic approximation in $L_p[-1,1]$ $(1\le p\le\infty)$, Mathematica Balkanica (new series) 2 (1988) 101-113.

[11] Стечкин, С.Б., О порядке наилучших приближений непрерывных функций, Известия АН СССР, сер. матем. **15** (1951) 219-242.

[12] Тиман, А.Ф., Теория приближения функций действительного переменного (Физматгиз, Москва, 1960)

[13] Тиман, М.Ф., Обратные теоремы конструктивной теории функций в пространствах L_p, Матем. сборник **46** (1958) 125-132.

Approximation, Optimization and Computing:
Theory and Applications, A.G. Law and C.L. Wang (eds.)
Elsevier Science Publishers B.V. (North-Holland)

A PIVOTING SCHEME FOR THE INTERVAL GAUSS-SEIDEL METHOD: NUMERICAL EXPERIMENTS

Chen-Yi Hu and R. Baker Kearfott

Department of Mathematics, University of Southwestern Louisiana
U.S.L. Box 4-1010, Lafayette, LA 70504

Abstract. Interval Newton methods in conjunction with generalized bisection form the basis of algorithms which find all real roots within a specified box $\mathbf{X} \subset \mathbf{R}^n$ of a system of nonlinear equations $F(X) = 0$ *with mathematical certainty*, even in finite precision arithmetic. The practicality and efficiency of such methods is, in general, dependent upon *preconditioning* a certain interval linear system of equations. Here, we present the results of numerical experiments for such a *pivoting* preconditioner we are developing.

1. INTRODUCTION AND ALGORITHMS

The general problem we address is:

(1.1) Find, with certainty, approximations to all solutions of the nonlinear system
$$F(X) = (f_1(x_1, x_2, \ldots, x_n), \ldots, f_n(x_1, x_2, \ldots, x_n)) = 0,$$
where bounds $\underline{x}_i$ and $\overline{x}_i$ are known such that
$$\underline{x}_i \le x_i \le \overline{x}_i \text{ for } 1 \le i \le n.$$

We write $X = (x_1, x_2, \ldots, x_n)$, and we denote the box given by the inequalities on the variables x_i by $\mathbf{B}$.

A successful approach to this problem is generalized bisection in conjunction with interval Newton methods, as described in [4], [8] or numerous other works. For an introduction to the interval arithmetic underlying these methods, see [1], [9], the recent review [6], etc. Also, the book [11] will contain an overview of interval methods for linear and nonlinear systems of equations.

In these methods, we first transform $F(X) = 0$ to the linear interval system

$$\mathbf{F}'(\mathbf{X}_k)(\tilde{\mathbf{X}}_k - X_k) = -F(X_k), \tag{1.2}$$

where $\mathbf{F}'(\mathbf{X}_k)$ is a suitable (such as an elementwise) interval extension of the Jacobian matrix over the box $\mathbf{X}_k$ (with $\mathbf{X}_0 = \mathbf{B}$), and where $X_k \in \mathbf{X}_k$ represents a *predictor* or *initial guess* point. (Consult [1], [6], [9], [11], [12], etc. for information on interval extensions.) If we formally solve (1.2) using interval arithmetic, the resulting box $\tilde{\mathbf{X}}_k$, which actually just satisfies

$$\mathbf{F}'(\mathbf{X}_k)(\tilde{\mathbf{X}}_k - X_k) \supset -F(X_k), \tag{1.2(b)}$$

will contain all solutions to all systems

$$A(X - X_k) = -F(X_k),$$

for $A \in \mathbf{F}'(\mathbf{X}_k)$. Also, for suitable interval extensions $\mathbf{F}'(\mathbf{X}_k)$, the mean value theorem implies that $\mathbf{X}_k$ will contain all solutions to $F(X) = 0$. We then define the next iterate $\mathbf{X}_{k+1}$ by

$$\mathbf{X}_{k+1} = \mathbf{X}_k \cap \tilde{\mathbf{X}}_k. \tag{1.3}$$

This scheme is termed an *interval Newton method*.

If the coordinate intervals of $\mathbf{X}_{k+1}$ are not smaller than those of $\mathbf{X}_k$, then we may bisect one of these intervals to form two new boxes; we then continue the iteration with one of these boxes, and put the other one on a stack for later consideration. The following fact (from [10]) allows such a composite generalized bisection algorithm to compute all solutions to (1.1) *with mathematical certainty*. For many methods of solving (1.2),

(1.4) if $\tilde{\mathbf{X}}_k \subset \mathbf{X}_k$, then the system of equations in (1.1) has a unique solution in $\mathbf{X}_k$. Conversely, if $\tilde{\mathbf{X}}_k \cap \mathbf{X}_k = \emptyset$ then there are no solutions of the system in (1.1) in $\mathbf{X}_k$.

We give complete details of the overall generalized bisection algorithm in [5]. Here, we are interested in the fact that the efficiency of generalized bisection depends on the way we find the solution bounds $\tilde{\mathbf{X}}_k$ to (1.2).

In [2], we derived a pivoting scheme for the *interval Gauss-Seidel* method for computing $\tilde{\mathbf{X}}_k$. We review that scheme and report computational results here. For proofs and additional details, see [2].

We use the following notation. We write

$$\mathbf{X} = (\mathbf{x}_1, \mathbf{x}_2, \ldots, \mathbf{x}_n)$$

for $\mathbf{X}_k$ and we let $\mathbf{A}_{i,j}$ be the interval in the i-th row and j-th column of $\mathbf{A} = \mathbf{F}'(\mathbf{X})$. We denote the components of F as boldface intervals, since they must be evaluated in interval arithmetic with directed roundings, so that we have

$F(X_k) = F = (\mathbf{f}_1, \mathbf{f}_2, \ldots, \mathbf{f}_n)$, and $X_k = (x_1, x_2, \ldots, x_n)$. In this notation, (1.2) becomes

$$\mathbf{A}(\tilde{\mathbf{X}}_k - X_k) = -F. \tag{1.5}$$

We generally *precondition* (1.5) by a scalar (i.e. non-interval) matrix Y to obtain

$$Y\mathbf{A}(\tilde{\mathbf{X}}_k - X_k) = -YF. \tag{1.6}$$

Let $Y_i = (y_1, y_2, \ldots, y_n)$ denote the i-th row of the preconditioner, let

$$\mathbf{k}_i = Y_i F,$$

and let

$$\begin{aligned} Y_i\mathbf{A} = \mathbf{G}_i &= (\mathbf{G}_{i,1}, \mathbf{G}_{i,2}, \ldots, \mathbf{G}_{i,n}) \\ &= ([\underline{g}_{i,1}, \overline{g}_{i,1}], [\underline{g}_{i,2}, \overline{g}_{i,2}], \ldots, [\underline{g}_{i,n}, \overline{g}_{i,n}]). \end{aligned}$$

Then the *preconditioned interval Gauss-Seidel algorithm* is

ALGORITHM 1. *(Preconditioned version of interval Gauss-Seidel) Do the following for $i = 1$ to n.*

1. *(Update a coordinate.)*
   (a) *Compute Y_i.*
   (b) *Compute $\mathbf{k}_i$ and $\mathbf{G}_i$.*
   (c) *Compute*

$$\tilde{\mathbf{x}}_i = x_i - \frac{\left[\mathbf{k}_i + \sum_{\substack{j=1 \\ j\neq i}}^{n} \mathbf{G}_{i,j}(\mathbf{x}_j - x_j)\right]}{\mathbf{G}_{i,i}}. \tag{1.7}$$

   using interval arithmetic.
2. *(The new box is empty.) If $\tilde{\mathbf{x}}_i \cap \mathbf{x}_i = \emptyset$, then signal that there is no root of F in $\mathbf{X}$, and continue the generalized bisection algorithm.*
3. *(The new box is non-empty; prepare for the next coordinate.)*
   (a) *Replace $\mathbf{x}_i$ by $\mathbf{x}_i \cap \tilde{\mathbf{x}}_i$.*
   (b) *Possibly re-evaluate $\mathbf{F}'(\mathbf{X}_k)$ to replace $\mathbf{A}$ by an interval matrix whose corresponding widths are smaller.*

We will denote the *width* of the interval $\mathbf{x}_i = [\underline{x}_i, \overline{x}_i]$ by $w(\mathbf{x}_i) = \overline{x}_i - \underline{x}_i$. Similarly, we will denote the *lower bound* for the box

$$\mathbf{X} = ([\underline{x}_1, \overline{x}_1], [\underline{x}_2, \overline{x}_2], \ldots, [\underline{x}_n, \overline{x}_n]).$$

by $\mathbf{X}\downarrow = (\underline{x}_1, \underline{x}_2, \ldots, \underline{x}_n)$ and the *upper bound* for the box by $\mathbf{X}\uparrow = (\overline{x}_1, \overline{x}_2, \ldots, \overline{x}_n)$. We will speak of the *diameter* of the box $\mathbf{X}$ as $\|\mathbf{X}\uparrow - \mathbf{X}\downarrow\|_2$.

The following theorem shows us that preconditioning does not affect the reliability of the overall generalized bisection scheme.

THEOREM 2. *Let $\mathbf{X}^+ = (\mathbf{x}_1^+, \mathbf{x}_2^+, \ldots, \mathbf{x}_n^+)$ denote the new (possibly altered and possibly empty) box which Algorithm 1 returns, and refer to the $\mathbf{X}$ entering Algorithm 1 as simply $\mathbf{X} = (\mathbf{x}_1, \mathbf{x}_2, \ldots, \mathbf{x}_n)$. Then any roots of F in $\mathbf{X}$ must also be in the new $\mathbf{X}^+$.*

See [5] for a proof of Theorem 2.

The pivoting preconditioner is based on the following characterization.

THEOREM 3. *In (1.7), if $\mathbf{k}_i \in -\sum_{\substack{j=1 \\ j\neq i}}^{n} \mathbf{G}_{i,j}(\mathbf{x}_j - x_j)$ and $0 \notin \mathbf{G}_{i,i}$, then*

$$w(\tilde{\mathbf{x}}_i) = \sum_{\substack{j=1 \\ j\neq i \\ \sigma_j=1}}^{n} \max\{|\underline{g}_{i,j}|, |\overline{g}_{i,j}|\} w(\mathbf{x}_j^1) + \sum_{\substack{j=1 \\ j\neq i \\ \sigma_j=0}}^{n} \lambda_j^* w(\mathbf{G}_{i,j}) w(\mathbf{x}_j^1),$$

where

$$\lambda_j^* = \begin{cases} \dfrac{\max\{(x_j - \underline{x}_j), (\overline{x}_j - x_j)\}}{w(\mathbf{x}_j^1)} & \text{if } w(\mathbf{x}_j^1) \neq 0, \\ 0 & \text{otherwise,} \end{cases}$$

where

$$\sigma_j = \begin{cases} 0 & \text{if } |\underline{g}_{i,j}|\frac{1}{\lambda_j^*} - 1 < |\overline{g}_{i,j}| < |\underline{g}_{i,j}|\frac{\lambda_j^*}{1-\lambda_j^*} \\ & \text{and } 0 \in \mathbf{G}_{i,j} = [\underline{g}_{i,j}, \overline{g}_{i,j}], \\ 1 & \text{otherwise,} \end{cases}$$

and where

$$\mathbf{x}_j^1 = \begin{cases} \mathbf{x}_j \cap \tilde{\mathbf{x}}_j & \text{if } j \leq i, \\ \mathbf{x}_j & \text{if } j > i. \end{cases}$$

The idea in [2] and here is to develop preconditioners which, though not necessarily giving a minimal $w(\mathbf{x}_i)$, require no numerical linear algebra and only $O(n^2)$ arithmetic operations, and are potentially effective on many problems. See [2] and [5] for *rationale* and details. We review the resulting preconditioner here.

Noting that terms in the sums in Theorem 3 for which $w(\mathbf{x}_j^1) = 0$ are irrelevant, we make

DEFINITION 4.

$$\mathcal{N}_{\text{col}} = \left\{ j \in \{1, 2, \ldots, i-1, i+1, \ldots, n\} \mid w(\mathbf{x}_j^1) \neq 0 \right\}.$$

Also, because of considerations explained in [2], our preconditioner algorithm makes use of

DEFINITION 5. $\mathcal{N}_{\text{row}} = \{j \in \{1, 2, \ldots, n\} \mid 0 \notin \mathbf{A}_{j,i}\}$.

Our pivoting preconditioner is based on selecting a row of the interval Jacobian matrix $\mathbf{A}$ to use as the preconditioned row $\mathbf{G}_i$; i.e., on selecting Y_i to be a unit vector. It is optimal in the sense of

DEFINITION 3.5. *We say that Y_i is pivoting first optimal if Y_i yields a $\mathbf{G}_i$ which minimizes $w(\tilde{\mathbf{x}}_i)$, where $\tilde{\mathbf{x}}_i$ is as in (1.7), subject to the condition that Y_i be a unit vector.*

We use the following lemma to determine which column indices m give pivoting first optimal preconditioner.

LEMMA 6. *Suppose there exists a pivoting first optimal preconditioner row $Y_i = e_m^T$. Suppose further that, as in Theorem 3, $\mathbf{k}_i \in -\sum_{\substack{j=1 \\ j\neq i}}^{n} \mathbf{A}_{m,j}(\mathbf{x}_j - x_j)$ and $0 \notin \mathbf{A}_{m,i}$, where $\mathbf{k}_i = \mathbf{f}_m$ is the m-th component of the function value $F(X_k)$.*

Then

$$w(\tilde{\mathbf{x}}_i) = \left[\sum_{\substack{j\in\mathcal{N}_{col}\\ \sigma_j=1}} \max\{|\underline{a}_{m,j}|, |\overline{a}_{m,j}|\} w(\mathbf{x}_j^1) + \sum_{\substack{j\in\mathcal{N}_{col}\\ \sigma_j=0}} \lambda_j^* w(\mathbf{A}_{m,j}) w(\mathbf{x}_j^1) \right] / \min\{|\underline{a}_{m,i}|, |\overline{a}_{m,i}|\},$$

where

$$\lambda_j^* = \begin{cases} \frac{\max\{(x_j - \underline{x}_j),(\overline{x}_j - x_j)\}}{w(\mathbf{x}_j^1)} & \text{if } w(\mathbf{x}_j^1) \neq 0 \\ 0 & \text{otherwise,} \end{cases}$$

where

$$\sigma_j = \begin{cases} 0 & \text{if } |\underline{a}_{m,j}|(1/\lambda_j^* - 1) < |\overline{a}_{m,j}| < \frac{|\underline{a}_{m,j}|\lambda_j^*}{1-\lambda_j^*} \\ & \text{and } 0 \in \mathbf{A}_{m,j} = [\underline{a}_{m,j}, \overline{a}_{m,j}] \\ 1 & \text{otherwise,} \end{cases}$$

and where

$$\mathbf{x}_j^1 = \begin{cases} \mathbf{x}_j \cap \tilde{\mathbf{x}}_j & \text{if } j \leq i \\ \mathbf{x}_j & \text{if } j > i. \end{cases}$$

Lemma 6 follows from Theorem 3; see [2], and leads to

ALGORITHM 7. *(Determination of the column index for the i-th first optimal pivoting preconditioner row)*

1. *Check* $w(\mathbf{x}_j^1)$ *for* $j = 1, 2, \ldots, i-1, i+1, \ldots, n$ *to find* $\mathcal{N}_{col}$.
2. *Check the* i*-th column of the interval Jacobian matrix* $\mathbf{A}$ *to determine* $\mathcal{N}_{row}$.
3. *Pick* $m_0 \in \mathcal{N}_{row}$ *which minimizes* $w(\tilde{\mathbf{x}}_i)$ *in Lemma 3.8 over all* $m \in \mathcal{N}_{row}$.
4. *Choose* $Y_i = e_{m_0}^T$ *in Step 1(a) of Algorithm 1.2, where* $e_{m_0}^T$ *is that unit row vector with 1 in the* m_0*-th coordinate and 0 in the other coordinates.*

2. ALTERNATE ALGORITHMS AND COST

The pivoting preconditioner may not result in an optimally small $w(\tilde{\mathbf{x}}_i \cap \mathbf{x}_i)$ in Algorithm 1; see [2]. However, it can sometimes result in reduction of $w(\mathbf{x}_i)$ with less arithmetic operations than other preconditioners. Here, we compare the number of arithmetic operations to complete Algorithm 1, when various different procedures are used to compute Y_i. In particular, we will compare the following schemes:

1) the pivoting preconditioner (Algorithm 7);
2) the minimum width linear programming preconditioner;
3) the inverse midpoint preconditioner;
4) solving for each variable in each equation.

The linear programming preconditioner is explained in [5]. It gives a *minimal* $w(\tilde{\mathbf{x}}_i)$ for each i, but is expensive to obtain. We are unsure of the exact operations count, but on two variable dimension test problems, the total work for the generalized bisection method (and possibly also for Algorithm 1.2) seems to increase like $O(n^5)$.

The inverse midpoint preconditioner is commonly used in Algorithm 1. In it, we take Y_i to be the i-th row of the inverse of the n by n non-interval matrix formed by taking the midpoint of each entry of the n by n interval matrix $\mathbf{F}'(\mathbf{X}_k)$. This preconditioner has some nice properties, but is not always appropriate, as explained in [5].

We use the following modification of the interval Gauss-Seidel algorithm when we solve for each coordinate.

ALGORITHM 8. *(Solve for each variable in each equation.) Let* $\mathbf{A} = \mathbf{F}'(\mathbf{X})$. *Do the following for* **for** $m = 1$ **to** n *and for* $i = 1$ *to* n.

1. *Compute*

$$\tilde{\mathbf{x}}_i = x_i - \left[\mathbf{f}_m + \sum_{\substack{j=1\\ j\neq m}}^{n} \mathbf{A}_{m,j}(\mathbf{x}_j - x_j) \right] \Big/ \mathbf{A}_{m,i}$$

using interval arithmetic.

2. *If* $\tilde{\mathbf{x}}_i \cap \mathbf{x}_i = \emptyset$, *then signal that there is no root of F in* $\mathbf{X}$, *and continue the generalized bisection algorithm.*
3. *(Prepare for the next coordinate.)*
   (a) *Replace* $\mathbf{x}_i$ *by* $\mathbf{x}_i \cap \tilde{\mathbf{x}}_i$.
   (b) *Possibly re-evaluate* $\mathbf{F}'(\mathbf{X}_k)$ *to replace* $\mathbf{A}$ *by an interval matrix whose corresponding widths are smaller.*

In our operations counts here, we assume a dense interval Jacobian matrix.

We give summary of the arithmetic complexity of the above algorithms, in terms of both non-interval and interval operations, in Table 1. Precise values appear in [2], and are derived in [3]. Values for sparse Jacobian matrices also appear in those two places. We emphasize that the values in Table 1 represent the order of operations to complete an entire n steps of Algorithm 1.

Table 1

Type of op.	Pivoting	Each variable	inverse midpoint
interval ×	$O(n^2)$	$O(n^2)$	$O(n^2)$
interval ÷	$O(n)$	$O(n^2)$	$O(n)$
interval +	$O(n^2)$	$O(n^3)$	$O(n^3)$
usual ×	$O(n^2)$	0	$O(n^3)$
usual ÷	$O(n^2)$	0	$O(n^2)$
usual +	$O(n^2)$	$O(n^2)$	$O(n^3)$

An alternative to the above three preconditioners is to apply no preconditioner at all, i.e. to always take $m_0 = i$. In that case, there are no scalar arithmetic operations, and the numbers of interval arithmetic operations are the same as for the pivoting preconditioner.

3. EXPERIMENTAL RESULTS

In this section, we report results of some preliminary

experiments on the three preconditioners mentioned in Section 2 and on the interval Gauss-Seidel method with no preconditioner at all (i.e. with Y equal to the identity matrix). The computation involves repeated execution of Algorithm 1 for each box $\mathbf{X}$ produced from the initial box via bisection, until either one of the possibilities in (1.4) happens or else until the maximum width $w(\mathbf{x}_i)$ is smaller than a fixed tolerance (10^{-5} here). See [4], [5], and [7] for details of the overall algorithm and of the way the interval Gauss-Seidel method is combined with generalized bisection.

Here, we examine a variable dimension test problem, which can be solved with no preconditioner at all. This allows us to measure the overhead costs of the preconditioner schemes in a relatively simple way. A more thorough analysis of the types of problems for which the pivoting preconditioner is appropriate appears in [2].

We report total virtual CPU times on an IBM 3090, with a code similar to that in [7], in Table 2. We report the total number of boxes considered in Table 3 and the total number of interval Jacobian evaluations in Table 4; these are measures of the effectiveness of a preconditioner which are independent of the cost to obtain the preconditioner. (For this test problem, no preconditioner and the pivoting preconditioner seem to give the same row indices.) In each case, the first column gives the order of the problem.

The algorithm was unable to finish in eleven CPU hours for the entries in the table marked with asterisks.

Table 2 – CPU times

n	Pivoting	Each variable	inverse midpoint	none
5	0.61	0.50	1.13	0.40
10	4.79	4.51	121.74	1.75
15	21.39	21.73	*	6.46
20	48.45	69.80	*	17.79
25	91.23	144.63	*	34.84
30	149.46	264.20	*	61.21
35	225.47	441.42	*	98.47
40	320.51	680.28	*	148.62

Table 3 – Number of boxes

n	Pivoting	Each variable	inverse midpoint	none
5	45	47	56	43
10	71	73	589	67
15	113	103	*	97
20	141	139	*	133
25	171	169	*	163
30	201	199	*	193
35	231	229	*	223
40	261	259	*	253

Table 4 – Number Jacobian matrix evaluations

n	Pivoting	Each variable	inverse midpoint	none
5	52	52	29	56
10	82	82	294	86
15	126	116	*	120
20	154	154	*	158
25	184	184	*	188
30	214	214	*	218
35	244	244	*	248
40	274	274	*	278

REFERENCES

[1] Alefeld, G., and Herzberger, J., "Introduction to Interval Computations", (Academic Press, New York, 1983).

[2] Hu, C.-Y., and Kearfott, R. B., *A Width Characterization, Efficiency, and a Pivoting Strategy for the Interval Gauss-Seidel Method*, Math. Comp., submitted (1989).

[3] Hu, C.-Y., "Preconditioners for Interval Newton Methods", Ph.D. dissertation, The University of Southwestern Louisiana (expected in 1990).

[4] Kearfott, R. B., *Some Tests of Generalized Bisection*, ACM Trans. Math. Software **13** 3, 197–220 (1987).

[5] Kearfott, R. B., *Preconditioners for the interval Gauss-Seidel method*, SIAM J. Numer. Anal., to appear.

[6] Kearfott, R. B. *Interval Arithmetic Techniques in the Computational Solution of Nonlinear Systems of Equations: Introduction, Examples, and Comparisons*, to appear in the proceedings of the 1988 AMS/SIAM Summer Seminar.

[7] Kearfott, R. B., and Novoa, M, *INTBIS, an Portable Interval Newton / Bisection Package*, to appear in ACM Trans. Math. Software.

[8] Moore, R. E., and Jones, S. T. *Safe Starting Regions for Iterative Methods*, SIAM J. Numer. Anal. **14** 6, 1051–1065 (1977).

[9] Moore, R. E., "Methods and Applications of Interval Analysis" (SIAM, Philadelphia, 1979).

[10] Neumaier, A. *Interval iteration for zeros of systems of equations*, BIT **25** 1, 256–273 (1985).

[11] Neumaier, A., "Interval Methods for Systems of Equations", Cambridge University Press, in press.

[12] Ratschek, H., and Rokne, J. "Computer Methods for the Range of Functions", (Horwood, Chichester, England, 1984).

Approximation, Optimization and Computing:
Theory and Applications, A.G. Law and C.L. Wang (eds.)
Elsevier Science Publishers B.V. (North-Holland)

RATIONAL APPROXIMATION BY THE METHOD OF NUMERICAL INTEGRATION

Jiang Yong

Dept. of Applied Mathematics, East China Institute of Technology
Nanjing, P.R.CHINA

A class of rational approximation is presented by the numerical integration. The method has been proved successful with the following properties. 1. The rational approximation is uniformly convergent and has the structure of fraction in the lowest terms; 2. The algorithmic complexity is low and the accuracy is high; 3. The number of nodes can be increased steadily so that the error can be controlled automatically; 4. The rational approximation has the same pole and its degree as the original function.

1. INTRODUCTION

Apart from some theoretical problems indicated by P.Turan [1], every rational approximation and its algorithm so far has more or less following problems in its theory, practical computation and application:

(1). algorithmic complexity is high, for instance, it often causes solving system of equations ;
(2). it is not alway convergent (especially for interpolation problem);
(3). it may add new pole, or lose old pole, or change the multiplicity of pole;
(4). the structure of fraction is not lowest term, which may bring difficulties in the application;
(5). the nodes can not be increased gradually to control automatically the error; and so on.

For example, all the alorithms have a problem closely related to the practice which has become a dilemma in the practical application of all the numerical rational approximation method, that is, the rational approximation function obtained may has different analytic properties to the original function in the region (mainly in the pole and its degree). This makes the rational approximation unexceptale [2], which causes worries to those who use the approximation function.

This paper gives a numerical rational approximation method which avoids the defects mentioned above, and its application in the numerical inversion of Laplace transform.

2. OUTLINE OF THE METHOD

An analytic function in multiply connected domain can be expressed by

$$H(z)=\frac{1}{2\pi i}\int_{c}\frac{H(t)}{t-z}dt \quad (Cauchy),$$

here contour C take the positive direction along the border of the region. A serise of approximation to H(z) are obtained by applying the different method of numerical integration to the above equality.

It is very exciting that the approximations can be expressed in rational functions by certain technique, therfore, it is practically the rational approximation to H(z) with the following properties:

(1). The ratoinal approximation to H(z) can be expressed by explicit formula, so the approximation realized by directly computing the function values;
(2). The rational approximation is the sum of fraction in the lowest terms, which provides a lot of convenience in the practical applications;
(3). The rational approximation has the same pole and its degree as the original function in the corresponding region;
(4). The algorithm has low complexity and high accuracy, and is easy to program;
(5). The number of nodes can be added gradually so that the error can be controlled automatically;
(6). The extraplation technique can be introduced to speed up the uniform convergence; and so on.

3. THE MAIN RESULTS

The followings are some main results.

Theorem 1. Assume that H(z) is analytic in multiply connected region D containing closed disk $|z| \leqslant R$, and continous in D, then $\tilde{H}_N(z)$ is a rational approximation to H(z):

$$H(z) \sim \tilde{H}_N(z) = \frac{r}{2N}\sum_{k=1}^{2N}\frac{a(\theta_k)z + b(\theta_k)}{z^2 + c(\theta_k)z + r^2} + i\frac{r}{2N}\sum_{k=1}^{2N}\frac{d(\theta_k)z + e(\theta_k)}{z^2 + c(\theta_k)z + r^2} \quad |z| < r,\ r \leqslant R$$

here

$$\begin{cases} a(\theta_k) = ImH(re^{i\theta_k})sin\theta_k - ReH(re^{i\theta_k})cos\theta_k \\ b(\theta_k) = rReH(re^{i\theta_k}) \\ c(\theta_k) = -2rcos\theta_k \\ d(\theta_k) = -ImH(re^{i\theta_k})cos\theta_k - ReH(re^{i\theta_k})sin\theta_k \\ e(\theta_k) = ImH(re^{i\theta_k}) \\ \theta_k = k\pi/N \qquad k = 1,2,\cdots,2N \end{cases}$$

Corollary. Assume that f(x) is a real function on [−R,R]. If its expansion to complex plane f(z) is analytic in a multiply connected region D containing $|z| \leqslant R$, then has rational approximation $\tilde{f}_N(x)$ in (−R,R):

$$f(x) \sim \tilde{f}_N(x) = \frac{r}{2N}\sum_{k=1}^{2N}\frac{a_k x + b_k}{x^2 + c_k x + r^2}, \quad x \in (-r,r),\ r \leqslant R$$

where

$$\begin{cases} a_k = Imf(re^{i\theta_k})sin\theta_k - Ref(re^{i\theta_k})cos\theta_k \\ b_k = rRef(re^{i\theta_k}) \\ c_k = -2rcos\theta_k \end{cases}$$

Theorem 2. Under the conditions of Theorem 1, $\tilde{H}_N(z)$ converges uniformly to H(z) in inner close set of $|z| < r$,and the truncation error $\tilde{R}_N(z,r)$ satisfies

$$|\tilde{R}_N(z,r)| \leqslant \frac{C_j}{N^{2j+1}}, \quad |z| < r, r \leqslant R$$

in which

$$C_j = 2rm_j(z,r)\zeta(2j+1),$$

$$M_j(z,r) = max_{0 \leqslant \theta \leqslant 2\pi}\left|\left(\frac{H(re^{i\theta})}{r - ze^{-i\theta_k}}\right)_\theta^{(2j+1)}\right|$$

are independable on N. ζ(2j+1) is Riemann–Zeta function, j is positive integer which make

$$\left(\frac{H(re^{i\theta})}{r - ze^{-i\theta}}\right)_\theta^{(2j+1)}$$

continuous.

Theorem 3. Let H(z) is analytic in $|z| < r$, then $\tilde{H}_N(z)$ analytic.

4. PROOFS OF THE THEOREMS

First, the proof of theorem 1 is given, which also shows the construction of the rational approximation method in this paper.

Proof of theorem 1. By Cauchy theorem and the analytic of function H(z) in D, the equality

$$H(z) = \frac{1}{2\pi i}\int_c \frac{H(t)}{t - z}dt, \quad z \in D$$

holds, where C has the positive direction. Furthermore,

$$H(z) = \frac{1}{2\pi i}\int_{|t| = r}\frac{H(t)}{t - z}dt, \quad |z| < r,\ r \leqslant R$$

where r is a constant and the integration is anticlockwise. Let $t = re^{i\theta}$, then

$$H(z) = \frac{1}{2\pi i}\int_0^{2\pi}\frac{H(re^{i\theta})}{re^{i\theta} - z}ire^{i\theta}d\theta = \frac{r}{2\pi}\int_0^{2\pi}\frac{H(re^{i\theta})}{r - ze^{-i\theta}}d\theta$$

Let

$$u(\theta,r) = ReH(re^{i\theta}),$$
$$v(\theta,r) = ImH(re^{i\theta}),$$

then

$$\begin{aligned} H(z) &= \frac{r}{2\pi}\int_0^{2\pi}\frac{H(re^{i\theta})(r - ze^{i\theta})}{r^2 - 2zcos\theta + z^2}d\theta \\ &= \frac{r}{2\pi}\int_0^{2\pi}\frac{(v(\theta,r)sin\theta - u(\theta,r)cos\theta)z + ru(\theta,r)}{r^2 - 2zrcos\theta + z^2}d\theta \\ &+ i\frac{r}{2\pi}\int_0^{2\pi}\frac{-(v(\theta,r)cos\theta + u(\theta,r)sin\theta)z + rv(\theta,r)}{r^2 - 2zrcos\theta + z^2}d\theta. \end{aligned}$$

By numerical integration formula (rectangle formula is used here with node $\theta_k = k\pi/N$, $k = 1,2,\ldots,2N$),

$$\begin{aligned} H(z) &= \frac{r}{2\pi}\sum_{k=1}^{2N}\frac{(v(\theta_k,r)sin\theta_k - u(\theta_k,r)cos\theta_k)z + u(\theta_k,r)}{r^2 - 2zrcos\theta_k + z^2}\frac{\pi}{N} \\ &+ i\frac{r}{2\pi}\sum_{k=1}^{2N}\frac{-(v(\theta_k,r)cos\theta_k + u(\theta_k,r)sin\theta_k)z + rv(\theta_k,r)}{r^2 - 2zrcos\theta + z^2}\frac{\pi}{N} \\ &+ \tilde{R}_N(z,r) \\ &= \tilde{H}_N(z) + \tilde{R}_N(z,r). \end{aligned}$$

So the rational approximation to H(z) is obtained :

$$H(z) \sim \tilde{H}_N(z) = \frac{r}{2N}\sum_{k=1}^{2N}\frac{a(\theta_k)z + b(\theta_k)}{z^2 + c(\theta_k)z + r^2} + i\frac{r}{2N}\sum_{k=1}^{2N}\frac{d(\theta_k)z + e(\theta_k)}{z^2 + c(\theta_k)z + r^2}, \quad |z| < r,\ r \leqslant R.$$

The proofs of theorem 2 and theorem 3 are omited.

5. NUMERICAL EXAMPLE

The rational approximation to e^x on [−1,1] has been discussed especially in recent years, for which D.J.Newman,et.al [3,4,5] have done a mount of work. Here only the computational result is presented by the method developed in this paper, not all the studies in detail.

From the corollary of theorem 1, the rational approximation to

$$f(x) = e^x \qquad x \in [-1,1]$$

is

$$\tilde{f}_N(x) = \frac{r}{N}\sum_{k=1}^{N}\frac{a(\theta_k)x + b(\theta_k)}{x^2 + c(\theta_k)x + r^2} \qquad x \in [-1,1],\ r > 1$$

where

$$\begin{cases} a(\theta_k) = e^{r\cos\theta_k}[\sin(r\sin\theta_k) - \cos(r\sin\theta_k)\sin\theta_k] \\ b(\theta_k) = re^{r\cos\theta_k}\cos(r\sin\theta_k) \\ c(\theta_k) = -2r\cos\theta_k \\ \theta_k = k\pi/N \qquad k = 1,2,\cdots,N. \end{cases}$$

Take N = 20, r = 10, the approximated value, exact value and the errorare as follow: (be omited, the maximun < 10^{-4}).

6. THE INVERSION OF LAPLACE TRANSFORM

Because of the properties of the rational approximation method in this paper, the numerical inverison method of Laplace transformation based this rational approximation avoids a great many defects, and has the characteristics that the half division algorithm and extrapolation technique can be used to speed up the computation, and the error can be controled automatically.

The main results in this paper are as follows.

Theorem 4. Assume that F(p) is analytic in

$$D = \{p \| p - p_0| \geqslant R\},$$

then there is a rational approximation $\tilde{F}_N(p)$ to F(p)

$$F(p) \sim \tilde{F}_N(p)$$

$$= \frac{r}{2N}\sum_{k=1}^{2N}\frac{A_R(p-p_0)B_k}{(p - p_0 + R\cos\theta_R/r) + (R\sin\theta_k/r)}$$

$$+ i\frac{r}{2N}\sum_{k=1}^{2N}\frac{D_k(p-p_0)+E_k}{(p - p_0 + R\cos\theta^k/r) + (R\sin\theta_k/r)}$$

$$|p - p_0| > R/r,\ r \leqslant 1$$

in which

$$\begin{cases} A_k = -R[ReF(p_k)\cos\theta_k + ImF(p_k)\sin\theta_k]/r^2 \\ B_k = -R^2 ReF(p_k)/r^3 \\ D_k = -R[ReF(p_k)\sin\theta_k + ImF(p_k)\cos\theta_k]/r^2 \\ E = -R^2 ImF(p_k)/r^3 \\ p_k = p_0 - R\cos\theta_k/r + iR\sin\theta_k/r \\ \theta_k = k\pi/N \qquad k = 1,2,\cdots,2N \end{cases}$$

Theorem 5. Assume that F(p) is generating function and analytic in sufficiently large region

$$D = \{P \| p - p_0| \geqslant R\} \quad ,$$

then there is a approximating function $\tilde{f}_N$(t) to the determining function f(t) which is a real function,

$$f(t) \sim \tilde{f}_N(t)$$

$$= \frac{1}{2N}\sum_{k=1}^{2N} e^{(p_0 - R\cos\theta_k/r)t} \cdot$$

$$\left[A_k\cos\left(\frac{R}{r}\sin\theta_k \cdot t\right) + B_k\sin\left(\frac{R}{r}\sin\theta_k \cdot t\right)\right]$$

where

$$\begin{cases} A_k = -R[ReF(p_k)\cos\theta_k + ImF(p_k)\sin\theta_k]/r \\ B_k = R[ImF(p_k)\cos\theta_k - ReF(p_k)\sin\theta_k]/r \end{cases}$$

$$k = 1,2,\cdots,2N.$$

Other theorems about convergence, etc. are omitted here (see [6]).

ACKNOWLEDGEMENTS

The auther is indebted to Prof. Wang Qiu and Dai Yu for interest in this work.

REFERENCES

[1] Turan, P., J.Approx.Th., 29(1980)1, 23.
[2] Baker, Jr.G.A. and Graves–Mooris, P., Pade Approximation Addison –Wesley, 1981).
[3] Newman, D.J., J.Approx.Th., 40(1984)2, 111.
[4] Braess, D., J.Approx.Th., 40(1984)4, 375.
[5] Trefethen, L.N., J.Approx.Th., 40(1984)4, 380.
[6] Jiang, Y., in print.
[7] Davies, B. and Martin, B., J.Comp.Appl.Math., 2(1978).
[8] Longman, I.M. and Shair, M., Geophys.J.R.Astr.Soc., 25(1971), 299.
[9] Luke, Y.L., J.Franklin.Inst., 305(1978)5, 259.
[10] Ртбов, В.М., Вестн.Ленингр.Унта. 1984.no1, 42.

Approximation, Optimization and Computing:
Theory and Applications, A.G. Law and C.L. Wang (eds.)
Elsevier Science Publishers B.V. (North-Holland)

RELIABILITY ANALYSIS USING DATAFLOW GRAPH MODELS AND APPROXIMATE SOLUTIONS

Krishna M. Kavi
Tsair-Chin Lin

Department of Computer Science Engineering
The University of Texas at Arlington
Arlington, Texas 76019, USA

Abstract

Traditional approaches to reliability analysis become intractable when dealing with distributed systems, networks and software. New approaches based on Petri-nets, dataflow graphs, simulations and approximations are now used in such cases. In order to extend the utility of Petri-nets and dataflow graphs, we present a decomposition technique that can be used to partition a large system into smaller subsystems, such that the system reliability can be obtained (at least approximately) from the subsystem reliabilities. The decomposition reduces the computational complexity of analysis significantly.

Key words. Stochastic dataflow graph models, Reliability, mean-time-between-failure, Imperfect coverage, Blocking.

1. Introduction

Traditional approaches to reliability analysis of complex systems such as distributed processing systems become intractable. Stochastic Petri nets and simulations have been used in such cases. In this paper we will describe how stochastic dataflow graph models can also be used for the reliability analysis of complex systems. Our approach will be illustrated using hypercube interconnection networks. The reliability in the presence of imperfect coverage will also be derived. The reliability of more complex systems including distributed processing systems, algorithms, and programs can also be obtained using stochastic dataflow graphs. We wanted to make the dataflow graph modeling technique very clear and hence used a simple example. In addition, the most important contribution of the paper is to show how a dataflow graph can be decomposed into subgraphs such that the computational complexity of the analysis is reduced. The resulting solution is, admittedly, only an approximation. But, we feel that the accuracy lost due to the approximation is less significant than the loss of accuracy due to numerical approximations resulting from the use of computerized tools. The decomposition approach can also be used with stochastic dataflow graphs.

Dataflow graph models are similar to Petri nets in terms of their modeling capabilities. However, dataflow graphs have several advantages over Petri net models. A detailed description of these advantages can be found in [7]. Moreover, dataflow graphs are hierarchical in that, subsystems (or subgraphs) can be analyzed using either dataflow techniques, or other methods such as stochastic Petri nets.

The authors are with the department of Computer Science Engineering, UTA, PO Box 19015, Arlington, Texas 76019, USA.
This research is supported in part by the State of Texas Coordinating Board on Higher Education, under the Advanced Research Project, Grant #1770

2. Stochastic Dataflow Graphs

Our model is based on the dataflow graphs used by Dennis [2, 4]. Dataflow graphs consists of two types of nodes called **actors** and **links.** Actors represent operations and links represent place holders as data (tokens) flows from actor to actor via edges. More formally,

$G = < A \cup L, E, S, T>$

A is the set of actors; L is the set of links

$E \subseteq (A \times L) \cup (L \times A)$, is the set of edges

$S = \{l \in L \mid (a,l) \notin E, \forall a \in A\}$ is the starting set of links

$T = \{l \in L \mid (l, a) \notin E, \forall a \in A\}$ is the terminating set of links

Let I(a), $a \in A$ ($I(l)$, $l \in L$) and O(a), $a \in A$, ($O(l)$, $l \in L$) denote the set of input and output links (actors) of actor a (link l), respectively. In our model, $|I(a)|$ and $|O(a)|$ must be nonzero while $|I(l)|$ and $|O(l)|$ are at most one. Starting set (terminating set) represent the external inputs (external outputs) of the dataflow graph.

A **marking** of a dataflow graph denotes the presence or absence of tokens in links. M: L -->{0,1,...,k}. We presently allow k = 1, permitting an alternative definition, $M \subseteq L$. A marking is distinguished as an initial or terminal marking if $M \subseteq S$ or $M \subseteq T$. Associated with each actor are input and output **firing semantic sets** denoting which links enable the actor and which receive tokens when the actor fires. These are denoted as F_1 and F_2 respectively.

$F_1(a,M) \subseteq I(a)$; $F_2(a,M) \subseteq O(a)$

The selection at decision points are represented nondeterministically by using probability mass functions with F_1 and F_2. Conditional probabilities can also be defined on firing semantic sets when the selection depends on the firing semantic sets of other actors. At present, we restrict the firing sets such that $|F_1(a,M)| = 1$ or $F_1(a,M) = I(a)$, but not both. A similar restriction applies to F_2. These restrictions do not result in any loss of generality for the intended applications. These restrictions give us four different types of dataflow actors (Fig. 1).

The firing of an actor results in a new marking and is indicated by M ---> M' where tokens from input firing semantic set are removed and tokens are added to output semantic set. The markings defined with dataflow graphs can be mapped into states of an equivalent homogeneous Markov process. The probabilities defined with the firing semantic sets of actors will translate into state transition probabilities. Failure rates associated with actors will be used in obtaining the transition rates of the corresponding Markov process.

A continuous time **stochastic dataflow graph** (SDFG) is a dataflow graph with two sets of rates $\lambda = (\lambda_1,$

$\lambda_2, ..., \lambda_n)$ and $\mu = (\mu_1, \mu_2, ..., \mu_n)$ representing respectively the failure and service rates of actors in the dataflow graph.

SDFG = < A L, E, S, T, λ, μ >

3. Reliability Analysis of Interconnection Networks

In this section we will use SDFG's to compute the time dependent reliability indexes of fault-tolerant interconnection networks. In our analysis we first translate the interconnection network into SDFG; the SDFG is then mapped to a Markov process. For each switch in the interconnection network there is one actor in the dataflow graph. In this paper we will analyze the reliability between a pair of terminals in an 8-processor. Multiterminal reliability can be computed by constructing SDFG's for all terminal pairs. The complexity of analyzing a single-terminal pair depends on the number of states in the Markov process. There is one state for each link in the SDFG. In a hypercube with N processors, since there are $\log_2$ (N) edges for each processor, the number of states in the corresponding Markov process is proportional to **O**(N * $\log_2$ (N)).

3.1. Reliability Analysis of hypercube.

Consider the 8-processor hypercube shown in Fig. 1. We will compute the reliability for the terminal pair 0 and 7. The communication paths between these terminals are represented by the dataflow graph shown in Fig. 2. For the purpose of this analysis, we will assume that there is continuous demand for communication between the terminals 0 and 7 (as shown by the connection between actors a_7 and a_0). It should also be noted that a "+" between edges in the dataflow graph indicates a choice and hence probabilities associated with the paths.

In this analysis we assume that for each terminal pair, there are priorities associated with paths. For example, we treat the right-side path as the primary choice, and thus, the path with l_2 (output of a_0) will be used only when actor a_1 fails. In general, any probability assignment with the output firing semantic sets $F_2(a,M)$ for the dataflow actors can be used. Blocking phenomenon present in interconnection networks can also be represented by associating probabilities with output firing semantic sets of actors to reflect blocking probabilities. Reliability in the presence of imperfect coverage can also be handled using stochastic dataflow graphs. Even though not shown explicitly, each dataflow actor has an output link l_f and a token is placed in l_f when the actor fails, leading to a failure marking M_f.

3.2. Mapping of SDFG into Markov Process.

The fourteen markings (M_0, M_1, ..., M_{12}, M_f) identified in Table 1 can be mapped into fourteen states (m_0, m_1, ..., m_{12}, m_f) of an equivalent Markov process. Table 2 shows the transition rate matrix for the Markov process corresponding to the dataflow graph of Fig. 2.

This analysis of interconnection networks does not consider failures of communication links. Dataflow graphs can, however, model such failures; each communication channel is represented as a dataflow actor along with the associated failure rate.

The Markov process shown in Table 2 can be solved using Chapman-Kolmogorov equations [1, 9]. Let $P_{mi}(t)$ denote the probability associated with the Markov state m_i. Then the reliability of the dataflow graph of Fig. 2 is given by $R[G(t)] = 1 - P_{mf}(t) = \Sigma P_{mi}(t)$

The mean-time-between-failures of the dataflow graph can be expressed as

$$\text{MTBF} = \int_0^\infty R[G(t)]\, dt$$

Since the actors in the dataflow graph of Fig. 2 represent switches of the hypercube, we assume that $\lambda_0 = \lambda_1 = ... = \lambda_7 = \lambda$ and $\mu_0 = \mu_1 = ... = \mu_7 = \mu$. Now, solving the Markov process, we obtain

$$\text{MTBF} = \frac{[\lambda^6+4\mu^6+23\lambda\mu^5+53\lambda^2\mu^4+60\lambda^3\mu^3+33\lambda^4\mu^2+9\lambda^5\mu]}{[\lambda^7+2\lambda\mu^6+12\lambda^2\mu^5+29\lambda^3\mu^4+35\lambda^4\mu^3+21\lambda^5\mu^2+7\lambda^6\mu]}$$

In general, to solve a Markov process of k states we have to solve a system of k linear equations of the type A X = b. Hence the complexity of the computation is **O** (k^3). For the single terminal analysis in a hypercube of N processors, the complexity of computation would be **O**((N $\log_2$ N$)^3$).

3.4. Complexity Reduction using Graph Decomposition.

As discussed earlier, the complexity (the size of the state space) for a hypercube single-terminal pair reliability analysis is **O**((N $\log_2$ N$)^3$), where N is the number of processors in the hypercube. And the analysis of all terminal pairs leads to a complexity of **O**($N^5 \log_2^3 N$), since there are N^2 terminal pairs. This complexity can be reduced by decomposing the dataflow graph of Fig. 2 into smaller subgraphs.

In our previous research [6], we have shown how a free-choice Petri net can be decomposed into strongly connected marked graphs. Using such decompositions we have obtained an approximate value for the expected value of performance indexes such as mean-time-to-an-event. This work was based on the work of Hack [3], and Ramamoorthy [8]. Since we also have shown the isomorphism between dataflow graphs and Petri nets [6], we can apply similar decompositions and approximations with dataflow graph models. For the purpose of this paper, we will introduce Marked graphs, informally, and in terms of dataflow graphs. For a more formal treatment (with respect to Petri nets) the reader is referred to [6]. A marked graph is a dataflow graph where there are no selective actors.

Thus, selective actors (where there is a choice associated with the output firing semantic set), provide the basis for our decomposition. Each subgraph thus obtained will be a (strongly connected) Marked graph. The original dataflow graph can be analyzed (albeit, approximately) by combining the analyses of these subgraphs. The dataflow graph of Fig. 2 is decomposed into six subgraphs as shown in Fig. 3.

Because of the priorities associated with paths (see Section 3.1), the MTBF analysis of subgraph G-1 will require the failure rates of the subgraphs G-II and G-IV; the analysis of G-II requires the failure rates of G-III and G-V; the analysis of G-IV will require the failure rate of G-VI. It is obvious that the failure behavior of a subgraph cannot be described by an exponential function, even when individual nodes have exponential failure rates. However, for the purpose of this paper, we will approximate the failure rate of the subgraphs G-II, G-III, G-IV, G-V, G-VI with an exponential value obtained from the MTBF for the subgraph. The following equations will show the result of the decomposition.

$$\text{MTBF} = \frac{[\{(\lambda_{II}+\mu_{II})(\lambda_{IV}+\mu_{IV})(\lambda^3+4\mu^3+4\lambda^2\mu+6\lambda\mu^2)\} + \{\lambda\mu(\lambda+\mu)^2(\lambda_{IV}+\mu_{IV})\} + \{\lambda\mu^2(\lambda+\mu)(\lambda_{II}+\mu_{II})\}]}{[\{(\lambda_{II}+\mu_{II})(\lambda_{IV}+\mu_{IV})(\lambda^4+4\lambda^3\mu+6\lambda^2\mu^2+4\lambda\mu^3)\} - \{\lambda\mu\mu_{II}(\lambda+\mu)^2(\lambda_{IV}+\mu_{IV})\} - \{\lambda\mu^2\mu_{IV}(\lambda+\mu)(\lambda_{II}+\mu_{II})\}]}$$

Where

$$\lambda_{II} = \frac{[\{(\lambda_{III}+\mu_{III})(\lambda_V+\mu_V)(\lambda^4+4\lambda^3\mu+6\lambda^2\mu^2+4\lambda\mu^3)\} - \{\lambda\mu\mu_{III}(\lambda+\mu)^2(\lambda_V+\mu_V)\}-\{\lambda\mu^2\mu_V(\lambda+\mu)(\lambda_{III}+\mu_{III})\}]}{[\{(\lambda_{III}+\mu_{III})(\lambda_V+\mu_V)(\lambda^4+4\mu^3+4\lambda^2\mu+6\lambda\mu^2)\} + \{\lambda\mu(\lambda+\mu)^2(\lambda_V+\mu_V)\} + \{\lambda\mu^2(\lambda+\mu)(\lambda_{III}+\mu_{III})\}]}$$

$$\lambda_{III} = \frac{[\{(\lambda_{VI}+\mu_{VI})(\lambda^4+4\lambda^3\mu+6\lambda^2\mu^2+4\lambda\mu^3)\}-\{\lambda\mu^3\mu_{VI}\}]}{[\{(\lambda_{VI}+\mu_{VI})(\lambda^3+3\mu^3+4\lambda^2\mu+6\lambda\mu^2)\} + \{\lambda\mu^3\}]}$$

$$\lambda_{IV} = \lambda_V = \lambda_{VI} = \frac{[\lambda^4+4\lambda^3\mu+6\lambda^2\mu^2+4\lambda\mu^3]}{[\lambda^3+4\mu^3+4\lambda^2\mu+6\lambda\mu^2]}$$

3.5. Complexity Analysis

The number of marked graphs (subgraphs based on our decomposition) that results from a dataflow graph depends on the number of selector actors and the number of selections made at each selector actor. In a hypercube, this is given by n!, where $n = \log_2 N$. Once again, the complexity of analyzing a subgraph depends on the number of states in the Markov process for the subgraph. Looking at the subgraphs for a hypercube (Fig. 3), the maximum number of states in its corresponding Markov process is equal to n+1. Thus, the computational complexity for each subgraph is $\mathbf{O}((n+1)^3)$. Our decomposition technique, thus, reduces the computational complexity of analyzing the reliability of a terminal pair in a hypercube with N processors from $\mathbf{O}(N^5 \log_2^3 N)$ to $\mathbf{O}(n!\,(n+1)^3)$. Furthermore, since the equations for the decomposed graphs involve fewer variables than the equations for the complete dataflow graph, the numerical errors (and the possibility of error accumulation) are smaller.

3.6. Error Analysis and Bounds for MTBF.

In most electronic components, the service rates of components are very large and failure rates are very small. Our approximate values for the MTBF (for a hypercube) approaches the exact value as m --> ∞ (or 1 << m). Table 5 shows the percentage error due to our approximations.

Table 3. Percentage Error Due to Decomposition Method

λ	Percent Error
10^{-9}	$1.875 * 10^{-9}$
10^{-8}	$1.875 * 10^{-8}$
10^{-7}	$1.8752 * 10^{-7}$
10^{-6}	$1.877 * 10^{-6}$
10^{-5}	$1.892 * 10^{-5}$
10^{-4}	$2.047 * 10^{-4}$
10^{-3}	$3.462 * 10^{-3}$

In a general approach to this decomposition we may consider two different structures for analysis at two different levels: 1) a cycle consisting of nodes with exponential failure and service distributions; 2) a network consisting of these cycles. In (1), using the usual Markov process analysis techniques, we can easily show the following.

Let the service rates and failure rates at the nodes be $(\mu_1, \mu_1, \ldots, \mu_n)$ and $(\alpha_1, \alpha_2, \ldots, \alpha_n)$ respectively. Let f(t) be the density for successful completion and g(t) be the density for failure. Then we get

$$f(t) = \sum_{r=1}^{n} \frac{\alpha_r}{\mu_r} \prod_{i=1}^{r} \left(\frac{\mu_i}{\mu_i+\alpha_i}\right) [h_i(t)]^*$$

$$g(t) = \prod_{i=1}^{r} \left(\frac{\mu_i}{\mu_i+\alpha_i}\right) [h_i(t)]^*$$

where $h_i(t) = (\mu_i+\alpha_i)\, e^{-(\mu_i+\sigma_i)t}$ and * denotes convolution.

For analysis at the network level, consider a simple two-node network with service in k exponential phases at each node with rates km_1 and km_2 respectively. Also let ka_1 and ka_2 be the failure rates for each phase. Now, using the usual Markov process techniques and assuming that service is given alternatively by the two nodes we get

$$\text{MTBF} = \frac{1}{k[1-\delta_1^k\delta_2^k]}[\frac{1-\delta_1^k}{\alpha_1} + \frac{\delta_1^k(1-\delta_1^k)}{\alpha_2}$$

where $\delta_i = [\mu_i / (\mu_i + \alpha_i)]$ for $i = 1, 2$.
From this we find that $(\text{MTBF})_{k=1} > (\text{MTBF})_{k=2}$. We believe that this relationship is more general and therefore MTBF with a single exponential phase for each mode provides an upper bound for MTBF for the network.

4. Conclusions

In this paper we have shown that stochastic dataflow graph models can be used for the reliability analysis of complex computer systems by using the MTBF analysis of interconnection networks. Imperfect coverage, blocking and link failures can also be modeled using dataflow graphs. The major contribution of this paper is the graph decomposition technique that can reduce the computational complexity of the analysis. Although this approach yields only approximate values for mean-time-between-failures, we feel that the accuracy lost is less significant than the loss of accuracy due to numerical approximations resulting from the use of computerized tools. Moreover, since the equations for the decomposed graphs involve fewer variables than the equations for the complete dataflow graph, the numerical errors (and the possibility of error accumulation) are smaller. As computer systems and networks become more complex, it becomes necessary to find ways to reduce the analysis complexities. We feel that our approach addresses this problem well.

5. References

[1]. U.N. Bhat. ***Elements of applied stochastic processes***, 2nd Ed., John Wiley, New York, 1984.

[2]. J.B. Dennis. "First version of data flow procedural language", ***Lecture notes in computer science***, Vol. 19, Springer-verlag, 1974.

[3]. M.H.T. Hack "Analysis of production schemata by Petri nets", MIT Tech. Rept. 84, Project MAC, 1972, 112 pp.

[4]. K.M. Kavi, B.P. Buckles and U.N. Bhat. "A formal definition of dataflow graph models", ***IEEE Tran. on Comp.*** Nov. 1986, pp 940-948.

[5]. K.M. Kavi, R.M. Boyd and S.R. Amble. "DFDLS: A dataflow language based on graphical structure and a dataflow simulator", ***Tech. Rept.*** CSE TR 86-002, Dept. of CSE, UTA, Arlington, TX 76019.

[6]. K.M. Kavi, B.P. Buckles and U.N. Bhat. "Isomorphisms between Petri nets and dataflow graphs", ***IEEE Trans. on SE,*** Oct. 1987, pp 1127-1134.

[7]. K.M. Kavi. "A comparison of dataflow graph model with other models of concurrency", (in preparation)

[8]. C.V. Ramamoorthy and G.S. Ho. "Performance of asynchronous concurrent systems using Petri nets", *IEEE Tr. SE,* Sept. 1980, pp 440-449.

[9]. K.S. Trivedi. ***Probability & statistics with reliability, queuing, and computer science applications.***, Prentice-Hall. 1982.

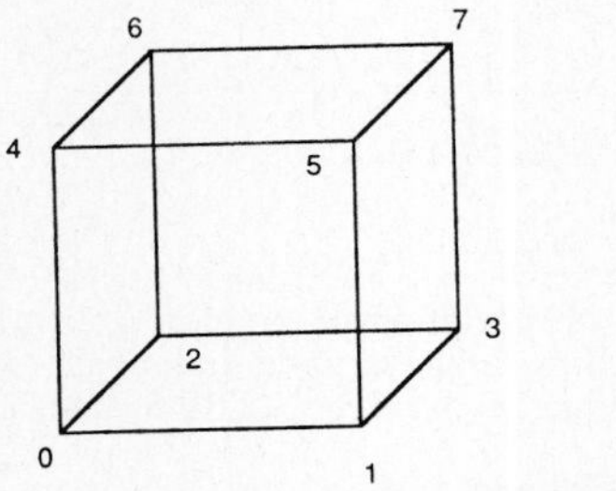

Fig. 1. An Eight Node hypercube

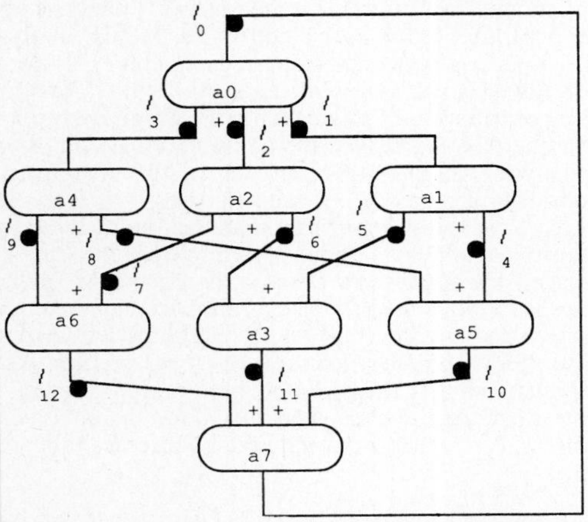

Fig. 2 Dataflow graph of paths between terminal pair (0,7)

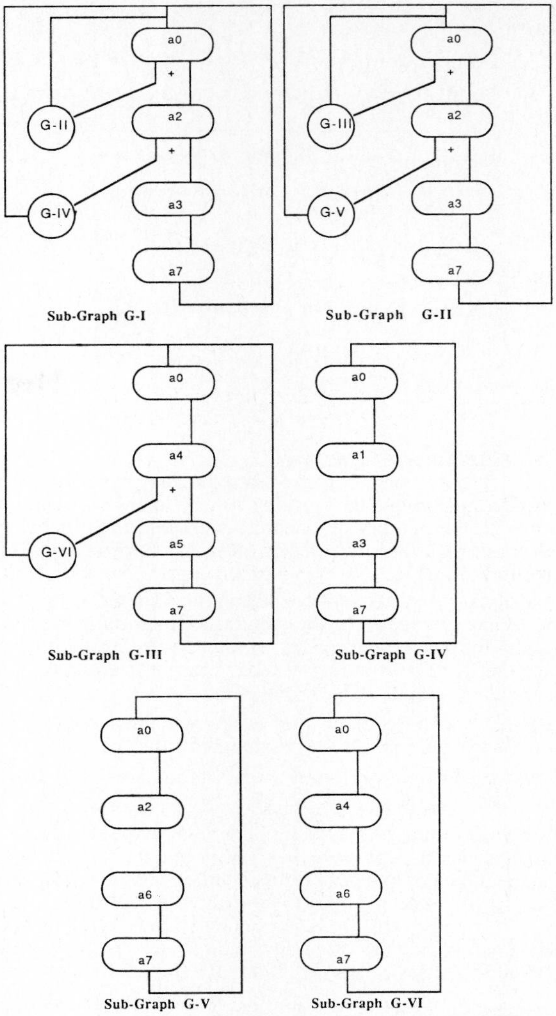

Fig. 3. Decomposition of Dataflow Graph of Fig. 2

Table 1. Markings

$M_0 = \{\ell_0\}$
$M_1 = \{\ell_1\}$
$M_2 = \{\ell_2\}$
$M_3 = \{\ell_3\}$
$M_4 = \{\ell_4\}$
$M_5 = \{\ell_5\}$
$M_6 = \{\ell_6\}$
$M_7 = \{\ell_7\}$
$M_8 = \{\ell_8\}$
$M_9 = \{\ell_9\}$
$M_{10} = \{\ell_{10}\}$
$M_{11} = \{\ell_{11}\}$
$M_{12} = \{\ell_{12}\}$
$M_f = \{\ell_f\}$

Table 2. Transition rate Matrix

	m_0	m_1	m_2	m_3	m_4	m_5	m_6	m_7	m_8	m_9	m_{10}	m_{11}	m_{12}	m_f
m_0	$-(\lambda_0+\mu_0)$	μ_0												λ_0
m_1		$-(\lambda_1+\mu_1)$	λ_1	μ_1										
m_2			$-(\lambda_2+\mu_2)$	λ_2			μ_2							
m_3				$-(\lambda_4+\mu_4)$					μ_4					λ_4
m_4					$-(\lambda_5+\mu_5)$	λ_5					μ_5			
m_5						$-(\lambda_3+\mu_3)$						μ_3		λ_3
m_6							$-(\lambda_3+\mu_3)$	λ_3				μ_3		
m_7								$-(\lambda_6+\mu_6)$					μ_6	λ_6
m_8									$-(\lambda_5+\mu_5)$	λ_5	μ_5			
m_9										$-(\lambda_6+\mu_6)$			μ_6	λ_6
m_{10}	μ_7										$-(\lambda_7+\mu_7)$			λ_7
m_{11}	μ_7											$-(\lambda_7+\mu_7)$		λ_7
m_{12}	μ_7												$-(\lambda_7+\mu_7)$	λ_7
m_f														1

Approximation, Optimization and Computing:
Theory and Applications, A.G. Law and C.L. Wang (eds.)
Elsevier Science Publishers B.V. (North-Holland)

COMPUTATION OF FILTERS BY DISCRETE APPROXIMATION

H. KOREZLIOGLU (1) and W.J. RUNGGALDIER (2)

(1) Ecole Nationale Supérieure des Télécommunications, 46 rue Barrault, 75634 Paris Cedex 13, France,
(2) Dipartimento di Matematica Pura ed Applicata Università di Padova, Via Belzoni , 35131 Padova, Italy.

Various computable approximations for solving nonlinear filtering problems are presented. The methods are partly extensions of previous works by the authors, in the sense that the model, considered here, covers also the case of nonwhite Gaussian noises and allows some correlation between signal and noise processes.

1. INTRODUCTION

The purpose of this paper is to present approximation methods for solving general nonlinear filtering problems given by a model of the following type (see e.g. [1])

$$(1.1)\quad x_t = x_0 + \int_0^t f(s,x_s,y_s)ds + \int_0^t g(s,x_s,y_s)db_s + \int_0^t \theta(s,x_s,y_s)\, dy_s$$

$$(1.2)\quad z_t = z_0 + \int_0^t a_s z_s\, ds + \int_0^t K_s\, dv_s$$

$$(1.3)\quad w_t = \int_0^t l_s z_s\, ds + v_t$$

$$(1.4)\quad y_t = \int_0^t h(s,x_s,y_s)\, ds + w_t$$

where (x_t), $x_t \in \mathbb{R}^q$, represents the signal process and (y_t), $y_t \in \mathbb{R}^q$, are the observations ; (z_t), $z_t \in \mathbb{R}^m$, is the state process for the observation noise (w_t) with z_0 a centered Gaussian r.v. The processes (b_t), $b_t \in \mathbb{R}^r$, and (v_t), $v_t \in \mathbb{R}^p$, are independent Brownian motions and (x_0, z_0, b, v) are mutually independent. Notice that (1.2) and (1.3) are a Markovian realization of (w_t) ; a large class of gaussian processes have this representation.

The filtering problem consists in computing, for each t in some interval [o,T], the conditional expectation of some function $F(x_t, z_t)$ of the state (x_t, z_t), given the observations y in the interval [o,t]. More formally, denoting by Y_t the completed σ-field generated by $\{y_s, s \leq t\}$, the problem consists in computing

$$(2)\quad \pi_t(F) := E\{F(x_t, z_t) / Y_t\}$$

It is well known that there are very few cases in which an expression of the form of (2) can be explicitly computed. This fact constitutes the main motivation for the elaboration of approximation schemes allowing a numerical solution.

We propose here an approximation procedure for problem (1), (2), that is based on previous works by the authors ([2], [3], [4]) and at the same time constitutes an extension of these works in the sense that it covers many cases where the observation noise is non-white and where the state process and the observation noise are not independent.

We shall make the following assumptions on model (1), (2) :

H.1 : The functions f,g, θ, h satisfy the linear growth and Lipschitz conditions in all their variables.

H.2 : The time functions a, l, K are bounded with K also Lipschitz-continuous

H.3 : The functions g and θ are bounded.

H.4 : The initial value z_0 is a centered gaussian r.v. and x_0 is such that $E\{e^{\varepsilon |x_0|^2}\} < \infty$ for some $\varepsilon > 0$.

2. FILTER REPRESENTATION BY THE REFERENCE PROBABILITY METHOD

The most suitable tool for performing the approximations that we are after is the representation of the filter (2) by the so-called reference probability method involving an absolutely continous transformation of measures. More precisely, denoting by (Ω, F, P) the underlying probability space for model (1), (2) and letting

$$(3)\quad L_t := \exp\{-\int_0^t H_s dy_s + \tfrac{1}{2}\int_0^t |H_s|^2 ds\}$$

where $H_s := h(s, x_s, y_s) + l_s z_s$, under our assumptions, we have that for the measure P, the process (L_t) is a martingale with mean 1 (see e.g. [1 ; & 6.2]). This allows us to define on (Ω, F) a new probability measure P_0 by

$$(3.b)\quad dP_0 = L_T\, dP$$

The space (Ω, F, P_0) is called the reference probability space. It turns out that on this space the state (x_t, z_t) and the observations y_t have the representation

$$(4.a)\quad x_t = x_0 + \int_0^t f(s, x_s, y_s)\, ds + \int_0^t g(s, x_s, y_s)\, db_s + \int_0^t \theta(s, x_s, y_s)\, dv_s^{\circ}$$

$$(4.b)\quad z_t = z_0 + \int_0^t [(a_s - K_s l_s) z_s - K_s h(s, x_s, y_s)]\, ds + \int_0^t K_s\, dv_s^{\circ}$$

$$(4.c)\quad y_t = v_t^{\circ}$$

where (b_t) and (v_t^o) are independent Brownian motions and (x_o, z_o, b, v^o) are mutually independent.

Furthermore, letting E_o denote the expectation under P_o, the filter (2) can be expressed as

(5.a) $\pi_t(F) = \sigma_t(F)/\sigma_t(1)$

where

(5.b) $\sigma_t(F) = E_o\{L_t^{-1} F(x_t, z_t) / Y_t\}$

Notice that by (3.a) and (4) the expression $L_t^{-1} F(x_t, z_t)$ on the right in (5.b) can be considered as a functional ψ_t of x_o, z_o and of the processes (b_t), (y_t). Notice furthermore from (4) that, under the measure P_o, the process $y_t = v_t^o$ is independent of the other sources of randomness, namely x_o, z_o and the Brownian motion (b_t). It follows that the conditional expectation at the right hand side of (5.b) can be interpreted as an ordinary expectation, with respect to the joint distribution of x_o, z_o and (b_t), of the functional ψ_t in which the values of the process y_t have been fixed equal to their observed values.

3. APPROXIMATION

The starting point for our approximation of $\pi_t(F)$ or, equivalently $\sigma_t(F)$, is to suppose that the observations are known only by their samples $\{y_o = o, y_\delta, \ldots, y_{n\delta};\ n \leqq T/\delta\}$ where δ is the length of the sampling period. We shall denote by [a/b] the integer part of a/b for positive a and b and write $\phi_n^\delta = \phi(n\delta) - \phi((n-1)\delta)$ with $\phi(-\delta) = o$ for a function ϕ on [o, T] ; we shall also denote by Y_t^δ the σ-field generated by $\{y_n^\delta ; n \leqq [t/\delta]\}$. Concerning the state process (x_t, z_t) we then consider the Euler approximation of (4.a), (4.b) :

$$(6)\quad \hat{x}_{n\delta} = \hat{x}_{(n-1)\delta} + f[(n-1)\delta, \hat{x}_{(n-1)\delta}, y_{(n-1)\delta}]\,\delta + g[(n-1)\delta, \hat{x}_{(n-1)\delta}, y_{(n-1)\delta}]\, b_n^\delta + \theta[(n-1)\delta, \hat{x}_{(n-1)\delta}, y_{(n-1)\delta}]\, y_n^\delta$$

$$(7)\quad \hat{z}_{n\delta} = \hat{z}_{(n-1)\delta} + \{a_{(n-1)\delta}\hat{z}_{(n-1)\delta} - K_{(n-1)\delta}[h((n-1)\delta, \hat{x}_{(n-1)\delta}, y_{(n-1)\delta}) + l_{(n-1)\delta}\, z_{(n-1)\delta}]\}\delta + K_{(n-1)\delta}\, y_n^\delta$$

where (see comments below (5.b)) y_n^δ are given by their observed values. Furthermore, taking $\hat{x}_o = x_o$, $\hat{z}_o = z_o$, define the continuous-time interpolated processes $\hat{x}_t$, $\hat{z}_t$ by

$$(8)\quad \hat{x}_t = \sum_{n=o}^{N} \hat{x}_{n\delta}\, 1_{[n\delta, (n+1)\delta]}(t)$$

$$(9)\quad \hat{z}_t = \sum_{n=o}^{N} \hat{z}_{n\delta}\, 1_{[n\delta, (n+1)\delta]}(t)$$

Finally, we approximate the martingale L_t by

$$(10.a)\quad \hat{L}_t = \exp \sum_{1}^{[t/\delta]} (-\hat{H}_k\, y_k^\delta + (\delta/2)\,|\hat{H}_k|^2)$$

where

$$(10.b)\quad \hat{H}_k = h((k-1)\delta, \hat{x}_{(k-1)\delta}, y_{(k-1)\delta}) + l_{(k-1)\delta}\,\hat{z}_{(k-1)\delta}$$

We now define, as our approximation to the filter (5) the following

(11.a) $\hat{\pi}_t(F) = \hat{\sigma}_t(F)/\hat{\sigma}_t(1)$

where, for $n\delta \leqq t < (n+1)\delta$

(11.b) $\hat{\sigma}_t(F) = E_o\{L_{n\delta}^{-1} F(\hat{x}_{n\delta}, \hat{z}_{n\delta}) \,|\, Y_{n\delta}^\delta\}$

and which, using again the reference probability approach, can be interpreted as optimal filter for a discrete-time approximation to model (1), (2).

In order that (11) can be regarded as acceptable approximation to (5), two things have now to be shown :

i) When $\delta \to o$, the approximation $\hat{\pi}_t(F)$ converges in some sense to $\pi_t(F)$;

ii) $\hat{\pi}_t(F)$ can indeed be computed.

Concerning point i), it is possible to adapt the method of [3] to our case to obtain that, if F is Lipschitz with Lipschitz constant L_F and bounded with sup-norm $\|F\|$, then there exists a positive K such that for all $t \in [o,T]$

(12) $E\{|\pi_t(F) - \hat{\pi}_t(F)|\} \leqq (\|F\| + L_F)\, K\sqrt{\delta}$

We remark that, if we had a model in which the observation noise was independent of the state process, then a relation of the type (12) could have been obtained along the lines of [4] requiring F to be only locally Lipschitz. Always under the assumption of independence of state process and observation noise, but requiring additional regularity of the functions in the model, the bound in (12) could be improved to a bound of order δ along the lines of [5]. Point ii) is the subject of the next section

4. RECURSIVE COMPUTATION OF THE APPROXIMATING FILTER

To actually compute in a recursive manner $\hat{\pi}_t(F)$ or, equivalently, $\hat{\sigma}_t(F)$, there are essentially two possiblities :

4.1 Monte-Carlo simulation

Since $\hat{\pi}_t(F)$ can be considered as an optimal filter for a discrete-time model, simulation may become feasible. In fact, by the comment following (5.b) the conditional expectation on the right in (11.b) is an ordinary expectation with respect to the joint distribution of the sequence $(\hat{x}_{i\delta}, \hat{z}_{i\delta})$, $i \leqq n$, defined in (6), (7) for a given sequence of observed values of (y_i^δ), $i \leqq n$. As a consequence, by simulating various realizations of $(\hat{x}_{n\delta}, \hat{z}_{n\delta})$ according to (6) and (7) for given values of (y_n^δ), we can recursively compute $\hat{\sigma}_t(F)$ via Monte-Carlo simulation.

4.2 Further approximations

Let us first give a more convernient recursive representation of $\hat{\sigma}_t(F)$ in (11.b). We have in fact

(13.a) $\hat{\sigma}_{n\delta}(F) = \hat{\sigma}_{(n-1)\delta}(\xi(.,F))$

where the kernel ξ is defined by

(13.b) $\xi(\hat{x}_{(n-1)\delta}, \hat{z}_{(n-1)\delta}; F) =$
$\exp(\hat{H}_n y_n^\delta - (\delta/2)|\hat{H}_n|^2) E_{P_0}[F(\hat{x}_{n\delta}, \hat{z}_{n\delta}) | \hat{x}_{(n-1)\delta}, \hat{z}_{(n-1)\delta}]$

with the conditional expectation computed for fixed (y_n^δ).

An equivalent and more explicit way is as follows (see e.g. [6]) :

Let p_0 denote the distribution of $(\hat{x}_0, \hat{z}_0)$; let $P(\hat{x}_{(n-1)\delta}, \hat{z}_{(n-1)\delta}, d\hat{x}_{n\delta}, d\hat{z}_{n\delta}; y^\delta)$ be the conditional probability kernel for the process $(\hat{x}_{n\delta}, \hat{z}_{n\delta})$ for fixed (y_n^δ) ; let $\Lambda_n(\hat{x}_{n\delta}, \hat{z}_{n\delta}, y_n^\delta) = \exp[\hat{H}_n y_n^\delta - (\delta/2)|\hat{H}_n|^2]$ where $\hat{H}_n$ is as in (10.b) ; then

(14.a) $\hat{\sigma}_t(F) = \int F(\hat{x}_{n\delta}, \hat{z}_{n\delta}) q_n(d\hat{x}_{n\delta}, d\hat{z}_{n\delta}; y^\delta)$

where $q_n(.\,; y^\delta)$ is a sequence of (random) measures defined by (let B be a borel set in $\mathbb{R}^q \times \mathbb{R}^m$)

$$(14.b)\quad \begin{cases} q_0(B) = \int_B p_0(d\hat{x}_0, d\hat{z}_0) \\ q_n(B; y^\delta) = \int_B 1_B(\hat{x}_{n\delta}; \hat{z}_{n\delta}) \Lambda_n(\hat{x}_{n\delta}, \hat{z}_{n\delta}; y_n^\delta)\,. \end{cases}$$
$$.\int P(\hat{x}_{(n-1)\delta}, \hat{z}_{(n-1)\delta}; d\hat{x}_{n\delta}, d\hat{z}_{n\delta}; y^\delta)\, q(d\hat{x}_{(n-1)\delta}, d\hat{z}_{(n-1)\delta}; y^\delta)$$

with $1_B(.)$ denoting the indicator function of B. Notice that $q_n(B, y^\delta)$ represents an unnormalized conditional distribution for $(\hat{x}_{n\delta}, \hat{z}_{n\delta}) \in B$, given $Y_{n\delta}^\delta$.

In order to make the computations in (13.b) and (14.b) numerically feasible, it is convenient to proceed to a further approximation which may consist in a spatial discretization (quantization) of $(\hat{x}_{n\delta}, \hat{z}_{n\delta})$. For this purpose one can simply adapt one of the procedures in [3] or [6] which allows the unnormalized conditional distribution q_n to be replaced by a finite-dimensional vector of unormalized conditional probability with the kernel P becoming a matrix, while the ensuing filter approximation error can again be bounded by a quantity of order $\sqrt{\delta}$.

If, for fixed values of y_k^δ, $(k \leq n)$, the quantity $\hat{L}_{n\delta}^{-1} F(\hat{x}_{n\delta}, \hat{z}_{n\delta})$ is a continuous and bounded function of $(\hat{x}_0, \hat{z}_0, \ldots \hat{x}_{n\delta}, \hat{z}_{n\delta})$, we also have the following additional possibility (see e.g. [7], [8]) : Take any discretization procedure leading to a sequence of finite-state Markov chains $(\bar{x}_{n\delta}, \bar{z}_{n\delta})$ that, for fixed values of y_k^δ $(k \leq n)$, converges weakly to $(\hat{x}_{n\delta}, \hat{z}_{n\delta})$. The computations in both (13) and (14) then becomes feasible as they involve the finite-dimensional transition probability matrix of the chain $(\bar{x}_{n\delta}, \bar{z}_{n\delta})$, and it follows immediately that the filter approximation error can be made arbitrarily small, though no explicit bound can be given in this case. Notice also that the unnormalized conditional distribution q_n then becomes again a finite-dimentional vector, this time representing the unnormalized conditional probabilities of $(\bar{x}_{n\delta}, \bar{z}_{n\delta})$, given $Y_{n\delta}^\delta$. Notice finally that this additional possibility could have been applied directly to (5.b) involving time and spacial discretisations (for details see [7]).

REFERENCES

[1] Liptser, R.S. and Shiryaev, A.N., Statistics of Random Processes (Springer, New York, 1977).

[2] Korezlioglu, H. and Mazziotto, G. , Approximations of Nonlinear Filters by Periodic Sampling and Quantization, in : Bensoussan, A. and Lions, J.L. (eds.), Analysis and Optimization of Systems, Part 1 (Springer, LN in Control and Info. Sc. Vol 62, 1984) pp 553-567.

[3] Korezlioglu, H., Computation of Filters by Sampling and Quantization, Center for Stochastic Processes, University of North Carolina, Technical Report n°208, Sep. 1987.

[4] Di Masi, G.B., Pratelli, M. and Runggaldier, W.J., An approximation for the Nonlinear Filtering Problem with Error Bound, Stochastics 14 (1985), 247.

[5] Picard, J., Approximation of Nonlinear Filtering Problems and Order of Convergence, in : Korezlioglu, H., Mazziotto, G. and Szpirglas, J. (eds), Filtering and Control of Random Processes (Springer, LN in Control of Info. Sc. Vol 61, 1984) pp 219-236.

[6] Bensoussan, A. and Runggaldier, W., An Approximation Method for Stochastic Control Problems with Partial Observation of the State, Acta Applicandae Mathematicae 10 (1987), 145.

[7] Kushner, H.J., Probability Methods for Approximations in Stochastic Control and for Elliptic Equations (Academic Press, New York, 1977).

[8] Di Masi, G.B. and Runggaldier, W.Y., On Approximation Methods for Nonlinear Filtering, in : Mitter, S.K. and Moro, A. (eds), Nonlinear Filtering and Stochastic Control (Springer, LN in Mathematics Vol 972,1983).

Approximation, Optimization and Computing:
Theory and Applications, A.G. Law and C.L. Wang (eds.)
Elsevier Science Publishers B.V. (North-Holland)

$L^1(\mathbf{R})$-OPTIMAL RECOVERY ON SOME DIFFERENTIABLE FUNCTION CLASSES

CHUN LI

Institute of Mathematics, Academia Sinica, Beijing, China

In this paper we study the optimal recovery problem for some differentiable function classes in $L^1(\mathbf{R})$ by using the information of function values with infinite cardinality. The optimal sampling points are determined equidistantly in a suitable set and the optimal algorithm is found to be cardinal $\mathcal{L}$-spline interpolation.

1. INTRODUCTION

Let r be a positive integer and $p_r(x)$ be a polynomial of degree r with following form:

$$p_r(x) = \prod_{j=1}^{r}(x - t_j), \qquad (1.1)$$

where $t_j \in \mathbf{R}, j = 1, \cdots, r$. Set

$$\begin{aligned}\mathcal{M}_1^r := \; & \{f \in L^1(\mathbf{R}) : f^{(r-1)} \text{is absolutely locally} \\ & \text{continuous on } \mathbf{R} \text{ and } \|p_r(D)f\|_1 \le 1\},\end{aligned} \qquad (1.2)$$

where $D := \frac{d}{dx}, \|h\|_1 := \int_{\mathbf{R}} |h(x)|dx$; and

$$\begin{aligned}\Theta := \; & \{\xi = \{\xi_j\}_{j\in\mathbf{Z}} : \xi_j < \xi_{j+1}, \quad \forall j \in \mathbf{Z}, \\ & \liminf_{a\to+\infty} \frac{\text{card }(\xi \cap [-a, a])}{2a} \le 1\},\end{aligned} \qquad (1.3)$$

where card (S) is the cardinality of the set S. For each $\xi = \{\xi_j\}_{j\in\mathbf{Z}} \in \Theta$, let $I_\xi : f \mapsto \{f(\xi_j)\}_{j\in\mathbf{Z}}$ be the information operator from $\mathcal{M}_1^r$ into l^∞. We consider the following optimal recovery problem in the sense of Micchelli and Rivlin [5]

$$E(\mathcal{M}_1^r) := \inf_{\xi\in\Theta} \inf_{A} \sup_{f\in\mathcal{M}_1^r} \|f - A(I_\xi(f))\|_1, \qquad (1.4)$$

where A runs through any mapping from $I_\xi(\mathcal{M}_1^r)$ into $L^1(\mathbf{R})$. Sometimes A is also called an algorithm. Our problems are as follows: (1) It is required to give an exact expression of $E(\mathcal{M}_1^r)$, which means that we should find an explicit formula for $E(\mathcal{M}_1^r)$; (2) How to determine the optimal set of sampling points $\xi^* \in \Theta$ and the optimal algorithm A^* for which we have

$$E(\mathcal{M}_1^r) = \inf_{A} \sup_{f\in\mathcal{M}_1^r} \|f - A(I_{\xi^*}(f))\|_1 = \sup_{f\in\mathcal{M}_1^r} \|f - A^*(I_{\xi^*}(f))\|_1.$$

In case of $L^\infty(\mathbf{R})$, Prof. Sun [2] first raised the optimal recovery problem of this type and solved it. In what follows we will solve above problems completely.

2. RESULTS AND PROOFS

By $[t_1, \cdots, t_r]f(\cdot)$ we denote the $(r-1)^{th}$ order divided difference of function f on points $t_1, \cdots, t_r$. We define

$$\begin{aligned}A_r(x, -1) &:= [0, t_1, \cdots, t_r]\frac{e^{\cdot x}}{e^{\cdot} + 1}, \\ A_r^*(x, -1) &:= [0, -t_1, \cdots, -t_r]\frac{e^{\cdot x}}{e^{\cdot} + 1}\end{aligned} \qquad (2.1)$$

on [0,1] and extend $A_r^*(x, -1)$ to the entire real line as follows:

$$A_r^*(x + 1, -1) = -A_r^*(x, -1), \quad \forall x \in \mathbf{R}. \qquad (2.2)$$

Set

$$\mathcal{E}_r^*(x) := 2A_r^*(x, -1). \qquad (2.3)$$

Then $p_r^*(D)\mathcal{E}_r^*(x) = 1$, $\forall x \in (0, 1)$, where $p_r^*(D) := (-1)^r \cdot p_r(-D)$ is the adjoint differential operator of $p_r(D)$. $\mathcal{E}_r^*(x)$ is called Euler spline with respect to $p_r^*(D)$. The following facts are well known[4]. There exists an unique $\alpha \in [0, 1)$ such that $A_r(\alpha, -1) = 0$. For this α there exists an unique $L(x) \in \mathcal{S}_{r-1}$ such that $L(j+\alpha) = \delta_{j,0}, \forall j \in \mathbf{Z}$, where $\mathcal{S}_{r-1} := \{s \in C^{r-2}(\mathbf{R}) : p_r(D)s(x) = 0, \; \forall x \in (j, j+1), \; \forall j \in \mathbf{Z}\}$ is the space of cardinal $\mathcal{L}$-spline. For any bounded data $\eta := \{\eta_j\}_{j\in\mathbf{Z}}$ we put

$$s_{r-1}(\eta; x) := \sum_{j\in\mathbf{Z}} \eta_j L(x - j + \alpha). \qquad (2.4)$$

Then $s_{r-1}(\eta; j) = \eta_j, \; \forall j \in \mathbf{Z}$. Let $\varsigma := \{j\}_{j\in\mathbf{Z}}$ and $\|\cdot\|_\infty$ be the max norm on $\mathbf{R}$. The following theorem is the main result of this paper.

Theorem. Let r be a positive integer. Then

$$\begin{aligned}E(\mathcal{M}_1^r) &= \inf_{A} \sup_{f\in\mathcal{M}_1^r} \|f - A(I_\varsigma(f))\|_1 \\ &= \sup_{f\in\mathcal{M}_1^r} \|f - s_{r-1}(I_\varsigma(f))\|_1 = \|\mathcal{E}_r^*(\cdot)\|_\infty.\end{aligned} \qquad (2.5)$$

Remark. Equalities (2.5) show that $\varsigma = \{j\}_{j\in\mathbf{Z}}$ is an optimal set of sampling points and the cardinal interpolant s_{r-1} is an optimal algorithm.

To prove the theorem we need the following lemmas.

Lemma 1. $\sup_{f\in\mathcal{M}_1^r} \|f - s_{r-1}(I_\varsigma(f))\|_1 \le \|\mathcal{E}_r^*(\cdot)\|_\infty$.

When $r \ge 3$, this lemma can be found in [6]. When $r = 1$ and $r = 2$, we can apply the similar methods employed in [6] to get the results we need. For abbreviation we omit the details. Lemma 1 shows that s_{r-1} maps $I_\varsigma(\mathcal{M}_1^r)$ into $L^1(\mathbf{R})$. To estimate $E(\mathcal{M}_1^r)$ from below we have to make some preparations. Let $I := [c,d]$ be an interval, $\xi = \{\xi_j\}_{j\in\mathbf{Z}} \in \Theta$ and $\Delta := \xi \cap I$. Set

$$\begin{aligned} S_{r-1}^*(\Delta) := \; & \{s \in C^{r-2}(I) : \; p_r^*(D)s(x) = 0, \\ & \forall x \in (\xi_j, \xi_{j+1}), \; \forall [\xi_j, \xi_{j+1}] \cap I \ne \phi\}, \end{aligned} \tag{2.6}$$

$$\begin{aligned} \mathcal{M}_1^r(\Delta)_0 := \; & \{f : f^{(r-1)} \text{ abs. cont. on } I, f(\xi_j) = 0, \\ & \forall \xi_j \in \Delta, \; f^{(i)}(c) = f^{(i)}(d) = 0, \\ & i = 0, \cdots, r-1, \; \|p_r(D)f\|_{L^1(I)} \le 1\}, \end{aligned} \tag{2.7}$$

where $\|h\|_{L^1(I)} := \int_I |h(x)|dx$. $S_{r-1}^*(\Delta)$ is just the space of spline corresponding to $p_r^*(D)$ with simple knots at Δ.

Lemma 2.

$$\begin{aligned} & \{p_r(D)f : f \in \mathcal{M}_1^r(\Delta)_0\} \\ = \; & \{\psi : \psi \perp S_{r-1}^*(\Delta) \text{ and } \|\psi\|_{L^1(I)} \le 1\}, \end{aligned}$$

where $\psi \perp S_{r-1}^*(\Delta)$ means that $\int_I \psi(x)s(x)dx = 0$, $\forall s \in S_{r-1}^*(\Delta)$.

Proof. Put $B := \{p_r(D)f : \; f \in \mathcal{M}_1^r(\Delta)_0\}$ and $C := \{\psi : \; \psi \perp S_{r-1}^*(\Delta)$ and $\|\psi\|_{L^1(I)} \le 1\}$. Let $\psi \in B$. Then $\exists f \in \mathcal{M}_1^r(\Delta)_0$ such that $\psi = p_r(D)f$ and $\|\psi\|_{L^1(I)} = \|p_r(D)f\|_{L^1(I)} \le 1$. Select an arbitrary $s \in S_{r-1}^*(\Delta)$. Observing that $s \in C^{r-2}(I)$ and $f^{(i)}(c) = f^{(i)}(d) = 0$, $i = 0, \cdots, r-1$ and using integration by parts, we have

$$\begin{aligned} & \int_I \psi(x)s(x)dx \\ = \; & \int_I s(x)p_r(D)f(x)dx \\ = \; & \int_I s(x)\prod_{j=1}^{r}(D - t_j)f(x)dx \\ = \; & (-1)^{r-1}\int_I \prod_{j=1}^{r-1}(D + t_j)s(x)(D - t_r)f(x)dx \\ = \; & (-1)^{r-1}\left[\int_c^{\xi_\mu} + \sum_{k=\mu}^{\nu-1}\int_{\xi_k}^{\xi_{k+1}} + \int_{\xi_\nu}^{d}\right] \\ & \cdot \prod_{j=1}^{r-1}(D + t_j)s(x)(D - t_r)f(x)dx, \end{aligned}$$

where $\xi_{\mu-1} \le c < \xi_\mu < \xi_\nu < d \le \xi_{\nu+1}$. Since $f(\xi_k) = 0$ and $p_r^*(D)s(x) = 0$, $\forall x \in (\xi_k, \xi_{k+1}), k = \mu, \cdots, \nu$, we obtain

$$\begin{aligned} & \int_{\xi_k}^{\xi_{k+1}} \prod_{j=1}^{r-1}(D + t_j)s(x)(D - t_r)f(x)dx \\ = \; & -\int_{\xi_k}^{\xi_{k+1}} f(x)p_r^*(D)s(x)dx \\ & + f(x)\prod_{j=1}^{r-1}(D + t_j)s(x)\Big|_{x=\xi_k}^{x=\xi_{k+1}} = 0, \end{aligned}$$

$k = \mu, \cdots, \nu - 1$. Similarly, $\int_c^{\xi_\mu} = \int_{\xi_\nu}^{d} = 0$.

Thus $\int_I \psi(x)s(x)dx = 0$ and $\psi \in C$. Hence $B \subseteq C$.

Otherwise, let $\psi \in C$. By the theory of ordinary differential equation we can find a function f for which $f^{(r-1)}$ is absolutely continuous on I such that $p_r(D)f = \psi$ and $f^{(i)}(c) = 0, i = 0, \cdots, r-1$. Therefore $\|p_r(D)f\|_{L^1(I)} = \|\psi\|_{L^1(I)} \le 1$. Since $\psi \perp S_{r-1}^*(\Delta)$, i.e., $\int_I s(x)p_r(D)f(x)dx = 0, \forall s \in S_{r-1}^*(\Delta)$, for each suitable choice of $s \in S_{r-1}^*(\Delta)$, using integration by parts again, we can conclude that $f^{(i)}(d) = 0$, $i = 0, \cdots, r-1$ and $f(\xi_k) = 0$, $\forall \xi_k \in \Delta$. Thus $f \in \mathcal{M}_1^r(\Delta)_0$ and $\psi = p_r(D)f \in B$. Therefore $B \supseteq C$ and $B = C$. This proves the lemma. ▯

Let

$$\begin{aligned} e(\mathcal{M}_1^r) & := \inf_{\xi\in\Theta} e(\mathcal{M}_1^r; \xi) \\ & := \inf_{\xi\in\Theta} \sup\{\|f\|_1 : f \in \mathcal{M}_1^r, \; I_\xi(f) = 0\}. \end{aligned} \tag{2.8}$$

It is known[5] that

$$E(\mathcal{M}_1^r) \ge e(\mathcal{M}_1^r). \tag{2.9}$$

Lemma 3. $e(\mathcal{M}_1^r) \ge \|\mathcal{E}_r^*(\cdot)\|_\infty$.

Proof. Let $0 < \varepsilon < 1$ and $\xi = \{\xi_j\}_{j\in\mathbf{Z}} \in \Theta$. Since $\liminf_{a\to+\infty} \frac{card(\xi \cap [-a,a])}{2a} \le 1$ we can find a sequence $\{I_k\}_{k=1}^\infty := \{[-a_k, a_k]\}_{k=1}^\infty$ of finite interval with $\lim_{k\to\infty} a_k = +\infty$ such that

$$n_k := \text{card } (\xi \cap I_k) \le 2(1+\varepsilon)a_k, \tag{2.10}$$

$k = 1, 2, \cdots$. It is obvious that

$$\begin{aligned} & e(\mathcal{M}_1^r; \xi) = \sup\{\|f\|_1 : \; f \in \mathcal{M}_1^r, \; I_\xi(f) = 0\} \\ \ge \; & \sup\{\|f\|_1 : \; f \in \mathcal{M}_1^r, \; I_\xi(f) = 0, \; f(x) = 0, \; \forall x \in \mathbf{R}/I_k, \\ & \quad f^{(i)}(-a_k) = f^{(i)}(a_k) = 0, \; i = 0, 1, \cdots, r-1\} \\ = \; & \sup\{\|f\|_{L^1(I_k)} : \; f \in \mathcal{M}_1^r(\Delta_k)_0\} \end{aligned} \tag{2.11}$$

where $\Delta_k := \xi \cap I_k$ and $\mathcal{M}_1^r(\Delta_k)_0$ is defined as (2.7). Let $S_{r-1}^*(\Delta_k)$ be given as (2.6) and

$$\begin{aligned} \mathcal{M}_\infty^{r,*}(I_k) := \; & \{f : f^{(r-1)} \text{ abs. cont. on } I_k \\ & \text{and } \|p_r^*(D)f\|_{L^\infty(I_k)} \le 1\}. \end{aligned} \tag{2.12}$$

Then according to Lemma 2 and the dual theorem[1] of the best approximation, we have

$$\sup\{\|f\|_{L^1(I_k)} : f \in \mathcal{M}_1^r(\Delta_k)_0\}$$

$$= \sup\left\{\int_{I_k} \varphi(x)f(x)dx : f \in \mathcal{M}_1^r(\Delta_k)_0, \|\varphi\|_{L^\infty(I_k)} \le 1\right\}$$

$$= \sup\left\{\int_{I_k} f(x)p_r^*(D)g(x)dx : f \in \mathcal{M}_1^r(\Delta_k)_0, g \in \mathcal{M}_\infty^{r,*}(I_k)\right\}$$

$$= \sup\left\{\int_{I_k} g(x)p_r(D)f(x)dx : f \in \mathcal{M}_1^r(\Delta_k)_0, g \in \mathcal{M}_\infty^{r,*}(I_k)\right\}$$

$$= \sup\left\{\int_{I_k} g(x)\psi(x)dx : g \in \mathcal{M}_\infty^{r,*}(I_k), \psi \perp S_{r-1}^*(\Delta_k), \|\psi\|_{L^1(I_k)} \le 1\right\}$$

$$= \sup_{g\in\mathcal{M}_\infty^{r,*}(I_k)} \inf_{s\in S_{r-1}^*(\Delta_k)} \|g-s\|_{L^\infty(I_k)}. \tag{2.13}$$

By $d_n(K;X)$ we denote the Kolomogorov n-width[1] of K in X. Observe that $\dim S_{r-1}^*(\Delta_k) \le n_k + r \le [2(1+\varepsilon)a_k] + r \le 2[([2(1+\varepsilon)a_k]+r+1)/2] = 2N_k$, where $[x]$ is the largest integer which is not greater than x and

$$N_k := [([2(1+\varepsilon)a_k]+r+1)/2]. \tag{2.14}$$

From the properties of Kolomogorov n-width we obtain

$$\begin{aligned}
&\sup_{g\in\mathcal{M}_\infty^{r,*}(I_k)} \inf_{s\in S_{r-1}^*(\Delta_k)} \|g-s\|_{L^\infty(I_k)}\\
\ge\ & d_{n_k+r}(\mathcal{M}_\infty^{r,*}(I_k); L^\infty(I_k))\\
\ge\ & d_{2N_k}(\mathcal{M}_\infty^{r,*}(I_k); L^\infty(I_k)) \qquad (2.15)\\
\ge\ & d_{2N_k}(\widetilde{\mathcal{M}}_\infty^{r,*}(I_k); L^\infty(I_k))\\
=\ & \|\varphi_{N_k}\|_\infty, \quad k = 1,2,\cdots.
\end{aligned}$$

in view of Sun and Huang [3] for the last equality, where $\widetilde{\mathcal{M}}_\infty^{r,*}(I_k) := \{f : f \in \mathcal{M}_\infty^{r,*}(I_k), f^{(i)}(-a_k) = f^{(i)}(a_k), i = 0,\cdots,r-1\}$ and φ_{N_k} satisfies that $\varphi_{N_k} \in C^{r-1}(\mathbf{R})$, $\varphi_{N_k}(\cdot + a_k/N_k) = -\varphi_{N_k}(\cdot)$ and $p_r^*(D)\varphi_{N_k}(x) = 1$, $\forall x \in (0, a_k/N_k)$. Set $y := \frac{N_k}{a_k}x$, $\psi_{N_k}(y) := \left(\frac{N_k}{a_k}\right)^r \varphi_{N_k}\left(\frac{N_k}{a_k}y\right) = \left(\frac{N_k}{a_k}\right)^r \varphi_{N_k}(x)$ and $p_{r,k}^*\left(\frac{d}{dy}\right) := \prod_{j=1}^r \left(\frac{d}{dy} + \frac{a_k}{N_k}t_j\right)$. Then $\psi_{N_k}(\cdot+1) = -\psi_{N_k}(\cdot)$, $p_{r,k}^*\left(\frac{d}{dy}\right)\psi_{N_k}(y) = 1$, $\forall y \in (0,1)$ and $\psi_{N_k} \in C^{r-1}(\mathbf{R})$. Hence $\psi_{N_k}(y)$ is the Euler spline with respect to $p_{r,k}^*\left(\frac{d}{dy}\right)$. Noticing $N_k/a_k \to 1+\varepsilon$ as $k \to \infty$, we have

$$p_{r,k}^*\left(\frac{d}{dy}\right) \to \prod_{j=1}^r \left(\frac{d}{dy} + \frac{t_j}{1+\varepsilon}\right) =: p_{r,\varepsilon}^*\left(\frac{d}{dy}\right)$$

and $\psi_{N_k}(y) \to \mathcal{E}_{r,\varepsilon}^*(y)$ uniformly as $k \to \infty$, where $\mathcal{E}_{r,\varepsilon}^*(y)$ is the Euler spline corresponding to $p_{r,\varepsilon}^*\left(\frac{d}{dy}\right)$. Thus, from (2.11), (2.13) and (2.15) it follows that

$$e(\mathcal{M}_1^r;\xi) \ge \|\varphi_{N_k}\|_\infty = \left(\frac{a_k}{N_k}\right)^r \|\psi_{N_k}\|_\infty,$$

$k = 1,2,\cdots$. Let $k \to \infty$ in above inequality,

$$e(\mathcal{M}_1^r;\xi) \ge \frac{1}{(1+\varepsilon)^r}\|\mathcal{E}_{r,\varepsilon}^*(\cdot)\|_\infty.$$

Let $\varepsilon \to 0^+$, since $\mathcal{E}_{r,\varepsilon}^*(\cdot) \to \mathcal{E}_r^*(\cdot)$ uniformly, we conclude that

$$e(\mathcal{M}_1^r;\xi) \ge \|\mathcal{E}_r^*(\cdot)\|_\infty.$$

Since $\xi \in \Theta$ is arbitrary,

$$e(\mathcal{M}_1^r) = \inf_{\xi\in\Theta} e(\mathcal{M}_1^r;\xi) \ge \|\mathcal{E}_r^*(\cdot)\|_\infty.$$

This proves Lemma 3. ▯

Proof of Theorem. By Lemma 1, Lemma 3, (2.9) and the definition of $E(\mathcal{M}_1^r)$, we obtain

$$\begin{aligned}
&\|\mathcal{E}_r^*(\cdot)\|_\infty \le e(\mathcal{M}_1^r) \le E(\mathcal{M}_1^r)\\
\le\ & \inf_A \sup_{f\in\mathcal{M}_1^r} \|f - A(I_\xi(f))\|_1\\
\le\ & \sup_{f\in\mathcal{M}_1^r} \|f - s_{r-1}(I_\xi(f))\|_1 \le \|\mathcal{E}_r^*(\cdot)\|_\infty.
\end{aligned}$$

Thus, all inequal signs in above inequalities should be equal signs. This proves the theorem. ▯

Corollary. $e(\mathcal{M}_1^r) = \|\mathcal{E}_r^*(\cdot)\|_\infty$.

ACKNOWLEDGEMENTS.

The author is grateful to Prof. H.L.Chen and Prof. Y.S.Sun for their valueable guidance and helpful discussion.

REFERENCES

[1] N.P.Korneicuk, Extremal Problems in Approximation Theory (in Russian), HAYKA, Moscow, 1976.

[2] Sun Yongsheng, On optimal interpolation for a differentiable function class(I), Approx. Theory & its Appl., Vol.2, No.4, Dec., 1986, 49—54.

[3] Sun Yongsheng & Huang Daren, On n-width of generalized Bernoulli kernel, Approx. Theory & its Appl., Vol.1, No.2, Mar., 1985, 83-92.

[4] C.A.Micchelli, Cardinal $\mathcal{L}$-spline, in "Studies in Spline Functions and Approximation Theory", Ed. by S.Karlin etc., Acad. Press,Inc., 1976.

[5] C.A.Micchelli & T.J.Rivlin, A survey of optimal recovery, in "Optimal Estimation in Approximation Theory", Ed. by C.A.Micchelli etc., Plenum Press, N. Y., 1977.

[6] Li Chun, On some extremal problems of Cardinal $\mathcal{L}$-spline (in Chinese), to appear.

Approximation, Optimization and Computing:
Theory and Applications, A.G. Law and C.L. Wang (eds.)
Elsevier Science Publishers B.V. (North-Holland)

SOME RESULTS ON PADÉ APPROXIMANTS

Li Jia – Liang

Department of Mathematics, Wuhan University, Wuhan, P. R. China

This paper is divided into two parts, in the first part, we investigate the existence of Padé approximation of a formal random power series. In the second part, we investigate the best approximation of a formal power series in Orlicz space.

1. ON THE EXISTENCE OF RANDOM PADÉ APPROXIMATIONS

1.1. Introduction

In this section we investigate a formal power series

$$f(w,z) = a_0\xi_0(w) + a_1\xi_1(w)z + \cdots \qquad (1.1)$$

where a_i $(i = 0,1,2,\ldots)$ are arbitrary real numbers which are not all zero, and $\xi_i(w)$ are continuous random variables defined in a probability space $(\Omega, \aleph, P)$. We prove that there exists almost sure in Ω arbitrary Padé approximants of $f(w,z)$.

First of all, we give some definitions and theorems which are known to us.

Definition 1.1. Let $(\Omega, \aleph, P)$ be a probability space and ξ be a random variable in $(\Omega, \aleph, P)$. If the distribution function of ξ is continuous, we call ξ a continuous random variable.

Theorem A. A suffcient and necessary condition for a random variable ξ in $(\Omega, \aleph, P)$ to be continuous is that the probability of ξ in an arbitrary single set is zero.

Definition 1.2. Let

$$f(z) = a_0 + a_1 z + \cdots \qquad (1.2)$$

and

$$C(L/M) = det \begin{pmatrix} a_{L-M+1} & \cdots & a_L \\ \cdots & \cdots & \cdots \\ a_L & \cdots & a_{L+M-1} \end{pmatrix}$$

we call $C(L/M)$ a C – determinant of $f(z)$. A table

c(0/0),	c(1/0),	c(2/0),	⋯
c(0/1),	c(1/1),	c(2/1),	⋯
⋮	⋮	⋮	⋱

is called the C – table. If all elements in C – table are not zero, then the C – table is said to be normal.

Theorem B. A sufficient and necessary condition for an arbitrary Padé approximant of $f(z)$ to exist is that the C – table of $f(z)$ is normal.

1.2. Main Results

Lemma 1.1. Let $(\Omega, \aleph, P)$ be a probability space, let $\xi_1, \ldots, \xi_n$ be n independent continuous random variables, and let $g(x_1, \ldots, x_n)$ be a polynomial of degree m in n elements, $g(x_1, \ldots, x_n) \neq 0$. Suppose $\eta = g(\xi_1, \ldots, \xi_n)$, then η is also a continuous random variable.

Proof: By theorem in [1], η is a random variable in $(\Omega, \aleph, P)$. $\forall A \in \Re(R)$

$$\begin{aligned} P_\eta(A) &= P(g(\xi_1, \ldots, \xi_n) \in A) \\ &= P_{(\xi_1,\ldots,\xi_n)}\{(x_1, \ldots, x_n) \in g^{-1}(A)\} \\ &= \int_{\{(x_1,\ldots,x_n)\in g^{-1}(A)\}} dP_{(\xi_1,\ldots,\xi_n)}(x_1, \ldots, x_n) \\ &= \int_{\{g(x_1,\ldots,x_n)\in A\}} dP_{\xi_1}(x_1) \cdots dP_{\xi_n}(x_n) \end{aligned}$$

where $P_{\xi_i}(x_1)$ are probability distributions of ξ_i $(i = 1, 2, \ldots, n)$, $P_{(\xi_1,\ldots,\xi_n)}(x_1, \ldots, x_n)$ is joint probability distribution of $(\xi_1, \ldots, \xi_n)$. Let $F_\eta(y)$ be a distribution function of η, we shall prove that $F_\eta(y)$ is continuous.

$\forall y_n \searrow 0, y_0 \in R$

$$\{\eta = y_0\} = \cap_{n=1}^{\infty}(y_0 \le \eta < y_0 + y_n)$$

$$(y_0 \le \eta < y_0 + y_n) \subset (y_0 \le \eta < y_0 + y_{n-1})$$

By theorem of continuity

$$P(\eta = y_0) = \lim_{n\to\infty} P(y_0 \le \eta < y_0 + y_n)$$

Hence

$$\lim_{y\to 0+0}(F_\eta(y + y_0) - F_\eta(y_0)) = P(\eta = y_0)$$

* This research was supported by National Science Foundation Grant.

By (1.3) and Fubini's theorem

$$P(\eta = y_0) = \int (dP_{\xi_2}(x_2)\cdots dP_{\xi_n}(x_n)) \int_{\{g(x_1,\ldots,x_n)=y_0\}} dP_{\xi_1}(x_1)$$

Since ξ_1 is a continuous random variable, $g(x_1,\ldots,x_n) - y_0$ is a polynomial of x_1 of degree less than or equal to m, $g(x_1,\ldots,x_n) - y_0$ has at most m real roots. Hence

$$\int_{\{g(x_1,\ldots,x_n)=y_0\}} dP_{\xi_1}(x_1) = 0.$$

It follows that

$$\int (dP_{\xi_2}(x_2)\cdots dP_{\xi_n}(x_n)) \int_{\{g(x_1,\ldots,x_n)=y_0\}} dP_{\xi_1}(x_1) = 0.$$

We prove that $F_\eta(y)$ is right continuous, and $F_\eta(y)$ is also left continuous, so that $F_\eta(y)$ is continuous. By defintion 1.1, lemma 1.1 is proved.

Theorem 1.1. Let

$$f(w,z) = a_0\xi_0(w) + a_1\xi_1(w)z + \cdots$$

be a formal power series defined by (1.1), then there exists almost sure in $(\Omega, \aleph, P)$ arbitrary Padé approximants of $f(z,w)$. We omit the proof.

2. PADÉ APPROXIMANTS AS LIMITS OF RATIONAL FUNCTION OF BEST APPROXIMATION IN ORLICZ SPACE

2.1 Introduction

We shall call a rational function of type (n,v) provided it can be writen in the form

$$\frac{s_1 + s_2z + \cdots + s_nz^{n-1}}{t_1 + t_2z + \cdots + t_vz^{v-1}} \sum |\, t_k \,| \neq 0 \qquad (2.1)$$

The Padé approximant to a given analytic function $f(z)$ is a rational function $P_{nv}(z)$ of type (n,v) with contact of the highest order at the origin to $f(z)$.

$$f(z) = a_0 + a_1z + \ldots + a_{n+v}z^{n+v} + O(z^{n+v+1}),$$

$$a_0 \neq 0, \qquad (2.2)$$

It is shown in [2] that provided a certain determinant of a_k is not zero, the rational function $R_{nv}(\epsilon,z)$ of type (n,v) of best approximant of $f(z)$ in the Chebyshev sense on the disc$\delta = \{|\, z \,| \leq \epsilon\}$ as $\epsilon \to 0$ approaches as a limit the function $P_{nv}(z)$ on any closed set within which $P_{nv}(z)$ is analytic. In this part we prove a analogous theorem in Orlicz space.

2.2. Notations and Results of Orlicz space [4]

Definition 2.2.1. A function $M(u)$ is called N – function if it can be written in the form $M(u) = \int_0^{|u|} P(t)dt$, where $P(t)$ is increasing right continuous function, and $P(t)$ is positive if $t > 0$ and $P(0) = 0$, $P(\infty) = \lim_{t\to\infty} P(t) = \infty$. Let $Q(s) = SUP_{P(t)\leq s}{}^t N(v) = \int_0^{|v|} Q(s)ds$. We call $N(v)$ complementary N – function.

Definition 2.2.2. Let $M(u)$ a N – function and $L_M(G)$ denote the class of all real value functions $U(z)$ in G which satisfy the following condition

$$\rho(U,M) = \int_G M[U(z)]d\sigma(z) < \infty$$

where G is a closed bounded set in R^n.

Definition 2.2.3. Let $M(u), N(v)$ be complementary N – functions, let $L^*_M(G)$ denote the class of functions $U(z)$ which for all $V(z) \in L_N$ satisfies condition

$$(U,V) = \int_G U(z)V(z)d\sigma(z) < \infty$$

we call $L^*_M(G)$ Orlicz space. $\forall U(z) \in L^*_M(G)$, we denote norm of $U(z)$ as follows

$$\|\, U(z) \,\|_M = SUP_{\rho(V,N)\leq 1} \left| \int_G U(z)V(z)d\sigma(z) \right|.$$

2.3 Main Results

Theorem 2.3.1. Let the function $f(z)$ defined by (2.2) be analytic at $z = 0$, and for $\epsilon(\geq 0)$ sufficiently small and fixed n and v, let $R_{nv}(\epsilon,z)$ denote the rational function of type (n,v) of best approximation to $f(z)$ in Orlicz space $L^*_M(G)$, where $L^*_M(G)$ is defined by N – functions $M(z)$ and $N(z)$, and δ denote the disc: $\{|\, z \,| \leq \epsilon\}$, $N(v)$ satisfy the following condition

$$N(|\, V \,|)/ |\, V \,|^2 \leq 1/\pi, as \ |\, V \,| \ large\ enough. \qquad (2.3)$$

Suppose we have

$$H_v(a_{n-v+1}) = det \begin{pmatrix} a_{n-v+1} & a_{n-v+2} & \cdots & a_n \\ a_{n-v+2} & a_{n-v+3} & \cdots & a_{n+1} \\ \cdots & \cdots & \cdots & \cdots \\ a_n & a_{n+1} & \cdots & a_{n+v-1} \end{pmatrix}$$

$$\neq 0. \qquad (2.4)$$

then as ϵ approaches zero $R_{nv}(\epsilon,z)$ approaches the Padé function $P_{nv}(z)$ of (2.2) uniformly on any closed bounded set containing no pole of $P_{nv}(z)$.

We omit the proof of the theorem.

Remark. There exists N – function which satisfies condition (2.3).

For example:

$M(u) = e^{|u|} - |\, u \,| - 1$

$N(v) = (1 + |\, v \,|)ln(1 + |\, v \,|) - |\, v \,|$

It is easy to verify by definition that $M(u)$ and $N(v)$ are complementary N – functions and

$$N(V)/ |\, V \,|^2 = \{(1 + |\, V \,|)ln(1 + |\, V \,|) - |\, V \,|\}/ |\, V \,|^2 \to 0$$

as $|\, V \,| \to \infty$.

ACKNOWLEDGEMENTS

We should like to thank professor Yu Chia-Yung and Wen Zhi-Ying who instructed and helped me in the preparation of this paper.

REFERENCES

1. P. L. Mayer, Introdutory Probability and Statistical Applications, 1972.
2. J. L. Walsh, Padé Approximants as Limits of Rational Function of Best Approximation, J. Math. Mech. 13 (1964), 305 – 312.
3. J. L. Walsh, On Approximation to an Analytic function by Rational functions of the Approximation, Math. Zeit. 38 (1934), 163 –176.
4. M. A. Krasnoselsky & Ya. B. Rutickiĭ , Convex function and the Orlicz Space, Scientific Press. Peking, 1962.
5. George A. Baker, Jr. and Peter Graves-Morris, Padé Approximants, Part 1: Basic Theory. Addison-Wesley Publishing Company. London, Amsterdam, etc. 1981.
6. A. L. Levin and D. S. Lubinsky, Best Rational Approximations of Entire Functions whose Maclaurin Series Coefficients Decrease Rapidly and Smoothly. Trans. Amer. Math. Soc. 293 (1986), 533 – 545.

Approximation, Optimization and Computing:
Theory and Applications, A.G. Law and C.L. Wang (eds.)
Elsevier Science Publishers B.V. (North-Holland)

RS METHOD FOR SOLVING INTEGRAL EQUATIONS AND ITS ERROR ESTIMATE

Yuesheng Li and **Hongyang Chao**

Zhongshan university, Guangzhou, China

In this paper, we will discuss a new method to solve the Fredholm integral equations of the first kind. An error estimate of the method will be presented, and a criterion to choose the regularized parameter α will be given.

1. Introduction

Solving *Fredholm integral equations of the first kind* is an important problem which involves a lot of areas such as CT technique, geophsical exploration and so on. Many authors have discussed the problem and presented various numerical methods, for example, B-G method [1], Phillips smoothing method , Tikhonov regularization method [4-5] ect. Here, we are interested in a variant of Tikhonov regularization—*Regularization-spline method* (RS method).

RS method was derived by the first author of this paper. He modified the regularization functional posed by Tikhonov, and revealed that the solution of the variational problem for the modified functional is a spline function defined by some differential operator. Moreover, a new numerical method has been constructed by the characteristic of spline function. Following is the main idea of the RS method.

Consider the Fredholm integral equation of the first kind

$$\int_a^b K(x,t)u(t)dt=f(x), \quad c\le x\le d. \qquad (1.1)$$

Using a numerical integral formula

$$\int_a^b g(t)dt\approx \sum_{j=1}^n a_j g(t_j), \quad a\le t_1\le\ldots\le t_n\le b,$$

yields

$$\int_a^b K(x,t)u(t)dt\approx \sum_{j=1}^n a_j K(x,t_j)u(t_j).$$

Given x_i and $f(x_i)$ with x_i satisfying $c\le x_1\le\ldots\le x_n\le d$, and $\alpha>0$, we construct a functional

$$\mathfrak{J}_\alpha[u]= \sum_{i=1}^l \Big(\sum_{j=1}^n a_j K(x_i,t_j)u(t_j)-f(x_i)\Big)^2\rho_i + \alpha \int_a^b (u^{(m)}(t))^2dt, \qquad (1.2)$$

where $\rho_1,\ldots,\rho_l$ are weight parameters. We will approximate the solution of equation (1.1) by the minimizer of the functional $\mathfrak{J}_\alpha[u]$. A necessary condition for u(t) minimizing the $\mathfrak{J}_\alpha[u]$ is vanishing of its variation, $\delta\mathfrak{J}_\alpha[u]=0$. With Dirac-$\delta$ function, we can present

$$\frac{1}{2}\,\delta\mathfrak{J}[u]=\alpha\int_a^b(-1)^m u^{(2m)}(t)\delta u(t)dt- \int_a^b \sum_{j=1}^n \alpha\beta_j\,\delta(t-t_j)\delta u(t)dt+ \qquad (1.3)$$
$$+\alpha\sum_{\upsilon=0}^{m-1}(-1)^\upsilon u^{(m+\upsilon)}(t)\delta u^{(m-\upsilon-1)}(t)\Big|_a^b,$$

where

$$\alpha\beta_j=f_j-\sum_{k=1}^n K^TK_{j,k}u(t_k). \qquad (1.4)$$

$$K^TK_{j,k}= \sum_{i=1}^l \rho_i a_j a_k K(x_i,t_j)K(x_i,x_k),$$

$$f_j= \sum_{i=1}^l \rho_i a_j f(x_i)K(x_i,t_j). \qquad (1.5)$$

Putting $\delta\mathfrak{J}_\alpha[u]=0$ leads to the generalized Euler differential equation in the sense of distribution

$$(-1)^m\frac{d^{2m}}{dt^{2m}}u(t) = \sum_{j=1}^n \beta_j\,\delta(t-t_j). \qquad (1.6)$$

and the natural boundary conditions

$$u^{(m+\upsilon)}(a)=u^{(m+\upsilon)}(b)=0, \quad \upsilon=0,1,\ldots,m-1. \qquad (1.7)$$

From (1.6) we know that the minimizer of functional (1.2), if exists, is a spline function of degree 2m-1 with knots $\{t_j\}_1^n$. It is so called *Regularization-spline solution* (RS solution) of equation (1.1). If m=1 or 2, the solution is just the wellknown spline of degree 1 or 3. According to [6], this spline can be expressed as

$$u(t)= \sum_{i=1}^{2m} c_i\varphi_i(t)+ \sum_{j=1}^n \beta_j G(t,t_j), \qquad (1.8)$$

where $\varphi_1(t),\ldots,\varphi_{2m}(t)$ form fundamental system of solutions of equation

$$\frac{d^{2m}}{dt^{2m}} u(t) = 0, \qquad (1.9)$$

and $G(t,\tau)$ satisfies

$$(-1)^m \frac{d^{2m}}{dt^{2m}} G(t,\tau) = \delta(t-\tau). \qquad (1.10)$$

How to find the unknowns $c_1,\ldots,c_{2m}$; $\beta_1,\ldots,\beta_n$ is showed in [6], and the numerical examples in [6] indicate the good approximation and numerical stability of the algorithm.

Following [3], let $\mathbf{H}=L^2(a,b)$, $\mathbf{F}=L^2(c,d)$, operators $\mathcal{A}:\mathbf{H}\to\mathbf{F}$ and $\mathcal{L}:\mathbf{H}\to\mathbf{H}$ are respectively defined as

$$\mathcal{A}u(x)=\int_a^b K(x,t)u(t)dt, \quad \forall u\in\mathbf{H}, \qquad (1.11)$$

$$\mathcal{L}u = \frac{d^m}{dt^m} u, \qquad \forall u\in\mathbf{H}, \qquad (1.12)$$

here $K\in L^2(\Omega)$, $\Omega=[c,d]\times[a,b]$, $m\geq 0$. Assuming the set $\mathbf{D}=\mathbf{H}^m(a,b)$ is the domain of operator $\mathcal{L}$, $f\in\mathbf{F}$, we consider the variational problem: find $\hat{u}\in\mathbf{D}$ such that

$$\|\mathcal{L}\hat{u}\|_H=\inf\{\|\mathcal{L}u\|_H;\ u\in\mathbf{U}_A\} \qquad (1.13)$$

where

$$\mathbf{U}_A= \{u\in\mathbf{D};\ \|\mathcal{A}u-f\|_F= \mu_A\},$$

$$\mu_A= \inf\{\ \|\mathcal{A}u-f\|_F\ ;\ u\in\mathbf{D}\ \}.$$

Suppose again $\mathfrak{J}_\alpha[u]$ is defined as (1.2) and consider the variational problem: finding $\tilde{u}_\alpha\in\mathbf{D}$ such that

$$\mathfrak{J}_\alpha[\tilde{u}_\alpha]=\inf\{\mathfrak{J}_\alpha[u];\ u\in D\}. \qquad (1.14)$$

The nature of RS method is using the exact solution of problem (1.14) as a approximate solution of problem (1.13). In the paper, we will, from the point of view above, discuss the error estimate of the RS method and the criterion of choosing regularized parameter α.

2. Nature of the operator $\mathcal{A}$ and $\mathcal{L}$

It is easy to point out that the range of operator $\mathcal{L}$ is closed and $\mathbf{D}$ is convex. From direct provement, we can get

Lemma 2.1. The operator $\mathcal{A}$ and $\mathcal{L}$ are jointly closed on $\mathbf{D}$, that is, for any $\{u_j\}\subset\mathbf{D}$ which satisfy

$$\lim_{j\to\infty} u_j =u_0 \text{ (in H)}, \quad \lim_{j\to\infty} \mathcal{A}u_j=f_0 \text{ (in F)},$$
$$\lim_{j\to\infty} \mathcal{L}u_j=g_0 \text{ (in H)}, \qquad (2.1)$$

we have that $u_0\in\mathbf{D}$, $\mathcal{A}u_0=f_0$ and $\mathcal{L}u_0=g_0$. #

Lemma 2.2. If $\mathbf{N}(\mathcal{A})\cap\mathbf{N}(\mathcal{L})=\{0\}$, there exists $\beta>0$ such that for any $u\in\mathbf{N}(\mathcal{L})$

$$\|\mathcal{A}u\|_F\geq\beta\|u\|_H$$

where $\mathbf{N}(\mathcal{A})$ and $\mathbf{N}(\mathcal{L})$ are respectively the null spaces of $\mathcal{A}$ and $\mathcal{L}$.

Proof. If (2.2) is not ture, there are $\varepsilon_n\to 0$, and $\{u_j\}\subseteq\mathbf{N}(\mathcal{L})$ where $\|u_j\|_H=1$ $(n=1,2,\ldots)$ such that

$$\|\mathcal{A}u_j\|_F< \varepsilon_n,$$

thus,

$$\lim_{j\to\infty} \mathcal{A}u_j=0.$$

Since $\mathbf{N}(\mathcal{L})$ is a finite dementional subspace, the set of polynomial of degree m-1, there is a convergent subsequence $\{u_{nj}\}$ and an element $u_0\in\mathbf{N}(\mathcal{L})$, such that

$$\lim_{j\to\infty} u_{nj} = u_0 \quad \text{(in H)}.$$

However, from the continuity of the operator $\mathcal{A}$,

$$\mathcal{A}u_0= \lim_{j\to\infty} \mathcal{A}\, u_{nj} =0.$$

So, $u_0\in\mathbf{N}(\mathcal{A})\cap\mathbf{N}(\mathcal{L})=\{0\}$. It contradicts $\|u_0\|_H=\lim_{j\to\infty}\|u_j\|_H=1$. The contradiction implies (2.2) to be ture.#

Lemma 2.3 [2]. If the following conditions are satisfied:
(i) $\mathbf{N}(\mathcal{A})\cap\mathbf{N}(\mathcal{L}) = \{0\}$;
(ii) $\mathbf{R}(\mathcal{L})$, range of $\mathcal{L}$, is closed;
(iii) there is a const $\beta>0$ such that

$$\|\mathcal{A}u\|_F \geq \beta\|u\|_H, \qquad \forall\ u\in\mathbf{N}(\mathcal{L}),$$

then, there is a const $\gamma>0$ such that

$$\|\mathcal{A}u\|_F^2+\|\mathcal{L}u\|_H^2\geq\gamma^2(\|u\|_H^2+\|\mathcal{L}u\|_H^2)=\gamma^2\|u\|_m^2,$$
$$\forall\ u\in\mathbf{D}. \qquad (2.3)\ \#$$

Toeorem 2.1. Suppose $\mathbf{N}(\mathcal{A})\cap\mathbf{N}(\mathcal{L})=\{0\}$, then, the solution of the basic problem (1.13) exists and is unique if $f\in\mathbf{R}(\bar{\mathcal{A}})+\mathbf{R}(\bar{\mathcal{A}})^\perp$, where $\bar{\mathcal{A}}=\mathcal{A}|_D$.

Proof. Since $f\in\mathbf{R}(\bar{\mathcal{A}})+\mathbf{R}(\bar{\mathcal{A}})^\perp$, the set $\mathbf{U}_A$ is nonempty. From Lemma 2.1-2.3 we can know that all the conditions of Theorem 1 in [3] are satisfied. Thus the solution of (1.13) exists and is unique.#

3. Error estimate (I)

Consider the formula of numerical integration

$$\int_a^b g(t)dt \approx \sum_{j=1}^n a_j g(t_j) \qquad (3.1)$$

here $\sum_{j=1}^n a_j=b-a$, $a_j\geq 0$, $j=1,\ldots,n$. Using (3.1), the summing operator $\mathcal{A}_\tau:\mathbf{H}\to\mathbf{F}$

$$\mathcal{A}_\tau u(x)=\sum_{j=1}^n a_j K(x,t_j)u(t_j) \qquad (3.2)$$

can be regarded as an approximation of the integral operator $\mathcal{A}$.

In this section, we will discuss the error caused by substituting summing operator $\mathcal{A}_\tau$ for integral operator $\mathcal{A}$ and the introduction of regularized parameter α. Suppose

$$\|\mathcal{A}u-\mathcal{A}_\tau u\|_F\leq \tau\|u\|_m. \qquad (3.3)$$

It is easy to see that the hypothesis

(3.3) is resonable. For example, when (3.1) is a composite trapezoid formula, then (3.3) holds if m>1, where

$$\tau=(\Delta t)^2\{ \int_c^d \|K(x,.)\|^2_{H^2(a,b)}dx \}^{1/2}$$

and Δt is the maximum mesh size.

Lemma 3.1. If $K\in\mathbf{C}(\Omega)$, the operators $\mathcal{L}$ and $\mathcal{A}_\tau$ are jointly closed on the set **D**.

Lemma 3.2. If $\mathbf{N}(\mathcal{A})\cap\mathbf{N}(\mathcal{L})=\{0\}$, when τ is small enough there is a const $\bar{\gamma}>0$ such that

$$\|\mathcal{A}_\tau u\|^2_F+\|\mathcal{L}u\|^2_H\geq\bar{\gamma}^2\|u\|^2_m, \quad \forall u\in\mathbf{D}. \qquad (3.4)$$

Proof. According to Lemma 2.2, the inequality (2.3) holds. So, it follows from (3.3) that

$$\|\mathcal{A}u-\mathcal{A}_\tau u\|_F\leq \tau\|u\|_m\leq\frac{\tau}{\gamma}\ (\|\mathcal{A}u\|_F+\|\mathcal{L}u\|_H)$$

or

$$(1-\tau/\gamma)(\|\mathcal{A}u\|_F+\|Lu\|_H)\leq \|\mathcal{A}_\tau u\|_F+\|\mathcal{L}u\|_H$$

$$\leq(1+\tau/\gamma)(\|\mathcal{A}u\|_F+\|\mathcal{L}u\|_H) .$$

If τ is small enough so that $1-\tau/\gamma\geq 1/2$,

$$\|\mathcal{A}_\tau u\|^2_F+\|\mathcal{L}u\|^2_H\geq \frac{1}{2}(\|\mathcal{A}_\tau u\|_F+\|\mathcal{L}u\|_H)^2$$

$$\geq \frac{1}{4}(\|\mathcal{A}u\|_F+\|\mathcal{L}\|_H)^2\geq \frac{1}{4}\ \gamma^2\|u\|^2_m.$$

Thus, (3.4) can be achieved if $\bar{\gamma}=\gamma/2$.#

Suppose $\alpha>0$, and functional

$$J_\alpha[\mathcal{A}_\tau;u]=\|\mathcal{A}_\tau u-f\|^2_F+\alpha\|\mathcal{L}u\|^2_H .$$

Consider the auxiliary variational problem: Find $\bar{u}_\alpha\in\mathbf{D}$, such that

$$J_\alpha[\mathcal{A}_\tau;\bar{u}_\alpha]= \inf_{u\in D} \{J_\alpha[\mathcal{A}_\tau;u]\}. \qquad (3.5)$$

It follows from Lemma 3.1-3.2 and Theorem 2 in [3] that the solution of problem (3.5) exists and is unique under the condition of Lemma 2.2. And, from Theorem 7, Theorem 8 and Theorem 11 in [3] we can obtain

Lemma 3.3. Let $\hat{u}$ is the solution of problems (1.13) and $\bar{u}_\alpha$ is that of(3.5). If $\hat{u}\in\mathbf{H}^{2m}(a,b)$, we have

$$\|\hat{u}-\bar{u}_\alpha\|_m\leq C_1(\alpha^s+\frac{\tau}{\alpha}) \qquad (3.6)$$

here s=1/2 if $d^{2m}\hat{u}/dt^{2m}\in\mathbf{R}(\mathcal{A}^*)$ and s=1 if $d^{2m}\hat{u}/dt^{2m}\in\mathbf{R}(\mathcal{A}^*\mathcal{A})$, and C_1 is a const independent of α and τ. #

4. Error estimate (II)

In the practical problem, the right side of equation (1.1) can be only obtained in a series of observations $f(x_1),\ldots,f(x_l)$. So, it is necessary to make further discretization .

Consider the following discrete integral equations

$$\int_a^b K(x_i,t)u(t)dt=f(x_i), \quad i=1,\ldots,l, \qquad (4.1)$$

where $c=x_1<\ldots<x_l=d$. Define an operator $\mathcal{B}$: $\mathbf{H}\rightarrow\mathbf{R}^l$

$$\mathcal{B}u=(\mathcal{B}_1u,\ldots,\mathcal{B}_lu), \quad \text{for any } u\in\mathbf{H}$$

where $\mathcal{B}_j:\mathbf{H}\rightarrow\mathbf{R}$ (**R** is the real number set)

$$\mathcal{B}_ju=\int_a^b K(x_j,t)u(t)dt, \quad \forall\ u\in\mathbf{H},\ j=1,\ldots,l.$$

Let operator $\mathcal{P}$ to be an interpolating operator such that for any $f\in\mathbf{F}$

$$\mathcal{P}f=(f(x_1),\ldots,f(x_l)) \in \mathbf{R}^l.$$

Thus, the equation (4.1) is equal to

$$\mathcal{B}u=\mathcal{P}f. \qquad (4.2)$$

Suppose the inner product in $\mathbf{R}^l$ is:

$$(\eta,\xi)_*=\sum_{i=1}^{l}\rho_i\eta_i\xi_i,\quad \|\xi\|^2_*=(\xi,\xi)_*,\quad \forall\eta,\xi\in\mathbf{R}^l$$

where the coefficients $\rho_1,\ldots,\rho_l$ are

$$\rho_1=\frac{x_2-x_1}{2},\quad \rho_i=\frac{x_{i+1}-x_{i-1}}{2},\quad i=2,\ldots,l-1,$$

$$\rho_l=\frac{x_l-x_{l-1}}{2}.$$

It is easy to see that $\|\xi\|_*$ is an equivalent norm in $\mathbf{R}^l$. Let

$$h=\max_{1\leq i\leq l-1}|x_{i+1}-x_i|,$$

it follows from Euler-Maclaulin summation formula that for any $f,g\in\mathbf{F}$, if both f and g belong in $\mathbf{H}^2(c,d)$, we have

$$(f,g)_F=(\mathcal{P}f,\mathcal{P}g)_*+O(h^2)$$

where

$$|O(h^2)|\leq h^2\|f\|_{H^2(c,d)}\|g\|_{H^2(c,d)}. \qquad (4.4)$$

Lemma 4.1. If $K\in\mathbf{C}(\Omega)$, the operators $\mathcal{B}$ and $\mathcal{L}$ are jointly closed on the set **D**.

Using Lemma 2.3 and discussing as Lemma 2.2 we can get

Lemma 4.2 If $\mathbf{N}(\mathcal{B})\cap\mathbf{N}(\mathcal{L})=\{0\}$, there exists a const $\mu>0$, such that

$$\|\mathcal{B}u\|^2_*+\|\mathcal{L}u\|^2_H\geq \mu^2(\|u\|^2_H+\|\mathcal{L}u\|^2_H)=\mu^2\|u\|^2_m,$$

$$\forall\ u\in\mathbf{D}, \qquad (4.5)$$

Remark. Condition $\mathbf{N}(\mathcal{B})\cap\mathbf{N}(\mathcal{L})=\{0\}$ includes $\mathbf{N}(\mathcal{A})\cap\mathbf{N}(\mathcal{L})=\{0\}$.

let operator $\mathcal{B}_\tau$ to be an approximation of the operator $\mathcal{B}$, where

$$\mathcal{B}_\tau u=(\mathcal{B}_{1,\tau}u,\ldots,\mathcal{B}_{l,\tau}u), \quad i=1,\ldots,l,$$

$$\mathcal{B}_{i,\tau}u=\sum_{j=1}^{n}a_jK(x_i,t_j)u(t_j), \quad i=1,\ldots,l,$$

and a_j ($j=1,\ldots,n$) are difined as §3. Suppose

$$\|\mathcal{B}u-\mathcal{B}_\tau u\|_*\leq \tau\|u\|_m, \qquad \forall\ u\in\mathbf{D},$$

and τ is the same as that in inequality (3.3). Discussing as §3, we have

Lemma 4.3. If $K \in \mathbf{C}(\Omega)$ and $m>0$, the operators $\mathcal{B}_\tau$ and $\mathcal{L}$ are jointly closed on the set $\mathbf{D}$.

Lemma 4.4. If $\mathbf{N}(\mathcal{B}) \cap \mathbf{N}(\mathcal{L}) = \{0\}$, when τ is small enough there exists a const $\bar{\mu}>0$, such that

$$\|\mathcal{B}_\tau u\|_*^2 + \|\mathcal{L}u\|_H^2 \geq \bar{\mu}^2 \|u\|_m^2, \ \forall\ u \in \mathbf{D}. \quad (4.6)$$

Let $\alpha>0$, a parameterized functional is difined as

$$\mathfrak{J}_\alpha[\mathcal{B}_\tau; u] = \|\mathcal{B}_\tau u - \mathcal{P}f\|_*^2 + \alpha\|\mathcal{L}u\|_H^2 \quad (4.7)$$

Consider following variational problem: Find $\tilde{u}_\alpha \in \mathbf{D}$, such that

$$\mathfrak{J}_\alpha[\mathcal{B}_\tau; \tilde{u}_\alpha] = \inf\{\mathfrak{J}_\alpha[\mathcal{B}_\tau; u];\ u \in \mathbf{D}\}. \quad (4.8)$$

Since $\mathfrak{J}_\alpha[\mathcal{B}_\tau; u]$ is the same as the functional $\mathfrak{J}_\alpha[u]$ difined in §1, the solution of problem (4.8) is just the RS solution. It follows from the discussion similar to that in the problem (3.5) that the solution of problem (4.8) exists and is unique if the conditions of Lemma 4.2 hold.

Next, we will estimate the error between the solution $\tilde{u}_\alpha$ of problem (4.8) and the basic solution $\hat{u}$. By the variation principle, the solutions u_α and $\bar{u}_\alpha$ of (3.5) must respectively satisfy

$$(\mathcal{B}_\tau\tilde{u}_\alpha - \mathcal{P}f, \mathcal{B}_\tau(w-\tilde{u}_\alpha))_* + \alpha(\mathcal{L}\tilde{u}_\alpha, \mathcal{L}(w-\tilde{u}_\alpha))_H = 0, \quad \forall w \in \mathbf{D}, \quad (4.9)$$

$$(\mathcal{A}_\tau\bar{u}_\alpha - f, \mathcal{A}_\tau(v-\bar{u}_\alpha))_F + \alpha(\mathcal{L}\bar{u}_\alpha, \mathcal{L}(v-\bar{u}_\alpha))_H = 0, \quad \forall v \in \mathbf{D}. \quad (4.10)$$

Since $\mathcal{P}(\mathcal{A}_\tau\bar{u}_\alpha) = \mathcal{B}_\tau\bar{u}_\alpha$, from equality (4.3), (4.10) is equivalent to

$$(\mathcal{B}_\tau\bar{u}_\alpha - \mathcal{P}f, \mathcal{B}_\tau(v-\bar{u}_\alpha))_* + \alpha(\mathcal{L}\bar{u}_\alpha, \mathcal{L}(v-\bar{u}_\alpha))_H = O(h^2). \quad (4.11)$$

Letting $z_\alpha = \tilde{u}_\alpha - \bar{u}_\alpha$, and assumming $w=\bar{u}_\alpha$ in (4.9), $v=\tilde{u}_\alpha$ in (4.11) and furthermore, adding both equalities obtained, we get

$$\|\mathcal{B}_\tau z_\alpha\|_*^2 + \alpha\|\mathcal{L}z_\alpha\|_H^2 = O(h^2) \quad (4.12)$$

If function $K \in \mathbf{C}^2(\Omega)$ and $m \geq 1$, it follows from (4.4) that

$$|O(h^2)| \leq h^2 \|\mathcal{A}_\tau\bar{u}_\alpha - f\|_{H^2(c,d)} \|\mathcal{A}_\tau z_\alpha\|_{H^2(c,d)}. \quad (4.13)$$

After some manipulation, we obtain

$$\|\mathcal{A}_\tau z_\alpha\|_{H^2(c,d)}^2 \leq C_2 \|z_\alpha\|_m^2, \quad (4.14)$$

where C_2 is a const independent of z_α. From the definition of $\bar{u}_\alpha$ and inequalities (3.6), we also have

$$\|\mathcal{A}_\tau\bar{u}_\alpha\|_{H^2(c,d)} \leq C_2 \|\bar{u}_\alpha\|_m \leq \text{const}(1+\alpha^s+\frac{\tau}{\alpha})$$

Thus,

$$\|\mathcal{A}_\tau\bar{u}_\alpha - f\|_{H^2(c,d)} \leq \text{const}(1+\alpha^s+\frac{\tau}{\alpha}), \quad s=1/2 \text{ or } 1. \quad (4.15)$$

Substituting (4.13)-(4.15) into (4.12), we get

$$\|\mathcal{B}_\tau z_\alpha\|_*^2 \leq \text{const}\ h^2(1+\alpha^s+\frac{\tau}{\alpha})\|z_\alpha\|_m,$$

$$\|\mathcal{L}z_\alpha\|_H^2 \leq \text{const}\frac{h^2}{\alpha}(1+\alpha^s+\frac{\tau}{\alpha})\|z_\alpha\|_m.$$

It follows from (4.6) again that

$$\|z_\alpha\|_m^2 \leq (\|\mathcal{B}_\tau z_\alpha\|_*^2 + \|\mathcal{L}z_\alpha\|_H^2)/\bar{\mu}$$
$$\leq \text{const}\ h^2(1+1/\alpha)(1+\alpha^s+\tau/\alpha)\|z_\alpha\|_m,$$

that is,

$$\|z_\alpha\|_m \leq \text{const}\ h^2(1+\frac{1}{\alpha})(1+\alpha^s+\frac{\tau}{\alpha}), \quad (4.16)$$

And, from (3.6)-(3.7) we can achieve

$$\|\tilde{u}_\alpha - \hat{u}\|_m \leq \text{const}[\alpha^s+\frac{\tau}{\alpha}+h^2(1+\frac{1}{\alpha})(1+\alpha^s+\frac{\tau}{\alpha})], \quad s=1/2 \text{ or } 1. \quad (4.17)$$

Theorem 4.1. Let that the conditions of Lemma 3.3 hold, functions $K \in \mathbf{C}^2(\Omega)$ and $f \in \mathbf{H}^2(c,d)$. Then, if $\mathbf{N}(\mathcal{B}) \cap \mathbf{N}(\mathcal{L}) = \{0\}$, $m>0$, there is relation (4.17) between RS solution $\tilde{u}_\alpha$ and basic solution $\hat{u}$. #

In fact, a criterion to choose the regularized parameter α has yet been given by (4.17), i.e.

$$\lim_{(\alpha,\tau,h)\to 0} [(\tau+h^2)/\alpha] = 0.$$

Following is an "optimal" choice

$$\alpha = O(h^{2/(s+1)} + \tau^{1/(s+1)}), \quad s=1/2 \text{ or } 1.$$

Under this case

$$\|\hat{u} - \tilde{u}_\alpha\|_m = O(\tau+h^2)^{s/(s+1)}. \#$$

Reference

[1] Backs, G. and Gilbert, F., Numerical applications of a formalism for geophysical inverse problems, Geophys. J. Roy. Astron., 13(1967), 247.

[2] Locker, J. and Prenter, P.M., Regularization with differential operators I: General theory, J. Math. Anal. Appl., 74(1980), 504-529.

[3] Morozov, V.A., Methods for solving incorrecctly posed problems, Springer-Verlag, New York, Berlin, Heidelberg, tokyo, 1984.

[4] Tikhonov, A.N., On the solution of ill-posed problem and the method of regularization, Soviet Math. Dokl., 4(1963), 1035-1038.

[5] Tikhonov, A.N., On the regularization of ill-posed problems, Soviet Math. Dokl., 4(1963), 1624-1627.

[6] Li, Y.S., Rugularization-spline method for solving integral equations of first kind, CAT #54 (1987), Texas A & M Univ.

Approximation, Optimization and Computing:
Theory and Applications, A.G. Law and C.L. Wang (eds.)
Elsevier Science Publishers B.V. (North-Holland)

ON CHEBYSHEV APPROXIMATION IN SEVERAL VARIABLES (III)

X. Z. Liang & X. M. Han

Jilin University, China

In this paper, we give a new proof of the characterization theorem of multivariate Chebyshev approximation and propose a new theorem which characterizes the strong unicity of the best approximation .

1. INTRODUCTION

In multivariate Chebyshev approximation theory, the theorem which characterizes the best approximation polynomials is very important. The first proof of the theorem was given by Zukhovitskii [1] in 1956, who made use of functional analysis. In 1961, using the properties of convex sets, Rivlin and Shapiro [2] gave another proof. In this paper, taking advantage of Zorn's lemma and Cantor's theorem in the set theory, we give a simple proof of the theorem. In addition, making use of the concept of uniqueness set of sign-alternate points (USSAP) to be defined, we prove a theorem which characterizes the strong unicity of multivariate Chebyshev approximation polynomial.

2. NEW PROOF OF CHARACTERIZATION THEOREM

Let C(D) denote the space of continuous, real-valued functions on the compact metric space D endowed with the uniform norm and let H be a subspace of C(D) with dimension k . A function $f \in C(D)$ is given. The Chebyshev approximation problem is to find $p^* \in H$ such that

$$\|p^*-f\| \leq \|p-f\| \text{ for every } p \in H, \quad (1)$$

where p^* is called the Chebyshev approximation polynomial of f from H. For $p \in H$ the sets

$$A_p(f)=\{x \in D \mid p(x)-f(x)= \|p-f\|\}$$

and

$$B_p(f)=\{x \in D \mid p(x)-f(x)=- \|p-f\|\}$$

are called the positive e-point set and the negative e-point set respectively.

A couple of point sets $\{A,B\}$, $(A,B \subset D)$, is called isolable with respect to H , if there exists a $q \in H, q \neq 0$, such that

$$\begin{cases} q(x) \geq 0 & \text{for every } x \in A \\ q(x) \leq 0 & \text{for every } x \in B \end{cases} \quad (2)$$

A couple of point sets$\{A,B\}$, $(A,B \subset D)$, is called strictly isolable with respect to H,if strict inequalities hold in (2). The following theorem is well known:

Theorem(Kolmogorov). p is the Chebyshev approximation polynomial of $f \in C(D)$ from H, if and only if $\{A_p(f),B_p(f)\}$ is not strictly isolable with respect to H.

Now we introduce some conceptions.

Definition 1. Let $A,B \subset D$ be two disjoint compact sets. The couple of point sets $\{A,B\}$ is called a critical set for H, if $\{A,B\}$ is not strictly isolable with respect to H and if every couple of point sets $\{A',B'\}$, $A' \subset A$ and $B' \subset B$ being two compact sets and $\{A',B'\} \neq \{A,B\}$, is strictly isolable with respect to H. Furthermore, $\{A,B\}$ is called a set of sign-alternate points for H (an SSAP for H), if $\{A,B\}$ is a critical point set for H and if A and B are two finite point sets.

Theorem (Cantor). Let $S_1 \supset S_2 \supset \dots \; S_\alpha \supset \dots$ all be nonempty compact subsets of a compact metric space D, Then $S = \bigcap_\alpha S_\alpha$ still is a nonempty compact subset of D.

Using above concept and Cantor's theorem, we prove the following two lemmas and then get the characterization theorem.

Lemma 1. p is a Chebyshev approximation polynomial of $f \in C(D)$ from H, if and only if $\{A_p(f), B_p(f)\}$ contains a critical point set for H .

Proof. By Kolmogorov's theorem the "if" part is clear. We need only prove the " only if " part. Suppose $p \in H$ is a Chebyshev approximation polynomial of f . Then both A_p and B_p are compact sets, and $\{A_p, B_p\}$ is not strictly isolable with respect to H . Let Π be the collection of all the couples of compact sets of points which are subsets of the couple $\{A_p, B_p\}$ and are not isolable with respect to H. For $\{A_1, B_1\}$ and $\{A_2, B_2\} \in \Pi$, we define a relation $\{A_1, B_1\} \geqslant \{A_2, B_2\}$ if $A_1 \subset A_2$ and $B_1 \subset B_2$. So we define a partial ordering relation on the collection Π, and Π is a partially ordered set . By Cantor's theorem every chain $\{\{A_\alpha, B_\alpha\} | \alpha \in \Lambda\}$ of Π has a nonempty compact intersection $\{A_0, B_0\} = \bigcap_\alpha \{A_\alpha, B_\alpha\}$. It can be shown that $\{A_0, B_0\}$ is not strictly isolable with respect to H . Otherwise , there exists a $p \in H$ such that $p(x) > 0$ for every $x \in A_0$ and $p(x) < 0$ for every $x \in B_0$. Let $A' = \{x \in D;\ p(x) \leqslant 0\}$ and $B' = \{x \in D; p(x) \geqslant 0\}$. Then $A_0 \cap A' = B_0 \cap B' = \phi$ is a empty set. However, By Cantor's theorem , $(A_0 \cap A') \cup (B_0 \cap B') = \bigcap_\alpha ((A_\alpha \cap A') \cup (B_\alpha \cap B'))$ is a nonempty set. This is a contradiction. Therefore every chain of Π has a supremum. By Zorn's lemma Π has a maximal element (A^*, B^*), which is the critical point set we look for .

Lemma 2. Every critical point set for H is a finite point set .

Proof. Suppose $\{M, N\}$ is a critical point set for H. If $M \cup N$ is a infinite point set, we will lead to a contradiction . Without loss of generality, we suppose M is infinite and x_0 is an accumulation point of M . Set

$U_n(x_0) = \{x \in D;\ \|x - x_0\| < 1/n\}, n = 1, 2, \ldots$

We know that $\{M \setminus U_n(x_0), N\}$ is strictly isolable with respect to H . suppose $q_{1,n}(x) = \langle c_{1,n}, \Phi(x)\rangle \in H$ is the strictly isolating polynomial for it , where

$c_{1,n} \in R^k ,\quad \|c_{1,n}\| = 1,$

and $\Phi(x) = (p_1(x), \ldots, p_k(x))$ is a base of H. From the sequence $\{c_{1,n}\}_{n=1}^{\infty}$ we can choose a subsequence (denoted still by $\{c_{1,n}\}_{n=1}^{\infty}$) which convergences to a unit vector $c_1 \in R^k$. It is easy to see that $q_1(x) = \langle c_1, \Phi(x)\rangle$ isolates $\{M, N\}$ and $q_1(x_0) = 0$. We can conclude that the curve $q_1(x) = 0$ go through all the points of $M \cup N$. If not, without loss of generality, we suppose that there is a point $x_1 \in M$ such that $x_1 \neq x_0$ and $q_1(x_1) > 0$. Then for a sufficiently small $\varepsilon > 0$, we can find a function $r(x) \in H$ which strictly isolates $\{M \setminus V_\varepsilon(x_1), N\}$, where $V_\varepsilon(x_1) = \{x \in D;\ \|x - x_1\| < \varepsilon\}$. For a sufficiently small $\lambda > 0$, $q_1(x) + \lambda r(x)$ would strictly isolates $\{M, N\}$, this is a contradiction.

Set

$$c'_{2,n} = c_{1,n} - \langle c_1, c_{1,n}\rangle c_1 ,$$
$$c_{2,n} = c'_{2,n} / \|c'_{2,n}\| .$$

It can be seen that $c_{2,n} \neq 0$, $\langle c_{2,n}, c_1\rangle = 0$, and $q_{2,n}(x) = \langle c_{2,n}, \Phi(x)\rangle$ strictly isolates $\{M \setminus U_n(x_0), N\}$. We can choose a subsequence of $\{c_{2,n}\}_{n=1}^{\infty}$ (denoting the subsequence by $\{c_{2,n}\}$ still) which convergences to a vector $c_2 \in R^k$.

$c_{2,n} \to c_2\ (n \to \infty)$, and $\|c_2\| = 1$.

Similarly, $q_2(x) = \langle c_2, \Phi(x)\rangle$ isolates $\{M, N\}$, all the points of $M \cup N$ are on the curve $q_2(x) = 0$.

Set

$$c'_{3,n} = c_{2,n} - \langle c_{2,n}, c_2\rangle c_2 ,$$
$$c_{3,n} = c'_{3,n} / \|c'_{3,n}\| .$$

Repeating the above process we can find a vector $c_3 \in R^k$ such that
$\|c_3\| = 1$, and $< c_3, c_1 > = < c_3, c_2 > = 0$.
And so on, finally we can get k+1 unit vectors in R^k which are orthogonal one another. This is impossible.

Theorem 1. p is a Chebyshev approximation polynomial of $f \in C(D)$ from H, if and only if $\{A_p(f), B_p(f)\}$ contains a set of sign-alternate points for H (an SSAP for H).

Proof. It can be proved that every finite critical point set for H is a set of sign-alternate points for H. So by Lemma 1 and Lemma 2 we can draw the conclusoin of Theorem 1 immediately.

3. UNIQUENESS AND STRONG UNICITY

From now on, we discuss the uniqueness and the strong unicity of multivariate Chebyshev approximation problem.

Definitoin 2. We say that $p^* \in H$ is a unique best approximation to $f \in C(D)$ from H, if

$$\|p-f\| > \|p^*-f\|, \quad \forall p \in H, \ p \neq p^*. \quad (3)$$

We say that $p^* \in H$ is a strongly unique best approximation to $f \in C(D)$ from H, if there is a constant $\gamma = \gamma(f) > 0$ such that

$$\|p-f\| \geq \|p^*-f\| + \gamma \|p-p^*\|, \forall p \in H. \quad (4)$$

Definition 3. Let $A, B \subset D$ be two disjoint finite sets. The couple of point sets $\{A, B\}$ is called a uniqueness set of sign-alternate points for H (a USSAP for H), if $\{A, B\}$ is not isolable with respect to H and if every couple of point sets $\{A', B'\}$, $A' \subset A$ and $B' \subset B$ being two finite sets and $\{A', B'\} \neq \{A, B\}$, is isolable with respect to H.

Lemma 3. Let $A, B \subset D$ be two disjoint compact sets, and $\{A, B\}$ be not isolable with respect to H. Then every point in $A \cup B$ is in some set of sign-alternate points contained in $\{A, B\}$.

Proof. Let $\Phi(x) = (p_1(x), \ldots, p_k(x))$ be a base of H. $C = A \cup B$. Define $\Psi(x) = \Phi(x)$ for $x \in A$ and $\Psi(x) = -\Phi(x)$ for $x \in B$. Consider the following subsets of R^k:
$A' = \{\Psi(x);\ x \in A\}$, $B' = \{\Psi(x);\ x \in B\}$,
$C' = A' \cup B'$, and
$Co(C')$ = the convex hull of C'.
We can prove that the origin point $O = (o, \ldots, 0)$ of R^k is in the interior of $Co(C')$. Otherwise, by the separation theorem of convex sets [5], we know that there is a vector $c' \in R^k (c' \neq 0)$ such that
$< c', y > \geq 0$, for every $y \in Co(C')$.
Hence $< c', \Psi(x) > \geq 0, \forall x \in C$.
From it we see that $\{A, B\}$ is isolable and get a contradiction. Thus we may choose a small real $\varepsilon > 0$ such that the ball $V'_\varepsilon(o) = \{y \in R^k;\ \|y\| \leq \varepsilon\}$ is contained in $Co(C')$.

For any $x_0 \in C$, without loss of generality, say $x_0 \in A$, we write $y_0 = \Psi(x_0)$. Then for a sufficiently small $\alpha_0 > 0$, we have
$-\alpha_0 y_0 \in V'_\varepsilon(o) \subset Co(C')$.
By Caratheodory Theorem (See[6]) we can find n vectors $y_1 = \Psi(x_1), \ldots, y_n = \Psi(x_n) \in C'$ and n positive real numbers $\alpha_1, \ldots, \alpha_n$ $(n \leq k+1)$ satisfying

$$\Sigma^n_{i=1} \alpha_0 \alpha_i = 1, \ -\alpha_0 y_0 = \Sigma^n_{i=1} \alpha_0 \alpha_i y_i.$$

Hence

$$-y_0 = \Sigma^n_{i=1} \alpha_i y_i = \Sigma^n_{i=1} \alpha_i \Psi(x_i). \quad (5)$$

If $y_1, \ldots, y_n$ is not linearly independent, then there exist $\lambda_1, \ldots, \lambda_n$, not all zero, such that

$$0 = \Sigma^n_{i=1} \lambda_i y_i. \quad (6)$$

Multiply (6) by an appropriate number and add it to (5), we can make some coefficients of y_i's vanish and keep the other coefficients of y_i's still positive. Delete the y_i's whose coefficients are zeros. Repeat the above procedure if the remaining y_i's are not independent. Finally we get linearly independent vectors $y_{i1}, \ldots, y_{im} \in C'$ such that

$$-y_0 = \Sigma^m_{j=1} \lambda_{ij} y_{ij} = \Sigma^m_{j=1} \lambda_{ij} \Psi(x_{ij}),$$

$\lambda_{i1}, \lambda_{i2}, \dots, \lambda_{im}$ being positive numbers. It is clear that the points $x_0, x_{i1}, \dots x_{im} \in C$ can constitute an SSAP for H, which completes the proof.

Making use of Lemma 3, we can prove the following theorem:

Theorem 2. Let $A, B \subset D$ be two disjoint compact sets. Then the following statements are equivalent:

(i) The couple of point sets $\{A,B\}$ is not isolable with respect to H.

(ii) $\{A,B\}$ contains a USSAP for H.

(iii) $f \in C(D)$, $p \in H$, $A_p(f)=A$, $B_p(f)=B$ imply that p is the unique best approximation to $f \in C(D)$ from H.

Lastly, making use of the concept of USSAP, we give a characterization of the strong unicity of multivariate Chebyshev approximation polynomial.

Theorem 3. $p^* \in H$ is a strongly unique best approximation to $f \in C(D)$ from H, if and only if $\{A_{p^*}(f), B_{p^*}(f)\}$ contains a USSAP for H.

Proof. Sufficiency. Suppose that $\{A_{p^*}(f), B_{p^*}(f)\}$ contains a USSAP $\{A,B\}$ for H. $C=A \cup B$. Then, by Theorem 1, p^* is a best approximation to f from H. From Definition 3 we know that the couple of sets $\{A,B\}$ is not isolable with respect to H. So for every $q \in H$ and $q \not\equiv 0$ we have

$$M(q) = \max_{x \in C} \operatorname{sign}(p^*(x)-f(x))q(x) > 0.$$

Since the set $\{q \in H, \|q\|=1\}$ is a compact subset of C(D), it is clear that

$$\gamma = \min_{\|q\|=1} M(q) > 0.$$

Thus, for any $p \in H$ we have

$$\begin{aligned} \|p-f\| &> \max_{x \in C} |p(x)-f(x)| \\ &\geq \max_{x \in C} \operatorname{Sign}(p^*(x)-f(x))((p^*-f)+(p-p^*)) \\ &= \|p^*-f\| + \max_{x \in C} \operatorname{Sign}(p^*(x)-f(x))(p(x)-p^*(x)) \\ &\geq \|p^*-f\| + \gamma\|p-p^*\|. \qquad (7) \end{aligned}$$

This indicates that p^* is a strongly unique best approximation to f from H.

Necessity. Suppose that p^* is the strongly unique best approximation to f from H. Then there exists a constant $\gamma = \gamma(f) > 0$ such that the relation (4) holds. By Theorem 2, we can prove that the couple $\{A_{p^*}(f), B_{p^*}(f)\}$ contains a USSAP for H. Suppose the couple is isolable with respect to H. Then there exists $q \in H$, $q \not\equiv 0$, such that $q(x) \geq 0$ if $x \in A_{p^*}(f)$; $q(x) \leq 0$ if $x \in B_{p^*}(f)$. Choose a number $\varepsilon_1 > 0$ such that

$$\varepsilon_1 < \gamma\|q\| . \qquad (8)$$

We can easily find two open subsets A_1 and B_1 of D satisfying

(i) $A_1 \supset A_{p^*}(f)$, $B_1 \supset B_{p^*}(f)$;

(ii) $q(x) > -\varepsilon_1$ if $x \in A_1$, $q(x) < \varepsilon_1$ if $x \in B_1$;

(iii) $p^*(x)-f(x) > 0$ if $x \in A_1$, $p^*(x)-f(x) < 0$ if $x \in B_1$.

Set $C=A_1 \cup B_1$, $F=D \setminus C$. It is easy to see that there exists a number $\varepsilon_2 > 0$ such that $|p^*(x)-f(x)| \leq \|p^*-f\| - \varepsilon_2$, $\forall x \in F$. Let $\lambda_0 = \operatorname{Min}\{\frac{1}{2}\|p^*-f\|/\|q\|, \varepsilon_2/\|q\|\}$. Then for any λ satisfying $0 < \lambda < \lambda_0$ we have

$$|(p^*-f-q)(x)| \leq \begin{cases} \|p^*-f\| - \varepsilon_2 + \|q\| & \text{if } x \in F \\ \|p^*-f\| + \lambda\varepsilon_1 & \text{if } x \in C . \end{cases}$$

Hence for $\lambda \in (0, \lambda_0)$ we have

$$\|p^*-f\| + \lambda\varepsilon_1 \geq \|p^*-f-q\| .$$

In addition, by the relation (4), we have

$$\|p^*-f-\lambda q\| \geq \|p^*-f\| + \lambda\gamma\|q\|$$

So we get $\varepsilon_1 \geq \|q\|$, which contradicts (8).

REFERENCES

[1] Zukhovitskii, S.I., Uspehi, 11(1956)125.
[2] Rivlin, T.J., and S hapiro, H.S., J. SIAM, 9(1961) 670.
[3] Liang, X.Z., Acta Scientiarum Naturalium Universitatis Jilinensis, (I), 1 (1965)1; (II), 2(1979) 9.
[4] Lorentz, G.G., Approximation of Functions (Holt, R.W., New York, 1966).
[5] Rockfellar, R.T., Convex Analysis (Princeton Univ. Press, Princeton, 1970).
[6] Cheney, E.W., Introduction to Approximation Theory (McGraw Hill, N.Y., 1966).
[7] Bartelt, M.W. and McLaughlin, H.W., J. Approx. Theory 9 (1973) 255.
[8] Bartelt, M.W. and Schmidt, D., J. Approx. Theory 40 (1984) 202.
[9] Wang, R.H. and Liang, X.Z., Approximation of Function in Several Variables, (Science Press, Beijing, 1988).

Approximation, Optimization and Computing:
Theory and Applications, A.G. Law and C.L. Wang (eds.)
Elsevier Science Publishers B.V. (North-Holland)

ASYMPTOTIC SOLUTIONS OF THE VECTOR NONLINEAR BOUNDARY VALUE PROBLEMS

Liu Guang-Xu

Department of Mathematics
Nankai University
Tianjin, P.R. China

In this paper, the objective is to give sufficient conditions for the existence of solution of the nonlinear boundary value problems. And we employ these results to consider the existence and asymptotic behavior, as $\mu \to o$, of the solution of singularly perturbed system of higher order.

1. DIFFERENTIAL INEQUALITIES

We consider the boundary value problem of the form

$$y'' = h(t,y,y'), \quad a<t<b \qquad (1.1)$$
$$y(a) = A, \quad y(b) = B$$

where y, h, A and B are n-vectors, the function h is continuous on $[a,b] \times R^n \times R^n$.

To simplify the exposition and to avoid repetition, let us agree upon the following notational conventions.

(1). All vectors are taken to be in R^n;

(2). Vectorial inequalities will be used with the understanding that the same inequalities hold between their respective components, for example, $\alpha(t) \leqslant y(t) \leqslant \beta(t)$ will mean that $\alpha_i(t) \leqslant y_i(t) \leqslant \beta_i(t)$, $i=1,2,\ldots,n$. The absolute value of a vector will appear in the same componentwise sense;

(3). The double subscripted vector $y_{\alpha i}$ equals $(y_1, \ldots, y_{i-1}, \alpha_i, y_{i+1}, \ldots, y_n)$.

DEFINITION 1 Let $\alpha(t)$ and $\beta(t)$ be of class $C^2[a,b]$ with $\alpha(t) \leqslant \beta(t)$. Suppose there exists a positive, nondecreasing function $\phi = \phi(t)$ of $C[o, \infty)$ which satisfies

$$\int_{\lambda_i}^{\infty} \frac{sds}{\phi_i(s)} > \max_{[a,b]} \beta_i(t) - \max_{[a,b]} \alpha_i(t) \qquad (1.2)$$

where

$$\lambda_i = \frac{1}{b-a} \max [|\alpha_i(a) - \beta_i(b)|, |\alpha_i(b) - \beta_i(a)|]$$

Further suppose that

$$|h_i(t,y,y')| < \phi_i(|y'_i|) \qquad (1.3)$$

for $a \leqslant t \leqslant b$, $\alpha(t) \leqslant y \leqslant \beta(t)$ and y' in R^n. Then function h is said to satisfy a componentwise Nagumo condition on [a,b] with respect to (α, β, ϕ).

DEFINITION 2 Let $\alpha(t)$ and $\beta(t)$, $a \leqslant t \leqslant b$, be two functions. If

(1). function pairs (α_i, β_i), $i = 1, \ldots, n$, are of piecewise-C^2 functions on [a,b], that is, there are n partitions $[t_i^j]_{j=0}^{l_i}$ of [a,b] with $a = t_i^o < t_i^1 < \cdots < t_i^{l_i} = b$ such that on each subinterval $[t_i^{j-1}, t_i^j]$, $i = 1, \ldots, n$; $j = 1, \ldots, l_i$, functions α_i and β_i are twice continuously differentiable;

(2). for each t in (a,b), $\alpha(t) \leqslant \beta(t)$, while at the endpoints $\alpha(a) \leqslant A \leqslant \alpha(a)$ and $\alpha(b) \leqslant B \leqslant \beta(b)$;

(3). for each t in (a,b), $D_L \alpha_i(t) \leqslant D_R \alpha_i(t)$ and $D_L \beta_i(t) \geqslant D_R \beta_i(t)$, $i = 1, 2, \ldots, n$, where D_L and D_R denot lefthand and righthand differentiation, respectively;

(4). on each subinterval (t_i^{j-1}, t_i^j), $j = 1, \cdots l_i$,

$$\alpha''_i \geqslant h_i(t, y_{\alpha i}, y'_{\alpha i}) \qquad (1.4)$$

and

$$\beta''_i \leqslant h_i(t, y_{\beta i}, y'_{\beta i}) \qquad (1.5)$$

provided $\alpha_k(t) \leqslant y_k \leqslant \beta_k(t)$ and $-\infty < y'_k < \infty$ $(k \neq i)$.

Then the function pair $(\alpha(t), \beta(t))$ is called the bounding function pair of BVP (1.1).

THEOREM 1 Assume that

(1). There is modified bounding function pair ($\alpha(t)$, $\beta(t)$) of BVP (1.1);

(2). $H(t, y, y')$ is modification of $h(t, y, y')$ associated with the triple ($\alpha(t)$, $\beta(t)$, c), where $c = (c_1, c_2, \ldots, c_n) > \max_{[a,b]} [|\alpha'(t)|, |\beta'(t)|]$[1].

Then the BVP

$$y'' = H(t, y, y') \qquad y(a) = A,\ y(b) = B \tag{1.1$_H$}$$

has a solution $y = y(t)$ satisfying $\alpha(t) \leqslant y(t) \leqslant \beta(t)$ on [a,b].

PROOF Because $\alpha(t)$ and $\beta(t)$ are of piecewise C^2 function on [a,b], so $|\alpha_i(t)|$ and $|\beta_i(t)|$ are consider to be $\max [|D_L\alpha_i(t_i^j)|, D_R\alpha_i(t_i^j)|]$ and $\max [|D_L\beta_i(t_i^j)|, |D_R\beta_i(t_i^j)|]$, respectively, at the partition points $[t_i^j]$. From the definition of modification it follows that $H(t,y,y')$ is continuous and bounded on $[a,b] \times R^n \times R^n$. By Scorza and Dragoni existence theorem[2] the modified BVP $(1.1)_H$ has a solution $y = y(t) \in C^2[a,b]$. Thus we only need to show that $\alpha(t) \leqslant y(t) \leqslant \beta(t)$ on [a,b]. We shall show only that $\alpha(t) \leqslant y(t)$ since the arguments for $y(t) \leqslant \beta(t)$ are essentially the same. Assume that there exists an index k, $1 \leqslant k \leqslant n$, and a point t^* of [a,b] such that $\alpha_k(t^*) > y_k(t^*)$. By conditon (2) of Definition 2, $t^* \in (a,b)$, and so the function $a_k(t) - y_k(t)$ has a positive maximum at a point t_o (a,b). If t_o is not one of the partition points t_k^j, $j=0,\ldots,l_k$, then it follows that $y_k'(t_o) = \alpha_k'(t_o)$ and $y_k'(t_o) < c_k$. Hence

$$\alpha_k''(t_o) - y_k''(t_o) \geqslant \frac{\alpha_k(t_o) - y_k(t_o)}{1 + y_k^2(t_o)} > 0$$

Therefore we have

$$\alpha_k''(t_o) - y_k''(t_o) > 0$$

which is impossible at a maximum of $\alpha_k(t) - y_k(t)$.

If $t_o = t_k^j$, for some j, is a partition point, then $D_L(\alpha_k(t_o) - y_k(t_o)) \geqslant 0$ and $D_R(\alpha_k(t_o) - y_k(t_o)) \leqslant 0$. And it follows that

$$D_L\alpha_k(t_o) = D_R\alpha_k(t_o) = \alpha_k'(t_o)$$

and

$$\alpha_k'(t_o) = y_k'(t_o)$$

Examining the lefthand derivatives, we have

$$D_L\alpha_k'(t_o) - D_L y_k'(t_o) \leqslant 0$$

One, similarly, may verify that $y(t) \leqslant \beta(t)$. We conclude that $\alpha(t) \leqslant y(t) \leqslant \beta(t)$ on [a,b].

THeorem 1 gurantees the existence of a solution $y = y(t)$ of modified BVP $(1.1)_H$ which satisfies the inequalities $\alpha(t) \leqslant y(t) \leqslant \beta(t)$ on [a,b]. Thus this solution $y = y(t)$ of $(1.1)_H$ is also a solution of the BVP

$$y'' = h(t, y, y^{*\prime})$$

$$y(a) = A,\ y(b) = B$$

where

$$y_j^{*\prime} = c_j, \text{ if } y_j' > c_j$$

$$y_j^{*\prime} = y_j', \text{ if } |y_j'| < c_j$$

$$y_j^{*\prime} = -c_j, \text{ if } y_j' < -c_j$$

$$j = 1, \ldots, n.$$

If we want to show that $y = y(t)$ is also a solution of the original BVP (1.1), we must show that $y^{*\prime}(t) = y'(t)$. In other words, we must show that $|y'(t)| \leqslant c$, $t \in [a,b]$. We shall point out that the componentwise Nagumo condition is used to guarantee $|y'(t)| \leqslant c$.

THEOREM 2 Assume that $h = h(t,y,y')$ satisfies componentwise Nagumo condition on [a,b] with respect to the triple $(\alpha(t), \beta(t), \phi(t))$. Then, for any solution $y = y(t) \in C^2[a,b]$ of BVP (1.1) with $\alpha(t) \leqslant y(t) \leqslant \beta(t)$ on [a,b], there is a constant vector $c = (c_1, \ldots, c_n) > 0$ depending only on $\alpha(t)$, $\beta(t)$ and ϕ such that $|y'(t)| \leqslant c$ on [a,b].

PROOF This proof is analogous to[3]. Since function $h(t, y, y')$ satisfies componentwise Nagumo condition on [a,b] with respect to the triple ($\alpha(t)$, $\beta(t)$, $\phi(t)$), then

$$\int_{\lambda_i}^{\infty} \frac{s\,ds}{\phi_i(s)} > \max_{[a,b]} \beta_i(t) - \min_{[a,b]} \alpha_i(t) \qquad i = 1, \ldots, n$$

where

$$\lambda_i = \frac{1}{b-a} \max[|\alpha_i(a) - \beta_i(b)|, |\alpha_i(b) - \beta_i(a)|]$$

we choose constant vector $c = (c_1, \ldots, c_n) > \lambda = (\lambda_1, \ldots, \lambda_n) > 0$ such that

$$\int_{\lambda_i}^{c_i} \frac{s\,ds}{\phi_i(s)} > \max_{[a,b]} \beta_i(t) - \min_{[a,b]} \alpha_i(t) \tag{1.6}$$

Suppose Theorem 2 is not true, that is $|y'(t)| \not\leqslant c$. Then there is an index k, $1 \leqslant k \leqslant n$, and a point $t^* \in (a,b)$ such that $|y_k'(t^*)| > c_k$. If $\xi \in (a,b)$ is such that $(b-a)y_k'(\xi) = y_k(b) - y_k(a)$, then by (1.6) $y_k'(\xi) \leqslant \lambda_k < c_k$. Then there is an interval $[c,d] \subset [a,b]$ such that one of the following four cases obtains:

Case 1. $y_k'(c) = \lambda_k$, $y_k'(d) = c_k$,

$\lambda_k < y_k'(t) < c_k$

Case 2. $y_k'(c) = c_k$, $y_k'(d) = \lambda_k$,

$\lambda_k < y_k'(t) < c_k$

Case 3. $y_k'(c) = -\lambda_k$, $y_k'(d) = -c_k$,

$-c_k < y_k'(t) < -\lambda_k$

Case 4. $y_k'(c) = -c_k$, $y_k'(d) = -\lambda_k$

$-c_k < y_k'(t) < -\lambda_k$,

$t \in (c,d)$

Let us consider the first case, then on [c,d],

$$|y_k''(t)|\ y_k'(t) \leqslant \phi_k(|y_k'(t)|)y_k'(t)$$

and

$$\left|\int_c^d \frac{y_k''(t)y_k'(t)dt}{\phi_k(|y_k'(t)|)}\right| \leqslant \max_{[a,b]} \beta_k(t) - \min_{[a,b]} \alpha_k(t) \qquad (1.7)$$

Making the chang of variable $s = y_k'(t)$, by relation (1.6), we have

$$\left|\int_c^d \frac{y_k''(t)y_k'(t)dt}{\phi_k(|y_k'(t)|)}\right| > \max_{[a,b]} \beta_k(t) - \min_{[a,b]} \alpha_k(t) \qquad (1.8)$$

This leads to the contradiction between (1.7) and (1.8). Other cases can be dealt with in a similar way and we coclude that $|y'(t)| \leqslant c$ on [a,b].

THEOREM 3 Assume that

(1). BVP (1.1) has a bounding function pair $(\alpha(t), \beta(t))$ of piecewise-C^2 on [a,b];

(2). Function $h(t,y,y')$ satisfies componentwise Nagumo condition on [a,b] with respect to the triple $(\alpha(t), \beta(t), \phi(t))$. THen BVP (1.1) has a solution $y = y(t)$ of class $C^2[a,b]$ with $\alpha(t) \leqslant y(t) \leqslant \beta(t)$ for $t \in [a,b]$.

PROOF Because $h(t,y,y')$ satisfies componentwise Nagumo condition, then, by Theorem 2, there is a constant vector $c = (c_1,\dots,c_n) > 0$ scuh that $|y'(t)| \leqslant c$ for any C^2 - solution of BVP (1.1) with $\alpha(t) \leqslant y(t) \leqslant \beta(t)$ on [a,b].Let $H = H(t,y,y')$ be the modification of $h(t,y,y')$ associated with the triple $(\alpha(t), \beta(t), N)$ where $N>0$ is chosen so that

$$N_i \geqslant \max_{[a,b]} [c_i, \max_{[a,b]} |\alpha_i'(t)|, \max_{[a,b]} |\beta_i'(t)|]$$

Then by Theorem 1 the modified BVP

$$y'' = H(t, y, y') \qquad y(a) = A,\ Y(b) = B \qquad (1.1)_H$$

has a solution $y = y(t)$ with $\alpha(t) \leqslant y(t) \leqslant \beta(t)$ on [a,b]. Hence, by the construction of function $H(t,y,y')$, this solution of $(1.1)_H$ is also a solution of the BVP

$$y'' = h(t,y,y^{*\prime}) \qquad y(a) = A,\ y(b) = B \qquad (1.1)_*$$

We are now ready to show that $y^{*\prime}$ can be replaced by y' in $(1.1)_*$, and therefore the solution $y = y(t)$ of BVP $(1.1)_H$ is also a solution of the original BVP (1.1).

First, because $H(t,y,y')$ is the modification of $h(t,y,y')$ associated with the triple $(\alpha(t), \beta(t), N)$, by definitions of $y^{*\prime}(t)$ and N, $|y^{*\prime}(t)| \leqslant N$. It remains to show that $|y^{*\prime}(t)| \leqslant N$. If $|y'(t)| \not\leqslant N$, then there is an index k, $1 \leqslant k \leqslant n$, and a point $t^* \in (a,b)$ such that $|y_k'(t^*)| > N_k$, and proceeding as in the proof of Theorem 2. Hence, we conclude that $|y'(t)| \leqslant N$ on [a,b]. This implies that

$$H(t,y(t),y'(t)) = h(t,y(t),y^{*\prime}(t))$$

$$= h(t,y(t),y'(t)), \quad t \in [a,b]$$

That is the solution $y = y(t)$ of modified BVP $(1.1)_H$ and is actually a solution of the original BVP (1.1).

2. SINGULARLY PERTURBED SYSTEM OF HIGHER ORDER

We now employ the Theorem 3 to consider the existence and asymptotic behavior, as $\mu \to 0$, of the solution of singularly perturbed system of higher order

$$\mu y^{(n)} = f(t,y,y',\dots,y^{(n-1)},\mu) \quad 0<t<1$$
$$y^{(j)}(0,\mu) = A_j(\mu), \quad 0 \leqslant j \leqslant n-2 \qquad (2.1)$$
$$y^{(n-2)}(1,\mu) = B(\mu)$$

where μ is a small positive parameter and y, A_j, B, f are m-vectors.

We define

$$D_i = [0 \leqslant t \leqslant 1,\ |y_i^{(j)} - u_i^{(j)}| \leqslant d_i(t),$$
$$0 \leqslant j \leqslant n-2,\ |y_i^{(n-1)}| < \infty,\ 0 \leqslant \mu \leqslant \mu_1]$$

Here $d_i(t)$ is a smooth positive function such that

$$|B^i(\mu) - u_i^{(n-2)}(1)| \leqslant d_i(t) \leqslant$$
$$|B^i(\mu) - u_i^{(n-2)}(1)| + \eta,\ t \in [1-\delta/2, 1]$$
$$d_i(t) \leqslant \eta,\quad t \in [0, 1-\delta]$$

where B^i are components of B, and η, δ are positive constants.

The following result can be proved using the Theorem 3.

THEOREM 4 Assume that

(1). The reduced problem

$$f(t, u, u', \ldots, u^{(n-1)}, 0) = 0$$
$$u^{(j)}(0) = A_j(0),$$
$$j = 0, 1, \ldots, n-2$$

has a solution $u = u(t) = (u_1(t), \ldots, u_m(t)) \in C^{(n)}$ on [0,1], such that

$$f^i(t, y_{ui}, y'_{u'i}, \ldots, y_{u(n-1)_i}^{(n-1)}, 0) = 0$$
$$u_i^{(j)}(0) = A_j^i(0)$$
$$0 \leqslant j \leqslant n-2$$

where

$$A_j^i(\mu) = A_j^i(0) + 0(\mu)$$
$$B^i(\mu) = B^i(0) + 0(\mu),\quad i = 1, \ldots, m$$

(2). $f^i(t, y, y', \ldots, y^{(n-1)}, \mu)$, $i = 1, \ldots, m$, are continuous with respect to t and continuously differentiable with respect to y, y', ..., $y^{(n-1)}$ in $D = \cup D_i$.

(3). $f^i(t, y, y', \ldots, y^{(n-1)}, \mu)$ satisfy Nagumo codition with respect to $y^{(n-1)}$, and

$$f^i(t, y_{ui}, \ldots, y_{u(n-1)i}^{(n-1)}, \mu) = 0(\mu)$$

(4). there exist positive constants L_j^i, j = 0, 1, ..., n-2, and K_i, i = 1, ..., m, such that

$$\left|\frac{\partial f^i}{\partial y_i^{(j)}}\right| < L_j^i, \quad \frac{\partial f^i}{\partial y_i^{(n-1)}} \geqslant K_i > 0$$

Then there exist an $\mu_o > 0$ such that for $0 < \mu < \mu_o$ the vector problem (2.1) has a solution $y = y(t,\mu)$ satisfying

$$y_i^{(j)}(t,\mu) = u_i^{(j)}(t) + 0(\mu),\quad j = 0, \ldots, n-3$$
$$y_i^{(n-2)}(t,\mu) = u_i^{(n-2)}(t) +$$
$$|B^i(\mu) - u_i^{(n-2)}(1)| \exp[-\lambda_{1i}(t-1)] + 0(\mu)$$
$$0 \leqslant t \leqslant 1$$

where λ_{1i} is negative real root of the following equation

$$\mu\lambda^n + K_i\lambda^{n-1} + L_{n-2}^i\lambda^{n-2} + \cdots +$$
$$(-1)^{n+1}L_1^i\lambda + (-1)^n L_o^i = 0$$

REFERENCES

[1] Liu Guang-Xu, On Singularly Perturbed Quasilinear Systems, Applied Mathematics and Mechanics Vol.8 No.11, Nov. (1987) pp. 1027-1036.

[2] Scorza-Dragoni, Sur Problema dei Valori ai Limiti per i Systemi di Equazioni Differenziali del Secondo Ordine, Boll. Un. Mat. Ital., 14(1935), pp.225-230.

[3] Jackson, L.K., Subfunctions and Second-Order Ordinary Differential Inequalities, Adv. in Math. 2(1968), pp.308-363.

Approximation, Optimization and Computing:
Theory and Applications, A.G. Law and C.L. Wang (eds.)
Elsevier Science Publishers B.V. (North-Holland)

SOME PROBLEMS IN PANSYSTEMS APPROXIMATION THEORY

Liu Huaijun

(Wuhan University)

Abstract

In this paper, we have presented the concepts, framework and idea of pansystems approximation theory, and have obtained a series of theorems about a kind of special pansystems approximation.

Pansystems theory is a transfield multilayer network–like methodological research with emphasis on pansystems, i.e., generalized systems, relations, symmetry etc.. The pansystems approximation theory (PAT)is concerning the generalized approximation for the pansystems within the framework of pansystems theory, where the generalized approximation includes mainly the so–called 15–transformation–simulations, namely, confinement, extension, projection, embodiment (inversion of projection), quotientization, productization (=inversion of quotientization), microscopy (=confinement * embodiment * productization), bird's–eye–view (=extension * projection * quotientization), epitome (= confinement * projection * quotientization), extended–embodiment (=extension * embodiment * productization), explicit–transformation (=projection * embodiment), implicit–transformation (=embodiment * projection), synergy–transformation (=extension * implicit–transformation * confinement), quasi–transformation (=epitome * extended–embodiment), panproduct (=quotient–epitome of direct–product), and related induced simulations.

The research of PAT shows that the generalized approximation of the majority of objects, structures, things in mathematics and various disciplines can be realized through the use of pansystems or the compositions and embodiments of the so–called 12–pansystems–relations (PR): macro–microscopy, whole–parts, body–shadow, observo–control, cause–effect, motion–rest, shengke (synergy–conflict, etc.), pan–order (generalized–order), simulation (modelling), clustering–discoupling, difference–identity, series–parallel.

The pansystems theory itself can be considered as a sort of generalized approximation to the whole knowledge–method system. Certain relatively universal concepts, principles, methods, theorems, models and techniques developed by pansystems theory can be also considered as some special PAT research on certain method–technique topics. Related contents are connected with recognition, control, modelling, games, design, strategy, simplification, detection, education, creation, tolerance, synergy and certain topics of the true, the good and the beautiful (see [1]–[6]).

At present, we consider a special pansystems approximation. Let G be an arbitrary set, P(G) is the power set of G. $I=\{(x,x)|x\in G\}$, then $I\in P(G^2)$.
Suppose $f,g\in P(G^2)$, let $f^{-1}=\{(y,x)|(x,y)\in f\}$, and define the composition operation:

$$f * g=\{(x,y)|(x,t)\in f,(t,y)\in g,t\in G\}$$

The symbols E[A], $E_s[A]$ are used to denote the classes of equivalent relations and tolerant relations about A respectively.

If $\delta\in E_s[A]$, the totality of subsets A_i which is tolerant with δ is called the quotient system of A about δ, and is denoted by

$$A/\delta=\{A_i\}.$$

The quotientization f_δ and panderivative f'_δ are defined as[8].

$$f_\delta=\{(x,A_i)|x\in A_i\}\subset A\times(A/\delta),$$

and

$$f'_\delta=f'_\delta(\theta)=f_\delta^{-1} * \theta * f_\delta\subset(A/\delta)^2,\theta\subset A^2.$$

The pansystems operators are defined as certain transformations which map things, structures to pansystems relations, especially to tolerant–equivalent relations. The important typical types are the so–called ε_i–operators and δ_i–operators which are defined as follows:

$$\varepsilon_1(\theta)=\theta\cup\theta^{-1}\cup I,\qquad \varepsilon_2(\theta)=\varepsilon_1(\theta\cap\theta^{-1}),$$
$$\varepsilon_3(\theta)=\varepsilon_1(\theta^t\cap\theta^{-t}),\qquad \varepsilon_4(\theta)=\varepsilon_1(\theta * \theta^{-1}),$$
$$\varepsilon_5(\theta)=\varepsilon_1(\theta^{-1} * \theta),$$

$$\varepsilon_{5+i}(\theta) = \varepsilon_1(A^2-\theta), \qquad i=1,2,3.$$
$$\delta_j(\theta) = [\varepsilon_j(\theta)]^t, \qquad j=1,2,...,8.$$

where $\theta \subset A^2$, $\theta^t = \theta \cup \theta^{(2)} \cup \theta^{(3)} \cup ...$,

$$\theta^{(2)} = \theta * \theta, \qquad \theta^{(n)} = \theta^{(n-i)} * \theta.$$

Obviously,

$$\varepsilon_i(\theta) \in E_s[A), \qquad \delta_i(\theta) \in E[A).$$

We can obtain the following results:

Theorem 1 $f'_\delta(\theta) = \{(A_i,A_j) | x \in A_i, y \in A_j, (x,y) \in \theta\}$, $A_i,A_j \in A/\delta$, $\theta \subset A^2$.

Theorem 2

$f'_\delta(\theta) = \{(A_i,A_j) | (f_\delta * A_i) \times (f_\delta * A_j) \cap \theta \neq \phi\}$.

Theorem 3 $f'_\delta(\theta) = \{(A_i, A_j) | (x,y) \in \theta, x * f_\delta = A_i, y * f_\delta = A_j\} = \{(x * f_\delta, y * f_\delta) | (x, y) \in \theta\} \subset (A/\delta)^2$.

Theorem 4 f'_δ makes the reflexivity, symmetry and tolerance conservative. If $\theta \supset \delta$ are all equivalent relations, then $f'_\delta(\theta) \in E[A/\delta)$.

Theorem 5 If $C \subset A$, $\delta = \varepsilon_1(C^2)$, then $A/\delta = \{C\} \cup \{x | x \in C\}$. In other words, A/δ is just to pointize or black–boxize the subset C.

$$\varepsilon_i(C^2) = \delta_j(C^2), \qquad i,j=1,2,3.$$

If $(x,y) \in C^2 \cap \theta \neq \phi$, then $(\{C\},\{C\}) \in f'_\delta(\theta)$,
if $(x,y) \in C \times (A-C) \cap \theta$, then $(\{C\},y) \in f'_\delta(\theta)$,
if $(x,y) \in (A-C) \times C \cap \theta$, then $(x,\{C\}) \in f'_\delta(\theta)$,
if $(x,y) \in (A-C)^2 \cap \theta$, then $(x,y) \in f'_\delta(\theta)$.

Theorem 6 If $C \subset A$, then $\varepsilon_1((A-C)^2) = (A-C)^2 \cup I \subset \varepsilon_6(C^2) = \varepsilon_1(A^2-C^2) = (A^2-C^2) \cup I$. But $A/\varepsilon_1((A-C)^2) = A/\varepsilon_6(C^2) = \{\{x\} | x \in C\} \cup \{A-C\}$, quotientization f_δ makes (A–C) black–boxized and C white–boxized simultaneously.

Similarly, we can obtain the corresponding results about $f'_\delta(\theta)$ as in theorem 5.

Sometimes panderivative plays the role which makes certain generalized softwares black–boxized.

Theorem 7 If $\delta = \delta_1(\theta)$, then θ is disjointed among A/δ and

$$f'_\delta(\theta) \subset I(A/\delta) = \{(A_i,A_i) | A_i \in A/\delta\}.$$

Theorem 8 If $\delta = \delta_3(\theta)$, then A/δ is of θ–isolation or unidirectional connection and $f'_\delta(\theta)$ is of antisymmetry:

$$f'_\delta(\theta) \cap [f'_\delta(\theta)]^{-1} \subset I(A/\delta).$$

Theorem 9 If $\delta = \varepsilon_1(\theta)$ or $\delta = \varepsilon_2(\theta)$, $A_i \in A/\delta$ does not be reduce to a single element, or $(A_i,A_i) \in \theta$, then $(A_i,A_i) \in f'_\delta(\theta)$. In other words, $f'_\delta(\theta)$ is of near–reflexivity.

Theorem 10 $f'_\delta(\theta^{-1}) = (f'_\delta(\theta))^{-1}$. If $\theta \subset \theta_1$, then $f'_\delta(\theta) \subset f'_\delta(\theta_1)$, $f'_\delta(I(A)) = \{(A_i,A_j) | A_i \cap A_j \neq \phi\}$.

Theorem 11 If $\delta = \delta_6(\theta)$ and $A_i \in A/\delta$ does not reduce into a single element subset, then

$$(A_i,A_i) \bar{\in} f'_\delta(\theta)$$

or $f'_\delta(\theta)$ is of approximate non–reflexivity.

It can be seen from the above that the essential significance of panderivative lies in its transformation of relations from those among elements to those among subsets.

Recently, Wu Xuemou presented a conjecture as follows:

Conjecture For $\delta = \delta_6(\theta)$, if $A_i,A_j \in A/\delta$ do not reduce into single element subsets $i \neq j$, then

$$(A_i,A_j) \in f'_\delta(\theta).$$

That is to say, $f'_\delta(\theta)$ is of approximate binary relation.

Reference

[1] Wu Xuemou, Pansystems Methodology 100, Science Exploration, 3(1986).

[2] Wu Xuemou, Approximation Transformation Theory and the Pansystems Concepts in Mathematics, Hunan Sci.Tech. Press. 1984.

[3] Wu Xuemou, The Pansystems View and Pansystems Poems 100, Guizhou Science, 2(1988).

[4] Wu Xuemou, Pansystems Methodology: Concepts, Theorems and Applications (Ⅰ)–(Ⅷ) (English), Sci. Expl., 1,2,4(1982), 1,4(1983), 1,4(1984), 1(1985).

[5] Wu Xuemou, Pansystems Methodology and its Applications: Cybernetics, Epistemology, Shengkeology and Sociology (English), Sci. Expl.3(1986).

[6] Wu Xuemou, Pansystems Methodology of Ecology, Medicine and Diagnostics (Ⅰ)–(Ⅳ), Expl. Nature, 2,3(1983), 1(1984), 2(1985).

[7] Liu Huaijun, Construction of a class of Analytic Functions, Journal of Math. 1(1981).

[8] Liu Huaijun, Some Mathematical Models of Pansystems Approximation and Pansystems Quotientization, Guizhou Sci., 2(1988).

Approximation, Optimization and Computing:
Theory and Applications, A.G. Law and C.L. Wang (eds.)
Elsevier Science Publishers B.V. (North-Holland)

AN EXTENSION OF A STRONG LAW OF LARGE NUMBERS OF MARKOV CHAINS

Liu Wen

Hebei Institute of Technology, Tianjin
The People's Republic of China

The strong law of large numbers for transition probabilities of denumerable homogeneous Markov chains is generalized to the nonhomogeneous case. A new analytical approach in proving strong law of large numbers of Markov chains is presented, the crucial part of which is to construct the Markov chain with the given initial distribution and transition matrix in the probability spacs $([0,1), \beta, \mu)$ where β is the class of Borel sets and μ is the Lebesgue measure, and then prove the existence of convergence a.e. by using Lebesgue differentiation theorem on monotone functions.

The purpose of this paper is to generalize a strong law of large numbers for transition probabilities of denumerable homogeneous Markov chains to the nonhomogeneous case (cf. Chung[1], p.92).

Let $\{X_n, n\geqslant 0\}$ be a nonhomogeneous Markov chains with state space $E=\{1,2,3,\cdots\}$, initial distribution

$$q(1),\ q(2), \cdots,\ q(n), \cdots \tag{1}$$

and one-step transition matrix

$$p(n)=\begin{pmatrix} p(n,1,1) & p(n,1,2) & p(n,1,3)\cdots \\ p(n,2,1) & p(n,2,2) & p(n,2,3)\cdots \\ p(n,3,1) & p(n,3,2) & p(n,3,3)\cdots \\ \cdots & \cdots & \cdots\ \cdots \end{pmatrix}, n=0,1,2,\cdots \tag{2}$$

where $p(n,i,j)=P\{X_{n+1}=j \mid X_n=1\}$. Assume $k, l\in E$, and let $S(k,n)$ be the number of occurrence of k in the partial sequence $X_0, X_1, \cdots, X_{n-1}$, $A(k,l,n)$ the number of occurrence of couple (k,l) in partial sequence

$$(X_0, X_1), (X_1, X_2), \cdots, (X_{n-1}, X_n).$$

THEOREM. Assume $0<p(n,k,l)<1$ for all $n\geqslant 0$. If there exist nonnegative real numbers a and b such that

$$a\leqslant \liminf_{n\to\infty} p(n,k,l)\leqslant \limsup_{n\to\infty} p(n,k,l)\leqslant b \tag{3}$$

then

$$a\leqslant \liminf_{n\to\infty}\frac{A(k,l,n)}{S(k,n)}\leqslant \limsup_{n\to\infty}\frac{A(k,l,n)}{S(k,n)}\leqslant b \tag{4}$$

for almost every ω in D_k where

$$D_k=\{\omega;\ \lim S(k,n)=\infty\}$$

Proof. Throught out this paper, we shall deal with the underlying probability space $([0,1)\ \beta, \mu)$, where β is the class of Borel sets and μ is the Lebesgue measure. We first prove that a Markov chain with initial distribution(1) and one-step transition matrix (2) can be constructed in this probability space.

Let

$$q(n_1),\ q(n_2),\ \cdots,\ q(n_i),\ \cdots$$

be the positive terms of (1), where $n_1<n_2<\cdots$.

Divide $[0,1)$ into countablely many right-semi-open intervals d_{x_0} $(x_0=n_1, n_2, \cdots)$ according to the ratio $q(n_1):q(n_2):\cdots$, that is,

$$d_{n_1}=[0, q(n_1)],\ d_{n_2}=[q(n_1), q(n_1+q(n_2)], \cdots$$

These intervals will be called the 0-th order intervals. Proceeding inductively, suppose the n-th order d-intervals $d_{x_0\cdots x_n}$ have been defined. Then divide the right-semi-open interval $d_{x_0\cdots x_n}$ into countablely many right-semi-open intervals

$$d_{x_0\cdots x_n x_{n+1}}\quad (x_{n+1}=m_i\ ,\ i=1,2,3,\cdots)$$

according to the ratio $p(n,x_n,m_1,):p(n,x_n,m_2):\cdots$, where

$$p(n,x_n,m_i),\ i=1,2,3,\cdots,\ m_1<m_2<\cdots$$

are the positive elements of the x_n-th row of the matrix $P(n)$. In this way the (n+1) st order d-intervals are created. It is easy to see that

$$\mu(d_{x_0\cdots x_n})=(d_{x_0})\prod_{m=0}^{n-1} p(m, x_m, x_{m+1}). \tag{5}$$

Define, for $n\geqslant 0$, a random variable $X_n: [0,1)\to E$ as follows:

$$X_n(\omega)=X_n\ ,\ \text{if}\ \omega\in d_{x_0\cdots x_n} \tag{6}$$

It is clear that

$$\mu(X_0=x_0, \cdots, X_n=x_n)=\mu(d_{x_0\cdots x_n}) =q(x_0)\prod_{m=0}^{n-1} p(m, x_m, x_{m+1}) \tag{7}$$

so $\{X_n, n\geqslant 0\}$ is a Markov chain with initial distribution (1) and one-step transition matrix (2). Assume

$$R(n)=\begin{pmatrix} r(n,1,1) & r(n,1,2) & r(n,1,3)\cdots \\ r(n,2,1) & r(n,2,2) & r(n,2,2)\cdots \\ r(n,3,1) & r(n,3,2) & r(n,3,3)\cdots \\ \cdots & \cdots & \cdots\ \cdots \end{pmatrix},\ n=0,1,2\cdots$$

is another one-step transition matrix, where $r(n,i,j)>0$ if and only if $p(n,i,j)>0$. Using the initial distribution (1) in the same procedure as above, except with $R(n)$ in place of $P(n)$, we construct a new collection of intervals $\Delta_{x_0\cdots x_n}$. It is clear that

$$\Delta_{x_0}=d_{x_0}\qquad (x_0=n_1, n_2, n_3, \cdots)$$

and

$$\mu(\Delta_{x_0\cdots x_n})=q(x_0)\prod_{m=0}^{n-1} r(m, x_m, x_{m+1}) \tag{8}$$

Let $d^-_{x_0\cdots x_n}$ and $d^+_{x_0\cdots x_n}$ be, respectively, the left and right end-points of $d_{x_0\cdots x_n}$; define $\Delta^-_{x_0\cdots x_n}$ and $\Delta^+_{x_0\cdots x_n}$ similarly. Let Q be the set of end-points of all d-intervals. Now we define a function $f: [0,1)\rightarrow[0,1)$ as follows,

$$f(d^-_{x_0\cdots x_n})=\Delta^-_{x_0\cdots x_n},\quad f(d^+_{x_0\cdots x_n})=\Delta^+_{x_0\cdots x_n}, \tag{9}$$

$$f(\omega)=\sup\{f(t), t\in Q\cap[0,\omega)\},\quad \omega\in[0,1)-Q \tag{10}$$

It is easy to see that f is increasing on $[0,1)$. Let $\omega\in[0,1)$ and $d_{x_0\cdots x_n}$ be the n-th order interval containing ω. From (7)-(9) it follows that

$$\frac{f(d^+_{x_0\cdots x_n})-f(d^-_{x_0\cdots x_n})}{d^+_{x_0\cdots x_n}-d^-_{x_0\cdots x_n}}=\frac{\Delta^+_{x_0\cdots x_n}-\Delta^-_{x_0\cdots x_n}}{d^+_{x_0\cdots x_n}-d^-_{x_0\cdots x_n}}$$

$$=\frac{\mu(\Delta_{x_0\cdots x_n})}{\mu(d_{x_0\cdots x_n})}=\prod_{m=0}^{n-1}\frac{r(m, x_m, x_{m+1})}{p(m, x_m, x_{m+1})}=t_n(\omega) \tag{11}$$

Given any positive real number λ, we can choose $r(m,k,l)$ such that

$$\frac{r(m,k,l)(1-p(m,k,l))}{p(m,k,l)(1-r(m,k,l))}=\lambda \tag{12}$$

It f llows that

$$r(m,k,l)=\frac{p(m,k,l)}{1+(\lambda-1)p(m,k,l)}, \tag{13}$$

$$\frac{1-r(m,k,l)}{1-p(m,k,l)}=\frac{1}{1+(\lambda-1)p(m,k,l)} \tag{14}$$

Let

$$r(m,k,j)=\frac{1-r(m,k,l)p(m,k,j)}{1-p(m,k,l)},\ j\neq k, \tag{15}$$

$$r(m,i,j)=p(m,i,j),\quad i\neq k \tag{16}$$

Now define, for $n\geq 1$,

$$I_i(x)=\begin{cases}1, & \text{if } x=i;\\ 0, & \text{if } x\neq i.\end{cases}$$

In view of (12) and (14), we have

$$\frac{r(m, x_m, x_{m+1})}{p(m, x_m, x_{m+1})}=\left[\frac{r(m,k,l)(1-p(m,k,l))}{p(m,k,l)(1-r(m,k,l))}\right]^{I_k(x_m)I_l(x_{m+1})}\cdot\left[\frac{1-r(m,k,l)}{1-p(m,k,l)}\right]^{I_k(x_m)}$$

$$=\lambda^{I_k(x_m)I_k(x_{m+1})}\left[\frac{1}{1+(\lambda-1)p(m,k,l)}\right]^{I_k(x_m)} \tag{17}$$

It is clear that

$$\sum_{m=0}^{n-1} I_k(x_m)\, I_l(x_{m+1})=A(k,l,n), \tag{18}$$

$$\sum_{m=0}^{n-1} I_k(x_m)=S(k,n) \tag{19}$$

Let f be the function defined by (9) and (10), and denote $t_n(\omega)$ in (11) by $t_n(\lambda,\omega)$ when $r(m,i,j)$ are given by (13), (15) and (16). In view of (11), (18) and (17), we heve

$$t_n(\lambda,\omega)=\lambda^{A(k,l,n)}\prod_{m=0}^{n-1}\left(\frac{1}{1+(\lambda-1)p(m,k,l)}\right)^{I_k(x_m)} \tag{20}$$

Letting $H(\lambda)$ be the set of points at which f is differentiable, $\mu(H(\lambda))=1$ (cf. [2], p406). Let $\omega\in H(\lambda)$, and assume $d_{x_0\cdots x_n}$ is the n-th order d-interval containing ω. By (11), if $\lim_{n\to\infty}\mu(d_{x_0\cdots x_n})=0$, then (cf. [3], p.345)

$$\lim_{n\to\infty} t_n(\lambda,\omega)=\lim_{n\to\infty}\frac{f_\lambda(d^+_{x_0\cdots x_n})-f_\lambda(d^-_{x_0\cdots x_n})}{d^+_{x_0\cdots x_n}-d^-_{x_0\cdots x_n}}=f'_\lambda(\omega)<\infty; \tag{21}$$

if $\lim_{n\to\infty}(d_{x_0\cdots x_n})=\delta>0$

$$\lim_{n\to\infty} t_n(\lambda,\omega)=\lim_{n\to\infty}\mu(\Delta_{x_0\cdots x_n})/\delta<\infty \tag{22}$$

In riew of (21) and (22)

$$\limsup_{n\to\infty}[t_n(\lambda,\omega)]^{1/s(k,n)}\leq 1,\ \omega\in H(\lambda)\cap D_k \tag{23}$$

It follows that

$$\limsup_{n\to\infty}\frac{1}{S(k,n)}\ln t_n(\lambda,\omega)\leq 0,\quad \omega\in H(\lambda)\cap D_k \tag{24}$$

There is no harm in assuming $b<1$. Given any $\varepsilon\in(0, 1-b)$, in view of (3), there exists positive integer N such that

$$p(n,k,l)\leq b+\varepsilon<1,\quad n\geq N. \tag{25}$$

Let $\lambda>1$. In view of (22), (25) and (19),

$$t_n(\lambda,\omega)=\lambda^{A(k,l,\omega)}\prod_{m=0}^{N}\left[\frac{1+(\lambda-1)(b+\varepsilon)}{1+(\lambda-1)p(m,k,l)}\right]^{X_k(X_m)}$$

$$\cdot\prod_{m=0}^{N}\left[\frac{1}{1+(\lambda-1)(b+\varepsilon)}\right]^{X_k(X_m)}\prod_{m=N+1}^{n-1}\left[\frac{1}{1+(\lambda-1)p(m,k,l)}\right]^{X_k(X_m)}$$

$$\geq\lambda^{A(k,l,n)}\left[\frac{1}{1+(\lambda-1)(b+\varepsilon)}\right]^{S(k,n)}\left[\frac{1+(\lambda-1)(b+\varepsilon)}{\lambda}\right]^{N} \tag{26}$$

Therefore, by (24) and (26),

$$\limsup_{n\to\infty}\frac{A(k,l,n)}{S(k,n)}\leq\frac{\operatorname{Ln}(1+(\lambda-1)(b+\varepsilon))}{\operatorname{Ln}\lambda},\ \omega\in H(\lambda)\cap D_k \tag{27}$$

Let $\varepsilon\to 0$, then, in view of (27),

$$\limsup_{n\to\infty}\frac{A(k,l,n)}{S(k,n)}\ \frac{\ln(1+(\lambda-1)b}{\ln\lambda},\ \omega\in H(\lambda)\cap D_k \tag{28}$$

Now choose real sequence $\{\lambda_i\}$ such that $\lambda_i>1$, and $\lambda_i\to 1$ as $i\to\infty$. Let $H^*=\bigcap_{i=1}^{\infty}H(\lambda_i)$. In view of (28),

$$\limsup_{n\to\infty}\frac{A(k,l,n)}{S(k,n)}\leq\frac{\ln(1+(\lambda_i-1)b)}{\ln\lambda_i},\quad H^*\cap D_k \tag{29}$$

It follows from (29) that

$$\liminf_{n\to\infty}\frac{A(k,l,n)}{S(k,n)}\geq b,\ \omega\in H^*\cap D_k \tag{30}$$

since $\lim_{\lambda\to 1+0}\frac{\ln(1+(\lambda-1)b)}{\ln\lambda}=b$

Choose positive real sequence $\{\lambda'_i\}$ such that $\lambda'_i<1$, and $\lambda'_i\to 1$ as $i\to\infty$. Let $H_*=\bigcap_{i=1}^{\infty}H(\lambda'_i)$. In the similar procedure as above, it can be shown that

$$\liminf_{n\to\infty}\frac{A(k,l,n)}{S(k,n)}\geq a,\quad \omega\in H_*\cap D_k \tag{31}$$

Finally, in view of (30) and (31), (4) holds if $\omega \in H^* \cap H_* \cap D_k$. Since $\mu(H^* \cap H_* \cap D_k) = \mu(D_k)$, the theorem follows.

REFERENCES

[1] K.L.Chung, Markov Chains with Sfationary Transition probabilities, Springer-Verlag, New york, 1967.
[2] A.E. Taylor, General Theory of Functions and Integration Blaisdell, New york, 1965.
[3] T.H.Hildebrandt, Introduction to the Theory of Integration, Academic Press, New York, 1963.

Approximation, Optimization and Computing:
Theory and Applications, A.G. Law and C.L. Wang (eds.)
Elsevier Science Publishers B.V. (North-Holland)

ON RELATIVE PSEUDO-CENTRE AND PSEUDO-RADIUS

Jianke Lu (Chien-Ke Lu)

Dept. of Math., Wuhan Univ. , Wuhan, Hubei 430072, China (PRC)

ABSTRACT: The author defined the pseudo-centre y_E and pseudo- radius r_E of a set E in R^n and gave a general method for finding them (cf. [1,2]). Here we extend these concepts to those relative to a hyperplane π^m in $R^n(1 \leq m < n)$. We also describe a general method for finding them.

1. THE RELATIVE PSEUDO-CENTRE AND PSEUDO-RADIUS

Let E be a compact set in R^n.

DEFINITION. The pseudo-radius of E relative to π^m is defined as

$$r_E(\pi^m) = \min_{y\in\pi^m} \max_{x\in E} |y - x| \tag{1}$$

and the point $y_E(\pi^m)$ which realizes this max-minium is called the pseudo-centre of E relative to π^m.

The radius of the smallest n-ball $K_E(\pi^m)$ with centre on π^m and covering E is actually $r_E(\pi^m)$, and its centre, $y_E(\pi^m)$. Thus, they are unique and are the same if E replaced by its convex closure and even by the boundary of the latter. The following properties are obvious.

(i) $E' \subset E$ implies $r_{E'}(\pi^m) \leq r_E(\pi^m)$.
(ii) $2r_E(\pi^m) \leq d_E$ (the diameter of E).
(iii) if $E \subset \pi^m$, then $y_E(\pi^m) = y_E$ and $r_E(\pi^m) = r_E$ in R^n.
(iv) If E contains a region G in R^n or in some hyperplane of lower dimension, then $y_E(\pi^m)$ and $r_E(\pi^m)$ remain unchanged when Int$G$ is taken away from E.

Therefore, we always assume E is not a subset of π^m.

2. ESTIMATES OF APPROXIMATION

Let E and E' be two arbitrary compact sets in R^n. We define

$$\delta_E(E') = \max_{x'\in E'} \min_{x\in E} |x - x'|. \tag{2}$$

Then

$$\delta(E, E') = \max\{\delta_{E'}(E), \delta_E(E')\} \tag{3}$$

is the well known Hausdorff measure of E and E'. [3] It is obvious $E' \subset E$ implies $\delta_E(E') = 0$ and $\delta_E(E, E') = 0$ means $E = E'$. We have

THEOREM 1. If $\delta(E, E') \leq \varepsilon$, then

$$|r_E(\pi^m) - r_{E'}(\pi^m)| \leq \varepsilon \tag{4}$$

and

$$| y_E(\pi^m) - y_{E'}(\pi^m) | \leq \min\{\sqrt{2\varepsilon(2r + \varepsilon)}, 2\sqrt{\varepsilon(r' + \varepsilon)}\} \tag{5}$$

where

$$r = \max\{r_E(\pi^m), r_{E'}(\pi^m)\},$$
$$r' = \min\{r_E(\pi^m), r_{E'}(\pi^m)\}. \tag{6}$$

PROOF. Put

$$f_E(y) = \max_{x\in E} | y - x | . \tag{7}$$

Since $\delta_E(E') \leq \varepsilon$, so we have, for any $x' \in E', \min |x - x'| \leq \varepsilon$. Thereby, for arbitrarily fixed $x' \in E'$, there exists $x_0 \in E$ such that $|x_0 - x'| \leq \varepsilon$. Then, for any $y \in R^n$, We have

$$|y - x'| \leq |y - x_0| + |x_0 - x'| \leq |y - x_0| + \varepsilon \leq f_E(y) + \varepsilon.$$

And then, by (7),

$$f_{E'}(y) \leq f_E(y) + \varepsilon. \tag{8}$$

Since $r_{E'}(\pi^m) = \min_{y\in\pi^m} f_{E'}(y)$, hence we have $r_{E'}(\pi^m) \leq f_E(y) + \varepsilon$. Take the minimum of the right-hand number for $y \in \pi^m$, we get

$$r_{E'}(\pi^m) \le r_E(\pi^m) + \varepsilon. \tag{9}$$

Changing E and E' in position, we obtain (4).

In order to prove (5), we may assume

$$r_{E'}(\pi^m) \le r_E(\pi^m) \text{ and } y_E(\pi^m) \ne y_{E'}(\pi^m).$$

Starting from $y_E(\pi^m)$, we draw a half-ray l passing through $y_{E'}(\pi^m)$. Then $l \subset \pi^m$. There exists a unique $c \in l$ such that

$$|y_E(\pi^m) - c| = \sqrt{[r_E(\pi^m) + \varepsilon]^2 - r_{E'}(\pi^m)^2}. \tag{10}$$

We want to prove $y_{E'}(\pi^m)$ is situated on the line-segment $[y_E(\pi^m), c]$.

Suppose it is not the case. We shall denote the n-ball with centre at y and radius r by $K(y,r)$. Since $K_E(\pi^m)$ covers E and $\delta(E,E') \le \varepsilon$, hence it also covers E'. On the other hand, since $K_{E'}(\pi^m)$ covers E', we then have

$$E' \subset K(y_E(\pi^m), r_E(\pi^m) + \varepsilon) \cap K_{E'}(\pi^m).$$

The set in the right-hand member is in the interior of $K(c, r_{E'}(\pi^m))$ by supposition and therefore it also covers E'. Thus, $r_{E'}(\pi^m)$ is not the radius of the smallest n-ball with centre on π^m covering E', which is a contradiction. Hence, by (10), we get

$$|y_E(\pi^m) - y_{E'}(\pi^m)| \le |y(\pi^m) - c| = \sqrt{(r+\varepsilon)^2 - r'^2} \tag{11}$$

which implies (6) since $0 < r - r' \le \varepsilon$.

If $E' \subset E$, (11) may be improved. In fact,we have

THEOREM 2. If $E' \subset E$ and $\delta(E,E')(= \delta_{E'}(E)) \le \varepsilon$, then

$$|y_E(\pi^m) - y_{E'}(\pi^m)| \le \min\{\sqrt{2\varepsilon r_E(\pi^m)},$$
$$\sqrt{\varepsilon(2r_{E'}(\pi^m) + \varepsilon)}\}. \tag{12}$$

PROOF. Since $E' \subset E$, then $E' \subset K_E(\pi^m)$. Therefore, $K(y_E(\pi^m), r_E(\pi^m)+\varepsilon)$ in the proof of Th.1 may be replaced by $K_E(\pi^m)$ and thence (11) may be improved as

$$|y_E(\pi^m) - y_{E'}(\pi^m)| \le \sqrt{(r^2 - r'^2)}$$
$$\le \sqrt{\varepsilon(r + r')}$$

which follows (12) immediately.

3. METHOD OF DETERMINATION

If E is a finite set, we may determine $K_E(\pi^m)$ by a finite number of procedures. We note that it is the same when E is replaced by its convex closure, namely, a polyhedron, and then by its vertices (by property (iv)). Thus, we may always assume E is the set of vertices of a convex polyhedron.

We establish the following

LEMMA. If E is the set of vertices of a convex polyhedron in R^n, then $K_E(\pi^m)$ is the smallest n-ball covering E with centre on π^m which contains least number of points of E lying on its boundary.

PROOF. Evidently, for $K_E(\pi^m)$, there must exist points on E lying on its boundary.

Consider an n-ball $K = K(C,r)$ covering E and having k points $p_1, \cdots, p_k \in E \cap \partial K$. If K is not the $K_E(\pi^m)$, then there exists point $C' \in \pi^m, |P_j - C'| < r = |P_j - C|$, sufficiently close to C such that $E \setminus \cup_j P_j \subset \text{Int} K'$ where $K' = K(C', r')$, $r' = \max |P_j - C'|$. Thus, we obtain an n-ball $K(C', r')(C' \in \pi^m, r' < r)$ which contains less or equal number of points of E on its boundary. This proves the lemma.

By this lemma, a method for determining $K_E(\pi^m)$ may be illustrated (assuming E is not in π^m). First, we take any point $x_0 \in E$ lies farthest from π^m and project it on π^m with projection y_0. If $K(y_0, |x_0 - y_0|)$ covers E, then it is $K_E(\pi^m)$. If it is not the case, then we take arbitrarily a pair of points $x_1, x_2 \in E$ (except those the straight line passing through x_1, x_2 orthogonal to π^m) and construct the smallest n-ball with centre $y_1 \in \pi^m$ whose boundary passes throuhgh x_1, x_2; y_1 must lie on the projection line π^1 (to π^m) of the line passing through x_1, x_2 and so the problem is reduced to that in R^2 for the set $\{x_1, x_2\}$ relative to π^1. If $K(y_1, y - x_1)$ covers E, then it is $K_E(\pi^m)$. If it is not the case for any such pair, then we take arbitrarily a triple of points $x_1, x_2, x_3 \in E$ (except those spanning a plane orthogonal to π^m) and construct the smallest n-ball with centre $y_2 \in \pi^m$ whose boundary passes through x_1, x_2, x_3; y_2 must lie on the projection plane π^2 (to π^m) of the plane spanned by x_1, x_2, x_3 and so the problem is reduced to that in R^3 for the set $\{x_1, x_2, x_3\}$ relative to π^2. If $K(y_2, x_1)$ covers E, then it is $K_E(\pi^m)$. If it is not the case for any such triple, we then continue the above process. Then, either our problem is solved after certain steps of the process, or the process continues until to a problem in R^{m+1} for an arbitrary set $\{x_1, \cdots, x_{m+1}\} \subset E$ relative to π^m. Evidently, the n-ball with centre on π^m whose boundary passes through $x_1, \cdots, x_{m+1}$ uniquely exists. Among such n-balls there must exists one covering E with smallest radius. This is just the $K_E(\pi^m)$.

Now we consider the case E being an infinite set. Take an ε−net E_ε in E so that $\delta(E, E_\varepsilon) \leq \varepsilon$ for any $\varepsilon > 0$. E_ε is a finite set. Then we may determine $y_\varepsilon(\pi^m)$ and $r_\varepsilon(\pi^m)$ for E_ε as shown above. By Th.2, we know that

$$|r_E(\pi^m) - r_\varepsilon(\pi^m)| \leq \varepsilon,$$

$$|y_E(\pi^m) - y_\varepsilon(\pi^m)| \leq \sqrt{\varepsilon(2r_\varepsilon(\pi^m) + \varepsilon)}.$$

Therefore, if we take a sequence $\varepsilon_k \to 0$, then we may conclude that $y_{\varepsilon_k}(\pi^m) \to y_E(\pi^m)$ and $r_{\varepsilon_k}(\pi^m) \to r_E(\pi^m)$ when $k \to \infty$.

REFERENCES

[1] Lu Jianke, Pseudo-centre and pseudo-radius of a point set, *Math. in Practice and Theory,* 1986, No. 2, 69-73 (in *Chinese*).

[2] Lu Jianke, The general method for finding the pseudo-centre of a point set, *Math. in Practice and Theory,* 1988, No. 1, 75-78 (in *Chinese*).

[3] Federer H., Geometric Measure Theory, *Springer-Verlag, New York,* 1969.

Approximation, Optimization and Computing:
Theory and Applications, A.G. Law and C.L. Wang (eds.)
Elsevier Science Publishers B.V. (North-Holland)

ON L_1-MEAN APPROXIMATION TO CONVEX FUNCTIONS BY POSITIVE LINEAR OPERATORS

Lu Xu-guang

Department of Applied Mathematics, Tsinghua University, Beijing, China

1. INTRODUCTION

Let E be a k-dimensional compact convex set in $\mathbf{R}^k$ and let $C(E)$ denote the space of real-valued continuous functions defined on E with the norm $\| f \|_E = max\{| f(x) | : x \in E\}$. Define the projection set $E(j)$ of E for $k \geq 2$ by $E(j) := \{(x_1,\cdots,x_{k-1}) \in \mathbf{R}^{k-1} : \exists t \in \mathbf{R}^1$ such that $(x_1 \cdots, x_{j-1}, t, x_j, \cdots, x_{k-1}) \in E.\}$. $\Omega(f,E) := max\{f(x) : x \in E\} - min\{f(x) : x \in E\}$; $e_j(x) := x_j, x = (x_1,\cdots,x_k) \in \mathbf{R}^k; e_1(x) = x, x \in \mathbf{R}^1$; $< x,y > := x_1y_1 + \cdots + x_ky_k,\ x,y \in \mathbf{R}^k$; $D_j = \partial/\partial x_j, j = 1,2,\cdots,k$; $T = T_k := \{(x_1,\cdots,x_k) \in \mathbf{R}^k : x_j \geq 0, j = 1,2,\cdots,k;\ x_1 + \cdots + x_k \leq 1\}, Q := [0,1]^k$; $m(S)$ is the Lebesgue measure of S. For a function $f \in C(E)$, (i) f is convex on E iff $f(\frac{1}{2}x+\frac{1}{2}y) \leq \frac{1}{2}f(x)+\frac{1}{2}f(y), x,y \in E$; (ii) $f \in C^1(E)$ iff $\exists$ an open set $G \supset E$ such that $f \in C^1(G)$. The following Bernstein type polynomial positive linear operators are well known:

$$B_nf(x) = \sum_{|\nu|\leq n} P_{n,\nu}(x)f(\frac{\nu}{n}), \quad n \in \mathbf{N}$$

$$M_nf(x) = \frac{(n+k)!}{n!} \sum_{|\nu|\leq n} P_{n,\nu}(x) \int_T P_{n,\nu}(t)f(t)dt, \ n \in \mathbf{N}$$

$$B_nf(x) = \sum_{\nu=0}^{n} P_{n,\nu}(x)f(\frac{\nu_1}{n_1},\dots,\frac{\nu_k}{n_k}), \ n = (n_1,\dots,n_k) \in \mathbf{N}^k$$

$$M_nf(x) = \prod_{j=1}^{k}(n_j+1)\cdot \sum_{\nu=0}^{n} P_{n,\nu}(x) \int_Q P_{n,\nu}(t)f(t)dt,$$

$$n = (n_1,\dots,n_k) \in \mathbf{N}^k$$

$$K_nf(x) = \prod_{j=1}^{k}(n_j+1)\cdot \sum_{\nu=0}^{n} P_{n,\nu}(x) \int_{I_{n,\nu}} f(t)dt,$$

$$n = (n_1,\dots,n_k) \in \mathbf{N}^k$$

where

$$n \in \mathbf{N}, P_{n,\nu}(x) = \frac{n!}{\nu!(n-|\nu|)!}x^\nu(1-|x|)^{n-|\nu|}; \nu = (\nu_1,\cdots,$$

$\nu_k)$ is multi-index; $\nu! = \nu_1!\nu_2!\cdots\nu_k!$; $|\nu| = \nu_1 + \cdots + \nu_k$;

$$x^\nu = x_1^{\nu_1}\, x_2^{\nu_2} \cdots x_k^{\nu_k};\ |x| = x_1 + x_2 + \cdots + x_k;$$

$$n = (n_1,\dots,n_k) \in \mathbf{N}^k, P_{n,\nu}(x) = \frac{n!}{\nu!(n-\nu)!}x^\nu(1-x)^{n-\nu},$$

$$\mathbf{1} = (1,\dots,1); I_{n,\nu} = \prod_{j=1}^{k}[\frac{\nu_j}{n_{j+1}}, \frac{\nu_{j+1}}{n_{j+1}}]; \sum_{\nu=0}^{n} = \sum_{\nu_1=0}^{n_1}\cdots\sum_{\nu_k=0}^{n_k}.$$

Consider the general case, let $L : C(E) \to C(E)$ be a positive linear operator with $L1 = 1$. By Lemma 1 of this paper, we have, for any convex function $f \in C(E), Lf(x) \geq f(\sigma(x))$, where $\sigma(x) = (Le_1(x), Le_2(x), \dots, Le_k(x), x \in E$. Especially, if $\sigma(x) \equiv x, (e.g.\ L = B_n)$, then

$$\int_E |Lf(x) - f(x)|dx = \int_E Lf(x)dx - \int_E f(x)dx$$

This fact makes us to believe that the positive linear operators are more suitable for L_1-mean approximation to convex functions. Another related facts (see Lemma 3 and Theorem 2 in this paper) show that even if $\sigma(x) \not\equiv x$, the L_1-mean approximation by such operators can also achieve their best orders.

2.MAIN RESULTS AND LEMMAS

THEOREM 1. *Let L: $C(E) \to C(E)$ be a positive linear operator with $L1 = 1$. Then $\forall$ convex function $f \in C(E)$,*

$$\int_E |Lf(x) - f(x)|dx \leq \int_E Lf(x)dx - \int_E f(x)dx$$

$$+\begin{cases} 4\Omega(f,E)\cdot \sum_{j=1}^{k} m(E(j))\cdot \| Le_j - e_j \|_E, & k \geq 2 \\ 4\Omega(f,E)\cdot \| Le_1 - e_1 \|_E, & k = 1 \end{cases} \quad (1)$$

THEOREM 2. *Let convex function $f \in C(T)$ or $C(Q)$, then we have respectively*

$$\int_T |B_nf(x) - f(x)|dx \leq \frac{2k^2}{k!}\Omega(f,T)\cdot \frac{1}{n+1}, \quad n \in \mathbf{N}, \quad (2)$$

$$\int_T |M_nf(x) - f(x)|dx \leq \frac{4k^3}{k!}\Omega(f,T)\cdot \frac{1}{n+k+1}, \quad n \in \mathbf{N}, \quad (3)$$

$$\int_Q |B_nf(x) - f(x)|dx \leq 2\Omega(f,Q)\cdot \sum_{j=1}^{k}\frac{1}{n_j+1}, \quad (4)$$

$$n = (n_1,\cdots,n_k) \in \mathbf{N}^k,$$

$$\int_Q |M_nf(x) - f(x)|dx \leq 4\Omega(f,Q)\cdot \sum_{j=1}^{k}\frac{1}{n_j+2}, \quad (5)$$

$$n = (n_1,\cdots,n_k) \in \mathbf{N}^k,$$

$$\int_Q |K_nf(x) - f(x)|dx \leq 2\Omega(f,Q)\cdot \sum_{j=1}^{k}\frac{1}{n_j+1}, \quad (6)$$

$$n = (n_1,\cdots,n_k) \in \mathbf{N}^k.$$

THEOREM 3. *Let $L : C[0,1] \to C[0,1]$ be a positive linear operator with $L1 = 1$, $Le_1 = e_1$. Then $\forall$ canvex function $f \in C[0,1]$,*

$$\int_0^1 |Lf(x) - f(x)|dx \le 24\Omega(f,[0,1]) \cdot \delta \log(1+\delta^{-1}), \quad (7)$$

where

$$\delta = \sup_{0<x<1} \frac{Le_1^2(x) - x^2}{x(1-x)}.$$

LEMMA 1.*If Λ is a positive linear functional on $C(E)$ with $\Lambda 1 = 1$, and $f \in C(E)$ is convex. Then $\sigma := (\Lambda e_1, \cdots, \Lambda e_k) \in E$ and $\Lambda f \ge f$.*

Proof. Put $F = \{(x,u) \in \mathbf{R}^{k+1} : x \in E, u \ge f(x)\}$, then F is a closed convex set.Suppose that $(\sigma, \Lambda f) \notin F$, then there exists a point $(x_0, u_0) \in F$ such that $\forall (x,u) \in F$, $< (x,u) - (x_0,u_0), (\sigma, \Lambda f) - (x_0, u_0) > \le 0$. Since $(x, f(x)) \in F$, we obtain, for $g(x) := < (x, f(x) - (x_0,u_0), (\sigma, \Lambda f) - (x_0,u_0) >$, $0 \ge \Lambda g = < (\sigma, \Lambda f) - (x_0,u_0), (\sigma,\Lambda f) - (x_0,u_0) > > 0$. This contradiction shows that $(\sigma, \Lambda f) \in F$ and $\sigma \in E$.

LEMMA 2[1].*If $f \in C(E)$ is convex, then there exists a sequence $\{P_n\}$ of polynomials such that each polynomial P_n is convex on E and $\| P_n - f \|_E \to 0$ $(n \to \infty)$.*

LEMMA 3.*Suppose that $f \in C^1(E)$ and f is convex on E, then*

$$\int_E |D_j f(x)|dx$$

$$\le \begin{cases} 2m(E(j)) \cdot \Omega(f,E), & j = 1,2,\cdots,k; k \ge 2 \\ 2\Omega(f,E) & ,k=1 \end{cases} \quad (8)$$

Proof. For the case $E = [a,b]$(*i.e.* $k = 1$), we may assume that $f'(c) = 0$ for some $c \in (a,b)$. Since f' is an increasing function,we get

$$\int_a^b |f'(x)|dx = \int_a^c -f'(x)dx + \int_c^b f'(x)dx$$

$$\le 2\Omega(f,[a,b]). \quad (9)$$

For the case $k \ge 2$, without loss of generality,let $j = 1$. Define,for $y \in E(1)$, $\varphi(y) = sup\{t \in \mathbf{R}^1 : (t,y) \in E\}$, $\psi(y) = inf\{t \in \mathbf{R}^1 : (t,y) \in E\}$. Since E is a compact convex set,we have $\forall y \in E(1), [\psi(y), \varphi(y)] \times E(1) \subset E$. Thus by Fubini's Theorem and the inequality (9),

$$\int_E |D_1 f(x)|dx = \int_{E(1)} (\int_{\psi(y)}^{\varphi(y)} |D_1 f(t,y)|dt)dy$$

$$\le \int_{E(1)} \{\sup_{y\in E(1)} 2\Omega(f(\cdot,y),[\psi(y),\varphi(y)])\}dy$$

$$\le 2\Omega(f,E) \cdot m(E(1)).$$

3. PROOFS OF THEOREMS

Proof of Theorem 1. By Lemma 2 we can suppose that $f \in C^1(E)$ since $\| LP_n - Lf \|_E \le \| P_n - f \|_E \to 0$ and $\Omega(P_n, E) \to \Omega(f,E)$ as $n \to \infty$. Lemma 1 gives $Lf(x) \ge f(\sigma(x)), x \in E$, and so

$$\int_E |Lf(x) - f(x)|dx$$

$$\le \int_E |Lf(x) - f(\sigma(x))|dx + \int_E |f(\sigma(x)) - f(x)|dx$$

$$= \int_E Lf(x)dx - \int_E f(x)dx + 2\int_{E(f \ge f\circ\sigma)} [f(x) - f(\sigma(x))]dx$$

Using the property of convex functions:

$$f(y) \ge f(x) + < \nabla f(x), y - x >$$

we have

$$f(x) - f(\sigma(x)) \le < \nabla f(x), x - \sigma(x) >$$

$$\le \sum_{j=1}^k |D_j f(x)| \cdot |x_j - Le_j(x)|, x \in E.$$

$$\int_{E(f\ge f\circ\sigma)} [f(x) - f(\sigma(x))]dx$$

$$\le \sum_{j=1}^k (\int_E |D_j f(x)|dx) \cdot \| Le_j - e_j \|_E .$$

Therefore by Lemma 3,the estimate (1) holds.

Proof of Theorem 2. Since we have the following equalities which are easy to verify:

$$M_n 1 = 1, \quad M_n e_j - e_j = \frac{1-(k+1)e_j}{n+k+1},$$

$$\int_T M_n f(x)dx = \int_T f(x)dx, \ f \in C(T), \ n \in \mathbf{N};$$

$$M_n = 1, \quad M_n e_j - e_j = \frac{1-2e_j}{n_j+2},$$

$$\int_Q M_n f(x)dx = \int_Q f(x)dx, \ f \in C(Q), \ n \in \mathbf{N}^k;$$

$$K_n 1 = 1, \quad K_n e_j = e_j = \frac{1/2 - e_j}{n_j+1},$$

$$\int_Q K_n f(x)dx = \int_Q f(x)dx, \ f \in C(Q), \ n \in \mathbf{N}^k$$

$$(j = 1,2,\cdots,k; \quad k \ge 1)$$

$$m(T(j)) = \frac{1}{(k-1)!}, \ m(Q(j)) = 1,$$

$$(j = 1,2,\cdots,k; k \ge 2)$$

by Theorem 1,the estimate (3),(5) and (6) hold. For the opertors $B_n (n \in \mathbf{N}$ or $n \in \mathbf{N}^k)$, we know that $B_n 1 = 1$ $B_n e_j = e_j$, $j = 1,2,\cdots,k$. Thus Lemma 1 implies that for convex function $f \in C(T)$ or $C(Q), B_n f \ge$ f on T or on Q respectively.We prove the estimates (2),(4) by induction on the dimension k. Let k=1 and let convex function $f \in C^1[0,1]$. Then

$$\int_0^1 |B_nf(x)-f(x)|dx = \int_0^1 B_nf(x)dx - \int_0^1 f(x)$$

$$= \sum_{s=0}^{n} \frac{1}{n+1} f(\frac{s}{n}) - \sum_{s=0}^{n} \int_{\frac{s}{n+1}}^{\frac{s+1}{n+1}} f(x)dx$$

$$= \sum_{s=0}^{n} \int_{\frac{s}{n+1}}^{\frac{s+1}{n+1}} (\int_x^{\frac{s}{n}} f'(t)dt)dx \le \sum_{s=0}^{n} \int_{\frac{s}{n+1}}^{\frac{s+1}{n+1}} (\int_{\frac{s}{n+1}}^{\frac{s+1}{n+1}} |f'(t)|dt)dx$$

$$= \frac{1}{n+1}\int_0^1 |f'(t)|dt \le 2\Omega(f,[0,1])\cdot\frac{1}{n+1}$$

(by Lemma 3).

Lemma 2 implies that the same estimate also holds for any convex function $f \in C[0,1]$, and therefore (2),(4) hold for $k=1$. Assume that (2),(4) hold for $k \ge 1$. Write $T = T_{k+1}$ and let convex function $f \in C(T)$. Define $P:[0,1]\times T_k \to T: P(t,u) = (tu_1,\cdots,tu_k,\ t(1-|u|))$. Then $T = P([0,1]\times T_k)$, $|detJP(t,u)| = t^k$. Write

$$f_u(t) := f(P(t,u)), f_{n,s}(u) := f(P(\frac{s}{n},u)) = f_u(\frac{s}{n})$$

$s = 1,2,\cdots,n$. Then we have the following relation

$$B_nf(x)-f(x) = B_nf_u(t) - f_u(t)$$

$$+\sum_{s=1}^{n} P_{n,s}(t)[B_sf_{n,s}(u) - f_{u,s}(u)], \quad x = P(t,u).$$

This gives

$$\int_T |B_nf(x)-f(x)|dx$$

$$= \int_{[0,1]\times T_k} |B_nf(P(t,u)) - f(P(t,u))|t^k dtdu$$

$$\le \int_{[0,1]\times T_k} |B_nf_u(t) - f_u(t)|t^k dtdu$$

$$+\sum_{s=1}^{n}\int_{[0,1]\times T_k} P_{n,s}(t)|B_sf_{n,s}(u) - f_{u,s}(u)|t^k dtdu$$

$$= I_o + \sum_{s=1}^{n} I_s. \tag{10}$$

Obviously, $f_u(t)$ and $f_{n,s}(u)$ are continuous convex functions of $t \in [0,1]$ and of $u \in T_k$ respectively.By induction hypothesis we obtain

$$I_o \le \int_{T_k}(\int_0^1 |B_nf_u(t) - f(u)|dt)du$$

$$\le \int_{T_k}[\sup_{u\in T_k} 2\Omega(f_u,[0,1])\cdot\frac{1}{n+1}]du$$

$$\le \frac{1}{k!}2\Omega(f,T)\cdot\frac{1}{n+1}. \tag{11}$$

$$I_s \le \int_0^1 t^k P_{n,s}(t)(\int_{T_k} |B_sf_{n,s}(u) - f_{n,s}(u)|du)dt$$

$$\le 2\frac{k^2}{k!}\Omega(f_{n,s},T_k)\cdot\frac{1}{s+1}\cdot\int_0^1 t^k P_{n,s}(t)dt$$

$$\le 2\frac{k^2}{k!}\Omega(f,T)\cdot\frac{1}{n+1}\cdot\int_0^1 t^{k-1}P_{n+1,s+1}(t)dt,$$

$$\sum_{s=1}^{n} I_s \le 2\frac{k_2}{k!}\Omega(f,T)\cdot\frac{1}{n+1}\cdot\int_0^1 t^{k-1}\sum_{s=1}^{n} P_{n+1,s+1}(t)dt$$

$$\le 2\frac{k^2}{k!}\Omega(f,T)\cdot\frac{1}{n+1}\cdot\int_0^1 t^{k-1}dt$$

$$= 2\frac{k^2}{k!}\Omega(f,T)\cdot\frac{1}{n+1}\cdot\frac{1}{k}. \tag{12}$$

Therefore by (10)-(12), (2) also holds for the case of $k+1$. Now,let convex function $f \in C(Q), Q = [0,1]^{k+1}$. To prove (4),let $n = (n_1,\cdots,n_k,n_{k+1})$, $m = (n_1,n_2,\cdots,n_k,)$ and define $f_y(t) := f(y,t), y \in \tilde{Q} := [0,1]^k, t \in [0,1]$; $f_{r,s}(y) := f(y,\frac{s}{r}) := f_y(\frac{s}{r})$, where $r := n_{k+1}, 0 \le s \le r$. Then for $x = (y,t) \in Q$, we have

$$B_nf(x) - f(x) = \sum_{s=0}^{r} P_{r,s}(t)[B_mf_{r,s}(y) - f_{r,s}(y)]$$

$$+B_rf_y(t) - f_y(t),$$

$$\int_Q |B_nf(x) - f(x)|dx$$

$$\le \sum_{s=0}^{r}\int_0^1 P_{r,s}(t)(\int_{\tilde{Q}} |B_mf_{r,s}(y) - f_{r,s}(y)|dy)dt$$

$$+\int_{\tilde{Q}}(\int_0^1 |B_rf_y(t) - f_y(t)|dt)dy := J_1 + J_2 \tag{13}$$

By induction hypothesis,we obtain

$$\int_{\tilde{Q}} |B_mf_{r,s}(y) - f_{r,s}(y)|dy$$

$$\le 2\Omega(f_{r,s},\tilde{Q})\cdot\sum_{j=1}^{k}\frac{1}{n_j+1} \le 2\Omega(f,Q)\cdot\sum_{j=1}^{k}\frac{1}{n_j+1},$$

$$\int_0^1 |B_rf_y(t) - f_y(t)|dt \le 2\Omega(f_y,[0,1])\cdot\frac{1}{r+1}$$

$$\le 2\Omega(f,Q)\cdot\frac{1}{r+1} = 2\Omega(f,Q)\cdot\frac{1}{n_{k+1}+1},$$

and so

$$J_1 \le 2\Omega(f,Q)\cdot\sum_{j=1}^{k}\frac{1}{n_j+1}\cdot\int_0^1\sum_{s=0}^{r} P_{r,s}(t)dt$$

$$= 2\Omega(f,Q)\cdot\sum_{j=1}^{k}\frac{1}{n_j+1}, \quad J_2 \le 2\Omega(f,Q)\cdot\frac{1}{n_{k+1}+1}.$$

Hence by (13), the estimate (4) holds again for the case of k+1.

Proof of Theroem 3. It is easy to see that $0 \le \delta \le 1$ and if $\delta = 0$ then $Lf = f$ for all $f \in C[0,1]$. In this case the estimate (7) clearly holds if we define $\delta\log(1+\delta^{-1}) = 0$ for $\delta = 0$. Suppose that $\delta > 0$. Choose $n \in \mathbf{N}$ such that $n \le \delta^{-1} < n+1$. By Lemma 1 and (2) we have $Lf \ge f, B_nf \ge f$

on [0,1] and

$$\int_0^1 |Lf(x) - fx|dx = \int_0^1 Lf(x)dx - \int_0^1 f(x)dx$$

$$\leq \int_0^1 [LB_nf(x) - B_nf(x)]dx + \int_0^1 [B_nf(x) - f(x)]dx$$

$$\leq \int_0^1 L[B_nf - B_nf(x) - B_n'f(x)(e_1 - x)](x)dx$$

$$+2\Omega(f,[0,1]) \cdot \frac{1}{n+1}. \tag{14}$$

Using the representation of $h \in C^2(\mathbf{R}^1)$

$$h(t) - h(x) - h'(x)(t-x) = \int_x^t h''(\tau)(t-\tau)d\tau,$$

and the inequality

$$\frac{|t-\tau|}{\tau(1-\tau)} \leq \frac{|t-x|}{x(1-x)},$$

where $x \in (0,1), t \in [0,1], t \neq x, \tau \in (t,x) or\ \tau \in (x,t)$,we have

$$|h(t) - h(x) - h'(x)(t-x)|$$

$$\leq \frac{(t-x)^2}{x(1-x)} \cdot \frac{1}{|t-x|} \left| \int_x^t |h''(\tau)|\tau(1-\tau)d\tau \right|.$$

This implies

$$|h(t) - h(x) - h'(x)(t-x)| \leq \frac{(t-x)^2}{x(1-x)} M(\Phi(h))(x), \tag{15}$$

$t \in [0,1], x \in (0,1)$; where

$$\Phi(h)(x) := x(1-x)h''(x)\chi_{[0,1]}(x),$$

$M(g)$ is the maximal function of $g \in L_1(\mathbf{R}^1) : M(g)(x)$

$$= sup\{\frac{1}{|I|}\int_I |g(t)|dt : I\ is\ an\ open\ interval,\ x \in I\}.$$

From [2], the weak-type estimate

$$m(\{x \in \mathbf{R}^1 : M(g)(x) > \alpha\}) \leq \frac{6}{\alpha}\int_{|g|>\frac{\alpha}{2}} |g(t)|dt, \quad (\alpha > 0)$$

gives the following estimate

$$\int_0^1 M(g)(x)dx \leq 1 + 6\int_0^1 |g(x)|\log(1+6|g(x))|dx$$

provided supp$g \subset [0,1]$. Thus , by (14),(15) and $B_n''f \geq 0$ on [0,1] (since the Berntein operator B_n is of preserving convexity),we obtain

$$\int_0^1 |Lf(x) - fx|dx$$

$$\leq \int_0^1 \frac{L(e_1-x)^2(x)}{x(1-x)} \cdot M(\Phi(B_nf))(x)dx + 2\Omega(f,[0,1]) \cdot \delta$$

$$\leq [1 + 6\int_0^1 x(1-x)B_n''f(x)\log(1+6x(1-x)B_n''f(x))dx$$

$$+2\Omega(f,[0,1])] \cdot \delta := H_n \cdot \delta$$

Now we suppose that $\Omega(f,[0,1]) \leq 1/6$. Then

$$x(1-x)B_n''f(x) \leq \frac{1}{4}n^2 \cdot 2\Omega(f,[0,1]) \leq \frac{1}{12} \cdot n^2,$$

$$\int_0^1 x(1-x)B_n''f(x)dx = -\int_0^1 B_n''f(x)(1-2x)dx$$

$$\leq \int_0^1 |B_n''f(x)|dx \leq 2\Omega(B_nf,[0,1]) \leq 2\Omega(f,[0,1]) \leq \frac{1}{3},$$

$$H_n \leq 1 + 6\log(1+\frac{1}{2}n^2) \cdot \frac{1}{3} + \frac{1}{3}$$

$$\leq \frac{4}{3} - 2\log 2 + 4\log(1+n) \leq 4\log(1+\delta^{-1})$$

and therefore

$$\int_0^1 |Lf(x) - fx|dx \leq 4\delta\log(1+\delta^{-1}).$$

Replacing f by $(6\Omega(f,[0,1]))^{-1}f$, we then obtain the final estimate (7).

4. REMARK

As a consequence,Theorem 2 gives the Chebyshev type inequalities. For instance, for Bernstein operator B_n and convex function $f \in C(T)$, we obtain the weak estimate

$$m(\{x \in T : |B_nf(x) - f(x)| > \varepsilon^{-1} \cdot 2k^2\Omega(f,T)\frac{1}{n+1}\})$$

$$\leq \varepsilon \cdot m(T), \quad \varepsilon > 0, n \in \mathbf{N}.$$

Combining this with the uniform estimate [3]

$$\| B_nf - f \|_T \leq C_k \cdot \max_{o \leq \alpha \leq 1} (\frac{1}{n})^{\frac{1-\alpha}{2}} \omega(f, n^{-\alpha})$$

We can conclude that the Bernstein type polynomial positive linear operators not only have some good properties in preserving convexity as we have already known [4],but,in convergence,they are also good operators for approximating continuous convex functions.
In Theorem 3,whether the factor $\log(1+\delta^{-1})$ can be taken off or not remains open to investigate.

REFERENCES

[1] Lu Xu-guang,On Multivariate Polynomials Convex Approximation,in print.
[2] Stein,E.M., Singular Integrals and Differentiability Properties of Functions (Princeton Univ. Press,Princeton.N.J. 1970)
[3] Lu Xu-guang, Mathematica Numerica Sinica 10 (1988) 398-407.
[4] Chang Geng-zhe,and P.J. Davis, Jour. Approx.Theory,40(1984) 11-28.

Approximation, Optimization and Computing:
Theory and Applications, A.G. Law and C.L. Wang (eds.)
Elsevier Science Publishers B.V. (North-Holland)

ON THE WIDTH OF CLASS OF PERIODIC FUNCTIONS OF TWO VARIABLES

Jun-Bo Luo

Department of Mathematics, Liaoning University, China.

1. Introduction

Let Lp ($1\leq p\leq\infty$) be the Banach space of functions of two variables with period 2π in each Variable and the norm is defined by

$\|f\|_p=$

$$\begin{cases}\left(\frac{1}{(2\pi)^2}\int_{-\pi}^{\pi}\int_{-\pi}^{\pi}|f(x,y)|^p dX\,dy\right)^{\frac{1}{p}} & \text{if } 1\leq p<\infty,\\ \operatorname{esssup}|f(x,y)|, & \text{if } p=\infty\end{cases}$$

Suppose $\sum_{k,j} C_{kj}e^{i(kx+jy)}$ is the Fourier series of $f(x,y)\in L_1$ and $r\geq 0$ a real number.

If $\sum_{k,j}(-1)^r(k^2+j^2)^r C_{kj}e^{i(kx+jy)}$ is the Fourier series of some function $\phi(x,y)\in L_1$, we denote

$\phi(x,y)$ by $\triangle^r f(x,y)$.

When r is positive integer, the symbol "$\triangle$" denotes the operator $\frac{\partial^2}{\partial x^2}+\frac{\partial^2}{\partial y^2}$.

We define the class of functions

$$\triangle_p^r=\left\{f(x,y)\in L_1;\quad \|\triangle^r f\|_p\leq 1\right\}$$

The N-Width $d_N(\triangle_p^r;L_q)$ in the sense of Kolmogoroff of the class $\triangle_p^r$ in L_q metric is defined by

$$d_N(\triangle_p^r;\ L_q)=\inf_{M_N\subset L_q}\ \sup_{f\in\triangle_p^r}\ \inf_{u\in M_N}\|f-u\|_q,$$

where the left infimum is taken over all N-dimensional subspaces M_N of L_q.

The purpose of the present paper is to discuss the degree of $d_N(\triangle_p^r;L_q)$ for $1<p$, $q<\infty$, the main result is the following theorem.

Theorem.

$d_N(\triangle_p^r;L_q)\sim N^{-r}$, if $1<q\leq p<\infty$ and $r>0$ (1)

$d_N(\triangle_p^r;L_q)\sim N^{-r}$, if $2\leq p<q<\infty$ and $r>\frac{1}{2}$ (2)

$d_N(\triangle_p^r;L_q)\sim N^{-r+\frac{1}{p}-\frac{1}{q}}$, if $1<p<q\leq 2$

and $r>\frac{1}{p}-\frac{1}{q}$ (3)

$d_N(\triangle_p^r;L_q)\sim N^{-r+\frac{1}{p}-\frac{1}{2}}$, if $1<p\leq 2\leq q$

and $r>\frac{1}{p}$ (4)

Here and throughout this paper by symbols "$\sim$" "$\ll$" "and" "$\gg$" we denote equality and inequalities of degree, respectively.

The result $d_N(\triangle_p^r;\ L_p)\sim N^{-r}$ ($1\leq P\leq\infty$) is obtained by Mitjagin B.C. [1] by means of the Fourier multipler theory.

2. The approximation of $\triangle_p^r$ by trigonometric polynomials.

By T_{nm} we denote the space of trigonometric polynomials $t_{nm}(x,y)$ of degree$\leq(n,m)$ and by T_R denote the space of trigonometric polynomials $t_{nm}(x,y)$, in which n, m satisfy $n^2+m^2\leq R^2$.

For $f\in\triangle_p^r$ and class $\triangle_p^r$ we define

$$E_{nm}(f)_p=\inf\{\|f-t_{nm}\|_p;\ t_{nm}\in T_{nm}\}$$

$$E_R(f)_p=\inf\{\|f-t_{nm}\|_p;\ t_{nm}\in T_R\}$$

and $E_R(\triangle_p^r)_q=\sup_{f\in\triangle_p^r}E_R(f)_q$

For each $f(x,y)\in\triangle_p^r$ we have the following integral representation:

$$f(x,y)=\frac{C_{00}(f)}{4}+$$

$$\frac{1}{\pi^2}\int_{-\pi}^{\pi}\int_{-\pi}^{\pi}\phi(t,z)\,D_r(x-t,y-z)\,dtdz$$

$$=\frac{C_{00}}{4}+\phi*D_r \tag{5}$$

where $\phi=\triangle^r f$.

$$D_r(x,y)=\sum_{k=0}^{\infty}\sum_{j=0}^{\infty}{}' \frac{(-1)^r \cos kx\cdot\cos jy}{(k^2+j^2)^r} \quad (6)$$

The prime " ' " denotes that the term is deleted for $k=j=0$ and multiplied by factor $\frac{1}{2}$ for $k=0$ or $j=0$.

To obtain the degree of $E_R(\triangle_p^r)_p$, first we discuss the approximation of the kernel $D_r(x,y)$ in (5) by trigonometric polynomials.

Proposition 1. If $1<P\leqslant\infty$, $r-1+\frac{1}{p}>0$, then

$$E_{nm}(D_r)_p\leqslant \| D_r-S_{nm}(D_r)\|_p << (nm)^{1-\frac{1}{p}}(n^2+m^2)^{-r} + n^{-2(r-1+\frac{1}{p})}+m^{-2(r-1+\frac{1}{p})} \quad (7)$$

where $S_{nm}(D_r)$ is the partial sums of the Fourier series of D_r.

In fact, write

$$D_{nm}^{(r)}(x,y)=D_r(x,y)-S_{nm}(D_r;x,y)=\left\{\sum_{k=0}^{n}\sum_{j=m+1}^{\infty}+\sum_{k=n+1}^{\infty}\sum_{j=0}^{m}+\sum_{k=n+1}^{\infty}\sum_{j=m+1}^{\infty}\right\}\frac{(-1)^r\cos kx\cos jy}{(k^2+j^2)^r}=\Sigma_1+\Sigma_2+\Sigma_3, \text{ say.}$$

On each Σ_i, using Abelian transformation (partial summation) twice and on account of

$\| D_k(t)\|_p=\left\|\frac{\sin(k+\frac{1}{2})t}{2\sin\frac{t}{2}}\right\|_p\leqslant k^{1-\frac{1}{p}}$, we get

$$\|\Sigma_i\|_p<<(nm)^{1-\frac{1}{p}}(n^2+m^2)^{-r}+n^{-2(r-1+\frac{1}{p})}+m^{-2(r-1+\frac{1}{p})},$$

$i=1,2,3.$

Proposition 2. If $1\leqslant q\leqslant\infty$, $r-1+\frac{1}{q}>0$, then

$$E_R(D_r)_q<<R^{-2(r-1+\frac{1}{q})} \quad (8)$$

If $1<q\leqslant\infty$, on acount of $T_n\subset T_{nn}\subset T_{\sqrt{2}n}$, (7) implies (8). So we need only to prove that (8) is corret for $q=1$.

Let us define the fuction $\psi(u)$ such that

$$\psi(u)=\begin{cases}1, & \text{for } 0\leqslant u\leqslant\frac{1}{2}\\ 0, & \text{for } 1\leqslant u<\infty\end{cases}$$

and $\psi(u)$ is infinite differentiable.

Putting $t_R^*(x,y)=\sum_{k=0}^{\infty}\sum_{j=0}^{\infty}(-1)^r\psi\left(\frac{(k^2+j^2)^r}{R^{2r}}\right)$

$$(k^2+j^2)^{-r}\cos kx\cos jy \quad (9)$$

we have

$$E_R(D_r)_1\leqslant \| D_r(x,y)-t_R^*(x,y)\|_1 = \left\|\sum_{k=0}^{\infty}\sum_{j=0}^{\infty}a_{kj}\cos kx\cos jy\right\|_1 \quad (10)$$

where $a_{kj}=\left\{1-\psi\left(\frac{(k^2+j^2)^r}{R^{2r}}\right)\right\}(k^2+j^2)^{-r}$

Carriing out Abelian transformation four times on the right hand side of (10) and on account of

$\| F_k(x)\|_1=\|\sum_{j=0}^{k}D_j(x)\|_1<<k+1$, we have

$$E_R(D_r)_1<<\sum_{k=0}^{\infty}\sum_{j=0}^{\infty}(k+1)(j+1)|\triangle_{kj}^{22}a_{kj}| \quad (11)$$

where $\triangle_{kj}^{22}a_{kj}=\triangle_k^2(\triangle_j^2a_{kj})$, $\triangle_j^2a_{kj}=a_{kj}-2a_{kj+1}+a_{kj+2}$

Through the estimate of partial derivatives of the function

$$\left\{1-\psi\left(\frac{(u^2+v^2)^r}{R^{2r}}\right)\right\}(u^2+v^2)^{-r}$$

we get the following estimate:

$$|\triangle_{kj}^{22}a_{kj}|\begin{cases}<<(k^2+j^2)^{-r-2} & \text{if } (k^2+j^2)^rR^{-2r}\geqslant\frac{1}{2},\\ =0, & \text{if } (k^2+j^2)^rR^{-2r}<\frac{1}{2}.\end{cases}$$

Subtituting this estimate into (11), we obtain

$$E_R(D_r)_1<<R^{-2r}$$

Proposition 3. $E_R(\triangle_p^r)_p\sim R^{-2r}$ for $1\leqslant p\leqslant\infty$.

Proof. For any $f\in\triangle_p^r$, put

$t_R(f;x,y)=\frac{C_{00}(f)}{4}+\phi*t_R^*$, where $\phi=\triangle^r f$ and t_R^* is defined in (9). Using the generalized Minkowski inequality (when $p=\infty$ directely) we get $\| f-t_R(f)\|_p<<\| D_r-t_R^*\|_1\cdot\|\phi\|_p<<R^{-2r}$

The lower bound of $E_R(\triangle_p^r)_p$ follows from the result of the fuctions of one variable.

3. Some lemmas

In order to prove the theorem, We need some lemmas.

Lemma 1. For each trigonometric polynomial $t_{nm}\in T_{nm}$, holds the following inequality

$$\|\triangle^r t_{nm}\|_p\leqslant(n^2+m^2)^r\| t_{nm}\|_p \quad (1\leqslant p\leqslant\infty)$$

This lemma is proved in [1] for $r=1$, but that proof is still correct for $r\geqslant 0$.

let

$\rho_{\mu\nu}=\{(k,j);\ 2^{\mu-1}\leqslant|k|<2^{\mu},\ 2^{\nu-1}\leqslant|j|<2^{\nu}\}$

and $\delta_{\mu\nu}(f;x,y)=\sum_{(k,j)\in\rho_{\mu\nu}}c_{kj}(f)e^{i(kx+jy)}$,

Lemma 2. (see [4]) For $1<p<\infty$,

$$\|f\|_p \sim \|(\sum_{\mu,\nu} |\delta_{\mu\nu}(f;x,y)|^2)^{\frac{1}{2}}\|_p$$

Lemma 3 (see [4] P139). For $1<q\leq 2$,

$$\|f\|_q \gg \{\sum_{\mu,\nu} \|\delta_{\mu\nu}(f)\|_2^q\, 2^{(\mu+\nu)(\frac{q}{2}-1)}\}^{\frac{1}{q}}$$

By lemma 3, we can obtain

Lemma 4. For $1<P\leq 2$ and any r,

$$\|\triangle^r f\|_p \gg \|\triangle^{r-\frac{1}{p}+\frac{1}{2}} f\|_2$$

Let $F_{\mu\nu}=\text{span}\{e^{i(kx+jy)};\ (k,j)\in\rho_{\mu\nu}\}$. For every $f\in F_{\mu\nu}$, write

$$f(x,y)=\{\sum_{k,j>0}+\sum_{k>0,j<0}+\sum_{k<0,j>0}+\sum_{k,j<0}\}C_{kj}(f)e^{i(kx+jy)}$$

$=\sum_{i=1}^{4} f_i(x,y)$, say. let $\tau(f)=$

$$\{\{f_1(\tau_{nm})\},\{f_2(\tau_{nm})\},\{f_3(\tau_{nm})\},\{f_4(\tau_{nm})\}\}$$

$$=\{f^{(i)}\}_{i=1}^{2^{\mu+\nu}},$$

where $\tau_{nm}=(2\pi n\cdot 2^{-\mu+1},\ 2\pi m\cdot 2^{-\nu+1})$.

$$n=1,2,\cdots,2^{\mu-1},\quad m=1,2,\cdots,2^{\nu-1}.$$

Lemma 5 (see [3]). The mapping $f\mapsto\tau(f)$ is a one-to-one mapping between spaces $F_{\mu\nu}$ and $R^{2^{\mu+\nu}}$ and

$$\|f\|_p\sim 2^{-\frac{\mu+\nu}{p}}(\sum_{i=1}^{2^{\mu+\nu}}|f^{(i)}|^p)^{\frac{1}{p}}=2^{-\frac{\mu+\nu}{p}}\|\tau(f)\|_{l_p}$$

for $1<p<\infty$.

Lemma 6 (see [3]).

$$d_n(B_p^m;l_q^m)\sim\begin{cases}(m-n)^{\frac{1}{q}-\frac{1}{p}}, & \text{if } 1\leq q\leq p\leq\infty\\ \min\{1;m^{\frac{1}{q}}n^{-\frac{1}{2}}\}^{\beta}, & \text{if } 2\leq p<q<\infty\end{cases}$$

where $n<m$, $\beta=\frac{\frac{1}{p}-\frac{1}{q}}{\frac{1}{2}-\frac{1}{q}}$ and

$$B_p^m=\{x=(x_1,\cdots,x_m)\in R^m;\ \|X\|_{l_p}\leq 1\}$$

4. The proof of the theorem.

Now we come to the proof of the theorem. First we prove the estimate (1)

Since $\|\phi\|_q\leq\|\phi\|_p$ for $q\leq p$

$\triangle_p^r\subset\triangle_q^r$ and $d_N(\triangle_p^r;L_q)\leq d_N(\triangle_q^r;L_q)$

$\ll E_N(\triangle_q^r)_q\ll N^{-r}$

by proposition 3.

For given N, choose positive integer s such that $2^{2s}>2N$. By Lemma2 We have

$$d_N(\triangle_p^r;L_q)\gg d_N(\triangle_p^r\cap F_{ss};L_q)\geq d_N(\triangle_p^r\cap F_{ss};L_q\cap F_{ss})$$

According to lemma 1 and 5, for every $f\in F_{ss}$,

$$\|\triangle^r f\|_p\ll 2^{2rs}\|f\|_p\ll 2^{2s(r-\frac{1}{p})}\|\tau(f)\|_{l_p},$$

so if $f\in F_{ss}$ and $\|\tau(f)\|_{l_p}\ll 2^{2s(\frac{1}{p}-r)}$,

then $f\in\triangle_p^r\cap F_{ss}$,

On the other hand, for any $g\in F_{ss}$,

$$\|g\|_q\gg 2^{-\frac{2s}{q}}\|\tau(g)\|_{l_q}$$

By lemma 6, we get

$$d_N(\triangle_p^r;L_q)\geq d_N(\triangle_p^r\cap F_{ss};L_q\cap F_{ss})\gg$$

$$2^{2s(\frac{1}{p}-\frac{1}{q}-r)}d_N(B_p^{4^s};l_q^{4^s})$$

$$\gg 2^{2rs(\frac{1}{p}-\frac{1}{q}-r)}(2s-N)^{(\frac{1}{q}-\frac{1}{p})}\gg N^{-r}$$

The relation (1) is then proved.

Secondly, We prove (2). Suppose $2\leq p<q<\infty$, $r>\frac{1}{2}$. The lower bound estimate of (2) follows from the relation (1):

$$d_N(\triangle_q^r;L_q)\geq d_N(\triangle_q^r;L_p)\gg N^{-r}.$$

On account of $d_N(\triangle_p^r;L_q)\leq d_N(\triangle_2^r;L_q)$, to complete the proof of (2), we need only to prove that $d_N(\triangle_2^r;L_q)\ll N^{-r}$

For given N, choose positive integer s such that $4^{4s}\sim N$

Let $S_{lm}=\{(\mu,\nu);\ 4^l\leq 4^{\mu+\nu}<4^{4(l+1)}$ and $4(l+m)\leq\mu+\nu<4(l+m+1)\}$

for $m=0,1,\cdots,l$ and $l=0,1,2,\cdots$.

Obviously, $|s_{lm}|\overset{\triangle}{=}\text{card}\ s_{lm}=32$ for $1\leq m\leq l-1$, $|S_{l0}|=20$, $|S_{ll}|=28$ and

$$\|S_{lm}\|\overset{\triangle}{=}\text{card}\ \{(k,j)\in\rho_{\mu\nu},(\mu,\nu)\in S_{lm}\}\sim 2^{4(l+m)}$$

Let us put

$$N_{lm}=\begin{cases}\|S_{lm}\|, & \text{for } l\leq s\\ \min\{|S_{lm}|2^{4(2s+\lambda s-2\lambda l+\lambda m)},\ \|S_{lm}\|\}, & \text{for } l>s\end{cases}$$

Where $\lambda>0$ is so small that $4r-2-\lambda>0$. We have

$$\sum_{l,m}N_{lm}\sim\sum_{l\leq s}\sum_{m=0}^{l}2^{4(l+m)}+$$

$$\sum_{l>s}\sum_{m=0}^{l}2^{4(2s+\lambda s-2\lambda l+\lambda m)}\sim 4^{4s}\sim N$$

let $E_{lm}=\text{span}\{e^{i(kx+jy)};\ (k,j)\in\rho_{\mu\nu}$ and $(\mu,\nu)\in S_{lm}\}$

According to lemma 5 for every $f\in E_{lm}\cap\triangle_2^r$ we have

$1\geq\|\triangle^r f\|_2=$

$$\{\sum_{(\mu,\nu)\in S_{lm}}\sum_{(k,j)\in\rho_{\mu\nu}}(k^2+j^2)^{2r}C_{kj}^2(f)\}^{\frac{1}{2}}$$

$$\gg 4^{4lr}(\sum_{(\mu,\nu)\in S_{lm}}\|\delta_{\mu\nu}(f)\|_2^2)^{\frac{1}{2}}$$

$$\sim 4^{4lr}\Big(\sum_{(\mu,v)\in S_{lm}} 2^{-(\mu+v)}\|\tau(\delta_{\mu v}(f))\|^2_{l_2}\Big)^{\frac{1}{2}}$$

$$\gg 4^{4lr}2^{-2(l+m)}\|\tau(f)\|_{l_2}$$

On the other hand, by lemma 2, 5 and the Hölder inequality, we get

$$\|f\|_q \sim \|(\sum_{(\mu,v)\in S_{lm}}|\delta_{\mu v}(f;x,y)|^2)^{\frac{1}{2}}\|_q \leqslant$$

$$\|(\sum_{(\mu,v)\in S_{lm}}1)^{\frac{1}{2}-\frac{1}{q}}(\sum_{(\mu,v)\in S_{lm}}|\delta_{\mu v}(f;x,y)|^q)^{\frac{1}{q}}\|_q$$

$$\ll\Big(\sum_{(\mu,v)\in S_{lm}}\|\delta_{\mu v}(f)\|^q_q\Big)^{\frac{1}{q}}$$

$$\sim(\sum_{(\mu,v)\in S_{lm}}2^{-(\mu+v)}\|\tau(\delta_{\mu v}(f)\|^q_{l_q})^{\frac{1}{q}}$$

$$\ll 2^{-\frac{4(l+m)}{q}}\|\tau(f)\|_{l_q}$$

Therefore by lemma 2 and 6 we obtain

$$d_N(\triangle_2^r;L_q)\ll\sum_{l,m}d_{N_{lm}}(\triangle_2^r\cap E_{lm};L_q\cap E_{lm})\ll$$

$$\sum_{l>s}\sum_{m=0}^{l}4^{-4lr}\cdot 2^{2(l+m)}\cdot 2^{-\frac{4(l+m)}{q}}\cdot d_{N_{lm}}(B_2^{\|S_{lm}\|}\ l_q^{\|S_{lm}\|})$$

$$\ll\sum_{l>s}4^{-4lr}\sum_{m=0}^{l}2^{2(l+m)}\cdot 2^{-\frac{4(l+m)}{q}}\|S_{lm}\|^{\frac{1}{q}}N_{lm}^{\frac{1}{2}}$$

$$\ll\sum_{l>s}4^{-4lr}\sum_{m=0}^{l}2^{2(l+m)}2^{-2(2s+\lambda s+2\lambda l+\lambda m)}\sim N^{-r}$$

The proof of relation (2) is completed. Now we prove (3)

Suppose $1<p\leqslant q\leqslant 2$, $r\geqslant\frac{1}{p}-\frac{1}{q}$. For any $f\in\triangle_p^r$, by proposition 1 and Young inequality, we get

$$\|f-S_{nn}(f)\|_q=\|\phi * D_{nn}^{(r)}\|_q$$

$$\leqslant\|\phi\|_p\|D_{nn}^{(r)}\|_\lambda\ll n^{-2(r+\frac{1}{q}-\frac{1}{p})}$$

where $\frac{1}{\lambda}=1+\frac{1}{q}-\frac{1}{p}$. So $d_N(\triangle_p^r;L_q)\ll N^{-(r+\frac{1}{q}-\frac{1}{p})}$.

For given N, choose s such that $2^{2s-2}\geqslant 2N$. The following conclusion is proved in [2] (see P178):

Suppose L_N is any N-dimention subspace of Fss, then there exists $g^*\in F_{ss}$ such that:

(i) $\|g^*\|_2\ll 2^s$ and

(ii) $\|g^*-u^*\|_2=\min_{u\in L_N}\|g^*-u\|_2\gg 2^s$

Putting $g=2^{-2s(r+1-\frac{1}{p})}g^*$ and according to the Nikoliski inequality $(\|\delta_{\mu v}(f)\|_p \ll 2^{(\mu+v)(\frac{1}{2}-\frac{1}{p})}\|\delta_{\mu v}(f)\|_2,)$ and lemma 1, we have

$$\|\triangle^r g\|_p\ll 2^{-2s(r+1-\frac{1}{p})}\cdot 2^{2sr}\|g^*\|_p\ll 2^{-2s(1-\frac{1}{p})}\cdot$$

$2^{-2s(\frac{1}{2}-\frac{1}{p})}\|g\|_2\ll 1$, so $g\in\triangle_p^r\cap F_{ss}$

By lemma 3

$$\sup_{f\in\triangle_p^r\cap F_{ss}}\inf_{u\in L_N}\|f-u\|_q\geqslant\inf_{u\in L_N}\|g-u\|_q$$

$$=2^{-2s(r+1-\frac{1}{p})}\|g^*-u^*\|_q$$

$$\gg 2^{-2s(r+1-\frac{1}{p})}\cdot 2^{2s(\frac{1}{2}-\frac{1}{q})}\|g^*-u^*\|_2\gg N^{-(r+\frac{1}{q}-\frac{1}{p})}$$

Since L_N is a arbitrary N-dimension subspace of F_{ss},

$$d_N(\triangle_p^r;L_q)\gg d_N(\triangle_p^r\cap F_{ss};L_q\cap F_{ss})\gg N^{-(r+\frac{1}{q}-\frac{1}{p})}$$

Hence (3) is proved.

Finally we prove the relation (4).

Suppose $1<p\leqslant 2\leqslant q<\infty$ and $r>\frac{1}{p}$.

According to the estimate (3), it is easy to see that $d_N(\triangle_p^r;L_q)\gg d_N(\triangle_p^r;L_2)\geqslant N^{-(r-\frac{1}{p}+\frac{1}{2})}$

By lemma 4, $\triangle_p^r\subset\triangle_2^{r-\frac{1}{p}+\frac{1}{2}}$

so $d_N(\triangle_p^r;L_q)\leqslant d_N(\triangle_2^{r-\frac{1}{p}+\frac{1}{2}};L_q)\ll N^{-(r-\frac{1}{p}+\frac{1}{2})}$

The estimate (4) is proved and the proof of the theorem is completed.

References

[1] Mitjagin, B.S., Mat. Sbornik (N.S.), 58 (100)(1962) 394-414.

[2] Temljakov, V. N., Izv. Akad. Nauk SSSR, 46(1982) No1, 171-186.

[3] Galeev, M., Izv. Akad. Nauk SSSR, 49 (1985)No5, 917-934.

[4] Temljakov, V. N. Izv. Akad. Nauk SSSR, 50(1986), No1, 137-155.

Approximation, Optimization and Computing:
Theory and Applications, A.G. Law and C.L. Wang (eds.)
Elsevier Science Publishers B.V. (North-Holland)

ERROR ESTIMATION FOR THE SHOOTING METHOD IN NONLINEAR BOUNDARY VALUE PROBLEMS§

P. MARZULLI
Istituto di Matematiche Applicate, Facoltà di Ingegneria, Università di Pisa
Via Bonanno 25B, I-56126 Pisa (Italy)

For a nonlinear boundary value problem with linear boundary conditions, solved by a shooting method, it is shown that whenever an error estimate with $O(h^p)$ accuracy is available for the related initial value problem, we can estimate the error for the given problem with an almost equal accuracy.

1. INTRODUCTION

We consider the nonlinear boundary value problem (BVP)

(1.1) $y'=f(x,y),\ Ay(a)+By(b)=\alpha,$

$(x,y)\in S=\{(x,y)\,|\,a\leq x\leq b,\ y\in R^m\},$

where $f,\ \alpha\in R^m$ and $A,\ B\in R^{m\times m}$.

In order to solve (1.1) by the simple shooting method, we seek a solution $y(x;s)$ for the initial value problem (IVP)

(1.2) $y'=f(x,y),\ y(a)=s,$

with a starting vector $s=s^*$, such that $y(x)=y(x;s^*)$ is a solution of (1.1) too, this implies that s^* is a solution of

(1.3) $As+By(b;s)-\alpha=0.$

Our goal is to suggest an algorithm for estimating the global discretization error

(1.4) $E(x_n)=y_n-y(x_n)=y_n-y(x_n;s^*),\ n=0,1,\ldots,N,$

where y_n is an approximation of $y(x_n)=y(x_n;s^*)$ at $x_n=a+nh$, $n=0,1,\ldots,N$, $h=(b-a)/N$.

Usually one assumes $y_n=y_n^{(k)}$, where $y_n^{(k)}$, $n=0,1,\ldots,N$, is a numerical solution of the IVP

(1.5) $y'=f(x,y),\ y(a)=s^{(k)},$

and $s^{(k)}$ is a numerical approximation of s^*.

Supposing to apply a p-th order integrator to (1.5), let $e(x_n)$ be the global discretization error

(1.6) $e(x_n)=y_n^{(k)}-y(x_n;s^{(k)})=O(h^p),$

then, using (1.4) and (1.6), we can write

(1.7) $E(x_n)=e(x_n)-(y(x_n;s^*)-y(x_n;s^{(k)})).$

We shall show that, if an IVP global error estimation procedure is available for $e(x_n)$, it can be used to determine an almost equally good estimate for the whole $E(x_n)$, provided $s^{(k)}$ is sufficiently close to s^*. An analogous result for the case of a linear BVP has been given in [1] by Marzulli and Gheri.

2. THE ESTIMATION OF $E(x_n)$

We premise a well known theorem:

THEOREM 2.1. *If the Jacobian f_y exists on S and is continuous and bounded there, then the solution $y(x;s)$ of (1.2) is differentiable with respect to s and the derivative is given by the matrix*

(2.1) $\partial y(x;s)/\partial s=Z(x;s)$

where Z is the solution of the matrix IVP

(2.2) $\partial Z(x;s)/\partial x=f_y(x,y(x;s))Z(x;s),\ Z(a;s)=I$

(I=the identity matrix).

PROOF. See Coddington and Levinson [2], Chapt. 1, theor. 7.2.

§Work supported in part by the Italian Ministero della Pubblica Istruzione.

If there exists $Z(x;s)$, we assume that $s^{(k)}$ can be obtained from (1.3) through the Newton-Raphson's algorithm

$$(2.3)\quad s^{(i+1)}-s^{(i)}=-(A+BZ_N^{(i)})^{-1}(As^{(i)}+By_N^{(i)}-\alpha),$$

where $Z_n^{(i)}$, $n=0,1,\ldots,N$, is a numerical solution of (2.2) corresponding to a numerical solution of (1.5) for k=i. An estimate of $E(x_n)$ is now furnished by the following theorem.

THEOREM 2.2. *Assume that the derivative (2.1) exists for $s=s^{(k)}$, where $s^{(k)}$ is given by (2.3) with k such that $(s^*-s^{(k)})=O(h^p)$; let the matrix $A+BZ(b;s^{(k)})$ be nonsingular; let $y_n^{(k)}$ and $Z_n^{(k)}$ be numerical solutions of (1.2) and (2.2) respectively, for $s=s^{(k)}$ and accuracy $O(h^p)$ guaranteing the nonsingularity of $A+BZ_N^{(k)}$; then $E(x_n)$ is given by*

$$(2.4)\quad E(x_n)=e(x_n)-Z_n^{(k)}(A+BZ_N^{(k)})^{-1}(Be(x_n)-(As^{(k)}+By_N^{(k)}-\alpha))+O(h^{2p}).$$

PROOF. From (1.7), applying the Taylor formula for differentiable operators, we have

$$(2.5)\quad E(x_n)=e(x_n)-Z(x_n;s^{(k)})(s^*-s^{(k)})+O(h^{2p}),$$

on the other hand s^* satisfies

$$As^*+By(b;s^*)-\alpha=0,$$

while from (2.3) with i=k one obtains

$$(2.6)\quad As^{(k)}+By_N^{(k)}-\alpha=-(A+BZ_N^{(k)})(s^{(k+1)}-s^{(k)})$$

then, by subtraction, we have

$$A(s^*-s^{(k)})-BE(x_N)=(A+BZ_N^{(k)})(s^{(k+1)}-s^{(k)}),$$

into which we substitute (2.5) with n=N obtaining

$$(A+BZ(x_N;s^{(k)}))(s^*-s^{(k)})=Be(x_N)+(A+BZ_N^{(k)})(s^{(k+1)}-s^{(k)}).$$

Then we have

$$s^*-s^{(k)}=(A+BZ(x_N;s^{(k)}))^{-1}(Be(x_N)+(A+BZ_N^{(k)})(s^{(k+1)}-s^{(k)}));$$

as, for the Banach lemma (cfr. Keller [3] pag.2), it is $\|(A+BZ(x_N;s^{(k)}))^{-1}-(A+BZ_N^{(k)})^{-1}\|=O(h^p)$ and having chosen k such that $s^{(k+1)}-s^{(k)}=O(h^p)$, it follows

$$s^*-s^{(k)}=(A+BZ(x_N;s^{(k)}))^{-1}Be(x_N)+(s^{(k+1)}-s^{(k)})+O(h^{2p}).$$

Substituting back into (2.5) one obtains

$$E(x_n)=e(x_n)-Z(x_n;s^{(k)})(A+BZ(x_N;s^{(k)}))^{-1}Be(x_N)-Z(x_n;s^{(k)})(s^{(k+1)}-s^{(k)})+O(h^{2p});$$

recalling that $e(x_n)=O(h^p)$ and using again the Banach lemma, we get

$$E(x_n)=e(x_n)-Z_n^{(k)}(A+BZ_N^{(k)})^{-1}Be(x_N)-Z_n^{(k)}(s^{(k+1)}-s^{(k)})+O(h^{2p}),$$

then, expressing $(s^{(k+1)}-s^{(k)})$ by means of (2.6), we reach the equality (2.4).

We note that if an estimate $e_n\cong e(x_n)$, $n=0,1,\ldots,N$, is known and k is such that $\|s^{(k+1)}-s^{(k)}\|\ll h^p$, the theorem 2.2. suggests a practical estimate of $E(x_n)$ given by

$$(2.7)\quad E_n=e_n-Z_n^{(k)}v$$

where v satisfies the linear system

$$(A+BZ_N^{(k)})v=Be_N.$$

If one uses a modified form of the Newton's formula, it is possible to avoid the numerical solution of the m initial value problems (2.2), by replacing in (2.3) the approximate derivative matrix $Z_N^{(i)}$ with a matrix $\Delta_N^{(i)}$ approximating the difference quotient matrix $\Delta y(x_N;s^{(i)})$, whose r-th column is assumed to be of the form

$$(y(x_N;\Delta s_r^{(i)})-y(x_N;s^{(i)}))/\delta s_r^{(i)},$$

where we have introduced the incremented vectors $\Delta s_r^{(i)}=(s_1^{(i)},s_2^{(i)},\ldots,s_r^{(i)}+\delta s_r^{(i)},s_{r+1}^{(i)},\ldots,s_m^{(i)})^T$, $r=1,\ldots,m$, and the scalar increments $\delta s_r^{(i)}$ are supposed to be "sufficiently" small. The calculation of $\Delta_N^{(i)}$ of course will require that $y(x_N;s^{(i)})$ and $y(x_N;\Delta s_r^{(i)})$ be approximated through the solution of the corresponding initial value problems. When this modified Newton method is used, the right-hand side of the equality (2.7) can be approximated by changing $Z_n^{(k)}$ and $Z_N^{(k)}$ into $\Delta_n^{(k)}$ and $\Delta_N^{(k)}$ respectively, provided the hypothesis of the theorem 2.2 are satisfied.

3. SECOND ORDER PROBLEMS

In the particular case of a BVP with separated boundary conditions, of the form

(3.1) $y''=f(x,y)$, $y(a)=\alpha_1$, $y(b)=\alpha_2$,

where y, f, α_1, α_2 R^m, an expression of $E(x_n)$ equivalent to (2.4) can be derived with a slight modification. Similarly to (1.2) we consider the IVP

(3.2) $y''=f(x,y)$, $y(a)=\alpha_1$, $y'(a)=s$,

and seek the solution $y(x;s^*)$ of (3.2) with $s=s^*$ such that s^* is a solution of the nonlinear equation

(3.3) $y(b;s)-\alpha_2=0$.

It is easy to verify, using the theorem 2.1, that $\partial y(x;s)/\partial s=Z(x;s)$ is the solution matrix of the IVP

(3.4) $Z''=f_y(x,y(x;s))Z$, $Z(a)=0$, $Z'(a)=I$,

which formally can be obtained setting in (3.2) $y(x)=y(x;s)$ and then differentiating the resulting identities with respect to s. We assume that an approximation $s^{(k)}$ of s^* can be obtained from (3.3) using the Newton-Raphson's iterations

(3.5) $s^{(i+1)}-s^{(i)}=-(Z_N^{(i)})^{-1}(y_N^{(i)}-\alpha_2)$,

$i=0,1,\dots,k-1$, where $Z_n^{(i)}$, $n=0,1,\dots,N$, is a numerical solution of (3.4) corresponding to a numerical solution $y_n^{(i)}$, $n=0,1,\dots,N$, of (3.2) for $s=s^{(i)}$. The following theorem is nothing more than theorem 2.2 for a problem of the form (3.1).

THEOREM 3.1. *Let the problem (3.2) be such that the derivative $\partial y(x;s)/\partial s=Z(x;s)$ exists for $s=s^{(k)}$, where $s^{(k)}$ is given by (3.5) and one has $(s^*-s^{(k)})=O(h^p)$; let the matrix $Z(b;s^{(k)})$ be nonsingular and the approximations $y_n^{(k)}$ and $Z_n^{(k)}$ have accuracy $O(h^p)$ such that $Z_N^{(k)}$ is nonsingular; then the error $E(x_n)=y_n^{(k)}-y(x_n;s^*)$ is given by*

(3.6) $E(x_n)=e(x_n)-Z_n^{(k)}(Z_N^{(k)})^{-1}(e(x_n)-(y_N^{(k)}-\alpha_2))+O(h^{2p})$

where $e(x_n)=y_n^{(k)}-y(x_n;s^{(k)})$.

PROOF. From $E(x_n)=e(x_n)-(y(x_n;s^*)-y(x_n;s^{(k)}))$ we have

(3.7) $E(x_n)=e(x_n)-Z(x_n;s^{(k)})(s^*-s^{(k)})+O(h^{2p})$,

on the other hand from the equalities

$$y(b;s^*)-y(b;s^{(k)})=\alpha_2-y(b;s^{(k)})=Z(b;s^{(k)})(s^*-s^{(k)})+O(h^{2p}),$$

one obtains

$$s^*-s^{(k)}=(Z(b;s^{(k)}))^{-1}(\alpha_2-y(b;s^{(k)}))+O(h^{2p})$$

which, setting $y(b;s^{(k)})=y_N^{(k)}-e(x_N)$, yelds

$$s^*-s^{(k)}=(Z(b;s^{(k)}))^{-1}(e(x_N)-(y_N^{(k)}-\alpha_2))+O(h^{2p});$$

substituting in (3.7) we have

$$E(x_n)=e(x_n)-Z(x_n;s^{(k)})(Z(b;s^{(k)}))^{-1}(e(x_N)-(y_N^{(k)}-\alpha_2))+O(h^{2p}),$$

from which applicating the Banach lemma we also get that (3.6) holds.

If an estimate e_n of $e(x_n)$ is available and k is "sufficiently" large, the theorem 3.1 suggests the following practical estimation similar to (2.7):

(3.8) $E_n=e_n-Z_n^{(k)}w$,

where w is the solution of the linear system $Z_N^{(k)}w=e_N$. An approximation of the right-hand side of (3.8) can be obtained using in place of $Z_n^{(k)}$ and $Z_N^{(k)}$ the corresponding approximate quotient matrices $\Delta_n^{(k)}$ and $\Delta_N^{(k)}$.

4. EXAMPLE

Among the numerical examples we have considered,

we present the following one which is a modified form of a BVP given by Stoer and Bulirsh in [4], pg. 471:

$$\begin{cases} y_1'=y_2 \\ y_2'=\frac{3}{2}y_1^2 \end{cases},$$

$$\begin{pmatrix}1 & 0\\ 0 & 0\end{pmatrix}\begin{pmatrix}y_1(0)\\ y_2(0)\end{pmatrix}+\begin{pmatrix}0 & 0\\ 1 & 0\end{pmatrix}\begin{pmatrix}y_1(1)\\ y_2(1)\end{pmatrix}=\begin{pmatrix}4\\ 1\end{pmatrix}.$$

For the exact solution $y_1(x)=4/(1+x)^2$ and $y_2(x)=-8/(1+x)^2$, numerical values have been calculated with three different values of $h=1/N$. Using as IVP integrator the Euler's method, the thrue $E(x_n)$ has been compared with the estimate (2.7) of E_n at $x_n=0.5$, giving:

h	1/10	1/100	1/500
$E(x_n)$	-0.1435D+00	-0.5642D-02	-0.1108D-02
	0.4529D+00	0.8172D-02	0.1562D-02
E_n	-0.1455D+00	-0.5711D-02	-0.1111D-02
	0.4616D+00	0.8258D-02	0.1566D-02

Employing as IVP integrator the 3-stage Runge-Kutta method $y_{n+1}=y_n+hf(y_n+\frac{1}{2}hf(y_n+\frac{1}{2}hf(y_n)))$, and using again the estimate (2.7) gives:

h	1/10	1/100	1/500
$E(x_n)$	-0.2493D-01	-0.3935D-03	-0.2299D-04
	0.7870D-01	0.1226D-02	0.7147D-04
E_n	-0.2709D-01	-0.4521D-03	-0.2770D-04
	0.8596D-01	0.1411D-02	0.8614D-04

Identical results have been obtained for the same problem in its original form $y''=(3/2)y^2$, $y(0)=4$, $y(1)=1$, employing the same IVP integrators as above but then the (3.8) estimate. In both cases the estimate e_n has been furnished by the numerical solution of the corresponding principal error equation (see Stetter [5]).

REFERENCES

[1] Marzulli,P. and Gheri, G., Estimation of the global discretization error in shooting methods for linear boundry value problems, J. Comput. Appl. Math., 27 (1989), in print.

[2] Coddington, E.A. and Levinson, N., Theory of Ordinary Differential Equations (Mc Graw-Hill, New-York, 1955).

[3] Keller, H.B., Numerical Solution of Two Point Boundary Value Problems (SIAM Regional Conference Series in Applied Mathematics 24, Philadelphia, 1976).

[4] Stoer, J. and Bulirsch, R., Introduction to Numerical Analysis (Springer-Verlag, New-York, Heidelberg, Berlin, 1980).

[5] Stetter, H. J., Local estimation of the global discretization error, SIAM J. Numer. Anal.,

Approximation, Optimization and Computing:
Theory and Applications, A.G. Law and C.L. Wang (eds.)
Elsevier Science Publishers B.V. (North-Holland)

DIRECT THEOREMS FOR BEST APPROXIMATION IN BANACH SPACES

Toshihiko NISHISHIRAHO

Department of Mathematics, University of the Ryukyus
Nishihara-Cho, Okinawa 903-01, Japan

Quantitative estimates for best approximation in Banach spaces are obtained by using a modulus of continuity and by taking higher order moments.

1. INTRODUCTION

Let X be a Banach space with norm $||\cdot||_X$ and let $\mathbf{Z}$ denote the set of all integers.

Let $\{f_j, f_j^*\}_{j\in\mathbf{Z}}$ be a fundamental, total biorthogonal system. That is, it is a system satisfying the following conditions:

(A) $\{f_j\}_{j\in\mathbf{Z}}$ and $\{f_j^*\}_{j\in\mathbf{Z}}$ are sequences of elements in X and X*, respectively, where X* denotes the dual space of X.

(B) $f_j^*(f_m) = \delta_{jm}$ for all j, m $\in \mathbf{Z}$, where δ_{jm} is the Kronecker's symbol.

(C) $f_j^*(f) = 0$ for all $j \in \mathbf{Z}$ implies $f = 0$.

(D) The linear subspace of X spanned by $\{f_j : j \in \mathbf{Z}\}$ is dense in X.

Let $\mathbb{N}$ denote the set of all non-negative integers. For each $n \in \mathbb{N}$, Π_n stands for the linear subspace of X spanned by $\{f_j : |j| \leq n\}$ and for any $f \in X$, we define

$$E_n(X;f) = \inf\{||f - g||_X : g \in \Pi_n\},$$

which is called the best approximation of degree n to f. Then we have

$$E_0(X;f) \geq E_1(X;f) \geq \cdots \geq E_n(X;f) \geq \cdots,$$

and Condition (D) implies that

$$E_n(X;f) \to 0 \quad \text{as} \quad n \to \infty$$

whenever f belongs to X.

The purpose of this paper is to establish direct theorems on estimating the order of magnitude of $E_n(X;f)$, i.e., we should like to relate the rapidity with which $E_n(X;f)$ approaches to zero to certain smoothness properties of f. This is achieved by making use of a study of convolution operators as well as multiplier operators in connection with Fourier series expansions with respect to the system $\{f_j, f_j^*\}_{j\in\mathbf{Z}}$ (cf. [2], [3], [4]).

2. MODULI OF CONTINUITY

Let B[X] denote the Banach algebra of all bounded linear operators of X into itself and s stands for the set of all sequences $\alpha = \{\alpha_j\}_{j\in\mathbf{Z}}$ of scalars. To any $f \in X$ one may associate its (formal) Fourier series expansion (with respect to the system $\{f_j, f_j^*\}_{j\in\mathbf{Z}}$)

$$f \sim \sum_{j=-\infty}^{\infty} f_j^*(f) f_j$$

(cf. [6]). An operator $T \in B[X]$ is called a multiplier operator if there exists a sequence $\alpha \in s$ such that for every $f \in X$,

$$T(f) \sim \sum_{j=-\infty}^{\infty} \alpha_j f_j^*(f) f_j,$$

and the following notation is used:

$$T \sim \sum_{j=-\infty}^{\infty} \alpha_j P_j,$$

where each P_j is the projection operator in B[X] defined by $P_j(f) = f_j^*(f) f_j$ for all $f \in X$. (cf. [2], [3]).

Let $\mathbb{R}$ denote the set of all real numbers and let $\{T_t : t \in \mathbb{R}\}$ be a family of isometric multiplier operators in B[X] having the expansions

$$T_t \sim \sum_{j=-\infty}^{\infty} \exp(-ijt) P_j \qquad (t \in \mathbb{R}).$$

Then $\{T_t: t \in \mathbb{R}\}$ becomes a strongly continuous group of operators in B[X] and its infinitesimal generator G with domain D(G) satisfies

$$G(f) \sim \sum_{j=-\infty}^{\infty} (-ij)P_j(f)$$

for every $f \in D(G)$, and if, with the n-th Cesàro mean operator

$$\sigma_n = \sum_{j=-n}^{n} (1 - |j|/(n+1))P_j,$$

the sequence $\{\sigma_n\}_{n\in\mathbb{N}}$ is uniformly bounded, then $D(G) = \{f \in X: g \sim \Sigma_{j=-\infty}^{\infty}(-ij)P_j(f) \text{ for some } g \in X\}$ (cf. [2; Proposition 2]).

For any $f \in X$ and $\delta \geq 0$, we define

$$\omega(X;f,\delta) = \sup\{\|T_t(f) - f\|_X: |t| \leq \delta\}$$

and

$$\omega^*(X;f,\delta) = \sup\{\|T_t(f) + T_{-t}(f) - 2f\|_X : 0 \leq t \leq \delta\},$$

which are called the modulus of continuity of f and the generalized modulus of continuity of f associated with the family $\{T_t: t \in \mathbb{R}\}$. These quantities have the following properties (cf. [2; Lemmas 2 and 5]): Let $f \in X$.

(a) $\omega(X;f,\cdot)$ and $\omega^*(X;f,\cdot)$ are non-decreasing functions on $[0,\infty)$ with $\omega(X;f,0) = \omega^*(X;f,0) = 0$.

(b) $\lim_{\delta\to 0+0} \omega(X;f,\delta) = 0$; $\lim_{\delta\to 0+0} \omega^*(X;f,\delta) = 0$.

(c) $\omega^*(X;f,\delta) \leq 2\omega(X;f,\delta) \leq 4\|f\|_X \quad (\delta \geq 0)$.

(d) $\omega(X;f,\xi\delta) \leq (1+\xi)\omega(X;f,\delta) \quad (\xi, \delta \geq 0)$.

(e) $\omega^*(X;f,\xi\delta) \leq (1+\xi)^2\omega^*(X;f,\delta) \quad (\xi, \delta \geq 0)$.

Let $M > 0$ and $\beta > 0$. An element $f \in X$ is said to satisfy the Lipschitz condition with constant M and exponent β, or to belong to the class $\mathrm{lip}_M(X;\beta)$ if $\omega(X;f,\delta) \leq M\delta^\beta$ for all $\delta \geq 0$. Also, an element $f \in X$ is said to satisfy the generalized Lipschitz condition with constant M and exponent β, or to belong to the class $\mathrm{Lip}_M^*(X;\beta)$ if $\omega^*(X;f,\delta) \leq M\delta^\beta$ for all $\delta \geq 0$.

3. MAIN RESULTS

For each $n \in \mathbb{N}$, we denote by H_n the set of all non-negative, even trigonometric polynomials k of degree not exceeding n of the form

$$k(t) = 1 + 2\sum_{j=1}^{n} a_j \cos jt \qquad (t \in \mathbb{R}).$$

For any $k \in H_n$ and $p > 0$, we define

$$m(k;p) = (1/2\pi)\int_{-\pi}^{\pi} |t|^p k(t)dt,$$

which is called the p-th moment of k.

THEOREM 1. For all $f \in X$ and all $n \in \mathbb{N}$,

$$E_n(X;f) \leq \inf\{(1 + \min\{\delta^{-p}m(k;p), \delta^{-1}(m(k;p))^{1/p}\})\omega(X;f,\delta): \delta > 0,\ p \geq 1,\ k \in H_n\}. \tag{1}$$

THEOREM 2. For all $f \in X$ and all $n \in \mathbb{N}$,

$$E_n(X;f) \leq \inf\{(1/2 + m(k;1)/\delta + m(k;p)/(2\delta^2))\omega^*(X;f,\delta): \delta > 0,\ k \in H_n\}. \tag{2}$$

THEOREM 3. For all $f \in D(G)$ and all $n \in \mathbb{N}$,

$$E_n(X;f) \leq \inf\{(m(k;1) + m(k;p+1)/(\delta^p(p+1))) \times \omega(X;G(f),\delta): \delta > 0,\ p \geq 1,\ k \in H_n\}. \tag{3}$$

THEOREM 4. For all $f \in \mathrm{Lip}_M^*(X;\beta)$ and all $n \in \mathbb{N}$,

$$E_n(X;f) \leq (M/2)\inf\{m(k;\beta): k \in H_n\}.$$

These theorems can be proved by using the techniques and the results of our previous work [4] (cf. [2], [3]). Denoting by $e_n(X;f)$, $e_n^*(X;f)$ and $d_n(X;f)$ the quantities of the right-hand sides of the inequalities (1), (2) and (3), respectively, we have:

$$e_n(X;f) \leq \inf\{(1 + \min\{\delta^{-2}(\pi^2/2)(1 - k\hat{\ }(1)), \delta^{-1}(2^{-1/2}\pi)(1 - k\hat{\ }(1))^{1/2}\}) \times \omega(X;f,\delta): \delta > 0,\ k \in H_n\}, \tag{4}$$

where

$$k\hat{\ }(1) = a_1 = (1/2\pi)\int_{-\pi}^{\pi} k(t)\cos t\,dt;$$

$$e_n^*(X;f) \leq \inf\{(1/2 + (2^{-1/2}\pi)(1 - k\hat{\ }(1))^{1/2}/\delta + (\pi^2/2)(1 - k\hat{\ }(1))/(2\delta^2)) \times \omega^*(X;f,\delta): \delta > 0,\ k \in H_n\}; \tag{5}$$

$$d_n(X;f) \leq \inf\{((2^{-1/2}\pi)(1 - k\hat{\ }(1))^{1/2} + (\pi^2/4)(1 - k\hat{\ }(1))/\delta) \times \omega(X;G(f),\delta): \delta > 0,\ k \in H_n\}. \tag{6}$$

Consider the Fejér-Korovkin kernel k_n given by

$$k_n(t) = \Lambda_n \left|\sum_{m=0}^{n} \lambda_{n,m} e^{imt}\right|^2 \qquad (n \in \mathbb{N},\ t \in \mathbb{R}),$$

where

$$\lambda_{n,m} = \sin((m+1)\pi/(n+2)) \qquad (m = 0, 1, \cdots, n)$$

and

$$\Lambda_n = (\lambda_{n,0}^2 + \lambda_{n,1}^2 + \cdots + \lambda_{n,n}^2)^{-1}.$$

Then we have

$$1 - k_n^{\wedge}(1) = 1 - \cos(\pi/(n+2)) < \pi^2/(2(n+2)^2),$$

and so Theorems 1, 2 and 3 and inequalities (4), (5) and (6) yield the following estimates:

$$E_n(X;f) \leqq e_n(X;f) \leqq (1+\pi^2/2)\omega(X;f,1/(n+2)) \qquad (f \in X,\ n \in \mathbb{N});$$

$$E_n(X;f) \leqq e_n^*(X;f) \leqq (1/2)(1+\pi^2/2)^2\omega^*(X;f,1/(n+2)) \qquad (f \in X,\ n \in \mathbb{N});$$

$$E_n(X;f) \leqq d_n(X;f) \leqq (\pi^2/2)(1+\pi^2/4)(1/(n+2))\omega(X;G(f),1/(n+2)) \qquad (f \in D(G),\ n \in \mathbb{N}).$$

In particular,

$$E_n(X;f) \leqq e_n(X;f) \leqq M(1+\pi^2/2)(n+2)^{-\beta} \qquad (f \in \mathrm{Lip}_M(X;\beta),\ n \in \mathbb{N});$$

$$E_n(X;f) \leqq e_n^*(X;f) \leqq (M/2)(1+\pi^2/2)^2(n+2)^{-\beta} \qquad (f \in \mathrm{Lip}_M^*(X;\beta),\ n \in \mathbb{N});$$

$$E_n(X;f) \leqq d_n(X;f) \leqq (M\pi^2/2)(1+\pi^2/4)(n+2)^{-(\beta+1)} \qquad (G(f) \in \mathrm{Lip}_M(X;\beta),\ n \in \mathbb{N}).$$

REMARK. Let F_n be the Fejér kernel, i.e.,

$$F_n(t) = 1 + 2\sum_{j=1}^{n}(1 - j/(n+1))\cos jt \qquad (n \in \mathbb{N},\ t \in \mathbb{R}).$$

Then for all $n \in \mathbb{N}$, we have:

$$m(F_n;\beta) \leqq (2^{\beta}\pi/(1-\beta^2))(n+1)^{-\beta} \qquad (0 < \beta < 1);$$

$$m(F_n;\beta) \leqq \pi(1/2 + \log(\pi/2) + \log(n+1))/(n+1) \qquad (\beta = 1);$$

$$m(F_n;\beta) \leqq \pi^{\beta}(\beta-1)^{-1}(n+1)^{-1} \qquad (\beta > 1).$$

Therefore, by Theorem 4, the following inequalities hold for all $f \in \mathrm{Lip}_M^*(X;\beta)$ and all $n \in \mathbb{N}$:

$$E_n(X;f) \leqq 2^{\beta-1}M\pi(1-\beta^2)^{-1}(n+1)^{-\beta} \qquad (0 < \beta < 1);$$

$$E_n(X;f) \leqq 2^{-1}M\pi(1/2 + \log(\pi/2) + \log(n+1))(n+1)^{-1} \qquad (\beta = 1);$$

$$E_n(X;f) \leqq 2^{-1}M\pi^{\beta}(\beta-1)^{-1}(n+1)^{-1} \qquad (\beta > 1).$$

4. APPLICATIONS

Let $L_{2\pi}^1$ denote the Banach space of all 2π-periodic, Lebesgue measurable functions f on $\mathbb{R}$ with norm

$$\|f\|_1 = (1/2\pi)\int_{-\pi}^{\pi}|f(t)|dt < \infty.$$

A linear subspace X of $L_{2\pi}^1$ is called a homogeneous Banach space if it is a Banach space with norm $\|\cdot\|_X$ which satisfies the following conditions (cf. [1], [2], [5]):

(H-1) X is continuously embedded in $L_{2\pi}^1$, i.e., there exists a constant $C > 0$ such that

$$\|f\|_1 \leqq C\|f\|_X$$

for all $f \in X$.

(H-2) Let $\{T_t: t \in \mathbb{R}\}$ be the family of translation operators defined by

$$T_t(f)(u) = f(u-t) \qquad (f \in X,\ t,\ u \in \mathbb{R}).$$

Then for each $t \in \mathbb{R}$, T_t is an isometric operator in $B[X]$.

(H-3) For each $f \in X$, the mapping $t \to T_t(f)$ is strongly continuous on $\mathbb{R}$.

Now let X be a homogeneous Banach space with norm $\|\cdot\|_X$, and we define

$$f_j(t) = e^{ijt}, \quad f_j^*(f) = f^{\wedge}(j) \qquad (j \in \mathbb{Z},\ t \in \mathbb{R},\ f \in X),$$

where $f^{\wedge}(j)$ stands for the j-th Fourier coefficient of f. Then $\{f_j, f_j^*\}_{j\in\mathbb{Z}}$ is a fundamental, total, biorthogonal system on account of the fact that

$$\lim_{n\to\infty}\|\sigma_n(f) - f\|_X = 0$$

for every $f \in X$ and the uniqueness property of Fourier coefficients. Also, we have

$$T_t(f) \sim \sum_{j=-\infty}^{\infty}\exp(-ijt)f_j^*(f)f_j \qquad (f \in X,\ t \in \mathbb{R}).$$

Obviously, Π_n consists of all trigonometric polynomials of degree not exceeding n, and we have

$$\omega(X;f,\delta) = \sup\{\|f(\cdot - t) - f(\cdot)\|_X : |t| \leqq \delta\}$$

and

$$\omega^*(X;f,\delta) = \sup\{\|f(\cdot + t) + f(\cdot - t) - 2f(\cdot)\|_X : 0 \leqq t \leqq \delta\}.$$

Consequently, all the results obtained in the preceding section are applied in the above situation. In particular, we obtain the theorems of Jackson type in homogeneous Banach spaces of which very special cases are $C_{2\pi}$, the Banach space of all 2π-periodic continuous functions f with the norm

$$\|f\|_\infty = \sup\{|f(t)|: |t| \leqq \pi\}$$

and $L^p_{2\pi}$, $1 \leqq p < \infty$, the Banach space of all 2π-periodic, p-th power Lebesgue integrable functions f with the norm

$$\|f\|_p = ((1/2\pi)\int_{-\pi}^{\pi}|f(t)|^p dt)^{1/p}$$

(cf. [5; Theorem 9.3.3.1]). Further details and related results will appear elsewhere.

REFERENCES

[1] Katznelson, Y., An Introduction to Harmonic Analysis (John Wiley, New York, 1968).

[2] Nishishiraho, T., Tôhoku Math. J. 33(1981) 109-126.

[3] Nishishiraho, T., Tôhoku Math. J. 34(1982) 23-42.

[4] Nishishiraho, T., The Degree of approximation by Convolution Operators in Banach Spaces, to appear in Approximation Theory VI (Proc. Int. Sympos. College Station, Texas 1989; ed. by Chui, C. K., Schumaker, L. L. and Ward, J. D., Academic Press, New York, 1989).

[5] Shapiro, H.S., Topics in Approximation Theory, Lecture Notes in Mathematics, Vol. 187 (Springer-Verlag, Berlin-Heidelberg, New York, 1971).

[6] Singer, I., Bases in Banach Spaces I (Springer-Verlag, Berlin-Heidelberg, New York, 1970).

Approximation, Optimization and Computing:
Theory and Applications, A.G. Law and C.L. Wang (eds.)
Elsevier Science Publishers B.V. (North-Holland)

Converse Theorems for Approximation on Discrete Sets

Gerhard SCHMEISSER

Mathematisches Institut, Universität Erlangen-Nürnberg, Bismarckstraße 1$^1/_2$, D–8520 Erlangen, FEDERAL REPUBLIC OF GERMANY

Converse theorems describe regularity properties of a function in terms of the speed of convergence of its best approximations on an interval. We extend some of these results to approximation on sequences of discrete sets. As approximating functions we consider polynomials, trigonometric polynomials and entire functions of exponential type.

1. INTRODUCTION

Denote by $E_n^\star(f)$ the error of best uniform approximation of a 2π-periodic function f by trigonometric polynomials of degree at most n . It is well known [5, Chap. 4, Theorem 9] that for $k \in \mathbf{N}_0$ and $\alpha \in (0,1)$ we have $E_n^\star(f) = O(n^{-k-\alpha})$ as $n \to \infty$ if and only if $f \in C^k(\mathbf{R})$ and $f^{(k)} \in \mathrm{Lip}\,\alpha$. This result is also contained [7, Sec. 2.6.21] in the following Theorem A. Denote by $A_\sigma(f)$ the error of best uniform approximation of a bounded function f on $\mathbf{R}$ by entire functions of exponential type σ and let us write $\mathrm{UCB}(\mathbf{R})$ for the class of uniformly continuous and bounded functions on $\mathbf{R}$. We then have

Theorem A [7, Sec. 5.1.4, 6.1.6]. *Let f be a bounded function on $\mathbf{R}$ and let $k \in \mathbf{N}_0$, $\alpha \in (0,1)$. Then*

$$(1) \qquad A_\sigma(f) = O(\sigma^{-k-\alpha}) \qquad as \qquad \sigma \to \infty$$

if and only if f is k times differentiable and

$$f^{(k)} \in \mathrm{UCB}(\mathbf{R}) \cap \mathrm{Lip}\,\alpha.$$

For approximation by polynomials the situation is more complicated. There are essentially two different ways to characterize the approximability of a function f on an interval in terms of regularity properties. One either works with a pointwise concept of approximability which increases precision towards the endpoints of the interval or uses a concept of regularity which relaxes towards the endpoints. The first way was followed by Timan, Dzjadyk and Teljakovskiĭ (see [5, Chap. 5], [6]) the second one by Ditzian, Ivanov and Totik (see [3]). Let us mention a result of the latter kind.

Denote by $E_n(f)$ the error of best uniform approximation by polynomials of degree at most n on $[-1,1]$. Furthermore, for $\varphi(x) := \sqrt{1-x^2}$ we define

$$\Delta_{h\varphi}^1 f(x) := f(x + h\varphi(x)) - f(x - h\varphi(x))$$

and recurrently

$$\Delta_{h\varphi}^j f(x) := \Delta_{h\varphi}^{j-1} f(x + h\varphi(x)) - \Delta_{h\varphi}^{j-1} f(x - h\varphi(x)).$$

Theorem B [3, Sec. 7.2]. *Let f be a bounded function on $[-1,1]$ and let $k \in \mathbf{N}_0$, $\alpha \in (0,1)$. Then*

$$(2) \qquad E_n(f) = O(n^{-k-\alpha}) \qquad as \qquad n \to \infty$$

if and only if

$$\| \Delta_{h\varphi}^{k+1} f \|_{L^\infty[-1,1]} = O(h^{k+\alpha}) \qquad as \qquad h \to \infty.$$

By computation the best approximation can be achieved on a discrete set of points only. We may therefore ask for sequences of discrete sets on which a prescribed rate of convergence of the form (1) or (2) implies already the same regularity properties of f as in the above theorems. The purpose of this paper is to study this question.

2. STATEMENT OF RESULTS

Notation 1. For positive real numbers δ , L and τ we denote by $\Lambda(\tau) := \Lambda(\delta, L, \tau)$ a set $\{x_\nu \in \mathbf{R} : \nu \in \mathbf{Z}\}$ of points such that

$$x_{\nu+1} - x_\nu \geq 2\pi \frac{\delta}{\tau} \qquad \text{and} \qquad \left| x_\nu - \frac{\pi\nu}{\tau} \right| \leq \pi \frac{L}{\tau}.$$

For fixed δ and L we write $\Lambda_\mu(\tau) := \Lambda(\delta, L, \tau 2^\mu)$ if we have a sequence of such sets with the property that

$$\Lambda(\delta, L, \tau 2^\mu) \subset \Lambda(\delta, L, \tau 2^{\mu+1})$$

for $\mu \in \mathbf{N}_0$. An example is given by the sets $\Lambda_\mu(\tau) := \{\frac{\pi\nu}{\tau 2^\mu} : \nu \in \mathbf{Z}\}$ with $\delta \in (0, \frac{1}{2}]$ and $L > 0$.

Concerning Theorem A we are now ready to state

Theorem 1. *Let $f \in C(\mathbf{R})$ be a bounded function on $\mathbf{R}$ and let $k \in \mathbf{N}_0$, $\alpha \in (0,1)$ and $\sigma > 0$. Suppose there exists a sequence of entire functions $g_{\sigma 2^\mu}$ of exponential type $\sigma 2^\mu$ such that*

$$(3) \qquad \sup_{x \in \Lambda_{\mu+1}(\tau)} | f(x) - g_{\sigma 2^\mu}(x) | \leq c\,(\sigma 2^\mu)^{-k-\alpha}, \qquad \mu \in \mathbf{N}_0$$

with constants $c > 0$ and $\tau > \sigma$. Then f is k-times differentiable and $f^{(k)} \in \mathrm{UCB}(\mathbf{R}) \cap \mathrm{Lip}\,\alpha$.

Notation 2. For positive real numbers δ, L and $n \in \mathbf{N}$ we denote by $\Omega(n) := \Omega(\delta, L, n)$ a set

$$\{x_\nu \in [-\pi, \pi] : \nu = 1-n, 2-n, ..., n\}$$

of $2n$ points such that

$$x_{\nu+1} - x_\nu \geq 2\pi\frac{\delta}{n} \qquad \text{and} \qquad \left| x_\nu - \frac{\pi\nu}{n} \right| < \pi\frac{L}{n}$$

for $\nu = 1-n, 2-n, ..., n$, where $x_{n+1} := 2\pi + x_{1-n}$.

For fixed δ and L we write $\Omega_\mu(n) := \Omega(\delta, L, n2^\mu)$ if we have a sequence of such sets with the property that

$$\Omega(\delta, L, n2^\mu) \subset \Omega(\delta, L, n2^{\mu+1})$$

for $\mu \in \mathbf{N}_0$. An example is given by the sets

$$\Omega_\mu(n) := \left\{ \frac{\pi\nu}{n2^\mu} : \nu = 1-n2^\mu, 2-n2^\mu, ..., n2^\mu \right\}$$

with $\delta \in (0, \frac{1}{2}]$ and $L > 0$.

From now on we denote by $C_{2\pi}(\mathbf{R})$ the class of all continuous 2π-periodic functions on $\mathbf{R}$. Since every trigonometric polynomial of degree at most n is an entire function of exponential type n we obtain as a consequence of Theorem 1

Corollary 1. *Let $f \in C_{2\pi}(\mathbf{R})$ and let $k \in \mathbf{N}_0$, $\alpha \in (0,1)$ and $n \in \mathbf{N}$. Suppose there exists a sequence of trigonometric polynomials t_{n2^μ} of degree at most $n2^\mu$ such that*

$$\max_{x \in \Omega_{\mu+1}(n+1)} |f(x) - t_{n2^\mu}(x)| \leq c\,(n2^\mu)^{-k-\alpha}, \qquad \mu \in \mathbf{N}_0$$

with a constant $c > 0$. Then $f \in C^k(\mathbf{R})$ and $f^{(k)} \in \mathrm{Lip}\,\alpha$.

Notation 3. For positive real numbers δ, L and an integer $n \geq 2$ we denote by $\Xi(n) := \Xi(\delta, L, n)$ a set of $n-1$ points $x_\nu = \cos\theta_\nu$ $(\nu = 1, 2, ..., n-1)$ such that $0 < \theta_1 < ... < \theta_{n-1} < \pi$,

$$\theta_{\nu+1} - \theta_\nu \geq 2\pi\frac{\delta}{n-1} \qquad \text{and} \qquad \left| \theta_\nu - \frac{\pi\nu}{n-1} \right| \leq \frac{\pi L}{n-1}$$

for $\nu = 0, 1, ..., n-1$, where $\theta_0 := -\theta_1$ and $\theta_n := 2\pi - \theta_{n-1}$.

For fixed δ and L we write $\Xi_\mu(n) := \Xi(\delta, L, n2^\mu)$ if we have a sequence of such sets with the property that

$$\Xi(\delta, L, n2^\mu) \subset \Xi(\delta, L, n2^{\mu+1})$$

for $\mu \in \mathbf{N}_0$. Obviously, for $\delta \in (0, 1/2]$ and $L > 0$ the zeros of the Chebyshev polynomial of the second kind $U_{n2^\mu - 1}(x)$ form such sets $\Xi_\mu(n)$.

Concerning Theorem B we state

Theorem 2. *Let $f \in C[-1,1]$ and let $k \in \mathbf{N}_0$, $\alpha \in (0,1)$ and $n \in \mathbf{N}$. Suppose there exists a sequence of polynomials P_{n2^μ} of degree at most $n2^\mu$ such that*

$$\max_{x \in \Xi_{\mu+1}(n+2)} |f(x) - P_{n2^\mu}(x)| \leq c\,(n2^\mu)^{-k-\alpha}, \qquad \mu \in \mathbf{N}_0$$

with a constant $c > 0$. Then

$$\| \Delta_{h\varphi}^{k+1} f \|_{L^\infty[-1,1]} = O(h^{k+\alpha}) \quad \textit{as} \quad h \to \infty.$$

Our considerations also give an answer to the interesting question for which discrete subsets S of an interval the best approximation on S has already the order of best approximation on the whole interval (see [2, Chap. 3, Sec. 7]).

Proposition 1. *For $f \in \mathrm{UCB}(\mathbf{R})$ and $\lambda > 1$ let g_σ be an entire function of exponential type $\sigma > 0$ such that*

$$\sup_{x \in \Lambda(\lambda\sigma)} |f(x) - g_\sigma(x)|$$

becomes minimal. Then

$$\sup_{x \in \mathbf{R}} |f(x) - g_\sigma(x)| \leq c\,A_\sigma(f)$$

with a constant c not depending on σ and f.

Proposition 2. *For $f \in C_{2\pi}(\mathbf{R})$ and $\lambda > 1$ let t_n be a trigonometric polynomial of degree at most n such that for an integer $n' \geq \lambda n$*

$$\max_{x \in \Omega(n')} |f(x) - t_n(x)|$$

becomes minimal. Then

$$\max_{x \in \mathbf{R}} |f(x) - t_n(x)| \leq c\,E_n^*(f)$$

with a constant c not depending on n and f.

Proposition 3. *For $f \in C[-1,1]$ and $\lambda > 1$ let P_n be a polynomial of degree at most n such that for an integer $n' \geq \lambda n$*

$$\max_{x \in \Xi(n'+1)} |f(x) - P_n(x)|$$

becomes minimal. Then

$$\max_{-1 \leq x \leq 1} |f(x) - P_n(x)| \leq c\,E_n(f)$$

with a constant c not depending on n and f.

Sharpness of the Results

The conclusions concerning the regularity of the function f are all best possible as follows from Theorems A and B.

The assumption that f is *continuous* can not be replaced by *bounded* as in Theorems A and B. In fact, the union U of all discrete sets of a sequence appearing in the statements is countable and so the function f zero on U and one elsewhere is a counterexample.

One may now ask if the "size" of the discrete sets can be diminished. Since the operator of Lagrange interpolation by entire functions of exponential type, trigonometric polynomials or algebraic polynomials is known to be unbounded, it is clear that our discrete sets must be "larger" than the sets guaranteeing unique interpolation. Hence in Propositions 1-3 the condition $\lambda > 1$ can not be replaced by $\lambda \geq 1$. In Theorem 1, Corollary 1 and Theorem 2, however, we approximate on discrete sets having more than twice the "size" of a set of unique interpolation. We found it convenient to design the discrete sets this way in order to have short proofs which show the essential ideas easily.

Given any $\lambda > 1$ we may consider a sequence of sets $\Lambda_\mu(\tau) := \Lambda(\delta, L, \tau\lambda^{\mu/2})$ with the property that

$$\Lambda(\delta, L, \tau\lambda^{\mu/2}) \subset \Lambda(\delta, L, \tau\lambda^{(\mu+1)/2}).$$

Then the proof of Theorem 1 will show that (3) may be replaced by

$$\sup_{x\in\Lambda_{\mu+1}(\tau)} |\, f(x) - g_{\sigma\lambda^{\mu/2}}(x)\,| \leq c\left(\sigma\lambda^{\mu/2}\right)^{-k-\alpha}.$$

For $\tau \in (\sigma, \sqrt{\lambda}\sigma)$ we see that $\Lambda_{\mu+1}(\tau)$ exceeds the "size" (here the density of points) of a set of unique interpolation by entire functions of exponential type $\sigma\lambda^{\mu/2}$ by a factor less than λ. Analogous improvements are possible in Corollary 1 and Theorem 2, if we take the integer part whenever $\lambda^{\mu/2}$ is not an integer. This was mentioned to us by Professor G.G. Lorentz. Unfortunately, the statements get a heavy form by such modifications.

3. Lemmas

We denote by K_1 and K_2 positive constants which may depend on δ, L and λ, but do not depend on x, θ, σ, m, n and the functions in consideration.

Lemma 1. *Let g be an entire function of exponential type $\sigma > 0$ and let $\Lambda(\lambda\sigma)$, where $\lambda > 1$, be a set of points as specified under Notation 1. If $|\,g(x)\,| \leq M$ for $x \in \Lambda(\lambda\sigma)$, then $|\,g(x)\,| \leq K_1 M$ for all $x \in \mathbf{R}$.*

Proof. Apply [1, Theorem 10.5.3] to $f(z) = g(\frac{\pi}{\lambda\sigma}z)$.

For the special case of a trigonometric polynomial we immediately deduce from Lemma 1 the following

Lemma 2. *Let t be a trigonometric polynomial of degree at most n and let $\Omega(m)$, where m is an integer such that $m \geq \lambda n$ with $\lambda > 1$, be a set of $2m$ points as specified under Notation 2. If $|\,t(x)\,| \leq M$ for $x \in \Omega(m)$, then $|\,t(x)\,| \leq K_1 M$ for all $x \in \mathbf{R}$.*

Defining $\theta_{-\nu+1} := -\theta_\nu$, the points θ_ν specified under Notation 3 form a set $\Omega(n-1)$ with L replaced by $L+1$. Now, applying Lemma 2 to $t(\theta) := P(\cos\theta)$ we obtain

Lemma 3. *Let P be a polynomial of degree at most n and let $\Xi(m+1)$, where m is an integer such that $m \geq \lambda n$ with $\lambda > 1$, be a set of m points as specified under Notation 3. If $|\,P(x)\,| \leq M$ for $x \in \Xi(m+1)$, then $|\,P(x)\,| \leq K_2 M$ for all $x \in [-1, 1]$.*

A complete characterization of the sets $S_n \subset [-1, 1]$ with the property that boundedness of a polynomial of degree at most n on S_n implies boundedness on the whole interval $[-1, 1]$, with a bound not depending on n, was given by Erdös [4]. We also mention that Cheney [2, Chap. 3, Sec. 7] presents a result similar to Lemma 3 but with stronger restrictions on the sets corresponding to $\Xi(m+1)$.

4. Proofs of the Results

We now denote by c_1, c_2 etc. appropriate positive numbers which do not depend on x, μ, i and j.

Proof of Theorem 1. The entire function $G_\mu := g_{\sigma 2^\mu} - g_{\sigma 2^{\mu-1}}$ is of exponential type $\sigma 2^\mu$ and

$$|\,G_\mu(x)\,| \leq |\,f(x) - g_{\sigma 2^\mu}(x)\,| + |\,f(x) - g_{\sigma 2^{\mu-1}}(x)\,|.$$

The first term on the right is bounded by $c\,(\sigma 2^\mu)^{-k-\alpha}$ for $x \in \Lambda_{\mu+1}(\tau)$, the second one by $c\,(\sigma 2^{\mu-1})^{-k-\alpha}$ for $x \in \Lambda_\mu(\tau)$. Since $\Lambda_\mu(\tau) \subset \Lambda_{\mu+1}(\tau)$ we obtain with the help of Lemma 1 that

$$(4) \qquad |\,G_\mu(x)\,| \leq c_1\,(\sigma 2^\mu)^{-k-\alpha} \quad \text{for} \quad x \in \mathbf{R},\ \mu \in \mathbf{N}.$$

Hence for any two integers $0 < i < j$ and all $x \in \mathbf{R}$

$$(5) \qquad \begin{aligned} &|\,f(x) - g_{\sigma 2^i}(x) - (f(x) - g_{\sigma 2^j}(x))\,| \\ &= \left|\sum_{\mu=i+1}^{j} G_\mu(x)\right| \leq c_2\left(\sigma 2^i\right)^{-k-\alpha}. \end{aligned}$$

This shows that $\psi_i := f - g_{\sigma 2^i}$ converges to a continuous function ψ, say, as $i \to \infty$. From (3) we conclude that $\psi(x) = 0$ on all the sets $\Lambda_\mu(\tau)$ ($\mu = 0, 1, \ldots$) and so $\psi(x)$ must be identically zero. Thus we obtain

$$(6) \quad f(x) = g_{\sigma 2^i}(x) + \sum_{\mu=i+1}^{\infty} G_\mu(x) \quad \text{for } x \in \mathbf{R},\ i \in \mathbf{N}_0.$$

Furthermore (5) and (6) yield

$$A_{\sigma 2^i}(f) \leq c_2\left(\sigma 2^i\right)^{-k-\alpha} \quad \text{for all} \quad i \in \mathbf{N}_0.$$

For every large real number ϱ we may choose i such that $\sigma 2^i \leq \varrho < \sigma 2^{i+1}$ and obtain

$$A_\varrho(f) \leq c_2\left(\sigma 2^i\right)^{-k-\alpha} = O(\varrho^{-k-\alpha}) \quad \text{as} \quad \varrho \to \infty.$$

Now the proof is completed by applying Theorem A.

Proof of Theorem 2. Using Lemma 3 and Theorem B instead of Lemma 1 and Theorem A the proof becomes completely analogous to the one of Theorem 1.

Proof of Proposition 1. Let g_σ and $g_\sigma^\star$ be entire functions of exponential type σ which furnish a best uniform approximation of f on $\Lambda(\lambda\sigma)$ and $\mathbf{R}$, respectively. Clearly,

$$\sup_{x\in\Lambda(\lambda\sigma)} |\,f(x) - g_\sigma(x)\,| \leq A_\sigma(f)$$

and so

$$\begin{aligned} &\sup_{x\in\Lambda(\lambda\sigma)} |\,g_\sigma(x) - g_\sigma^\star(x)\,| \\ &\leq \sup_{x\in\Lambda(\lambda\sigma)} |\,f(x) - g_\sigma(x)\,| + \sup_{x\in\mathbf{R}} |\,f(x) - g_\sigma^\star(x)\,| \\ &\leq 2\,A_\sigma(f). \end{aligned}$$

Since $g_\sigma - g_\sigma^\star$ is an entire function of exponential type σ it follows from Lemma 1 that

$$\sup_{x\in\mathbf{R}} |g_\sigma(x) - g_\sigma^\star(x)| \le 2\,K_1 A_\sigma(f).$$

Hence

$$\begin{aligned}
&\sup_{x\in\mathbf{R}} |f(x) - g_\sigma(x)| \\
&\quad\le \sup_{x\in\mathbf{R}} |f(x) - g_\sigma^\star(x)| + \sup_{x\in\mathbf{R}} |g_\sigma(x) - g_\sigma^\star(x)| \\
&\quad\le (1 + 2K_1)\, A_\sigma(f).
\end{aligned}$$

The proofs of Proposition 2 and 3 are completely analogous.

REFERENCES

[1] Boas, R.P., Jr., Entire Functions (Academic Press, New York, 1954).

[2] Cheney, E.W., Introduction to Approximation Theory (McGraw Hill, New York, 1966).

[3] Ditzian, Z. and Totik, V., Moduli of Smoothness (Springer, New York, 1987).

[4] Erdös, P., J. d'Analyse Math. 19 (1967) 135-148.

[5] Lorentz, G. G., Approximation of Functions (Holt, Rinehart and Winston, New York, 1966).

[6] Teljakovskiĭ, S.A., Amer. Math. Soc. Translations, Ser. 2, 77 (1966) 163-177.

[7] Timan, A.F., Theory of Approximation of Functions of a Real Variable (Pergamon Press, Oxford, 1963).

Approximation, Optimization and Computing:
Theory and Applications, A.G. Law and C.L. Wang (eds.)
Elsevier Science Publishers B.V. (North-Holland)

GEOMETRIC CONTINUITY ON CURVES AND SURFACES

Xi-Quan Shi and Ren-hong Wang

Inst. of Math., Dalian Univ. of Technology, Dalian 116024, China

ABSTRACT: A kind of definition on GC^n joint of curves and surfaces which depends only on their geometric properties is presented, and at the same time, several necessary and sufficient conditions on the GC^n joint of curves and surfaces are obtained.

KEYWORDS: GC^n joint, Generalized GC^n joint, Curvature, Deflection.

1. GEOMETRIC CONTINUITY FOR CURVES

Up to now, the point of departure of defining n-th order geometric continuity (GC^n) of curves and surfaces is transformation. In this paper, we present a kind of definition on GC^n which depends only on the geometric properties of curves and surfaces.

For two regular vectorvalued functions $F_1(t)$ and $G_1(s)$ ($t, s \in [0,1]$) which are assumed having analytic components for simplicity, we have

DEFINITION 1. $F_1(t)$ and $G_1(s)$ are called to be *a GC^n joint* if

$$F_1(0) = G_1(1),$$
$$F_i'(0) = \alpha G_i'(1), \quad \text{for } 1 \le i \le n,$$

hold, where $\alpha > 0$ is a real constant, $F_i(t) = F_{i-1}'(t)/|F_{i-1}'(t)|$, and $G_i(s) = G_{i-1}'(s)/|G_{i-1}'(s)|$, and if $F_{i-1}'(t_0) = 0, F_i(t_0)$ will be definied by

$$F_i(t_0) = \lim_{t \to t_0} F_i(t).$$

THEOREM 1. $F_1(t)$ and $G_1(s)$ form a $GC^i (1 \le i \le 3)$ joint if and only if the first $i+1$ of the following equalities

$$F_1(0) = G_1(1),$$
$$F_1'(0) = \alpha G_1'(1),$$
$$F_1''(0) = \alpha^2 G_1''(1) + \beta\alpha^2 G_1'(1),$$
$$F_1^{(3)}(0) = \alpha^3 G_1^{(3)}(1) + \gamma\alpha^3 G_1''(1) + \theta\alpha^3 G_1'(1)$$

hold, where β, γ and θ are real constants .

It is perhaps interesting to note that the conatant γ given here is free. According to other definitions, however, γ has to be 3β .

THROREM 2. (i). If $F_1(t)$ and $G_1(s)$ form a GC^2 joint, then their curvatures at the common point are the same.
(ii). If $F_1(t)$ and $G_1(s)$ form a GC^3 joint, then their curvatures and deflections at the common point are the same, respectively .

DEFINITION 2. Let $F_i(t)$ and $G_i(s) (i \ge 1)$ be the same as in Definition 1. Then $F_1(t)$ and $G_1(s)$ are called *a generalized GC^n joint* if

$$F_1(0) = G_1(1) ,$$
$$F_i'(0) = \alpha_i G_i'(1)$$

hold for $1 \le i \le n$, where $\alpha_i > 0$ are real constants .

THEOREM 3. $F_1(t)$ and $G_1(s)$ form a generalized GC^i joint, if and only if the first $i+1$ of the following equalities

$$F_1(0) = G_1(1),$$
$$F_1'(0) = \alpha_1 G_1'(1),$$
$$F_1''(0) = (G_1''(1) + \beta G_1'(1))\alpha_1\alpha_2,$$
$$F_1^{(3)}(0) = (G_1^{(3)}(1) + \gamma G_1''(1) + \theta G_1'(1))\alpha_1\alpha_2\alpha_3$$

hold, where $1 \le i \le 3$.

THEOREM 4. (i). If $F_1(t)$ and $G_1(s)$ form a generalized GC^2 joint, and their curvatures coincide at the common point, then

$$\alpha_1 = \alpha_2.$$

(ii). If $F_1(t)$ and $G_1(s)$ become a generalized GC^3 joint, and their curvatures and deflections coincide at the common point, respectively, then

$$\alpha_1 = \alpha_2 = \alpha_3.$$

Theorem 3 and Theorem 4 show that Definition 1 is reasonable .

THEOREM 5. The elements $k'_{F_1}(0), k_{F_1}(0), \alpha, \beta$ and γ are related by the following equality

$$\alpha^{-1}k'_{F_1}(0) = \frac{(G'_1(1)\times G_1^{(3)}(1), G'_1(1)\times G''_1(1))}{|G'_1(1)|^3 \cdot |G'_1(1)\times G''_1(1)|}$$

$$+(\gamma - 3\beta - 3\frac{(G'_1(1), G''_1(1))}{|G'_1(1)|^2})k_{F_1}(0).$$

2. GEOMETRIC CONTINUITY FOR SURFACES

Let $F(u,v)$ and $G(s,t)$ be two regular vectorvalued functions which describe two sufficiently smooth surface patches (cf. Fig.1), and let their common boundary curve be given by

$$F(0,v) = G(1,t),\ 0 \le v,\ t = f(v) \le 1,$$

where $f{:}[0,1] \mapsto [0,1]$ is a sufficiently smooth one-to-one mapping, and $f'(v) > 0$ for $0 \le v \le 1$.

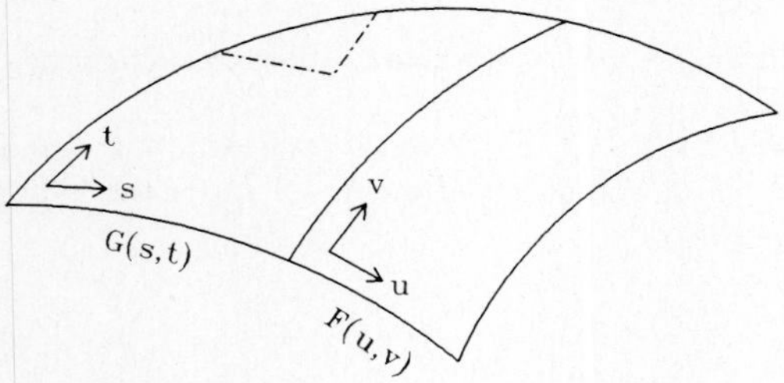

Figure 1

It is well known that $F(u,v)$ and $G(s,t)$ will be a GC^1 joint if both patches have coincident tangent planes at every point of their common boundary .

Therefore the relation

$$G_s(1,t) = a(v)F_u(0,v) + b(v)F_v(0,v)$$

with some functions $a(v) > 0$ and $b(v)$ ensures the GC^1 joint, where $t = f(v)$.

DEFINITION 3. $F(u,v)$ and $G(s,t)$ are called *a GC^n joint* if

$$\frac{\partial^{m_1}}{\partial t^{m_1}} \cdot \frac{\partial^{m_2}}{\partial s^{m_2}} G(1,t)$$

$$= \frac{\partial^{m_1}}{\partial t^{m_1}} (a(v)\frac{\partial}{\partial u} + b(v)\frac{\partial}{\partial v})^{m_2} F(0,v)$$

hold for $0 \le v, t = f(v) \le 1$ and $0 \le m_1 + m_2 \le n$, where

$$(a(v)\frac{\partial}{\partial u} + b(v)\frac{\partial}{\partial v})^m F(0,v)$$

$$= a(v)\frac{\partial}{\partial u}(a(v)\frac{\partial}{\partial u} + b(v)\frac{\partial}{\partial v})^{m-1} F(0,v)$$

$$+ b(v)\frac{\partial}{\partial v}(a(v)\frac{\partial}{\partial u} + b(v)\frac{\partial}{\partial v})^{m-1} F(0,v),$$

and

$$\frac{\partial^m}{\partial t^m} P(v) = \frac{1}{f'(v)} \frac{1}{\partial v}(\frac{\partial^{m-1}}{\partial t^{m-1}} P(v)),\ m \ge 1.$$

THEOREM 6. If $F(u,v)$ and $G(s,t)$ form a GC^2 joint, then their Gaussian curvatures and mean curvatures coincide at every point of their common boundary .

The method used in this section is also valid if $F(u,v)$ and/or $G(s,t)$ are triangular patches, and the relevant results can be obtained, too .

Approximation, Optimization and Computing:
Theory and Applications, A.G. Law and C.L. Wang (eds.)
Elsevier Science Publishers B.V. (North-Holland)

CHARACTERIZATION OF WEIGHTS IN WEIGHTED TRIGONOMETRIC APPROXIMATION

Shi, Xian – Liang

Department of Mathematics, Hangzhou University, Hangzhou, Zhejiang, people's Republic of China

In this paper the author has characterized weights in weighted trigonometric approximation of periodic integrable functions.

1. INTRODUCTION

The problem of characterization of weights in weighted rational approximation of piecewice smooth functions was introduced and studied in our two papers [1] and [2]. The aim of this paper is to characterize weights in weighted trigonometric approximation of periodic integrable functions.

2. DIRECT AND CONVERSE THEOREMS

Denote by $\prod_n$ the collection of trigonometric polynomials of degree n. Let U(x) be a periodic weight function with period 2π. For $1 \leqslant p \leqslant \infty$, denote the $L^p(U)$ norm of f by

$$\|f\|_{L^p(u)} = \left(\int_0^{2\pi} |f(x)U(x)|^p dx \right)^{1/p}.$$

The "distance" of f from $\prod_n$ is defined by

$$E_n(f)_{L^p(U)} = \inf_{T_n \in \prod_n} \|f - T_n\|_{L^p(U)}.$$

Denote by $\omega_k(f, t)_{L^p(U)}$ the modulus of smoothness of degree k for f in $L^p(U)$, i. e.

$$\omega_k(f, t)_{L^p(U)} = \sup_{0 \leqslant |h| \leqslant t} \|\Delta_h^k f(\cdot)\|_{L^p(U)},$$

where $\Delta_h^k = \Delta_h^{k-1}\Delta_h^1$ and $\Delta_h^1 f(x) = f(x+h) - f(x)$.
It is well known that if $U(x) = 1$ then we have the following direct estimate

$$E_n(f)_{L^p(U)} \leqslant C_1 \omega_k \left(f, \frac{1}{n+1}\right)_{L^p(U)}, \tag{1}$$

and the converse estimate

$$\omega_k\left(f, \frac{1}{n+1}\right)_{L^p(U)} \leqslant \frac{C_2}{(n+1)^k} \sum_{j=0}^{n} (j+1)^{k-1} E_j(f)_{L^p(U)}, \tag{2}$$

where C_1 and C_2 are constants independent on n and f.
Naturally, we will propose the following problems.
PROBLEM 1. How to characterize U for which the inequality (1) holds?
PROBLEM 2. How to characterize U for which the inequality (2) holds?
The following two theorems answer these two problems.

THEOREM 1. Let $1 \leqslant p \leqslant \infty$ and k be a positive integer. Suppose U and V are two given weights. Then there exists a constant C such that

$$E_n(f)_{L^p(U)} \leqslant C\, \omega_k\left(f, \frac{1}{n+1}\right)_{L^p(V)}$$

holds for any f and n, if and only if there exists a constant B such that $U(x) \leqslant BV(x)$, a. e.
THEOREM 2. Let k be a positive integer, and U and V two given weights. Suppose $1 \leqslant p \leqslant \infty$. Then there exists a constant C such that for any f and n holds

$$\omega_k\left(f, \frac{1}{n+1}\right)_{L^p(U)} \leqslant \frac{C}{(n+1)^k} \sum_{j=0}^{n} (j+1)^{k-1} E_j(f)_{L^p(V)}$$

if and only if there exists a constant B such that for any interval I, $|I| \leqslant 2\pi$, holds

$$\left(\frac{1}{|I|}\int_I U^p(x)\,dx\right)^{1/p} \left(\frac{1}{|I|}\int_I \left(\frac{1}{V(x)}\right)^{p'} dx\right)^{1/p'} \leqslant B, \tag{3}$$

where $1/p + 1/p' = 1$.
Next we discuss the approximation problem in weghted BMO spaces.
The BMO space was introduced by John and Nirenberg in [3], and the weighted BMO spaces were introduced by Muchenhoupt and Wheeden in [4].
For any interval I and $f \in L_{2\pi}$ denote $f_I = |I|^{-1}\int_I f(u)\,du$ and for a weight U write $U(E) = \int_E U(x)\,dx$, where E is any measurable set. A function f is called a BMO_U function if it satisfies the condition

$$\|f\|_{BMO_U} := \sup \frac{1}{U(I)} \int_I |f(x) - f_I|\,dx < \infty,$$

where "sup" is taken over all intervals I with $|I| \leqslant 2\pi$.
We have the following
THEOREM 3. Let U(x) be any weight and k a positive integer. Then there exists a constant C such that for any f and n holds

$$\omega_k\left(f, \frac{1}{n+1}\right)_{BMO_U} \leqslant \frac{C}{(n+1)^k} \sum_{j=0}^{n} (j+1)^{k-1} E_j(f)_{BMO_U},$$

if and only if U satisfies the following conditions :
i) for any interval I, $|I| \leqslant 2\pi$, holds

$$\left(\frac{1}{|I|}\int_I U(x)\,dx\right)\exp\left(\frac{1}{|I|}\int_I \log\frac{1}{U(x)}\,dx\right) \leqslant B < \infty$$

ii) for any interval I, $|I| \leqslant \pi$, holds

$$|I|\int_{|I|\leqslant t<\pi} \frac{U(x_I+t)+U(x_I-t)}{t^2}\,dt \leqslant \frac{B}{|I|}\int_I U(x)\,dx,$$

where x_I is the center of I.

In order to prove these theorems we generalize the Bernstein's inequality on trigonometric polynomials.

LEMMA 1. Let $1 \leqslant p \leqslant \infty$. Suppose that $U(x)$ and $V(x)$ are two given weights satisfying the condition (3) of the theorem 2. Then for any $T_n \in \prod_n$ holds

$$\|T_n'\|_{L^p(U)} + \|\widetilde{T}_n'\|_{L^p(U)} \leqslant Cn\|T_n\|_{L^p(V)},$$

where C is a constant, independent on n and T_n, and $\widetilde{T}_n$ denotes the conjugate function of T_n.

LEMMA 2. Let $U(x)$ be a weight satisfying the conditions i) and ii) of Theorem 3. Then for any $T_n \in \prod_n$ holds

$$\|T_n'\|_{BMO_U} + \|\widetilde{T}_n'\|_{BMO_U} \leqslant Cn\|T_n\|_{BMO_U},$$

where C is a constant independent on n and T_n.

3. LEBESGUE CONSTANTS

In this section we consider the Lebesgue constants of some linear means.

Suppose that $f(z) = \sum_{k=0}^{\infty} b_k z^k$ is an analytic function on $\{z: |z| \leqslant 1\}$ and $f = C$. Let $A = (a_{mk})$ be a matrix of summability defined by

$$a_{00} = 1,\ a_{0k} = 0\ \ (k \geqslant 1),$$

$$\prod_{j=1}^{m} \frac{f(z)+d_j}{f(1)+d_j} = \sum_{k=0}^{\infty} a_{mk} z^k \quad (m = 1,2,\cdots).$$

This method of summability is called the (f, d_n) method of summability, which was introduced by Smith [5]. If $b_k \geqslant 0$ and $d_j \geqslant 0$, then the method (f, d_n) is called to be non-negative. It is known that (see[5]) a non-negative (f, d_n) method of summability is regular, if and only if

$$H_m = \sum_{j=1}^{m} \frac{1}{f(1)+d_j} \to \infty \quad (m \to \infty).$$

Let $g \in L_{2\pi}$ and

$$S[g] = \frac{a_0}{2} + \sum_{k=1}^{\infty} a_k \cos kx + b_k \sin kx$$

be its Fourier series. Denote by $s_k(g, x)$ its k-th partial sum and

$$\sigma_m(g, x) = \sum_{k=0}^{\infty} a_{mk} s_k(g, x)$$

its m-th (f, d_n) mean. We will consider the norm of operator σ_m in the space $L^p(U)$, i. e. the quantities

$$L_m(A, U)_p = \sup_{\|\sigma_m(g,\cdot)\|_{L^p(U)} \leqslant 1} \|g\|_{L^p(U)},$$

where $m = 0,1,2,\cdots$. It is known that (see [6]), if the method (f, d_n) is non-negative and regular, then

$$L_m(A, U)_p \sim \begin{cases} C_1, & \text{when } 1<p<\infty, \\ C_2 \ln(m+2), & \text{when } p=1 \text{ or } \infty \end{cases}$$

We discuss the following problem.

PROBLEM 3. How to characterize $U(x)$ for which hold the following estimations

$$L_m(A, U)_p = \begin{cases} O(1), & \text{when } 1<p<\infty \\ O(\ln(m+2)), & \text{when } p=1 \text{ or } \infty\ ? \end{cases}$$

In order to solve this problem we introduce some conditions.

DEFINITION 1. Let $1 \leqslant p < \infty$. A weight $U(x)$ is called to satisfy A_p condition, if for any interval I, $|I| \leqslant 2\pi$, holds

$$\left(\frac{1}{|I|}\int_I U(x)^p dx\right)^{1/p}\left(\frac{1}{|I|}\int_I U(x)^{-p'} dx\right)^{1/p'} \leqslant C,$$

where $1/p + 1/p' = 1$

Define

$$M_{\log} g(x) = \sup_{\pi \geqslant r>0}\ \overline{\lim_{\varepsilon \to 0^+}} \frac{1}{2\int_{\varepsilon}^{r} \frac{du}{u}} \int_{\varepsilon \leqslant |u| \leqslant r} \frac{|g(x+u)|}{|u|}\,du$$

DEFINITION 2. A weight $U(x)$ is called to satisfy $A_{\log,1}$ condition if

$$M_{\log} U(x) \leqslant C\,U(x), \text{ a. e.}$$

We have the following

THEOREM 4. Let the (f, d_n) wethod be non-negative and regular.

i) If $1<p<\infty$, then a necessary and sufficient condition for

$$L_m(A, U)_p = O(1)$$

is that the weight $U(x)$ satisfies the A_p condition.

ii) If $p=1$, then a necessary and sufficient condition for

$$L_m(A, U)_p = O(\log(m+1)) \tag{4}$$

is that the weight $U(x)$ satisfies the $A_{\log,1}$ condition.

iii) If $p = \infty$, then a necessary and sufficient condition for (4) is that the function $1/U(x)$ satisfies the $A_{\log,1}$ condition.

REFERENCES

[1] Chui, C. K. and Shi, X. L., J. Approximation Theory, 54 (1988) pp. 180-195.

[2] Chui, C. K. and Shi, X. L., J. Approximation Theory, 54 (1988) pp. 196–209.

[3] John, F. and Nirenberg, L., Comm. Pure App. Math., 14 (1961) pp. 415–426.

[4] Muckenhoupt, B. and Wheeden, R. L., Studia Math., 54 (1976) pp. 221–237.

[5] Smith, G., Canad. J. Math., 17 (1965) pp. 506–526.

[6] Shi X. L., Chinese Math. Ann., 3(1982) pp. 365–374.

Approximation, Optimization and Computing:
Theory and Applications, A.G. Law and C.L. Wang (eds.)
Elsevier Science Publishers B.V. (North-Holland)

THE CONVERGENCE CRITERIA AND CRITICAL ORDER OF HERMITE–FEJÉR TYPE INTERPOLATION

YING GUANG SHI

Computing Center, Chinese Academy of Sciences, P.O.Box 2719, Beijing 100080, China

1. INTRODUCTION

Let $X := \{x_{Nk}, y_{Nj} : \ k = 1,2,\cdots,n; \ j = 1,2,\cdots,m(n \geq 1, m \geq 0); N := n+m = 1,2,\cdots\}$ be a matrix of nodes satisfying

$$-1 \leq x_{N1} < x_{N2} < \cdots < x_{Nn} \leq 1,$$

$$-1 \leq y_{N1} < y_{N2} < \cdots < y_{Nm} \leq 1$$

and

$$x_{Nk} \neq y_{Nj}, \quad k = 1,2,\cdots,n, \quad j = 1,2,\cdots,m.$$

In the sequel for simplicity we usually omit the superfluous notations. Define the following $2n+m$ uniquely determined polynomials of degree $\leq N_1 := 2n+m-1$

$$A_i(x_k) = \delta_{ik}, \quad A_i(y_j) = 0, \quad A_i'(x_k) = 0,$$
$$i,k = 1,\cdots,n, \quad j = 1,\cdots,m;$$
$$A_i^*(x_k) = 0, \quad A_i^*(y_j) = \delta_{ij}, \quad A_i^{*'}(x_k) = 0,$$
$$k = 1,\cdots,n, \quad i,j = 1,\cdots,m;$$
$$B_i(x_k) = 0, \quad B_i(y_j) = 0, \quad B_i'(x_k) = \delta_{ik},$$
$$i,k = 1,\cdots,n, \quad j = 1,\cdots,m.$$

Then by [1] for $f \in C$ (the space of all continuous functions on $[-1,1]$) the uniquely determined polynomial $S_N(f,x)$ of degree $\leq N_1$ satisfying that

$$\left.\begin{array}{l} S_N(f,x_k) = f(x_k), S_N'(f,x_k) = 0, \quad k = 1,2,\cdots,n; \\ S_N(f,y_j) = f(y_j), \quad j = 1,2,\cdots,m \end{array}\right\} \quad (1)$$

has the form ($\sum_{j=1}^{m} := 0$ for $m = 0$)

$$S_N(f,x) = \sum_{k=1}^{n} f(x_k)A_k(x) + \sum_{j=1}^{m} f(y_j)A_j^*(x). \quad (2)$$

Meanwhile

$$B_k(x) = \frac{(x-x_k)l_k^2(x)W(x)}{W(x_k)}, k = 1,2,\cdots,n, \quad (3)$$

where

$$W(x) = \begin{cases} 1, & \text{for } m = 0 \\ (x-y_1)(x-y_2)\cdots(x-y_m), & \text{for } m > 0 \end{cases}$$

$$w(x) = (x-x_1)(x-x_2)\cdots(x-x_n),$$

$$l_k(x) = \frac{w(x)}{(x-x_k)w'(x_k)}, k = 1,2,\cdots,n.$$

In Section 2 we give a very interesting result: Under some assumptions there exists an integer r ($\leq min\{n,m\}$, sometimes, $= 2$) such that if

$$\lim_{N\to\infty} \|S_N(x^i,x) - x^i\| = 0, i = 1,2,\cdots,r \quad (4)$$

where $\|\cdot\|$ stands for the Chebyshev norm, then

$$\lim_{N\to\infty} \|S_N(x^i,x) - x^i\| = 0, i = 1,2,\cdots,r,r+1,\cdots \quad (5)$$

In Section 3 we prove that the critical order of $S_N(f,x)$ is only $\frac{1}{N}$ if m is fixed and

$$y_{N1}, y_{N2}, \cdots, y_{Nm} \notin [x_{N1}, x_{Nn}], N = 1,2,\cdots \quad (6)$$

2. CONVERGENCE CRITERIA

Define by σ_N the number of sign changes in the sequence

$$W(x_{N1}), W(x_{N2}), \cdots, W(x_{Nn}).$$

Our first main result is

THEOREM 1. *If*

$$r := 2 + \sup_N \sigma_N < \infty,$$

then (4) implies (5).

Using the Banach theorem [2,p.216] we can immediately obtain the following convergence criterion.

THEOREM 2. *Under the assumptions of Theorem 1 the following statements are equivalent :*

(a) $S_N(f,\cdot)$ converges for each $f \in C$, i.e.,

$$\lim_{N\to\infty} \|S_N(f,\cdot) - f\| = 0, \quad \forall f \in C; \quad (7)$$

(b) (4) holds and

$$\left\| \sum_{k=1}^{n} |A_k| + \sum_{j=1}^{m} |A_j^*| \right\| = O(1) \quad (8)$$

(c) (8) holds and

$$\lim_{N\to\infty} \left\| \sum_{k=1}^{n} x_k^i B_k \right\| = 0, i = 0,1,\cdots,r-1. \quad (9)$$

For the special and important case when (6) holds it is easy to see that in this case $\sigma_N = 0$ and hence $r = 2$. For convenience of use we state the theorem for this case.

THEOREM 3. *Let (6) hold. Then*
(a) If

$$\lim_{N\to\infty} \|S_N(x^i, x) - x^i\| = 0, i = 1, 2, \tag{10}$$

then (5) is valid;
(b) (7) is equivalent to (8) and (10);
(c) (7) is equivalent to (8) and

$$\lim_{N\to\infty} \left\|\sum_{k=1}^{n} x_k^i B_k\right\| = 0, i = 0, 1. \tag{11}$$

3. CRITICAL ORDER

THEOREM 4. *Let m be fixed. Then for any sequence of positive numbers $e = \{e_n\}$ and for any matrix X there exist sets*

$$I_n := I_n(e, X)$$
$$= (\bigcup_{k=1}^{n} [x_k - c_k, x_k + d_k]) \bigcup (\bigcup_{j=1}^{m} [y_j - c'_j, y_j + d'_j])$$

with $c_k, d_k, c'_j, d'_j > 0$ such that $|I_n|$ (:=the measure of I_n) $\leq e_n$ and

$$\sum_{k=1}^{n} |(x - x_k)B_k(x)| \geq \frac{ce_n^{m+2}}{n}$$

for all $x \in [-1, 1] \setminus I_n$ and $n = 1, 2, \cdots$, where

$$c = 48^{-2}(8m)^{-m}.$$

Moreover, the order $\frac{1}{n}$ is the best possible.

To prove this theorem we need a related result [3, Theorem 1] which is as follows.

PROPOSITION. *Let $t \geq 1$ be an integer. Then for any sequence of positive numbers $e = \{e_n\}$ and for any matrix X with $m = 0$ there exist sets*

$$I'_n := I'_n(e, t, X) = \bigcup_{k=1}^{n} [x_k - c_k, x_k + d_k]$$

such that $|I'_n| \leq e_n$ and

$$\sum_{k=1}^{n} |(x - x_k)l_k(x)|^t \geq \frac{c'e_n^t}{n^{t-1}}$$

for all $x \in [-1, 1] \setminus I'_n$ and $n = 1, 2, \cdots$, where $c' = 24^{-t}$. Moreover, the order n^{1-t} is the best possible.

From the above theorem we may deduce an important result which states that the critical order of Hermite-Fejér type interpolation is $\frac{1}{N}$ if m is fixed and (6) is valid. Precisely we have

THEOREM 5. *If m is fixed and (6) is valid, then for any matrix X either*

$$\|S_N(x, x) - x\| \neq o(\frac{1}{N})$$

or

$$\|S_N(x^2, x) - x^2\| \neq o(\frac{1}{N}).$$

REFERENCES

[1] Vértesi, P., Hermite-Fejér Type Interpolation. III, Acta Math. Acad. Sci. Hungar., 34(1979), 67-84.
[2] Cheney, E.W., Introduction to Approximation Theory, McGraw-Hill, New York, 1966.
[3] Shi, Y.G., On Critical Order of Hermite-Fejér Type Interpolation, J. Approximation Theory, in print.

Approximation, Optimization and Computing:
Theory and Applications, A.G. Law and C.L. Wang (eds.)
Elsevier Science Publishers B.V. (North-Holland)

On Queueing Approximation Formulas for Mean Waiting Time[1]

Howard A. Sholl and Yiping Ding

Department of Computer Science and Engineering, and
Taylor L. Booth Center For Computer Applications and Research, University of Connecticut, Storrs, CT 06268, USA.

Abstract

A mean waiting time approximation formula (X-Y formula) with two parameters X and Y for the GI/G/1 queueing system is presented in this paper. By changing these two parameters, the formula covers all the mean waiting time formulas we know so far. In view of the mathematical difficulties already encountered with the GI/G/1 queue, it seems unlikely to have a simple mean waiting time formula which works for all the possible distribution functions of the interarrival and service times in terms of the mean and variance of the functions. Idle time distributions play an important role here. Since for a broad subclass of GI/G/1 queues, the ratio of the first two moments of the idle time distribution $(\overline{I^2})/(2\overline{I})$ is bounded in terms of the first two moments of the interarrival time distribution, it is possible to choose X and Y so that one can find a mean waiting time approximation formula which is exact for the M/G/1 queue and also exact/approximate for other types of single server queueing models (assuming an exact/approximation mean waiting time formula for that type of model is known). This allows one to develop customized analytic expressions to match specific classes of arrival and service distributions.

I. Introduction

Performance is one of the key factors that must be taken into account in the design, development, configuration, and tuning of a computer system. Computer systems have become so complex that intuition alone is not enough to predict their performance, and simulation models are difficult and costly to construct, validate, and run. Because of that mathematical modeling plays an important role.

A queueing model is a mathematic model used to represent a physical system in which there is contention for resources. The purpose of using this model is to predict the performance of the system so that we can control, design, and use the system effectively [Sholl 87] . To predict the system performance means to estimate performance parameters. One of the most important performance parameters is the mean system waiting time. Of course, the variance of the waiting time is also very important [Ding 87]. But in this paper, we only consider the mean response time.

A large number of papers have been written about queue waiting time. Two well-known results are the Pollaczek-Khinchine "equilibrium" theory for M/G/1 (P-K mean waiting time formula) and Lindley integral equation of waiting time distribution for GI/G/1 [Klein 75]. Although both results are exact solutions to the corresponding problems, the former is correct only for M/G/1, the latter is difficult to solve in general. So attempts were made, starting in 1962 with J. F. C. Kingman [King 62] , to break away from exact analytic techniques. There have been two such kinds of analytic researches. One is to find bounds (inequalities) for the mean waiting time in certain classes of GI/G/1 queues [King 70] [Mars 68] [Ott 87] [Rosb 87]. The objective of inequality techniques is to eliminate the need for detailed distribution analysis by providing upper and lower bounds for such entities as the mean waiting time and the probability of no wait. This technique has also been used for the waiting time distribution functions [Berg 76]. The other research direction is to find simple form approximation functions for waiting time distribution functions for a GI/G/1 queue [Fred 82].

Little work has been done on approximations for mean waiting time, although it is very useful from practical point of view. Notable exceptions are Kingman's formula [King62], which in fact is the upper bound of the mean waiting time, Marchal's formula [Marc74] , and Allen-Cunneen's formula [Alle78, 80]. In this paper, we will present a mean waiting time approximation formula with two parameters. This formula not only covers all the mean waiting time formulas we know so far, but also opens the door for some new approximation formulas. II. Background: Waiting Time Formulas for Single Server Queueing Models

II-1. M/G/1 Queueing Model

For M/G/1, we have the following famous P-K mean-value formula [Klein 76]:

$$W_{P\text{-}K} = \frac{\rho(1 + C_b^2)}{2(1-\rho)\mu} \qquad (1)$$

where
$\rho = \lambda/\mu$ is traffic intensity;
λ is mean arrival rate;
μ is mean service rate;
C_b^2 is squared coefficient of variation for service time.

Formula (1) states that mean waiting time increases as C_b^2, the coefficient of variation of the service time, and/or ρ, increase. If $C_b^2 = 0$, then service time is deterministic (M/D/1). It is quite obvious that term $\rho C_b^2 / 2(1-\rho)\mu$ is the contribution to the mean waiting time due to the randomness of the service times. We will see in section II-4 that $\rho C_b^2 / 2(1-\rho)\mu$ is also the upper bound for D/G/1.

II-2. GI/M/1 Queueing Model

Compared to the M/G/1 queueing model few analytical results are available for the GI/M/1 queueing model. Its mean waiting time is [Klein 76]:

$$W = \frac{\sigma}{\mu(1-\sigma)} \qquad (2)$$

where σ is the unique root of

[1] This research was partially supported by the National Science Foundation under Grant No. CCR-8701839 and partially by Dapco Industries and the Connecticut Department of Higher Education.

$$\sigma = A^*(\mu - \mu\sigma),\ 0 < \sigma < 1 \ . \tag{3}$$

Where $A^*(s)$ is the Laplace transform associated with a(t), the pdf for interarrival time.

Thus from (2) it is straight forward to derive the well-known mean waiting time formula for M/M/1:

$$W = \frac{\rho}{\mu(1-\rho)} \ . \tag{4}$$

While it would be nice to have explicit expressions for σ, it has only been obtained in special cases. For a given interarrival time distribution function, we must solve (3) in order to get σ. If the interarrival time distribution function is unknown, we cannot get any useful, explicit expression of mean waiting time for GI/M/1.

[Lips 87] shows that when ρ is close to one, σ can be expanded in a series depending on ever increasing moments:

$$\sigma(\rho) = 1 - \sigma'(1)(1-\rho)$$

where

$$\sigma'(1) = \frac{2/\lambda^2}{(\sigma_a^2 + 1/\lambda^2)} = \frac{2}{1 + C_a^2} \ ; \tag{6}$$

and σ_a^2 is a variance of interarrival time and C_a^2 is a squared coefficient of variation of interarrival time.

In (5), if we consider the first two terms only, we have

$$\sigma(\rho) \cong = \frac{C_a^2 + 2\rho - 1}{C_a^2 + 1} \ .$$

Thus we can get a mean waiting time approximation formula for GI/M/1:

$$W \cong \frac{C_a^2 + 2\rho - 1}{2\mu(1-\rho)} \equiv W_{GM} \ . \tag{7}$$

We can treat (7) as a heavy-traffic approximation formula for GI/M/1.

Note that when GI = M, formula (7) becomes formula (4), the mean waiting time formula for M/M/1. Later in section II-4, we will show that this approximation formula actually implies the following assumption. Given it has been a time t since the last customer arrived, the probability that a customer arrives in the next small interval δ is independent from t : a memoryless assumption.

II-3. GI/G/1 Queueing Model

So far we have considered queueing models that represent the contention for a single server either under the assumption that the interarrival time has an exponential distribution or under the assumption that the service time has an exponential distribution. When both the interarrival time and the service time have a general distribution there are very few results available. Even in some very simple cases for which explicit analytical solutions can be found, these solution are often too complicated to be of practical use. The following result exists for GI/G/1 [Mars 68]:

$$W = \frac{\sigma_a^2 + \sigma_b^2 + (\frac{1}{\lambda})^2(1-\rho)^2}{2(\frac{1}{\lambda})(1-\rho)} - \frac{\overline{I^2}}{2\overline{I}} \tag{8}$$

Where I is the duration of the idle period.

When $\rho \to 1$, we have the following heavy-traffic approximation formula for GI/G/1 [King 62]:

$$W_K = \frac{\sigma_a^2 + \sigma_b^2}{2(\frac{1}{\lambda})(1-\rho)} \tag{9}$$

As a matter of fact, the heavy-traffic approximation to the mean waiting time forms a strict upper bound for the mean waiting time in any GI/G/1 queue [Klein 76].

Since we do not have any precise formula in terms of the first few moments for GI/G/1, approximation formulas have come to play a role:

Marchal has proposed that the upper bound above be scaled down so that it is exact for M/G/1:

$$W_M = \frac{1 + C_b^2}{(\frac{1}{\rho})^2 + C_b^2}\left[\frac{\sigma_a^2 + \sigma_b^2}{2(\frac{1}{\lambda})(1-\rho)}\right] \tag{10}$$

Then John Cunneen and Arnold Allen found their approximation formula for GI/G/1 [Allen 78, 80]:

$$W_{A\text{-}C} = \frac{\rho(C_a^2 + C_b^2)}{2(1-\rho)\mu} \tag{11}$$

By comparing (1) and (11), we can see that what Allen and Cunneen really did was changing the "1" in the P-K formula into C_a^2

What is the relationship between W_K and $W_{A\text{-}C}$? If we transform the coefficient of variations in $W_{A\text{-}C}$ into variances, then we get

$$W_{A\text{-}C} = \frac{\lambda(\rho^2\sigma_a^2 + \sigma_b^2)}{2(1-\rho)} \ . \tag{12}$$

It is easy to see from (9) and (12) that the difference between W_K and $W_{A\text{-}C}$ is the coefficient of σ_a^2 (the ρ^2 factor).

II-4. IFR(DFR)/G/1 Queueing Model

From formula (8) we can see that the mean waiting time depends on only the first two moments of the interarrival, service, and idle distributions. In general these idle period moments are difficult to calculate. Any closer estimate or sharper bounds obtained for

$\overline{I^2}/2\overline{I}$, the mean residual life time of the idle time I, will directly help us to get better approximation formulas or bounds for the mean waiting time. In this section, we will discuss two subclasses of GI/G/1 model: IFR/G/1 and DFR/G/1. They are defined as [Mars68]:

Definition 1. A nondiscrete distribution F has Increasing (Decreasing) Failure Rate, denoted IFR (DFR), if and only if, for any $\delta > 0$,

$$\frac{F(t+\delta) - F(t)}{F^c(t)}$$

is nondecreasing (nonincreasing) for all $t \geq 0$, where $F^c(t) \equiv 1 - F(t) > 0$.

The physical interpretation of IFR(DFR) arrival distribution is: given it has been a time t since the last customer arrived, the probability that a new customer arrives in the next small interval δ is increasing (decreasing) in time t. For IFR(DFR), we have the following theorem [Mars68]:

Theorem 1. For IFR(DFR)/G/1 queues,

$$\frac{\overline{I^2}}{2\overline{I}} \leq (\geq) \frac{(C_a^2 + 1)}{2\lambda} \ .$$

From Theorem 1 and formula (8), we can get the following inequalities:

$$\frac{C^2_a + \rho^2 C^2_b}{2\lambda(1-\rho)} - \frac{(C^2_a + \rho)}{2\lambda} \le W_{IFR/G/1}$$

and

$$W_{IFR/G/1} \le \frac{C^2_a + \rho^2 C^2_b}{2\lambda(1-\rho)}$$

For the constant interarrival time (D/G/1), the probability that a new customer arrives in the next small interval δ is increasing in time t. So D/G/1 queues are a special subclass of IFR/G/1 queues. In the case of D/G/1, $C^2_a = 0$, we have

$$\frac{\rho C^2_b}{2\mu(1-\rho)} - \frac{1}{2\mu} \le W_{D/G/1} \le \frac{\rho C^2_b}{2\mu(1-\rho)} \quad . \tag{13}$$

As we mentioned in section II-1, the upper bound of $W_{D/G/1}$ is the contribution to the mean waiting time for M/G/1 queues due to the randomness of the service times.

Since the exponential distribution has a constant failure rate, it is both IFR and DFR [Kauf77]. If we start with (8) and assume that the interarrival time is independent from t (exponential), then

$$W_{14} = \frac{C^2_a + \rho C^2_b + \rho - 1}{2\mu(1-\rho)} \quad . \tag{14}$$

Formula (14) has some interesting properties:
* It is a lower bound of $W_{IFR/G/1}$. If $C^2_a = 0$ (D/G/1) then

$$W_{14} = \frac{\rho C^2_b}{2\mu(1-\rho)} - \frac{1}{2\mu} \quad ,$$

which is the lower bound of $W_{D/G/1}$.
* If $C^2_a = 1$ (M/G/1) then

$$W_{14} = \frac{\rho(C^2_b + 1)}{2\mu(1-\rho)} = W_{P-K} \quad .$$

* If $C^2_b = 1$ (GI/M/1) then

$$W_{14} = \frac{C^2_a + 2\rho - 1}{2\mu(1-\rho)} = W_{GM} \quad .$$

III. A General Waiting Time Formula for GI/G/1

In section II we introduced three approximation formulas for GI/G/1, namely Kingman's heavy traffic approximation formula (9), Marchal's formula (10) and Allen-Cunneen approximation formula (11)/(12). By comparing all these approximation formulas, we can see that the basis for them is the formula (9), the upper bound for mean waiting time in GI/G/1 queue. So it is very natural to introduce the following general waiting time formula (X-Y formula):

$$W_{X\text{-}Y} = \frac{\lambda(X\sigma_a^2 + Y\sigma_b^2)}{2(1-\rho)} \quad . \tag{15}$$

Since we also want the general formula be precise for M/G/1, we constrain

$$W_{X\text{-}Y} = W_{P\text{-}K} \quad .$$

That is

$$\frac{\lambda(X\sigma_a^2 + Y\sigma_b^2)}{2(1-\rho)} = \frac{\rho(1 + C_b^2)}{2(1-\rho)\mu}$$

$$\lambda(X\frac{1}{\lambda^2} + Y\sigma_b^2) = \frac{\lambda(1 + \sigma_b^2\mu^2)}{\mu^2} \quad ,$$

where $C_a^2 = 1$.

Finally, we obtain the condition:

$$\frac{1}{\lambda^2}X + Y\sigma_b^2 = \frac{1}{\mu^2} + \sigma_b^2 \quad . \tag{16}$$

Formula (15) can produce an infinite number of approximation formulas for GI/G/1. As long as X and Y satisfy the condition (16), these approximation formulas are precise for M/G/1. As matter of fact, all the formulas we discussed in the section II are special cases of this general formula:

(i). Let $X = Y = \dfrac{1 + C_b^2}{\frac{1}{\rho^2} + C_b^2} = A$, we get Marchal's formula (10). That is:

$$W_M = W_{A\text{-}A} = W_{X\text{-}Y}|_{X=A,\, Y=A}$$

(ii). Let $X = \rho^2$, $Y = 1$, we get Allen-Cunneen's formula (11). That is:

$$W_{A\text{-}C} = W_{X\text{-}Y}|_{X=\rho^2,\, Y=1}$$

(iii). Let $X = \dfrac{1}{\mu^2\sigma_a^2}$, $Y = 1$, we get the P-K formula (1). That is:

$$W_{P\text{-}K} = W_{X\text{-}Y}|_{X=\frac{1}{\mu^2\sigma_a^2},\, Y=1}$$

(iv). Let $X = 1$, $Y = 1$, we get Kingman's formula (9). That is:

$$W_K = W_{X\text{-}Y}|_{X=1,\, Y=1}$$

Note that $X = 1$ and $Y = 1$ do not satisfy condition (16), if $\lambda \neq \mu$, so in general, Kingman's formula is not exact for M/G/1. In fact, it is not exact even for M/M/1.

IV. Some New Approximation Formulas

Based on the general approximation formula presented above, we derive some new waiting time formulas.

Formula W_1

From (ii) in section III we know that if $X = \rho^2$ and $Y = 1$ we get the Allen-Cunneen formula. Now we let $X = 1$. Solving (16) we get

$$X = 1 \ ,$$

$$Y = \frac{1}{\sigma_b^2}[\frac{1}{\mu^2} - \frac{1}{\lambda^2} + \sigma_b^2] \ .$$

Thus formula W_1 is

$$W_1 = \frac{\lambda(\sigma_a^2 + \sigma_b^2 + \frac{1}{\mu^2} - \frac{1}{\lambda^2})}{2(1-\rho)} \tag{17}$$

For M/G/1, $C_a^2 = 1$, and $W_1 = W_{P\text{-}K}$. So it is exact for M/G/1 and is always less than W_K

Formula W_2

Solving the following equations:

$$\frac{1}{\lambda^2}X + Y\sigma_b^2 = \frac{1}{\mu^2} + \sigma_b^2 \tag{16}$$

and

$$W_{X\text{-}Y} = W_{GM} \quad ,$$

or using the GM heavy traffic approximation (7)

$$\frac{\lambda(X\sigma_a^2 + Y\sigma_b^2)}{2(1-\rho)} = \frac{C_a^2 + 2\rho - 1}{2\mu(1-\rho)} \quad ,$$

that is

$$\frac{1}{\lambda^2}X + \frac{C_b^2}{\mu^2}Y = \frac{1}{\mu^2} + \frac{C_b^2}{\mu^2}$$

$$\frac{C_a^2}{\lambda^2}X + \frac{1}{\mu^2}Y = \frac{C_a^2 + 2\rho - 1}{\lambda\mu},$$

we can get

$$X = \frac{\lambda^2(1 + C_b^2) - \lambda\mu C_b^2(C_a^2 + 2\rho - 1)}{\mu^2(1 - C_a^2 C_b^2)}$$

and

$$Y = \frac{\mu(C_a^2 + 2\rho - 1) - \lambda C_a^2(1 + C_b^2)}{\lambda(1 - C_a^2 C_b^2)}$$

Thus

$$W_2 = \frac{\lambda(\frac{C_a^2}{\lambda^2}X + \frac{C_b^2}{\mu^2}Y)}{2(1-\rho)}$$

$$= \frac{\lambda C_a^2(1 + C_b^2)(1 - C_b^2) + \mu C_b^2(C_a^2 + 2\rho - 1)(1 - C_a^2)}{2(1-\rho)\mu^2(1 - C_a^2 C_b^2)} \quad (18)$$

Formula W_2 generalizes the P-K formula for M/G/1 and the heavy-traffic approximation formula for GI/M/1. So now we have one formula for both M/G/1 and GI/M/1 cases, and this formula is exact for M/G/1.

W_2 has the following properties:

i) For M/G/1, since $C_a^2 = 1$, we have

$$W_2 = \frac{\lambda(1 + C_b^2)}{2(1-\rho)\mu^2} = \frac{\rho(1 + C_b^2)}{2(1-\rho)\mu} = W_{P\text{-}K.}$$

ii) For GI/M/1, we have

$$W_2 = \frac{(C_a^2 + 2\rho - 1)}{2(1-\rho)\mu} = W_{GM}.$$

iii) For D/G/1, we have

$$W_2 = \frac{C_b^2(2\rho - 1)}{2(1-\rho)\mu} = \frac{\rho C_b^2}{2(1-\rho)\mu} - \frac{C_b^2}{2\mu}.$$

If $0 \le C_b^2 \le 1$, W_2 is between lower bound and upper bound of $W_{D/G/1}$. In this case,

$$W_{A-C} = \frac{\rho C_b^2}{2(1-\rho)\mu}$$

which is the upper bound of $W_{D/G/1}$

iv) For D/M/1, we have

$$W_2 = \frac{\rho}{2(1-\rho)\mu} - \frac{1}{2\mu},$$

$$W_{A-C} = \frac{\rho}{2(1-\rho)\mu}.$$

They are lower and upper bounds of $W_{D/M/1}$ respectively.

v) For GI/D/1, we have

$$W_2 = \frac{\rho C_a^2}{2(1-\rho)\mu} = W_{A-C}.$$

Note: If $C_a^2 = C_b^2$ and $\rho \to 1$, then we have

$$W_2 \to W_{A-C}$$

or

$$W_2 \to W_K.$$

VI. Summary

In this paper, we present a mean waiting time approximation formula with parameters X and Y (X-Y formula) for the GI/G/1 queueing system. By choosing X and Y, one can find a mean waiting time approximation formula which is exact for M/G/1 queue and exact or approximate for other type of single server queueing model, depending on whether we know an exact or approximation mean waiting time formula for that type of model. This allows one to develop customized analytic expressions to match specific classes of arrival and service distributions. As an example, we derive a new mean waiting time approximation formula W_2, which is exact for M/G/1, and a heavy traffic approximation for GI/M/1. Since W_2 is based on a heavy traffic approximation formula for GI/M/1, a better approximation mean waiting time formula for GI/M/1 may yield a better approximation mean waiting time formula for GI/G/1.

References

[Allen 78] A. O. Allen, " Probability , Statistics, and Queueing Theory with Computer Science Applications ", Academic Press, New York, 1978.

[Allen 80] A. O. Allen, " Queueing Models of Computer Systems ", IEEE, April, 1980.

[Ding 87] Y. P. Ding, " Optimal Performance Control in Real Time Operating Systems ", Master's Thesis, Department of Computer Science and Engineering, and Computer Application and Research Center, University of Connecticut, Dec. 1987.

[Fred 82] A. A. Fredericks, "A Class of Approximations for the Waiting Time Distribution in a GI/G/1 Queueing System", Bell System Technical Journal, Vol. 61, No. 3, March 1982.

[King 62] J. F. C. Kingman, "Some inequalities for the Queue GI/G/1", Biometrika 49, 315-324, 1962.

[King 70] J. F. C. Kingman, " Inequalities in the theory of Queues", J. Roy. Stat. Soc. B 59, 102-110.

[Klein 75] L. Kleinrock, " Queueing Systems Volume 1: Computer Applications ", John Wiley, New York, 1975.

[Klein 76] L. Kleinrock, " Queueing Systems Volume 2: Computer Applications ", John Wiley, New York, 1976.

[Lips 87] L. Lipsky, " The Approximation of Density Functions for Use in Queueing Theory ", Research Proposal, Department of Computer Science and Engineering, University of Connecticut, November, 1987.

[Marc74] W. G. Marchal, "Some Simple Bounds and Approximations in Queueing", Technical Memorandum Serial T-294, Institute for Management Science and Engineering, The George Washington University, Jan. 1974.

[Mars 68] K. T. Marshall, "Some Inequalities in Queueing", Opens. Res. 16, 651-665, 1968.

[Ott 87] T. J. Ott, "Simple Inequalities for the D/G/1 Queue", Oper. Res., Vol. 35, No. 4, 1987.

[Rosb 87] Z. Rosber, " Bounds on the expected waiting time in a GI/G/1 Queue : Upgrading for Low Traffic intensity", J. Appl. Prob. 24, 749-757, 1987.

[Sholl 87 *a*] H. A. Sholl, Y. P. Ding, " Performance Parameter Control in Real Time Operating Systems ", Fourth Workshop on Real-Time Operating Systems, Cambridge, Ma., July, 1987.

Approximation, Optimization and Computing:
Theory and Applications, A.G. Law and C.L. Wang (eds.)
Elsevier Science Publishers B.V. (North-Holland)

AN ESTIMATE OF THE RATE OF CONVERGENCE OF FOURIER SERIES OF A FUNCTION OF WIENER'S CLASS*

RAFAT N. SIDDIQI

Department of Mathematics, Kuwait University, Kuwait, Arabian Gulf

Abstract. In this paper we give an estimate of the rate of convergences of Fourier series of a function of Wiener's class which is strictly larger class than the class of functions of bounded variation. This extends a theorem of R. Bojanic [1]. We further study the estimate of rate of convergence of the conjugate Fourier series of a function of Wiener's class and get another extension of a theorem of W.H. Young [6].

1. INTRODUCTION

Let f be a 2π-periodic function defined on $[0,2\pi]$. We set

$$\overset{b}{\underset{a}{V_p}}(f) = \sup \left\{ \sum_{i=1}^{n} |f(t_i) - f(t_{i-1})|^p \right\}^{1/p} \quad (1 < p < \infty),$$

where suprema has been taken with respect to all partitions $P: a = t_o < t_1 < t_2 < \ldots < t_n = b$ of any segment $[a,b]$ contained in $[0,2\pi]$. We call $\overset{b}{\underset{a}{V_p}}(f)$ the p-th total variation of f on $[a,b]$. If we denote p-th total variation of f on $[0,2\pi]$ by $V_p(f)$, then we can define Wiener's class simply by

$$V_p = \{f: V_p(f) < \infty\} \quad . \tag{1}$$

It is clear that V_1 is an ordinary class of functions of bounded variation, introduced by Jordan. The class V_p was first introduced by N. Wiener [5]. He [5] showed that the functions of class V_p could only have simple discontinuities. We note [3] that

$$V_{p_1} \subset V_{p_2} \quad (1 \le p_1 \le p_2 < \infty) \tag{2}$$

is a strict inclusion. Hence for an arbitrary $1 \le p < \infty$, Wiener's class V_p is strictly larger than the class V_1 .

2. RATE OF CONVERGENCE OF FOURIER SERIES

Let $f \in V_p$ $(1 \le p < \infty)$ and let

$$S(f) = \frac{1}{2} a_o + \sum_{n=1}^{\infty} (a_n \cos nx + b_n \sin nx) \tag{3}$$

be its Fourier series. Wiener [5] proved the following theorem.

Theorem A.

If $f \in V_p$ $(1 \le p < \infty)$ then $S(f)$ converges almost everywhere in $[0,2\pi]$.

Recently R. Bojanic [1] gave an estimate of the rate of convergence of Fourier series of functions of bounded variation in the following form:

Theorem B

If $f \in V_1$, then

$$\left|S_n(x) - \frac{1}{2}\{f(x+0) + f(x-0)\}\right| \le \frac{3}{n} \sum_{k=1}^{n} \overset{\pi/k}{\underset{o}{V_1}}(g_x(t)) \qquad |x| \le \pi ,$$

where $\overset{k}{\underset{o}{V_1}}(g_x(t))$ is the first total variation of

$$g_x(t) = f(x+t) + f(x-t) - f(x+0) - f(x-0)$$

in $[0,k]$ and $S_n(x)$ is the n-th partial sum of $S(f)$.

The main object of this paper is to obtain the rate of convergence of Fourier series of $f \in V_p (1 \le p < \infty)$ which is strictly larger class than the class V_1 . To be precise, first we prove the following theorem.

Theorem 1

If $f \in V_p$ $(1 \le p < \infty)$, then

*This research work was supported by the grant No.SM054 of Research Council of Kuwait University.

$$|S_n(x) - \tfrac{1}{2}\{f(x+0) + f(x-0)\}| \le \frac{3M}{n} \sum_{k=1}^{n} \overset{\pi/k}{\underset{o}{V_p}} (g_x(t))$$

for $|x| \le \pi$ where M is a positive real number.

3. PROOF

Since we can write

$$S_n(x) - \frac{1}{2}\{f(x+0) + f(x-0)\}$$

$$= \pi^{-1} \int_0^{\pi} \frac{\sin nt}{t} g_x(t)\,dt + o(1)$$

$$= \pi^{-1} \sum_{k=0}^{n-1} \int_{\frac{k\pi}{n}}^{\frac{(k+1)\pi}{n}} \frac{\sin nt}{t} g_x(t)\,dt + o(1).$$

By change of variable the above expression can be written into

$$= (\pi)^{-1} \sum_{k=0}^{n} \int_{\frac{(k-1)\pi}{n}}^{\frac{k\pi}{n}} \left(\frac{g_x(t + 2k\pi/n)}{t + 2k\pi/n} - \frac{g_x(t + (2k+1)(\pi/n))}{t + (2k+1)(\pi/n)} \right) \sin nt\,dt + o(1)$$

$$= (\pi)^{-1} \int_0^{\frac{\pi}{n}} \left[\sum_{k=1}^{[n/2]} \left(\frac{g_x(t + 2k\pi/n)}{t + 2k\pi/n} - \frac{g_x(t + (2k+1)(\pi/n))}{t + (2k+1)(\pi/n)} \right) \right] \sin nt + o(1)$$

$$= (\pi)^{-1} \int_0^{\frac{\pi}{n}} \left[\sum_{k=1}^{[\varepsilon n]} + \sum_{[\varepsilon n]+1}^{[n/2]} \right] \sin nt\,dt + o(1)$$

$$= I_n(\varepsilon) + J_n(\varepsilon) + o(1) \tag{4}$$

Now we consider

$$I_n(\varepsilon) = (\pi)^{-1} \int_0^{\frac{\pi}{n}} \left[\sum_{k=1}^{[\varepsilon n]} \left(\frac{g_x(t + 2k\pi/n)}{t + 2k\pi/n} - \frac{g_x(t + (2k+1)(\pi/n))}{t + (2k+1)(\pi/n)} \right) \right] \sin nt\,dt$$

$$\le (\pi)^{-1} \int_0^{\frac{\pi}{n}} \left[\sum_{k=1}^{[\varepsilon n]} \frac{g_x(t+2k\pi/n) - g_x(t + (2k+1)(\pi/n))}{t + 2k\pi/n} \right] \sin nt\,dt +$$

$$+ \frac{1}{n} \int_0^{\frac{\pi}{n}} \left[\sum_{k=1}^{[\varepsilon n]} \frac{g_x(t + (2k+1)(\pi/n))}{(t+2k\pi/n)(t+(2k+1)(\pi/n))} \right] \sin nt\,dt$$

$$= I_{n_1}(\varepsilon) + I_{n_2}(\varepsilon) . \tag{5}$$

Applying Hölder's inequality on the sum of integrand, we obtain,

$$I_{n_1}(\varepsilon) \le (\pi)^{-1} \left[\int_0^{\frac{\pi}{n}} \left| \sum_{k=1}^{[\varepsilon n]} |g_x(t + 2k\pi/n) - g_x(t + (2k+1)(\pi/n))|^p \right]^{1/p} \times \left[\left(\sum_{k=1}^{[\varepsilon n]} \left|\frac{1}{2k\pi/n}\right|^q \right)^{1/q} \right] \sin nt\,dt$$

$$\le (\pi)^{-1} \int_0^{\frac{\pi}{n}} \overset{3\varepsilon\pi}{\underset{o}{V_p}} (g_x(t) \left(\sum_{k=1}^{\varepsilon n} \left|\frac{1}{2k\pi/n}\right|^q \right)^{1/q} \sin nt\,dt \tag{6}$$

$$\le \frac{M_1}{n} \overset{3\varepsilon\pi}{\underset{o}{V_p}} (g_x(t)) \tag{7}$$

because the series of right hand side of (6) converges for $q > 1$, hence we can find a positive number M_1 satisfying the above inequality (7). And also

$$I_{n_2}(\varepsilon) =$$

$$\frac{1}{n} \int_0^{\frac{\pi}{n}} \sum_{k=1}^{[\varepsilon n]} \left| \frac{g_x(t + (2k+1)(\pi/n)}{(t+2k\pi/n)(t+(2k+1)(\pi/n)} \right| \sin nt\,dt$$

$$\le \frac{1}{n} \sup_{0 \le t \le 3\varepsilon\pi} \{g_x(t)\} \le \frac{1}{n} \overset{3\varepsilon\pi}{\underset{o}{V_p}} (g_x(t))\, M_2 \tag{8}$$

where M_2 is a positive real number. Hence from (7) and (8), we obtain

$$I_n(\varepsilon) \le \frac{M_1}{n} \overset{3\varepsilon\pi}{\underset{o}{V_p}} (g_x(t)) + \frac{M_2}{n} \overset{3\varepsilon\pi}{\underset{o}{V_p}} g_x(\varepsilon)$$

$$\le \frac{M_3}{n} \overset{3\varepsilon\pi}{\underset{o}{V_p}} (g_x(t)) \tag{9}$$

where $M_3 = M_1 + M_2$. Similarly we can prove that

$$|J_n(\varepsilon)| \le \frac{1}{n} \overset{\pi}{\underset{o}{V_p}} (g_x(t) + \frac{1}{n} \le \frac{2}{n} \overset{\pi}{\underset{o}{V_p}} (g_x(t)). \tag{10}$$

Collecting the terms of (9) and (10), we obtain

$$|S_n(x) - \frac{1}{2}\{f(x+0) + f(x-0)\}|$$

$$\leq \frac{M_3}{n} \overset{3\varepsilon\pi}{\underset{o}{V_p}} (g_x(t)) + \frac{2}{n} \overset{\pi}{\underset{o}{V_p}} (g_x(t))$$

$$\leq \frac{3M}{n} \sum_{k=1}^{n} \overset{\pi/k}{\underset{o}{V_p}} (g_x(t))$$

where M is a positive real number. This completes the proof of theorem 1 .

Since $\overset{t}{\underset{o}{V_p}} (g_x(t))$ is continuous when $f(x)$ is continuous, it follows that

$$\frac{1}{n} \sum_{k=1}^{n} \overset{\pi/k}{\underset{o}{V_p}} (g_x(t)) \to 0 \quad \text{as} \quad n \to \infty .$$

Thus we deduce the following result which is generalized version of Theorem A due to Wiener [5] (cf. Zygmund [7] p.59) .

Corollary 1

If $f \in V_p$ $(1 \leq p < \infty)$ then the Fourier series of f converges to $\frac{1}{2}[f(x+0) + f(x-0)]$ at every $x \in [0,2\pi]$. In particular, $S(f)$ converges to $f(x)$ at every point x of continuity of f of V_p $(1 \leq p < \infty)$.

4. RATE OF CONVERGENCE OF THE CONJUGATE FOURIER SERIES

Let $f \in V_p$ $(1 \leq p < \infty)$ and let the series conjugate to (3) is given by

$$\sum_{n=1}^{\infty} (a_n \sin nx - b_n \cos nx) . \tag{11}$$

We denote by $\overline{S_n(x)}$ the n-th partial sums of the conjugate Fourier series (11). We also denote by

$$\psi_x(t) = \frac{1}{2}\{f(x+t) - f(x-t)\},$$

$$\bar{f}(x) = \lim_{\varepsilon\to 0^+} \bar{f}(x,\varepsilon) = \lim_{\varepsilon\to 0^+} \left\{\frac{-2}{\pi}\int_{\varepsilon}^{\pi} \frac{\psi_x(t)}{2\tan t/2}\, dt\right\} \tag{12}$$

and

$$\bar{D}_n(t) = \sum_{k=1}^{n} \sin kt = \frac{\cos t/2 - \cos(n+\frac{1}{2})t}{2 \sin t/2}$$

There is following well known theorem due to W.H. Young [6] (cf. Zygmund [7] p.59) on convergence of the conjugate Fourier series of a function of bounded variation.

Theorem C

If $f \in V_1$, then the conjugate series (11) converges to $\bar{f}(x)$ at every $x \in [0,2\pi]$, provided that $\bar{f}(x)$ exists.

There is a classical result of Privalov (cf. Zygmund [7] Chapter IV, Theorem 3.1) which states that $\bar{f}(x)$ exists a.e. for any $f \in L^1$.

The main object of this section is to extend Theorem C into strictly larger class V_p for every p. To be precise, we give the simple proof of the following theorem in which we obtain the estimate of the rate of convergence of the conjugate Fourier series of a function of V_p .

Theorem 2

If $f \in V_p$ $(1 \leq p < \infty)$, then

$$|\bar{S}_n(x) - \bar{f}(x, \frac{\pi}{n})| \leq \frac{M}{n} \sum_{k=1}^{n} \overset{\pi/k}{\underset{o}{V_p}} (\psi_x)$$

where $\overset{\pi/k}{\underset{o}{V_p}} (\psi_x)$ denotes the total p-th variation of ψ_x on the interval $(0, \frac{\pi}{k}) \subset (0,\pi)$ and M is an absolute constant independent of n .

5. PROOF

Since we can write

$$\overline{S_n}(x) = -\frac{2}{\pi}\int_0^{\pi} \psi_x(t)\, \overline{D_n}(t)\, dt , \tag{13}$$

hence

$$\overline{S_n}(x) - \bar{f}(x,\pi/n)$$

$$= -\frac{2}{\pi}\int_0^{\pi} \frac{\psi_x(t)}{2\sin t/2} (\cos t/2 - \cos(n+\frac{1}{2})t)$$

$$+ \frac{2}{\pi}\int_{\pi/n}^{\pi} \frac{\psi_x(t)}{2\tan t/2}\, dt$$

$$= -\frac{2}{\pi}\int_0^{\pi/n} \frac{\psi_x(t)}{2\sin t/2} (\cos t/2 - \cos(n+\frac{1}{2})t)dt$$

$$+ \frac{2}{\pi}\int_{\pi/n}^{\pi} \frac{\psi_x(t)}{2\sin t/2} \cos(n+\tfrac{1}{2})t\, dt = I_1 + I_2 , \text{ say.}$$

Using Hölder's inequality (cf. Rudin [2]), we get

$$|I_1| = \left|\frac{2}{\pi}\int_0^{\pi/n} \psi_x(t)\ \overline{D_n}(t)dt\right|$$

$$\le \frac{2}{\pi}\left\{\left(\int_0^{\pi/n} |\psi_x(t)-\psi_x(0)|^p\right)^{1/p}\left(\int_0^{\pi/n} |\overline{D_n}(t)|^q\right)^{1/q}\right. \tag{14}$$

$$\le \frac{2}{\pi}\left(\int_0^{\pi/n}\left[\overset{\pi/n}{\underset{o}{V_p}}(\psi_x)\right]^p dt\right)^{1/p}\left(\int_0^{\pi/n} n^q\, dt\right)^{1/q}$$

$$\le \frac{2}{n}\ \overset{\pi/n}{\underset{o}{V_p}}(\psi_x)\ .$$

Now consider the integral

$$I_2 = \frac{2}{\pi}\int_{\pi/n}^{\pi} \frac{\psi_x(t)\cos(n+\tfrac{1}{2})t}{2\sin t/2}\, dt$$

$$= -\frac{2}{\pi}\int_{\pi/n}^{\pi} \psi_x(t)\ dg(t) \tag{15}$$

where

$$g(t) = \int_t^{\pi} \frac{\cos(n+\tfrac{1}{2})x}{2\sin x/2}\, dx\ . \tag{16}$$

Hence using integration by parts, we get

$$|I_2| \le \frac{2}{\pi}|\psi_x(\tfrac{\pi}{n})\ g(\tfrac{\pi}{n})| + \frac{2}{\pi}\int_{\pi/n}^{\pi} |g(t)\ d\psi_x(t)|. \tag{17}$$

But by using the second mean value theorem, we get

$$g(t) = \frac{1}{2\sin t/2}\int_t^{\xi}\cos(n+\tfrac{1}{2})x\ dx\ , \quad (t<\xi<\pi),$$

hence

$$|g(t)| \le \frac{2}{n}\,\frac{1}{2\sin t/2} \le \frac{\pi}{nt}\ . \tag{18}$$

From (17) and (18) we obtain

$$|I_2| \le \frac{2}{n}\ \overset{\pi/n}{\underset{o}{V_p}}(\psi_x) + \frac{2}{n}\int_{\pi/n}^{\pi}\frac{|d\psi_x(t)|}{t}\ . \tag{19}$$

But $\overset{\pi/n}{\underset{o}{V_p}}(\psi_x)$ tends to zero as $n\to\infty$, and by using Hölder's inequality (cf. Rudin [2]), we get

$$\left|\int_{\pi/n}^{\pi}\frac{|d\psi_x(t)|}{t}\right|$$

$$\le \left(\int_{\pi/n}^{\pi}|\psi_x(t)-\psi_x(0)|^p\right)^{1/p}\left(\int_{\pi/n}^{\pi}\frac{dt}{t^q}\right)^{1/q} \tag{20}$$

where $\frac{1}{p}+\frac{1}{q}=1$. Since $|\psi_x(t)-\psi_x(0)|^p$ is majorised by $[V_p(\psi_x)]^p$ and the integral $\int_{\pi/n}^{\pi} t^{-q}\, dt$ is convergent for $q>1$, hence we can find a positive number M independent of n such that

$$I_2 \le \frac{M}{n}\sum_{k=1}^{n}\ \overset{\pi/k}{\underset{o}{V_p}}(\psi_x)\ .$$

This completes the proof of Theorem 2.

Since the estimates of the items of I_1 and of I_2 in the proof of Theorem 2 tend to zero as $n\to\infty$, we deduce the following Corollary 2 as an extension of Theorem C into Wiener's class V_p .

Corollary 2

If $f\in V_p$ $(1\le p<\infty)$, then the conjugate series (11) converges to $f(x)$ at every $x\in[0,2\pi]$, provided that $\bar{f}(x)$ exists.

6. ACKNOWLEDGEMENTS

I like to thank the Research Council of Kuwait University for granting me Research Project SM054 under which this paper was completed.

REFERENCES

[1] R. Bojanic, An estimate of the rate of convergence of Fourier series of a function of bounded variation, Publ.Inst. Math. (Beograd), 26(1979), 57-60 .

[2] W. Rudin, Fourier Analysis on groups, New York, Interscience Pulbishers, 1962.

[3] R.N. Siddiqi, On jump of function of higher variation, Rev. Tec. Ing. Univ. Zulia, Vol.10 (1) (1987), 75-80.

[4] R.N. Siddiqi, Generalized absolute continuity of a function of Wiener's class, Bull. Austral. Math. Soc. Vol.22 (1980), 253-258.

[5] N. Wiener, The quadratic variation of a function and its Fourier coefficients, Massachusetts J. Math., Vol.3 (1924), 72-94.

[6] W.H. Young, Konvergenzbedingungen fur die verwandte Reichc einer Fourier Schen Reich Munch. Sitzungsberichte, 41(1911), 361-371.

[7] A. Zygmund, Trigonometric series, Vol.I, Cambridge University Press (1959).

Department of Mathematics
Kuwait University,Kuwait.

Approximation, Optimization and Computing:
Theory and Applications, A.G. Law and C.L. Wang (eds.)
Elsevier Science Publishers B.V. (North-Holland)

A REMARK ON THE APPROXIMATION IN THE LOCALLY CONVEX SPACE

Wenhua Song

Inst. of Math., Dalian Univ. of Technology, Dalian 116024, China

ABSTRACT: In this paper, we improve the theorem of [1], and give two applications of this improvement. In these applications, the subset need not be f-proximinal.

1. INTRODUCTION

Let X and Y be a pair of linear spaces put in duality by a separating bilinear form $<,>$, and equipped with locally convex topologies compatible with the pairing. A topology on X is said to be *a compatible topology* if it is a separated locally convex topology for which continuous linear functionals on X are precicely of the form

$$< ,y >: Y \to < x,y >,\ for\ x\ in\ X.$$

Suppose f is a continuous convex function defined on X and satisfies $f(0) = 0$. Given a nonempty subset U of X and x in X. We denote $f_U(x) = inf\{f(x-u); u \in U\}$ and $P_f(x) = \{u \in U; f(x-u) = f_U(x)\}$. The set-valued mapping $P_f : x \to P_f(x)$ is called to be *f-projection supported on U*. U is called to be *f-proximal* (resp. *f-Chebyshev*) if $P_f(x) \neq \phi$ (resp. $P_f(x)$ is a singleton set) for each x in X.

In thix paper, we improve the theorem of [1] and obtain some results by this improvement. In section 2, we get certain result similar to the proposition of [1] without the proximinality of U. In section 3, we give two applications of the theorem 2.3.

2. THE MAIN THEOREM

Let f be a continuous convex function and $f(0) = 0$. For r in R, $r > 0$, let $S_r = \{x \in X; f(x) \leq r\}$. Then S_r is a convex absorbing subset containing the origin 0 in its interior. Let

$$P_r(x) = inf\{t > 0; x \in t \cdot S_r\}, x \in X$$

denote the Minkowski guage of S_r. Then P_r is a non-negative continuous sub-linear functional.

LEMMA 2.1. (P. Govindarajulu, [2]) Let f be a non-negative continuous convex function on X and $f(0) = 0$, satisfying the following condition:

There exists a continuous bijection $\psi : R_+ \to R_+ (R_+ = [0, \infty))$ such that:

$$(*) \quad f(rx) = \psi(r)f(x), for\ r > 0, x \in X.$$

Then one has $S_r = (1/\lambda)S_{r\psi(\lambda)}$ and $P_r = P_{r\psi(\lambda)}$, for $r, \lambda > 0$.

REMARK: By lemma 2.1, we have $P_{P_r}(x) = P_{P_\lambda}(x)$ for each x in X and any $r, \lambda > 0$.

LEMMA 2.2. We assume that f satisfies the conditions of lemma 2.1. Let r be a positive constant. Then we have

1) $f(x) = r \iff P_r(x) = 1$.

2) For each $\varepsilon > 0$, there exists a $\delta > 0$ such that

a) $f(x) \leq r - \varepsilon \iff P_r(x) \leq 1 - \delta$,

b) $f(x) \geq r + \varepsilon \iff P_r(x) \geq 1 + \delta$.

PROOF: 1) is the lemma 3.3 of [2].

$a)$ of 2). By assumption, ψ is a strictly increasing function and $\psi(1) = 1$ (except for the trivial case in which $f \equiv 0$). We can assume $0 < \varepsilon < r$. Then we have $\psi^{-1}(r/(r-\varepsilon)) > 1$. Let $\delta = 1 - 1/\psi^{-1}(r/(r-\varepsilon))$. If $f(x) \leq r - \varepsilon$, we get $f(\psi^{-1}(r/(r-\varepsilon))x) = (r/(r-\varepsilon))f(x) \leq r$. So we have $P_r(\psi^{-1}(r/(r-\varepsilon))x) = \psi^{-1}(r/(r-\varepsilon))P_r(x) \leq 1$. Hence $P_r(x) \leq 1/\psi^{-1}(r/(r-\varepsilon)) = 1 - \delta$.

According to the lemma 3.3 of [2], the inversion is obvious.

The proof of $b)$ is similar to that of $a)$.

THEOREM 2.3. Let f be a function satisfying the conditions of lemma 2.1. Then for each x in X and each $r > 0$, we have $P_f(x) = P_{P_r}(x)$.

REMARK: Since D.V.Pai and P.Govindarajulu [1] have shown that, if U is f-proximinal, one has $P_f(x) = P_{P_r}(X)$, we need only to prove $P_f(x) = \phi$ iff $P_{P_r}(x) = \phi$.

PROOF: We only prove that, if $y \in P_{P_r(x)}$, then $y \in P_f(x)$. Let $r_\circ = f_U(x)$. Since $y \in P_{P_{r_\circ}}(x)$, we have $P_{r_\circ}(x-y) = P_{r_\circ,U}(x)$. We must prove $P_{r_\circ}(x-y) = 1$. If $P_{r_\circ}(x-y) > 1$, for $\varepsilon = (P_{r_\circ}(x-y) - 1)/2$, using lemma 2.2, there exists a $\delta > 0$ such that $f(x-y) \geq r_\circ + \delta$. Since $f_U(x) = r_\circ$, there exists a z in U such that $f(x-z) < r_\circ + \delta$. Hence $P_{r_\circ}(x-z) \leq 1+\varepsilon < P_{r_\circ}(x-y)$. This is contradictory to $y \in P_{P_{r_\circ}}(x)$. If $P_{r_\circ}(x-y) < 1$, then we have $f(x-y) < r_\circ$. This is contradictory to $f_U(x) = r_\circ$. Hence $P_{r_\circ}(x-y) = 1$. By lemma 2.2, we have $f(x-y) = r_\circ$. So we get $y \in P_f(x)$.

3. THE APPLICATIONS OF THEOREM 2.3

In this section, we give two applications of the theorem 2.3, in which it is not necessary to know whether the subset of X is f-proximinal or not.

THEOREM 3.1. Let f satisfy the conditions of lemma 2.1, and U be a convex closed subset of X. Suppose $x \in X$ and $y \in P_f(x)$. Let $z_t = y + t(x-y)$, $t > 0$. Then we have $y \in P_f(z_t)$ for $t \in (0,1]$.

PROOF: By theorem 2.3, it is sufficient to prove $y \in P_{P_r}(z_t)$. Since P_r is a Minkowski functional, the proof is similar to the normed case (see [4]).

In the following theorem, we generalize Kolmogorov's criterion on the normed space to the locally convex space.

THEOREM 3.2. Let U be a convex subset of X and x in X, $u_\circ$ in U. Suppose f satisfies the conditions of lemma 2.1 and $r = f(x-u_\circ)$. Then the following statements are equivalent:

1) $u_\circ \in P_f(x)$.

2) There exists a g in X' with the following properties:
  a) $\mathrm{Reg}(y) \leq r \cdot P_r(y)$, for each $y \in X$
  b) $\mathrm{Reg}(x-u_\circ) = r$
  c) $\mathrm{Reg}(u-u_\circ) \leq 0$, for each $u \in U$.

PROOF: 1) $\Longrightarrow$ 2). Applying theorem 2.3, we have $u_\circ \in P_{P_r}(x)$ and $P_r(x-u_\circ) = 1$. Since f is convex and continuous, $V = \{y;\ f(y) < r\}$ is a nonempty open convex subset of X and $V \cap (x-U) = \phi$. Using the separability theorem, there exists a $g_\circ$ in $X'\backslash\{0\}$ and a constant c such that, for v in V and u in U, $Reg_\circ(v) \leq c \leq Reg_\circ(x-u)$. Since $x-u_\circ \in \overline{V} \cap (x-U)$, we obtain that $Reg_\circ(x-u_\circ) = c$.

If $c \leq 0$, since V is an open subset containing the origin 0, we get $g_\circ = 0$. It is contradictary to $g_\circ \in X'\backslash\{0\}$. So $c > 0$. Let $g = r \cdot g_\circ/c$. It is obvious that *b*) is true. For u in U, since $Reg(x-u) \geq Reg(x-u_\circ) = r$, we have $Reg(u-u_\circ) \leq 0$, namely *c*).

For y in X, if $P_r(y) = 0$, then for any $t > 0$, $P_r(ty) = 0$. Hence $Reg(ty) = t \cdot Reg(y) \leq r$. So $Reg(y) \leq 0 = r \cdot R_r(y)$. If $P_r(y) \neq 0$, then $P_r(y/P_r(y)) = 1$. Hence $(1/P_r(y)) \cdot Reg(y) \leq r$, namely $Reg(y) \leq r \cdot P_r(y)$.

2) $\Longrightarrow$ 1). Using the assumption, we have $r = Reg(x-u_\circ) \leq Reg(x-u) \leq r \cdot P_r(x-u)$. Then $P_r(x-u) \geq 1$. By lemma 2.2, we get $f(x-u) \geq r = f(x-u_\circ)$. So $u_\circ \in P_f(x)$.

REFERENCES

[1] D.V.Pal and P.Govindarajulu, On set-valued f-projections and f-farthest point mappings, *J. Approx. Theory*, 42 (1984) 4-13.
[2] P.Govindarajulu, On simultaneous approximation in locally convex spaces, *Indian J. Pure Appl.* 16(6) (1985) 617-626.
[3] S.P.Singh, Some results on best approximations in locally convex spaces, *J. Approx. Theory*, 28 (1980) 329-332.
[4] D.Braess,H Nonlinear approximation theory,*Springer-Verlag, Berlin, Heidelberg, New York, London, Paris, Tokyo*, 1986.

Approximation, Optimization and Computing:
Theory and Applications, A.G. Law and C.L. Wang (eds.)
Elsevier Science Publishers B.V. (North-Holland)

APPROXIMATION THEORY AND HARMONIC ANALYSIS ON LOCALLY COMPACT GROUPS

Su Weiyi

Mathematics Department, Nanjing University, Nanjing, 210008, PRC

This note is a survey of current main development in the area of Approximation Theory and Harmonic Analysis on Locally Compact Groups and Local Fields. As we know, this area has strong theoretical basis, delicate techniques and a bright applicable future.

1. Introduction

The structures of locally compact groups are quite different from that of $\mathbb{R}^n$.

Definition 1. Suppose that G is a locally compact Abelian topological group which contains a strictly decreasing sequence of open compact subgroups $\{G_n\}_{n\in\mathbb{Z}}$ such that

(1) $\bigcup_{n=-\infty}^{\infty} G_n=G$ and $\bigcap_{n=-\infty}^{\infty} G_n=\{0\}$,

(2) $M:=\sup\{order(G_n/G_{n+1}): n\in\mathbb{Z}\}<\infty$.

Then G is called a locally compact group[11].

Such groups are totally disconnected, locally compact, non-discreted. As a special case of such groups, we have so called local fields[18]. A local field is a totally disconnected, non-discrete, locally compact and complete topological field, denoted by K. A non-archimedean norm $|\cdot|$ endowed on K is a mapping from K into $\mathbb{R}^+$ such that

(a) $|x|=0$ iff $x=0$;

(b) $|xy|=|x||y|$;

(c) $|x+y| \le \max\{|x|, |y|\}$.

Let $q=p^c$, $p\geq 2$ be a prime integer, $c\in\mathbb{N}$. Then the set $\mathfrak{B}^k=\{x\in K: |x|\le q^{-k}\}$ can be regarded as G_k, $k\in\mathbb{Z}$. We note that:

(A) the range of the non-archimedean norm $|\cdot|$ on K is the set $\{q^k: k\in\mathbb{Z}\}$.

(B) the position of two balls in K only has two cases: disjoint each other, or one is contained in the other one.

(C) the dimension of K is zero, i.e., the family of all sets that are both open and closed is an open basis for the topology of K.

The structures of G and K are so interesting that many mathematicians pay great attention to the study of these groups and fields, and obtain a lot of valuable results (see [2],[3],[5],[12],[20]). In fact, there are certain background and important motivation of our studing. For example, a. The Dyadic Analysis is used in the area of computer science and signal analysis as well as in many applied areas. b. Walsh Analysis and its applications are developed, especially dyadic derivatives introduced by P.L.Butzer in 1973[1], and p-adic derivatives introduced by Zheng Weixing in 1979[22]; several approximate identity kernels defined by Su Weiyi[14],[16] and Zheng Weixing[22],[23]; the Jackson and Bernstein theorems for Walsh System are proved[27]; etc. c. Some properties of approximation operators on K may be used in the area of scientific investigation or application (see [17],[18],[24],[26]). Moreover, lots of open problems are presented in these areas.

Further notations are needed in this note. Let Γ denote the dual group of G. For each $n\in\mathbb{Z}$, let Γ_n be the annihilator of G_n, that is,

$\Gamma_n = \{\xi\in\Gamma: \xi(x)=1 \text{ for all } x\in G_n\}$.

We may choose a Haar measure μ on G and λ on Γ such that $\mu(G_0)=\lambda(\Gamma_0)=1$, then $\mu(G_n)=(\lambda(\Gamma_n))^{-1}$, $n\in\mathbb{Z}$. Set $m_n=\lambda(\Gamma_n)$. Now if G is a local field K, by the dual theory, Γ is topologically isomorphic to K. Hence Γ_n is isomorphic to $\mathfrak{B}^{-n}$. Choose a $\chi\in\Gamma$, called a character of K, such that it is non-trivial on $\mathfrak{B}^{-1}$ but trivial on $\mathfrak{B}^0$. Then we have $\Gamma\simeq K$: $\chi_\lambda\to\lambda$, $\lambda\in K$.

---

* Project supported by the National Natural Science Foundation of China.

2. Main Results

(1) Define derivatives for functions on K.

C.W.Onneweer defined the fractional, pointwise and strong derivatives for f on p-series and p-adic number fields, two special cases of local fields, in 1977-1980[6],[7]. Zheng Weixing suggested the other definitions of pointwise and strong derivatives for f on p-series and p-adic number fields[26], studied some properties of approximation operators by using his derivatives. Recently, we have given a modified definition of Zheng's derivative for general local fields (see [28]).

Deninition 2. Suppose that $f:K\to\mathbf{C}$ is a given function. For any $N\in\mathbf{N}$, set

$$\Delta_N f(x) = $$

$$= \sum_{j=-N}^{N} q^{-N-j+1} \sum_{l=0}^{q^N-1} \sum_{\nu=1}^{p-1} exp(-2\pi i\nu l p^{-1})\cdot f(x+l\mathfrak{p}^{-j}),$$

where $\mathfrak{p}$ is a prime element in K with $|\mathfrak{p}|=q^{-1}$. If the limit $lim_N\Delta_N f(x)$ exists at $x\in K$, denoted by $f^{<1>}(x)$, then it is said to be a pointwise (p) type derivative at x. A strong derivative $D^{<1>}f$ is defined as an $L^r(K)$-sense limit of $\Delta_N f(x)$, $1\le r\le\infty$.

We also defined the strong integral of f[16], denoted by $I^{<1>}$, and proved many similar properties like those in the classical case for this kind of derivatives and integrals (see [15],[28]). Moreover, the direct and inverse theorems on the best approximation for p-series field and p-adic number field are established by virtue of Onneweer's derivative[21].

(2) Construct several important approximate identity kernels on K.

The Radial approximate identity kernels (see [14]):
Assume that $\omega\in L^1(K)$ satisfies
a. ω is radial, i.e., $\omega(x)=\omega(|x|)$, $x\in K$,
b. there is an $\alpha>0$ such that $lim_{t\to 0}(\omega(t)-1)|t|^{-\alpha} = c \ne 0$.
If there exists $w\in L^1(K)$, such that $w\hat{}=\omega$, then we call w a radial approximate identity kernel.

The Poisson type kernels (see [17]):
The following R_y is called the kernel of Poisson type

$$R_y(x) = $$

$$=|y|^{-1}\Phi_0(xy^{-1})\{ 1+\sum_{h=1}^{m} c_h \chi_{xy^{-1}}(\alpha_h\mathfrak{p}^{-1})\}$$

with $y\in K$, $c_h\in\mathbf{C}$, $\alpha\in\{\varepsilon_0,\varepsilon_1,\cdots,\varepsilon_{q-2}\}$, $h=,\cdots,m$, and $m\in\{1,2,\cdots,q-2\}$; Φ_0 is the characteristic function of the $\mathfrak{B}^0=\{x\in K:|x|\le 1\}$; χ is a character of K which is trivial on $\mathfrak{B}^0$ but nontrivial on $\mathfrak{B}^{-1}$; $\{\varepsilon_0,\varepsilon_1,\cdots,\varepsilon_{q-2}\}$ is the full set representatives of $\mathfrak{B}^0$ in $\mathfrak{B}^{-1}$.

We have also constructed the product type kernels (see References in [16]); a class of approximate identity kernels $K_\omega(t)$ (see [24], [26]); the de la Vallée Poussin kernels (see [4]), etc. All of these kernels have many interesting properties.

(3) Investigate function spaces on G or K.

In this sub-section, we assume that $\alpha\in\mathbb{R}$, and $0<r,s\le\infty$.

The Homogeneous Besov spaces $B(\alpha,r,s)$, (see [11]):

$$B(\alpha,r,s) = \{ f\in Z'(G): \|f\|_{B(\alpha,r,s)} = \{ \sum_{n=-\infty}^{\infty} ((m_n)^{\alpha}\|f*f_n\|_r)^s \}^{1/s} <\infty \}$$

with the usual modification if $s=\infty$; here $m_n=\mu(G_n)^{-1}$, μ is a Haar measure on G; $\phi_n=\Delta_n-\Delta_{n-1}$, $\Delta_n(x)=m_n\Phi_{G_n}(x)$, and Φ_{G_n} is the characteristic function of G_n; $Z'(G)$ is the distribution space of the space $Z(G) = \{ \psi\in S(G): \psi\hat{}(0)=\int_G\psi(t)d\mu(t)=0 \}$, where $S(G)$ is the set of functions on G that have compact support and are constant on the cosets of some G_n in G. We have proved the relationship between $B(\alpha,r,s)$ and its dual space $B(\alpha,r,s)^*$:
a. $B(\alpha,r,s)^* \simeq B(-\alpha,r',s')$, $1\le r<\infty$.
b. $B(\alpha,r,s)^* \simeq B(-\alpha-1+1/r,r',s')$, $0<r<1$.

The regular function spaces $A(\alpha,r,s)$, (see [11]):

$$A(\alpha,r,s) = \{ u\in\mathfrak{R}(G\times\mathbf{Z}): \|u\|_{A(\alpha,r,s)} = \{ \sum_{n=-\infty}^{\infty} ((m_n)^{-\alpha}\|u(\cdot,n)\|_r)^s\}^{1/s} <\infty \}$$

with the usual modification if $s=\infty$; here $\mathfrak{R}(G\times\mathbf{Z})$ is the regular function space on $G\times\mathbf{Z}$ (for definition, see [11],[19]). The regularization of $f\in S'(G)$ is defined as $F(x,n)=f*\Delta_n(x)$, $x\in G$, $n\in\mathbf{Z}$. One can see that in [11] and [19], a regularization $F(x,n)$ of f is a regular function. Every regular function $u(x,n)$ is the regularization of some $f\in S'(G)$, and each $f\in S'(G)$ is the boundary value of a regular function u in the sence that $lim_k u(\cdot,k)=f$ in $S'(G)$, moreover, u is the regularization of its boundary value f. It is very similar to that of a harmonic function on the upper halfspace of $\mathbb{R}^{n+1}$.

The Atomic decomposition spaces $AD(\alpha,r,s)$, (see [4]):
If $f:G\to\mathbb{C}$ has an atomic decomposition

$$f = \sum_{j=-\infty}^{\infty} \sum_{l=0}^{\infty} \lambda_{lj}a_{lj}, \quad \lambda_{lj}\in\mathbb{C},$$

where $\|\lambda\|_{rs} = \{\sum_j(\sum_l|\lambda_{lj}|^r)^{s/r}\}^{1/s}<\infty$, a_{lj} is an (α,∞)-atom with $supp(a_{lj}) \subset z_{lj}+G_{j-1}$, then we say that $f\in AD(\alpha,r,s)$. Here we decompose G as $G=\cup_l G_{ln}$, $n\in\mathbb{Z}$, and $G_{ln} = z_{ln}+G_n$, $z_{ln}\in G$ to be chosen so that $G_{ln} \cap G_{kn}=\emptyset$ for $l\neq k$, and $G_{0n}=G_n$. Then we have
a. $B(\alpha,r,s) \subset AD(-\alpha+1/r,r,s)$, $\alpha\in\mathbb{R}$,
b. $B(\alpha,r,s) \supset AD(-\alpha+1/r,r,s)$, $\alpha<0$.
It implies that $B(\alpha,r,s)$ can be characterized by $AD(-\alpha+1/r,r,s)$ for $\alpha<0$.

The mean oscillation spaces $MO(\alpha,r,s)$ are introduced in [11]. And we have for $\alpha>0$
a. $B(\alpha,r,s) \subset MO(\alpha-1/r,r,s)$, $1\leq r,s\leq\infty$,
b. $B(\alpha,r,s) \supset MO(\alpha-1/r,r,s)$, $0\leq r,s\leq\infty$.
Thus, $B(\alpha,r,s)$ can be described by the spaces $MO(\alpha-1/r,r,s)$, $\alpha>0$.

The other function spaces, for example, the generalized Lipschitz spaces, the Herz spaces and their Fourier transforms (see [8],[9]), multipliers on weighted L_p-spaces (see [10]), etc., are studied too. It is clear that all of the above spaces will play the important role in the study of Function Space Theory on G.

(4) Study pseudo-differential operators in the Besov spaces $B(\alpha,r,s)$ on K.

We introduce two function classes on G or K: basic class and symbol class.

The Basic class $\mathfrak{S}(G)$ (see [13]):
If a function $\phi:G\to\mathbb{C}$ satisfies
a. for any $N\in\mathbb{N}$, there is a constant $c_N>0$, such that

$$|\phi(x)| \leq c_N\langle x\rangle^{-N}, \quad x\in G,$$

where $\langle x\rangle = max(1,|x|)$,
b. for any $(\mu,N)\in\mathbb{P}\times\mathbb{N}$, $\mathbb{P}=\{0,1,\cdots\}$, there is a constant $c_{\mu N}>0$, such that for $|y|<1$

$$|\Delta_y\phi(x)| \leq c_{\mu N}|y|^{\mu}\langle x\rangle^{-N}, \quad x\in G.$$

where $\Delta_y\phi(x)=\phi(x+y)-\phi(x)$. Then we say $\phi\in\mathfrak{S}(G)$. With certain semi-norms, $\mathfrak{S}(G)$ is a Fréchet space.

The Symbol class $S^m_{\rho,\delta}$ (see [13],[18]):
Let $m\in\mathbb{R}$, $\rho,\delta\geq0$. If a function $\sigma(x,\xi)$ on $G\times\Gamma$ satisfies
a. there is a constant $c>0$, such that

$$|\sigma(x,\xi)| \leq c\langle\xi\rangle^m, \quad \xi\in\Gamma,$$

b. for any $(\mu,\nu)\in\mathbb{P}\times\mathbb{P}$, there is a constant $c_{\mu\nu}>0$, such that

$$|\Delta_y\Delta_\zeta\sigma(x,\xi)| \leq c_{\mu\nu}|y|^{\mu}|\zeta|^{\nu}\langle\xi\rangle^{m+\delta\mu-\rho\nu},$$

where $y\in G$, $\zeta\in\Gamma$ with $|\zeta| < \langle\xi\rangle$. Moreover,

$$|\Delta_y\sigma(x,\xi)| \leq c_\mu|y|^{\mu}\langle\xi\rangle^{m+\delta\mu},$$

and

$$|\Delta_\zeta\sigma(x,\xi)| \leq c_\nu|\zeta|^{\nu}\langle\xi\rangle^{m-\rho\nu}.$$

Then we say $\sigma\in S^m_{\rho,\delta}$. This symbol class is a Fréchet space with certain semi-norms. The corresponding pseudo-differential operator is defined as follows.

$$T_\sigma f(x) = \int_\Gamma\{\int_G\sigma(x,\xi)f(x)\bar{\chi}_\xi(t-x)dt\}d\xi,$$

where $f\in\mathfrak{S}(G)$.
We have proved that the L^r-boundedness and $B(\alpha,r,s)$-boundedness of $T_\sigma f$[18]:

a. If $\sigma\in S^m_{\rho,\delta}$ with $m+3(1-\rho)<0$ and $m<0$, then for $f\in L^r(K)$, $1\leq r<\infty$, we have

$$\|T_\sigma f\|_r \leq c\|f\|_r,$$

$c>0$ is a constant depending on q, m, ρ.

b. If $\sigma\in S^m_{\rho,\delta}$ with $m+([\alpha]+1)\delta+3(1-\rho)<0$, $[\alpha]$ is the integer part of α. Then for $f\in B(\alpha,r,s)$, $1\leq r,s<\infty$, we have

$$\|T_\sigma f\|_{B(\alpha,r,s)} \leq c\|f\|_{B(\alpha,r,s)}.$$

We also have unified the definitions of derivatives for a function on K introduced in the section 2,(1), and have got many interesting properties.

REFERENCES

[1] Butzer,P.L. and Wagner,H.J., Walsh-Fourier Series and the Concept of a Derivative, Applicable Anal., 3(1973), 29-46.
[2] Chao,J.A. and Taibleson,M.H., Generalized Conjugate System on Local Fields, Studia Math., T.LXIV.(1979), 213-225.
[3] Daly,J. and Phillips,K., On Singular Integrals, Multipliers, H^1 and Fourier Series ---a Local Phenomenon, Math. Ann., 265 (1983), 181-219.
[4] Jiang Huikun, The Kernels of de la Vallée Poussin Type on p-adic Fields, to appear.

[5] Ombe,H., Besov-type spaces on certain groups, Ph.D. Thesis, Univ. of New Mexico, 1984.

[6] Onneweer,C.W., Differentiation on a p-adic or p-series Field, Linear Spaces and Approximation, Birkhäuser Verlag Basel, 1978, 187-198.

[7] Onneweer,C,W., Fractional Differentiation and Lipschitz spaces on Local Fields, Trans. Amer. Math. Soc., 258(1980), 155-165.

[8] Onneweer,C.W., The Fourier Transform of Generalized Lipschitz Spaces on Certain Groups, National Univ. of Singapore, Lecture Notes 16, 1982.

[9] Onneweer,C.W., The Fourier Transform of Herz Spaces on Certain Groups, Monatsh. f. Math., 97(1984), 297-310.

[10] Onneweer,C.W., Multipliers on Weighted L_p-spaces Over Certain Totally Disconnected groups, Trans. Amer. Math. Soc., 288(1985), 347-362.

[11] Onneweer,C.W. and Su Weiyi, Homogeneous Besov Spaces on Locally Compact Vilenkin Groups, Studia Math. (in print).

[12] Phillips,K., Distributional and Operational Integral Homogeneity over Local Fields, Math. Ann., 242(1979), 69-84.

[13] Saloff-Coste,L., Opérateurs pseudo-Différentiels sur certains groupes totalement discontinus, Studia Math. T.IXXXIII(1986), 205-228.

[14] Su Weiyi, The Kernels of Abel-Poisson Type on Walsh System, Chinese Ann. of Math. 2(1981), (English Issue), 81-92.

[15] Su Weiyi, The Derivatives and Integrals on Local Fields, J. of Nanjing Univ. Math. Biquarterly, 2:1(1985), 32-40.

[16] Su Weiyi, The Approximation Identity Kernels of Product Type for the Walsh System, J. of Approx. Theory, 47:4(1986), 284-301.

[17] Su Weiyi, The Kernels of Poisson Type on Local fields, Scientia Sinica, 16:6A(1988), 641-653.

[18] Su Weiyi, Pseudo-Differential Operators in Besov Spaces over Local Fields, Approx. Theory Appl., 4:2(1988), 119-129.

[19] Taibleson,M.H., Fourier Analysis on Local Fields, Princeton Univ. Press, Princeton, 1975.

[20] Taibleson,M.H., The Failure of Even Conjugate Characterizations of H^1 on Local fields, Pacific J. of Math. 74:2(1978), 501-506.

[21] Xiao Changbai, On Best Approximation over Local Fields, to appear.

[22] Zheng Weixing, Generalized Walsh Transform and an Extreme problem (Chinese), Acta Math. Sinica, 22:3(1979), 362-374.

[23] Zheng Weixing, Approximate Identity Kernels on Walsh System (Chinese), Chinese Ann. of Math., 4A(2),(1983), 177-184.

[24] Zheng Weixing, A class of Approximation Identity Kernels, Approx. Theory Appl., 1:1(1984), 65-76.

[25] Zheng Weixing, Derivatves and Approximation Theorems on Local Fields, Rocky Mountain J. of Math., 15:4 (1985), 803-817.

[26] Zheng Weixing, Further on a Class of Approximation Identity Operators on Local Fields, Scientia Sinica, 30:9A(1987), 641-653.

[27] Zheng Weixing and Su Weiyi, Walsh Analysis and Approximation Operators (Chinese), Adv. in Math.(Beijing), 12:2(1983), 81-93.

[28] Zheng Weixing, Su Weiyi and Jiang Huikun, A Note to the concept of Derivatives on Local Fields, to appear.

Approximation, Optimization and Computing:
Theory and Applications, A.G. Law and C.L. Wang (eds.)
Elsevier Science Publishers B.V. (North-Holland)

A FE–SPLITTING–UP METHOD AND ITS APPLICATION TO DISTRIBUTED PARAMETER IN PARABOLIC EQUATIONS

Xue–Cheng TAI*), Pekka NEITTAANMÄKI

University of Jyväskylä, Department of Mathematics, 40100 Jyväskylä, Finland

1. INTRODUCTION

In this paper, we present a new computing technique for estimating the spatially dependent parameter $a(x)$ in the parabolic equation

$$\begin{cases} \dfrac{\partial u}{\partial t} = \nabla \cdot (a(x)\nabla u) + f(x,t) \\ \qquad x \in \Omega \subset \mathbf{R}^n \quad t \in [0,T] \\ u(x,0) = u_0(x)\ ,\ x \in \Omega \\ u|_{\partial\Omega \times [0,T]} = 0 \end{cases} \tag{1.1}$$

This kind of problem arises in a lot of practical applications. For example, underground water exploration, petroleum reservoirs, heat conduct media, dispersion in rivers and a lot of other applications.

For the computing of such an identification problem, one always uses the output–least square method to estimate the unknown parameter. Because this problem is ill–posed, a certain regularization method has been introduced and successively used as in [**2,7**]. Another alternative is to apply finite dimensional approximation. Especially good results are obtained for 1–D problems by this approximation technique [**1**].

In this paper, we do some further research into the finite dimensional approximation technique for both 1–D and 2–D problems. In [**6**] for elliptic problems, we proved that if one discretizes $a(x)$ and $u(x,t)$ by applying the finite element spaces S_h^r, S_h^{r+1} respectively and then identify $a(x)$ in some set of S_h^r, the obtained solution a^h satisfies the error estimate

$$\left(\int_\Omega |a - a_h|^2 \left(|u'|^2 + h|u''|^2 + h^2|u^{(3)}|^2 + \cdots + h^r|u^{(r+1)}|^2 \right) dx \right)^{\frac{1}{2}} \le C \left(h^{r+1} + \frac{\varepsilon}{h} \right) . \tag{1.2}$$

Here r denotes the order of the finite element space, h is the finite element mesh size and ε is the observation error. According to this estimate, we can see that in the application of FEM, a small h is not prefered, but a higher order of r is prefered. By using a lower order finite element space very poor results may be produced, especially, when the derivative of the given function u vanishes at some points. So we try to use higher order FE–spaces for our identification problems. In one–dimensional cases it is easy to handle higher order FE–spaces, but for multidimensional problems it is quite tedious to construct higher order FE–spaces, which also satisfy the boundary conditions. Moreover the implementation of a parameter identification problem by the FE–method with higher order elements is quite complicated. The condition number and the bandwidth of the occuring matrices increases if one applies high order elements. This causes troubles in solving algebraic equations. This is the reason for proposing in the first part of our paper [**5**] the splitting–up method for the problem (1.1). This method reduces the multidimensional problem (1.1) into a series of 1–D problems, enabling us to solve the multidimensional problem using only a 1–D finite

*)Permanent address: Institute of Systems Science, Academia Sinica, Beijing 100080, P.R. China

element method. Because only a 1–D FE–method is necessary, we have a lot of possibilities to find convenient higher order FE–spaces. For example, we can use any order spline functions, h–p–version FE–spaces etc. In the numerical tests examined below, only cubic spline functions are tested for the identification of one– and two–dimensional problems. It seems that in the literature only a few successful examples have been shown for multidimensional problems [**2,7**].

2. PARAMETER IDENTIFICATION PROBLEM

We consider the problem in which the observation z of the state u in (1.1) is known and we try to recover parameter $a(x)$ through this observation. A widely used technique for determining a is to solve the following problem:

$$\underset{a\in \mathrm{M}}{\text{Minimize}}\left\{ \mathcal{J}(a) = \int_0^T \int_\Omega |u(x,t) - z(x,t)|^2 dx dt \right\} \tag{P}$$

subject to u satisfying (1.1) .

Here $z(x,t)$ is the given observation and M is the set of admissible controls containing some priori information about $a(x)$.

We should in practice use an iterative method for solving (P), typically a gradient method. By applying the standard definition of the Gâteux derivative, we have:

THEOREM 2.1. *$\mathcal{J}(a)$ is infinitely G–differentiable in $L^2(\Omega)$ and*

$$\mathcal{J}'(a) = 2\int_0^T \nabla u \cdot \nabla p\, dt \ , \tag{2.1}$$

where p is the solution of adjoint problem:

$$\begin{cases} \dfrac{\partial p}{\partial t} = -\nabla\cdot(a(x)\nabla p) + \\ \qquad (u(x,t) - z(x,t)) \\ p(x,T) = 0 \qquad x \in \Omega \\ p|_{\partial\Omega\times[0,T]} = 0 \qquad t \in [0,T] \ . \end{cases} \tag{2.2}$$

Moreover, if M is a compact subset in $L^2(\Omega)$, then (P) at least has one solution.

3. THE DISCRETIZED PROBLEM

In order to identify $a(x)$ through (P), we have to determine $a(x)$ approximately and also compute u approximately by some numerical methods. A common approach is to give some finite element approximations for both $a(x)$ and $u(x,t)$.

We take two finite element spaces T^{h_1} and S^{h_2} that are spanned by some finite element basis

$$T^{h_1} = \text{span}\,\{\varphi_1, \varphi_2, \ldots, \varphi_N\} \tag{3.1}$$

$$S^{h_2} = \text{span}\,\{\psi_1, \psi_2, \ldots, \psi_M\} \tag{3.2}$$

and let

$$U = \left\{ (a_1, a_2, \ldots, a_N) \in \mathbf{R}^N \mid a_h(x) = \sum_{i=1}^N a_i\varphi_i(x) \in M \cap T^{h_1} \right\} .$$

In order to determine a_h we solve the following problem:

$$\underset{(a_1,a_2,\ldots,a_N)\in U}{\text{Minimize}} \left\{ \mathcal{J}_h(a_1, a_2, \ldots, a_N) = \int_0^T \int_\Omega |u_h(x,t) - z(x,t)|^2\, dx dt \right\} \tag{P$_\mathrm{h}$}$$

subject to $u_h \in S^{h_2}$ satisfying the semidiscrete problem

$$\begin{cases} \displaystyle\int_\Omega \frac{du_h}{dt} v_h\, dx + \int_\Omega a_h \nabla u_h \cdot \\ \displaystyle\nabla v_h\, dx = \int_\Omega f v_h\, dx \ \forall v_h \in S^{h_2} \\ u_h(x,0) = u_0^h(x) \ . \end{cases} \tag{3.3}$$

Here $u_0^h(x)$ is some approximate function of $u_0(x)$ in S^{h_2}. In real computing, the constraints from U are quite complicated to implement, so we always take them to be the "box" conditions.

Similar to Theorem 2.1 we can also have:

THEOREM 3.1. *$\mathcal{J}_h(a_1, a_2, \ldots, a_N)$ is infinitely differentiable and*

$$\frac{\partial \mathcal{J}_h}{\partial a_i} = 2\int_0^T \int_\Omega \nabla u_h \cdot \nabla p_h \varphi_i \, dxdt \, .$$

Here $p_h \in S^{h_2}$ is the solution of

$$\begin{cases} \displaystyle\int_\Omega \frac{dp_h}{dt} v_h \, dx = \int_\Omega a_h \nabla p_h \cdot \\ \displaystyle\nabla v_h \, dx + \int_\Omega (u_h - z) v_h \, dx \\ \qquad \forall v_h \in S^{h_2} \\ p_h(x,T) = 0 \end{cases} \tag{3.4}$$

and if U is a closed bounded set in $\mathbf{R}^N$, $(\mathrm{P_h})$ has at least one solution.

4. NUMERICAL EXPERIMENTS FOR 1–D PROBLEMS

In this section, we give some numerical experiments for the following 1–D model:

$$\begin{cases} \displaystyle\frac{\partial u}{\partial t} = \frac{\partial}{\partial x}\left(a(x)\frac{\partial u}{\partial x}\right) + f(x,t) \\ \qquad (x,t) \in (0,1) \times (0,1) \\ u_0(x,0) = u_0(x) \, , \quad x \in (0,1) \\ u(0,t) = u(1,t) = 0 \qquad t \in (0,1) \, . \end{cases} \tag{4.1}$$

We choose f such that we know $a(x)$ and $u(x,t)$. Let $z(x,t) = u(x,t)$, i.e., no observation error is added. Then we try to recover $a(x)$ from $z(x,t)$ by solving $(\mathrm{P_h})$. In the computing, we take a uniform mesh size for both T^{h_1} and S^{h_2}. We take T^{h_1} to be the cubic B–spline function and S^{h_2} to be the B–spline functions which satisfy the boundary conditions as in [**5**, §3]. (3.3) and (3.4) are solved by the implicite Euler scheme as in [**5**, §3]. In the computing, double precision is required, but no regularization is used. As in [**4**, p. 282, Corollary 2.1], we also always keep dim $T^{h_1} \leq$ dim S^{h_2} in the computing, so correspondingly we use a different mesh size for T^{h_1} and S^{h_2}. The constraint set U is simply taken to be a "box" $U = [0.01, 100]^N$. In the following examples, $a_h(x)$ denotes the computed value and $a(x)$ is the exact value.

EXAMPLE 4.1. It was pointed out in [**3**] that if $|u_x| = 0$ at some points, then we cannot identify $a(x)$ very well at those points. Our computing shows that by applying the spline approximation, we can still get a good result. Here we suppose $a(x) = 1 + x$, $u(x,t) = \sin(\pi x)e^{-t}$ (u_0, f we get from (4.1)). The initial guess $a_i \equiv 1.1$, dim $T^{h_1} = 10$, dim $S^{h_2} = 11$. The accuracy of the approximation can be seen in Table 4.2.

x	$a_h(x)$	$a(x)$
0.0	1.0025	1.0000
1/7	1.1453	1.1428
2/7	1.2882	1.2857
3/7	1.4371	1.4286
4/7	1.5739	1.5714
5/7	1.7168	1.7143
6/7	1.8600	1.8571
1.0	2.0025	2.0000

Table 4.2.

This is the same example as in [**3**], but we have tried to identify $a(x)$ in $\frac{\partial}{\partial x} \cdot (a\frac{\partial}{\partial x}u) = f$ through the equation

$$v_t = \frac{\partial}{\partial x} \cdot (a\frac{\partial}{\partial x}v) + e^{-t}(f-u)$$

by solving $(\mathrm{P_h})$. Here $v = e^{-t}u$.

We may note, in the above example, $|u_x| = 0$ at $x = 1/2$, but $|u_{xx}| \neq 0$. So the computing accuracy is almost uneffected there. However in the following example, when higher order derivatives have also vanished, the accuracy is effected.

EXAMPLE 4.2. Here the test functions are changed to $u(x,t) = 4x^4(x-1)^4(x-1/2)^4(1-t)^4, a \equiv 1$, initial guess $a_i \equiv 1.5$. The computed result is:

x	$a_h(x)$	$a(x)$
0.0	1.0719	1.0000
1/7	0.9981	1.0000
2/7	1.0039	1.0000
3/7	0.9847	1.0000
4/7	0.9847	1.0000
5/7	1.0039	1.0000
6/7	0.9981	1.0000
1.0	1.0719	1.0000

Table 4.3.

5. NUMERICAL EXPERIMENT FOR 2–D PROBLEMS

From the computing of the previous section, the spline approximation can produce very good results for identification problems. However, for multidimensional problems it is very complicated to construct higher order FE–spaces, which satisfy the boundary conditions, for solving (3.3) and (3.4). So our splitting–up method has a very good advantage on this point. Namely, we can avoid the difficulties in constructing multi-dimensional FE–spaces, because we compute the multidimensional problem by a series of one–dimensional problems.

In the following, we have done some numerical computing for the following model:

$$\begin{cases} \frac{\partial u}{\partial t} = \nabla \cdot (a(x_1, x_2)\nabla u) + f(x_1, x_2, t) \\ \quad (x_1, x_2, t) \in \Omega \times (0, T) \ , \\ u(x_1, x_2, 0) = u_0(x_1, x_2) \quad (x_1, x_2) \in \Omega \ , \\ u|_{\partial\Omega \times [0,T]} = 0 \end{cases} \tag{5.1}$$

where $\Omega = (0,1) \times (0,1), T = 1$. Again we choose a known $a(x_1, x_2)$, $u(x_1, x_2)$, and take the observation from the known function u. Then we try to recover a from this observation. The observation is taken to be the value of u at the following points:

$$\begin{cases} z_{ijk} = (x_i, x_j, t_k) \quad \text{with} \\ x_i = 0,1i, \ i = 2, \ldots, 7, \ x_j = 0,1j, \\ j = 2, \ldots, 7, \\ t_k = 0,1k \ , \quad k = 1, \ldots, 10. \end{cases} \tag{5.2}$$

and

$$\mathcal{J}(a) = \sum_{i=2}^{7} \sum_{j=2}^{7} \sum_{k=1}^{10} |u(x_{1i}, x_{2j}, t_k) - z_{ijk}|^2 \ .$$

In the following, we compute u_h by our splitting–up method. Because no boundary condition is needed for $a(x_1, x_2)$, we can simply take T^{h_1} to be the product of the cubic spline functions with uniform mesh, i.e.

$$a_h(x_1, x_2) = \sum_{i=1}^{N_{x_1}} \sum_{j=1}^{N_{x_2}} a_{ij}\varphi_i(x_1)\varphi_j(x_2). \tag{5.4}$$

Here N_{x_1} is the dimension of splines in the x_1–direction, N_{x_2} is that in the x_2–direction and $\{\varphi_i(x_1)\}, \{\varphi_j(x_2)\}$ are the splines in these directions respectively.

For the details of the computing of u_h, see the ideas in §3 and §4 of [**5**]. By the splitting–up method, only a 1–D finite element space is needed. We take it to be the 1–D B–cubic splines [**5**, §3] with uniform mesh size h_2. The time interval $[0, 1]$ is divided into M elements. In solving ($\mathrm{P_h}$), we use the quasi–Newton method with finite difference gradient. We take $N_{x_1} = N_{x_2} = 10$ in (5.4). This means $h_1 = \frac{1}{7}$. For the splitting–up method, we take $h_2 = 1/10 = 0.1$. In the example $a_h(x_1, x_2)$ is the computed solution. Double precision is used in the computing.

EXAMPLE 5.1. We choose the known u and a as: $u(x_1, x_2, t) = 4x_1x_2(1 - x_1)(1 - x_2)(1 - t)$, $a(x_1, x_2) \equiv 1$. The initial guess for a_{ij} is $\{a_{ij} \equiv 1.1\}_{i,j=1}^{10}$. Let the time step $\tau = 1/100$. The identified parameter is shown in the following table.

$a_h(x_1, x_2)$

$x_2^{x_1}$	0.0	1/7	2/7	3/7	4/7	5/7	6/7	1.0
0.0	1.09	1.07	1.04	1.02	1.02	1.04	1.07	1.09
1/7	1.07	1.04	1.01	0.98	0.98	1.01	1.04	1.07
2/7	1.05	1.02	1.01	1.02	1.02	1.01	1.02	1.05
3/7	1.04	1.01	1.03	1.05	1.05	1.03	1.01	1.04
4/7	1.04	1.01	1.03	1.05	1.05	1.03	1.01	1.04
5/7	1.05	1.02	1.01	1.02	1.02	1.01	1.02	1.05
6/7	1.07	1.04	1.01	0.98	0.98	1.01	1.04	1.07
1.0	1.09	1.07	1.04	1.02	1.02	1.04	1.07	1.09

Table 5.1.

REFERENCES

[1] Banks, H.T. and Lam, P.D., *Estimation of variable coefficients in parabolic distributed systems*, IEEE Trans. Automat. Contr. **AC-30** (1985), 386–398.

[2] Kravaris, C. and Seinfeld, J.H., *Identification of spatially varying parameter in distributed parameter systems by discrete regularization*, J. Math. Anal. Appl. **119** (1986), 128–152.

[3] Kunish, K., *Inherent identifiability of parameters in elliptic differential equations*, J. Math. Anal. Appl. **132** (1988), 453–472.

[4] Kunish, K. and White, L., *Identifiability under approximation for an elliptic boundary value problem*, SIAM J. Contr. Optim. **25** (1987), 279–297.

[5] Tai, X.–C. and Neittaanmäki, P., *A parallel FE–splitting–up method for solving a class of parabolic partial differential equations*, to appear in Proc. of IFIP WG2.5 and WOCO 6 "Symposium on Scientific Software" in Beijing, (1989).

[6] Tai, X.–C. and Neittaanmäki, P., *Error estimates for the numerical identification of distributed parameters*, to appear in Proc. of IFIP–TC7 –Conference on "Control Theory of Distributed Parameter Systems and Applications" in Shanghai, (1989).

[7] Yu, W.H. and Seinfeld, J.H., *Identification of parabolic distributed parameter systems by regularization with differential operators*, J. Math. Anal. Appl. **132**, (1988), 365–387.

Approximation, Optimization and Computing:
Theory and Applications, A.G. Law and C.L. Wang (eds.)
Elsevier Science Publishers B.V. (North-Holland)

KOROVKIN TYPE THEOREM ON C[0,1]

Sin-Ei TAKAHASI*

Department of Basic Technology, Yamagata University, Yonezawa 992, Japan

A class of operators on C[0,1] which satisfy a Korovkin-Wulbert type theorem is introduced. It is shown that some one-dimensional operators and the compositions of homomorphisms and isometric multipliers on C[0,1] are such examples.

1. INTRODUCTION

P. P. Korovkin [1] proved the following approximation theorem : If $\{T_\lambda\}$ is a net of nonnegative linear operators on C[0,1] such that $\lim ||T_\lambda|| = 1$ and $\lim ||T_\lambda x^n - x^n|| = 0$ (n = 0, 1, 2), then $\lim ||T_\lambda f - f|| = 0$ for all $f \in C[0,1]$. After that, it was proved by D. E. Wulbert [5] that the nonnegativity assumption on the operators could be dropped. Then given a subset F of C[0,1], it will be valuable to investigate a class of operators T on C[0,1] such that if $\{T_\lambda\}$ is a net of bounded linear operators on C[0,1] with $\lim ||T_\lambda|| = ||T||$ and $\lim ||T_\lambda f - Tf|| = 0$ for all $f \in F$, then $\lim ||T_\lambda f - Tf|| = 0$ for all $f \in C[0,1]$. We call such T a Korovkin-Wulbert (KW) operator on C[0,1] for F. Of course the Korovkin-Wulbert theorem asserts that the identity operator on C[0,1] is a KW-operator for $\{1, x, x^2\}$. Also any operator on C[0,1] is a KW-operator for $\{1, x, x^2, \ldots\}$ by the Weierstrass approximation theorem for algebraic polynomials. In this note we will investigate a class of KW-operators on C[0,1] for $\{1, x, x^2\}$.

2. RESULTS

Let M[0,1] be the space of all bounded Borel measures on [0,1]. Then a measure in M[0,1] can be regarded as a linear functional on C[0,1]. If $g \in C[0,1]$ and $\mu \in M[0,1]$, we denote by $g \boxtimes \mu$ the linear operator on C[0,1] defined by the relation : $(g \boxtimes \mu)f = \mu(f)g$, $f \in C[0,1]$. In this setting our main result is the following

Theorem. (i) If T is a homomorphism on C[0,1] and S is an isometric multiplier on C[0,1], then ST is a KW-operator on C[0,1] for $\{1, x, x^2\}$.
(ii) If u is a function in C[0,1] such that $|u(t)|$ = constant ($t \in [0,1]$), then $u \boxtimes \delta_t$ ($t \in [0,1]$), $u \boxtimes (\delta_0 - \delta_t)$ ($t \in [1/2,1]$) and $u \boxtimes (\delta_1 - \delta_t)$ ($t \in [0,1/2]$) are KW-operators on C[0,1] for $\{1, x, x^2\}$, where δ_t denotes the Dirac measure consentrated at $t \in [0,1]$.

Let G be the semigroup generated by all homomorphisms and isometric multipliers on C[0,1]. Note that any operator in G is of form ST, where S is an isometric multiplier and T is a homomorphism. Then (i) asserts that all operators in the semigroup G are KW for $\{1, x, x^2\}$. Note also that

$$(u \boxtimes (\delta_a - \delta_b))(v \boxtimes (\delta_t - \delta_s)) = (v(a) - v(b))u \boxtimes (\delta_t - \delta_s)$$

for all $u, v \in C[0,1]$ and $a, b, t, s \in [0,1]$. Then the set of all one-dimensional operators given in (ii) makes a semigroup. Now we shall prove the main theorem by solving some moment problem inspired by [4].

3. LEMMAS

Given a subset F of C[0,1], let $M_{ue}(F)$ be the set of all $\mu \in M[0,1]$ such that if ν is a measure in M[0,1] with $||\nu|| \leq ||\mu||$ and $\nu(f) = \mu(f)$ for all $f \in F$, then $\nu = \mu$. For an operator T on C[0,1], let us denote by T* the adjoint operator of T. Under these notations we have the follow-

*Research partially supported by the Grant-in-Aid for Scientific Research C-635400092 from the Education Ministry of Japan.

ing lemma which follows from the method of L. C. Krutz [2].

Lemma 1. Let F be a subset of C[0,1] and let T be an operator on C[0,1] such that $||T|| = ||T^*\delta_t||$ and $T^*\delta_t \in M_{ue}(F)$ for all $t \in [0,1]$. Then T is a KW-operator for F.

Proof. Let $\{T_\lambda : \lambda \in \Lambda\}$ be a net of operators on C[0,1] such that $\lim ||T_\lambda|| = ||T||$ and $\lim ||T_\lambda f - Tf|| = 0$ for all $f \in F$. We have to show that $\lim ||T_\lambda f - Tf|| = 0$ for all $f \in C[0,1]$. We can assume without loss of generality that $||T_\lambda|| = ||T|| \neq 0$ for all $\lambda \in \Lambda$. Suppose that there exists $f_0 \in C[0,1]$ with $\lim ||T_\lambda f_0 - Tf_0|| \neq 0$. Then there exists $\rho > 0$ such that for any $\lambda \in \Lambda$, there is $\alpha_\lambda \in \Lambda$ and $t_\lambda \in [0,1]$ with $\alpha_\lambda \geqq \lambda$ and

(1) $|(T_{\alpha_\lambda} f_0)(t_\lambda) - (Tf_0)(t_\lambda)| \geqq \rho.$

We can assume that the net $\{t_\lambda\}$ converges to a point $t \in [0,1]$. Hence we have

(2) $\lim \mu_\lambda(f) = (T^*\delta_t)(f) \quad (f \in F),$

where $\mu_\lambda = (T_{\alpha_\lambda})^*\delta_{t_\lambda} \quad (\lambda \in \Lambda)$. Therefore we have weak*-lim $\mu_\lambda = T^*\delta_t$. In fact, assume that there exists a weak*-neighbourhood U of $T^*\delta_t$ such that

(3) for any $\lambda \in \Lambda$, there is $\beta(\lambda) \in \Lambda$ with $\beta(\lambda) \geqq \lambda$ and $\mu_{\beta(\lambda)} \notin U$.

Observe that $||\mu_\lambda|| \leqq ||T||$ for all $\lambda \in \Lambda$, so that we can assume that the net $\{\mu_{\beta(\lambda)}\}$ converges to a measure $\mu \in M[0,1]$ in the weak*-topology. Of course $||\mu|| \leqq ||T||$. Since $\{\mu_{\beta(\lambda)}\}$ is a subnet of $\{\mu_\lambda\}$, it follows from (2) that $\mu(f) = (T^*\delta_t)f$ for all $f \in F$. Recall that $||T|| = ||T^*\delta_t||$, so $||\mu|| \leqq ||T^*\delta_t||$. Then $\mu = T^*\delta_t$ by the assumption : $T^*\delta_t \in M_{ue}(F)$. Hence we have weak*-lim $\mu_{\beta(\lambda)} = T^*\delta_t$. This contradicts (3). Thus weak*-lim $\mu_\lambda = T^*\delta_t$ was proved. In particular, $\lim (T_{\alpha_\lambda} f_0)(t_\lambda) = (Tf_0)(t) = \lim (Tf_0)(t_\lambda)$. This contradicts (1). Q. E. D.

Let X be a norm space and X* its dual space. For subset E of X*, let us call $f \in X^*$ a minimal functional of E if $f \in E$ and $||f|| = \inf \{||g|| : g \in E\}$. Also for $x_1,\dots, x_n \in X$ and $c_1,\dots, c_n \in \mathbb{C}$, the complex numbers, set

$$H^*_X(x_1,\dots, x_n;c_1,\dots, c_n) = \{f \in X^* : f(x_i) = c_i \ (i=1,\dots, n)\}.$$

Then we have the following lemma whose proof is straightforward from the Hahn-Banach extension theorem (cf. [3, 4]).

Lemma 2. If $x_1,\dots, x_n$ in X are linearly independent, then there exists a minimal functional of $H^*_X(x_1,\dots, x_n;c_1,\dots, c_n)$ and

$$\inf \{||g|| : g \in H^*_X(x_1,\dots, x_n;c_1,\dots, c_n)\} = \sup \{|\textstyle\sum_{i=1}^n \alpha_i c_i|/||\sum_{i=1}^n \alpha_i x_i|| : \alpha_1,\dots, \alpha_n \in \mathbb{C}\}.$$

In the next lemma, minimal functionals of some important moment problems on C[0,1] is given by applying the above lemma.

Lemma 3. Let $\mu \in M[0,1]$ and $c \in [0,1]$. Then

(i) if $1/2 < c \leqq 1$, $\mu(1) = 0$ and $\mu((x - c)^2) = c^2$, then $||\mu|| \geqq 2$ and $||\mu|| = 2$ iff $\mu = \delta_0 - \delta_c$,

(ii) if $0 \leqq c < 1/2$, $\mu(1) = 0$ and $\mu((x - c)^2) = (1 - c)^2$, then $||\mu|| \geqq 2$ and $||\mu|| = 2$ iff $\mu = \delta_1 - \delta_c$,

and

(iii) if $\mu(1) = 0$ and $\mu((x - 1/2)^2) = 1/4$, then $||\mu|| \geqq 2$ and $||\mu|| = 2$ iff $\mu = a\delta_0 + (1 - a)\delta_1 - \delta_{1/2}$ for some $a \in [0,1]$.

Proof. We only show the assertion (i). The assertions (ii) and (iii) can be proved by the similar method and so we shall omit those proofs. Let $1/2 < c \leqq 1$ and let $\mu \in M[0,1]$ with $\mu(1) = 0$ and $\mu((x - c)^2) = c^2$. By Lemma 2,

$$\begin{aligned} ||\mu|| &\geqq \sup \{||c^2/(z + (x - c)^2)||_\infty : z \in \mathbb{C}\} \\ &= c^2/\inf \{||z + (x - c)^2||_\infty : z \in \mathbb{C}\} \\ &= c^2/(c^2/2) \\ &= 2, \end{aligned}$$

where $|| \ ||_\infty$ denotes the usal supremum norm on C[0,1]. Now suppose that μ is a real measure with $||\mu|| = 2$ and let $\mu = \mu^+ - \mu^-$ be the Hahn decomposition of μ. Then $\mu^+(1) = \mu^-(1)$ and hence $||\mu^+|| = ||\mu^-|| = 1$. Since $(x - c)^2 \leqq c^2$, it follows that

$$\begin{aligned} c^2 + \mu^-((x - c)^2) &= \mu((x - c)^2) + \mu^-((x - c)^2) \\ &= \mu^+((x - c)^2) \\ &\leqq c^2. \end{aligned}$$

Then $\mu^-((x - c)^2) = 0$ and hence $\mathrm{supp}(\mu^-) = \{c\}$. Consequently $\mu^- = \delta_c$. Furthermore we have that

$$c^2 \leqq c^2 + \mu^+(c^2 - (x - c)^2)$$

$= -\mu(c^2 - (x - c)^2) + \mu^+(c^2 - (x - c)^2)$

$= \mu^-(c^2 - (x - c)^2)$

$= c^2.$

Then $\mu^+(c^2 - (x - c)^2) = 0$ and hence $\mathrm{supp}(\mu^+) = \{0\}$. Consequently we have $\mu^+ = \delta_0$. In other words $\mu = \delta_0 - \delta_c$. We next show the complex case. Let μ_1 and μ_2 be the real part and the imaginary part of μ, respectively. Then $||\mu_1|| \leq 2$, $\mu_1(1) = \mu_2(1) = 0$, $\mu_1((x - c)^2) = c^2$ and $\mu_2((x - c)^2) = 0$. Hence μ_1 must be equal to $\delta_0 - \delta_c$ by the real case. Since $||\mu|| = 2$ and $\mu_1 = \delta_0 - \delta_c$, it follows that $\mu_2(f) = 0$ for all $f \in C[0,1]$ such that $f(0) = 1$, $f(c) = -1$ and $-1 \leq f \leq 1$. Set

$$f_{n,m}(t) = 2c^{-(n + m)}|t - c|^n|t^2 - c|^m - 1$$

for each $t \in [0,1]$ and nonnegative integers n, m with $n + m \geq 1$. Then $\mu_2(f_{n,m}) = 0$ and hence

$$\mu_2(|x - c|^n|x^2 - c|^m) = 0$$

for all n, m = 0, 1, 2,.... Note that $\{|x - c|^n, |x^2 - c|^m : n, m = 0, 1, 2,...\}$ separates the points of [0,1]. Therefore we have by the Stone-Weierstrass theorem that $\mu_2 = 0$ and hence $\mu = \delta_0 - \delta_c$. Q. E. D.

Corollary 4. $\{\delta_t : t \in [0,1]\}$, $\{\delta_0 - \delta_t : t \in [1/2,1]\}$, $\{\delta_1 - \delta_t : t \in [0,1/2]\}$ are contained in $M_{ue}(\{1,x,x^2\})$.

Proof. We first show that $\delta_{1/2} \in M_{ue}(\{1,x,x^2\})$. Let $\nu \in M[0,1]$ be such that $\nu(1) = 1$, $\nu(x) = 1/2$, $\nu(x^2) = 1/4$ and $||\nu|| \leq 1$. Set $\mu = \delta_0 - \nu$. Then $||\mu|| \leq 2$, $\mu(1) = 0$ and $\mu((x - 1/2)^2) = 1/4$ and so, by Lemma 3 - (iii), $\mu = a\delta_0 + (1 - a)\delta_1 - \delta_{1/2}$ for some $a \in [0,1]$. Hence $\mu(x) = 1 - a - 1/2 = 1/2 - a$. On the other hand, $\mu(x) = \delta_0(x) - \nu(x) = -1/2$. Then we have $a = 1$, so that $\mu = \delta_0 - \delta_{1/2}$, hence $\nu = \delta_{1/2}$. In other words, $\delta_{1/2} \in M_{ue}(\{1,x,x^2\})$. Similarly, it follows from Lemma 3 - (i) that $\delta_t \in M_{ue}(\{1,x,x^2\})$ for each $t \in (1/2,1]$. Also it follows from Lemma 3 - (ii) that $\delta_t \in M_{ue}(\{1,x,x^2\})$ for each $t \in [0, 1/2)$. We next show that $\delta_0 - \delta_t \in M_{ue}(\{1,x,x^2\})$ for each $t \in (1/2,1]$. To do this, let $t \in (1/2, 1]$ be fixed and let $\nu \in M[0,1]$ be such that $\nu(1) = 0$, $\nu(x) = -t$, $\nu(x^2) = -t^2$ and $||\nu|| \leq 2$. Then we have $\nu((x - t)^2) = t^2$ and hence $\nu = \delta_0 - \delta_t$ by Lemma 3 - (i). In other words, $\delta_0 - \delta_t \in M_{ue}(\{1,x,x^2\})$. Similarly, it follows from Lemma 3 - (ii) that $\delta_1 - \delta_t \in M_{ue}(\{1,x,x^2\})$ for each $t \in [0,1/2)$. Q. E. D.

4. PROOF OF MAIN THEOREM

(i) Let T be a nonzero homomorphism on C[0,1]. Then $T(1) = 1$ and hence $T^*\delta_t$ is a nonzero multiplicative linear functional on C[0,1] for each $t \in [0,1]$. So there exists a continuous function ϕ of [0,1] into itself such that $T^*\delta_t = \delta_{\phi(t)}$ for all $t \in C[0,1]$. Let S be an isometric multiplier on C[0,1]. Then there exists $u \in C[0,1]$ such that $|u(t)| = 1$ for all $t \in [0,1]$ and $S(f) = uf$ for all $f \in C[0,1]$. Therefore we have

$$(ST)^*\delta_t = T^*(u(t)\delta_t) = u(t)T^*\delta_t$$
$$= u(t)\delta_{\phi(t)}$$

for all $t \in [0,1]$. Hence for each $t \in [0,1]$, $(ST)^*\delta_t$ must belong to $M_{ue}(\{1,x,x^2\})$. Moreover we have

$$||(ST)^*\delta_t|| = |u(t)|\ ||\delta_{\phi(t)}|| = 1$$
$$\geq ||ST|| \geq ||(ST)^*\delta_t||$$

for all $t \in [0,1]$. Then, by Lemma 1, ST is a KW-operator for $\{1, x, x^2\}$. Also the zero operator is usually KW.

(ii) Let F be any subset of C[0,1] and let $\mu \in M_{ue}(F)$ and $u \in C[0,1]$ with $|u(t)| = r$ (nonzero constant) for all $t \in [0,1]$. Then $(r^{-1}u \otimes \mu)^*\delta_t = r^{-1}u(t)\mu \in M_{ue}(F)$ for all $t \in [0,1]$. Moreover

$$||r^{-1}u \otimes \mu|| = ||\mu|| = |r^{-1}u(t)|\ ||\mu||$$
$$= ||(r^{-1}u \otimes \mu)^*\delta_t||$$

for all $t \in [0,1]$. In particular, put $F = \{1, x, x^2\}$. Then we obtain the desired result by Corollary 4 and Lemma 1. Q. E. D.

Note. The theorem of Korovkin for positive linear functionals follows directly from our main theorem. In fact, let $t \in [0,1]$ be fixed and let $\{\mu_\lambda\}$ be a net of positive linear functionals on C[0,1] such that $\lim \mu_\lambda(x^n) = t^n$ (n = 0, 1, 2). Set $T_\lambda = 1 \otimes \mu_\lambda$. Then $\lim ||T_\lambda|| = \lim \mu_\lambda(1) = 1 = ||1 \otimes \delta_t||$ and $\lim ||T_\lambda(x^n) - (1 \otimes \delta_t)(x^n)|| = \lim |\mu_\lambda(x^n - t^n| = 0$ for each

$n = 0, 1, 2$. Since $1 \otimes \delta_t$ is a KW-operator on $C[0,1]$ for $\{1, x, x^2\}$, it follows from our theorem that $\lim ||T_\lambda(f) - (1 \otimes \delta_t)(f)|| = 0$ and so $\lim |\mu_\lambda(f) - f(t)| = 0$ for all $f \in C[0,1]$.

REFERENCES

[1] Korovkin P. P., Linear Operators and Approximation Theory (Hindustan Pub. Delhi, India, 1960).

[2] Krutz L. C., Unique Hahn-Banach Extension and Korovkin's Theorem, Proc. Amer. Math. Soc., 47(1975) pp. 413-416.

[3] Larsen R., Functional Analysis an Introduction, Marcel Dekker, Inc. New York, 1973.

[4] Takahasi S.-E., Hatori O., Fujii M. and Fujii J.-I., A Note on Mendelsohn's Problem Involving Moments, preprint.

[5] Wulbert D. E., Convergence of Operators and Korovkin's Theorem, J. Approximation Theory, 1(1968) pp. 381-390.

Approximation, Optimization and Computing:
Theory and Applications, A.G. Law and C.L. Wang (eds.)
Elsevier Science Publishers B.V. (North-Holland)

DIRECT NUMERICAL INTEGRATION OF STRUCTURAL DYNAMIC EQUATIONS USING HIGHER ORDER PADE APPROXIMATIONS*

Qingsheng Tao

Dept. of Mechanics, Zhejiang University, Hangzhou, PRC

Higher order Padé approximant methods are used to develop a family of single step, higher order integration schemes for solving structural dynamic equations, based on a generalized equivalent first-order system and a consistent approximation to forcing term. Constructions of the algorithms are aided by computer algebra system REDUCE. Numerical comparisons with some well-known integration schemes showed the efficiency of the proposed method.

1. INTRODUCTION

A wide variety of methods have been presented [1,2] for solving second-order systems of differential equations such as arise in structural and mechanical vibration analysis:

$$M\ddot{x}(t) + C\dot{x}(t) + Kx(t) = f(t) \qquad t > 0 \qquad (1)$$

with initial conditions

$$x(0) = X_0 , \qquad \dot{x}(0) = V_0$$

where mass matrix M, damping matrix C and stiffness matrix K are constant, square and of dimension N; f is the vector of applied loads and x is the displacement vector.

D. M. Trujillo presented [3] several integration formulas stemming from the Padé exponential matrix approximations such as Padé (1,1) and Padé (2,2), but could not give explicit integration formulas from Padé (3,3), Padé (4,4) and higher order Padé approximations. In this paper, we extend the derivations to obtain whole families of methods using any order of Padé aproximations based on a generalized equivalent first-order system. After showing that the forcing term approximated by a set of piecewice linear segments in [3] gives inconsistent low order of accuracy if using higher order Padé approximations, we present consistent approximations to forcing term by using Obrechkoff method [4] or multiderivative method for the generalized equivalent first order system.

In the paper, a lot of tedious matrix manipulation have been done with the aids of REDUCE[5], a computer algebra system, which shows great potentials for computer aided construction of numerical algorithms. Two simple models are tested numerically to show good accuracy of the proposed algorithms based on Padé (3,3).

2. PADE APPROXIMATIONS FOR SECOND-ORDER SYSTEM

The second-order dynamic equation (1) can be written as following generalized equivalent q-set first-order system :

$$\dot{Y}(t) = HY(t) + G(t), \qquad t > 0, \qquad Y(0) = Y_0 \qquad (2)$$

where the matrix H is constant, square and of dimension $qN \times qN$, the vectors Y and G are of dimension qN, defined respectively as for $q \geq 2$:

$$Y = \{ \; x \;\; \dot{x} \;\; \ddot{x} \;\; \cdots \;\; \overset{q-1}{x} \; \}^T \qquad (3)$$

$$G = \{ \; 0 \;\; 0 \;\; \cdots \;\; 0 \;\; (M^{-1}\overset{q-2}{f}) \; \}^T \qquad (4)$$

$$H = \begin{bmatrix} 0 & I & 0 & 0 & \cdots & 0 & 0 & 0 \\ 0 & 0 & I & 0 & \cdots & 0 & 0 & 0 \\ \vdots & \vdots & \vdots & \vdots & & \vdots & \vdots & \vdots \\ 0 & 0 & 0 & 0 & \cdots & 0 & I & 0 \\ 0 & 0 & 0 & 0 & \cdots & 0 & 0 & I \\ 0 & 0 & 0 & 0 & \cdots & 0 & -M^{-1}K & -M^{-1}C \end{bmatrix} \qquad (5)$$

The analytical solution of equation (2) has the relation:

$$Y(t+h) = e^{Hh}Y(t) + e^{Hh}e^{Ht}\int_t^{t+h} e^{-Hs}G(s)ds \qquad (6)$$

*This work was partially supported by Natural Science Foundation of Zhejiang Province, PRC.

Using Padé (S,T) exponential matrix approximant[4] with $q = max\{S,T\}$:

$$e^{Hh} \approx \left(\sum_{i=0}^{T} b_i (Hh)^i\right)^{-1} \left(\sum_{i=0}^{S} a_i (Hh)^i\right) \tag{7}$$

we have a family of formulas for direct integration of equation (2) as follows

$$\sum_{i=0}^{T} b_i (Hh)^i Y_{n+1} = \sum_{i=0}^{S} a_i (Hh)^i Y_n \tag{8}$$

with $G = 0$ for simplicity.

To eliminate the terms such as $M^{-1}K$, $M^{-1}KM^{-1}C$ and so on, we may use the corresponding premultiplier matrix R_m ($m = q$) defined recursively as

$$R_2 = \begin{bmatrix} M & 0 \\ C & M \end{bmatrix}$$

$$R_3 = \begin{bmatrix} M & 0 & 0 \\ C & M & 0 \\ K & C & M \end{bmatrix} = \left[\begin{array}{cc|c} \multicolumn{2}{c|}{R_2} & 0 \\ --- & + & 0 \\ K & C & M \end{array}\right]$$

$$R_4 = \begin{bmatrix} M & 0 & 0 & 0 \\ C & M & 0 & 0 \\ K & C & M & 0 \\ 0 & K & C & M \end{bmatrix} = \left[\begin{array}{ccc|c} \multicolumn{3}{c|}{R_3} & 0 \\ & & & 0 \\ -- & -- & + & 0 \\ 0 & K & C & M \end{array}\right]$$

for $m \geq 4$

$$R_m = \left[\begin{array}{cccccc|c} R_{m-1} & & & & & & 0 \\ & & & & & & \vdots \\ --- & --- & --- & --- & + & & 0 \\ 0 & \cdots & 0 & K & C & & M \end{array}\right] \tag{9}$$

As an example, using Padé (3,3) exponential matrix approximant

$$e^{Hh} \approx Q^{-1}P \tag{10}$$

where

$$Q = I - \frac{Hh}{2} + \frac{(Hh)^2}{10} - \frac{(Hh)^3}{120}$$

$$P = I + \frac{Hh}{2} + \frac{(Hh)^2}{10} + \frac{(Hh)^3}{120}$$

a new integration formula can be derived:

$$A_0 Y_{n+1} = A_1 Y_n \tag{11}$$

$$Y = \{\ x\ \ \dot{x}\ \ \ddot{x}\ \}^T$$

where

$$A_0 = \begin{bmatrix} M & h(-\frac{1}{2}M + \frac{1}{120}h^2K) & h^2(\frac{1}{10}M + \frac{1}{120}hC) \\ C & M - \frac{1}{2}hC - \frac{1}{10}h^2K & h(-\frac{1}{2}M + \frac{1}{120}h^2K) \\ K & C & M \end{bmatrix}$$

$$A_1 = \begin{bmatrix} M & h(\frac{1}{2}M - \frac{1}{120}h^2K) & h^2(\frac{1}{10}M - \frac{1}{120}hC) \\ C & M + \frac{1}{2}hC - \frac{1}{10}h^2K & h(\frac{1}{2}M - \frac{1}{120}h^2K) \\ K & C & M \end{bmatrix}$$

If we had defined H as

$$H = \begin{bmatrix} 0 & I \\ -M^{-1}K & -M^{-1}C \end{bmatrix} \tag{12}$$

we would have failed to give the explicit formula (11) as [3].

3.APPROXIMATIONS TO THE FORCING TERM

Based on a representation of the forcing function by a set of piecewice linear segments, D. M. Trujillo presented a approximation formula[3] to equation (6):

$$Y(t+h) = e^{Hh}Y(t) + (e^{Hh} - I)H^{-1}G(t+h)$$
$$+ [e^{Hh}h - (e^{Hh} - I)H^{-1}]H^{-1}[G(t) - G(t+h)]/h \tag{13}$$

By using Taylor series expansion, we can expand partially equation (6) as

$$Y(t+h) = e^{Hh}Y(t) + hG(t) + \frac{h^2}{2}[HG(t) + \dot{G}(t)]$$
$$+ \cdots + \frac{h^i}{i!}\sum_{j=1}^{i} H^{i-j} \overset{j-1}{G}(t) + \cdots \tag{14}$$

and expand equation (13) similarly as

$$Y(t+h) = e^{Hh}Y(t) + hG(t) + \frac{h^2}{2}[HG(t) + \dot{G}(t)]$$
$$+ h^3[\frac{1}{6}H^2G(t) + \frac{1}{6}H\dot{G}(t) + \frac{1}{4}\ddot{G}(t)] + \cdots \tag{15}$$

Comparing (15) with (14) shows that the algorithm (13) is at most second order of accuracy no matter how high order of Padé exponential matrix approximant is used. The integration algorithm with consistent order of accuracy with the order of Padé approximant can be obtained by introducing a multiderivative form[4] of equation (8):

$$\sum_{i=0}^{T} b_i Y_{n+1}^{(i)} = \sum_{i=0}^{S} a_i Y_n^{(i)} \tag{16}$$

Substitution of equation (2) into equation (16) results in

$$\sum_{i=0}^{T} b_i (Hh)^i Y_{n+1} = \sum_{i=0}^{S} a_i (Hh)^i Y_n + V \qquad (17)$$

where V comes from the contribution of forcing function

$$V = \sum_{i=0}^{S} a_i h^i \sum_{j=1}^{i} H^{i-j} \overset{j-1}{G}_{\ n} - \sum_{i=0}^{T} b_i h^i \sum_{j=1}^{i} H^{i-j} \overset{j-1}{G}_{\ n+1} \qquad (18)$$

Obviously equation (17) becomes equation (8) when $G = 0$.

Also take Padé (3,3) approximant as an example. With premultilier matrix R_3, equation (11) can be obtained with one more term V added as

$$V = \{\ V_1 \quad V_2 \quad V_3\ \}^T \qquad (19)$$

$$V_1 = \frac{h^3}{120}(\dot{f}_{n+1} + \dot{f}_n)$$

$$V_2 = \frac{h^2}{10}(-\dot{f}_{n+1} + \dot{f}_n) + \frac{h^3}{120}(\ddot{f}_{n+1} + \ddot{f}_n)$$

$$V_3 = \frac{h}{2}(\dot{f}_{n+1} + \dot{f}_n) + \frac{h^2}{10}(-\ddot{f}_{n+1} + \ddot{f}_n) + \frac{h^3}{120}(\dddot{f}_{n+1} + \dddot{f}_n)$$

In deriving previous algorithms, we have assumed that the derivatives of the forcing function can be computed. For the cases when those derivatives are not readily available we can express the derivatives by multistep values of f_{n+1-j} for $j = 0, 1, \cdots, (S + T - 1)$ by finite difference method.

4.CONSTRUCTION OF ALGORITHM WITH REDUCE

Construction of the integration algorithm (17) involves tedious matrix manipulation, such as H^2,H^3,HG,$H\dot{G}$, and $R_m H$,$R_m H^2$ and so on, which can be done by computer algebra system [6] such as MACSYMA, REDUCE, SMP, muMATH, MAPLE,$\cdots$. With REDUCE system a general program has been developed to generate exlicit form of formula (17) premultiplied by R_m and to output FORTRAN subroutines for algorithms of direct numerical integration.

5. NUMERICAL EXPERIMENTATION

The numerical schemes proposed were applied to two test examples. The accuracy of the results from the Padé (3,3), equation (11) and (19), was compared with those from some well-known integration techniques.

As a first example, we consider a simple oscillator described as

$$\ddot{x} + x = 2, \quad x(0) = 3, \quad \dot{x}(0) = 0 \qquad (20)$$

with the exact solution:

$$x(t) = 2 + \cos t$$

The results from the proposed Padé (3,3) method with those obtained by Brusa-Nigro method[7] (third order accurate) and Gellert[8] or Serbin[9] method (fourth order accurate) are listed in global relative error in Table I, which shows the best global accuracy of the proposed method among the numerical schemes considered.

TABLE I Results for example 1

Method	Global error at $t = 50h$	
	$h = \pi/2$	$h = \pi/20$
Brusa-Nigro	0.10×10	0.37×10^{-3}
Gellert or Serbin	0.15	0.72×10^{-5}
Proposed Padé (3,3)	0.52×10^{-4}	0.11×10^{-8}

For second example, we consider the following dynamic equations from Bathe's book[10]

$$M\ddot{x} + Kx = f$$

where

$$M = \begin{bmatrix} 2 & 0 \\ 0 & 1 \end{bmatrix}, \quad K = \begin{bmatrix} 6 & -2 \\ -2 & 4 \end{bmatrix}$$

$$f = \begin{Bmatrix} 0 \\ 10 \end{Bmatrix}, \quad x(0) = \begin{Bmatrix} 0 \\ 0 \end{Bmatrix}, \quad \dot{x}(0) = \begin{Bmatrix} 0 \\ 0 \end{Bmatrix}$$

The numerical results from the proposed Padé (3,3) method are added to the table in [10] for comparison as shown in Table II where $h = 0.28$.

TABLE II Results for example 2

Method		$t = h$	$t = 2h$	$t = 5h$	$t = 10h$
Exact	x_1	0.003	0.038	0.996	2.806
Solution	x_2	0.382	1.41	5.00	2.806
Central	x_1	0	0.0307	1.02	2.77
Difference	x_2	0.392	1.45	5.02	2.78
Wilson-θ	x_1	0.00605	0.0525	0.952	2.82
($\theta = 1.4$)	x_2	0.366	1.34	4.88	3.06
Newmark	x_1	0.00673	0.0505	0.961	2.85
(Trapezoidal)	x_2	0.364	1.35	4.95	2.90
Proposed	x_1	0.003	0.038	0.996	2.806
Padé (3,3)	x_2	0.382	1.41	5.00	2.806

An examination of Table II shows that the dynamic response calculated by the Padé (3,3) algorithm are in excellent agreement with the exact solution.

6. DISCUSSIONS

With the unified method proposed in the paper, any order of Padé approximations can be used directly to structural dynamic equations, and the analysis of accuracy and stability of algorithms based on Padé appoximations to first-order system can also be used to select a method for any particular structural dynamic problem. The following results concerning the Padé (S,T) approximations, with the maximum possible order $S+T$, are known:[4] The corresponding method to Padé (S,T) approximation is A-stable if $T=S$; A_0-stable if $T \geq S$; L-stable if $T=S+1$ or $T=S+2$. For example, the Padé (3,3) approximant, with sixth order of accuracy, is A-stable.

Some well-known integration formulas are related to those from Padé approximations, as shown in [11].

In general, the higher order approximations do require more computer storage and computations per integration step. However, the disadvanteges with large numbers of degrees-of-freedom can be minimized by employing the conjugate gradient method as done in [12] for Trujillo's algorithms based on Padé (1,2) and Padé (2,2).

ACKNOWLEDGEMENT

I would like to thank Professor T. J. R. Hughes of Stanford University for his suggestions and help on this work.

REFERENCES

[1] Hughes, T. J. R., ' Analysis of transient algorithms with particular reference to stability behavior', in Computational Methods for Transient Analysis (eds. T. Belytschko and T.J.R. Hughes), North-Holland, New York, chapter 2, 1983.

[2] Tao, Qingsheng, 'Unified multistep and multiset algorithms for dynamic problems', Engineer Thesis, Stanford University, 1984.

[3] Trujillo, D. M., ' The direct numerical integration of linear matrix differential equations using Padé approximations', Int J. Numer. Methods Eng., 9, 259-270, 1975.

[4] Lambert, J. D., Computational Methods in Ordinary Differential Equations, Wiley, New York, 1973.

[5] Hearn, A. C., REDUCE User's Manual, Version 3.0, Rand Corp., Santa Monica, California, U. S. A., 1983.

[6] van Hulzen, J. A. and Calmet, J., Computer algebra systems, in Computer Algebra- Symbolic and Algebraic Computation (eds. B. Buchberger, G. E. Collins, R. Loos), Springer-Verlag, 1983.

[7] Brusa, I. and Nigro, I., ' A one-step method for direct integration of structural dynamic equations', Int J. Numer. Methods Eng., 15, 685-699, 1980.

[8] Gellert, M., ' A new algorithm for integration of dynamic systems', Computers Struct., 9, 401-408, 1978.

[9] Serbin, S. M.,' On a fourth order unconditionally stable scheme for damped second order systems', Com. Meths. Appl. Mech. Eng., 23, 333-340, 1980.

[10] Bathe, K. J., Finite Element Procedures in Engineering Analysis, Prentice-Hall, Inc.,1982.

[11] Thomas, R. M. and Gladwell, I., ' The methods of Gellert and of Brusa and Nigro are Padé approximant methods', Int J. Numer. Methods Eng., 20, 1307-1322, 1984.

[12] Carter, A. L., Shiflett, G. R. and Laub, A. J., ' The solution of higher order integration formulae for dynamic response equations by the conjugate gradient method', Int J. Numer. Methods Eng., 20, 339-351, 1984.

Approximation, Optimization and Computing:
Theory and Applications, A.G. Law and C.L. Wang (eds.)
Elsevier Science Publishers B.V. (North-Holland)

ON HERMITE INTERPOLATION IN ROOTS OF UNITY*

Tian-Liang Tu

Department of Basic Sciences, Yellow-River University, Zhengzhou, Henan province, P.R. China

In this paper the divergence and mean convergence of Hermite interpolation in $D=\{|z|\leq 1\}$ are condidered.

1. Divergence

Let $A^k(\bar{D})$ denote the space of funtions which are analytic in $D=\{|z|<1\}$ and its kth deviva-tives are continuous on $\bar{D}$. For the system of nodes $T=\{z_k=e^{2k\pi i/(n+1)}\}_{k=0}^n$, $n\in N$, consider the Hermite interpolation operator

$H_{2n+1}: A^1(\bar{D})\to \Pi_{2n+1}$

$$H_{2n+1}(f,z)=\sum_{k=0}^{n}[1-\frac{\omega''(z_k)}{\omega'(z_k)}(z-z_k)]l_k^2(z)f(z_k)$$

$$+\sum_{k=0}^{n}(z-z_k)l_k^2(z)f'(z_k)$$

Define

$$\|f\|_k := \max_{0\leq l\leq k}\{\sup_{z\in\bar{D}}|f^{(1)}(z)|\},$$

$$\|H\|_{2n+1}\|_k = \sup_{\substack{f\in A^k(\bar{D})\\ \|f\|_k=1}} \|H_{2n+1}f\|_k$$

Lemma 1.1 (Clunie - Mason)[1]. Let $z_1,\ldots, z_n$ be distinct points in complex plane. Define

$$\phi_k(z,\xi) = (z-\xi)(z\bar{z}_k-\xi z_k) \quad (1\leq k\leq n)$$

and put

$$R_{k,n}(z)=\sum_{\substack{j=1\\ j=k}}^{n}\{\frac{\phi_k(z,z_j)}{k(z_k,z_j)}\}^2, \quad W_k(z)=(\frac{1+zz_k}{1+|z_k|})^{m_k}$$

where $m_k\in N$ will be specified later, and set

$$P_k(z) = W_k(z)R_{k,n}(z).$$

Then, suppose that for some k, $z_k\in\partial D$. Given $K>0$, $\varepsilon>0$, for all large m_k there is a $\delta>0$ such that, if $\arg(z/z_k)=\theta(0\leq|\theta|\leq\pi)$, then

(a) $|P_k(z)| < 1-K\theta^2 \quad (|z|=1, |\theta|<\delta)$

(b) $|P_k(z)|<\varepsilon \quad (|z|=1, \delta\leq|\theta|\leq\pi)$.

(c) If $l\neq k$, $|P_l(z)|=O(\theta^2)$, $(|z|=1, \theta\to 0)$ uniformly with respect to the choice of m_l.

Remark. In lemma 1.1, let $m_1,\ldots,m_k$ be large enough, $|\sigma_k|\leq 1$, $k\in[1,n]$, and for z_0, $|z_0|=1$, set polynomial

$$Q(z) = \int_{z_0}^{z}\sum_{n=1}^{n}\sigma_kP_k(t)dt$$

then

1) $Q(z)\ A^1(\bar{D})$;

2) $Q'(z)=\sum_{k=1}^{n}\sigma_kP_k(z)$, and $Q'(z_k)=\sigma_k$.

3) $\|Q'\|<1$. In fact, (a) and (c) of lemma 1.1 cope with the behaviour of $Q'(z)$ on $\{|z|=1\}$ near to points $z_1,\ldots,z_k$, and (b) copes with the remainder of the unit circle.

4) Choosing $\frac{1}{128(n+1)^2}$, we set $\frac{1}{64(n+1)^3}$. Denote by Δ_k the arc with centre at z_k and $\text{mes}\Delta_k<\delta$. Let $A=\bigcup_{k=1}^{n}\Delta_k$, $B=\{|z|=1\}-A$, then

$$|Q(z)|\leq\int_{z_0}^{z}|\sum_{k=1}^{n}\sigma_kP_k(t)||dt|\leq\int_A+\int_B<\frac{1}{32(n+1)^2}$$

Unite this inequality and 3), we have $\|Q\|_1=1$.

One proves easily the following inequality.

Lemma 1.2. For the system of nodes $\{z_k=e^{2k\pi i/(n+1)}\}_{k=0}^{n}$, $n\in N$, we have

$$\sum_{k=0}^{n}|l_k^p(z)|\leq 3, \quad p\geq 2.$$

*Supported by the Science Fund of Henan province P.R. China

Theorem 1.1. For the system of nodes $\{z_k = e^{2k\pi i/(n+1)}\}_{k=0}^{n}$, $n \in N$, and $f \in A^1(\bar{D})$ with $\|f\|_1 = 1$, we have $\|H_{2n+1}\|_1 \geq \frac{2}{\pi} ln\, n - 4$.

Proof.

$$\|H_{2n+1}\|_1 \geq \|H_{2n+1}Q\|_1 \geq |(H_{2n+1}Q)'(e^{\pi i/(n+1)})| \quad (1.1)$$

$$(H_{2n+1}Q)'(z) = \sum_{k=0}^{n} \frac{-n}{z_k} l_k^2(z)Q(z_k) + \sum_{k=0}^{n} l_k^2(z)Q'(z_k)$$
$$+2\sum_{k=0}^{n}[1-\frac{n}{z_k}(z-z_k)]l_k(z)l_k'(z)Q(z_k)$$
$$+2\sum_{k=0}^{n}(z-z_k)l_k(z)l_k'(z)Q'(z_k) \quad (1.2)$$

Because

$$l_k'(z) = -\frac{1}{z-z_k}[z_k z^n - l_k(z)] ,$$

we have

$$2\sum_{k=0}^{n}(z-z_k)l_k(z)l_k'(z)Q'(z_k)$$
$$= 2\sum_{k=0}^{n} z_k z^n l_k(z)Q'(z_k) - 2\sum_{k=0}^{n} l_k^2(z)Q'(z_k) \quad (1.3)$$

and

$$2\sum_{k=0}^{n}[1-\frac{n}{z_k}(z-z_k)]l_k(z)l_k'(z)Q(z_k)$$
$$=2\sum_{k=0}^{n}\frac{z_k z^n}{z-z_k}l_k(z)Q(z_k) - 2\sum_{k=0}^{n}\frac{l_k^2(z)}{z-z_k}Q(z_k)$$
$$-2n\sum_{k=0}^{n} z^n l_k(z)Q(z_k) + 2\sum_{k=0}^{n}\frac{n}{z_k}l_k^2(z)Q(z_k) \quad (1.4)$$

Uniting (1.2)--(1.4),

$$(H_{2n+1}Q)'(e^{\pi i/(n+1)})$$
$$=\{2\sum_{k=0}^{n}\frac{z_k z^n}{z-z_k}l_k(z)Q(z_k) - 2\sum_{k=0}^{n}\frac{l_k^2(z)}{z-z_k}Q(z_k)$$
$$-2n\sum_{k=0}^{n} z^n l_k(z)Q(z_k) + \sum_{k=0}^{n}\frac{n}{z_k}l_k^2(z)Q(z_k)$$
$$+2z^n\sum_{k=0}^{n} z_k l_k(z)Q'(z_k) - \sum_{k=0}^{n} l_k^2(z)Q'(z_k)\}_{z=e^{\pi i/(n+1)}}$$
$$=: 2I_1 - 2I_2 - 2I_3 + I_4 + 2I_5 - I_6 \quad (1.5)$$

$$|I_1| < \frac{2}{n+1}\sum_{k=0}^{n}\frac{1}{|e^{\pi i/(n+1)} - z_k|^2}|Q(z_k)|$$
$$\leq 2(n+1)\|Q\| \quad (1.6)$$

Similarly,

$$|I_2| < \frac{3}{4}(n+1)\|Q\| \quad (1.7)$$

$$|I_3| < n\|Q\|\{\frac{1}{n} ln(n+1)+2\}^{[3,p.86]} \quad (1.8)$$

$$|I_4| \leq 3n\|Q\| \quad \text{(Lemma 1.2)} \quad (1.9)$$

Set

$$\sigma_k = \exp\{-i \arg\frac{z_k}{e^{\pi i/(n+1)} - z_k}\}$$

by virtue of lemma 1.1,

$$|I_5| = \sum_{k=0}^{n}\frac{2}{|e^{\pi i/(n+1)} - z_k|}\frac{1}{n+1}$$
$$\geq \frac{1}{\pi} ln(n+1)^{[3,p.86]} \quad (1.10)$$

$$|I_6| < 3 \quad \text{(Lemma 1.2)} \quad (1.11)$$

From (1.1) and (1.5--(1.11), we have

$$\|H_{2n+1}\|_1 \geq 2|I_5| - 2(|I_1| + |I_2| + \frac{1}{2}|I_4|) - |I_6|$$
$$\geq \frac{2}{\pi} ln(n+1) - 4 ,$$

the proof is completed. (cf. [2])

By virtue of Banach-Cteinhaus' theorem we have

Corollary. There is a function $f_0 \in A^1(\bar{D})$, such that $\|H_{2n+1}f_0 - f_0\|_1 \not\to 0$.

2. Mean convergence

It is natural to consider simultaneous approximation in $L^p(|z|=1)$ space by Hermite interpolation in the roots of unity because of the above corollary.

For $f \in A^k(\bar{D})$, define

$$\|f\|_{L^p} = \{\int_c |f(z)|^p |dz|\}^{\frac{1}{p}} , \quad p>0, \ c: |z|=1$$

$$\|f\|_{L^p,k} = \max_{0\leq 1\leq k}\{\|f^{(1)}\|_{L^p}\}, \quad p>0, \ k\text{--integer}.$$

Lemma 2.1.[4] Suppose $P(z) = c_0 + c_1 z + \ldots + c_n z^n$ is a polynomial of degree n, $z_k = e^{2k\pi i/(n+1)}$, $k \in [0,n]$, then

$$\{\int_0^{2}|P(e^{i\theta})|^q|d\theta|\}^{\frac{1}{q}} \le B_q\{\frac{2\pi}{n}\sum_{k=0}^{n}|P(z_k)|^q\}^{\frac{1}{q}}, \quad q>1.$$

Theorem 2.1. For the system of nodes $\{z_k=e^{2k\pi i/(n+1)}\}_{k=0}^{n}$, $n\in N$ and $f\in A^1(\bar{D})$ we have

$\|H_{2n+1}f - f\|_{L^{p,1}} \le C\omega(f',\frac{1}{n})$, $p>0$, C--const.

where $\omega(f',\frac{1}{n})$ is the modulus of continuity of $f'(z)$.

Proof. Without loss of generality, may set $p>1$ by means of Holder's inequality. It is enough to prove the following inequalities:

$$\{\int_c|H_{2n+1}(f,z)-f(z)|^p|dz|\}^{\frac{1}{p}} \le C_1\omega(f',\frac{1}{n}), \qquad (2.1)$$

$$\{\int_c|(H_{2n+1}f)'(z)-f'(z)|^p|dz|\}^{\frac{1}{p}} \le C_2\omega(f',\frac{1}{n}), \qquad (2.2)$$

where $p>1$, C_1 and C_2 are constants. To prove (2.1), suppose $P_{n-1}(z)$ is the best approximation polynomial of degree $(n-1)$ of $f'(z)$, set $P_n(z)=f(0)+\int_0^z P_{n-1}(t)dt$, $\Delta_n(z)=f(z)-P_n(z)$, then

$$\|P_n(z)-f(z)\| \le \int_0^z\|P_{n-1}(t)-f'(t)\|\,|dt| \le E_{n-1}(f').$$

$$\{\int_c|H_{2n+1}(f,z)-f(z)|^p dz|\}^{\frac{1}{p}}$$

$$\le\{\int_c|H_{2n+1}(\Delta_n,z)|^p|dz|\}^{\frac{1}{p}}+\{\int_c|\Delta_n(z)|^p|dz|\}^{\frac{1}{p}}$$

$$\le A+(2\ )^{\frac{1}{p}}E_{n-1}(f') \qquad (2.3)$$

where $A=\{\int_c|H_{2n+1}(\Delta_n,z)|^p|dz|\}^{\frac{1}{p}}$,

$$A\le\{\int_c|\sum_{k=0}^{n}l_k^2(z)\Delta_n(z_k)|^p|dz|\}^{\frac{1}{p}}$$

$$+\{\int_c|\sum_{k=0}^{n}\frac{n}{z_k}(z-z_k)l_k^2(z)\Delta_n(z_k)|^p|dz|\}^{\frac{1}{p}}$$

$$+\{\int_c|\sum_{k=0}^{n}(z-z_k)l_k^2(z)\Delta_n'(z_k)|^p|dz|\}^{\frac{1}{p}}$$

$$=:A_1+A_2+A_3 \qquad (2.4)$$

$$A_1\le 3E_{n-1}(f')(2\pi)^{\frac{1}{p}} \qquad \text{(Lemma 1.2,)} \qquad (2.5)$$

$$A_2\le 2B_p(2\pi)^{\frac{1}{p}}E_{n-1}(') \qquad \text{(Lemma 2.1,)} \qquad (2.6)$$

$$A_3\le\frac{2}{n+1}B_p(2\pi)^{\frac{1}{p}}E_{n-1}(f') \qquad (2.7)$$

Uniting (2.3)--(2.7), inequality (2.1) is proved.

It remains to prove (2.2). Consider

$$|H_{2n+1}'(f,z)-f'(z)|\le|H_{2n+1}'(\Delta_n,z)|+E_{n-1}(f') \qquad (2.8)$$

However

$$H_{2n+1}'(\Delta_n,z)$$

$$=\sum_{k=0}^{n}\frac{-n}{z_k}l_k^2(z)\int_z^{z_k}\Delta_n'(t)dt$$

$$+2\sum_{k=0}^{n}[1-\frac{n}{z_k}(z-z_k)]l_k(z)l_k'(z)\int_z^{z_k}\Delta_n'(t)dt$$

$$+2\sum_{k=0}^{n}(z-z_k)l_k(z)l_k'(z)\Delta_n'(z_k)+\sum_{k=0}^{n}l_k^2(z)\Delta_n'(z_k)$$

$$=:I_1+2I_2+2I_3+I_4 \qquad (2.9)$$

To estimate (2.9), first consider the fact: suppose z, $z_k\in\bar{D}$, and L is a straight-line: $L=z_k+te^{i\alpha}$, $0\le t\le|z-z_k|$, $\alpha=\arg(z-z_k)$. Let the integral be along the path L. Then

$$\int_{z_k}^{z}f'(\xi)d\xi=\int_0^{|z-z_k|}(\tilde{u}(t)+i\tilde{v}(t))e^{i\alpha}dt \quad (f'=u+iv)$$

$$=(z-z_k)\tilde{u}(\theta_{k,1})+i(z-z_k)\tilde{v}(\theta_{k,2}) \qquad (2.10)$$

where $0<\theta_{k,1}$, $\theta_{k,2}<|z-z_k|$.

Now we consider I_1 in (2.9). Set $\Delta_n'=u_n+iv_n$.

$$I_1=\sum_{k=0}^{n}\frac{n}{z_k}l_k^2(z)(z-z_k)[u_n(\theta_{k,1})+iv_n(\theta_{k,2})]$$

$$=\frac{n}{n+1}\omega(z)\sum_{k=0}^{n}l_k(z)u_n(\theta_{k,1})+i\sum_{k=0}^{n}l_k(z)v_n(\theta_{k,2}),$$

$$\{\int_c|I_1|^p|dz|\}^{\frac{1}{p}}\le C_4\omega(f',\frac{1}{n}) \qquad \text{(Lemma 2.1)} \qquad (2.11)$$

$$I_2=-\sum_{k=0}^{n}l_k(z)l_k'(z)\int_{z_k}^{z}\Delta_n'(\xi)d\xi$$

$$+\sum_{k=0}^{n}\frac{n}{z_k}(z-z_k)l_k(z)l_k'(z)\int_{z_k}^{z}\Delta_n'(\xi)d\xi$$

$$=:g_1+g_2. \qquad (2.12)$$

$$g_1=-\sum_{k=0}^{n}(z_kz^n-l_k(z))l_k(z)[u_n(\theta_{k,1})+iv_n(\theta_{k,2})]$$

by lemma 2.1 and lemma 1.2,

$$\{\int_c |g_1|^p dz\}^{\frac{1}{p}} \leq 2B_p(2\ )^{\frac{1}{p}}E_{n-1}(f')+6E_{n-1}(f')(2\pi)^{\frac{1}{p}}$$

$$\leq C_5\omega(f',\tfrac{1}{n}) \tag{2.13}$$

$$g_2=nz^n\sum_{k=0}^{n} l_k(z)\int_{z_k}^{z}\Delta_n'(\xi)d\xi-\sum_{k=0}^{n}\frac{n}{z_k}l_k^2(z)\int_{z_k}^{z}\Delta_n'(\xi)d\xi$$

$$=: g_{2,1}+g_{2,2}\ . \tag{2.14}$$

From (2.11),

$$\{\int_c |g_{2,2}|^p|dz|\}^{\frac{1}{p}} \leq C_4\omega(f',\tfrac{1}{n}) \tag{2.15}$$

To estimate $g_{2,1}$ suppose $Q_n(z)$ is the best approximation polynomial of degree n of f(z).

$$g_{2,1}=nz^n\sum_{k=0}^{n} l_k(z)[\Delta_n(z)-\Delta_n(z_k)]$$

$$=nz^n[\Delta_n(z)-ln(\Delta_n,z)]$$

$$=-nz^n[ln(f,z)-f(z)]$$

$$=-nz^n ln(f-Q_n,z)+nz^n[Q_n(z)-f(z)]$$

Therefore

$$\{\int_c |g_{2,1}|^p|dz|\}^{\frac{1}{p}} \leq nB_p(2\pi)^{\frac{1}{p}}E_n(f)+n(2\pi)^{\frac{1}{p}}E_n(f)$$

$$\leq C_6\omega(f',\tfrac{1}{n}) \tag{2.16}$$

From (2.13, (2.15) and (2.16),

$$\{\int_c |I_2|^p|dz|\}^{\frac{1}{p}} \leq C_7\omega(f',\tfrac{1}{n}) \tag{2.17}$$

$$I_3=\sum_{k=0}^{n}[z_k z^n-l_k(z)]l_k(z)\Delta_n'(z_k)$$

$$=z^n\sum_{k=0}^{n} l_k(z)z_k\Delta_n'(z_k)-\sum_{k=0}^{n} l_k^2(z)\Delta_n'(z_k)$$

therefore

$$\{\int_c |2I_3+I_4|^p|dz|\}^{\frac{1}{p}} \leq 2B_p(2\pi)^{\frac{1}{p}}E_{n-1}(f')$$

$$+3(2\pi)^{\frac{1}{p}}E_{n-1}(f')$$

$$\leq C_8\omega(f',\tfrac{1}{n}) \tag{2.18}$$

Uniting (2.11), (2.17) and (2.18), the inequality (2.2) follows.

The order of approximation in theorem 2.1 is sharp. For example[5],

$$f_0(z):=\sum_{k=0}^{+\infty} 2^{-\frac{k}{2}}(2^k+1)z^{2^k+1},\quad f_0\in A^1(\bar{D}),\quad f_0'\in Lip\tfrac{1}{2}.$$

For p=2, it follows from theorm 2.1 that

$$\|(H_{2\cdot 2^n+1}f_0)'-f_0'\|_{L^2} \leq C_9\, 2^{-\frac{n}{2}}.$$

On the other hand, the best approximation in space $L^2(|z|=1)$ of $f_0'(z)$ is exactly the following expression:

$$\{\int_c \Big|\sum_{k=n+2}^{+\infty} 2^{-\frac{k}{2}}z^{2^k}\Big|^2|dz|\}^{\frac{1}{2}}=\sqrt{\tfrac{\pi}{2}}\,2^{-\frac{n}{2}}\ .$$

Corollary. Suppose $f \in A^1(\bar{D})$. Then by Cauchy's integral formula, both differences $\{H_{2n+1}(f,z)-f(z)\}$ and $\{H_{2n+1}'(f,z)-f'(z)\}$ converge to zero at the same time uniformly on $|z|\leq r<1$.

ACKNOWLEDGEMENT

The author would like to thank Professor Mo Qi-wu for her kind help.

REFERENCES

[1] J. G. Clunie and J. C. Mason, J. Approx. Theory 41, 149-159 (1984)
[2] J. E. Deng. On Hermite Interpolation, J. Jiaozuo Mining Institute, 1989. No.2.
[3] V. I. Smirnov, N. A. Lebedev, Constructive theory of functions of a complex variable, M-L, 1964. (in Russian)
[4] A. Zygmund, Trigonometric series, Vol. II Cambridge 1959.
[5] X. C. Shen, L. F. Zhong, Kexue Tongbao, No. 11, 810-814 (1988)

Approximation, Optimization and Computing:
Theory and Applications, A.G. Law and C.L. Wang (eds.)
Elsevier Science Publishers B.V. (North-Holland)

A NOTE ON NON-NEGATIVE POLYNOMIAL INTERPOLATION*

Wang Quan-long and Wang Sen

Department of Mathematics, Shanxi University, Taiyuan, Shanxi, P. R. C.

1. Introduction

Let n be even and suppose f is non-negative and bounded on [0,1], and continuous on some open subset of [0,1]. Then there exist n+1 points $0 \leqslant x_1 < \cdots < x_{n+1} \leqslant 1$ such that the polynomial which interpolates f at these points is non-negative on [0,1], as was shown in [1]. If one drops the hypothesis that f is continuous, then there is neither proof of the above statement nor any counterexample of it. To attack this problem seems not easy even in the case n=2. This leads us to consider the situation when f is of bounded variation since it is of no less importance than continuous function in real analysis. We study that situation in this paper and claim that the above statement remains true if we suppose f is of bounded variation instead of being continuous.

2. Some basic facts

If f is a real-valued function on a set S, given n+1 distinct points $\{x_1, \cdots, x_{n+1}\} \subseteq S$, there exists a unique $p \in \Pi_n$, the polynomials of degree $\leqslant n$, such that $p(x_j) = f(x_j)$ for $j=1, \cdots, n+1$, we say that p is a Lagrange interpolant of f at $x_1, \cdots, x_{n+1}$. We use the notation $p = L(f; x_1, \cdots, x_{n+1})$ and p is given in Newton form by:

$$P(x) = f(x_1) + f[x_1, x_2](x-x_1) + \cdots + f[x_1, \cdots, x_{n+1}](x-x_1)\cdots(x-x_n),$$

where $f[x_1 \cdots x_{j+1})$ is the j-th order divided difference of f at $x_1, \cdots, x_{j+1}$, defined inductively by:

$$f[x_1, \cdots, x_{j+1}] = \frac{f[x_2, \cdots, x_{j+1}] - f[x_1, \cdots, x_j]}{x_{j+1} - x_1}. \quad (1)$$

We also make use of the Newton form of the error in polynomial interpolation:

$$f(x) - L(f; x_1, \cdots, x_n) = f[x, x_1, \cdots, x_n](x-x_1)\cdots(x-x_n)$$

and the expression of divided difference:

$$f[x, x_1, \cdots, x_n] = \sum_{j=1}^{n} \frac{f[x, x_j]}{\prod_{\substack{i=1 \\ i \neq j}}^{n} (x_j - x_i)}. \quad (2)$$

A function f is of bounded variation on [a, b] if and only if f is the difference of two increasing functions, say, u(x) and v(x), on [a, b]. Suppose $f(x) = u(x) - v(x)$, and let $A = \{x_1, x_2, \cdots\}$ be the set of the discontinuities of u(x) or v(x). Then we give

Difinition 1. A point $x \in [a, b] - A$ is called a good point of f if f is continuous at x and $f'(x)$ exists and finite. Otherwise x is called a bad point of f.

Remark 1. The set of bad points of f is at most countably infinite.

Following the difinitions in [1] we have

Difinition 2. Let f be bounded on [0,1]. A set $S = \{x_1, \cdots, x_m\} \subseteq [0,1]$ is called a I-type set for f if

$$f(x_j) > 0 \text{ for } j = 1, \cdots, m, \quad (3)$$

and

for each j, there exists $\delta_j > 0$ such that f is continuous on $(x_j - \delta_j, x_j + \delta_j) \cap [0,1]$.

S is called a II-type set for f if (3) holds and

for each j, there exists a $\delta_j > 0$ such that f is of bounded variation on $[x_j - \delta_j, x_j + \delta_j] \cap [0,1]$ and x_j is a good point of f.

Difinition 3. For the set S in difinition 2, n=2m, and for small h>0, define:

$$x_{m+j} = x_j + h, \text{ for } j = 1, \cdots, m. \quad (4)$$

* The research was supported by a gant from Shanxi Natural Science Fund Commission.

$$I_j(h)=(x_j-h,\ x_{m+j}+h)\cap[0,1],\ j=1\cdots,m.$$
$$I(h)=\bigcup_{j=1}^{m} I_j(h).$$
$$g(x)=f[x,x_1,\cdots,x_n],\text{ for } x\neq x_j \text{ for any } j. \quad (5)$$
$$s(x)=(x-x_1)(x-x_2)\cdots(x-x_n). \quad (6)$$
$$p(x)=L(f;x_1,\cdots,x_n)+c(h)s(x), \quad (7)$$

where $c(h)=\sup_{x\in I(h)^c=[0,1]-I(h)} g(x)$, (8)

and hence

$$p(x)\in\Pi_n \text{ and } p(x_j)=f(x_j), \text{ for } j=1,\cdots,n. \quad (9)$$
$$E(x)=f(x)-p(x)=s(x)(g(x)-c(h)). \quad (10)$$

3. Results

Theorem 1. Let f be bounded and non-negative on $[0,1]$ and let $S=\{x_1,\cdots,x_m\}$ be a I-type set or a II-type set for f. Let $n=2m$. Then there exist $n+1-m$ distinct points $\{x_{m+1},\cdots,x_{n+1}\}\subseteq[0,1]-S$ such that $L(f;x_1,\cdots,x_{n+1})$ is non-negative on $[0,1]$.

We shall give the details of the proof for II-type under the case $0<x_1<\cdots<x_m<1$, for I-type see [1].

First we prove

Lemma. $\lim_{h\to 0^+}\|E(x)\|=0$,

where $E(x)$ is defined by (10) and $\|E(x)\|=\sup_{x\in I(h)}|E(x)|$.

Proof. From (5) and (2),

$$g(x)=\sum_{j=1}^{n}\frac{f[x,x_j]}{\prod_{\substack{i=1\\ i\neq j}}^{n}(x_j-x_i)}.$$

By (4), we have

$$\|s(x)\|=\sup_{x\in I(h)}|s(x)|=0(h^2) \text{ as } h\to 0^+. \quad (12)$$

and

$$\frac{1}{\prod_{\substack{i=1\\ i\neq j}}^{n}(x_j-x_i)}=0(h^{-1}) \text{ as } h\to 0^+ \text{ for each}$$

$j=1,\cdots,n$. (13)

We can choose a fixed $\delta>0$ such that $f(x)>0$ and $f[x,x_j]$ is bounded on $\overline{I(\delta)}$ for $j=1,\cdots,m$, since each x_j is a good point of f. Notice that $f(x)$ is bounded, from (11), (1), and (13),

$$\sup_{x\in I(\delta)^c}|g(x)|=0(h^{-1}).$$

and for $h<\delta$, since $f[x,x_j]$ is bounded,

$$d(h)=\sup_{x\in I(\delta)-I(h)}|g(x)|=0(h^{-1}),$$

so from (12), we have

$$\lim_{h\to 0^+} c(h)\,\|s(x)\|=0. \quad (14)$$

To finish the proof of the lemma it remains to show

$$\lim_{h\to 0^+}\sup_{\substack{x\in I(h)\\ x\notin\{x_1,\cdots,x_n\}}}|s(x)g(x)|=0 \quad (15)$$

For any $x\in I_1(h)$, $x\neq x_1$ or x_{m+1}, since x_1 is good, $u(x)$ and $v(x)$ are increasing,

$$|s(x)\cdot\frac{f[x,x_1]}{\prod(x_1-x_i)}|\leq M|f(x)-f(x_1)|$$
$$\leq M(|u(x)-u(x_1)|+|v(x)-v(x_1)|)$$
$$\leq 2M\cdot\max\begin{Bmatrix}\max[u(x_1)-u(x_1-h),u(x_1+2h)-u(x_1)],\\ \max[v(x_1)-v(x_1-h),v(x_1+2h)-v(x_1)]\end{Bmatrix}, \quad (16)$$

where M is a constant independent of x and h. A similar estimate can hold for $s(x)\dfrac{f[x,x_{m+1}]}{\prod(x_{m+1}-x_i)}$, since, by remark 1, h can be chosen such that x_{m+1} is good. The other terms in (11) are $0(h^{-1})$, since $x\in I_1(h)$.

Similar estimates hold on $I_j(h)$ for $j>1$.

From all these estimates, notice that $u(x)$ and $v(x)$ are continuous at x_j, $j=1,\cdots,m$, it follows that (15) holds.

Proof of Theorem 1. First we claim

$$p(x)\geq 0 \text{ on } [0,1], \quad (17)$$

where $p(x)$ is defined by (7).

Let $\varepsilon=\inf_{x\in\overline{I(\delta)}} f(x)>0$, then for sufficient small $h_0<\delta$ and any $x\in I(h_0)$, by lemma,

$$E(x)\leq\sup_{x\in I(h_0)}|E(x)|<\varepsilon\leq\inf_{x\in I(h_0)} f(x)\leq f(x).$$

Thus by (10),

$$p(x)>0 \text{ for any } x\in I(h_0).$$

From (8) and (10), it follows that $E(x)\leq 0$ for any $x\in I(h_0)^c$, since $s(x)>0$ on $I(h_0)^c$. Notice that f is non-negative, so $E(x)\leq f(x)$ on $I(h_0)^c$, and again by (10),

$p(x) \geqslant 0$ on $I(h_0)^c$.

Now we determine the $(n+1)$-st interpolating point.

If there exists t in $I(h_0)^c$ such that $g(t) = c(h_0)$, then by taking $x_{n+1} = t$ and from (10) and (7), $p(x) = L(f; x_1, \cdots, x_{n+1})$. By (17) the theorem follows.

If $g(x) \neq c(h_0)$ for any $x \in I(h_0)^c$, then $E(x) < 0$ on $I(h_0)^c$ and hence $p(x)$ is positive on [0,1]. By difinition (8), there exist $\{t_k\} \subseteq I(h_0)^c$ with $g(t_k)$ converging to $c(h_0)$. Now let $p^*(x) = p(x) + (g(t_k) - c(h_0))s(x)$. Then $p^*(t_k) = f(t_k)$ and for large k, $p^*(x)$ is positive on [0,1].

Corollary. Let n be even and suppose f is non-negative and bounded on [0,1], and of bounded variation on some closed interval I of [0,1], then there exist $n+1$ points $0 \leqslant x_1 < \cdots < x_{n+1} \leqslant 1$ such that $L(f; x_1, \cdots, x_{n+1})$ is non-negative on [0,1].

Proof. If f has infinite zero points in I, then the result follows trivially. Otherwise, we can choose a II-type set for f from I, with $m = \frac{n}{2}$, the corollary follows from theorem 1.

From the theorem 1 and the properties of f of bounded variation, we can also establish the following

Theorem 2. Let f be of bounded variation and non-negative on [0,1]. Assume n is even, $n \geqslant 2$.

i). Suppose f is generalized convex of orded $n+1$ and 0 is a good point of f, with $f'(0) = 0$. Then there exists a non-negative $p \in \Pi_n$ on [0,1] which interpolates f at $n+1$ points in [0,1], 0 included, if and only if $f(0) > 0$.

ii). Suppose f is generalized convex of order $n+1$, and let $c \in (0,1)$ is a good point of f. Then there exists a non-negative $p \in \Pi_n$ on [0.1] which interpolates f at $n+1$ points in [0,1], c included, if and only if $f(c) > 0$.

Remark 2. Since $f \in C^1_{[0,1]}$ is of bounded variation, this theorem is a generalization of theorem 2 in [1]. and the positivity assumption on $f(x)$ at $\{x_1, \cdots, x_m\}$ is essential in general.

Remark 3. The proof of the lemma implies the importance of the fact that $f[x, x_i]$ is bounded on $I(\delta)$, for example, by (12) and (13), (16) can be replaced by a simple form

$$\left| s(x) \frac{f[x, x_i]}{\prod (x_i - x_j)} \right| \leqslant 4 |f[x, x_i]| h.$$

The condition for II-type can be weakened to the following

Theorem 1'. Let f be bounded and non-negative on [0,1], and $S = \{x_1, \cdots, x_m\} \subseteq [0,1]$ satisfies the conditions that $f(x_j) > 0$, and there exists $\delta_j > 0$ such that $f[x, x_j]$ is bounded on $(x_j - \delta_j, x_j + \delta_j) \cap [0,1]$, for $j = 1, \cdots, m$. Let $n = 2m$, then there exist $n+1-m$ distinct points $\{x_{m+1}, \cdots, x_{n+1}\} \subseteq S^c$ such that $L(f; x_1, \cdots, x_{n+1})$ is non-negative on [0,1].

We omit the proof, since it is more easy than that of theorem 1.

References

[1] A. L. Horwitz, A constructive approach to non-negative polynomial interpolation, Approx. Theory Appl. 3(1987), 25-36.

[2] J. Briggs and L. A. Rubel, On interpolation by non-negative polynomials, J. Approx. Theory 30(1980), 160-168.

[3] E. Isaacson and H. B. Keller, Analysis of Numerical Methods, Willey, New York, 1966.

Approximation, Optimization and Computing:
Theory and Applications, A.G. Law and C.L. Wang (eds.)
Elsevier Science Publishers B.V. (North-Holland)

$S^{\mu}_{\mu+1}$ SURFACE INTERPOLATIONS OVER TRIANGULATIONS

Ren-Hong Wang and Xi-Quan Shi

Inst. of Math., Dalian Univ. of Technology, Dalian 116024, China

ABSTRACT: A kind of surface interpolations is described. The domain is assumed to have been triangulated. The interpolant has local support, and is a piecewise polynomial of degree $\mu+1$ with global μ-smoothness, and reproduces polynomials of degrees up to $\mu+1$. The space $S^{\mu}_{\mu+1}$ given by this paper seems to be a natural generalization of the space of univariate splines of order $\mu+1$.

Let Δ be a triangulation consisting of triangles Δ_i, $i = 1,\cdots,T$. Denote by $S^{\mu}_k(\Delta)$ the bivariate spline space

$$S^{\mu}_k(\Delta) := \{s \in C^{\mu}(\Delta) \mid s\,|_{\Delta_i} \in P_k, \quad i = 1,\cdots,T.\},$$

where P_k is the collection of bivariate polynomials with real coefficients and total degree k.

According to the existence theorem on multivariate spline given in [2], in order to get the non-degenerate spline surface in $S^{\mu}_k(\Delta)$, k must be not less than $\mu+1$.

In[3], *Ženišek* obtained a kind of interpolation surfaces in $S^{\mu}_k(\Delta)$ when $k \geq 4\mu+1$. However, this kind of surfaces will be not convenient for practical applications because the degree $4\mu+1$ is too high.

The purpose of the present paper is to show a general method for constructing $S^{\mu}_{\mu+1}(\Delta^*)$ surface interpolations, where Δ^* is a certain subdivision of the triangulation Δ.

Given a triangle $\Delta_i \in \Delta$, we divide each of its three sides into $\mu+1$ equal segments respectively. Thus we get $3\mu+3$ boundary points of Δ_i. Take arbitrarily two boundary points which are not in the same side of Δ_i, and join them by a straight line. So we have constructed a subdivision of Δ_i. It will be called *a μ-local self adaptive partition of* Δ_i, and is denoted by $(\Delta_i)^{\mu}_{lsa}$. The union of all μ-local self adaptive partitions is called *$\mu-self$ adaptive partition of the triangulation* Δ, and is denoted by:

$$\Delta^{\mu}_{sa} := \cup_i(\Delta_i)^{\mu}_{lsa}.$$

It is obvious that the subdivision given by M.J.D.Powell and M.A.Sabin([4]) is the 1-self adaptive partition of a given triangulation.

Denote by V, E, and T the numbers of vertices, edges, and triangles of the triangulation Δ respectively.

M.J.D.Powell and M.A.Sabin ([4]) have proved the dimension of the spline space $S^1_2(\Delta^1_{sa})$:

$$dim S^1_2(\Delta^1_{sa}) = 3V + E, \tag{1}$$

and the following theorem.

THEOREM 1. Given $f_{i,j,t}, 0 \leq i+j \leq 1; t = 1,\cdots,V$; and $f_{n_k}, k = 1,\cdots,E$. Then there exists a unique $s \in S^1_2(\Delta'_{sa})$ for which

$$\frac{\partial^{i+j}}{\partial x^i \partial y^j} s\,|_{V_t} = f_{i,j,t}, 0 \leq i+j \leq 1, \quad t = 1,\cdots,V,$$

$$\frac{\partial s}{\partial n_k}|_{m_k} = f_{n_k}, \quad k = 1,\cdots,E, \tag{2}$$

where V_t is the i-th vertex of Δ , n_k is a normal direction of the k-th edge E_k , and m_k is the mid-point of E_k of Δ .

It is well known that the C^2 smoothness of surfaces would be useful, e.g. for the design of aircrafts, etc. For getting C^2 surface interpolations over any triangulation, the 2-self adaptive partition will be needed. In fact, we have:

THEOREM 2. Given $f_{i,j,t}$, $0 \leq i+j \leq 2, t = 1,\cdots,V$; f_{n_k}, $f_{n^2_{k_1}}$, $f_{n^2_{k_2}}$, $k = 1,\cdots,E$; and f_{cr}, $r = 1,\cdots,T$. Then there exists a unique $s \in S^2_3(\Delta^2_{sa})$ for which

$$\frac{\partial^{i+j}}{\partial x^i \partial y^j} s \mid_{V_t} = f_{i,j,t},$$
$$\frac{\partial s}{\partial n_k} \mid_{m_k} = f_{n_k},$$
$$\frac{\partial^2 s}{\partial n_k^2} \mid_{\nu_{kl}} = f_{n_{kl}^2}, s \mid_{cr} = f_{cr}, \tag{3}$$
$$0 \le i + j \le 2; \quad l = 1,2;$$
$$t = 1, \cdots, V; \quad k = 1, \cdots, E; \quad r = 1, \cdots, T,$$

where V_{k1} and V_{k2} are two trisecton points of the k-th edge of Δ ; cr is the barycenter of the r-th triangle of Δ; V_t, n_k, and m_k are the same as in Theorem 1.

COROLLARY 1.

$$dim S_3^2(\Delta_{sa}^2) = 6V + 3E + T. \tag{4}$$

By using the *"smoothing cofactor-conformality"* method given in [2],we can prove Theorem 2 and Corollary 1. Here is, however, the second proof that contains some useful lemmas.

LEMMA 1. A cubic polynomial $p(x)$ defined on the segment V_1V_2 can be represented by:

$$p(x) = f(V_1)u_1^3 + [3f(V_1) + D_{1,2}f(V_1)]u_1^2u_2$$
$$+[3f(V_1) + 2D_{1,2}f(V_1) + \frac{1}{2}D_{1,2}^2 f(V_1)]u_1u_2^2$$
$$+f(V_2)u_2^3, \tag{5}$$

where $D_{i,j}^k = \partial^k/\partial(V_j - V_i)^k$ denotes the k-th directional derivative in the directon $V_j - V_i$, (u_1, u_2) is the barycentric coordinates of x with respect to V_1V_2.

LEMMA 2. Let V_2, V_3 be two trisection points of V_1V_4. The cubic spline $s(x)$ which is defined on V_1V_4 and satisfies interpolation conditions:

$$D_{i,j}^k s(V_i) = D_{i,j}^k f(V_i), \quad k = 0,1,2; \quad i,j = 1,4 \tag{6}$$

can be obtained by using the following equations:

$$D_{3,4}s(V_3) = -D_{3,2}s(V_3)$$
$$= \frac{1}{45}[D_{1,4}f(V_1) + 4D_{4,1}f(V_4)],$$

$$s(V_2) = f(V_1) + \frac{2}{9}D_{1,4}f(V_1) + \frac{1}{54}D_{1,4}^2 f(V_1)$$
$$-\frac{1}{135}[4D_{1,4}f(V_1) + D_{4,1}f(V_4)], \tag{7}$$

$$s(V_3) = f(V_4) + \frac{2}{9}D_{4,1}f(V_4) + \frac{1}{54}D_{4,1}^2 f(V_4)$$
$$-\frac{1}{135}[D_{1,4}f(V_1) + 4D_{4,1}f(V_4)],$$

and by Lemma 1.

LEMMA 3. Suppose $V_i(= 1,2,3,4)$ are the same as in Lemma 2. The quadratic spline $s(x)$ which is defined on V_1V_4 and satisfies interpolation conditions

$$D_{i,j}^k s(V_i) = D_{i,j}^k f(V_i), \quad k = 0,1; \quad i,j = 1,4,$$

$$s(V_5) = f(V_5), \quad V_5 = (V_1 + V_2)/2 \tag{8}$$

can be obtained by using its definition and the relations

$$s(V_2) = \frac{2}{3}f(V_5) + \frac{1}{24}[10f(V_1) + \frac{5}{3}D_{1,4}f(V_1)$$
$$-2f(V_4) - \frac{1}{3}D_{4,1}f(V_4)],$$

$$s(V_3) = \frac{2}{3}f(V_5) + \frac{1}{24}[10f(V_4) + \frac{5}{3}D_{4,1}f(V_4)$$
$$-2f(V_1) - \frac{1}{3}D_{1,4}f(V_1)]. \tag{9}$$

LEMMA 4. The smoothing cofactor of any segment located on a straight line will be the same constant. If

$$q_3(V) = q_1(V) + \bar{C}_{2,1}L_{2,1}^3 + \bar{C}_{2,3}L_{2,3}^3$$

(cf. Fig. 1), then

$$\bar{C}_{2,1} = \frac{1}{18}[9s(V_7) + 5D_{7,5}s(V_7) + D_{7,5}^2 s(V_7)$$
$$-9s(V_5) - 4D_{5,7}s(V_5) - \frac{1}{2}D_{5,7}^2 s(V_5)],$$
$$\bar{C}_{2,3} = \frac{1}{18}[9s(V_5) + 5D_{5,7}s(V_5) + D_{5,7}^2 s(V_5)$$
$$-9s(V_7) - 4D_{7,5}s(V_7) - \frac{1}{2}D_{7,5}^2 s(V_7)], \tag{10}$$

where $L_{2,1} = 2\bar{u}_5 - \bar{u}_7, L_{2,3} = \bar{u}_5 - 2\bar{u}_7$,and $(\bar{u}_2, \bar{u}_5, \bar{u}_7)$ is the barycentric coordinates of V with respect to the triangle $V_2V_5V_7$.

LEMMA 5. Let

$$\bar{p}(V) = s(V_4)u_4^3 + [3s(V_4) + D_{4,7}s(V_4)]u_4^2u_7$$

$$+[3s(V_7) + D_{7,4}s(V_7)]u_4u_7^2 + s(V_7)u_7^3$$
$$+h_3[D_{n_{1,2}}s(V_4)u_4(2u_4 - u_7)$$
$$+4D_{n_{1,2}}s(V_{10})u_4u_7$$
$$+D_{n_{1,2}}s(V_7)u_7(2u_7 - u4)]u_0$$
$$+ \frac{1}{2}[D^2_{n_{1,2}}s(V_4)u_4 + D^2_{n_{1,2}}s(V_7)u_7](h_3u_0)^2,$$
$$p_0(V) = \bar{p}(V) + C_1u_0^3,$$
$$p_1 - p_0 = C_2(u_0 - u_4)^3, p_2 - p_1 = C_3(u_0 - u_7)^3,$$
$$p_3 - p_2 = \bar{C}_{2,1}(u_0 - 2u_4 - u_7)^3, \quad (11)$$
$$p - p_3 = \bar{C}_{1,2}(u_0 - u_4 - 2u_7)^3,$$

$$p - \bar{p} = C_1u_0^3 + C_2(u_0 - u_4)^3 + C_3(u_0 - u_7)^3$$
$$+\bar{C}_{3,1}(u_0 - 2u_4 - u_7)^3 + \bar{C}_{1,2}(u_0 - u_4 - 2u_7)^3.$$

Then

$$C_1 = 3p(V_0) + \frac{1}{3}(D_{0,4} + D_{0,7})p(V_0) - 3\bar{p}(V_0)$$
$$-\frac{1}{3}(D_{0,4} + D_{0,7})\bar{p}(V_0) + 2(\bar{C}_{1,2} + \bar{C}_{2,1})$$
$$C_2 = -p(V_0 - \frac{1}{3}D_{0,4}p(V_0) + \bar{p}(V_0)$$
$$+\frac{1}{3}D_{0,4}\bar{p}(V_0) - \bar{C}_{1,2} - 2\bar{C}_{2,1}, \quad (12)$$
$$C_3 = -p(V_0) - \frac{1}{3}D_{0,7}p(V_0) + \bar{p}(V_0)$$
$$+\frac{1}{3}D_{0,7}\bar{p}(V_0) - 2\bar{C}_{1,2} - \bar{C}_{3,1},$$

where $n_{1,2}$ is the normal direction of the edge V_1V_2, $V_{10} = (V_1 + V_2)/2$, $h_3 = 2area(V_0V_4V_7)/\mid V_4V_7\mid$, and (u_0, u_4, u_7) is the barycentric coordinates of V with respect to the triangle $V_0V_4V_7$. (cf. Fig. 1)

LEMMA 6. $D_{0,4}s(V_0)$ and $D_{0,7}s(V_0)$ can be represented by

$$D_{0,4}s(V_0) = D_{0,4}p(V_0) = \frac{1}{12}[2h_3D_{n_{1,2}}s(V_4)$$
$$+h_3^2D^2_{n_{1,2}}s(V_4) + 12\bar{C}_{2,1} - 2h_1D_{n_{2,3}}s(V_8)$$
$$-h_1^2D^2_{n_{2,3}}s(V_8) - 12\bar{C}_{2,3}],$$
$$D_{0,7}s(V_0) = D_{0,7}p(V_0) = \frac{1}{12}[2h_3D_{n_{1,2}}s(V_7)$$
$$-h_3^2D^2_{n_{1,2}}s(V_7) + 12\bar{C}_{1,2}$$
$$-2h_2D_{n_{3,1}}s(V_6) - h_2^2D^2_{n_{3,1}}s(V_6) - 12\bar{C}_{1,3}] \quad (13)$$

respectively, where $n_{1,2}, n_{2,3}, n_{3,1}, h_1$, and h_2 are similar to $n_{1,2}$ and h_3 given in Lemma 5.

Next we proceed to the proof of Theorem 2. It suffices to show that the spline $s \in S_3^2(\Delta^2_{sa})$ determined by the homogeneous interpolation conditions corresponding to (3) will be identically zero on Δ^2_{sa}.

According to lemma 1, 2, and 4, we can show $\bar{C}_{i,j} = 0$. Using Lemma 5 and 6, it will get $s \equiv 0$ on the triangle $V_1V_2V_3$. Thus s is identically zero on the triangulation Δ^2_{sa}, and the rest follows.

The explicit formula of the S_3^2 (Δ^2_{sa}) surface interpolation can be also obtained by using Lemma 1 to Lemma 6. The details of the formula may be omitted.

In the case of μ-self adaptive partition of Δ, we have

THEOREM 3. For given $f_{i,j,t}$, $0 \le i+j \le \mu$, $t = 1,\cdots,V$; $f(V^k_{l,r}), l = 1,\cdots,\mu, r = 1,\cdots,l, k = 1,\cdots,E$; $f(\lambda_{\tau\rho})$, $\tau = 1,\cdots,T$, $\rho = 1,\cdots,\mu(\mu-1)/2$, there exists a unique $s \in S^{\mu}_{\mu+1}(\Delta^{\mu}_{sa})$ for which

$$\frac{\partial^{i+j}}{\partial x^i\partial y^j}s\mid_{V_t} = f_{i,j,t},$$
$$0 \le i+j \le \mu, \quad t = 1,\cdots,V,$$

$$\frac{\partial^l}{\partial n^l_k}s\mid_{V^k_{l,r}} = f(V^k_{l,r}),$$
$$l = 1,\cdots,\mu, \quad r = 1,\cdots,l, \quad k = 1,\cdots,E \quad (14)$$

$$S(\lambda_{\tau\rho}) = f(\lambda_{\tau\rho}), \quad \tau = 1,\cdots,T,$$
$$\rho = 1,\cdots,\mu(\mu-1)/2,$$

where V_t and n_k are the same as in Theorem 1; V_1^k and V_2^k are two endpoints of the k-th edge of Δ; $V^k_{l,r}$ are equally spaced points of division of $V_1^kV_2^k$:

$$V^k_{l,r} := \frac{1}{l+1}[rV_1^k + (l-r+1)V_2^k];$$

and $\Lambda_\tau = \{\lambda_{\tau\rho}\mid_\rho = 1,\cdots,\mu(\mu-1)/2\}$ is a suitable set of points in the τ-th triangle Δ_τ of Δ (e.g. Λ_τ is not on a nontrivial algebraic curve of degree $\mu-2$ in $\Delta_\tau \subset \Delta$, cf. [1]).

Corollary 2.

$$dim S^{\mu}_{\mu+1}(\Delta^{\mu}_{sa}) = \frac{1}{2}(\mu+1)(\mu+2)V+$$
$$\frac{1}{2}\mu(\mu+1)E + \frac{1}{2}(\mu-1)\mu T. \quad (15)$$

The details of the proof on Theorem 3 are omitted.

It seems that the calculation of $s \in S_3^2(\Delta_{sa}^2)$ possesses some complications. However, it is easy to realize by using the above-mentioned Lemmas.

It is perhaps of interest to show the numbers of independent parameters of $s \in S_{\mu+1}^{\mu}$ and Ženišek's model $s \in S_{4\mu+1}^{\mu}$ respectively. In fact, we have (one triangle)

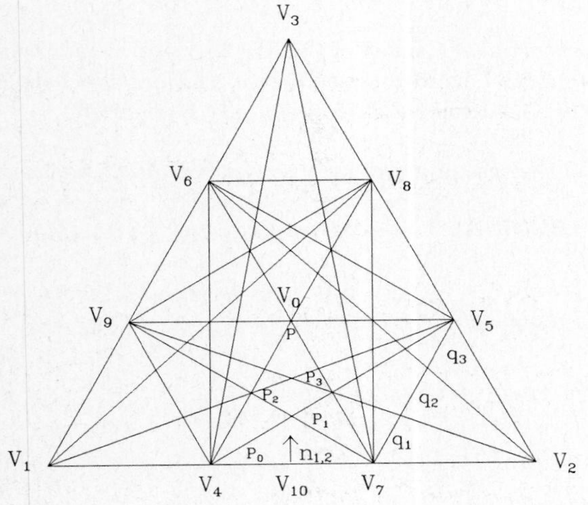

Figure 1

μ	$s \in S_{\mu+1}^{\mu}$	$s \in S_{4\mu+1}^{\mu}$ (Ženišek's model)
1	12	21
2	28	55
3	51	105
...	...	...
μ	$\frac{7}{2}\mu^2 + \frac{11}{2}\mu + 3$	$8\mu^2 + 6\mu + 3$

REFERENCES

[1] R.H.Wang, and X.Z.Liang, Approximation of Multivariate Functions, *Science Press*, 1988, *Beijing*.

[2] Ren-Hong Wang, The structural characterization and interpolation for multivariate splines, *Acta Math. Sinica*, 18(1975), 91-106.

[3] A. Ženišek, Interpolation polynomials on the triangle, *Numer. Math.*, 15(1970), 283-296.

[4] M.J.D.Powell and M.A.Sabin, Piecewise Quadratic Approximations on Triangles, *ACM Trans. Math. Software*, 3(1977), 316-325.

[5] X.Q.Shi, Higher-dimensional splines, *Ph.D. Thesis*, *Jilin University*, 1988.

Approximation, Optimization and Computing:
Theory and Applications, A.G. Law and C.L. Wang (eds.)
Elsevier Science Publishers B.V. (North-Holland)

ON ASYMPOTOTIC EXPANSION OF RAPIDLY OSCILLATORY INTEGRALS

J. N. Wei C. X. Huang and Z. D. Dong

Abstract. This paper represents an asymptotic expansion for oscillatory integrals of the form $\int_0^1 F(x, \langle\lambda\varphi(x)\rangle)dx$. some results obtained by us can be regarded as improvement and extension of mentioning in paper [3].

Up to now, many scholars in maths have worked in research of various approximation methods to compute integrals of oscillatory function.

Early in 1950, Erugin — Sobolov, Krylov, and Riekstens, respectively, present the asympototic expansions for following parameter integrals

$$I(\lambda) = \int_0^1 f(x)g\langle\lambda x\rangle dx \qquad (1)$$

where $\langle\lambda x\rangle$ denotes the fractional part of λx, λ being a large real parameter. The details may be found in reference [5].

In 1958, to offer an asymptotic expansion for the more general form oscillatory integrals

$$I_\lambda(F) = \int_0^1 F(x, \langle\lambda x\rangle)dx \qquad (2)$$

L. C. Hsu required the following results: if $F(x,y) \in C$ (the symbol C denotes the set of continuous functions on the square $[0,1]\times[0,1]$) has a continuous partial derivative with respect to x up to m—th order, then the asympotic expansion formula

$$I_\lambda(F) = \int_0^1\int_0^1 F(x,y)dxdy + \sum_{k=1}^{m-1} \frac{A_{k,\lambda}}{\lambda^k} + O(\frac{1}{\lambda^m}) \qquad (3)$$

holds for a large parameter λ, where $A_{k,\lambda}$ is defined by

$$A_{k,\lambda} = \frac{1}{K!}\int_0^1 [F_x^{(k-1)}(1,y)\bar{B}_k(y - \langle\lambda\rangle) - F_x^{(k-1)}(0,y)B_k(y)]dy \qquad (4)$$

in which $B_k(x)$ and $\bar{B}_k(x)$ denote the Bernoulli polynomial of degree k and the corresponding Bernoulli function with unit period respectively, and $\langle x\rangle = x - [x]$, where $[x]$ denotes the integral part of x.

Afterwards, L. C. Hsu, Y. S. Chou, X. H. Wang and others obtained a series of varied and perfect results on precise estimate for the error terms in (3) and on a general of the form $\int_0^1 x^{-\alpha}F(x, \langle\lambda x\rangle)dx$ $(0 < \alpha < 1)$, in which
$F(x,y)$ is periodic in y with unit period and λ is a large parameter, The details may be found in reference [4].

In 1978, both at Hangzhou Math Symposium and the Third National Maths Congress of China, L. C. Hsu proposed the following problem in his report on the appoximation theory; If $\varphi(x)$ is a continuous function on $[0, 1]$, and if $F(x, y)$ is a continuous function on the square $[0,1]\times[0,1]$, the integral

$$I_\lambda(F,\varphi) = \int_0^1 F(x, \langle\lambda\varphi(x)\rangle)dx \qquad (5)$$

will be the asymptotic expansion formula of $I_\lambda(F,\varphi)$, where parameter λ is large enough.

In the next year, X. L. Shi [3] first introduced the concept of the class H_c: for any $\varphi(x)$ in the class H_c: integral (5) is expressed in the following formula

$$I_\lambda(F,\varphi) = \int_0^1\int_0^1 F(x,y)dxdy + o(1) \qquad (\lambda \to \infty) \qquad (6)$$

where the function is $F(x,y) \in C$. Hence he also obtained the following results:

Theorem A: A real valued function $\varphi(x)$ defined on $[0,1]$ is of class H_c if and only if the equality

$$\lim_{\lambda\to\infty} \int_\alpha^\beta g(\langle\lambda\varphi(x)\rangle)dx = (\beta-\alpha)\int_0^1 g(x)dx \qquad (7)$$

holds for arbitrary real numbers α and β, $0 \leqslant \alpha < \beta \leqslant 1$, and any continuous piecewise linear function $g(x)$ defined on $[0,1]$.

+Mathematics Teaching Group, Wuhan University of Water Transportation Engineering, Wuhan, PEOPLES REPUBLIC OF CHINA

Obviously, according to theorem A, we can draw the conclusion: there exists a continuous and strictly increasing function $\varphi(x)$, which maps onto itself and is not of the class H_c, and, in the meanwhile, there exists another function $\varphi(x)$, which is constant on any subinterval $[\alpha,\beta]$ of the interval $[0,1]$ and is not of the class H_c. Such situation is rather difficult.

In order to establish the asymptotic expansion formula in form (3), Z. L. Shi introduced condition $K_m(\triangle)$, that is, if the function defined on $[0,1]$ satisfies condition $K_m(\triangle)$, then in each subinterval (x_i, x_{i+1}) of the partition $\triangle: 0=x_0<x_1<x_2<\cdots<x_i=1$, $\varphi(x)$ is continuous and strictly monotonic, while the inversel function of $\varphi(x)$ has continuous derivativese up to $(m+1)$—th order on the corresponding interval (each continuous derivative on point of any subinterval can be extened by right or left limit). Hence its formula which is similar to (3) has been established.

It is shown that if $\varphi(x)$ satisfies the condition $K_m(\triangle)$, then $\varphi'(x)\neq 0$. But the condition is too difficult to be satisfied. Besides, theorem A that represents the approximate property of the condition $K_m(\triangle)$ might be further regarded as the structure of function $\varphi(x)$.

To improve the above mentions. we have also introduced another piecewise linear function $\Gamma_\lambda(x)$ to approximate to function $\varphi(x)$ so that the following results can be obtained.

Theorem 1: A real valued continuous function $\varphi(x)$ defined on $[0,1]$ is of class H_c if and only if function $\varphi'(x)$ is not almost everywhere zero on a close interval $[0,1]$.

From theorem 1, it is easy to see that the condition $K_m(\triangle)$, i. e. $\varphi'(x)\neq 0$ has been relaxed and nevertheless the asymptotic expansion formula as (3) has been established. We have done a further study of theorem A and got some results that can be regarded as extension of theorem A. these results which will be mentioned later can not only reveal the property of the class H_c, but also make the application of it much easier. The main results of our study are stated as follows.

First, let $\varphi(x)$ be a continuous monotonic function, using Steklov function $\varphi_h(x)\,(h>0)$ to approximate to function $\varphi(x)$, where the so—called Steklov function $\varphi_h(x)$ is defined by

$$\varphi_h(x) = \frac{1}{2h}\int_{-h}^{h}\varphi(x)dx \qquad (8)$$

So the following theorem is derived by theorem 1.

Theorem 2: Let $\varphi(x)$ be a real function on close integral $[0,1]$. A necessary and sufficient condition for $\varphi(x)\in H_c$ is that there exists a inverse function $\overline{\varphi}(u)$ of $u=\varphi(x)$ such that $\overline{\varphi}(u)$ is an absolutely continuous function on corresponding interval.

Secondly, for a general continuous function $\varphi(x)$, according to the property of Lebesque measure and integral, the following two theorems are also true.

Theorem 3: A real valued function on a close interval $[0,1]$ is of the class H_c if and only if the equality

$$\lim_{\lambda\to\infty}\int_a^\beta g(\langle\lambda\varphi(x)\rangle)dx = (b-a)(\beta-\alpha) \qquad (9)$$

holds for any function which have the following form

$$g(x) = \begin{cases} 1 & x\in[\alpha,\beta], \\ 0 & elsewhere. \end{cases} \qquad (10)$$

and arbitrary real numbers $0\leqslant\alpha<\beta\leqslant 1$ and $0\leqslant a<b\leqslant 1$.

Throrem 4: Suppose that there is a real function $\varphi(x)$ defined on interval $[0,1]$. Then $\varphi(x)$ is of the class H_c if and only if the Lebesque measure of the inverse image $(\lambda\varphi(x))^{-1}(A)=E$ of the set $A=\bigcup_{k=-\infty}^{+\infty}[k+\alpha,k+\beta]$ on the interval $[a,b]$ satisfies

$$mE = (b-a)(\beta-\alpha)+o(1) \qquad (\lambda\to\infty) \qquad (11)$$

which holds for arbitrary real numbers $0\leqslant\alpha<\beta\leqslant 1$ and $0\leqslant a<b\leqslant 1$.

In fact, theorem 4 is another expression of theorem 3. But it is more intuitive than the previous one, from a geometric point of view. Theorem 4 describes in details that the inverse image $(\lambda\varphi(x))^{-1}(A)=E$ of the set A approches the uniform destribution as $\lambda\to\infty$. In other words, as parameter λ is large enough, function $\lambda\varphi(x)$ has almost the same linear degree at every point, i. e. the absolute values of gradient of tangent for every point are uniform.

Let $\varphi(x)$ be a non—negative integrable function on a

close interval [0,1]. We considered its distribution function $x=m(y)$, $m(y)=mE_y$, where the E_y is the set $\{x \mid x\in[a.b], \varphi(x)\geqslant y\}$. mE_y is Lebesque measure.

It follows that function $m(y)$ is a strictly decreasing continuous function, while function $\varphi(x)$ is continuous and not constant on any subinterval [a', b'] of the interval [a, b]. Hence its inverse function, i. e. a non—increasing re—ordered function $y=\tilde{\varphi}(x)$ is also a strictly decreasing continuous function on the interval [0, b—a]. The details may be found in reference [6].

According to theorem 4, the condition $\varphi(x)$ being of the class H_c depends on information given by the inverse image distribution, and a non—increasing re—order function $\tilde{\varphi}(x)$ of function $\varphi(x)$ keeps the amount of such information about $\varphi(x)$. Therefore, we immediately derive the following result from theorem 4.

Theorem 5: If $\varphi(x)$ is a real valued continuous function on the interval, then $\varphi(x)$ is of the class H_c if and only if the distribution function $x=m(y)$ of function $\psi(x)=|\varphi(x)|$ is an absolutely continuous function, or a non — increasing re — order function $\tilde{\psi}(x)$ is of the class H_c.

Therefore it is easy to obtain the following corollary.
Theorem 6: Let $\varphi(x)$ be a continuous function on a close interval [0,1]. Then $\varphi(x)\in H_c$ if and only if for any subset $E\subset[0,1]$ satisfies $m^*E>0$, and the following inequality

$$m^*\varphi(x)>0 \qquad (12)$$

holds.

Acknowledgment. The authors are much indebted to Prof. L. C. Hsu, Prof. R. H. Wang and Prof. Y. S. Zhou for their help, who gave our many valuable suggestions.

REFERENCES

[1] L. C. Hsu. "A reginement of the line integral approximation method and its application." Science Record N. S. Academic Sinica. No. 6. 1958 pp. 193—196.

[2] L. C. Hsu. "Annal. of Math. On Approximation Theory." Hang Zhou univ. (1980) (Chinese)

[3] X. L. Shi. "Note on a Problem of Asymptotic Expansion of Oscillatory Intograls" J. Hang Zhou univ. (Natureal Science) 3 (1979) No. 1. pp120—122 (Chinese)

[4] L. C. Hsu, Y. S. Chow and T. X. Hou "Selected Topic on Method unmerical integration ." Science Press. Anhui nhui (Chinese)

[5] E. Riekstens "On asymptotic expansion of some integrals involving a large parameter." Ucen. ap Leningrad Gos. univ. V41. (1961) pp. 5—23 (Russian)

[6] Colneachock. N. P. " Extremum problems for approximation of function classes" Science Press. Moscow. (1976) (Russian)

Approximation, Optimization and Computing:
Theory and Applications, A.G. Law and C.L. Wang (eds.)
Elsevier Science Publishers B.V. (North-Holland)

THE APPLICATION OF BIVARIATE CUBIC SPLINE

Zhen–xiang Xiong,

Dept. of Appl. Math. & Physics, Beijing Univ. of Aero. & Astro., Beijing, China

Xu–yang Li,

Computer Center, Beijing Institute of Economic Management, Beijing, China

This paper gives a practical example to show the application of our bivariate cubic spline for solving the linear partial differential equations with constant or variable coefficients. This method has only very simple calculation, and has high accuracy.

1. INTRODUCTION

There are plenty of applications of multivariate spline functions in many fields recently. In this paper, we calculate the numerical solutions of a certain kind of partial differential equations by using bivariate cubic spline functions.

The theory of multivariate spline functions is developed very fast recently. The bivariate interpolating splines developed by Z. X. Xiong have many fine properties. They have approximation order 4 and are very simple to calculate on a computer. Thus using them to find numerical solutions of partial differential equations is very significant.

2. PROBLEM

Given two metal disks of the same size and the same material. They are piled up coaxially. They have the radius $r_d = 10$ and the thickness $z_t = 0.5$. there are two electric poles pressed on both sides at the center of the disk. The distribution of electric field $\Phi(x,y,z)$ in the interior of the disks satisfies the Laplace equation:

$$\frac{\partial^2\Phi}{\partial x^2}+\frac{\partial^2\Phi}{\partial y^2}+\frac{\partial^2\Phi}{\partial z^2}=0$$

Change it into the cylindrical coordinate system, we have

$$\frac{\partial^2\Phi}{\partial r^2}+\frac{1}{r}\frac{\partial\Phi}{\partial r}+\frac{1}{r^2}\frac{\partial^2\Phi}{\partial\theta^2}+\frac{\partial^2\Phi}{\partial z^2}=0$$

Since the electric field is axially symmetric, it is independent on θ, hence we have

$$\frac{\partial^2\Phi}{\partial r^2}+\frac{1}{r}\frac{\partial\Phi}{\partial r}+\frac{\partial^2\Phi}{\partial z^2}=0 \qquad (1)$$

From the symmetry of the disks, we can only consider the shadow part of Fiqure 1.

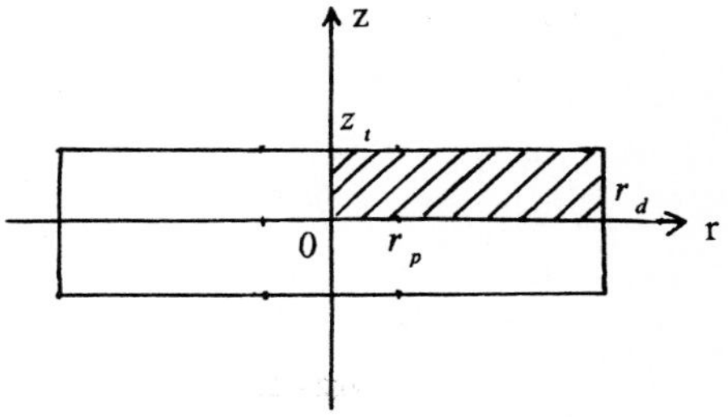

Figure 1

In this case we have the following boundary conditions:

$$\begin{cases}\Phi=\Phi_0, & r\leqslant r_p, \quad z=z_t;\\ \Phi=0, & r\leqslant r_p, \quad z=0;\\ \dfrac{\partial\Phi}{\partial n}=0, & \text{at the rest part of boundary}\\ & \text{(n is the direction of outer normal)}\end{cases} \qquad (2)$$

3. SOLUTION

Let the approximate solution of (1) with boundary conditions (2) be a bivariate spline function S(r,z). Divide the region

$$D=[0,\ r_d]\times[0,\ z_t]$$

into mn subrectangles by using partitions (see Figure 2):

$$r_0 = 0 < r_1 < \cdots < r_p < \cdots < r_m = r_d$$

$$z_0 = 0 < z_1 < \cdots < z_n = z_t$$

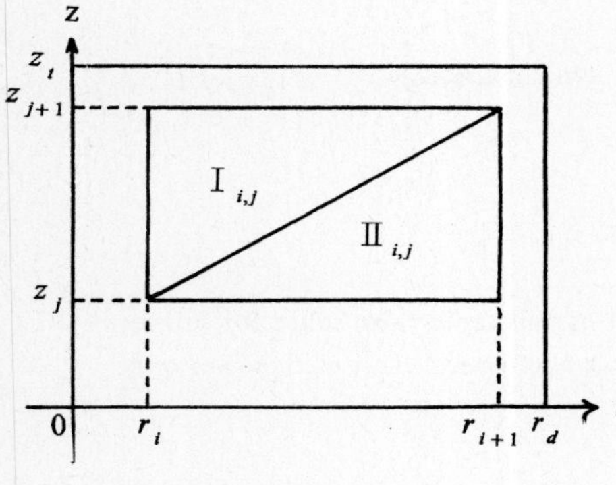

Figure 2

Separate each subrectangle into two right triangles $\mathrm{I}_{i,j}$ and $\mathrm{II}_{i,j}$ by the diagonal joining the two points (r_i, z_j) and (r_{i+1}, z_{j+1}). Then

$$S(x,z) = \begin{cases} \Phi_{i,j}(1-v) + \Phi_{i,j+1}(v-u) + \Phi_{i+1,j+1}u \\ - S_{02}(i,j)(v-u)(1-v)(2-v+u)k^2_{j+1}/6 \\ - S_{02}(i,j+1)(v-u)(1-v)(1+v+2u)k^2_{j+1}/6 \\ - S_{20}(i,j+1)u(v-u)(4-u-2v)h^2_{i+1}/6 \\ - S_{20}(i+1,j+1)u(v-u)(2-v+u)h^2_{i+1}/6 \\ - S_{11}(i,j+1)u(1-v)(v-u)h_{i+1}k_{j+1} \\ - D^2_{i,j}S(i,j)u(1-v)(2-2v+u)/6 \\ - D^2_{i,j}S(i+1,j+1)u(1-v)(1-v+2u)/6, \quad (r,z)\in \mathrm{I}_{i,j} \\ \Phi_{i,j}(1-u) + \Phi_{i+1,j}(u-v) + \Phi_{i+1,j+1}v \\ - S_{20}(i,j)(u-v)(1-u)(2-u+v)h^2_{i+1}/6 \\ - S_{20}(i+1,j)(u-v)(1-u)(1+u+2v)h^2_{i+1}/6 \\ - S_{02}(i+1,j)v(u-v)(4-u+2v)k^2_{j+1}/6 \\ - S_{02}(i+1,j+1)v(u-v)(2-u+v)k^2_{j+1}/6 \\ - S_{11}(i+1,j)v(1-u)(u-v)h_{i+1}k_{j+1} \\ - D^2_{i,j}S(i,j)v(1-u)(2-2u+v)/6 \\ - D^2_{i,j}S(i+1,j+1)v(1-u)(1-u+2v)/6, \quad (r,z)\in \mathrm{II}_{ij} \end{cases}$$

where $h_{i+1} = r_{r+1} - r_i, k_{j+1} = z_{j+1} - z_j$,

$$D^2_{i,j} = (h_{i+1}\frac{\partial}{\partial r} + k_{j+1}\frac{\partial}{\partial z})^{(2)}$$

$$u = (r - r_i)/h_{i+1}, \qquad v = (z - z_j)/k_{j+1},$$

and $S_{p,q}(i,j)$ denotes the value of $\dfrac{\partial^{p+q}S(r,z)}{\partial r^p \partial z^q}$ at the point (r_i, z_j), where $p+q=2$.

At the point (r_i, z_{j+1}), we have

$$\begin{cases} S_{20}(i,j+1) + \frac{1}{r_i}[\Phi_{i+1,j+1}/h_{i+1} - \Phi_{i,j+1}/h_{i+1} \\ - S_{20}(i+1,j+1)h_{i+1}/6 - S_{20}(i,j+1)h_{i+1}/3] \\ + S_{02}(i,j+1) = 0, \quad (i = 1,\cdots,m-1;\ j = 0,\cdots,n-1) \\ S_{20}(0,j+1) + S_{02}(0,j+1) = 0, \quad (j = 0,\cdots,n-1) \end{cases} \tag{3}$$

And at the point (r_{i+1}, z_j), we have

$$S_{20}(i+1,j) + \frac{1}{r_{i+1}}[-\Phi_{i,j}/h_{i+1} + \Phi_{i+1,j}/h_{i+1} + S_{20}(i,j)h_{i+1}/6$$
$$+ S_{20}(i+1,j)h_{i+1}/3] + S_{02}(i+1,j) = 0, \tag{4}$$

There left two points (0, 0) and (r_d, z_t) which are not considered, we have

$$S_{20}(0,0) + S_{02}(0,0) = 0 \tag{5}$$

and

$$S_{20}(m,n) + \frac{1}{r_d}[\Phi_{m,n}/h_m - \Phi_{n-1,n}/h_m + S_{20}(m,n)h_m/3$$
$$+ S_{20}(m-1,\ n)h_m/6] + S_{02}(m,\ n) = 0 \tag{6}$$

From [3] we have

$$k_{j+1}S_{02}(i,j-1) + 2(k_j + k_{j+1})S_{02}(i,j) + k_{j+1}S_{02}(i,j+1)$$
$$= 6[(\Phi_{i,j+1} - \Phi_{i,j})/k_{j+1} - (\Phi_{i,j} - \Phi_{i,j-1})/k_{j+1}], \tag{7}$$

$(\ i=0,\cdots,\ m;\ j=1,\cdots,n-1)$

$$h_iS_{20}(i-1,j) + 2(h_i + h_{i+1})S_{20}(i,j) + h_{i+1}S_{20}(i+1,j)$$
$$= 6[(\Phi_{i+1,j} - \Phi_{i,j})/h_{i+1} - (\Phi_{i,j} - \Phi_{i-1,j})/h_{i+1}], \tag{8}$$

$(\ j=0,\cdots,\ n;\ i=1,\cdots,m-1)$

The end conditions corresponding to (7) is

If $r_i \leqslant r_p$, then $\Phi_{i,0} = 0,\ \Phi_{i,n} = \Phi_0$, otherwise

$$\begin{cases} 2S_{02}(i,0) + S_{02}(i,1) + 6(\Phi_{i,0} - \Phi_{i,1})/k_1^2 = 0 \\ S_{02}(i,n-1) + 2S_{02}(i,n) + 6(\Phi_{i,0} - \Phi_{i,n-1})/k_n^2 = 0 \end{cases} \tag{9}$$

$(\ i=0,\cdots,m)$

Corresponding to (8), we have the follwoing end conditions

$$\begin{cases} 2S_{20}(0,j) + S_{20}(1,j) + 6(\Phi_{0,j} - \Phi_{1,j})/h_1^2 = 0 \\ S_{20}(n-1,j) + 2S_{20}(m,j) + 6(\Phi_{m,j} - \Phi_{m-1,j})/h_m^2 = 0 \end{cases} \tag{10}$$

$(\ j=0,\cdots,n)$

Equations (3)–(10) form a large sparse matrix equation, By solving this system of equations we can obtain all $\Phi_{i,j}$, $S_{02}(i,j)$ and $S_{20}(i,j)$. The values of $S_{11}(i,j)$ and $D^2_{i,j}S(i,j)$ can be determined by some reccurent formula in [3]. Thus the function S(r,z) is well defined. We divide the

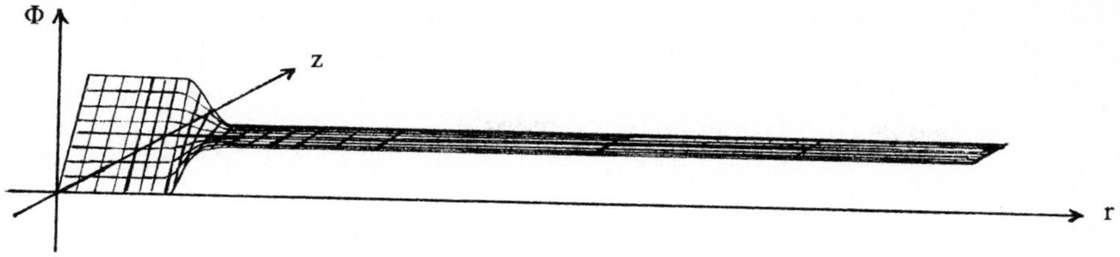

Figure 3

region [0,10]× [0,0.5] into 20× 8=160 subrectangles. The solution is shown as Figure 3. It has a high accuracy.

4. DISCUSSION

We can use the above method to find numerical solutions of linear partial differential equations of second order with variable coefficients. The advantages of this method are:

(1) The result is not only a system of approximate values at the mesh points but an approximate function on the whole region.

(2) The spline function has a high approximation order ana therefore the solution has a high accuracy.

(3) We can analyze other properties of the original equation by studying the spline function.

(4) The calculation is very simple. On VAX−11 / 750(with other users), we found the solution in less than two minutes.

REFERENCES

[1] de Boor, C., A practical guide to splines, Springer−Verlag, New York, 1978.

[2] Schumaker, L.L., SPLINE FUNCTIONS: Basic Theory, John Wiley & Sons, Inc. New York, 1981.

[3] Xiong, Z.X., Multivariate Interpolating Polynomials and Splines, to appear.

Approximation, Optimization and Computing:
Theory and Applications, A.G. Law and C.L. Wang (eds.)
Elsevier Science Publishers B.V. (North-Holland)

The Problem of Matrix Padé Approximation *

Xu Guo-liang †
Computing Center, Chinese Academy of Sciences, Beijing
A. Bultheel
Department of Computer Science, K.U.Leuven, Belgium

Abstract

Some proposals are made to give a general definition of matrix Padé approximants. Depending on the normalization of the denominator we define type I (constant term is the unit matrix) or type II (by conditions on the leading coefficient) approximants. Existence and uniqueness are considered, determinant expressions are given and relations among type I/II and left/right approximants are considered.

1. Introduction.

The problem of classical (scalar) Padé approximation is well established. However, when the given power series has matrix coefficients, it is not at all clear how the notion of Padé approximant should be generalized. Some papers have appeared on the case of a square matrix function where the scalar normality condition immediately generalizes to the matrix case (see e.g. [7-8]). The more general problem of Padé approximation in a non commutative algebra has been considered by a lot of authors. A. Draux has contributions in this area and has compiled a commented bibliography of about 300 references on this topic[5]. Also the vector case has attracted some attention in the literature. The papers of P.R. Graves-Morris (see e.g. [9]) are among the more recent ones. In some papers, the first author has considered a general matrix Padé problem for a rectangular matrix series, but a restricion is imposed on the sizes of column and row of the matrix series (see [2],[6]). There is also an extensive literature in linear system theory on the problem of minimal partial realization (see e.g. [10]). This problem is in a certain sense equivalent with a matrix approximation problem, but the given power series there is a series in z^{-1} while the numerator and denominator of the rational approximant are polynomials in z. Moreover, minimality is an important issue in those applications. This translates into a minimality of the degree of the determinant of the denominator. An effort has been done to translate these results into Padé results (see [3]). This resulted in a reformulation of the Padé problem as a minimal Padé problem, and it is the latter that is generalized to the matrix case. In this text we shall refer to it as the problem MPA(δ)(see [3]). This paper is a short written form of [4]. Due to the limitation of space, some results and proofs of the conclusions given here are omitted. The readers should consult reference [4] for details.

2. Definition of Matrix Padé Approximation.

Let $\mathbf{C}^{p\times m}$ be complex $p\times m$ matrices and $\mathbf{C}^{p\times m}[z]$ the polynomials with coefficients in $\mathbf{C}^{p\times m}$. Given $f(z)=\sum_{k=0}^{\infty} c_k z^k$, $c_k\in \mathbf{C}^{p\times m}$, we want to determine a rational approximant $N(z)M(z)^{-1}$ to f with residual $R(z)$ defined by

$$\begin{aligned} &R(z)=f(z)M(z)-N(z),\\ &M(z)\in \mathbf{C}^{m\times m}[z],\quad N(z)\in \mathbf{C}^{p\times m}[z]. \end{aligned} \tag{1}$$

This seems to be a fair proposal to generalize the definition of scalar Padé approximation(PA) to the matrix Padé approximation(MPA). However, there is still much choice on how to impose degree and order conditions. Whatever this choice is, the following principles seem natural: (i) When $p=m=1$, MPA should coincide with the PA. (ii) Solvability. The number of unknowns = the number of approximation conditions, so that the solution is unique in a generic situation. (iii) Nice properties of PAs should generalize to MPAs (e.g., invariance under linear fractional transformation).

The difficulties in defining the MPAs originate from:

(a) **Non-commutativity :** The MPA related to (1) is called a right MPA, but we could as well defined a left MPA $\tilde{M}(z)^{-1}(z)\tilde{N}(z)$ with $\tilde{N}(z)\in\mathbf{C}^{p\times m}[z], \tilde{M}(z)\in\mathbf{C}^{p\times m}[z]$ and residual $\tilde{R}(z)=\tilde{M}(z)f(z)-\tilde{N}(z)$.

(b) **Choice of degrees and orders :** The most general situation is to choose a degree for each entry of numerator and denominator and a order for each entry of the residual. Set

$$\begin{aligned} \mathbf{H}_k &:= \{p(z) : p(z)=\textstyle\sum_{i=0}^{k} a_i z^i,\quad a_i\in\mathbf{C}\},\\ \mathbf{E}_k &:= \{e(z) : e(z)=\textstyle\sum_{i=k+1}^{\infty} a_i z^i,\quad a_i\in\mathbf{C}\}, \end{aligned}$$

$\mathbf{Z}_+ := \{0,1,2,\cdots\}$, $\mathbf{Z}_+^{p\times m} := \{p\times m$ matrices with entries in $\mathbf{Z}_+\}$. For $V=(v_{ij})\in\mathbf{Z}_+^{p\times m}$, let

$$\begin{aligned} \mathbf{H}_V^{p\times m} &= \{P(z)=(p_{i,j}(z))_{i,j=1}^{p,m} : p_{ij}(z)\in\mathbf{H}_{v_{ij}}\},\\ \mathbf{E}_V^{p\times m} &= \{R(z)=(r_{i,j}(z))_{i,j=1}^{p,m} : r_{ij}(z)\in\mathbf{E}_{v_{ij}}\}. \end{aligned}$$

If V is not $p\times m$, it denotes row degree($p\times 1$) or column degree ($1\times m$) or matrix degree(1×1).

(c) **Normalization :** We shall consider two possibilities. (i) type I: $M(0)=I$ and type II: $M(z)$ is $C1$ canonical (see[3]).

---

*The work was done while the first author worked at the Department of Computer Science, K.U.Leuven

†Supported by National Science Foundation of China for Youth

Definition 2.1. *We say that $M(z) \in \mathbf{H}_V^{m\times m}$ is C1 canonical iff (i) There is a vector $U \in \mathbf{Z}_+^{m\times 1}$ such that $M(z) \in \mathbf{H}_U^{m\times m} \cap \mathbf{H}_{U^T}^{m\times m}$. (ii) The leading row coefficient matrix M^{hr} of $M(z)$ is the unit matrix. (iii) The leading column coefficient matrix M^{hc} of $M(z)$ is unit upper triangular.*

Now we are ready to give the following definitions of type I and type II right MPAs.

Definition 2.2. *Let $f(z) \in \mathbf{C}^{p\times m}[[z]]$ be a (formal) power series with coefficients in $\mathbf{C}^{p\times m}$ and let $V = (v_{ij}) \in \mathbf{Z}_+^{p\times m}, U = (u_{ij}) \in \mathbf{Z}_+^{m\times m}$ and $W = (w_{ij}) \in \mathbf{Z}_+^{p\times m}$ such that*

$$\sum_{k=1}^{p} w_{kj} = \sum_{k=1}^{p} v_{kj} + \sum_{k=1}^{m} u_{kj}, \quad j = 1,2,\ldots,m, \tag{2}$$

and $W - V \geq 0$. Then the right MPA problem of the first type is denoted by ${}^R(V,U,W)_I^f$ and consists in finding polynomials $N(z) \in \mathbf{H}_V^{p\times m}$ and $M(z) \in \mathbf{H}_U^{m\times m}$ such that $f(z)M(z) - N(z) \in \mathbf{E}_W^{p\times m}$ and $M(0) = I$.

Definition 2.3. *Let $f(z) \in \mathbf{C}^{p\times m}[[z]]$ be a (formal) power series with coefficients in $\mathbf{C}^{p\times m}$ and let $V = (v_{ij}) \in \mathbf{Z}_+^{p\times m}, U = (u_j) \in \mathbf{Z}_+^{m\times 1}$ and $W = (w_{ij}) \in \mathbf{Z}_+^{p\times m}$ such that*

$$\sum_{\substack{u_i > u_j \\ 1 \leq i \leq m}} u_j + \sum_{\substack{u_i \leq u_j \\ 1 \leq i \leq m}} u_i + \sum_{\substack{u_i > u_j \\ 1 \leq i < j}} 1 = \sum_{k=1}^{p} w_{kj} - \sum_{k=1}^{p} v_{kj}, \quad j = 1,2,\ldots,m, \tag{3}$$

and $W - V \geq 0$. Then the right MPA problem of the second type is denoted by ${}^R(V,U,W)_{II}^f$ and consists in finding polynomials $N(z) \in \mathbf{H}_V^{p\times m}$ and $M(z) \in \mathbf{H}_U^{m\times m}$ such that $f(z)M(z) - N(z) \in \mathbf{E}_W^{p\times m}$ and $M(z)$ is C1 canonical.

The solution set of ${}^R(V,U,W)_{I/II}^f$ is denoted by ${}^R[V,U,W]_{I/II}^f$

3. The Existence of MPA.

For $g(z) = \sum_{k=0}^{\infty} a_k z^k, \quad a_k \in \mathbf{C}^{s\times t}$, let

$$T_{mn}^l(g) = \begin{bmatrix} a_l & a_{l-1} & \cdots & a_{l-n+1} \\ a_{l+1} & a_l & \cdots & a_{l-n+2} \\ \cdots & \cdots & \cdots & \cdots \\ a_{l+m-1} & a_{l+m-2} & \cdots & a_{l+m-n} \end{bmatrix} \in \mathbf{C}^{ms\times nt}$$

be a block Toeplitz matrix. Let $f = (f_{ij}) \in \mathbf{C}^{p\times m}[[z]]$. For the problem ${}^R(V,U,W)_I^f$ we introduce the matrices $E = (e_{ij}) = W - V$ and

$${}^RH_I^j(V,U,W) = \begin{bmatrix} T_{e_{1j},u_{1j}}^{v_{1j}}(f_{11}) & \cdots & T_{e_{1j},u_{mj}}^{v_{1j}}(f_{1m}) \\ \cdots & \cdots & \cdots \\ T_{e_{pj},u_{1j}}^{v_{pj}}(f_{p1}) & \cdots & T_{e_{pj},u_{mj}}^{v_{pj}}(f_{pm}) \end{bmatrix},$$

$${}^RB_I^j(V,U,W) = \begin{bmatrix} T_{e_{1j},1}^{v_{1j}+1}(f_{1j}) \\ \vdots \\ T_{e_{pj},1}^{v_{pj}+1}(f_{pj}) \end{bmatrix}.$$

Then the problem of determining the j-th column of the denominator $M(z)$ in ${}^R(V,U,W)_I^f$ can be expressed by the equations ${}^RH_I^j(V,U,W)X = -\,{}^RB_I^j(V,U,W)$. If the j-th column $M_j(z)$ of $M(z)$ is determined, the j-th column $N_j(z)$ of $N(z)$ can be easily found from $N_j(z) = (fM_j)^{(V_j)}(z)$, where V_j is the degree vector of $N_j(z)$ and $(fM_j)^{(V_j)}$ is $f(z)M_j(z) \bmod z^{V_j}$. Therefore we have

Theorem 3.1. *The problem ${}^R(V,U,W)_I^f$ is solvable if and only if, for $j = 1,2,\ldots,m$,*

$$rank\ {}^RH_I^j(V,U,W) = rank \left[{}^RB_I^j(V,U,W) \ \ {}^RH_I^j(V,U,W) \right].$$

A similar conclusion holds for ${}^L(\tilde{V},\tilde{U},\tilde{W})_I^f$.

To the sequence $(u_i)_1^m$, we associate the matrices

$$C(U) = (c_{ij})_{i,j=1}^m, \quad T(U) = (t_{ij})_{i,j=1}^m, \quad S(U) = (s_{ij})_{i,j=1}^m,$$

where

$$c_{ij} = \begin{cases} 1, & \text{for } i \geq j, \text{ or } u_i \leq u_j \text{ and } i < j \\ 0, & \text{otherwise}, \end{cases}$$
$$t_{ij} = \min\{u_i, u_j\} - c_{ij}, \quad \text{and } s_{ij} = t_{ij} + 1.$$

Using these notations, the problem for determining the j-th column of the denominator $M(z)$ in the problem ${}^R(V,U,W)_{II}^f$ can be expressed by the following set of equations.

$${}^RH_{II}^j(V,U,W)X = -\,{}^RB_{II}^j(V,U,W),$$

where

$${}^RH_{II}^j(V,U,W) = \begin{bmatrix} T_{e_{1j},s_{1j}}^{v_{1j}+1}(f_{11}) & \cdots & T_{e_{1j},s_{mj}}^{v_{1j}+1}(f_{1m}) \\ \cdots & \cdots & \cdots \\ T_{e_{pj},s_{1j}}^{v_{pj}+1}(f_{p1}) & \cdots & T_{e_{pj},s_{mj}}^{v_{pj}+1}(f_{pm}) \end{bmatrix},$$

$${}^RB_{II}^j(V,U,W) = \begin{bmatrix} T_{e_{1j},1}^{v_{1j}-u_j+1}(f_{1j}) \\ \vdots \\ T_{e_{pj},1}^{v_{pj}-u_j+1}(f_{pj}) \end{bmatrix},$$

and $E = (e_{ij}) = W - V$. Therefore we have

Theorem 3.2. *The problem ${}^R(V,U,W)_{II}^f$ is solvable if and only if, for $j = 1,2,\ldots m$,*

$$rank\ {}^RH_{II}^j(V,U,W) = rank[\ {}^RB_{II}^j(V,U,W) \ \ {}^RH_{II}^j(V,U,W)].$$

A similar conclusion can be established for ${}^L(\tilde{V},\tilde{U},\tilde{W})_{II}^f$.

4. Determinant Expressions for MPAs.

For a given power series $g(z) = \sum_{i=0}^{\infty} a_i z^i, \quad a_i \in \mathbf{C}$, let

$$g^{(k)}(z) = \sum_{i=0}^{k} a_i z^i, \quad \bar{g}^{(k)}(z) = \sum_{i=k+1}^{\infty} a_i z^i, \quad a_i \in \mathbf{C}.$$

If $g(z) = (g_{ij}(z))_{i,j=1}^{s,t}$ and $V \in \mathbf{Z}_+^{s\times t}$, then we denote $(g_{ij}^{(v_{ij})}(z))_{i,j=1}^{s,t}$ and $(\bar{g}_{ij}^{(v_{ij})}(z))_{i,j=1}^{s,t}$ by $g^{(V)}(z)$ and $\bar{g}^{(V)}(z)$ respectively. The zero vector of k entries is denoted as : $\theta_k = [0,0,\ldots,0] \in \mathbf{C}^{1\times k}$. Then we have

Theorem 4.1. *Let $N(z)M(z)^{-1} \in {}^R[V,U,W]_I^f$. Then, if the matrix ${}^RH_I^j(V,U,W)$ is nonsingular,*

$$M_{ij}(z) = \frac{1}{\alpha_I^j} \det \begin{bmatrix} \delta_{ij} & \lambda_I(i,j) \\ {}^RB_I^j(V,U,W) & {}^RH_I^j(V,U,W) \end{bmatrix},$$

$$N_{ij}(z) = \frac{1}{\alpha_I^j} \det \begin{bmatrix} f_{ij}^{(v_{ij})}(z) & \omega_I(i,j) \\ {}^R B_I^j(V,U,W) & {}^R H_I^j(V,U,W) \end{bmatrix},$$

where $\alpha_I^j = \frac{1}{\det\ {}^R H_I^j(V,U,W)}$,

$$\begin{aligned} \lambda_I(i,j) &= [\theta_{u_{1j}}, \ldots, \theta_{u_{i-1,j}}\ z\ , \ldots, z^{u_{ij}},\ \theta_{u_{i+1,j}}\ , \ldots, \theta_{u_{m,j}}], \\ \omega_I(i,j) &= [\delta_I(i,j,1), \ldots, \delta_I(i,j,m)], \\ \delta_I(i,j,l) &= [z f_{il}^{(v_{ij}-1)}(z), \ldots, z^{u_{lj}} f_{il}^{(v_{ij}-u_{lj})}(z)]. \end{aligned}$$

Theorem 4.2. *Let $N(z)M(z)^{-1} \in {}^R[V,U,W]_{II}^f$. Then, if the matrix ${}^R H_{II}^j(V,U,W)$ is nonsingular,*

$$M_{ij}(z) = \frac{1}{\alpha_{II}^j} \det \begin{bmatrix} \delta_{ij} z^{s_{jj}} & \lambda_{II}(i,j) \\ {}^R B_{II}^j(V,U,W) & {}^R H_{II}^j(V,U,W) \end{bmatrix},$$

$$N_{ij}(z) = \frac{1}{\alpha_{II}^j} \det \begin{bmatrix} z^{s_{jj}} f_{ij}^{(v_{ij}-s_{jj})}(z) & \omega_{II}(i,j) \\ {}^R B_{II}^j(V,U,W) & {}^R H_{II}^j(V,U,W) \end{bmatrix},$$

where $\alpha_{II}^j = \frac{1}{\det\ {}^R H_{II}^j(V,U,W)}$,

$$\begin{aligned} &\lambda_{II}(i,j) = [\theta_{s_{1j}}, \ldots, \theta_{s_{i-1,j}}, 1,\ z\ , \ldots, z^{t_{ij}},\ \theta_{s_{i+1,j}}, \ldots, \theta_{s_{m,j}}], \\ &\omega_{II}(i,j) = [\delta_{II}(i,j,1), \ldots, \delta_{II}(i,j,m)], \\ &\delta_{II}(i,j,l) = [f_{il}^{(v_{ij})}(z), \ldots, z^{t_{lj}} f_{il}^{(v_{ij}-t_{lj})}(z)]. \end{aligned}$$

For the left MPA, similar formulas can be obtained.

5. Relations Among the Different MPAs.

Theorem 5.1. *Let $V,W \in \mathbf{Z}_+^{p\times 1}, U \in \mathbf{Z}_+^{m\times 1}$ be given. Then*
(i) if solvability equality (3) holds and

$$u_1 \geq u_2 \geq \cdots \geq u_m, \quad |u_i - u_j| \leq 1, \quad \forall(i,j), \qquad (4)$$

then $\{NM^{-1} \in {}^R[V,U,W]_{II}^f : M(0)$ is nonsingular$\} \subset {}^R[V,U,W]_I^f$.
(ii) if solvability equality (2) holds, then $\{NM^{-1} \in {}^R[V,U,W]_I^f : \exists\, Q \in \mathbf{C}^{m\times m}$ s.t. MQ is C1 canonical $\} \subset {}^R[V,U,W]_{II}^f$.

Theorem 5.2. *Let $V,W \in \mathbf{Z}_+^{p\times 1}$ and $U \in \mathbf{Z}_+$ be given such that solvability equality (2) holds (Note that in this case (2) is the same as (3)). For the given $f(z)$, define $g(z) = (g_{ij})_{i,j=1}^{p,m}$, with $g_{ij}(z) = z^{w_i} f_{ij}^{(w_i)}(z^{-1})$. Suppose furthermore that there exists a type II solution. Then we have*

$$\begin{aligned} &{}^R[V,U,W]_{II}^f(z) = f^{(W)}(z) \\ &-diag[z^{w_1}, \ldots, z^{w_p}]\ {}^R[U-1,U,W+U-V-1]_I^g(z^{-1}). \end{aligned}$$

Theorem 5.3. *Let $f(z) = \sum_{k=0}^{\infty} c_k z^k$ be given, $\omega \in \mathbf{Z}_+$, $\delta \in \mathbf{Z} \cap [-\omega, \omega]$. Let $U = (u_i)_{i=1}^m \in \mathbf{Z}^{1\times m}$ and $\tilde{U} = (\tilde{u}_i)_{i=1}^p \in \mathbf{Z}^{1\times p}$ be the Kronecker indices and dual Kronecker indices respectively associate with the sequence $\{c_\omega, c_{\omega-1}, \ldots, c_{-\delta+1}\}$. Then the following is true :*

(i) If $V = \omega - \tilde{U}$, $W = \omega$, then the solvability equality (2) holds for V, U and W. If NM^{-1} is a solution of problem MPA(δ), then $NM^{-1} \in {}^R[V,U,W]_I^f$, provided $M(0)$ is nonsingular.

(ii) For any $V,W \in \mathbf{Z}_+^{p\times 1}$, such that $V \geq \omega - \tilde{U}$, $W \leq \omega$, suppose that the solvability equality (3) holds for V,U,W. In that case the MPA(δ) solution NM^{-1} will be a type II MPA.

Because the type II MPAs and the solutions of the problem MPA(δ) both have C1 canonical normalization, one can ask if these two problems are equivalent in a certain sense. In the next theorem we shall show that this is true under a normality condition of the function.

In what follows $\lfloor a \rfloor$ will denote the largest integer not exceeding a.

Definition 5.1. *Let $f(z) = \sum_{k=0}^{\infty} c_k z^k$ be given. We say that $f(z)$ is normal if all the matrices H_k^ω are nonsingular for any $\omega \in \mathbf{Z}_+$ and $k > 0$, satisfying*

$$-\lfloor -k/m \rfloor \leq \omega, \quad -\lfloor -k/p \rfloor \leq \omega,$$

where H_k^ω is a $k \times k$ matrix defined by $H_k^\omega =$

$$\begin{bmatrix} c_\omega & \cdots & c_{\omega-q_1+1} & c_{\omega-q_1}(p,s) \\ \cdots & \cdots & \cdots & \cdots \\ c_{\omega-q_2+1} & \cdots & c_{\omega-q_1-q_2+2} & c_{\omega-q_1-q_2-1}(p,s) \\ c_{\omega-q_2}(t,m) & \cdots & c_{\omega-q_1-q_2+1}(t,m) & c_{\omega-q_1-q_2}(t,s) \end{bmatrix},$$

and

$$k = q_1 m + s, \quad s < m, \quad k = q_2 p + t, \quad t < p,$$

and $c_l(i,j)$ denotes a matrix formed from the first i rows and the first j columns of c_l.

Theorem 5.4. *For the given (ω, δ) as in the MPA(δ) problem, let $\lambda = \max\{0, -\delta+1\}$,*

$$u = \lfloor \frac{p(\omega - \lambda + 1) - m}{p+m} \rfloor, \quad d = \lfloor \frac{m(\omega - \lambda + 1) - p}{p+m} \rfloor,$$

$$s = \max\{0, p(\omega - \lambda + 1) - (u+1)(p+m)\},$$

$$t = \max\{0, m(\omega - \lambda + 1) - (d+1)(p+m)\}.$$

Assume $f(z)$ is normal. Then the problem MPA(δ) is equivalent to the problem ${}^R(V,U,W)_{II}^f$, where $V \in \mathbf{Z}_+^{p\times 1}$, $U \in \mathbf{Z}_+^{m\times 1}$, $W \in \mathbf{Z}_+$, are defined by $W = \omega$,

$$v_i = \omega - \tilde{u}_i, \quad i = 1,2,\ldots,p,$$

$$u_i = \begin{cases} u+2, & i = 1,2,\ldots,s, \\ u+1, & i = s+1,\ldots,m, \end{cases} \quad \text{and}$$

$$\tilde{u}_i = \begin{cases} d+2, & i = 1,2,\ldots,t, \\ d+1, & i = t+1,\ldots,p. \end{cases}$$

6. The Duality of MPA.

The duality considered here has to be understood in the following sense : To a given right MPA problem, we want to associate a left MPA problem which has the same solutions as the right one. The left problem will be called the dual problem of the right one and conversely.

Definition 6.1. *Consider the right and left MPA problems of type I : ${}^R(V,U,W)_I^f$ and ${}^L(\tilde{V},\tilde{U},\tilde{W})_I^f$. If for any $(N,M) \in \mathbf{H}_V \times \mathbf{H}_U$ and $(\tilde{N},\tilde{M}) \in \mathbf{H}_{\tilde{V}} \times \mathbf{H}_{\tilde{U}}$ satisfying*

$$f(z)M(z) - N(z) \in \mathbf{E}_W^{p\times m}, \quad \tilde{M}(z)f(z) - \tilde{N}(z) \in \mathbf{E}_{\tilde{W}}^{p\times m},$$

one has $\tilde{N}(z)M(z) = \tilde{M}(z)N(z)$, then we say that these first type left and right problems (and their solution sets) are each others dual.

Definition 6.2. *Consider the type II problems ${}^R(V,U,W)_{II}^f$ and ${}^L(\tilde{V},\tilde{U},\tilde{W})_{II}^f$. Define $U' = (u'_{ij}) \in \mathbf{Z}_+^{m\times m}$ and $\tilde{U}' = (\tilde{u}'_{ij}) \in \mathbf{Z}_+^{p\times p}$ by $u'_{ij} = \min\{u_i, u_j\}$, $\tilde{u}'_{ij} = \min\{\tilde{u}_i, \tilde{u}_j\}$. If for any $(N,M) \in \mathbf{H}_V \times \mathbf{H}_{U'}$ and $(\tilde{N},\tilde{M}) \in \mathbf{H}_{\tilde{V}} \times \mathbf{H}_{\tilde{U}'}$ satisfying*

$$f(z)M(z) - N(z) \in \mathbf{E}_W^{p\times m}, \quad \tilde{M}(z)f(z) - \tilde{N}(z) \in \mathbf{E}_{\tilde{W}}^{p\times m},$$

one has $\tilde{N}(z)M(z) = \tilde{M}(z)N(z)$, then we say that these second type left and right problems (and their solution sets) are dual.

Similarly, we can also define mixed dualilty. i.e., when the left and right problems are of different type.

Theorem 6.1. *Assume $V,W \in \mathbf{Z}_+^{p\times m}$, $U \in \mathbf{Z}_+^{m\times m}$, $\tilde{V},\tilde{W} \in \mathbf{Z}_+^{m\times p}$ and $\tilde{U} \in \mathbf{Z}_+^{p\times p}$. If relations*

1. $w_{ij} = \tilde{w}_{st} = w$, *for any (i,j) and (s,t),*
2. $u_{ij} = u_i, \quad j = 1,2,\ldots,m,$
3. $v_{ij} = v_i, \quad j = 1,2,\ldots,m,$
4. $\tilde{u}_{ij} = \tilde{u}_j, \quad i = 1,2,\ldots,p,$
5. $\tilde{v}_{ij} = \tilde{v}_j, \quad i = 1,2,\ldots,p,$
6. $u_i + \tilde{v}_i = w, \quad i = 1,2,\ldots,m,$
7. $v_i + \tilde{u}_i = w, \quad i = 1,2,\ldots,p,$

hold, then the problems ${}^R(V,U,W)_I^f$ and ${}^L(\tilde{V},\tilde{U},\tilde{W})_I^f$ are dual problems in the sense of definition 6.1.

If also (4) holds for U and also U is replaced by $\tilde{U}$, then type II problems ${}^R(V,U,W)_{II}^f$ and ${}^L(\tilde{V},\tilde{U},\tilde{W})_{II}^f$ are dual to each other in the sense of definition 6.2.

7. Uniqueness of MPA.

Let $V \in \mathbf{Z}_+^{p\times 1}$, $U \in \mathbf{Z}_+^{m\times 1}$ and $W \in \mathbf{Z}_+$, and assume relation (2) holds. Set further

$$\tilde{V} = (W-U)^T, \quad \tilde{U} = (W-V)^T \text{ and } \tilde{W} = W.$$

We introduce the following sets.

$$\begin{aligned} &{}^L R(\tilde{V},\tilde{U},\tilde{W})^f \\ &= \{(\tilde{N},\tilde{M}) \in \mathbf{H}_{\tilde{V}}^{p\times m} \times \mathbf{H}_{\tilde{U}}^{p\times p} \setminus \{0\} : \tilde{M}f - \tilde{N} \in \mathbf{E}_{\tilde{W}}^{p\times m}\}, \\ &{}^R R(V,U,W)^f \\ &= \{(N,M) \in \mathbf{H}_V^{p\times m} \times \mathbf{H}_U^{m\times m} \setminus \{0\} : fM - N \in \mathbf{E}_W^{p\times m}\}. \end{aligned}$$

Theorem 7.1. *Let $V \in \mathbf{Z}_+^{p\times 1}$ $U \in \mathbf{Z}_+^{m\times 1}$ and $W \in \mathbf{Z}_+$. Suppose $[V,U,W]_I^f \neq \emptyset$. Then the following statements are equivalent.*

(i) ${}^R[V,U,W]_I^f$ *is unique.*

(ii) ${}^L[W-U^T, W-V^T, W]_I^f \neq \emptyset$.

(iii) There exists a $(\tilde{N},\tilde{M}) \in {}^L R(W-U^T, W-V^T, W)^f$ such that $\det \tilde{M} \neq 0$.

(iv) For any $(N,M) \in {}^R R(V,U,W)^f$, we have ${}^R[V,U,W]_I^f M - N = \{0\}$.

For the problem of second type, we have

Corollary 7.2. *Let $V,W \in \mathbf{Z}_+^{p\times 1}$ and $W-V, U \in \mathbf{Z}_+$. Let $g(z) = (g_{ij})_{i,j=1}^{p,m}$, with $g_{ij}(z) = z^{w_i} f_{ij}^{(w_i)}(z^{-1})$. Suppose ${}^R[V,U,W]_{II}^f \neq \emptyset$. Then the following statements are equivalent.*

(i) ${}^R[V,U,W]_{II}^f$ *is unique.*

(ii) ${}^R[U-1, U, U+W-V-1]_I^g$ *is unique.*

(iii) ${}^L[W-V-1, W-V, U+W-V-1]_I^g \neq \emptyset$.

(iv) There exists a $(\tilde{N},\tilde{M}) \in {}^L R(W-V-1, W-V, U+W-V-1)^g$ such that $\det \tilde{M} \neq 0$.

(v) For any $(N,M) \in {}^R R(U-1, U, U+W-V-1)^g$, we have ${}^R[U-1, U, U+W-V-1]_I^g M - N = \{0\}$.

Corollary 7.3. *Let $V \in \mathbf{Z}_+^{p\times 1}$ $W \in \mathbf{Z}_+$, and $U \in \mathbf{Z}_+^{m\times 1}$ satisfy condition (4). If ${}^R[V,U,W]_I^f \neq \emptyset$ and ${}^R[V,U,W]_{II}^f \neq \emptyset$, then the uniqueness of ${}^R[V,U,W]_I^f$ implies the uniqueness of ${}^R[V,U,W]_{II}^f$*

Theorem 7.4. *Let $V \in \mathbf{Z}_+^{p\times 1}$ $W \in \mathbf{Z}_+$, $U \in \mathbf{Z}_+^{m\times 1}$ and let conditions (4) and $v_1 \le v_2 \le \cdots \le v_p$, $|v_p - v_1| \le 1$ be satisfied. If ${}^R[V,U,W]_{II}^f \neq \emptyset$, then the uniqueness of ${}^R[V,U,W]_{II}^f$ is equivalent to the existence of ${}^L[(W-U)^T, (W-V)^T, W]_{II}^f$.*

We should mention here that although the three results concerning the uniqueness of the second type MPA seem to be similar in forms, each one treats a different case. So they do not overlap.

References

[1] M. Van Barel, A.Bultheel, A Minimal partial algorithm for MIMO systems, I–V, 1988, Technical reports TW79, TW91, TW93, TW94, TW100, Department of Computer Science, K.U.Leuven.

[2] G.L. Xu, J.K.Li, Generalized Matrix Padé Approximants, submitted.

[3] A. Bultheel, M. Van Barel, A matrix Euclidean algorithm and matrix Padé approximations, Report TW 104, January 1988.

[4] G. L. Xu, A. Bultheel, The problem of Matrix Padé Approximations, Technical Report TW116, November 1988, Department of Computer Science, K.U.Leuven.

[5] A. Draux, Bibliography - Index - Report ANO - 145, Université de Sciences et techniques de Lille, November 1984.

[6] G.L.Xu, Existence and Uniqueness of Matrix Padé Approximants, J. Comp. Math. , to appear.

[7] A. Bultheel, Recursive Relations for Block Hankel and Toeplitz systems, Part I–II, J. Comp. Appl. Math., 10(1984), 301–354.

[8] A. Bultheel, Recursive algorithm for the matrix Padé tabe, Math. Comp., 35(1980), 875–892.

[9] P. R. Graves–Morris, Vector valued rational approximants II. IMA J. Numer. Anal,. 4(1984), 209–224.

[10] A. Bultheel, M. Van Barel, Padé techniques for model reduction in linear system theory: a survey, J. Comp. Appl. Math., 14(1986), 401–438.

Approximation, Optimization and Computing:
Theory and Applications, A.G. Law and C.L. Wang (eds.)
Elsevier Science Publishers B.V. (North-Holland)

INTERPOLATION THEOREMS

Yang Lihua

Department of Mathematics, Hunan Normal University, Changsha 410006, P. R. China.

In this paper we study interpolation theorems and extend the interpolation theorems of order r=2 established by V.Totik(see [1])to that of order r, any even number. By introducing the concept of weak monotone we constructed a briefer moduli of smoothness and treat different cases simultaneously which were done separately in [1]. the conditions of the theorems are also weakened.

1. INTRODUCTION

Let

$$X_p(a,b)=\begin{cases} L_p(a,b) & \text{if } 1\leqslant p<\infty \\ C(a,b) & \text{if } p=\infty \end{cases}$$

where Lp(a,b) is the space of all Lebesgue p-power integrable functions on (a,b) and C(a,b) all the bounded continous functions on (a,b). For $r\in N$ (all the natural numbers) and real-valued function ϕ on (a,b) write

$$W^r X_p(a,b)=\begin{cases} \{g\in L_p(a,b):\exists \tilde{g}(x)=g(x) \text{ a.e } \tilde{g}^{(r-1)} \text{is locally absolutely continuous and } \phi^r\tilde{g}^{(r)}\in L_p(a,b)\} & 1\leqslant p<\infty \\ \{g\in C(a,b):\ g^{(r-1)} \text{is locally absolutely continuous and } \phi^r g^{(r)}\in C(a,b)\} & p=\infty \end{cases}$$

We define the K-functional of $f\in X_p(a,b)$ by

$$K_r(t^r,f)=\inf_{g\in W^r X_p(a,b)}\{\|f-g\|_{X_p(a,b)}+t^r\|\phi^r g^{(r)}\|_{X_p(a,b)}\}$$

where

$$\|f\|_{X_p(a,b)}=\begin{cases} (\int_a^b |f(x)|^p dx)^{1/p} & \text{if } 1\leqslant p<\infty \\ \sup_{x\in(a,b)} |f(x)| & \text{if } p=\infty. \end{cases}$$

The concept of K-functional was introduced first by Peetre(see [2] [3]) in 1963 and 1964, and then studied by R.A.DeVore(see [4]), Z.Ditzian(see [5]), Zhou Xinlong(see [6]) and V.Totik(see [1]). the results of V.Totik are very good. For twice continuously differentiable function ϕ on (a,b) satisfying some conditions V.Totik constructed a kind of moduli of smoothness $\omega(f,t)$(see [1]) and obtained the following theorems.

Theorem A. Let $\phi, K_2(t^2,f)$ and $\omega(f,t)$ be as in [1]. There is a constant K independent of $f\in Lp(a,b)$ $(1\leqslant p<\infty)$ and $0<t\leqslant t_0$ such that

$$K^{-1}\omega(f,t)\leqslant K_2(t^2,f)\leqslant K\omega(f,t)$$

holds.

Theorem B. With the assumptions of theorem A let ϕ have limit zero at finite endpoints of (a,b) ((a,b)=(0,1) or (0,∞)) and let $\phi(x)/x$ be bounded at infinity ((a,b)=(0,∞) or (-∞,∞)). Then there is a constant K such that

$$K^{-1}V(f,t)\leqslant K_2(t^2,f)\leqslant K\int_0^t \frac{V(f,\tau)}{\tau}d\tau$$

holds for all $f\in Lp(a,b)$ and $0<t\leqslant t_0$, where

$$V(f,t)=\sup_{0<h\leqslant t}\|\Delta^2_{h\phi}f\|_{Lp(h^*,h^{**})}$$

The similar results in C[a,b] are discussed briefly in [1].

In this paper we construct an integral of one variable by using the Peano kernel to extend V.totik's work to that of any even order r.

2. THE INTERPOLATION THEOREMS ON Lp(a,b) $(1\leqslant P<\infty)$

We let (a,b)=(0,1),(0,∞) or (-∞,∞).

Definition 1. Let C be a positive number, ϕ be a veal-valued function on (a,b). If

$$\phi(x)\leqslant C\phi(y) \qquad (\text{or: } \phi(y)\leqslant C\phi(x))$$

holds for any $x,y\in(a,b)$ with $x\leqslant y$, ϕ is called C-weak increasing (or:decreasing). we also say briefly that ϕ is weak increasing (or: decreasing) if no necessity to mention C. Both weak increasing and weak decreasing are called weak monotone.

Let $r\in N$ be even, $\phi(x)$ be continusly differentiable for r-time on (a,b) satisfying:

1. there exists a constant $C>0$ such that

①. in a neighborhood of the endpoints $a=0,-\infty$ or $b=\infty$:

$$\begin{cases} C^{-1}\phi(y)\leqslant\phi(x)\leqslant C\phi(y) & \text{for } \frac{1}{2}\leqslant\left|\frac{x}{y}\right|\leqslant 2 \\ |\phi^{k-1}(x)\phi^{(k)}(x)|\leqslant C\left(\frac{\phi(X)}{|X|}\right)^k & (1\leqslant k\leqslant r) \end{cases}$$

②. in a neighborhood of the endpoint $b=1$:

$$\begin{cases} C^{-1}\phi(y)\leqslant\phi(x)\leqslant C\phi(y) & \text{for } \frac{1}{2}\leqslant\frac{1-x}{1-y}\leqslant 2 \\ |\phi^{k-1}(x)\phi^{(k)}(x)|\leqslant C\left(\frac{\phi(x)}{1-x}\right)^k & (1\leqslant k\leqslant r) \end{cases}$$

2.①. In a right neighborhood of $a=0$, $\phi(x)/x$ is bounded or; $\phi(x)/x$ is unbounded but $\phi(x)$ is weak increasing and $\phi(x)/x^{\delta}$ weak decreasing for some $0<\delta<1$;

②. In a left neighborhood of $b=1$, $\phi(x)/(1-x)$ is bounded or; $\phi(x)/(1-x)$ is unbounded but $\phi(x)$ weak decreasing and $\phi(x)/(1-x)^{\delta}$ weak increasing for some $0<\delta<1$;

③. at infinity $a=-\infty$(or: $b=\infty$), $\phi(x)/|x|$ is bounded or; $\phi(x)/|x|$ is unbounded but $\phi(x)/|x|$ weak decreasing (or: increasing).

Later we assume that the constant C in condition 1 is so large that all the weak monotones in condition 2 are C-weak monotones.

Let K denote positive constant, not necessarily the same at each occurants. Let Cr be positive constants depending only on r and ϕ, not necessarily the same at each occurants.

For positive even number r and small $h>0$ let

$$h^{*}=\begin{cases}\inf\{x:\ x-\frac{r}{2}h\phi(x)>0\} & \text{if } (a,b)=(0,1),(0,\infty) \\ \inf\{x:\ x+\frac{r}{2}h\phi(x)<0\} & \text{if } (a,b)=(-\infty,\infty)\end{cases}$$

$$h^{**}=\begin{cases}\sup\{x:\ x+\frac{r}{2}h\phi(x)<1\} & \text{if } (a,b)=(0,1) \\ \sup\{x:\ x-\frac{r}{2}h\phi(x)>0\} & \text{if } (a,b)=(0,\infty),(-\infty,\infty)\end{cases}$$

After these preliminaries we define, for $f\in Lp(a,b)$, small $t>0$ and a constant $C_1>C$ to be specified later,

$$\Omega(t)=\Omega(f,t)=\sup_{0<h\leqslant t}\|\Delta_h^r f\|_{Lp((C_1 h)^{*},(C_1 h)^{**})}$$

$$\Omega o(t)=\Omega o(f,t)=\sup_{0<h\leqslant t^{*}}\|\Delta_h^r f\|_{Lp(\frac{r}{2}h,\,2(C_1 h)^{*}+\frac{r}{2}h)}$$

$$\Omega_1(t)=\Omega_1(f,t)=\sup_{0<h\leqslant 1-t^{**}}\|\Delta_h^r f\|_{Lp(2(C_1 h)^{**}-1-\frac{r}{2}h,\,1-\frac{r}{2}h)}$$

$$\Omega_{\infty}(t)=\Omega_{\infty}(f,t)=\begin{cases}\|f\|_{Lp(\frac{(C_1 t)^{**}}{2},\,\infty)} & \text{if } (C_1 t)^{**}<\infty \\ 0 & \text{if } (C_1 t)^{**}=\infty\end{cases}$$

$$\Omega_{-\infty}(t)=\Omega_{-\infty}(f,t)=\begin{cases}\|f\|_{Lp(-\infty,\,2(C_1 t)^{*})} & \text{if } (C_1 t)^{*}>-\infty \\ 0 & \text{if } (C_1 t)^{*}=-\infty,\end{cases}$$

where the differences are symetric.

Now we define the moduli of smoothness for $f\in Lp(a,b)$

$$\omega_r(f,t)=\begin{cases}\Omega(t)+\Omega o(t)+\Omega_1(t) & (a,b)=(0,1) \\ \Omega(t)+\Omega o(t)+\Omega_{\infty}(t) & (a,b)=(0,\infty) \\ \Omega(t)+\Omega_{\infty}(t)+\Omega_{-\infty}(t) & (a,b)=(-\infty,\infty)\end{cases}$$

The following theorem 1 extends theorem A.

Theorem 1. Let $r\in N$ be even, $K_r(t,f)$ and $\omega_r(f,t)$ be as above. There is a constant $C_1\geqslant 2C>2$ not depending on f,t or p such that

$$K^{-1}\omega_r(f,t)\leqslant K_r(t^r,f)\leqslant K\omega_r(f,t)$$

holds for $f\in Lp(a,b)$ and small $t>0$, where K is a positive constant not depending on f or t.

Proof. Using the following lemmas we can prove the theorem. We omit the details here.

Lemma 1. For $1\leqslant s\leqslant r$ ther holds

$$\left(\frac{1}{\phi^r(x)}\right)^{(s)}=\frac{1}{\phi^{r+s}(x)}\sum{}^{*}C_{i_1,j_1,\dots,i_n,j_n}\cdot\phi^{(i_1-1)j_1+\dots+(i_n-1)j_n}\cdot[\phi^{(i_1)}(x)]^{j_1}\cdots[\phi^{(i_n)}(x)]^{j_n}=\frac{Ps(\phi,x)}{\phi^{r+s}(x)}=\frac{Ps(\phi)}{\phi^{r+s}}$$

and

$$\begin{cases}|Ps(\phi,x)|\leqslant Cr\left(\frac{\phi(x)}{|x|}\right)^{s} & \text{around } a=0,-\infty \text{ or } b=\infty \\ |Ps(\phi,x)|\leqslant Cr\left(\frac{\phi(x)}{1-x}\right)^{s} & \text{around } b=1\end{cases}$$

where $C_{i_1,j_1,\dots,i_n,j_n}$ depends only on $i_1,j_1,\dots,i_n,j_n$ and Σ^{*} is the sum over $i_1,j_1,\dots,i_n,j_n$ satisfying $i_1 j_1+\dots+i_n j_n=s$, $i_1,j_1,\dots,i_n j_n\geqslant 1$ and $n\in N$.

Furthermore, if we write $Po(\phi,x)=Po(\phi)\equiv 1$ the inequalities above also hold for $s=0$.

Lemma 2. For $0\leqslant j\leqslant r,\ j\neq\frac{r}{2}$ let

$$F_j(x)=\frac{1}{(r-1)!}\sum_{L=0}^{r}(-1)^{r-1}\binom{r}{L}\int_x^{x+(L-r/2)\frac{2j-r}{r}t\phi}$$

$$\left[x+\left(L-\frac{r}{2}\right)\frac{2j-r}{r}t\phi-\tau\right]^{r-1}f(\tau)d\tau.$$

Then for $0\leqslant k\leqslant r$ we have

$$F_j^{(k)}(x)=\sum_{L=0}^{r}(-1)^{r-L}\binom{r}{L}\Big\{\sum_{\substack{P=0\\ p\leqslant r-1}}^{k}\sum_{(Ki)}$$

$$C^{(p)}_{k_1,\dots,k_{n_p}}\cdot\left[1+\left(L-\frac{r}{2}\right)\frac{2j-r}{r}t\phi'(x)\right]^{k_1}$$

$$\cdot\prod_{m=2}^{n_p}\left[\left(L-\frac{r}{2}\right)\frac{2j-1}{r}t\phi^{(m)}(x)\right]^{Km}$$

$$\int_x^{x+(L-r/2)\frac{2j-r}{r}t\phi}\left[x+\left(L-\frac{r}{2}\right)\frac{2j-r}{r}t\,\cdot\right.$$

$$\left.\phi(X)-\tau\right]^{r-p-1}f(\tau)d\tau+C_kI_k\Big\}$$

($\sum_{(Ki)}$ means the sum over $k_1,\dots,k_{n_p}$

satisfying $\left\{\begin{array}{l}k_1+2k_2+\dots+n_pk_{n_p}=k\\ k_1+k_2+\dots+k_{n_p}=p\\ k_1,k_2,\dots,k_{n_p}\geqslant 0\\ n_p\leqslant r\end{array}\right.$)

where $C^{(p)}_{k_1,\dots,k_{n_p}}$ depends only on $k_1,\dots,k_{n_p}$, p and r, C_k depends on k and

$$I_k=\begin{cases}0 & \text{if } 0\leqslant k<r,\\ \left[1+\left(L-\frac{r}{2}\right)\frac{2j-r}{r}t\phi'(x)\right]^r f\left(x+\left(L-\frac{r}{2}\right)\frac{2j-r}{r}t\phi(x)\right) & \text{if } k=r.\end{cases}$$

Lemma 3. For $C_1\geqslant 2C>2$ and small $t>0$ there holds

$$\sup_{\substack{|h|\leqslant t\\ x\in((C_1t)^*,(C_1t)^{**})}}\frac{\phi(x)}{\phi(x+rh\phi(x)/2)}\leqslant C$$

Lemma 4. For small $t>0$ we have

$$\sup_{x\in((C_1t)^*,(C_1t)^{**})}|\phi^{k-1}(x)\phi^{(k)}(x)|$$

$$\leqslant C^{1+r}\left(\frac{1}{C_1t}\right)^k,\qquad(1\leqslant k\leqslant r)$$

We omit the proofs of lemma 1-4 here.

The moduli of smoothness above can be improved. If $\frac{\phi(x)}{x}$ is unbounded around $a=0$, for any $\xi\in(0,1)$ put

$$\Omega_0^*(f,t)=\sup_{0<h\leqslant t^*}\|\Delta_h^r f\|_{Lp(\frac{r}{2}h,\ \xi)};$$

and if $\frac{\phi(x)}{1-x}$ is unbounded around $b=1$, for any $\eta\in(0,1)$ put

$$\Omega_1^*(f,t)=\sup_{0<h\leqslant 1-t^{**}}\|\Delta_h^r f\|_{Lp(\eta,\,1-\frac{r}{2}h)}.$$

For small $h>0$ let

$$a\leqslant\alpha(h)\leqslant(C_1h)^*<(C_1h)^{**}\leqslant\beta(h)\leqslant b$$

satisfying $x\pm\frac{r}{2}h\phi(x)\in(a,b)$ and $1\pm\frac{r}{2}h\phi'(x)>K$ for all $x\in(\alpha(h),\beta(h))$, where K is a positive constant and C_1 is as in theorem 1. Put

$$\Omega^*(f,t)=\sup_{0<h\leqslant t}\|\Delta_{h\phi}^r f\|_{Lp(\alpha(h),\beta(h))}.$$

Now if we substitute $\Omega^*(f,t)$, $\Omega_0^*(f,t)$ and $\Omega_1^*(f,t)$ for $\Omega(f,t)$, $\Omega_0(f,t)$ and $\Omega_1(f,t)$ respectively in theorem 1 (or: substitute one or two of the three terms), the result of theorem 1 is still true.

In application it will be important to supplement theorem 1 with an estimate of $K_r(t^r,f)$ by the r-th difference $\Delta_{h\phi}^r f$ alone. Keeping the above notations let

$$V(t)=V(f,t)=\sup_{0<h\leqslant t}\|\Delta_{h\phi}^r f\|_{Lp(\alpha(h),\beta(h))}$$

and for this we can prove

Theoerm 2. with the assumptions of theorem 1 let $\frac{\phi(x)}{x}$ be bounded at infinity ($(a,b)=(0,\infty)$ or $(-\infty,\infty)$). Then there is a constant K such that

$$K^{-1}V(f,t)\leqslant K_r(t^r,f)\leqslant K\int_0^t\frac{V(f,\tau)}{\tau}d\tau$$

holds for all $f\in Lp(a,b)$ and sufficiently small $t>0$.

Specially, $K_r(t^r,f)=O(t^\alpha)$ and $V(f,t)=O(t^\alpha)$ are equivalent for $\alpha>0$.

3. THE INTERPOLATION THEOREMS ON C(a,b)

Let $\phi(x)$ be as in 2. We keep all the assumptions in 2 and strengthen the condition 2.③ by (see 2):

2③*. At infinity $a=-\infty(b=\infty)$ suppose $\frac{\phi(x)}{|x|}$ is bounded or; $\frac{\phi(x)}{|x|}$ is unbounded but $\frac{\phi(x)}{|x|^\sigma}$ is weak decreasing (increasing) for some $\sigma>1$. Let

$$\tilde{\Omega}_{\infty}(f,t)=\begin{cases}\sup\limits_{x,y\geq\frac{1}{2}(C_1 t)^{**}}|f(x)-f(y)| & \text{if } (C_1 t)^{**}<\infty \\ 0 & \text{if } (C_1 t)^{**}=\infty\end{cases}$$

$$\tilde{\Omega}_{-\infty}(f,t)=\begin{cases}\sup\limits_{x,y\leq 2(C_1 t)^{*}}|f(x)-f(y)| & \text{if } (C_1 t)^{*}>-\infty \\ 0 & \text{if } (C_1 t)^{*}=-\infty\end{cases}$$

and define

$$\omega_r(f,t)=\Omega(f,t)+\begin{cases}\Omega_0(f,t)+\Omega_1(f,t) & (a,b)=(0,1)\\ \Omega_0(f,t)+\tilde{\Omega}_{\infty}(f,t) & (a,b)=(0,\infty)\\ \tilde{\Omega}_{\infty}(f,t)+\tilde{\Omega}_{-\infty}(f,t) & (a,b)=(-\infty,\infty)\end{cases}$$

Then we have

Theorem 3. Let $K_r(t^r,f)$, $\omega_r(f,t)$ be as above. Then there exists $C_1>0$ such that

$$K^{-1}\omega_r(f,t)\leq K_r(t^r,f)\leq K\omega_r(f,t)$$

holds for $f\in Lp(a,b)$ and small $t>0$, where K is a positive constant not depending on f or t.

To simplify the moduli of smoothness we let $a\leq \alpha(h)\leq(C_1 h)^{*}<(C_1 h)^{**}\leq\beta(h)\leq b$ satisfying

(i) $x\pm\frac{r}{2}h\phi(x)\in(a,b)$ for $x\in(\alpha(h),\beta(h))$

(ii) If $a=-\infty$ and $\frac{\phi(x)}{|x|}$ is unbounded at infinity $a=-\infty$, then there is a constant $K>0$ such that $K^{-1}\leq|1\pm\frac{r}{2}h\frac{\phi(x)}{x}|\leq K$ for all $x\in(\alpha(h),(C_1 h)^{*})$;
If $b=\infty$ and $\frac{\phi(x)}{|x|}$ is unbounded at infinity $b=\infty$ then there is a constant $K>0$ such that $K^{-1}\leq|1\pm\frac{r}{2}h\frac{\phi(x)}{x}|\leq K$ for all $x\in((C_1 h)^{**},\beta(h))$.

We define

$$\Omega^{*}(f,t)=\sup_{0<h\leq t}\|\Delta^r_{h\phi}f\|_{C(\alpha(h),\beta(h))}$$

For any $\xi,\eta\in(0,1)$ if $\frac{\phi(x)}{x}$ is unbounded around $a=0$ we define

$$\Omega^{*}_0(f,t)=\sup_{0<h\leq t^{*}}\|\Delta^r_h f\|_{C(\frac{r}{2}h,\xi)}$$

and if $\frac{\phi(x)}{1-x}$ is unbounded around $b=1$ we define

$$\Omega^{*}_1(f,t)=\sup_{0<h\leq 1-t^{**}}\|\Delta^r_h f\|_{C(\eta,1-\frac{r}{2}h)}.$$

Similarly to the case of $Lp(a,b)$ if we substitute $\Omega^{*}(f,t)$, $\Omega^{*}_0(f,t)$ and $\Omega^{*}_1(f,t)$ for $\Omega(f,t)$, $\Omega_0(f,t)$ and $\Omega_1(f,t)$ respectively in theorem 3 (or: substitute one or two of the three), the result of theorem 3 is still true.

Keep the assumptions and notations above we let

$$V(t)=V(f,t)=\sup_{0<h\leq t}\|\Delta^r_{h\phi}f\|_{C(\alpha(h),\beta(h))}$$

Then similarly to theorem 2 we can prove

Theorem 4. Suppose $\frac{\phi(x)}{|x|}$ is bounded at infinity ($(a,b)=(0,\infty)$ or $(-\infty,\infty)$). Then there is a constant $K>0$ such that

$$K^{-1}V(f,t)\leq K_r(t^r,f)\leq K\int_0^t\frac{V(f,\tau)}{\tau}d\tau$$

holds for all $f\in C(a,b)$.

REMARKS

(a) For $r=2$, $\infty>p\geq 1$ $(a,b)=(0,\infty)$ and $\phi(x)=x^2$ let $f\in C^2[0,\infty)$ satisfying

(i) $f(x)=(\frac{1}{x\ln^2 x})^{1/p}$ for $x\geq 2$

(ii) there exists a constant $K>0$ such that $f(x)>K>0$ for all $x\in[0,2]$.

We can easily prove that $f\in Lp(0,\infty)$ and

$$\frac{\Omega_{\infty}(f,t)}{\Omega(f,t)}\longrightarrow\infty\qquad(t\to 0+)$$

which shows that $\Omega_{\infty}(f,t)$ can't be eliminated in general.

(b) When $(a,b)=(0,\infty)$ or $(-\infty,\infty)$ whether $\Omega(f,t)=O(t^{\alpha})$ and $\Omega_{\pm\infty}(f,t)=O(t^{\alpha})$ are equivalent remains a problem.

(c) Can we establish the similar theorems for odd r?

REFERENCE

[1]. V.Totik, Pacific.J.Math. 111,2(1984), 447-481.

[2]. Peetre.J., A Theory of Interpolation of Normed Space, Lecture Notes, Brazilia, 1963.

[3]. ------ "Espaces d'interpolation generalisations, applications", Rend. Sem. Math. Fis. Milano, 34, (1964), 133-164.

[4]. R.A.DeVore, "Degree of approximation" in Lorentz, Chui, and Schumaker (1976) PP. 117-161.

[5]. Z.Ditzian, Pacific.J.Math. Vol 90, No.2, (1980), 307-323.

[6]. Zhou Xinlong, On a Problem of Ditzian, J. of Hangzhou University, Vol 12, No.2, Apr.(1985), 178-182.

[7]. Larry.L.Schumaker, Spline Functions: Basic Theory.

Approximation, Optimization and Computing:
Theory and Applications, A.G. Law and C.L. Wang (eds.)
Elsevier Science Publishers B.V. (North-Holland)

ERROR ANALYSIS OF RECURRENCE TECHNIQUE FOR THE CALCULATION OF BESSEL FUNCTION $I_\nu(x)$

Toshio YOSHIDA
Chubu University, Kasugai, Aichi, Japan 487

The modified Bessel function $I_\nu(x)$ can be expanded in terms of $I_{\mu+2k}(x)$ $(k = 0, 1, \ldots)$. In this paper, we describe the error analysis of the approximation to $I_\nu(x)$ obtained by substituting approximations to $I_{\mu+2k}(x)$ $(\mu \geq 0)$ computed by recurrence technique into the truncated form of this expansion.

1. INTRODUCTION

The modified Bessel function of the first kind $I_\nu(x)$ can be expanded in terms of $I_{\mu+2k}(x)$ $(k = 0, 1, \ldots)$[1].

$$I_\nu(x) = \sum_{k=0}^{\infty} \rho_k I_{\mu+2k}(x) \tag{1}$$

where

$$\rho_k = \left(\frac{x}{2}\right)^{\nu-\mu} \frac{(-1)^k \Gamma(\mu+k)\Gamma(\nu+1-\mu)(\mu+2k)}{k!\ \Gamma(\nu+1-\mu-k)\Gamma(\nu+k+1)}. \tag{2}$$

From the above expansion, we can calculate $I_\nu(x)$ by using $I_{\mu+2k}(x)$ $(k = 0, 1, \ldots)$ for $\mu \geq 0$ as follows.

Consider a function $G_{\mu+n}(x)$, which obey the recurrence relation

$$G_{\mu-1}(x) = \frac{2\mu}{x} G_\mu(x) + G_{\mu+1}(x) \tag{3}$$

and is defined such that

$$G_{\mu+m+1}(x) = 0, \quad G_{\mu+m}(x) = \alpha \tag{4}$$

where m is an appropriately chosen positive even integer and α is an arbitrarily chosen small constant. By successive application of recurrence relation (3), we generate $G_{\mu+m-1}(x)$, $G_{\mu+m-2}(x), \ldots, G_\mu(x)$. Then for $n = 0, 1, \ldots, m/2$,

$$I_{\mu+2n}(x) \approx e^x G_{\mu+2n}(x) / \sum_{k=0}^{m} \epsilon_k G_{\mu+k}(x) \tag{5}$$

where

$$\epsilon_k = 2\left(\frac{x}{2}\right)^{-\mu} \frac{(\mu+k)\Gamma(\mu+1)\Gamma(2\mu+k)}{k!\ \Gamma(2\mu+1)}. \tag{6}$$

The efficient method for the calculation of $I_\mu(x)$ was studied by I.Ninomiya[2], S.Makinouchi[3] and T.Yoshida[4]. Ninomiya also gave the estimation of the error of the approximation.

Substituting Eq.(5) into the truncated form of Eq.(1), we can obtain an approximation to $I_\nu(x)$. This method is useful for computing both $I_\mu(x)$ and $I_\nu(x)$ $(\nu \neq \mu)$ at the same time and/or for computing $I_\nu(x)$ in the case of $\nu < 0$. In this paper, the error analysis of this approximation to $I_\nu(x)$ is described. Using the summation theorem[5] of the generalized hypergeometric series, a tedious manipulation of the expression leads to a relatively simple form for the error.

2. METHOD AND ERROR ANALYSIS

Let n be an integer and $\mu \geq 0$. The functions $I_{\mu+n}(x)$ and $\overline{K}_{\mu+n}(x) = (-1)^n K_{\mu+n}(x)$ ($K_\mu(x)$:the modified Bessel function of the second kind) both obey the same recurrence relation (3). Inversely the general solution of Eq.(3) is expressed in terms of

$$G_{\mu+n}(x) = \xi I_{\mu+n}(x) + \eta \overline{K}_{\mu+n}(x) \tag{7}$$

with arbitrary constants ξ and η. From Eq.(4),we can assume that

$$G_{\mu+m+1}(x) = \xi I_{\mu+m+1}(x) + \eta \overline{K}_{\mu+m+1}(x) = 0, \tag{8}$$

$$G_{\mu+m}(x) = \xi I_{\mu+m}(x) + \eta \overline{K}_{\mu+m}(x) = \alpha. \tag{9}$$

Eliminating η from Eqs.(7) and (8),we obtain

$$G_{\mu+n}(x) = \xi \left(I_{\mu+n}(x) - \frac{I_{\mu+m+1}(x)\overline{K}_{\mu+n}(x)}{\overline{K}_{\mu+m+1}(x)} \right). \tag{10}$$

From Eq.(10) and the relation

$$\sum_{k=0}^{\infty} \epsilon_k I_{\mu+k}(x) = e^x, \tag{11}$$

we obtain

$$\sum_{k=0}^{m} \epsilon_k \left(\frac{G_{\mu+k}(x)}{\xi} + \frac{I_{\mu+m+1}(x)\overline{K}_{\mu+k}(x)}{\overline{K}_{\mu+m+1}(x)} \right) + \sum_{k=m+1}^{\infty} \epsilon_k I_{\mu+k}(x) = e^x. \tag{12}$$

Eliminating ξ in Eqs.(10) and (12),

$$I_{\mu+n}(x) = \frac{e^x G_{\mu+n}(x)}{\sum_{k=0}^{m} \epsilon_k G_{\mu+k}(x)} \left\{ 1 - e^{-x} \left(\sum_{k=0}^{m} \epsilon_k \frac{I_{\mu+m+1}(x)\overline{K}_{\mu+k}(x)}{\overline{K}_{\mu+m+1}(x)} + \sum_{k=m+1}^{\infty} \epsilon_k I_{\mu+k}(x) \right) \right\} + \frac{I_{\mu+m+1}(x)\overline{K}_{\mu+n}(x)}{\overline{K}_{\mu+m+1}(x)}. \tag{13}$$

It is therefore found that the approximation to $I_\mu(x)$ with p significant digits is expressed using $G_{\mu+m-1}(x)$, $G_{\mu+m-2}(x), \ldots, G_\mu(x)$ generated by successive application of recurrence relation with initial values (4) as

$$I_{\mu+n}(x) \approx e^x G_{\mu+n}(x) / \sum_{k=0}^{m} \epsilon_k G_{\mu+k}(x) \tag{14}$$

if the following inequalities hold:

$$|\Phi_{\mu,m}(x)| < 0.5 \times 10^{-p} \tag{15}$$

and

$$\left| \frac{I_{\mu+m+1}(x) K_{\mu+n}(x)}{I_{\mu+n}(x) K_{\mu+m+1}(x)} \right| < 0.5 \times 10^{-p} \tag{16}$$

where

$$\Phi_{\mu,m}(x) = e^{-x} \left(\sum_{k=0}^{m} \epsilon_k \frac{I_{\mu+m+1}(x) \overline{K}_{\mu+k}(x)}{\overline{K}_{\mu+m+1}(x)} + \sum_{k=m+1}^{\infty} \epsilon_k I_{\mu+k}(x) \right). \tag{17}$$

For the computation of $I_\nu(x)$ which is a main object, from Eqs.(1) and (13), we obtain

$$I_\nu(x) = \sum_{k=0}^{\infty} \rho_k I_{\mu+2k}(x) = \frac{e^x \sum_{k=0}^{m/2} \rho_k G_{\mu+2k}(x)}{\sum_{k=0}^{m} \epsilon_k G_{\mu+k}(x)}$$

$$\times \left\{ 1 - e^{-x} \left(\sum_{k=0}^{m} \epsilon_k \frac{I_{\mu+m+1}(x) \overline{K}_{\mu+k}(x)}{\overline{K}_{\mu+m+1}(x)} + \sum_{k=m+1}^{\infty} \epsilon_k I_{\mu+k}(x) \right) \right\}$$

$$+ \frac{I_{\mu+m+1}(x)}{\overline{K}_{\mu+m+1}(x)} \sum_{k=0}^{m/2} \rho_k \overline{K}_{\mu+2k}(x) + \sum_{k=m/2+1}^{\infty} \rho_k I_{\mu+2k}(x) \tag{18}$$

Therefore the approximation to $I_\nu(x)$ with p significant digits is given as

$$I_\nu(x) \approx \frac{e^x \sum_{k=0}^{m/2} \rho_k G_{\mu+2k}(x)}{\sum_{k=0}^{m} \epsilon_k G_{\mu+k}(x)}, \tag{19}$$

if the following inequalities hold:

$$|\Phi_{\mu,m}(x)| < 0.5 \times 10^{-p} \tag{20}$$

and

$$|\Psi_{\nu,\mu,m}(x)| < 0.5 \times 10^{-p} \tag{21}$$

where

$$\Psi_{\nu,\mu,m}(x) = \frac{1}{I_\nu(x)} \left(\frac{I_{\mu+m+1}(x)}{\overline{K}_{\mu+m+1}(x)} \sum_{k=0}^{m/2} \rho_k \overline{K}_{\mu+2k}(x) + \sum_{k=m/2+1}^{\infty} \rho_k I_{\mu+2k}(x) \right). \tag{22}$$

Note that $\Phi_{\mu,m}(x)$ is independent of ν and the condition (15) in the computation of $I_\mu(x)$ is identical with the condition (20) in that of $I_\nu(x)$.

3. MODIFICATION OF EXPRESSION OF $\Phi_{\mu,m}(x)$

Let us rewrite the expression (17) of $\Phi_{\mu,m}(x)$.

$$\Phi_{\mu,m}(x) = e^{-x} \left(\sum_{k=0}^{m} \epsilon_k \frac{I_{\mu+m+1}(x) \overline{K}_{\mu+k}(x)}{\overline{K}_{\mu+m+1}(x)} + e^x - \sum_{k=0}^{m} \epsilon_k I_{\mu+k}(x) \right)$$

$$= \frac{e^x K_{\mu+m+1}(x) - \sum_{k=0}^{m} \epsilon_k x^{-1} \overline{R}_{m-k,\mu+k+1}(x)}{e^x K_{\mu+m+1}(x)} \tag{23}$$

where

$$\overline{R}_{m-k,\mu+k+1}(x) = x \left(I_{\mu+k}(x) K_{\mu+m+1}(x) + (-1)^{m+k+2} I_{\mu+m+1}(x) K_{\mu+k}(x) \right) \tag{24}$$

which is called the modified Lommel polynomial.

Now we rewrite the first term of the numerator of Eq.(23). Using the representation of $e^x K_\mu(x)$ in terms of the Kummer's confluent function[6],

$$e^x K_{\mu+m+1}(x) = \frac{1}{2} \sum_{k=0}^{2m+1} \frac{\Gamma(2\mu+2m+2-k)\Gamma(\mu+m+1-k)}{k!\,\Gamma(2\mu+2m+2-2k)} \left(\frac{x}{2}\right)^{-\mu-1-m+k}$$

$$+ \frac{\pi^{\frac{1}{2}}(-1)^{m+1}(2x)^{m+1}}{2 \sin \mu\pi} \left\{ \sum_{k=0}^{\infty} \frac{\Gamma(-\mu+m+k+\frac{3}{2})(2x)^{-\mu+k}}{(2m+k+2)!\,\Gamma(-2\mu+k+1)} \right.$$

$$\left. - \sum_{k=0}^{\infty} \frac{\Gamma(\mu+m+k+\frac{3}{2})(2x)^{\mu+k}}{k!\,\Gamma(2\mu+2m+k+3)} \right\}. \tag{25}$$

Next we can modify the second term of the numerator of Eq.(23) as follows.

$$\sum_{k=0}^{m} \epsilon_k x^{-1} \overline{R}_{m-k,\mu+k+1}(x) = \frac{\pi}{2 \sin \mu\pi}$$

$$\times \sum_{k=0}^{m} \epsilon_k \left(I_{\mu+k}(x) I_{-\mu-m-1}(x) - I_{\mu+m+1}(x) I_{-\mu-k}(x) \right)$$

$$= \frac{1}{2} \left(\frac{x}{2}\right)^{-m-1} \sum_{k=0}^{m} \epsilon_k \left(\frac{x}{2}\right)^k$$

$$\times \sum_{n=0}^{[(m-k)/2]} \frac{\Gamma(m-k+1-n)\Gamma(\mu+m+1-n)}{n!\,\Gamma(m-k+1-2n)\,\Gamma(\mu+k+1+n)} \left(\frac{x}{2}\right)^{2n}$$

$$= \frac{1}{2} \left(\frac{x}{2}\right)^{-m-1} \sum_{l=0}^{m} \left(\frac{x}{2}\right)^l \frac{1}{(m-l)!}$$

$$\times \sum_{n=0}^{[l/2]} \epsilon_{l-2n} \frac{(m-l+1+n)!\,\Gamma(\mu+m+1-n)}{n!\,\Gamma(\mu+l+1-n)}$$

$$= \left(\frac{x}{2}\right)^{-m-1-\mu} \frac{\Gamma(\mu+1)}{\Gamma(2\mu+1)} \sum_{l=0}^{m} \left(\frac{x}{2}\right)^l \frac{1}{(m-l)!}$$

$$\times \sum_{n=0}^{[l/2]} \frac{(\mu+l-2n)\Gamma(2\mu+l-2n)(m-l+n)!\,\Gamma(\mu+m+1-n)}{n!\,\Gamma(l-2n+1)\Gamma(\mu+l+1-n)}$$

$$= \left(\frac{x}{2}\right)^{-m-1-\mu} \frac{\Gamma(\mu+1)}{\Gamma(2\mu+1)} \sum_{l=0}^{m} \left(\frac{x}{2}\right)^l \frac{1}{(m-l)!}$$

$$\times \sum_{n=0}^{\infty} \frac{(\mu+l-2n)\Gamma(2\mu+l-2n)(m-l+n)!\,\Gamma(\mu+m+1-n)}{n!\,\Gamma(l-2n+1)\Gamma(\mu+l+1-n)}$$

$$= \left(\frac{x}{2}\right)^{-m-1-\mu} \frac{\Gamma(\mu+1)\Gamma(\mu+m+1)}{\Gamma(2\mu+1)} \sum_{l=0}^{m} \left(\frac{x}{2}\right)^l \frac{(\mu+l)\Gamma(2\mu+l)}{l!\,\Gamma(\mu+l+1)}$$

$$\times \sum_{n=0}^{\infty} \frac{\left(1-\frac{\mu+l}{2}\right)_n (m-l+1)_n \left(-\frac{l}{2}\right)_n \left(\frac{1}{2}-\frac{l}{2}\right)_n (-\mu-l)_n}{n!\,\left(-\frac{\mu+l}{2}\right)_n \left(\frac{1}{2}-\frac{2\mu+l}{2}\right)_n \left(1-\frac{2\mu+l}{2}\right)_n (-\mu-m)_n}. \tag{26}$$

Using the summation theorem[5] of the generalized hyphergeometric series

$${}_5F_4\left(a, 1+\frac{a}{2}, b, c, d; \frac{a}{2}, 1+a-b, 1+a-c, 1+a-d; 1\right)$$

$$= \frac{\Gamma(1+a-b)\Gamma(1+a-c)\Gamma(1+a-d)\Gamma(1+a-b-c-d)}{\Gamma(1+a)\Gamma(1+a-b-c)\Gamma(1+a-b-d)\Gamma(1+a-c-d)}, \tag{27}$$

we obtain

$$\sum_{k=0}^{m} \epsilon_k x^{-1} \overline{R}_{m-k,\mu+k+1}(x)$$
$$= \left(\frac{x}{2}\right)^{-m-1-\mu} \frac{\Gamma(\mu+1)\Gamma(\mu+m+1)\Gamma(-\mu-m)}{\Gamma(2\mu+1)\Gamma(\frac{1}{2}-\mu)}$$
$$\times \sum_{l=0}^{m} \left(\frac{x}{2}\right)^{l} \frac{\Gamma(2\mu+l)\Gamma\left(1-\mu-\frac{l}{2}\right)\Gamma\left(\frac{1}{2}-\mu-\frac{l}{2}\right)\Gamma\left(-\mu-m+l-\frac{1}{2}\right)}{l!\Gamma(\mu+l)\Gamma(1-\mu-l)\Gamma\left(-\mu-m+\frac{l}{2}\right)\Gamma\left(-\mu-m+\frac{l}{2}-\frac{1}{2}\right)}$$
$$= \frac{1}{2}\sum_{k=0}^{m} \frac{\Gamma(2\mu+2m+2-k)\Gamma(\mu+m+1-k)}{k!\ \Gamma(2\mu+2m+2-2k)} \left(\frac{x}{2}\right)^{-\mu-1-m+k} \tag{28}$$

By the aid of the summation theorem of the generalized hypergeometric series, double summation reduces to single summation. Thus it is found that this theorem plays a major role in simplification of the expression. Substituting Eqs.(25) and (28) into Eq.(23), we obtain

$$\Phi_{\mu,m}(x) = \left[\frac{1}{2}\sum_{k=m+1}^{2m+1} \frac{\Gamma(2\mu+2m+2-k)\Gamma(\mu+m+1-k)}{k!\ \Gamma(2\mu+2m+2-2k)} \left(\frac{x}{2}\right)^{-\mu-1-m+k}\right.$$
$$+ \frac{\pi^{\frac{1}{2}}(-1)^{m+1}(2x)^{m+1}}{2\sin\mu\pi} \left\{\sum_{k=0}^{\infty} \frac{\Gamma(-\mu+m+k+\frac{3}{2})(2x)^{-\mu+k}}{(2m+k+2)!\ \Gamma(-2\mu+k+1)}\right.$$
$$\left.\left. - \sum_{k=0}^{\infty} \frac{\Gamma(\mu+m+k+\frac{3}{2})(2x)^{\mu+k}}{k!\ \Gamma(2\mu+2m+k+3)}\right\}\right] / \{e^x K_{\mu+m+1}(x)\}. \tag{29}$$

In the above equation, if $\mu \ll m$, the second part in [ ] is negligibly small compared to the first part. Also in the first part, if x/m is small, the leading term (which corresponds to the case $k = m+1$) is dominant. Therefore we obtain the following useful estimation of $\Phi_{\mu,m}(x)$.

$$\Phi_{\mu,m}(x) \approx \frac{\Gamma(2\mu+m+1)\Gamma(\mu+1)}{(m+1)!\ \Gamma(2\mu+1)e^x K_{\mu+m+1}(x)} \left(\frac{x}{2}\right)^{-\mu} \tag{30}$$

The representation (29) and this estimation coincide with the result by I.Ninomiya[2]. He derived the same result by predicting the form of Eq.(28) and proving it by means of induction with a very tedious manipulation of the expression. Though our derivation is straightforward, we however need a tedious manipulation.

4. MODIFICATION OF EXPRESSION OF $\Psi_{\nu,\mu,m}(x)$

Now we shall consider the modification of the expression (22) of $\Psi_{\nu,\mu,m}(x)$.

$$\Psi_{\nu,\mu,m}(x) = \frac{1}{I_\nu(x)}\left(\frac{I_{\mu+m+1}(x)}{\overline{K}_{\mu+m+1}(x)}\sum_{k=0}^{m/2} \rho_k \overline{K}_{\mu+2k}(x) + I_\nu(x) - \sum_{k=0}^{m/2} \rho_k I_{\mu+2k}(x)\right)$$
$$= \frac{1}{I_\nu(x)\overline{K}_{\mu+m+1}(x)}\left\{I_\nu(x)\overline{K}_{\mu+m+1}(x) + \sum_{k=0}^{m/2} \rho_k \left(I_{\mu+m+1}(x)\overline{K}_{\mu+2k}(x) - \overline{K}_{\mu+m+1}(x) I_{\mu+2k}(x)\right)\right\}$$
$$= \frac{1}{I_\nu(x)K_{\mu+m+1}(x)}\left\{I_\nu(x)K_{\mu+m+1}(x) - \sum_{k=0}^{m/2} \rho_k x^{-1}\overline{R}_{m-2k,\mu+2k+1}(x)\right\} \tag{31}$$

Let us rewrite the first term in { } on the right hand of Eq.(31).

$$I_\nu(x)K_{\mu+m+1}(x) = \frac{\pi I_\nu(x)}{2\sin(\mu+m+1)\pi}\{I_{-\mu-m-1}(x) - I_{\mu+m+1}(x)\}$$
$$= \frac{(-1)^{m+1}\pi}{2\sin\mu\pi}\left(\frac{x}{2}\right)^{\nu}\sum_{i=0}^{\infty}\frac{(x/2)^{2i}}{i!\ \Gamma(\nu+i+1)}$$
$$\times\left\{\left(\frac{x}{2}\right)^{-\mu-m-1}\sum_{j=0}^{\infty}\frac{(x/2)^{2j}}{j!\Gamma(-\mu-m+j)} - \left(\frac{x}{2}\right)^{\mu+m+1}\sum_{j=0}^{\infty}\frac{(x/2)^{2j}}{j!\Gamma(\mu+m+2+j)}\right\}$$
$$= \frac{(-1)^{m+1}\pi}{2\sin\mu\pi}\left(\frac{x}{2}\right)^{\nu}\sum_{n=0}^{\infty}\left(\frac{x}{2}\right)^{2n}$$
$$\times\left\{\left(\frac{x}{2}\right)^{-\mu-m-1}\sum_{l=0}^{n}\frac{1}{l!\Gamma(n-l+1)\Gamma(-\mu-m+l)\Gamma(\nu+n-l+1)}\right.$$
$$\left. - \left(\frac{x}{2}\right)^{\mu+m+1}\sum_{l=0}^{n}\frac{1}{l!\Gamma(n-l+1)\Gamma(\mu+m+2+l)\Gamma(\nu+n-l+1)}\right\}$$
$$= \frac{(-1)^{m+1}\pi}{2\sin\mu\pi}\left(\frac{x}{2}\right)^{\nu}\sum_{n=0}^{\infty}\left(\frac{x}{2}\right)^{2n}$$
$$\times\left\{\left(\frac{x}{2}\right)^{-\mu-m-1}\frac{1}{n!\ \Gamma(-\mu-m)\Gamma(\nu+n+1)}\sum_{l=0}^{\infty}\frac{(-n)_l(-\nu-n)_l}{l!\ (-\mu-m)_l}\right.$$
$$\left. - \left(\frac{x}{2}\right)^{\mu+m+1}\frac{1}{n!\ \Gamma(\mu+m+2)\Gamma(\nu+n+1)}\sum_{l=0}^{\infty}\frac{(-n)_l(-\nu-n)_l}{l!\ (\mu+m+2)_l}\right\} \tag{32}$$

Using the relation

$${}_2F_1(a,b;c;1) = \frac{\Gamma(c)\Gamma(c-a-b)}{\Gamma(c-a)\Gamma(c-b)}, \tag{33}$$

we obtain

$$I_\nu(x)K_{\mu+m+1}(x) = \frac{(-1)^{m+1}\pi}{2\sin\mu\pi}\left(\frac{x}{2}\right)^{\nu}\sum_{n=0}^{\infty}\left(\frac{x}{2}\right)^{2n}$$
$$\times\left\{\left(\frac{x}{2}\right)^{-\mu-m-1}\frac{\Gamma(\nu-\mu-m+2n)}{n!\ \Gamma(-\mu-m+n)\Gamma(\nu-\mu-m+n)\Gamma(\nu+n+1)}\right.$$
$$\left. - \left(\frac{x}{2}\right)^{\mu+m+1}\frac{\Gamma(\nu+\mu+m+2n+2)}{n!\ \Gamma(\mu+m+n+2)\Gamma(\nu+\mu+m+n+2)\Gamma(\nu+n+1)}\right\}. \tag{34}$$

Next let us rewrite the second term of { } of Eq.(31).

$$\sum_{k=0}^{m/2} \rho_k x^{-1}\overline{R}_{m-2k,\mu+2k+1}(x)$$
$$= \sum_{k=0}^{m/2} \rho_k \frac{1}{x}\sum_{n=0}^{\frac{m-2k}{2}} \frac{(m-2k-n)!\Gamma(\mu+m-n+1)}{n!(m-2k-2n)!\Gamma(\mu+2k+n+1)}\left(\frac{x}{2}\right)^{-m+2k+2n}$$

$$=\frac{1}{x}\left(\frac{x}{2}\right)^{-m}\sum_{n=0}^{m/2}\left(\frac{x}{2}\right)^{2n}\sum_{i=0}^{n}\rho_i\frac{(m-n-i)!\Gamma(\mu+m-n+1+i)}{(m-2n)!\Gamma(n-i+1)\Gamma(\mu+n+1+i)}$$

$$=\frac{1}{x}\left(\frac{x}{2}\right)^{-m}\sum_{n=0}^{m/2}\left(\frac{x}{2}\right)^{2n}\sum_{i=0}^{\infty}\rho_i\frac{(m-n-i)!\Gamma(\mu+m-n+1+i)}{(m-2n)!\Gamma(n-i+1)\Gamma(\mu+n+1+i)}$$

$$=\frac{1}{2}\left(\frac{x}{2}\right)^{\nu-\mu-m-1}\sum_{n=0}^{m/2}\left(\frac{x}{2}\right)^{2n}\frac{\Gamma(\mu+1)\Gamma(m-n+1)\Gamma(\mu+m-n+1)}{n!(m-2n)!\Gamma(\nu+1)\Gamma(\mu+n+1)}$$

$$\times\sum_{i=0}^{\infty}\frac{(\mu)_i(\mu-\nu)_i\left(\frac{\mu}{2}+1\right)_i(\mu+m-n+1)_i(-n)_i}{i!\left(\frac{\mu}{2}\right)_i(n-m)_i(\nu+1)_i(\mu+n+1)_i}$$

$$=\frac{1}{2}\left(\frac{x}{2}\right)^{\nu-\mu-m-1}\sum_{n=0}^{m/2}\left(\frac{x}{2}\right)^{2n}\frac{(-1)^n\Gamma(\mu+m-n+1)\Gamma(\nu-\mu-m+2n)}{n!\,\Gamma(\nu-\mu-m+n)\Gamma(\nu+n+1)}$$

$$=\frac{(-1)^{m+1}\pi}{2\sin\mu\pi}\left(\frac{x}{2}\right)^{\nu-\mu-m-1}$$

$$\times\sum_{n=0}^{m/2}\left(\frac{x}{2}\right)^{2n}\frac{\Gamma(\nu-\mu-m+2n)}{n!\,\Gamma(-\mu-m+n)\Gamma(\nu-\mu-m+n)\Gamma(\nu+n+1)}.\quad(35)$$

Note that we used Eq.(27) in the above reduction.

Substituting Eqs.(34) and (35) into Eq.(31), we obtain

$$\Psi_{\nu,\mu,m}(x)=\frac{1}{I_\nu(x)K_{\mu+m+1}(x)}\left[\frac{(-1)^{m+1}\pi}{2\sin\mu\pi}\right.$$

$$\times\left\{\left(\frac{x}{2}\right)^{\nu-\mu-m-1}\sum_{n=\frac{m}{2}+1}^{\infty}\frac{\Gamma(\nu-\mu-m+2n)\,(x/2)^{2n}}{n!\Gamma(-\mu-m+n)\Gamma(\nu-\mu-m+n)\Gamma(\nu+n+1)}\right.$$

$$\left.\left.-\left(\frac{x}{2}\right)^{\nu+\mu+m+1}\sum_{n=0}^{\infty}\frac{\Gamma(\nu+\mu+m+2n+2)\,(x/2)^{2n}}{n!\Gamma(\mu+m+n+2)\Gamma(\nu+\mu+m+n+2)\Gamma(\nu+n+1)}\right\}\right]$$

$$=\frac{1}{I_\nu(x)K_{\mu+m+1}(x)}\left[\frac{(-1)^{m+1}\pi}{2\sin\mu\pi}\right.$$

$$\times\left\{\left(\frac{x}{2}\right)^{\nu-\mu-m-1}\sum_{n=\frac{m}{2}+1}^{m}\frac{\Gamma(\nu-\mu-m+2n)\,(x/2)^{2n}}{n!\Gamma(-\mu-m+n)\Gamma(\nu-\mu-m+n)\Gamma(\nu+n+1)}\right.$$

$$+\left(\frac{x}{2}\right)^{\nu-\mu-m-1}\sum_{n=m+1}^{\infty}\frac{\Gamma(\nu-\mu-m+2n)\,(x/2)^{2n}}{n!\Gamma(-\mu-m+n)\Gamma(\nu-\mu-m+n)\Gamma(\nu+n+1)}$$

$$\left.\left.-\left(\frac{x}{2}\right)^{\nu+\mu+m+1}\sum_{n=0}^{\infty}\frac{\Gamma(\nu+\mu+m+2n+2)\,(x/2)^{2n}}{n!\Gamma(\mu+m+n+2)\Gamma(\nu+\mu+m+n+2)\Gamma(\nu+n+1)}\right\}\right]\quad(36)$$

Thus we can obtain the expression of $\Psi_{\nu,\mu,m}(x)$.

$$\Psi_{\nu,\mu,m}(x)$$

$$=\left[\frac{1}{2}\left(\frac{x}{2}\right)^{\nu-\mu-m-1}\sum_{n=\frac{m}{2}+1}^{m}\left(\frac{x}{2}\right)^{2n}\frac{(-1)^n\Gamma(\mu+m-n+1)\Gamma(\nu-\mu-m+2n)}{n!\Gamma(\nu-\mu-m+n)\Gamma(\nu+n+1)}\right.$$

$$+\left\{\left(\frac{x}{2}\right)^{-\mu}\sum_{n=0}^{\infty}\frac{\Gamma(\nu-\mu+m+2n+2)\,(x/2)^{2n}}{(n+m+1)!\Gamma(-\mu+n+1)\Gamma(\nu-\mu+n+1)\Gamma(\nu+m+n+2)}\right.$$

$$\left.-\left(\frac{x}{2}\right)^{\mu}\sum_{n=0}^{\infty}\frac{\Gamma(\nu+\mu+m+2n+2)\,(x/2)^{2n}}{n!\Gamma(\mu+m+n+2)\Gamma(\nu+\mu+m+n+2)\Gamma(\nu+n+1)}\right\}$$

$$\left.\times\left(\frac{x}{2}\right)^{\nu+m+1}\frac{(-1)^{m+1}\pi}{2\sin\mu\pi}\right]/\{I_\nu(x)K_{\mu+m+1}(x)\}.\quad(37)$$

Consider the usual case $0<\nu-\mu+2<<m/2$. If $\nu-\mu$ is not close to an integer and $\mu<<m$ then the second part in [ ] of Eq.(37) is negligibly small compared to the first part. Then in the first part, if x/m^2 is small, the leading term ($n=\frac{m}{2}+1$) is dominant. Therefore we obtain the following estimation of $\Psi_{\nu,\mu,m}(x)$ in the case that $\nu-\mu$ is not close to an integer.

$$\Psi_{\nu,\mu,m}(x)\approx\frac{\frac{1}{2}(-1)^{\frac{m}{2}+1}(x/2)^{\nu-\mu+1}\Gamma(\mu+\frac{m}{2})\Gamma(\nu-\mu+2)}{(\frac{m}{2}+1)!\Gamma(\nu-\mu-\frac{m}{2}+1)\Gamma(\nu+\frac{m}{2}+2)I_\nu(x)K_{\mu+m+1}(x)}.\quad(38)$$

The first part in [ ] of Eq.(37) vanishes in the case of $\nu-\mu=-1$ or 0. It is found that if $\nu-\mu$ is an integer within $-1\le\nu-\mu<<m/2$ then $\Psi_{\nu,\mu,m}(x)$ is extremely small.

ACKNOWLEDGEMENT

I wish to thank Prof. I.Ninomiya for helpful comments.

REFERENCES

[1] Watson,G.N.,A Treatise on the Theory of Bessel Functions (Cambridge Univ. Press 1966) p.139.

[2] Ninomiya,I.,Computation of Bessel Functions by Recurrence, in : Numerical Method for a Computer (Baihukan,Tokyo,1967) pp.103-121(in Japanese).

[3] Yoshida,T. and Umeno,M.,Recurrence Techniques for the Calculation of Bessel Functions $I_n(z)$ with Complex Argument, Information Processing in Japan 13(1973) pp.100-104.

[4] Makinouchi,S.,Note on the Recurrence Techniques for the Calculation of Bessel Function $I_\nu(x)$, Information Processing Society of Japan, 5(1965) pp.247-252(in Japanese).

[5] Slater,L.J.,Generalized Hypergeometric Functions (Cambridge Univ. Press, 1966) p.56.

[6] Abramowitz,M. and Stegun,I.A.,Handbook of Mathematical Functions (Dover, New York, 1968) p.510.

Approximation, Optimization and Computing:
Theory and Applications, A.G. Law and C.L. Wang (eds.)
Elsevier Science Publishers B.V. (North-Holland)

THE BLENDING INTERPOLATION ON A CURVED TRIANGLE

Y.S.ZHOU & Z.P.SHENG

Department of Mathematics, Jilin University, China

In this paper, for the curved triangle with one curved side and two straight sides, we provide a method which can be used to construct many different schemes of C^0 or C^1 blending interpolation.

1. INTRODUCTION

In order to achive the modelling of free-form surface, in this paper, we consider the blending interpolation on a curved traingle. It has many applications. Examples are the disign of airplanes, automobiles, ships, and modelling the surface of the human heart and of scientific phenomena. Our idea is following: First, we transform the curved triangle into a straight triangle using a differentiable homeomorphic transformation. Then with the help of the blending interpolation schemes on the straight triangle, the scheme on the curved triangle will be obtained.

2. NOTATIONS

We assume T is a curved triangle; $V_i=(x_i, y_i)$, (i=1,2,3), are the vertices of T; Γ_i are the sides of T which are faced by the vertices V_i; and Γ_1 is a curved side, Γ_2 and Γ_3 are straight sides. Besides, the straight triangle with vertices V_i (i=1, 2,3) $\overline{T}$ is accompanying triangle of the T. $\overline{\Gamma}_i$ (i=1,2,3) are the sides of $\overline{T}$ which are faced by the vertices V_i (i=1,2,3) respectively. Obviosly, $\Gamma_i=\overline{\Gamma}_i$, (i=2,3). Let V=(x,y) be an arbitrary point on $T\setminus V_1$, and L be an arbitrary half line from V_1 to V. We always assume the cross point of L and Γ_1 is unique after.

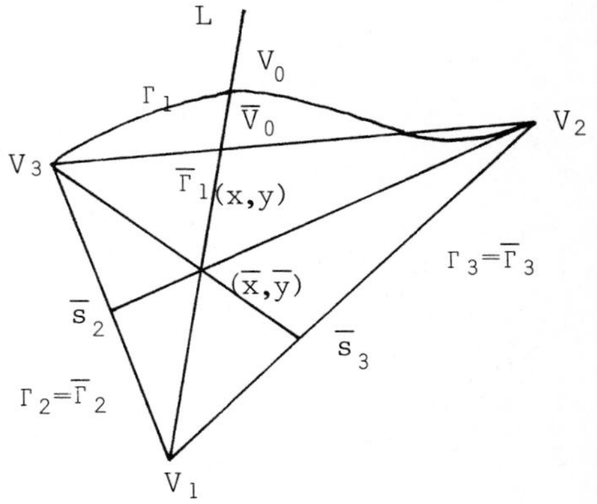

Figure 1

Also, let $V_0=(x_0(x,y), y_0(x,y))$ be the cross point of arbitrary half line L and Γ_1; and $\overline{V}_0=(\overline{x}_0(x,y), \overline{y}_0(x,y))$ be the cross point of L and $\overline{\Gamma}_1$. Now we introduce the following notations:

$$d_0=d_0(x,y)=\|V_0-V_1\|=((x_0-x_1)^2+(y_0-y_1)^2)^{\frac{1}{2}}$$

$$\overline{d}_0=\overline{d}_0(x,y)=\|\overline{V}_0-V_1\|=((\overline{x}_0-x_1)^2+(\overline{y}_0-y_1)^2)^{\frac{1}{2}}$$

$$d=d(x,y)=\|V-V_1\|=((x-x_1)^2+(y-y_1)^2)^{\frac{1}{2}}$$

$$\lambda=\lambda(x,y)=\overline{d}_0/d_0$$

$$\tau=\tau(x,y)=d/d_0$$

Finally, we let ∂A represent the boundary

of A.

3. Differentiable Homeomorphism

Let Γ_1 be defined by the equation $h(x,y)=0$. We consider the following transformation S_n(from T to $\overline{T}$):

$$\overline{x}=\overline{x}(x,y)=\Lambda_n x+(1-\Lambda_n)x_1$$

$$\overline{y}=\overline{y}(x,y)=\Lambda_n y+(1-\Lambda_n)y_1$$

where

$$\Lambda_n=\Lambda_n(x,y)=(\lambda-1)\tau^n+1, \quad n=0,1,2,\cdots$$

If $h(x,y)\in C(\Gamma_1)$, then $x_0, y_0, \overline{x}_0, \overline{y}_0$ are continuous and bounded on $T\setminus V_1$. If $h(x,y)\in C^n(\Gamma_1)$; any L is not a tangent line of the curved side ; and

$$\text{grad } h=(h'_x(x,y), h'_y(x,y))\neq 0$$

for $(x,y)\in\Gamma_1$, then it is easy to verify y_0, $x_0, \overline{y}_0, \overline{x}_0 \in C^n(T\setminus V_1)$, and all their n-th partial derivatives are same order with d^{-n} at point (x_1,y_1). Furthermore, we can know $\Lambda_n\in C^{n-1}(T)$ and all its n-th partial derivatives are continuous and bounded on $T\setminus V_1$.(n=1,2,$\cdots$).

Let

$$J=J(x,y)=\begin{bmatrix}\overline{x}'_x & \overline{x}'_y \\ \overline{y}'_x & \overline{y}'_y\end{bmatrix}, \quad \Delta=\det J$$

and, for simplicity, set

$$\Lambda=\Lambda_n,\ S=S_n,\quad n=1,2,\cdots$$

We have the following

Lemma 1. $\Delta=\Lambda((n+1)\Lambda-1)$. And therefore, when $\min\lambda>\frac{n}{n+1}$,

$$\Delta=\Delta(x,y)>0 \qquad \text{for} \quad (x,y)\in T.$$

Proof. According to the definition of partial derivatives, we see that

$$J(x_1,y_1)=\begin{bmatrix}1 & 0\\ 0 & 1\end{bmatrix}$$

Now, let $(x,y)\neq(x_1,y_1)$. Then it is easy to conclude that

$$\Delta=\Lambda(\Lambda+(x-x_1)\Lambda'_x+(y-y_1)\Lambda'_y)$$

$$\Lambda+(x-x_1)\Lambda'_x+(y-y_1)\Lambda'_y=\Lambda+\frac{\tau^n}{d_0^2}A$$

$$A=n(\lambda-1)d_0^2-((n+1)\lambda-n)\cdot$$

$$\cdot((x_0-x_1)A_1+(y_0-y_1)A_2)d+$$

$$+\lambda^{-1}((\overline{x}_0-x_1)B_1+(\overline{y}_0-y_1)B_2)d$$

$$A_1=\frac{x-x_1}{d}\frac{\partial x_0}{\partial x}+\frac{y-y_1}{d}\frac{\partial x_0}{\partial y}$$

$$A_2=\frac{x-x_1}{d}\frac{\partial y_0}{\partial x}+\frac{y-y_1}{d}\frac{\partial y_0}{\partial y}$$

$$B_1=\frac{x-x_1}{d}\frac{\partial \overline{x}_0}{\partial x}+\frac{y-y_1}{d}\frac{\partial \overline{x}_0}{\partial y}$$

$$B_2=\frac{x-x_1}{d}\frac{\partial \overline{y}_0}{\partial x}+\frac{y-y_1}{d}\frac{\partial \overline{y}_0}{\partial y}$$

Obviously, A_1 is the directional derivative of the function $x_0(x,y)$ along the direction L. But the function x_0 is a constant along the direction L, and so $A_1=0$. Analogically, $A_2=B_1=B_2=0$. Therefore

$$\Delta=\Lambda(\Lambda+n(\lambda-1)\tau^n)=\Lambda((n+1)\Lambda-n)$$

and

$$\Delta>0, \qquad \text{if} \quad \min\lambda>\frac{n}{n+1}. \qquad \#$$

Now, it is obvious to conclude the following result:

Theorem 1. Under the aforesaid conditions, S_n is a C^n differentiable homeomorphism, (n=0,1,2,$\cdots$).

From Lemma, we see that if n=1, then Γ_1 may be any curve lying $\Delta V_1V'_2V'_3$ with end points V_2 and V_3; and if n=2, then Γ_1 may be any curve lying $\Delta V_1\overline{V}_2\overline{V}_3$, etc.

(see Figure 2).

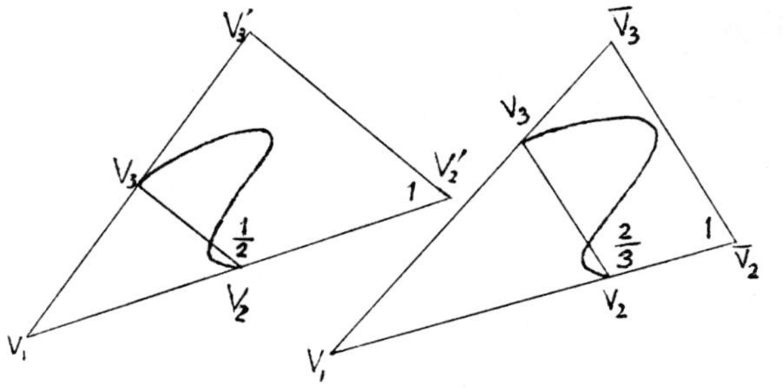

Figure 2

If Γ_1 is outside above scope, then we can take $\Lambda=(\lambda-\mu)\tau^n+\mu(n=1,2,\cdots)$ in transformation S, and to take available μ, for example

$$0<\mu<\min\lambda(1+\frac{1}{n})$$

4. GENERAL OPERATOR THEOREM

Let q_A represent a kind of differentiable property of a real-valued function on set A, Q_A represent the function class possessing the property q_A.

Assume that S is a homeomorphic transformation from T to $\overline{T}$. Besides, (1). if $S(\Gamma_i)=\overline{\Gamma}_i$ and $S(V_i)=V_i(i=1,2,3)$, S is designated as boundary preserving; (2). Let the properties q_T and $q_{\overline{T}}$ are corresponding each other by S (obviously, only one is free). If the components of the transformation $S:\overline{x}=\overline{x}(x,y)\in Q_T$ and $\overline{y}=\overline{y}(x,y)\in Q_T$, the components of the inverse transformation $S^{-1}:x=x(\overline{x},\overline{y})\in Q_{\overline{T}}$ and $y=y(\overline{x},\overline{y})\in Q_{\overline{T}}$, then S is designated as a Q differentiable homeomorphism. Obviously, this means , if $f\in Q_T$, then $fS^{-1}\in Q_{\overline{T}}$; if $\overline{f}\in Q_{\overline{T}}$, then $\overline{f}S\in Q_T$.

<u>Theorem 2.</u> Let $\overline{P}$ be a C^1 blending interpolation operator on $\overline{T}$:

$$(1) \qquad \overline{F}(\overline{x},\overline{y})=\overline{P}[\overline{f}](\overline{x},\overline{y}), \quad (\overline{x},\overline{y})\in\overline{T}$$

where $\overline{f}\in Q_{\overline{T}}$. And assume $S:T\to\overline{T}$ is a boundary preserving Q differentiable homeomorphism. Then for $f\in Q_T$, the function:

$$(2) \qquad F(x,y)=\overline{P}[fS^{-1}]S(x,y), \quad (x,y)\in T$$

interpolates f and all its partial derivatives on ∂T, and $F(x,y)\in C^1(T)$.

For this reason, define operator P:

$$(3) \qquad P[f]=\overline{P}[fS^{-1}]S, \qquad f\in Q_T$$

then P is a C^1 blending interpolation operator on T. Set $\overline{f}=fS^{-1}$. Then

$$(4) \qquad (\overline{f}'_{\overline{x}},\overline{f}'_{\overline{y}})=(f'_x,f'_y)\begin{bmatrix}\frac{\partial\overline{x}}{\partial x} & \frac{\partial\overline{x}}{\partial y}\\ \frac{\partial\overline{y}}{\partial x} & \frac{\partial\overline{y}}{\partial y}\end{bmatrix}^{-1} S^{-1}$$

<u>Proof.</u> $\overline{P}$ is a C^1 blending interpolation operator, hence $Q_{\overline{T}}\subset C^1(\overline{T})$. And S is a Q differentiable homeomorphism, therefore, S is a C^1 differentiable homeomorphism. Then we have

i) $\quad \overline{f}=fS^{-1}$, that is, $f(x,y)=\overline{f}(\overline{x},\overline{y})$

and hence

$$f'_x(x,y)=\overline{f}'_{\overline{x}}(\overline{x},\overline{y})\frac{\partial\overline{x}}{\partial x}+\overline{f}'_{\overline{y}}(\overline{x},\overline{y})\frac{\partial\overline{y}}{\partial x}$$

$$f'_y(x,y)=\overline{f}'_{\overline{x}}(\overline{x},\overline{y})\frac{\partial\overline{x}}{\partial y}+\overline{f}'_{\overline{y}}(\overline{x},\overline{y})\frac{\partial\overline{y}}{\partial y}$$

Furthermore we get

$$(\overline{f}'_{\overline{x}}(\overline{x},\overline{y}),\overline{f}'_{\overline{y}}(\overline{x},\overline{y}))$$

$$=(f'_x(x,y),f'_y(x,y))\begin{bmatrix}\frac{\partial\overline{x}}{\partial x} & \frac{\partial\overline{x}}{\partial y}\\ \frac{\partial\overline{y}}{\partial x} & \frac{\partial\overline{y}}{\partial y}\end{bmatrix}^{-1}$$

This is to say, (4) is true.

ii) $f\in Q_T$, so $\overline{f}=fS^{-1}\in Q_{\overline{T}}$. Thus $\overline{F}$ interpolates $\overline{f}$ and all its first partial derivatives on $\overline{T}$, and

$$F(x,y)=\overline{F}(\overline{x},\overline{y})=\overline{F}(\overline{x}(x,y),\overline{y}(x,y)),$$

therefore, $F(x,y)\in C^1(T)$. Then we have

$$F(x,y)|_{(x,y)\in\partial T} = \overline{F}(\overline{x},\overline{y})|_{(\overline{x},\overline{y})\in\partial\overline{T}}$$
$$= \overline{f}(\overline{x},\overline{y})|_{(\overline{x},\overline{y})\in\partial\overline{T}} = f(x,y)|_{(x,y)\in\partial T}$$

Furthermore, by (4), we have

$$(F'_x(x,y),F'_y(x,y))|_{(x,y)\in\partial T}$$
$$= (\overline{F}'_{\overline{x}}(\overline{x},\overline{y}),\overline{F}'_{\overline{y}}(\overline{x},\overline{y}))|_{(\overline{x},\overline{y})\in\partial\overline{T}}\begin{bmatrix}\frac{\partial\overline{x}}{\partial x} & \frac{\partial\overline{x}}{\partial y}\\ \frac{\partial\overline{y}}{\partial x} & \frac{\partial\overline{y}}{\partial y}\end{bmatrix}$$
$$= (\overline{f}'_{\overline{x}}(\overline{x},\overline{y}),\overline{f}'_{\overline{y}}(\overline{x},\overline{y}))|_{(\overline{x},\overline{y})\in\partial\overline{T}}\begin{bmatrix}\frac{\partial\overline{x}}{\partial x} & \frac{\partial\overline{x}}{\partial y}\\ \frac{\partial\overline{y}}{\partial x} & \frac{\partial\overline{y}}{\partial y}\end{bmatrix}$$
$$= (f'_x(x,y),f'_y(x,y))|_{(x,y)\in\partial T}$$

Until now, we have proved the function F interpolates to f and all its first partial derivatives on ∂T. #

5. TWO SCHEMES

Let

$$\overline{s}_i=\left(\frac{\overline{x}-b_ix_i}{1-b_i},\frac{\overline{y}-b_iy_i}{1-b_i}\right),\quad (i=1,2,3)$$

where $b_i=b_i(\overline{x},\overline{y}),(i=1,2,3)$, are the barycentric coordinates of a point $(\overline{x},\overline{y})$ on $\overline{T}$. And set $s_1=V_0$; $s_i=\overline{s}_i$, $(i=2,3)$;

$(\alpha_i,\beta_i)=\overline{s}_i-V_i,(i=1,2,3)$; $h(t)=t^2(3-2t)$; $\hat{h}(t)=t^2(t-1)$. Define the function class

$$C_A^2 = \{f;f\in C^1(A) \text{ and } f\in C^2(\partial A)\}$$

1°. C^0 scheme

Using the C^0 Nielson operator[1] and the homeomorphism S_0, then, for $f\in C(T)$, we can conclude that

$$P[f](x,y)=\sum_{i=1}^{3}(1-b_i)f(s_i)-\sum_{i=1}^{3}b_if(V_i)$$

is contained in C(T) and interpolates to f on ∂T, where

$$\overline{x}=\overline{x}(x,y)=\lambda x+(1-\lambda)x_1$$
$$\overline{y}=\overline{y}(x,y)=\lambda y+(1-\lambda)y_1$$

2°. C^1 scheme

Using the C^1 Nielson operator[1] (Little-Brown formula) and the C^2 differentiable homeomorphism S_2, then, for $f\in C_T^2$, we can conclude that

$$D[f] = \frac{b_2^2b_3^2D_1[f]+b_3^2b_1^2D_2[f]+b_1^2b_2^2D_3[f]}{b_2^2b_3^2+b_3^2b_1^2+b_1^2b_2^2}$$

is contained in $C^1(T)$ and interpolates to f and its first partial derivatives on ∂T, where

$$D_i[f]$$
$$=h(1-b_i)f(s_i)+\hat{h}(1-b_i)(A_if'_x(s_i)+B_if'_y(s_i))$$
$$+h(b_i)f(V_i)-\hat{h}(b_i)(A_if'_x(V_i)+B_if'_y(V_i))$$
$$A_i=\begin{vmatrix}\alpha_i & \beta_i\\ \overline{x}'_y & \overline{y}'_y\end{vmatrix}\Delta^{-1},B_i=\begin{vmatrix}\beta_i & \alpha_i\\ \overline{y}'_x & \overline{x}'_x\end{vmatrix}\Delta^{-1},(i=1,2,3)$$
$$\Delta=\Lambda(3\Lambda-2)$$

and

$$\overline{x}=\overline{x}(x,y)=\Lambda x+(1-\Lambda)x_1$$
$$\overline{y}=\overline{y}(x,y)=\Lambda y+(1-\Lambda)y_1$$
$$\Lambda=(\lambda-1)\tau^2+1.$$

REFERENCE

[1] G.M.Nielson, The side-vertex method for interpolation in triangles, Journal of Approximation Theory, 25(1979), 318-336.

II
OPTIMIZATION
Theory and Techniques

Approximation, Optimization and Computing:
Theory and Applications, A.G. Law and C.L. Wang (eds.)
Elsevier Science Publishers B.V. (North-Holland)

A MAXIMUM PRINCIPLE FOR CONTROL OF PIECEWISE DETERMINISTIC MARKOV PROCESSES

M.A.H. DEMPSTER and J.J. YE

Department of Mathematics, Statistics and Computing Science,
Dalhousie University, Halifax, Nova Scotia, Canada B3H 3J5

Piecewise deterministic Markov processes (PDPs) are continuous time homogeneous Markov processes whose trajectories are solutions of an ordinary differential equation with possible random jumps between the different integral curves. Both continuous deterministic motion and the random jumps of the processes are controlled in order to minimize the expected value of a performance criterion consisting of discounted running and boundary jump costs. In the first part of the paper we give a nonsmooth maximum principle for a deterministic control problem with boundary conditions which require the state of the system to stop at the boundary. In the second part by reducing the original problem to a family of deterministic control problems with boundary conditions parametrized by initial states, we derive a nonsmooth maximum principle for control of PDPs as a necessary condition for optimality.

1 Introduction

In this paper, our aim is to develop a nonsmooth maximum principle for control of piecewise deterministic Markov processes under weak assumptions. These processes, first introduced explicitly by Davis [4] in 1984, are a general family of stochastic models covering virtually almost all nondiffusion applications with pure jump processes and deterministic dynamical systems as special cases. Such processes provide a framework for studying optimization problems arising in queueing systems, inventory theory, resource allocation, capacity expansion problems and many other areas of operations research.

Now we give a precise definition for a PDP. Let $E \subset I\!R^n$ be the state domain defined by $E := \{x \in I\!R^n : \psi(x) < 0\}$ where ψ is some C^1 function such that $|\nabla \psi(x)| \geq 1$ for $x \in \partial E := \{x : \psi(x) = 0\}$. A PDP taking values in E is determined by its three "local characteristics ":

(i) a *vector field* $f : E \to I\!R^n$
(ii) a *jump rate* $\lambda : E \to I\!R_+$
(iii) a *transition measure* $Q : E \cup \partial E \to \mathcal{P}(E)$,

where $\mathcal{P}(E)$ denotes the set of probability measures on E.

A sample path x_t of the PDP starting at $x \in E$ is constructed as follows: the vector field f determines an *integral curve* $\phi(t,x)$ in E satisfying

$$\frac{\partial}{\partial t}\phi(t,x) = f(\phi(t,x))$$
$$\phi(0,x) = x$$

for $0 \leq t < T_1$, where T_1 is the first *jump time*, whose distribution is given by

$$P_x[\mathbf{T_1} > t] = \begin{cases} \exp[-\int_0^t \lambda(\phi(t,x))ds] & t < t^*(x) \\ 0 & t \geq t^*(x), \end{cases}$$

where $t^*(x) := \inf\{t > 0 : \phi(t,x) \in \partial E\}$ with the convention $\inf \emptyset := +\infty$. Having selected T_1 with this distribution we have $x_{T_1^-} = \phi(T_1, x)$ and the distribution $\mathbf{x_{T_1}}$ is given by

$$P_x[\mathbf{x_{T_1}} \in A | \mathbf{T_1} = T_1] = Q(A; \phi(T_1, x)).$$

The process now restarts at x_{T_1} according to the same recipe, giving a sequence of jump times $T_1, T_2, \ldots$, between which x_t follows the integral curves of f. We assume that for each $x \in E$, there exists $\varepsilon > 0$ such that $\int_0^\varepsilon \lambda(\Phi(t,x)) < \infty$ to ensure that jumps occur at isolated times. We also assume that $P_x[\mathbf{T_n} \uparrow \infty] = 1$ for all x. The reader is referred to Davis [4] for further details, general properties and examples of PDPs.

Control problems arise when the local characteristics f, λ and Q depend on a free *control parameter* u from a compact set U. The set of admissible controls may be different for interior and boundary states. We assume that $u \in U_\circ \subset I\!R^m$ if $x \in E$ and $u \in U_\partial \subset I\!R^l$ if $x \in \partial E$. Therefore, we shall distinguish the transition measure $Q_\circ(dy; x, u)$, for $x \in E$ and $u \in U_\circ$, describing jumps from *interior* points from $Q_\partial(dy; x, u)$, for $x \in \partial E$ and $u \in U_\partial$, describing jumps from *boundary* points. The following assumptions will be in force throughout:

(A1) $U_\circ$, U_∂ are compact

(A2) $f : E \times U_\circ \to I\!R^n$ is continuous and Lipschitz continuous in x uniformly in u

(A3) $\lambda : E \times U_\circ \to I\!R_+$ is bounded and continuous

(A4) $Q_\circ : E \times U_\circ \to \mathcal{P}(E)$
$Q_\partial : \partial E \times U_\partial \to \mathcal{P}(E)$ are continuous,

where $\mathcal{P}(E)$ denotes the set of probability measures on E with the topology of weak convergence.

The "usual" class of admissible controls in Markovian problems is that of state feedback controls $u_t = u(x_t)$. Unfortunately, if u is only a measurable function, the differential equation $\dot{x}_t = f(x_t, u(x_t))$ has in general no uniquely determined solution. Hence, following Vermes [6], we shall use a less restrictive class of "piecewise open loop" controls by adjoining to the process two extra states: z_t, the state at the last jump time, and τ_t, the time since the last jump. *Admissible controls* for interior points are now functions $u_\circ(z_t, \tau_t)$ so that the deterministic dynamics on the interval $t \in [T_i, T_{i+1})$ become

$$\dot{x}_t = f(x_t, u_\circ(z_t, \tau_t))$$

$$\dot{z}_t = 0 \quad z_{T_i} = x_{T_i}$$

$$\dot{\tau}_t = 1 \quad \tau_{T_i} = 0.$$

Note that (a) this system of equations has a unique solution for any measurable function $u_\circ : E \times I\!R_+ \to U_\circ$ by the Carathéodory theorem, (b) the augmented process $\tilde{x}_t = (x_t, z_t, \tau_t)$ is still a PDP, and (c) (z_t, τ_t) determines (x_t) so that controls of the above form still represent "complete observations".

Assume the following condition holds:

(A5) For any admissible control $u_\circ$ we have

$$P_x[\lim_n T_n = \infty] = 1 \text{ for all } x \in E.$$

The performance criterion includes a *running cost* component $l_\circ : E \times U_\circ \to I\!R_+$ and a *jump cost* $l_\partial : \partial E \times U_\partial \to I\!R_+$ incurred by *jumps* from the boundary.

We also make the following assumption:

(A6) $l_\circ : E \times U_\circ \to I\!R_+$ and $l_\partial : \partial E \times U_\partial \to I\!R_+$ are bounded continuous functions.

Let $\delta > 0$ be a *discount factor*. Then the optimal control problem for PDPs is to find an *optimal control* (i.e. a pair $u = (u_\circ, u_\partial)$ of measurable functions $u_\circ : E \times I\!R_+ \to U_\circ$ and $u_\partial : \partial E \to U_\partial$) which minimizes the *expected cost*

$$J_x(u) := E_x[\int_0^\infty e^{-\delta t} l_\circ(x_t, u_\circ(z_t, \tau_t))dt + \sum_i \exp(-\delta T_i) l_\partial(x_{T_i^-}, u_\partial(x_{T_i^-})) 1_{(x_{T_i^-} \in \partial E)}],$$

where E_x denotes the conditional expectation given the initial point x.

In a companion paper [5], we have shown that the optimal control for the PDP control problem is to choose after each jump a control function which is an optimal control in a deterministic optimal control problem with boundary conditions. Therefore it is obvious that a maximum principle for the control of PDPs will follow once the one for the control problem with boundary conditions is established. This deterministic control problem is however non-standard in that the terminal time t^* is not fixed, but instead is either the first time the trajectory reaches the boundary of the state space or $+\infty$. In the proof, we will consider separately the cases when t^* is finite and when t^* is $+\infty$. A nonsmooth maximum principle developed by Clarke [3] will be used when t^* is finite, while an infinite horizon nonsmooth maximum principle will be developed when t^* is infinite by using some results on differential inclusions of Aubin and Cellina [1].

2 A nonsmooth maximum principle for the control problem with boundary conditions

In this section, we consider the deterministic control problem with boundary conditions formulated as follows:

(P_z) minimize

$$J(z, u(\cdot)) := \int_0^{t^*} e^{-\Lambda^u(t)} f_\circ(x(t), u(t))dt + e^{-\Lambda^u(t^*)} F(x(t^*))$$

over the class Ω consisting of all pairs $(x(\cdot), u(\cdot))$

s.t. $u(\cdot) : [0, t^*] \to I\!R^m$ is measurable,

$u(t) \in U \subset I\!R^m$, compact

$\dot{x}(t) = f(x(t), u(t))$ a.e. $t \in [0, t^*]$

$x(0) = z \in E$,

where $t^* = \inf\{t \geq 0 : x(t) \in \partial E\} \leq +\infty$ is the first exit time of the trajectory $x(t)$ and $\Lambda^u(t) := \int_0^t \tilde{\lambda}(x(s), u(s))ds$.

To make sure that J is well defined, we assume that $\inf_{x,v} \tilde{\lambda}(x, v) > 0$. Thus even if t^* is $+\infty$, the integral converges and we agree that in this case the term $e^{-\Lambda^u(t^*)}F(x(t^*))$ vanishes.

We assume that all data $f(\cdot,\cdot)$, $f_\circ(\cdot,\cdot)$, $\tilde{\lambda}(\cdot,\cdot)$ and $F(\cdot)$ are bounded measurable and $f(\cdot, u)$, $f_\circ(\cdot, u)$, $\tilde{\lambda}(\cdot, u)$ and $F(\cdot)$ are Lipschitz continuous in E for every $u \in U$. We also assume that for any $\alpha > 0, x \in \partial E, v \in U_\circ$

$$f(x, v) \cdot n(x) \geq \alpha > 0 \qquad x \in \partial E,$$

where $n(x)$ is the unit outward normal to ∂E at the point $x \in \partial E$. This implies that ∂E is an *exit* boundary.

Define the *Hamiltonian function*

$$H(r, q, p, x, u) := \langle p, f(x, u)\rangle - r \cdot f_\circ(x, u) - q \cdot \tilde{\lambda}(x, u)$$

for r and $q \in I\!R$, $p \in I\!R^n$, $x \in I\!R^n$ and $u \in U$.

Theorem 1 (Maximum Principle)
Let $(x^*(\cdot), u^*(\cdot))$ be an optimal pair for the problem (P_z) and let t^* be the corresponding exit time. Then there exist:

(a) *absolutely continuous functions*

$$q : [0, t^*] \longmapsto I\!R \qquad p : [0, t^*] \longmapsto I\!R^n,$$

where q is actually a continuous, piecewise C^1 function, and

(b) *a scalar r, which is either equal to 0 or 1,*

such that the following relations hold:

(i) (q, p) *satisfies the "adjoint inclusion"*

$$-\dot{q}(t) = H(r, q(t), p(t), x^*(t), u^*(t)) - \langle p(t), f(x^*(t), u^*(t))\rangle \quad \text{a.e. } t \in [0, t^*] \tag{1}$$

$$
\begin{aligned}
-\dot{p}(t) \in\ & \partial_x f(x^*(t), u^*(t))^\top p(t) \\
& - q(t)\partial_x \tilde{\lambda}(x^*(t), u^*(t)) \\
& - r \cdot \partial_x f_\circ(x^*(t), u^*(t)) \\
& - \tilde{\lambda}(x^*(t), u^*(t))p(t) \\
& \qquad \text{a.e.} \quad t \in [0, t^*]
\end{aligned} \tag{2}
$$

(ii) *H is maximized at $u^*(t)$*

$$
\begin{aligned}
\max_{u\in U} & H(r, q(t), p(t), x^*(t), u) \\
&= H(r, q(t), p(t), x^*(t), u^*(t)) \\
&= 0 \qquad \text{a.e.} \quad t \in [0, t^*]
\end{aligned} \tag{3}
$$

(iii) *transversality condition, if $t^* < \infty$, then*

$$(q(t^*), p(t^*)) + r \cdot \xi \in (0, -C\partial d_E(x^*(t^*)) \tag{4}$$

for some nonnegative scalar C and an $n+1$ dimensional vector ξ with

$$\xi \in (F(x^*(t^*)), \partial F(x^*(t^*))$$

(iv) *if $t^* < \infty$, then*

$$\|p\| + \|q\| + r > 0, \tag{5}$$

where $^\top$ denotes the transpose, ∂ denotes the Clarke generalized derivative, ∂_x denotes the generalized partial derivative with respect to x (see [3]) and $\|\cdot\|$ is the supremum norm for the spaces of appropriate continuous functions on $[0, t^*]$

Proof. It is convenient to replace the exponential term in the cost by an extra differential relation

$$
\begin{aligned}
\dot{x}_0(t) &= -x_0(t)\tilde{\lambda}(x(t), u(t)) \\
x_0(0) &= 1.
\end{aligned}
$$

Problem (P_z) can be equivalently posed as follows:

$$
\begin{aligned}
(\overline{P}_z) \quad & \min \int_0^{t^*} x_0(t) f_\circ(x(t), u(t))dt + x_0(t^*)F(x(t^*)) \\
& \text{the class } \Omega \text{ of all pairs } (\overline{x}(\cdot), u(\cdot)) \text{ with} \\
& \overline{x}(\cdot) := (x_0(\cdot), x(\cdot)) \\
\text{s.t.} \quad & \tfrac{d}{dt}\overline{x}(t) = [-x_0(t)\tilde{\lambda}(x(t), u(t)), f(x(t), u(t))] \\
& \qquad \text{a.e. } t \in [0, t^*] \\
& \overline{x}(0) = (1, z) \\
& t^* := \inf\{t > 0 : x(t) \in \partial E\}.
\end{aligned}
$$

For an optimal pair $(x^*(t), u^*(t))$ in Ω we denote by $(\overline{x}^*(\cdot), u^*(\cdot))$ the corresponding solution for $(\overline{P}_z)$ in the class Ω. Now we divide the analysis into two cases:

(a) the exit time of the optimal trajectory $x^*(\cdot)$ is finite

(b) the exit time of the optimal trajectory $x^*(\cdot)$ is infinite.

Since the boundary of E is an exit boundary, it is obvious that $(\overline{x}^*(\cdot), u^*(\cdot))$ is an optimal solution of the problem:

$$
\begin{aligned}
(\overline{P}) \quad & \min \int_0^{t^*} x_0(t) f_\circ(x(t), u(t))dt + x_0(t^*)F(x(t^*)) \\
\text{s.t.} \quad & \tfrac{d}{dt}\overline{x}(t) = [-x_0(t)\tilde{\lambda}(x(t), u(t)), f(x(t), u(t))] \\
& \qquad \text{a.e. } t \in [0, t^*] \\
& x(t) \in T(x^*; \varepsilon) \\
& \overline{x}(0) = (1, z) \\
& (t^*, x(t^*)) \in M,
\end{aligned}
$$

where $M := [0, \infty) \times \partial E$ in case (a), $M := \{+\infty\} \times E$ in case (b), $T(x^*; \varepsilon)$ is the ε-tube about the optimal trajectory x^* defined by

$$T(x^*; \varepsilon) := \{v \in I\!R^n : \|x^*(t) - v\| < \varepsilon, t \geq 0\},$$

and $\varepsilon > 0$ is sufficiently small to ensure that $T(x^*(t); \varepsilon) \subset E$ for $t \in [0, t^*)$.

Case (a)

Here the time interval is finite and the endpoint constraint set $[0, \infty) \times \partial E$ is closed in $I\!R \times I\!R^n$. In this case the nonsmooth maximum principle developed by Clarke [3] is applicable. We refer to theorems 5.2.1 and 5.2.3 of Clarke and identify the data for $(\overline{P})$ with corresponding data in the theorems.

The Hamiltonian function for the problem $(\overline{P})$ is defined as follows:

$$\overline{H}(\overline{r}, \overline{q}, \overline{p}, \overline{x}, u) := \langle \overline{p}, f(x, u)\rangle - x_0\overline{q}\lambda(x, u) - \overline{r}x_0 f_\circ(x, u)$$

for $\overline{r}$ and $\overline{q} \in I\!R$, $\overline{p} \in I\!R^n$, $\overline{x} = (x_0, x) \in I\!R^{n+1}$ and $u \in U$.

Applying theorems 5.2.1 and 5.2.3 of Clarke [3], we have the following maximum principle for the problem $(\overline{P})$:

There exists a scalar $\overline{r}$ equal to 0 or 1, an absolutely continuous function $(\overline{q}(\cdot), \overline{p}(\cdot))$ on $[0, t^*]$ such that

(i) $(\overline{q}, \overline{p})$ satisfies the "adjoint inclusion"

$$-\frac{d}{dt}(\overline{q}(t), \overline{p}(t)) \in \partial_{\overline{x}}\overline{H}(\overline{r}, \overline{q}(t), \overline{p}(t), \overline{x}^*(t), u^*(t)) \qquad \text{a.e. } t \in [0, t^*] \tag{6}$$

(ii) $\overline{H}$ is maximized at $u^*(t)$:

$$
\begin{aligned}
\max_{u\in U} & \overline{H}(\overline{r}, \overline{q}, \overline{p}, \overline{x}^*(t), u) \\
&= \overline{H}(\overline{r}, \overline{q}, \overline{p}, \overline{x}^*(t), u^*(t)) \\
&= 0 \qquad \text{a.e. } t \in [0, t^*]
\end{aligned} \tag{7}
$$

(iii) the following transversality condition holds:

$$\overline{q}(t^*) = -\overline{r} \cdot F(x^*(t^*)) \tag{8}$$

$$\overline{p}(t^*) \in -\overline{r} \cdot x_0^*(t^*)\partial F(x^*(t^*)) - C \cdot \partial d_E(x^*(t^*)) \tag{9}$$

for some nonnegative scalar C

(iv)

$$\|\overline{p}\| + \|\overline{q}\| + \overline{r} > 0. \tag{10}$$

Now we need to rearrange the expressions so that we have a maximum principle for the problem (P_z).

Since $x_0^*(\cdot) \neq 0$, $p(\cdot) := \frac{\overline{p}(\cdot)}{x_0^*(\cdot)}$ is well defined. We also identify

$$r := \overline{r} \qquad \text{and} \qquad q(\cdot) := \overline{q}(\cdot) \tag{11}$$

to obtain the Hamiltonian H for the problem (P_z):

$$H(r,q,p,x,u) = \langle p, f(x,u)\rangle - q \cdot \widetilde{\lambda}(x,u) - r \cdot f_\circ(x,u).$$

It follows from the above definitions that (7) implies (3), (8) and (9) imply (4) and (10) implies (5).

By the finite sums formula (see [3]), we have

$$\begin{aligned}&\partial_{\overline{x}}\overline{H}(\overline{r},\overline{q}(t),\overline{p}(t),\overline{x}^*(t),u^*(t))\\ &\quad\subset \partial_{\overline{x}}\overline{H}_1(\overline{p}(t),\overline{x}^*(t),u^*(t)) + \partial_{\overline{x}}\overline{H}_2(\overline{r},\overline{q}(t),\overline{x}^*(t),u^*(t))\end{aligned}$$

where $\overline{H}_1(\overline{p},\overline{x},u) := \langle \overline{p}, f(x,u)\rangle$ and $\overline{H}_2(\overline{r},\overline{q},\overline{x},u) := -x_0(\overline{q}\widetilde{\lambda}(x,u) + \overline{r}f_\circ(x,u))$.

In general, there is no relationship between the generalized derivative and generalized partial derivative. However, since $\overline{H}_1$ as a function of $\overline{x}$ is independent of x_0 and $\overline{H}_2$ can be written in a form $x_0G(x)$ where $x_0 > 0$ and $G(x)$ is continuous, we can show that the following inclusions hold:

$$\begin{aligned}&\partial_{\overline{x}}\overline{H}_1 \subset \{0\} \times \partial_x\overline{H}_1\\ &\partial_{\overline{x}}\overline{H}_2 \subset \{-\overline{q}\widetilde{\lambda}(x,u) - \overline{r}f_\circ(x,u)\} \times \partial_x\overline{H}_2.\end{aligned}$$

Therefore (6) implies

$$-\frac{d}{dt}\overline{q}(t) = -\overline{q}(t)\lambda(x^*(t),u^*(t)) - \overline{r}f_\circ(x^*(t),u^*(t)) \qquad \text{a.e. } t \in [0,t^*]$$

and

$$\begin{aligned}-\frac{d}{dt}\overline{p}(t) &\in \partial_x\overline{H}_1(\overline{p}(t),\overline{x}^*(t),u^*(t))\\ &\quad + \partial_x\overline{H}_2(\overline{p}(t),\overline{x}^*(t),u^*(t)) \qquad \text{a.e. } t \in [0,t^*],\end{aligned}$$

from which we derive (1) and (2) by definition (11).

Case (b)

The situation here is not covered directly by the results employed in the previous case due to the fact that $t^* = +\infty$. For infinite horizon optimal control problems with *smooth* data, a maximum principle is developed in [2]. Here we develop a *maximum principle* for *nonsmooth* infinite horizon optimal control problems.

Take a strictly increasing sequence $\{t_i\}$ in $[0,\infty)$ such that $t_i \to \infty$ as $i \to \infty$.

A collection of deterministic problems $\{(\overline{P}_i)\}$ can be defined on $T(x^*;\varepsilon)$ as follows:

$(\overline{P}_i)$ min $\displaystyle\int_0^{t_i} x_0(t)f_\circ(x(t),u(t))dt$

over the class $\overline{\Omega}_i$ consisting of all pairs

$(\overline{x}(\cdot),u(\cdot))$ on $[0,t_i]$

$$\begin{aligned}\text{s.t.}\quad &\tfrac{d}{dt}\overline{x}(t) = [-x_0(t)\lambda(x(t),u(t)), f(x(t),u(t))] \qquad \text{a.e. } t \in [0,t^*]\\ &x(t) \in T(x^*;\varepsilon)\\ &\overline{x}(0) = (1,z)\\ &\overline{x}(t_i) = \overline{x}^*(t_i).\end{aligned}$$

By the standard argument of the principle of optimality, it is easy to show that $\overline{x}^*(t)$ restricted to $[0,t_i]$ is an optimal trajectory for $(\overline{P}_i)$.

Thus we can again use Clarke's nonsmooth maximum principle. By theorems 5.2.1 and 5.2.3 of [3], we conclude that for Problem $(\overline{P}_i)$ there exist absolutely continuous functions

$$\overline{q}_i : [0,t_i] \to I\!R \qquad \overline{p}_i : [0,t_i] \to I\!R^n$$

and a nonnegative scalar $\overline{r}_i$, such that

$$\begin{aligned}&-\frac{d}{dt}(\overline{q}_i(t),\overline{p}_i(t)) \in \partial_{\overline{x}}H(\overline{r},\overline{q}_i(t),\overline{p}_i(t),\overline{x}^*(t),u^*(t)) \qquad \text{a.e. } t \in [0,t_i]\\ &\max_{u\in U} H(\overline{r}_i,\overline{q}_i(t),\overline{p}_i(t),\overline{x}^*(t),u)\\ &\qquad = H(\overline{r}_i,\overline{q}_i(t),\overline{p}_i(t),\overline{x}^*(t),u^*(t))\\ &\qquad = 0 \qquad \text{a.e. } t \in [0,t_i]\\ &\|\overline{p}_i\| + \|\overline{q}_i\| + \overline{r}_i > 0.\end{aligned}$$

As in case (a), we can rearrange the expressions by redefining

$$r_i := \overline{r}_i, \quad q_i(\cdot) := \overline{q}_i(\cdot), \quad p_i(\cdot) = \frac{\overline{p}_i(\cdot)}{x_i^*(\cdot)}$$

and have

$$\begin{aligned}-\dot{q}_i(t) &= H(r_i,q_i(t),p_i(t),x^*(t),u^*(t))\\ &\quad -\langle p_i(t), f(x^*(t),u^*(t))\rangle \qquad \text{a.e. } t \in [0,t_i]\end{aligned} \tag{12}$$

$$\begin{aligned}-\dot{p}_i(t) &\in \partial_x f(x^*(t),u^*(t))^\top p_i(t)\\ &\quad - q_i(t)\partial_x\widetilde{\lambda}(x^*(t),u^*(t))\\ &\quad - r_i\partial_x f_\circ(x^*(t),u^*(t))\\ &\quad - \widetilde{\lambda}(x^*(t),u^*(t))p_i(t) \qquad \text{a.e. } t \in [0,t_i]\end{aligned} \tag{13}$$

$$\begin{aligned}&\max_{u\in U} H(r_i,q_i(t),p_i(t),x^*(t),u)\\ &\quad = H(r_i,q_i(t),p_i(t),x^*(t),u^*(t))\\ &\quad = 0 \qquad \text{a.e. } t \in [0,t_i]\end{aligned} \tag{14}$$

$$\|p_i\| + \|q_i\| + r_i > 0. \tag{15}$$

By normalization, condition (15) could be equivalently replaced by $r_i + \|p_i\| + \|q_i\| = 1$, so that $r_i + |p_i(0)| + |q_i(0)|$ is bounded. Hence by passing to an appropriate subsequence one may assume that

$$\lim_{i\to\infty} r_i = r, \quad \lim_{i\to\infty} p_i(0) = p(0), \quad \lim_{i\to\infty} q_i(0) = q(0)$$

exist.

Let $q : [0, \infty) \to I\!R$ be the unique continuous piecewise C^1 solution of the differential equation (1) with initial condition $q(0) = \lim_{i\to\infty} q_i(0)$. One has $\lim_{i\to\infty} q_i(t) = q(t)$ for $t > 0$, due to the continuous dependence of solutions of the differential system (1) with respect to the initial data.

Rewrite the differential equation (1) and the differential inclusion (2) as a differential inclusion in the form

$$\frac{d}{dt}(q(t), p(t)) \in F(t, (q(t), p(t))) \qquad \text{a.e. } t > 0 \tag{16}$$

where the set value map $F(t, x)$ can be shown to be convex, compact valued and Lipschitzian in the terminology of Aubin and Cellina [1]. By [1, Theorem 1, p. 121], for each $(q_i(\cdot), p_i(\cdot))$, we can associate a solution $y_i(\cdot)$ of the differential inclusion (2) (or equivalently $(q_i(\cdot), y_i(\cdot))$ of the differential inclusion (16)) such that $y_i(0) = p(0)$ and for $t > 0$

$$|y_i(t) - p_i(t)| \le \max\{\|p(0) - p_i(0)\|, \|q(0) - q_i(0)\|\} e^{\int_0^t k(s)ds},$$

where $k(s)$ is the Lipschitz constant of the set valued map $F(s, x)$. Since it may be shown that the minimum norm trajectory of (16) remains in a compact subset of $I\!R^n$, the set of all solutions of the differential inclusion (2) with initial data $p(0) = \lim_{i\to\infty} p_i(0)$ is compact in the topology of uniform convergency on compacta for $C([0, \infty); I\!R^{n+1})$ by [1, Theorem 1, p. 104]. It follows that there exists a solution $p(t)$ of the differential inclusion (2) satisfying $p(0) = \lim_{i\to\infty} p_i(0)$ such that $\lim_{i\to\infty} p_i(t) = p(t)$ for $t > 0$.

Taking limits in (14) we obtain (3) due to the linearity of H in r, q and p.

It is obvious that r can be taken as 0 or 1 according as it is originally 0 or positive. ■

3 The main result

Definition The *Hamiltonian function* for the PDP control problem is defined as follows:

$$\begin{aligned} H(r, q, p, x, u, [\theta]) := {} & \langle p, f(x, u)\rangle - r \cdot l_\circ(x, u) \\ & - r \cdot \lambda(x, u) \int_E \theta(y) Q_\circ(dy; x, u) \\ & - q(\lambda(x, u) + \delta) \end{aligned}$$

for r and $q \in I\!R$, $p \in I\!R^n$, $x \in I\!R^n$ and $u \in U_\circ$, $\theta \in C(E)$, the space of continuous functions on E.

Theorem 2 (A nonsmooth maximum principle for optimal control of PDPs)

In addition to A1–A6, assume the following conditions are met:

(a) *f is bounded, $l_\circ(\cdot, u)$, $\lambda(\cdot, u)$, $l_\partial(\cdot, u)$ are Lipschitz continuous on E uniformly in u and for all $\theta \in C(E)$, $x \longmapsto \int_E \theta(y) Q_\circ(dy; x, u)$ and $x \longmapsto \int_E \theta(y) Q_\partial(dy; x, u)$ are Lipschitz continuous on E uniformly in u.*

(b) *For any $x \in \partial E$ and $v \in U_\circ$, there exists $\alpha > 0$ such that*

$$f(x, v) \cdot n(x) \ge \alpha > 0,$$

where $n(x)$ is the unit outward normal to ∂E at the point $x \in \partial E$.

(c)
$$\begin{aligned} N_\theta(x) := \{ & (f(x, u), l_\circ(x, u) \\ & + \lambda(x, u) \int_E \theta(y) Q_\circ(dy; x, u), \lambda(x, u)) : u \in U_\circ\} \end{aligned}$$

is convex for any $x \in E$ and $\theta \in C(E)$.

Suppose $u = (u_\circ, u_\partial)$ is an optimal control for the PDP control problem. For any $z \in E$, let $x(\cdot)$ be the corresponding deterministic trajectory for the control function $u_\circ(z, \cdot)$, i.e. $x^*(\cdot)$ is the solution of the following differential system

$$\begin{cases} \dot{x}(t) = f(x(t), u_\circ(z, t)) & 0 \le t \le t^* \\ x(0) = z \end{cases}$$

where t^* is the corresponding exit time. Then for every $z \in E$ there exist

(1) *absolutely continuous functions*

$$q : [0, t^*] \longmapsto I\!R \qquad p : [0, t^*] \longmapsto I\!R^n$$

where q is actually a continuous, piecewise C^1 function

(2) *a scalar which is either equal to 0 or 1, such that the following relations hold:*

(i) *(q, p) satisfies the "adjoint inclusion"*

$$\begin{aligned} -\dot{q}(t) = {} & H(r, q(t), p(t), x(t), u_\circ(z, t), [J(u)]) \\ & - \langle p(t), f(x(t), u_\circ(z, t))\rangle \\ & \text{a.e. } t \in [0, t^*] \\ -\dot{p}(t) \in {} & \partial_x f(x(t), u_\circ(z, t))^{\mathsf{T}} p(t) \\ & - q(t) \partial_x \lambda(x(t), u_\circ(z, t)) \\ & + r \cdot \partial_x [l_\circ(x(t), u_\circ(z, t)) \\ & + \lambda(x(t), u_\circ(z, t)) \\ & \cdot \int_E J_y(u) Q_\circ(dy; x(t), u_\circ(z, t))] \\ & - [\lambda(x(t), u_\circ(z, t)) + \delta] p(t) \\ & \text{a.e. } t \in [0, t^*], \end{aligned}$$

where $J_y(u)$ is the cost corresponding to the optimal control u starting from interior state $y \in E$ and $J(u)$ is defined as the function $y \longmapsto J_y(u)$

(ii) *H is maximized at $u_\circ(z, t)$, i.e.*

$$\begin{aligned} & \max_{u \in U_\circ} H(r, q(t), p(t), x(t), u, [J(u)]) \\ & = H(r, q(t), p(t), x(t), u_\circ(z, t), [J(u)]) \\ & = 0 \qquad \text{a.e. } \; t \in [0, t^*] \end{aligned}$$

(iii) *transversality condition, if $t^* < \infty$, then*

$$(q(t^*), p(t^*)) + r \cdot \xi \in (0, -C \partial d_E(x(t^*)))$$

for some nonnegative scalar C and an $(n+1)$ dimensional vector ξ with

$$\xi \in (F_\partial(x(t^*)), \partial F_\partial(x(t^*)),$$

where

$$F_\partial(x) := \min_{u\in U_\partial}\{l_\partial(x,u) + \int_E J_y(u)Q_\partial(dy;x,u)\}$$

and $d_E(x) := \inf\{\|x-y\| : y \in E\}$ *is the distance function*

(iv) *if* $t^* < \infty$, *then*

$$\|p\| + \|q\| + r > 0.$$

Remark. For every $z \in E$ there is a multiplier function (q,p) which depends Borel measurably on z with respect to the topology of uniform convergence on compacta for $C([0,\infty); \mathbb{R}^{n+1})$. Hence we may consider the multiplier process $(\mathbf{q},\mathbf{p})$ corresponding to the PDP $\mathbf{z}$ as a random process.

Proof. In our paper [5], we have shown that a control $(u_\circ, u_\partial)$ is optimal if and only if for each $z \in E$, $u_\circ(z,\cdot)$ is an optimal control in the deterministic optimal control problem with boundary conditions (P_z) with the following data:

$$f_\circ(x,u) := l_\circ(x,u) + \int_E V(y)Q_\circ(dy;x,u)\cdot\lambda(x,u) \quad (17)$$

$$F(x) := \min_{v\in U_\partial}\{l_\partial(x,v) + \int_E V(y)Q_\partial(dy;x,v)\} \quad (18)$$

$$\tilde{\lambda}(x,u) := \lambda(x,u) + \delta. \quad (19)$$

Notice that we are dealing with a necessary condition. The value function $V(y)$ is known to be equal to the expected cost $J_y(u)$ and to be Lipschitz continuous on E. We subsititute $f_\circ(x,u)$,$F(x)$,$\tilde{\lambda}(x,u)$ defined by (17), (18) and (19) respectively in Theorem 1. The maximum principle for control of PDPs then follows in a straightforward manner. ∎

4 Extensions

If in addition the problem is subject to the *calmness condition* in the terminology of Clarke [3], it can be shown that the scalar r can be taken as 1.

By the multiplicative functional technique, these results can be extended to cost functionals with growth bounded, for example, exponentially.

Acknowledgement

The authors would like to thank Mark Davis for helpful discussions.

References

[1] Aubin, J. P. and Cellina, A., *Differential Inclusions*, Springer-Verlag, Berlin (1984).

[2] Carlson, D. A. and Haurie, A., *Infinite Horizon Optimal Control*, Springer-Verlag, Berlin (1987).

[3] Clarke, F. H., *Optimization and Nonsmooth Analysis*, Wiley, New York (1983).

[4] Davis, M. H. A., Piecewise-deterministic Markov processes: a general class of non-diffusion stochastic models, *J. Roy. Statist. Soc. Ser. B* **46** (1984) 353–388.

[5] Dempster, M. A. H. and Ye, J. J., A necessary and sufficent optimality condition for control of piecewise deterministic Markov processes, in preparation.

[6] Vermes, D., Optimal control of piecewise deterministic Markov processes, *Stochastics* **14** (1985) 165–208.

Approximation, Optimization and Computing:
Theory and Applications, A.G. Law and C.L. Wang (eds.)
Elsevier Science Publishers B.V. (North-Holland)

AN AXIOMATIC APPROACH TO THE DERIVATIVES OF NONSMOOTH FUNCTIONS

J.L. Dong and T. Yu

Department of Applied Mathematics
Jilin University of Technology, Changchun, CHINA

In this paper, we present an axiomatic definition of the generalized directional derivatives of nonsmooth functions on manifolds. Some properties of the generalized gradients are discussed.

1. INTRODUCTION

After Clarke's poineer work (see[1,2]), the generalized directional derivatives and the generalized gradients for nonsmooth functions have been developed, and applied to several areas of science. Generally speaking, investigators in this field usually focus their attention on giving various concrete definitions of the generalized directional derivatives and the generalized gradients for various kinds of nonsmooth functions (e.g. see [3]).

The purpose of this note is to formulate a general theory for nonsmooth analysis. The main motivations come from differential geometry.

2. DEFINITIONS AND THEOREMS

Let M be a n-dimensional differentiable manifold and B a space of functions on Ω (Ω is an open set of M). The following definitions show that the generalized directional derivatives posses a very clear geometric picture.

Definition 1. Suppose that $G_p: T_pM \times B \rightarrow R$ is a real-valued function. If G_p is both positively homogeneous and subadditive for two variables, and satisfies the following axioms:

1. $G_p(-v,f) = G_p(v,-f)$.

2. $G_p(v,fg) \leqslant G_p(v,g(p)f) + G_p(v,f(p)g)$.

Where $v \in T_pM$ and $f,g \in B$. Then we say that $G_p(v,f)$ is the generalized directional derivative of f in direction v at p.

Remarks. Where of course p is a point in Ω. In some cases, we assume that G_p is a continuous function.

Definition 2. Let $f \in B$, we define the generalized gradient of f as follows:

$$G(f)(p) = \{\xi \in T_p^*M \quad \xi(h) \leqslant G_p(h, f) \text{ for all } h \in T_pM\}.$$

Remark. T_pM and T_p^*M are tangent space and cotangent space of M at p, respectively.

With definition 1 and 2, the following theorems are obtained.

Theorem 1. Let $f \in B$, then $G(f)(p)$ is a nonempty, convex, compact subset of T_p^*M.

The proof of Theorem 1 is easily obtained by using the same argument in [3].

Theorem 2. For every $h \in T_pM$, one has

$$G_p(h,f) = \max\{\xi(h) \mid \xi \in G(f)(p)\}.$$

Proof. Let

$$g(p,h) = \max\{\xi(h) \mid \xi \in G(f)(p)\}.$$

By the definition of g, we know

$$g(p,h) \leqslant G_p(h,f).$$

Suppose that for some $v \in T_pM$

$$g(p,v) < G_p(v,f).$$

By the definition of G_p, we know there exists $\alpha \in T_p^*M$, such that

$$\alpha(v) = G_p(v,f),$$

$$\alpha(h) \leqslant G_p(h,f) \text{ for any } h \in T_pM.$$

It follows that $\alpha \in G(f)(p)$, whence $G_p(v,f) > \alpha(v) = G_p(v,f)$, a contradiction which establishes the theorem.

Before we discuss the properties of the generalized gradients, we now stop to investigate the meaning of $G(f)(p)$.

Theorem 3. Let (U, ϕ) be a coordinate system of M at p, and $x_i (i = 1,2,\ldots,n)$ the coordinate functions. Then for any $\xi \in G(f)(p)$, one has

$$\xi = \sum_i a_i dx_i,$$

where $a_i = \langle \xi, \partial/\partial x_i \rangle$.

Remark. The proof of Theorem 3 is very easy, therefore it is omitted.

As we know, for the smooth function f on the differentiable manifold M, we have the following conclusion(cf. [4]):

$$df = \sum_{k=1}^{n} \frac{\partial f}{\partial x_k} dx_k, \qquad (*)$$

df is the differential of f at p. Comparing the Theorem 3 with formula (*), we know that the generalized gradients are an extension of the classical differentials.

Turning to the properties of the generalized gradient $G(f)(p)$. For the sake of simplicity, we shall henceforth write $G(f)$ in place of $G(f)(p)$.

Theorem 4. For any scalar s, one has

$$G(sf) = sG(f).$$

Proof. Let us first consider $s \geqslant 0$. By the definition of G_p, it follows easily that $G(sf) = sG(f)$. It suffices now to prove the formula for $s = -1$. An element ζ belongs to $G(-f)$ iff $\zeta(h) \leqslant G_p(h,-f)$ for all h. By Axiom 1, this is equivalent to: $\zeta(h) \leqslant G_p(-h,f)$ for all h. Thus $\zeta \in G(-f)$ iff $\zeta \in -G(f)$.

Theorem 5. $G(f+g) \subset G(f)+G(g)$.

Theorem 6. $G(fg) \subset g(p)G(f) + f(p)G(g)$.

Corollary. For any scalar s_k, one has

$$G\left(\sum_{k=1}^{n} s_k f_k\right) \subset \sum_{k=1}^{n} s_k G(f_k).$$

The proof of Theorem 5 is quite analogous to the proof of Theorem 6, therefore we only prove Theorem 5.

Proof. Note that the support functions of the left- and right-hand sides are, respectively, $G_p(v,f+g)$ and $G_p(v,f)+G_p(v,g)$. By the definition 1, we know

$$G_p(v,f+g) \leqslant G_p(v,f) + G_p(v,g).$$

This proves the theorem.

Remark. The proof of Corollary follows by induction.

We conclude this section with the verification of the consistency between the generalized gradients and the classical differentials. Namely, we shall show that the generalized gradients that we defined in this section are an appropriate extension of the classical differentials. Now, we assume that B is a smooth functions space and G_p is a bilinear functional on $T_pM \times B$. In addition, the sign of equality holds in the axiom 2. Then, we have $G_p(v,\cdot) \in T_pM$. For any function f which belongs to B, a very elementary argument shows that $G(f)$ contains only one element, denoted by ζ, that is

$$G(f) = \{\zeta\}.$$

Since the differential of f (see [4]) belongs to T_p^*M, df is a 1-form, therefore we have following result:

$$df(v) = v(f) \text{ for all } v \in T_pM.$$

So that

$$\zeta = df.$$

That is

$$df = G(f).$$

The consistency is verified.

3. CONCLUSION

What has been presented here is an attempt to elaborate the theory of generalized derivatives for nonsmooth functions from a few general concepts. More thorough study for an axiomatic structure of nonsmooth analysis would be useful.

REFERENCES

[1] Clarke, F.H., Generalized gradients and applications, Trans. Am. Math. Soc. 205 (1975) 247-262.
[2] Clarke, F.H., A new approach to Lagrange Multipliers, Math. Op. Research 1 (1976) 165-174.
[3] Clarke, F.H., Optimization and Nonsmooth Analysis (John Wiley & sons, Inc. 1983)
[4] Warner, F., Foundations of differentiable manifolds and Lie groups (Springer-Verlag, 1983).

Approximation, Optimization and Computing:
Theory and Applications, A.G. Law and C.L. Wang (eds.)
Elsevier Science Publishers B.V. (North-Holland)

SHAPE OPTIMIZATION FOR GENERAL STRUCTURES USING MCADS

Gu Yuanxian and Cheng Gengdong

Research Institute of Engineering Mechanics
Dalian University of Technology
Dalian 116024, China

ABSTRACT. Based on the idea of integration of Finite Element Method (FEM), optimization, and Computer-Aided Design (CAD), a microcomputer based system, MCADS, has been developed for optimal design of general structures. Its featured approaches, i.e. semi-analytical sensitivity analysis, optimization-analysis modeling for shape design, application oriented user interfaces, and cooperation of automatic optimization and user intervention are presented in this paper.

1. INTRODUCTION

The integration of FEM, optimization, and CAD is an attractive research direction for both fields of computational mechanics and computer-aided engineering. It will make structural optimization suitable to environment of CAD and become a really applicable design tool like the state-of-art of FEM as an analysis tool. By means of this integration, more powerful CAD systems capable not only of graphics but of efficient analysis and optimization as well are being produced.

This integration needs multi-disciplinary researches and their cooperation to implement interfaces for linking of FEM-CAD, FEM-Optimization, and Optimization-CAD. Some important subjects, e.g. interfacing finite element modeling with geometric modeling [3], system developing using existing FEM and optimization packages as "black box", shape optimization coupling with CAD facilities [1], sensitivity analysis via the semi-analytical method [5] or variational method with boundary integral [2, 4], have caused research enthusiasm recently.

An integrated system, MCADS [7], has been developed on microcomputers for optimal design of general structures. The optimization with MCADS is same versatile as the common used FEM analysis due to the featured approaches given in following sections.

2. VERSATILE SENSITIVITY ANALYSIS VIA SEMI-ANALYTIC METHOD (SAM)

Sensitivity analysis is a fundamental procedure of the structural optimization, because almost all efficient optimization algorithms are based on the usage of sensitivity information. In order to make optimization practically applicable to general structures, a versatile approach of sensitivity analysis being suited to various type of finite elements, design variables, and state variables is necessary. Besides computational efficiency and accuracy, programming efficiency is the key of such an approach. In structural shape optimization, especially, the sensitivity formulae are related to the detailed structure and formulation of elements, the way of the change of boundary, mesh, and loads with shape variable, and the constraint functions such as smoothed node stress. These relations may be too complicated to be described explicitly. So that formulating and programming in the conventional analytical method is very consumptive. In contrast, two programming oriented approaches, Variational Design Sensitivity Analysis (VDSA) with boundary integral [3, 4] and Semi-Analytical Method (SAM) [5], are easy to be implemented due to making use of existing FEM programs and interfacing them with optimization algorithms.

The MCADS system takes a FEM program as a "black box" in which structural analysis follows

$$K\,U = P, \qquad R_j = R_j(X,U)\ (j=1,2,\ldots,m) \tag{1}$$

and sensitivity analysis via SAM follows

$$K\frac{\partial U}{\partial x_i} = \frac{\partial P}{\partial x_i} - \frac{\partial K}{\partial x_i}U \tag{2}$$

$$\frac{\partial R_j}{\partial x_i} = [R_j(X+\Delta X_i, U(X)+\frac{\partial U}{\partial x_i}\Delta X_i) - R_j(X,U)]/\Delta x_i \tag{3}$$

where K, U, P are stiffness matrix, displacement and load vector respectively, $R_j(X, U)$ is state variable. X is design variable vector, $\Delta X_i=(0,\ldots \Delta x_i,\ldots,0)$ is perturbation of X with a finite difference ΔX_i for i-th element. The calculations of sensitivities of K and P which respect to x_i are based on the element local differences:

$$\frac{\partial K}{\partial x_i} = \sum_{k=1}^{M}[K_k(X+\Delta X_i) - K_k(X)]/\Delta x_i \qquad (4)$$

$$\frac{\partial P}{\partial x_i} = \sum_{k=1}^{M}[P_k(X+\Delta X_i) - P_k(X)]/\Delta x_i \qquad (5)$$

Computations of Eq. (2)-(5) can be done easily by means of calling subroutines of FEM program treated as "black box".

An unified scheme was proposed in MCADS for sensitivity analysis of general cases: Giving a perturbation to X, updating finite element model with new design, computing new values of K_K, P_K, R_J and sensitivities from Eq. (2)-(5). this scheme is type of elements and design variables independent. there is no need to know computational details of elements and state variables. Only interfaces of subroutines calling are needed. The influence of load change related to shape has also been evaluated without difficulties encountered in analytic method. Therefore the sensitivity analysis of MCADS via SAM is suitable to shape and size optimization of general structures composed of multi-type elements, such as bar, beam, membrane, plate, shell, brick, axisymmetric brick and revolution shell, etc., and subject to various boundary conditions and loads.

The computational efficiency of SAM is same as that of analytic method. Because the computing effort involved in Eq. (3-5) is no more than that in calculating derivatives of K_K, P_K, and R_J in analytic method. About accuracy of SAM, however, some difficulties have been observed in shape variable case by recent researches. Our research and numerical tests [6] have shown that the error of sensitivity calculated by SAM will increase severely while mesh is refined for such elements their nodal displacements have nonuniform dimensions, e.g. beam, plate and shell element. While for other elements, their nodal displacements have the uniform dimensions, e.g. membrane, brick, bar, etc., there is no such problem. By making use of alternative forward/backward difference scheme suggested in [6], the sensitivity accuracy of MCADS has been improved remarkably and can be applied to structural shape optimization.

3. OPTIMIZATION-ANALYSIS MODELING APPROACH

It is important for optimal shape design, especially in the case of continuum structures, to build a design model with suitable description and modification of the shape information. Reasonable modeling approach should first construct design model with fewer design parameters so called the "natural variable", then generate FE model for analysis from the design model. Structural shape and FE model (mesh, loads, and boundary condition) will be updated by these natural variables within the design process.

Such an optimization-analysis modeling (OAM) approach based on the concepts of the natural variable and the design element has been developed in MCADS system by means of a mesh generator MESHG. Structure is divided into some regular shaped patches with edge curve/surface described in quadratic interpolation or spline. some interpolation parameters or coordinates of control-node of curve/surface are selected as design variables (natural variable). Patches near changeable boundary, called "design elements", as well as those parts of mesh and loads related to them are updated during iteration. This kind of model is convenient for continuum structures (composed of membrane, plate, shell, brick, and axisymmetric elements) subject to optimization of boundary shape or inner-geometry, such as distribution of thickness or material. The design model of truss/frame shape can be described in master-slave variable relation, i.e. the master variable which may be any control parameter of linear linking or user defined curve, is taken as the design variable, and the node coordinates as the slave variables.

Automatic generation of mesh, load, and connection data between models of design and analysis are finished by program MESHG. MESHG employes the mapping method for mesh generation of surface and volume of any complicated shape. So that patch division required in the design modeling can just be used as the mapped elements for meshing and re-meshing. The functions of re-meshing and load change are embedded into optimization iteration to implement SAM sensitivity analysis.

4. SHAPE OPTIMIZATION FACILITIES

The optimization of MCADS system is versatile for general structures with elements of membrane, plate, shell, brick, axisymmetric brick and shell, beam, bar and design variables of component sectional size and boundary shape. It is particularly applicable to optimum shape design of continuum structures by means of SAM and OAM approaches mentioned above. The constraints and/or objective include structural weight, stress, and displacement. The goal may be to minimize weight and/or stress, and to maximize or minimize stiffness. The load case, node, and element related to constraints can be assigned by user or selected by the program automatically according to design status.

The design operation of MCADS is composed of facilities of automatic optimization and user intervention. During iteration process, the user can interactively check and change design model according to his/her experience and information of design sensitivity. This update involves modification of variable values and constraint bounds, deletion or addition of constraints and/or design variables. User's interaction with system makes good use of their experience and creativity, and makes optimization more flexible. The optimization solver is an independent module in MCADS and is aimed at an algorithm library. It has employed algorithm of sequential quadratic/linear programming with some improvements, such as treatment for unfeasible design or multiple objectives with goal programming, approximate line search with Goldstein criterion, and adaptive move limit. The algorithm solver has been designed as a well tuned black box for users lacking optimization knowledge.

5. APPLICATION-ORIENTED USER INTERFACE

As a commercial CAD software package, an attractive feature of MCADS is that it is open to the user via some interfaces, such as the data interfaces for pre- and post-processing of FEM analysis, and for description of the design model. Especially it provides the application oriented subroutine interface for the user to define particular optimization problem itself. Because there are always special problems in practical engineering application which can not be preconceived and treated with definite formulas when programming. For instance, relationship between the natural variable and the interpolation curve/surface of the various boundary shape, calculations of parameter and stress of complicated section of components, etc. The procedures dealing with such application related problems are located in a few independent modules which opens to the user as the programming interface. So that the user can deal with special cases by means of modifying these procedures and replacing new formulas. And design model of optimization can be easily extended for complicated engineering applications. This application oriented interface has been used for computations of the beam section parameters, the stress on beam section, the stress at node with smoothing, the coordinate of control node of interpolation curve/surface, and the load changed with structural shape. The optimization of U-shaped bellow illustrated in section 6 was performed by adding new relations between the variables and the control nodes into only one module, no more programming is needed.

6. EXAMPLE ILLUSTRATION

The shape optimal design of bellow, used as equipment of pipe linkage and deformation compensation in industries, is illustrated here. Flexibility and strength of bellow, principle performances of structure, may be improved via optimization of wave shape (cross-section shape). Let us consider an U-shaped bellow, whose cross-section of a half wave is shown in Fig.1. Diameter D=40 cm, wall thickness t=0.8 cm, R=4.5 cm, H=9.8 cm. The

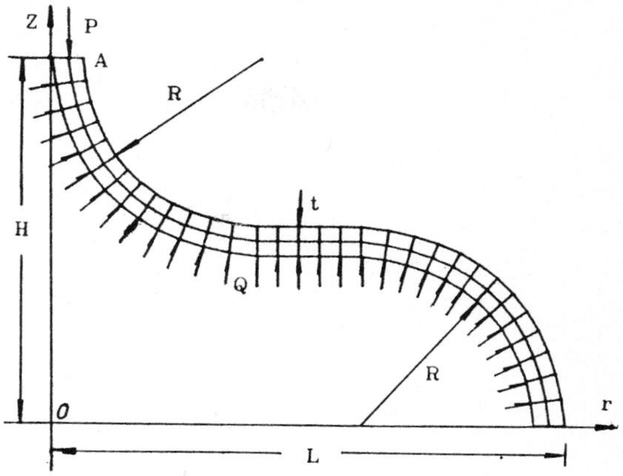

Fig.1

material is 16 MnR, E=1.93 10^6 Kg/cm^2, ν=0.3. The mesh of Axisymmetric brick element, used due to the axial symmetry of structure and load, is composed of 96 nodes and 62 elements. The load cases include: evenly distributed pressure Q=16 Kg/cm^2, and axial pressure P=125 Kg/cm which is the reaction of the neighbour waves. The contractive displacement at end is δ=-2.25 mm, and maximum stress is σ_θ=-35.2 Kg/cm^2 at point A.

The arcs of two ends and the middle straight part are divided into three design elements with 18 control nodes of quadratic interpolation of edges. In the first design model we select H and L (initial value is 13.3 cm) as design (natural) variables: H determines the radius R, L determines the length of the straight part, and they determine the coordinates of 18 control nodes. The mesh and the nodal load reduced from distributed pressure Q are updated while the shape chages. The variable bounds are: $8.0 \leq H \leq 12.0$ cm, $11.0 \leq L \leq 15.0$ cm. Two design schemes are considered: (1) To maximize bellow flexibility via increasing δ with the constraint of $\sigma_\theta \leq 35.0$ Kg/cm^2 for all nodes. (2) To minimize bellow stress with the constraint of $|\delta| \geq 2.25$ mm. After 10 and 15 iterations respectively, optimum designs are obtained. (1) δ=-2.71 mm (its absolute value increases 20%), and maximum stress σ_θ=-34.9 Kg/cm^2. H=10.5 cm, L=14.8 cm. (2) Maximum stress σ_θ=-28.7 Kg/cm^2 (its absolute value decreases 18%), and δ=-2.25 mm. H=12.0 cm, L=13.7 cm.

In the second design model we select L, t, and R as design variables with the bounds: $11.0 \leq L \leq 15.0$ cm, $0.6 \leq t \leq 1.0$ cm, $3.0 \leq R \leq 6.0$ cm. H keeps constant. All the other data and design elements are same as above. The same two design schemes are considered, better results are gained after 13 and 16 iterations: (1) δ=-3.41 mm (profit is 52%), maximum stress σ_θ=34.9

Kg/cm². L=14.84 cm, t=0.65 cm, R=5.83 cm. (2) Maximum stress σ_θ=26.6 Kg/cm² (decreases 24.4%), δ is still -2.25 mm. L=14.8 cm, t=0.83 cm, R=5.84 cm.

7. REFERENCES

[1] Atrek, E., et al (Eds.), "New directions in optimum structural design", John Wiley & Sons, 1984.

[2] Mota Soares (Eds.), "Computer aided optimal design", NATO/NASA/NSF/USAF advanced study institute, Portugal, 1986.

[3] Bennett, J.A. and Botkin, M.E. (Eds.), "The optimum shape, Automated structural design", Plenum press, 1986.

[4] Haug, E.J., Choi, K.K. and Komkov, V., "Design sensitivity analysis of structural system", Academic Press Inc., 1986.

[5] Cheng, G.D. and Liu, Y.W., "A new computation scheme for sensitivity analysis", Eng. Opt. Vol.12, no.3, pp. 219-235, 1987.

[6] Cheng, G.D., Gu, Y.X. and Chou, Y.Y., "Accuracy of semi-analytic sensitivity analysis", to appear in J. of Finite Elements in Analysis and Design.

[7] Gu, Y.X., "Researches on computer aided structural optimum design and development of MCADS system", Ph.D. thesis (in Chinese), Dalian University of Technology, China, 1988.

Approximation, Optimization and Computing:
Theory and Applications, A.G. Law and C.L. Wang (eds.)
Elsevier Science Publishers B.V. (North-Holland)

A NEW METHOD FOR LARGE SCALE SPARSE LINEAR PROGRAMMING

Yongchang JIAO and Kaizhou CHEN

Dept. of Applied Mathematics, Xidian University, Xi'an, Shaanxi, CHINA 710071

This paper presents a new method for LP, especially for large scale sparse problems. First we describe the partial relaxation multiplier (PRM) method, then we use it to transform LP into a series of quadratic programming (QP) problems with simple constraints, and we use the generalized conjugate gradient (GCG) method to solve these special QP problems. Convergence results for algorithms presented are given. Our method does not need to calculate any inverse matrix. It never changes the constrained matrix in LP, and zero elements in the matrix will neither be stored nor be used in calculation, so it can handle large scale sparse LP. Also we can find the original solution and the dual solution at the same time. Some numerical experiments show that our method is feasible and efficient.

1. INTRODUCTION

It is well–known that LP is an important branch in operations research. There are several fairly ripe methods, e.g. simplex method, to solve moderate size LP problems. Recently the algorithm research for large scale LP becomes a very important aspect for people to study.

In the recent years, a new approach for solving LP appears. Its main idea is that fairly ripe nonlinear programming methods are used to solve LP. This runs through the well–known Karmarkar's algorithm and its variants. But they need to calculate inverse matrices, this is their fatal weakness. In this paper we propose a new method for LP. It never needs to calculate any inverse matrix, so it has less calculation. Also it never changes the constrained matrix in LP and in fact only needs to use the matrix to form products with arbitrary vectors.so our method is suitable for sparse constrained matrices. In addition, we can obtain the original and the dual solutions at the same time. Some test results show our method is feasible and efficient.

2. THE PRM METHOD

Consider the following problem

$$\begin{cases} \min f(x) \\ s.t.\ h(x)=0 \\ x \geqslant 0, \qquad x \in R^n \end{cases} \tag{2.1}$$

where h(x) is $R^n \to R^l$, $h(x)=(h_1(x),\cdots,h_l(x))^T$. In order to deal with (2.1), we generalize the multiplier method which was proposed seperately by Hestenes and Powell in 1969 [1] [2].

Suppose that f, $h_1, \cdots h_l$ are twice continuously differentiable functions, that $x^* \geqslant 0$, and that there exist vectors $\lambda^* = (\lambda_1^*,\cdots\lambda_n^*)^T$ and $u^* = (u_1^*, \cdots, u_l^*)^T$ such that if we set

$$L(x, \lambda, u) = f(x) - \sum_{i=1}^{n} \lambda_i x_i - \sum_{j=1}^{l} u_j h_j(x),$$

then

$$\nabla_x L(x^*, \lambda^*, u^*) = 0 \tag{2.2}$$

$$h_j(x^*) = 0 \qquad j = 1,2,\cdots,l \tag{2.3}$$

$$\lambda_i^* x_i^* = 0 \qquad i = 1,2,\cdots,n \tag{2,4}$$

$$\lambda_i^* \geqslant 0 \tag{2.5}$$

hold, and for every $z{:}z \neq 0$ *and* $z \in \hat{Z}^1(x^*)$

$$z^T \nabla_x^2 L(x^*,\lambda^*,u^*) z > 0 \quad , \tag{2.6}$$

where $\hat{Z}^1(x^*) = \{z | z_i = 0,\ i \in \hat{I}(x^*);\ z_i \geqslant 0,\ i \in I(x^*)$ and $i \notin \hat{I}(x^*); z^T \nabla h_j(x^*) = 0,\ j = 1,2,\cdots,l\}$,

$$I(x^*) = \{i | x_i^* = 0, 1 \leqslant i \leqslant n\}$$

and

$$\hat{I}(x^*) = \{i | i \in I(x^*) \ \ and \ \ \lambda_i^* > 0\}.$$

We have the following result.

THEOREM 1. Suppose that x^*, λ^* and u^* satisfy (2.2) to (2.6) and that $x^* \geqslant 0$, then there exists a number $\sigma^* > 0$ such that x^* is a strict local minimum of problem

$$\begin{cases} \min\ f(x) - u^{*T} h(x) + \frac{1}{2}\sigma h(x)^T h(x) \\ s.t.\ x \geqslant 0 \end{cases} \tag{2.7}$$

for $\sigma > \sigma^*$. Conversely,if $h(\hat{x}) = 0$ and $\hat{x}$ is a minimum of problem

$$\begin{cases} \min\ f(x) - u_o^T h(x) + \frac{1}{2}\sigma h(x)^T h(x) \\ s.t.\ x \geqslant 0 \end{cases}$$

for some u_o and σ,then $\hat{x}$ is a minimum of (2.1),and u_o is the optimal Lagrange multiplier corresponding to $\hat{x}$ and the equality constraints in (2.1).

Based on Hestenes' method of multipliers, we give the following formula to update the Lagbange multiplier u:

$$u_{k+1} = u_k - \sigma_k \cdot h(x_k) \tag{2.8}$$

here x_k solves problem

$$\begin{cases} \min\ f(x) - u_k^T h(x) + \frac{1}{2}\sigma_k h(x)^T h(x) \\ s.t.\ x \geqslant 0 \end{cases} \tag{2.9}$$

Below we show the PRM method.

1° Select a sufficiently large scalar σ_1 and an initial constant vector $u_1 \in R^l$, set k = 1.

2° Let x_k solve (2.9). If $h(x_k) = 0$, then x_k solves (1.1), stop. Otherwise, goto 3°

3° Set $u_{k+1} = u_k - \sigma h(x_k)$,k=k+1, Let $\sigma_{k+1} \geqslant \sigma_k$, goto 2

Based on [3]、[4]、[5] we have given the convergence theorems for the method. Because these theorems are very complicated, we omit them here.

3. OUR METHOD AND ITS CONVERGENCE RESULT

Consider LP Problem

$$\begin{cases} \min\ c^T x \\ s.t.\ Ax = b \\ \quad\ x \geqslant 0 \end{cases} \tag{3.1}$$

where $c \in R^n$, $b \in R^m$ and $A \in R^{m \times n}$. If we use the PRM method to handle (3.1),there are some special properties .

THEOREM 2. Consider (3.1). Let x^*,u^* and λ^* satisfy $Ax^* = b, x^* > 0$, $C - A^T u^* - \lambda^* = 0$, $\lambda^{*T} x^* = 0$ and $\lambda_j^* \geqslant 0$ $(i = 1,2,\cdots,n)$, then for any positive scalar σ, x^* is a minimum of problem

$$\begin{cases} \min\ c^T x - u^{*T}(Ax - b) + \frac{1}{2}\sigma \|Ax - b\|^2 \\ s.t.\ x \geqslant 0. \end{cases} \tag{3.2}$$

Otherwise, if $Ax_o = b$, x_o is a minimum of problem

$$\begin{cases} \min\ c^T x - u_o^T(Ax - b) + \frac{1}{2}\sigma \|Ax - b\|^2 \\ s.t.\ x \geqslant 0 \end{cases} \tag{3.3}$$

for some u_o and σ, then x_o is a solution of (3.1), and u_o is its dual solution.

Let

$$d(u) = \min\{c^T x - u^T(Ax - b) | x \geqslant 0\},$$

$$L_\sigma(x.u) = c^T x - u^T(Ax - b) + \frac{1}{2}\sigma \|Ax - b\|^2,$$

we have following result.

THEOREM 3. Suppose that (3.1) has a nonempty and compact optimal solution set . Then for every scalar $\sigma > 0$ and $u \in R^m$, problem $\begin{cases} \min L_\sigma(x.u) \\ s.t.\ x \geqslant 0 \end{cases}$ has at least one solution and its optimal solution set is compact.

Below is our basic method.

Algorithm A.

1° select a suitable scalar $\sigma_1 > 0$ and an initial constant vector

$u_1 \in R^m$. Set k = 1 .

2° Let x_k solve problem

$$\begin{cases} \min\ L_{\sigma_k}(x,u_k) \\ s.t.\ x \geqslant 0 \end{cases} \tag{3.4}$$

If $Ax_k = b$. then x_k solves (3.1), and u_k solves its dual problem, stop; otherwise goto 3° .

3° Set $u_{k+1} = u_k - \sigma_k(Ax_k - b)$, k=k+1. Let $\sigma_{k+1} \geqslant \sigma_k$, goto 2° .

ASSUMPTION (A). (3.1) has a nonempty and bounded optimal solution set and a bounded dual solution set.

THEOREM 4. Let Assumption (A) hold, and $\{(x_k, u_k)\}$ be a sequence generated by Algorithm A. If $Ax_{k_o} = b$ for some integer $k_o > 1$, then x_{ko} is a optimal solution of (3.1), u_{ko} is its dual solution. Conversely, if $Ax_k \neq b$ for every $k \geqslant 1$, then the seguence has at least one limit point, and its every limit point $(\bar{x},\bar{u})$ is a primal—dual solution pair of (3.1).

Furthermore,

$$\lim_{k\to\infty} c^T x_k = \lim_{k\to\infty} d(u_k) = c^T x^*,\ \lim_{k\to\infty} \|Ax_k - b\| = 0.$$

We can prove further that the sequence $\{u_k\}$ generated by Algorithm A is convergent.

THEOREM 5. The sequence $\{u_k\}$ generated by Algorithm A converges to the dual solution of (3.1).

Based on some results in [6], we have proved these theorems. Because of the space limitation, the complicated proofs are omited.

The remaining problem is how to solve the special QP problem

$$(SQP)\begin{cases} \min\ L_\sigma(x,u) \\ s.t.\ x \geqslant 0 \end{cases}$$

for every $u \in R^m$ and $\sigma > 0$.

4. A METHOD FOR SOLVING SQP PROBLEM

Consider (SQP) problem. notice that the Hessian matrix of its objective function is $\sigma A^T A$. It is a positive semidefinite matrix rather than a positive definite one, which makes us to be difficult in implementing the conjugate gradient method reasonably. In order to surmount this difficulty, we transform (SQP) into a series of problems

$$\begin{cases} \min\ L_\sigma(x,u) + \frac{1}{2}q\|x - x^l\|^2 \\ s.t.\ x \geqslant 0 \end{cases} \tag{4.1}$$

where q is a positive scalar, x^l is the optimal estimate of (SQP) in the preceding iteration. Generally we choose q as a constant.

It is well–known that the conjugate gradient method has small storage and it can converge stably. Based on [7], we use the GCG method to solve (4.1).

Algorithm B

Initialization

- Select an $x^{(o)}$ such that $x^{(o)} \geqslant 0$, and set k=0.
- Set I={1,2,···,n} .

Outer Iteration

- Set k=k+1, $x^{(k)}=x^{(k-1)}$, $y^{(k)}=Hx^{(k)}+W$ (here H $=\sigma A^TA+qI, W=C-A^Tu-\sigma A^Tb-qx^l$), and $I_{k-1}=I$.
- If $I_k=I_{k-1}$, stop. The optimal solution of (4.1) has been found, i.e.$x^{l+1}=x^{(k)}$. Otherwise,set $I=I_k$ and begin the inner iteration.

1° Partition and rearrange the matrix system as

$$x^{(k)}\rightarrow\begin{bmatrix}x_I^{(k)}\\ x_J^{(k)}\end{bmatrix},\ W\rightarrow\begin{bmatrix}W_I\\ W_J\end{bmatrix},\ H\rightarrow\begin{bmatrix}H_{II} & H_{JI}^T\\ H_{JI} & H_{JJ}\end{bmatrix}$$

where J={1,2,···,n}∖I. Let the element number in J is s.Set q=0, $z^{(o)}=x_J^{(k)}, p^{(o)}=r^{(o)}=-W_J-H_{JJ}z^{(o)}$

2° Calculate

$$a_{cg}=\frac{(r^{(q)},r^{(q)})}{\left(p^{(q)},H_{JJ}p^{(q)}\right)} \tag{4.2}$$

$$a_{max}=\min_{\substack{1\leqslant j\leqslant s\\ p_j^{(q)}<0}}\frac{-z_j^{(q)}}{p_j^{(q)}} \tag{4.3}$$

$$a_q=\min(a_{cg},a_{max})$$

$$z^{(q+1)}=z^{(q)}+a_qp^{(q)}$$

$$r^{(q+1)}=r^{(q)}-a_qH_{JJ}p^{(q)}\ .$$

3° If $r^{(q+1)}=0$, set $x_J^{(k)}=z^{(q+1)}$ and restart the outer iteration. If $\{j|z_j^{(q+1)}=0,1\leqslant j\leqslant s\}=\Phi$, goto 4° ; otherwise, set $x_J^{(k)}=z^{(q+1)}$ and $I=\{i|x_i^{(k)}=0,1\leqslant i\leqslant n\}$.If $I=\{1,2,\cdots,n\}$,then restart the outer iteration; otherwise, restart the inner iteration.

4° Calculate

$$b_q=\frac{(r^{(q+1)},r^{(q+1)})}{(r^{(q)},r^{(q)})} \tag{4.4}$$

$$p^{(q+1)}=r^{(q+1)}+b_qp^{(q)}$$

Set q=q+1, goto 2° .

Based on Algorithm B, we describe a method for (SQP).

Algorithm C

1° Select an vector $x^o\geqslant 0$and a suitable scalar q> 0, Set l=0.

2° We choose x^l as the initial vector $x^{(o)}$, and use Algorithm B to find x^{l+1}.

3° Calculate $\varepsilon=\|X^{l+1}-X^l\|^2$. If $\varepsilon<\varepsilon_1$ (ε_1 is a small positive mumber we choose in advance), then we have found the approximative solution x^{l+1} of (SQP) ; otherwise, set l=l+1, goto 2°

The following theorems are concerned with the convevgence results for Algorithms B and C.

THROTRM 6. Select an initial vector $x^{(o)}\geqslant 0$ arbitrarily. The sequence $\{x^{(k)}\}$ generated by Algorithm B converges to the optimal solution of (4.1) in a finite number of iterations.

THEOREM 7. select an initial vector $x^o\geqslant 0$ arbitrarily. The sequence $\{x^l\}$ generated by Algorithm C converges to the optimal solution of (SQP).

5. COMPUTATIONAL CONSIDERATION AND SOME TEST RESULTS

we have presented our method above. It can handle large scale sparse LP. Because it never needs to calculate any inverse matrix, it will converge stably and it has less calculation. In addition, we can modify our method in order to improve its convergence rate. This remains to be studied further.

we also give some techniques to handle sparse matrix, and use them to our method. We have tested some examples in microcomputer IBM—AT and medium size computer IBM—4381. The test results are presented in Tables 1 and 2.

EXMPLE. Klee—Minty problem

$$\begin{cases}\min -x_d\\ s.t.\ x_1-r_1=\varepsilon\\ x_1+\sigma_1=1\\ x_j-\varepsilon x_{j-1}-r_j=0\\ x_j+\varepsilon x_{j-1}+s_j=1 \qquad j=2,3,\cdots,d\\ x_j\geqslant 0,\ r_j\geqslant 0,\ s_j\geqslant 0 \qquad j=1,2,\cdots,d\end{cases}$$

The problem has n=3d variables and m=2d equality constraints.

Its optimal value is ε^d-1, here ε is a constant and $0<\varepsilon<\frac{1}{2}$.

TABLE 1

cpu \ n / computer	30	90	300	360	498	999
IBM—4381	2.05s	8.44s	1m 41.37s	2m 1.99s	5m 58.22s	25m 4.05s

TABLE 2

cpu \ n / computer	30	60	90	120	150	180	210
IBM—AT	11s	1m 5s	2m 34s	4m 37s	8m 39s	11m 21s	15m 47s
	240	270	300	360	498	999*	
	24m 30s	32m 27s	36m 14s	18m 38s	36m 36s	12h50m 19s	

Note: h—hour, m—minute, s—second

* : using virtual memory

According to the partial comparison between our method and other methods, e.g. shang's method [8] and Diao's method [9], it is seen that our method is feasible and efficient.

REFERENCES

[1]. M.Avriel, Nonlinear Programming: Analysis and Methods, Prentice—Hall, Englewood Cliffs, New Jersey, 1976.

[2]. R. Fletcher, Practice Methods of optimization— vol.2, Constrained Optimization,John Wiley & Sons., Ltd., 1980.

[3]. D.P.Bertsekas, Multiplier Method, A Survey, Automatica, Vol.12 (1976), PP.131—145.

[4]. D.P.Bertsekas, On Penalty and Multiplier Method for Constrained Minimization, SIAM J.Control & Optimization, Vol.14(1976), PP.217—235.

[5]. B.T.Polyak, The Convergence Rate of the Penalty Function Methods, Z.Vycisl, Mat.iMat.FiZ., 11(1971), PP.3～11.

[6]. R.T.Rockafellar, Convex Analysis, Princeton University Press, Princeton, New Jersey, 1970.

[7]. D.P.O'leary, A Generalized Conjugate Gradient Algorithm for Solving a class of Quadratic Programming Problems, in Large Scale Matrix Problems, Ed. by Ake Bjorck, R.T.Plemmons and H.Schneider, Elsevior North Holland Inc., 1981.

[8]. Shang Yi, New Polynomial—Time Algorithm for Linear Programming—Saddle Surface Algorithm, J. of shenyang Institute of Chemical Technology,Vol.1 (1987), PP.1～16.

[9]. Diao Zaijun, A Modified Karmarkar's Algorithm, Applied Mathematics— A Journal of chinese Universities, Vol.3(1988), PP41～56.

Approximation, Optimization and Computing:
Theory and Applications, A.G. Law and C.L. Wang (eds.)
Elsevier Science Publishers B.V. (North-Holland)

NONLINEAR CONTROL THEORY

John Jones, Jr.

Air Force Institute of Technology, School of Engineering, Department of Mathematics and Computer Science, Wright-Patterson AFB, Ohio, 45433

The main purpose of this work is to show how several recent results of J. Jones, Jr. in the areas of mathematical modelling and computer simulation on large-scale supercomputers may apply to large multiparameter multidimensional nonlinear dynamical systems.

1. INTRODUCTION

Modelling and simulation of large-scale multidimensional multiparameter dynamical systems require the use of large-scale computers to generate feedback control laws, especially when model uncertainties exist. Now robust control theory is concerned with the problem of analyzing and synthesizing control systems that provide an acceptable level of performance where many model parameters or uncertainties may exist since mathematical models of physical systems are usually never exact due to the presence of such parameters.

The need to be able to design robust feedback control laws is very important in such systems. Usually a physical model will have significant structural information about the interconnection of components and subsystems, but less information concerning their integrated system performance. Hence, many variations of parameters must be carried out on supercomputers in order to determine the more significant and sensitive parameters which must be adjusted very rapidly to accomplish a desired level of performance.

The elements of the matrices, $E(\theta)$, $A(\theta)$, $C(\theta)$, $D(\theta)$ belong to the ring of polynomials $C[\theta]$ where $\theta = (\theta_1, \theta_2, \theta_3, \ldots, \theta_q)$ is a multiparameter and the coefficients of the polynomials belonging to $C[\theta]$ have elements belonging to the field C of complex numbers, or the elements of such multiparameter matrices may be of the form $f(\theta) = a(\theta)|b(\theta)$ where the polynomials $a(\theta)$, $b(\theta) \in C[\theta]$, for $b(\theta) \neq 0$. $E(\theta)$ may be a singular matrix for possible parameter values of θ. The parameter θ may also be a holomorphic function of a single complex variable z for z belonging to simply-connected bounded regions in the complex z-plane.

The basic questions of stabilization, as regards the overall dynamical system, needs to be treated in response to changing parameters in subsystems.

Fast numerical methods requiring parallel processing are necessary to compute adjustments as time t changes. Transfer function matrices, controllability matrices, observability matrices, feedback and control laws need to be recomputed as parameters change. Such matrices may be multiparameter matrices and may allow for improvement of control laws in cases where $E(\theta)$ may become a singular matrix and the dynamical system requires considerable fast changes in feedback control laws.

Use is made of recent results of J. Jones, Jr. concerning generalized inverses of such multiparameter matrices to aid in computer aided changes to carry out modelling and simulation and analysis of such dynamical systems. Some examples of nonlinear control theory problems are given which illustrate methods of approach to such difficult type problems in contrast to linear control theory.

2. EXAMPLE 1 (COUPLED RICCATI DIFFERENTIAL EQUATIONS)

The mathematical modelling and simulation of nonlinear dynamical systems and nonlinear control problems starts with the process of obtaining as much information as possible concerning properties of solutions of the basic nonlinear differential equations before the notions of stabilization, model reduction, controllability, observability, feedback control laws, etc. can be developed. It would be desirable to be able to solve such nonlinear dynamical equations explicitly first, but in general, this is a very difficult problem if not impossible in some cases.

Matrices of polynomials in several holomorphic functions of z having complex coefficients will be used throughout this paper to treat nonlinear problems. Use will be made of results of J. Jones, Jr. [1] concerning generalized inverses of such type matrices and results involving inverse dynamical systems of certain classes of nonlinear problems.

Consider the following coupled set of nonlinear Riccati type differential equations:

$$\begin{pmatrix} \dot{x}(t) \\ \dot{y}(t) \end{pmatrix} = \begin{pmatrix} -x(t) & x(t) \\ y(t) & \frac{2}{y(t)} - y(t) \end{pmatrix} \begin{pmatrix} x(t) \\ y(t) \end{pmatrix} \quad (1.1)$$

for $0 < t < \infty$ and $y(t) \neq 0$. The main objective of this example is to obtain an analytical solution if possible and to illustrate important approaches to such types of nonlinear problems. Equation (1.1) may be written as follows:

$$\text{(i)} \quad \dot{x}(t) = x(t)y(t) - x^2(t) \quad (1.1)'$$

$$\text{(ii)} \quad \dot{y}(t) = y(t)x(t) - y^2(t) + 2$$

for $0 < t < \infty$. Such type systems may have many solutions.

Case (i). Let $x(t) = t$, then equation (i) of (1.1)' above implies that $y(t) = t + \frac{1}{t}$ for $0 < t < \infty$. Using equation (ii) above for $x(t) = t$, then $\dot{y}(t) = 2 + ty(t) + y(t)(-1)y(t)$ is an ordinary nonlinear differential Riccati type equation of the form $\dot{y}(t) = A(t) + B(t)y(t) + y(t)C(t)y(t)$, where $A(t) = 2$, $B(t) = t$, $C(t) = -1$. For $y(t) \triangleq -[C(t)]^{-1}\frac{w'(t)}{w(t)}$, then $w(t)$ is a solution of the second order linear differential equation of the following form

$$w''(t) + \{-C(t)B(t)C^{-1}(t) - C^{1}(t)[C(t)]^{-1}\} \quad (1.2)$$

$$w'(t) + \{C(t)A(t)\}w(t) = 0$$

which reduces to the equation

$$w''(t) - tw'(t) - 2w(t) = 0 \quad (1.3)$$

which has a solution $w(t) = te^{t^2/2}$ for $0 < t < \infty$. Therefore for $x(t) = t$, $y(t) = w'(t)/w(t)$ a solution of (ii) above is of the form

$$y(t) = -[C(t)]^{-1}\frac{w'(t)}{w(t)} = \frac{w'(t)}{w(t)} \quad (1.4)$$

$$= \frac{e^{t^2/2} + t^2 e^{t^2/2}}{te^{t^2/2}} = \left(t + \frac{1}{t}\right)$$

for $0 < t < \infty$.

Case (ii). Let $y(t) = t + \frac{1}{t}$ in equation (i) of (1.1)' it follows that

$$\dot{x}(t) = 0 + \left(t + \frac{1}{t}\right) x(t) + x(t)(-1)x(t) \quad (1.5)$$

and

$$w''(t) + (-1)\left(t + \frac{1}{t}\right) w'(t) = 0 \quad (1.6)$$

where

$$w(t) = e^{t^2/2}; \; w'(t) = te^{t^2/2} \quad (1.7)$$

and

$$x(t) = \frac{w'(t)}{w(t)} = \frac{te^{t^2/2}}{e^{t^2/2}} = t \quad (1.8)$$

Therefore, a solution vector of the equation (1.1) is of the form

$$\begin{pmatrix} t \\ t + \frac{1}{t} \end{pmatrix}, \quad 0 < t < \infty \quad (1.9)$$

3. EXAMPLE (REDUCTION PROBLEM)

The following example concerns the reduction of a given class of nonlinear dynamical systems to one of simpler form. Consider the following coupled nonlinear set of ordinary differential equations:

$$\begin{pmatrix} \dot{x}(t) \\ \dot{y}(t) \\ \dot{z}(t) \end{pmatrix} = \begin{pmatrix} -x(t) & 0 & y(t) \\ 0 & -y(t) & x(t) \\ y(t) & 0 & -z(t) \end{pmatrix} \begin{pmatrix} x(t) \\ y(t) \\ z(t) \end{pmatrix} \quad (2.1)$$

The basic problem of reducing (2.1) to a simpler form arises in order to obtain more information concerning properties of solutions of (2.1) arises. Writing (2.1) in the following form:

$$\begin{cases} \dot{x}(t) = y(t)z(t) - x^2(t) \\ \dot{y}(t) = z(t)x(t) - y^2(t) \\ \dot{z}(t) = x(t)y(t) - z^2(t) \end{cases} \quad (2.1)'$$

First adding the equations of (2.1)'; next multiply the three equations of (2.1)' successively by $\{1,\omega,\omega^2\}$ respectively and adding the resulting equations, where ω is obtained by making use of $\omega^3 = 1$, $\omega^3 - 1 = 0$, $(\omega-1)(\omega^2+\omega+1) = 0$, $\omega = e^{2\pi i/3}$; and finally multiplying the equations of (2.1)' respectively by $\{1,\omega^2,\omega^4 = \omega\}$ and adding; the system (2.1) is transformed to the following system:

$$\begin{cases} \dot{u}(t) = \dot{x}(t) + \dot{y}(t) + \dot{z}(t) \\ \dot{v}(t) = \dot{x}(t) + \omega\dot{y}(t) + \omega^2\dot{z}(t) \\ \dot{w}(t) = \dot{x}(t) + \omega^2\dot{y}(t) + \omega\dot{z}(t) \end{cases} \quad (2.2)$$

$$= y(t)z(t)+z(t)x(t)+x(t)y(t)-x^2(t)$$
$$-y^2(t)-z^2(t)$$

$$= y(t)z(t)+\omega z(t)x(t)+\omega^2 x(t)y(t)-x^2(t)$$
$$-\omega y^2(t)-\omega^2 z^2(t)$$

$$= y(t)z(t)+\omega^2 z(t)x(t)+\omega x(t)y(t)-x^2(t)$$
$$-\omega^2 y^2(t)-\omega z^2(t)$$

where u(t), v(t), w(t) are defined by equations

$$\begin{pmatrix} u(t) \\ v(t) \\ w(t) \end{pmatrix} = \begin{pmatrix} 1 & 1 & 1 \\ 1 & \omega & \omega^2 \\ 1 & \omega^2 & \omega \end{pmatrix} \begin{pmatrix} x(t) \\ y(t) \\ z(t) \end{pmatrix} \; ; \tag{2.3}$$

$$\omega \overset{\Delta}{=} e^{2\pi i/3}$$

Now

$$\begin{pmatrix} x(t) \\ y(t) \\ v(t) \end{pmatrix} \tag{2.4}$$

$$= \frac{1}{3} \begin{pmatrix} 1 & 1 & 1 \\ 1 & e^{-2\pi i/3} & e^{-4\pi i/3} \\ 1 & e^{-4\pi i/3} & e^{-8\pi i/3} \end{pmatrix} \begin{pmatrix} u(t) \\ v(t) \\ w(t) \end{pmatrix}$$

Making use of (2.4) in (2.2) it follows that the reduced system is obtained:

$$\begin{pmatrix} \dot{u}(t) \\ \dot{v}(t) \\ \dot{w}(t) \end{pmatrix} = \begin{pmatrix} 0 & 0 & -v(t) \\ 0 & -u(t) & 0 \\ -w(t) & 0 & 0 \end{pmatrix} \begin{pmatrix} u(t) \\ v(t) \\ w(t) \end{pmatrix} \tag{2.5}$$

This system may be solved analytically for u(t), v(t), w(t) and then using (2.4) solutions of (2.1) are obtained. The reduction of the coupled system (2.1) may be carried out by use of a MACSYMA or symbolic type computer language. Solving (2.5) to get $\dot{u}(t) = k^2 w^2(t)$; $\dot{w}(t) = -w(t)u(t)$; and $v(t) = k^2 w(t)$, for k^2 an arbitrary constant.

Now

$$\begin{cases} u(t) = -c \tan c(t+b) \\ v(t) = ck \operatorname{Sec} c(t+b) \\ w(t) = \frac{c}{k} \operatorname{Sec} c(t+b) \end{cases} \tag{2.6}$$

and using (2.4) we obtain an explicit form of x(t), y(t), z(t) in terms of trignometric functions sec c(t+b), tan(t+b), for suitable constants, k^2, c, b.

In order to choose a matrix B, it is necessary to know when x(t), y(t), z(t) become unbounded or have peaks. Then after stabilization of the system (2.1), one proceeds to choose B such that possible controllability is obtained, and to choose a matrix C such that observability is possible, and finally a matrix D such that a feedback law exists.

4. EXAMPLE 3 (CONSTANT MATRIX CASE)

$$\begin{cases} \dot{x} = Ax + Bu \\ y = Cx + Du \end{cases} \tag{3.1}$$

where:

$$\begin{cases} \dot{x}(t) = \begin{pmatrix} 0 & 1 \\ -2 & -3 \end{pmatrix} x(t) + \begin{pmatrix} 0 & 1 \\ 1 & -2 \end{pmatrix} u(t) \\ y(t) = (1 \;\; 0)x(t) + (0 \;\; 1)u(t) \end{cases} \tag{3.2}$$

and the 1-generalized inverse D_1 of D is $\binom{k}{1}$ for the arbitrary k. A 1-generalized inverse system of (3.1) above is as follows:

$$\begin{cases} \dot{x}(t) = \begin{pmatrix} -1 & 1 \\ k & -3 \end{pmatrix} x(t) + \begin{pmatrix} 1 \\ k-2 \end{pmatrix} y(t) + \begin{pmatrix} 0 \\ g \end{pmatrix} \\ u(t) = \begin{pmatrix} -k & 0 \\ -1 & 0 \end{pmatrix} x(t) + \begin{pmatrix} k \\ 1 \end{pmatrix} y(t) + \begin{pmatrix} g \\ 0 \end{pmatrix} \end{cases} \tag{3.3}$$

let g = 0 and the state matrix of the 1-inverse system (3.2) of (3.1) has eigenvalues

$$\lambda_1 = -2 + \sqrt{1-k} \;\; ; \; \lambda_2 = -2 - \sqrt{1-k} \tag{3.4}$$

where k is an arbitrary scalar. Now for k = 0 or k = 1 the inverse system (3.3) is not <u>controllable</u> since the controllability matrix

$$C = \{(B_1 \vdots A_1 B_1)\} \tag{3.5}$$

$$= \left\{ \begin{pmatrix} 1 \\ k-2 \end{pmatrix} \vdots \begin{pmatrix} -1 & 1 \\ -k & -3 \end{pmatrix} \begin{pmatrix} 1 \\ k-2 \end{pmatrix} \right\}$$

$$= \left\{ \begin{pmatrix} 1 & \vdots & k-3 \\ k-2 & \vdots & -4k+6 \end{pmatrix} \right\}$$

has rank 1. However, for the observability matrix of the 1-inverse system:

$$O = \begin{pmatrix} C_1 \\ \cdots \\ C_1A_1 \end{pmatrix} = \begin{pmatrix} \begin{pmatrix} -k & 0 \\ -1 & 0 \end{pmatrix} \\ \cdots \\ \begin{pmatrix} -k & 0 \\ -1 & 0 \end{pmatrix}\begin{pmatrix} -1 & 1 \\ k & -3 \end{pmatrix} \end{pmatrix} \tag{3.6}$$

$$= \begin{pmatrix} \begin{pmatrix} -k & 0 \\ -1 & 0 \end{pmatrix} \\ \cdots \\ \begin{pmatrix} k & -k \\ 1 & -1 \end{pmatrix} \end{pmatrix}$$

for k any arbitrary scalar has rank 2 and the 1-generalized inverse system of (3.2) is observable for any k.

5. EXAMPLE 4 (TIME VARYING CASE)

$$\begin{cases} \dot{x}(t) = A(t)x(t) + B(t)u(t) \\ y(t) = C(t)x(t) + D(t)u(t) \end{cases} \tag{4.1}$$

where

$$\dot{x}(t) = \begin{pmatrix} \frac{t^3+t^2-2t-3}{(t+1)^2} & 4 \\ -1 & \frac{t^3+5t^2+6t+1}{(t+1)^2} \end{pmatrix} x(t) + \begin{pmatrix} \frac{t^2+t-1}{(t+1)} & 1 \\ -1 & \frac{t^2+3t+1}{(t+1)} \end{pmatrix} u(t) \tag{4.2}$$

$$y(t) = (1 \quad 0)\; x(t) + (0 \quad 1)\; u(t)$$

and the 1-generalized inverse of $D = (0 \quad 1)$ is of the form $D_1 = \begin{pmatrix} k \\ 1 \end{pmatrix}$ for arbitrary scalar k. A 1-generalized inverse system of (4.2) above is of the form:

$$\begin{cases} \dot{x}(t) = (A-BD_1C)x(t)+BD_1y(t)+B(I-D_1D)h \\ u(t) = (-D_1C)x(t)+(D_1)y(t)+(I-D_1D)h \end{cases} \tag{4.3}$$

where h is an arbitrary vector, $(DD_1D = D)$.

A 1-generalized inverse system of (4.2) is as follows:

$$\dot{x}(t) = \left\{ \begin{pmatrix} \frac{t^3+t^2-2t-3}{(t+1)^2} & 4 \\ -1 & \frac{t^3+5t^2+6t+1}{(t+1)^2} \end{pmatrix} - \begin{pmatrix} \frac{t^2+t-1}{(t+1)} & 1 \\ -1 & \frac{t^2+3t+1}{(t+1)} \end{pmatrix} \begin{pmatrix} k & 0 \\ 1 & 0 \end{pmatrix} \right\} x(t) \tag{4.4}$$

$$+ \begin{pmatrix} \frac{t^2+t-1}{(t+1)} & 1 \\ -1 & \frac{t^2+3t+1}{(t+1)} \end{pmatrix} \begin{pmatrix} k & 0 \\ 1 & 0 \end{pmatrix} \Bigg\} y(t)$$

$$+ \begin{pmatrix} \frac{t^2+t-1}{t+1} & 1 \\ -1 & \frac{t^2+3t+1}{t+1} \end{pmatrix}$$

$$\left(\begin{pmatrix} 1 & 0 \\ 0 & 1 \end{pmatrix} - \begin{pmatrix} 0 & k \\ 0 & 1 \end{pmatrix} \right) h$$

$$u(t) = -\left\{ \begin{pmatrix} k \\ 1 \end{pmatrix} (1 \quad 0) \right\} x(t) + \begin{pmatrix} k \\ 1 \end{pmatrix} y(t)$$

$$+ \begin{pmatrix} 1 & -k \\ 0 & 0 \end{pmatrix} h \;, \; \forall\, h$$

and

$$\dot{x}(t) = \left\{ \begin{pmatrix} \frac{t^3+t^2-2t-3}{(t+1)^2} & 4 \\ -1 & \frac{t^3+5t^2+6t+1}{(t+1)^2} \end{pmatrix} - \begin{pmatrix} k\left(\frac{t^2+t-1}{t+1}\right)+1 & 0 \\ -k+\frac{(t^2+3t+1)}{(t+1)} & 0 \end{pmatrix} \right\} x(t) \tag{4.5}$$

$$+\begin{pmatrix} \frac{k(t^2+t-1)}{(t+1)}+1 & 0 \\ -k+\frac{(t^2+3t+1)}{t+1} & 0 \end{pmatrix} y(t)$$

$$+\begin{pmatrix} \frac{t^2+t-1}{t+1} & 1 \\ -1 & \frac{t^2+3t+1}{t+1} \end{pmatrix}\begin{pmatrix} 1 & -k \\ 0 & 0 \end{pmatrix} h$$

$$u(t) = \begin{pmatrix} -k & 0 \\ -1 & 0 \end{pmatrix} x(t) + \begin{pmatrix} k \\ 1 \end{pmatrix} y(t)$$

$$+\begin{pmatrix} 1 & -k \\ 0 & 0 \end{pmatrix} h \ ; \ \forall\ h$$

The above 1-generalized inverse system may then be examined for its various properties for choices of the parameter k. Also B, C, D may be chosen to obtain desirable properties of this continuous dynamical system.

6. CHARACTERIZATIONS OF VARIOUS GENERALIZED INVERSES OF DYNAMICAL SYSTEMS

Consider the multiparameter multidimensional dynamical system of the following form:

$$\begin{cases} E(\theta)\dot{x} = A(\theta)x + B(\theta)u \\ y = C(\theta) + D(\theta)u \ , \ t \geq 0 \end{cases} \tag{5.1}$$

Let

$$S_1(\theta) = \begin{pmatrix} E_1(\theta) & B(\theta)D_1(\theta) \\ 0 & D_1(\theta) \end{pmatrix} \tag{5.2}$$

be associated with (6.1), where $E(\theta)E_1(\theta)E(\theta) = E(\theta)$ and $E(\theta)B(\theta)D_1(\theta) = B(\theta)D_1(\theta)$. Next let

$$S(\theta) = \begin{pmatrix} E(\theta) & -B(\theta) \\ 0 & D(\theta) \end{pmatrix} \tag{5.3}$$

also be associated with (6.1), then $S(\theta)S_1(\theta)S(\theta) = S(\theta)$ and $S_1(\theta)$ is a 1-generalized inverse of $S(\theta)$. If in addition to the above assumptions are made, namely

$$E_1(\theta)E(\theta)B(\theta)D_1(\theta) = E_1(\theta)B(\theta)D_1(\theta) \ ;$$
$$E_1(\theta)E(\theta)E_1(\theta) = E_1(\theta); \tag{5.4}$$

$$D_1(\theta)D(\theta)D_1(\theta) = D_1(\theta); \ E(\theta)E_1(\theta)A(\theta) = A(\theta);$$
$$D(\theta)D_1(\theta) = I$$

where $E_1(\theta)$, $D_1(\theta)$ are 2-generalized inverses of $E(\theta)$, $D(\theta)$ respectively, then $S_1(\theta)$ is a 2-generalized inverse of $S(\theta)$.

The system (6.1) may be written as follows:

$$\begin{pmatrix} E(\theta) & -B(\theta) \\ 0 & D(\theta) \end{pmatrix}\begin{pmatrix} \dot{x} \\ u \end{pmatrix} = \begin{pmatrix} A(\theta)x \\ y-C(\theta)x \end{pmatrix} \tag{5.5}$$

and has a solution $\begin{pmatrix} \dot{x} \\ u \end{pmatrix}$ if and only if

$$\begin{pmatrix} E(\theta) & -B(\theta) \\ 0 & D(\theta) \end{pmatrix}\begin{pmatrix} E(\theta) & -B(\theta) \\ 0 & D(\theta) \end{pmatrix}_1\begin{pmatrix} A(\theta)x \\ y-C(\theta)x \end{pmatrix} = \begin{pmatrix} A(\theta)x \\ y-C(\theta)x \end{pmatrix} \tag{5.6}$$

or written as follows:

$$S(\theta)S_1(\theta)\begin{pmatrix} A(\theta)x \\ y-C(\theta)x \end{pmatrix} = \begin{pmatrix} A(\theta)x \\ y-C(\theta)x \end{pmatrix} \tag{5.7}$$

and in which case the general solution of (6.5) for $\begin{pmatrix} \dot{x} \\ u \end{pmatrix}$ may be written as follows:

$$\begin{pmatrix} \dot{x} \\ u \end{pmatrix} = S_1(\theta)\begin{pmatrix} A(\theta)x \\ y-C(\theta)x \end{pmatrix} \tag{5.8}$$
$$+ (I-S_1(\theta)S(\theta))z(\theta), \ \forall\ z(\theta)$$

where $z(\theta)$ is of appropriate size.

7. SUMMARY

Other generalized inverse systems of (6.1) may be obtained which may possess more desirable properties for computing feedback matrices, controllability and observability properties, etc. of these inverse systems. Use is made of parallel processing to compute these various inverses of $S(\theta)$. A more detailed treatment of these various inverse systems will appear elsewhere.

REFERENCES

[1] Jones, J. Jr., Adaptive Feedback Control of Nonlinear Dynamical Systems, to appear.

Approximation, Optimization and Computing:
Theory and Applications, A.G. Law and C.L. Wang (eds.)
Elsevier Science Publishers B.V. (North-Holland)

ON THE OPTIMAL CONTROL SYNTHESIS FOR A CLASS OF NONLINEAR MECHANICAL SYSTEMS

P. Kiriazov, P. Marinov

Institute of Mechanics & Biomechanics, Bulgarian Academy of Sciences,
Acad. G. Bonchev Str., Bl. 4, 1113 Sofia, Bulgaria

Abstract. The subject of this work are nonlinear mechanical systems whose number of degrees of freedom is equal to the number of actuators. A decentralized controllability condition is imposed in the sense that the sign of any control input at the extremevalue is the same as the sign of the corresponding acceleration, irrespective of the other bounded control inputs. The performance index is a weighted minimum time-energy loss criterion. In the direct search optimization scheme adopted we employ appropriate spline approximations of the real optimal control laws via the Pontryagin's maximum principle. A sufficiently large set of feasible solution can be obtained among which a satisfactory suboptimal solution thus can be found.

1. INTRODUCTION

Many industrial robots, walking mashines and other mechanical devices are mechanical systems each degree of freedom of which being controlled individually. The dynamic performance of those systems is characterized mainly by the positioning accuracy, the movement execution time and the energy loss. These contradictory requirements are even more difficult to satisfy because of the existing couplings between the subsystems. To ensure existence of feasible solutions we assume a decentralized controllability condition in the sense that the sign of any control input (at least at extreme value) is the same as the sign of the corresponding acceleration, regardless of the other subsystems' influence. For the individually controlled mechanical systems to behave well such a condition is found to be a reasonable one [1,2,3].

An attempt to solve that complex highly nonlinear problem of the optimal control theory is presented in [4]. A suboptimal solution is obtained there utilizing control laws being optimal for one-degree-of-freedom mechanical systems.

The direct optimization approach is improved in the present work employing spline approximations of the control laws corresponding with the Pontryagin's maximum principle. The final switching times of the trial control functions are used in a natural way to solve the given TPBVP with values of all other describing parameters being fixed. Existence of a sufficiently large set of feasible solutions is guaranteed by the decentralized controllability condition and thus a satisfactory suboptimal solution can be obtained.

2. PROBLEM STATEMENT AND PONTRYAGIN'S MAXIMUM PRINCIPLE

Consider mechanical systems with dynamics described by the following coupled nonlinear differential equations:

$$\ddot{x}(t) = g(x(t),\dot{x}(t)) + f(x(t))u(t) \tag{1}$$

where: $x(t)\in R^n$, $u(t)\in R^n$, $g(.,.)\in C(R^n\times R^n;R^n)$
$f(.)\in C(R^n;R^{n\times n})$

According to [5] such systems with the number of degrees of freedom equal to the number of control functions are regarded as being "systems with full control".

Boundary conditions:

$$x(t^o) = x^o,\ \dot{x}(t^o) = 0 \text{ - initial state} \tag{2}$$

$$x(t^f) = x^f,\ \dot{x}(t^f) = 0 \text{ - final state,} \tag{3}$$

where t^f is not given in advance.

Performance index:

$$J = \int_{t^o}^{t^f} (1+1/2\alpha u^T(t)u(t))dt,\ \alpha > 0. \tag{4}$$

Control constraints:

$$|u_i(t)| \leq M_i \quad i = 1,\dots,n \leftrightarrow u\in U. \tag{5}$$

The problem is to find $u\in U$ that will drive the system (1) from the initial state (2) to the final state (3) so that the weighted time-energy loss functional (4) to be minimal and the constraints (5) to be satisfied.

According to [6,7] the problem so defined is a totally nonsingular control problem. Applying the Pontryagin's maximum principle in the latter work, the following optimal control laws have been derived:

$$u_i^*(t) = \begin{cases} +M_1, & p_i(t)/\alpha > M_i \\ p_i(t)/\alpha, & -M_i < p_i(t)/\alpha < M_i \\ -M_i, & p_i(t)/\alpha < -M_i \end{cases} \quad (6)$$

where $p_i(t)$ are the costate functions.

Generally speaking it does not make much sense to compute with great effort the exact solution of the mathematical model and to effectuate then this solution in an approximate way and with big efforts for the real system. In such cases the direct computation of such an "approximation" will be of a greater value.

3. DIRECT SEARCH APPROACH

Relying on the optimal control laws (6), we suggest the spline approximations as depicted in Fig. 1 to construct a set of trial control functions u_i, where:

$k_i = 1,\ldots,\bar{k}_i$ is the index of the switching times of function u_i and $L^{k_i} > 0$ are the corresponding slopes.

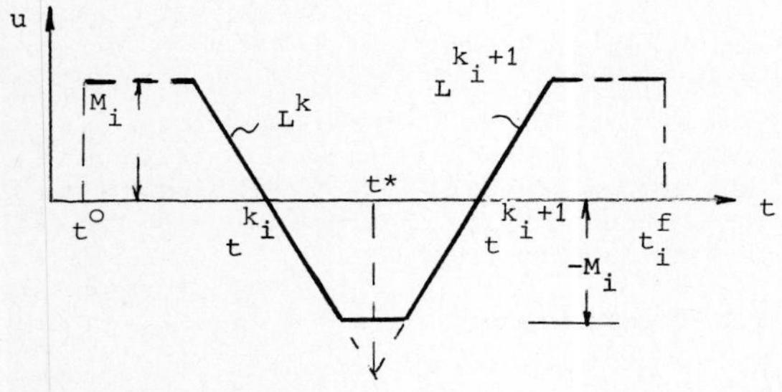

Fig. 1

The time interval (t^o, f_i^f) within the control function u_i is actuated, is devided by the switching times into $\bar{k}_i+1$ subintervals $(t^{k_i}; t^{k_i+1})$ where $t^{k_i}=t^o$ if $k_i=0$ and $t^{k_i+1}=t_i^f$ if $k_i=\bar{k}_i$.

On describing the control functions, it is necessary to define some other parameters. As shown in Fig. 1, the time t in an interval $(t^{k_i}; t^{k_i+1})$ is calculated through

$$t^* = (L^{k_i}.t^{k_i}+L^{k_i+1}.t^{k_i+1})/(L^{k_i}+L^{k_i+1}) \text{ when } k_i = 1,\ldots,\bar{k}_i-1. \quad (7)$$

We set $t^* = t^o$ if $k_i=0$ and $t^* = t^{\bar{k}_i}$ if $k_i = \bar{k}_i$.

We define, also, sign-switching functions $S_i(t)$ corresponding to control functions $u_i(t)$ as follows:

$$S_i(t) = S^{k_i} = \text{const.},\ t\epsilon(t^{k_i}; t^{k_i+1}), \quad (8)$$

where $S^{k_i+1} = -S^{k_i}$, $k_i = 0,\ldots,\bar{k}_i$ and $S_i(0) = \pm 1$.

The control function $u_i(t)$ on an interval $(t^{k_i}; t^{k_i+1})$ wher $k_i = 0,\ldots,\bar{k}_i$ is determined by

$$u_i(t) = \begin{cases} S_i(t)M_i(t), & |t-t^k| \geq M_i/L^k \\ S_i(t)|t-t^k|L^k, & |t-t^k| < M_i/L^k \end{cases} \quad (9)$$

where $k=k_i$ if $t \leqq t^*$ and $k=k_i+1$ if $t \geqq t^*$.

Therefore we need the following parameters to describe a trial control function:

- number $\bar{k}_i$ of the switching times (in practice $\bar{k}_i \leqq 2$)
- switching times t^{k_i}, $k_i = 1,\ldots,\bar{k}_i$
- initial value $S_i(0) = +1$ or -1
- slopes L^{k_i}, $k_i = 1,\ldots,\bar{k}_i$ (10)

In this way the optimal control laws can be parameterized by the maximum principle quite naturally. We devide the set of describing control parameters into two subsets. One of them consists of the final switching times $t^{\bar{k}_i}$, $i = 1,\ldots,n$, thus forming a vector $\bar{t}$. With fixed values of all the other control parameters their set denoted by p_{opt}, the n-dimensional vector $\bar{t}$ is assigned to solve the two-point boundary-value problem TPBVP (1÷3) and thus obtaining a feasible solution. And varying the values of the parameters p_{opt} we are looking for the optimum (the best suboptimal feasible solution as regards the performance index (4)).

The main point of our direct search approach is the solution of the TPBVP.

Denote by

$$x(t) = x(t,\bar{t},p_{opt}) \quad (11)$$

the dependence of motion on $\bar{t}$ and p_{opt}. We determine the final times of the particular coordinate motions $x_i(t)$ quite naturally:

$$f_i^f:\ f_i^f > f_i^{\bar{k}_i}\ \&\ \dot{x}_i(t_i^f,\bar{t},p_{opt}) = 0,\ i = 1,\ldots,n \quad (12)$$

Thus the final conditions $(3)_2$ for the velocities are satisfied with $t^f = \max t_i^f$, $i = 1,\ldots,n$. The correspondent reached positions at these final times are denoted by

$$F_i^{p_{opt}}(\bar{t}) = x_i(t_i^f, \bar{t}, p_{opt}), \quad i = 1,\ldots,n \qquad (13)$$

So, for the other final condition $(3)_1$ to be fulfilled, the missed distance from the required final position must be zero:

$$F_i^{p_{opt}}(\bar{t}) - x_i^f = 0, \quad i = 1,\ldots,n. \qquad (14)$$

This system of n equalities is regarded as being a system of n shooting equations with respect to the n-dimensional vector $\bar{t}$. So that the main steps of the optimization procedure we propose are the following:

1. Guess $\bar{k}_i$; t^{k_i}; L^{k_i}; $S_i(0)$; for $k_i = 1,\ldots,\bar{k}_i$ $i = 1,\ldots,n$
2. Perform a test movement. Check Eqs. (14): if "yes" Go to 4, if "no" Go to 3
3. Update $\bar{t}$. Go to 2
4. If the value of the performance index (4) is not satisfactory then update p_{opt} and Go to 2. Else Stop.

4. DISCUSSION

To succeed in carrying out the optimisation procedure above proposed, that is to succeed in finding sufficiently large number of feasible solutions, we need the following decentralized controllability (DC) conditions on (1) to be fulfilled:

$$|f_{ii}(x(t))M_i| > |g_i(x(t),\dot{x}(t))| + $$

$$+\Big|\sum_{j=1,j\neq i}^{n} f_{ij}(x(t))u_j\Big|, \quad i = 1,\ldots,n \qquad (15)$$

$\forall u \in U$, $\forall (x,\dot{x}) \in X \times V$ where $X \ni \overset{o}{x}, x^f$ and $V \ni 0$ are parallelipipeds in R^n.

In the presence of the DC-conditions (15) one can easily verify the following statements:

LEMMA 1: The final times t_i^f in (12) are finite and the reached positions $x_i(t_i^f)$ thus can be defined by (13) as function F_i^{opt} of $\bar{t} = (t^{\bar{k}_1},\ldots,t^{\bar{k}_n})$.

LEMMA 2: F_i ("opt" is omitted) are continuous functions of $\bar{t}$ [8,9].

LEMMA 3: There exist bounds t_i^- and t_i^+, $t_i^+ > t_i^-$, $i = 1,\ldots,n$ such that the function F considered on the cube $P = \{\bar{t} \mid t_i^- \leq t^{\bar{k}_i} < t_i^+\}$ has the following property: for any pair of points on the boundary of $P: \bar{t} \in bdP$, $\bar{t}_{sym} \in bdP$ symmetrical about the centre of the cube P there exists i:

$$(F_i(\bar{t}) - x_i^f)(F_i(\bar{t}_{sym}) - x_i^f) < 0 \qquad (16)$$

Now, we are in the position to apply a known T.1 asserting existence of solution of a nonlinear vector equation [10]:

THEOREM 1: (Borsuk-Ulam): Let $G \in C(P;R^n)$

$$G(z) = -G(z_{sym}) \quad \forall z, z_{sym} \in bdP \qquad (17)$$

then $\exists z^* \in P$: $G(z^*) = 0$.

Relying on Lemmas 1÷3 and Theorem 1 we can state the following theorem:

THEOREM 2: With p_{opt} fixed, there exists a solution of the shooting equations (14).

Some details about the proof of this theorem are given in [11].

Therefore the TPBVP (1÷3) is solvable employing the control laws described in section 3 (7÷10) and one can varying p_{opt} synthesize a satisfactory number of feasible solutions. Thus in the presence of the DC-conditions (15) a guarantee in the realization of our direct search optimization procedure can be given.

5. CONCLUSION

A problem on the optimal control of mechanical systems with individually controllable degrees of freedom is taken into consideration. The performance index is a weighted minimum time-energy loss criterion. A direct search optimization procedure is suggested involving control laws matching the Pontryagin's maximum principle. In other words, the suboptimal solution obtained by means of that procedure is an appropriate spline approximation to the optimal solution itself. To ensure the success of the optimization procedure an inherent decentralized controllability condition is assumed.

An other advantage of the method proposed is that the direct search procedure can be implemented on the mechanical system itself for a final adjustment of the control parameters. This is required in most cases when the dynamic model is not complete and exact.

REFERENCES

[1] Tourassis, V., and Ch. Neuman. - The inertial characteristics of dynamic robot models. - Mech. and Mach. Theory, 1985, Vol. 20, No 1, pp. 41- 52.

[2] Asada, H. - The kinematic design and mass redistribution of manipulator arms for decoupled and invariant inertia. - Proc. of the VIth CISM-IFToMM Symposium on Theory and Practice of Robots an Manipulators. Ed. by A. Morecki et al., 19870 Hermes

[3] Kiriazov, P., and P. Marinov - On the decoupled drive system design of industrial robots. - Theor. and Appl. Mech., 1987. Year XVIII, No 4, Bulg. Acad. of Sci., pp. 25-29

[4] Marinov, P., and P. Kiriazov. - A direct method for optimal control sythesis of manipulator point-to-point motion. - 9th Wotld Congr. IFAC. Ed. by J. Gertler et al., 1985. Vol. 1, pp. 453-457

[5] Sontag, E., and H. Sussman. - Time-optimal control of manipulators. - IEEE Conf. on Robotics and Automation, 1986. Vol. 3, pp. 1692-1697

[6] Jacobson, D.H., et al. - Computation of optimal singular controls IEEE Trans. Auto-Contr., 1970. Vol. AC-15, pp. 67-73

[7] Chen, Y., and A. Desrochers. - Time-optimal of two-degree of freedom robot arms. - IEEE Conf. on Robotics and Automation, 1988, pp. 1210-1216

[8] Rozenwasser, E., and R. Usoupov. - Sensitivity of Control Systems. - Nauka, 1981 (in Russian)

[9] Wen, J., and A. Desrochers. - Control system design for a robotic and loader. - Robotics and Automation. Laboratory Report No 25, 1984. Rensselaer Polytechnic Inst., Troy, N.Y.

[10] Todd, M., and A. Wright. - A variable-dimension symplicial algorithm for antipodal fixed-point theorems. - Numerical Functional Analysis and Optimization, 1980 (2), pp. 155-186.

[11] Kiriazov, P. On the Controllability of Some Nonlinear Dynamic Systems. Godishnik Vishite Technich. Zavedeniya. Prilozna Mathematika, 1988, in Bulgarian.

Approximation, Optimization and Computing:
Theory and Applications, A.G. Law and C.L. Wang (eds.)
Elsevier Science Publishers B.V. (North-Holland)

AN INFORMATIONAL ENTROPY APPROACH TO OPTIMIZATION

X.S. Li (Li Xingsi)* and A.B. Templeman**

* Research Institute of Engineering Mechanics, Dalian University of Technology, Dalian 116024, CHINA
** Department of Civil Engineering, The Univeristy of Liverpool, Liverpool PO Box 147, L69 3BX, UK

This paper presents a new solution method for constrained optimization problems, termed the informational entropy approach, which combines the concept of Shannon entropy with the surrogate duality theory. Some information-theoretic interpretations for the proposed method are provided. The optimum sizing problem of trusses is taken as an example for which an explicit surrogate dual problem can be derived such that it can be effectively solved by the present method.

1 INTRODUCTION

The geometric and deterministic point of view has played a large part in the development of constrained optimization methods. These conventional methods are usually devised from considering objective functions as hypersurfaces and constraints as deterministic boundaries. Certain search strategies are then used so as to minimize some objectives while keeping the constraint boundaries not to be crossed. Therms such as gradients, steepest descent and barriers all have topological associations. Thus, these methods have used calculated information (function values, gradients, etc.) in a geometric way, searching a deterministic topological domain for an optimum point. Information theory appears to be incompatible with this as it is essentially concerned with certain probabilities.

In contrast with these approaches, a different point of view is adopted in the present work. One of its objectives is to set an optimization process in a non-deterministic context without topological analogies and to use the calculated information in a non-geometrical, information-theoretic way to locate the optimum point. The idea behind this approach is based upon a speculation that an optimization procedure could be thought as a process in which messages are, alternatively, received and transmitted, as done in a communication system. The methods in information theory might then be useful in developing optimization methods.

2 ENTROPY AND MAXIMUM ENTROPY PRINCIPLE

One of the fundamental building blocks of modern information theory is the paper by Shannon (1948) in which a new mathematical model of communication systems was proposed and investigated. The most important innovation of this model was that it considered the components of a communication system as probabilistic entities. In his paper, Shannon proposed a quantitative measure of the amount of uncertainty about possible outcomes of a probabilistic experiment.

Consider a probabilistic experiment having n discrete possible outcomes $a_1, \ldots, a_n$ with respective discrete probabilities $p_1, \ldots, p_n$ that satisfy the following axiomatic conditions

$$p_i \geq 0 \quad \text{and} \quad \sum_{i=1}^{n} p_i = 1$$

In such an experiment, there, of course, exists an amout of uncertainty about a particular outcome that will occur if we perform this experiment. It can be seen that this amount of uncertainty, contained a priori by the probabilistic experiment, essentially depends on the probabilities of all possible outcomes of the experiment. For instance, if we have a probabilistic experiment having only two possible outcomes a_1 and a_2 with the probabilities of p_1 and p_2. Suppose that we assign two different sets of the probabilities, (p_1=0.5, p_2=0.5) and (p_1=0.95, p_2=0.05). It is obvious that the first case contains more uncertainty than that of the second since in the latter case the result of this experiment is "almost surely" a_1 whereas in the former case we cannot make any prediction on a particular outcome which will occur. This shows that a uniform probability distribution has a larger amout of uncertainty associated with it than a non-uniform distribution. This measure of uncertainty was then defined by Shanon in terms of the probabilities as

$$H(p) = -k \sum_{i=1}^{n} p_i \mathrm{Ln} p_i \tag{1}$$

where p_i represent the probabilities associated with each possible outcome, k is merely a positive constant depending upon a suitable choice for the units of measure and H is usually referred to as the Shannon or informational entropy.

The Shannon's entropy measure was an important step forward in that it allowed the amount of uncertainty in a probabilistic experiment to be quantified provided that the probabilities of all outcomes were known. The next important advance was made by Jaynes (1957) who realized that in many cases these probabilities could not be known in advance. In this case, Jaynes extended the use of the Shannon's entropy measure to calculate the unknown discrete probabilities from observable data on the experiment, and hence extended the role of the Shannon entropy from a simple measure to a crucial role in an inference process; that is, given a probabilistic process and observed aggregated data from this process, what does the Shannon's measure of uncertainty allow us to logically infer about the probability distributions underlying the process?

Suppose there exists an observable process in which a discrete random variable can take on any one of n values $x_1, \ldots, x_n$. Suppose also that as a result of observations on the process, it can be deduced that the outcomes satisfy certain aggregated functional relationships $g_1, \ldots, g_m$, such as mean value and variance, etc., where $m \ll n$. What can be deduced about the probabilities $p_1, \ldots, p_n$ of the random variable attaining the values $x_1, \ldots, x_n$? Clearly, there will be an infinite number of probability distributions that can satisfy the m observed functions $g_1, \ldots, g_m$. Which one is the best and should therefore be chosen? Obviously, we need some additional selection criteria. Jaynes provided us with such a criterion. In his paper, he wrote: in making inference on the basis of partial information we must use that probability distribution which has maximum entropy subject to whatever is known. This is the only unbiased assignment we can make; to use any other would amout to arbitrary assumptions of information which by hypothesis do not have. Jaynes thus claimed that among the infinite distributions, the one with the highest value of entropy should be chosen as this introduces the minimum artificial bias into the choice.

Mathematically, to maximize the entropy function H in (1) subject to the given information leads to an optimization problem:

$$\text{(E)}\ \max\quad H(p) = -k\sum_{i=1}^{n} p_i \operatorname{Ln} p_i \qquad (1)$$

$$\text{s.t.}\quad \sum_{i=1}^{n} p_i\, g_j(x_i) = E[g_j] \qquad j=1,\ldots,m \qquad (2)$$

$$\sum_{i=1}^{n} p_i = 1 \qquad (3)$$

$$p_i \geq 0 \qquad i=1,\ldots,n \qquad (4)$$

where E[.] denotes the expectation operator. The Jaynes' selection criterion is usually referred to as the maximum entropy principle. This criterion has a subjective character by nature, but it can be rather considered as the most "objectively" subjective one.

3 SURROGATE DUALITY APPROACH TO CONSTRAINED OPTIMIZATION

The constrained optimization problem to be considered here is defined as

$$\text{(P)}\quad \min\ f(x) \qquad (5)$$
$$\text{s.t.}\quad g_j(x) \leq 0 \qquad j=1,2,\ldots,m \qquad (6)$$

A surrogate constraint for Problem (P) is usually defined as a positive linear combination of the constraints (6), as suggested by Glover (1968), in the form:

$$g_s(x) = \sum_{j=1}^{m} \lambda_j\, g_j(x) \leq 0 \qquad (7)$$

where the weighting coefficients λ_j, j=1,..., m, are refered to as the surrogate multipliers that are usually normalized by requiring

$$\sum_{j=1}^{m} \lambda_j = 1 \qquad (8)$$

The solution x^* to (P) will then be sought by solving a sequence of surrogate problems (S), which consist of the minimization of f(x) subject to the single surrogate constraint (7). This approach assumes therefore that (P) and (S) are equivalent at the optimum point; specifically that a set of multipliers λ^* exists and can be found such that $x(\lambda^*)$ which solves (S) also solves (P). Previous researchers (See Glover (1968) and (1975), Greenberg and Pierskalla (1970)) have studied this assumption, establishing conditions on its validity.

It should be noted that a point which satisfies the original constraints (6) must satisfy the surrogate constraint (7) for any set of positive surrogate multipliers. Thus a surrogate problem (S) represents a relaxation of (P) so that the optimum objective value of a surrogate problem (S) for any fixed λ, denoted by $S(\lambda)$, cannot exceed the optimum objective $f(x^*)$ of (P). As the surrogate constraint (7) becomes an increasingly more "faithful" representation of the original constraints (6), $S(\lambda)$ will approach $f(x^*)$ more closely. Choices of the surrogate multipliers λ_j that improve the proximity of (S) to (P), i.e., that provide the greatest value of $S(\lambda)$ yield the strongest surrogate constraint, and motivate the definition of the surrogate dual formulation:

(SD) max $S(\lambda)$ (9)

$$\text{s.t.} \quad \sum_{j=1}^{m} \lambda_j = 1 \qquad (8)$$

$$\lambda_i \geq 0 \qquad j=1,\ldots,m \qquad (10)$$

where

$$S(\lambda)=\{\min f(x): \sum_{j=1}^{m} \lambda_j g_j(x) \leq 0, \ \sum_{j=1}^{m} \lambda_j = 1, \quad \lambda \geq 0\} \qquad (11)$$

which is a quasiconcave function as proved by Greenberg and Pierskalla (1970).

Using the surrogate dual formulation, we have transformed (P) into an equivalent problem (SD) for which the task is to seek a set of surrogate multipliers, λ^*, that maximizes $S(\lambda)$.

4 AN ENTROPY-BASED SOLUTION FOR PROBLEM (SD)

Concerning the solution of (SD), we propose that instead of directly solving (SD), we solve a modified problem in the form:

(SD_P) max $S_P(\lambda) = S(\lambda) + H(\lambda)/p$ (12)

$$\text{s.t.} \quad \sum_{j=1}^{m} \lambda_j = 1 \qquad (8)$$

$$\lambda_i \geq 0 \qquad j=1,\ldots,m \qquad (10)$$

where p is a positive controlling parameter and $H(\lambda)$ represents the entropy of surrogate multipliers. It might have been noted that to make the two problems (SD) and (SD_P) equivalent at the solution point of (SD), we need the parameter p to tend to infinity.

The idea of applying an axtra term $H(\lambda)/p$ to $S(\lambda)$ is to supply the solution of (SD) with an additional criterion, i.e., the maximum entropy principle. The definition of entropy is here based upon a probabilistic interpretation for the surrogate multipliers. As we can see from the construction of a surrogate constraint (7), the multipliers λ_j really play a part of probabilities if we regard the surrogate constraint (7) as an average or mean one of the original constraints (6). The maximization of $S_P(\lambda)$ thus implies a simultaneous maximization of both $S(\lambda)$ and $H(\lambda)$. Mathematically, the parameter p assumes the responsibility for balancing the two functions $S(\lambda)$ and $H(\lambda)$. An information-theoretic interpretation is that as iterations proceed, the uncertainty contained in the solution procedure should diminish. Therefore, p will be set as an increasing sequence of positive numbers. It is apparent that with p tending to infinity, (SD_P) will approach (SD).

It can be proved (See Li (1987)) that as p increases, $S(\lambda)$ will increase and $H(\lambda)$ decrease at the solution points of a series of problems (SD_P), monotonically. This ensures, on one hand, that the solution of (SD_P) will eventually approach that of (SD) when p tends to infinity. On the other hand, the decrease of $H(\lambda)$ simply implies the fact that the uncertainty about λ^* has diminished. This may become an evidence that the idea of information theory is indeed applicable to an optimization process.

The solution of (SD_P) can be found using stationarity conditions with respect to λ_j (j=1, ..., m) of its Lagrangean function:

$$L_P(\lambda,\alpha)=S(\lambda)-\frac{1}{p}\sum_{j=1}^{m}\lambda_j \mathrm{Ln}\,\lambda_j + \alpha\left(\sum_{j=1}^{m}\lambda_j - 1\right) \qquad (13)$$

from which we have

$$\frac{\partial L_P}{\partial \lambda_j} = S,j - \frac{1}{p}(1+\mathrm{Ln}\,\lambda_j) + \alpha = 0 \qquad j=1,\ldots,m \qquad (14)$$

where S,j represent partial derivatives of $S(\lambda)$ with respect to λ_j. From (14), we obtain

$$\lambda_j = \exp(pS,j + p\alpha - 1) \qquad j=1,\ldots,m \qquad (15)$$

which are then substituted into the normality condition (8), giving

$$\lambda_j = \exp(pS,j)/\sum_{j=1}^{m}\exp(pS,j) \qquad j=1,\ldots,m \qquad (16)$$

This formula provides a basis for the development of new numerical algorithms. It should be noted, however, that the equation (16) is merely a recursive formula because S,j on the right are also functions of λ_j. In mathematical terms, (16) is a one-point iteration formula for solving the system of nonlinear stationarity equations (See Traub (1964)). This means that the solution of (SD_P) has to be found by iteratively using (16).

5 AN EXPLICIT SURROGATE DUAL FOR THE OPTIMUM SIZING OF TRUSSES

In general, the surrogate dual function $S(\lambda)$ cannot be analytically obtained. For the optimum sizing problem of trusses, however, an explicit dual can be found by solving the corresponding surrogate problem. Therefore, it can be effectively solved by the entropy-based method proposed in the last section.

The optimum sizing of trusses is usually formulated, by using reciprocal variables $x_i=1/a_i$, as

$$(\text{Po}) \quad \min \quad W = \sum_{i=1}^{n} \rho_i l_i / x_i \qquad (17)$$

$$\text{s.t.} \quad g_j(x) = \sum_{i=1}^{n} c_{ij} x_i - 1 \leq 0 \qquad j=1,\ldots,m \tag{18}$$

$$g_{m+i}(x) = \bar{a}_i x_i - 1 \leq 0 \qquad i=1,\ldots,n \tag{19}$$

where ρ_i, l_i, $\bar{a}_i$ and x_i are the mass density, length, minimum size and reciprocal design variable of the ith bar, respectively; W, m and n are the structural weight and the numbers of constraints and design variables.

The corresponding surrogate problem for the above structural optimization problem takes a very simple form as

$$(\text{So}) \quad \min \quad W = \sum_{i=1}^{n} \rho_i l_i / x_i \tag{17}$$

$$\text{s.t.} \quad g_s(x) = \sum_{i=1}^{n} \zeta_i x_i - 1 \leq 0 \tag{20}$$

$$\text{where} \qquad \zeta_i = \sum_{j=1}^{m} c_{ij}\lambda_j + a_i \lambda_{m+i} \qquad i=1, \;\; ,n \tag{21}$$

Problem (So) is a convex programming and can be analytically solved by using Kuhn-Tucker conditions, which yield

$$x_i = \sqrt{\rho_i l_i / \zeta_i} \Big/ \sum_{i=1}^{n} \sqrt{\rho_i l_i \zeta_i} \qquad i=1,\ldots, n \tag{22}$$

Substituting x_i from (22) into (17) gives an explicit surrogate dual shown in (23).

The surrogate dual problem for (Po) is thus defined as

$$(\text{SDo}) \quad \max \quad S(\lambda) = \Big(\sum_{i=1}^{n} \sqrt{\rho_i l_i \zeta_i}\Big)^2 \tag{23}$$

$$\text{s.t.} \qquad \sum_{j=1}^{m+n} \lambda_j = 1 \tag{24}$$

$$\lambda_j \geq 0 \qquad j=1,\ldots,m+n \tag{25}$$

which takes a form that is exactly the same as that proposed by Templeman (1976) except for some symbolic changes. Nevertheless, the dual formulation presented herein has a rigorous mathematical basis in that the dual variables λ_j have a clear meaning of surrogate multipliers.

6 DISCUSSIONS AND CONCLUSIONS

The introduction of the Shannon's informational entropy into a constrained optimization method was motivated by the authors' appreciation that concepts and methods of the information theory might be useful in the systems other than a communication system as long as the "messages" are concerned in some way. On interpreting the surrogate multipliers λ_j as the probabilities of each constraint being active at the solution point, the maximum entropy principle was applied to establish our entropy-based method. The numerical algorithms may rest entirely on the recursive formula (16) which is extremely simple so that the present method is very easy to be implemented on a computer. Computational results show (See Li and Templeman (1988)) that this method has high accuracy and stability.

ACKNOWLEDGEMENT

This work is financially supported by National Science Foundation of China.

REFERENCES

Glover, F., 1968: Surrogate Constraints, Opns. Res. 16, 741-749.

Glover, F., 1975: Surrogate constraint duality in mathematical programming,. Opns. Res. 23, 434-451.

Greenberg, H.J. and Pierskalla, W.P., 1970: Surrogate mathematical programming, Opns. Res. 18, 924-939.

Jaynes, E.T., 1957: Information theory and statistical mechanics. Phys. Rev. 106, 620-630.

Li, X.S., 1987: Entropy and optimization. Ph.D thesis, University of liverpool, UK.

Li, X.S. and Templeman, A.B., 1988: Entropy-based optimum sizing of trusses. Civ. Engng Syst. 5, 121-128.

Shannon, C.E., 1948: The mathematical theory of communication. Bell Sys. Tech.J. 27, 279-428.

Templeman, A.B. and Li, X.S., 1987: A maximum entropy approach to constained nonlinear programming. Engng. Opt. 12, 191-205.

Templeman, A. B., 1976: A dual approach to optimum truss design. J. Struc. Mech. 4, 235-255.

Approximation, Optimization and Computing:
Theory and Applications, A.G. Law and C.L. Wang (eds.)
Elsevier Science Publishers B.V. (North-Holland)

First and Second Order Sensitivity Analysis of Shape Optimization Based on Variational method

H.Q.Liu R.W.Xia

Beijing University of Aeronautics and Astronautics Beijing , P.R.C.

The first and second order sensitivity analysis formulations in shape optimization based on the variational theory of a functional defined on a variable domain have been derived in this paper. In consideration of the character of practical engineering problems, sensitivity analysis is discussed for three different shape deformations respectively.

NOMENCLATURE

U	= displacement tensor
W	= adjoint displacement tensor
T	= boundary traction tensor
S	= boundary surface of domain Ω
S_u	= prescribed displacement part of S
S_t	= prescribed traction part of S
T°	= known boundary traction
U°	= known boundary displacement
X_n	= $X \cdot N$
N	= normal vector
" • "	= dot product
": "	= quadritic dot product
" × "	= vector product
$[ABC]$	= mixed product, A,B,C are vectors
∇	= $\partial / \partial x$ tensor operotor of rank one
∇_1	= $\partial / \partial z$ tensor operator of rank one
$\square$	= d / dx tenser operator of rank one
∇_2	= $\partial / \partial z$ tensor operator of second order
δ_{ij}	= kronecker delta
$\in$	= Permutation tensor
I	= unit matrix
δ^-	= pure behavior variation
δ_Ω	= pure shape variation
λ,μ	= elastic constant

INTRODUCTION

For shape optimization of continuum system, there are two kinds of approaches to calculate sensitivity derivatives. The first is based on differentiation of the discretized system[1][2] and the other on variation of the continuum system [3][4]. From the analysis standpoint, the former is approximate and the latter is accurate. Because the discretized model is strongly influenced by the shape changes in optimization process, it is difficult to assure the accuracy of the first approach. Therefore many researchers have paid more attentions to the second approach and developed material derivative method[5][6] and variational method[7] etc. In this paper, the sensitivity derivatives in shape optimization are formulated by use of the variational method for general, translation, rigid rotation and normal deformation cases. And it is limited to the 3–D linearly elastic system for which the concept of an adjoint system[8] can be easily applied in order to derive the pure shape variation. It is worth noting that the computational amount of the second order sensitivity analysis obtained in this paper is approximately the same order of magnitude as the first order sensitivity. This merit is very important in structrual optimization for improving both computational efficiency and convergence. Besides optimal design, the present method can be applied to other fields [8][9][10].

VARIATION AND QUADRATIC VARIATION REVIEW

For a functional defined on a variable domain

$$\psi = \int_\Delta F(X,Z(X),\nabla Z(X))d\Omega$$

its total variation is

$$\delta\psi = \delta^{(1)}\overline{\psi} + \int_S F\delta X_n ds + \delta^{(2)}\overline{\psi} + \int_S(\delta\overline{Z} \cdot \nabla_1 F + \delta\nabla\overline{Z}:\nabla_2 F)\delta x_n$$

$$+\frac{1}{2}\{\delta X\cdot\Box F\delta X+F(\nabla\cdot\delta X\delta X-\delta X\cdot\nabla\delta X)\}\cdot Nds \quad (1.1)$$

in which

$$\delta^{(1)}\bar{\psi}=\int_\Omega\delta\bar{Z}\cdot\nabla_1F+\delta\nabla\bar{Z}:\nabla_2Fds\Omega$$

$$\delta^{(2)}\bar{\psi}=\int_\Omega\frac{1}{2}(\delta\bar{Z}\cdot\nabla_1\nabla_1F\cdot\delta\bar{Z}+2\delta\bar{Z}\cdot\nabla_1\nabla_2F:\delta\nabla\bar{Z}+\delta\nabla\bar{Z}:\nabla_2\nabla_2F:\delta\nabla\bar{Z})d\Omega$$

Three cases of shape deformations are considered as follows.

1. For the thanslation case, that is

$\delta X=\delta A=const$

then, the variation is

$$\delta^{(1)}\psi=\delta^{(1)}\bar{\psi}+\int_sF\delta A_nds$$

$$\delta^{(2)}\psi=\delta^{(2)}\bar{\psi}+\int_s(\delta\bar{Z}\cdot\nabla_1F+\delta\nabla\bar{Z}:\nabla_2F)\delta A_n+\frac{1}{2}\delta A\cdot\Box F\delta A_nds \quad (1.2)$$

2. For the rigid rotation case, that is

$\delta X=X\times\delta\omega$

its variation is

$$\delta^{(1)}\psi=\delta^{(1)}\bar{\psi}+\int_SF[XN\delta\omega]ds$$

$$\delta^{(2)}\psi=\delta^{(2)}\bar{\psi}+\int_S(\delta\bar{Z}\cdot\nabla_1F+\delta\nabla\bar{Z}:\nabla_2F)[XN\delta\omega]+\frac{1}{2}\delta\omega\cdot\{(\Box F\cdot\epsilon\cdot X)(N\cdot\epsilon\cdot X)-F(NX-(N\cdot X)I)\}\cdot\delta\omega ds \quad (1.3)$$

Where $[XN\delta\omega]$ is mixed product, $\delta\omega$ is infinitesimal rotation vector;

3. For the normal deformation case, that is

$\delta X=\delta X_nN$

its variation is

$$\delta^{(1)}\psi=\delta^{(1)}\bar{\psi}+\int_sF\delta X_nds$$

$$\delta^{(2)}\psi=\delta^{(2)}\bar{\psi}+\int_s(\delta\bar{Z}\cdot\nabla_1F+\delta\nabla\bar{Z}:\nabla_2F)\delta X_n+\frac{1}{2}(N\cdot\Box F+FH)\delta X_n\delta X_nds \quad (1.4)$$

The formulations above are only limited to the functional defined on a variable domain. And for the functional defined on a variable boundary, the theory results will be published in other paper.

SENSITIVITY ANALYSIS BY MEANS OF ADJOINT SYSTEM EQUATION

An adjoint system method is generally applied to sizing design variable and shape optimization. Especially in shape optimization based on variational method, the pure shape variation can be obtained with the help of the adjoint system equation. Here only for linearly elastic problems,the first and second order sensitivity are discussed.

For the shape optimization, let the general form of the objective functional be

$$J=\int_\Omega F(X,U(X),\nabla U(X))d\Omega$$

and U should satisfy the state equation of linear elastic system

$$-\nabla\cdot(D:\nabla U)=f \quad \text{on }\Omega$$
$$(D:\nabla U)\cdot N=T^\circ \quad \text{on }S_t$$
$$U=U^\circ \quad \text{on }S_u$$

Where D is the tensor of elastic constants of an isotropic elastic body in Cartisian coordinates, the form is

$$D_{ijkl}=\lambda\delta_{ij}\delta_{kl}+\mu(\delta_{ik}\delta_{jl}+\delta_{il}\delta_{jk})$$

And for an arbitrary $W\in H^1$ and $W|_{Su}=0$, there is a bilinear form

$$\int_\Omega\nabla W:D:\nabla Ud\Omega=\int_\Omega f\cdot Wd\Omega+\int_{St}T^\circ\cdot Wds$$

In order to get the shape variation of an objective functional, let its adjoint system equation take the form

$$\int_\Omega\nabla W:D:\nabla\varphi d\Omega=\int_\Omega\varphi\cdot\nabla_1F+\nabla\varphi:\nabla_2Fd\Omega \qquad \forall\varphi\in H_0^1$$

then for an arbitrary $\delta\bar{U}$, it satisfies above formulation

$$\int_\Omega\nabla W:D:\delta\nabla\bar{U}d\Omega=\int_\Omega\delta\bar{U}\cdot\nabla_1F+\delta\nabla\bar{U}:\nabla_2Fd\Omega$$

In view of the first variation of the bilinear form of

state equation

$$\int_\Omega \nabla W{:}D{:}\delta\nabla \bar{U} d\Omega =$$
$$-\int_s (\nabla W{:}D{:}\nabla U)\delta X_n ds$$
$$+\int_S (f\cdot W)\delta X_n ds$$
$$+\int_{St} \Box\cdot(\delta X T^\circ \cdot W)$$
$$-(T^\circ \cdot W)(N\cdot\nabla\delta X\cdot N)ds$$

the following formulation is hold ,that is

$$\int_s \delta\bar{U}\cdot\nabla_1 F + \delta\nabla\bar{U}{:}\nabla_2 F d\Omega$$
$$=\int_S (f\cdot W - \nabla W{:}D{:}\nabla U)\delta X_n ds$$
$$+\int_{St} \Box\cdot(\delta X T^\circ \cdot W)$$
$$-(T^\circ \cdot W)N\cdot\nabla\delta X\cdot N)ds$$

And finally, the pure shape variation of its objective functional can be derived as follows :

$$\delta_\Omega^{(1)} J = \int_S (F + f\cdot W - \nabla W{:}D{:}\nabla U)\delta X_n ds$$
$$+\int_{St} \Box\cdot(\delta X T^\circ \cdot W)$$
$$-\ (T^\circ \cdot W)(N\cdot\nabla\delta X\cdot N)ds \qquad (1.5)$$

In practical engineering problems, sometimes only some special form occurs,and then three of the special deformations are discussed as follows respectively

1. For the translation case, that is

$$\delta X = \delta A = const$$

In view of Eq.(1.2) and Eq.(1.5), it follows that

$$\delta_\Omega^{(1)} J = \int_S (F + f\cdot W - \nabla W{:}D{:}\nabla U)\delta A_n ds$$
$$+\int_{St} \delta A\cdot\Box(T^\circ \cdot W)ds$$

2. For the rigid rotation case, that is

$$\delta X = X\times\delta\omega$$

and

$$\nabla\delta X = +\in\cdot\delta\omega$$

In view of Eq.(1.3) and Eq.(1.5), the first variation in rotation case

$$\delta_\Omega^{(1)} J = \int_S (F + f\cdot W$$
$$-\nabla W{:}D{:}\nabla U)[NX\delta\omega]ds$$
$$+\int_{St} \nabla(T^\circ \cdot W)\cdot(X\times\delta\omega)ds$$

3. For the normal deformation case, that is

$$\delta X = \delta X_n N$$

By use of Eq.(1.4), Eq.(1.5) and relations

$$\nabla N\cdot N = 0$$
$$-H = divN$$

where H is the Guass average curvature,then the first shape variation along normal direction

$$\delta_\Omega^{(1)} J = \int_S (F + f\cdot W - \nabla W{:}D{:}\nabla U)\delta X_n ds$$
$$+\int_{St} \{N\cdot\Box(T^\circ \cdot W)$$
$$-(T^\circ \cdot W)H\}\delta X_n ds$$

When only force boundary changes, the Scale Gradient is

$$G_n = F + f\cdot W - \nabla W{:}D{:}\nabla U$$
$$+N\cdot\Box(T^\circ \cdot W) - (T^\circ \cdot W)H$$

It is well known that the normal deformation δX_n play an important role in shape design perturbation and can be expressed in shape parameter. Therefore only the second order sensitivity analysis in the normal deformation is discussed. And then for the bilinear form of the state equation, its quadritic variation is

$$\int_S (\nabla W{:}D{:}\delta\nabla\bar{U})\delta X_n$$
$$+\frac{1}{2}\Box\cdot(N\nabla W{:}D{:}\nabla U)\delta X_n\delta X_n ds$$
$$=\int_S \frac{1}{2}\{\Box\cdot(Nf\cdot W)\}\delta X_n\delta X_n ds$$
$$+\int_{St}\frac{1}{2}\{N\cdot\Box\Box(T^\circ \cdot W)\cdot N$$
$$-2N\cdot\Box(T^\circ \cdot W)H$$
$$+(T^\circ \cdot W)(N\cdot\nabla N\cdot(\nabla N)^T\cdot N$$
$$+H^2 - \nabla N{:}\nabla N)\}\delta X_n\delta X_n ds$$

In view of above formulation and the differential form of its adjoint system equation which is the same as the adjoint system equation in the first sensitivity analysis, the couple term can be expressed approaximately as

$$\int_S (\delta\bar{U}\cdot\nabla_1 F + \delta\nabla\bar{U}{:}\nabla_2 F)\delta X_n ds$$
$$=\frac{1}{2}\int_s\{\Box\cdot(Nf\cdot W)$$
$$-\quad (N\nabla W{:}D{:}\nabla U)\}\delta X_n\delta X_n ds$$
$$+\frac{1}{2}\int_{St}\{N\cdot\Box\Box(T^\circ \cdot W)\cdot N$$
$$+(T^\circ \cdot W)(N\cdot\nabla N\cdot(\nabla N)^T\cdot N$$
$$+H^2 - \nabla N{:}\nabla N)$$
$$-2N\cdot\Box(T^\circ \cdot W)H\}\delta X_n\delta X_n ds$$

and then the second shape variation is derived from Eq(1.4)

$$\delta_{\Omega}^{(2)} J = \frac{1}{2}\int_S \{N \cdot \Box F - FH + \Box \cdot (Nf \cdot W) - \Box \cdot (N\nabla W{:}D{:}\nabla U)\}\delta X_n \delta X_n ds + \frac{1}{2}\int_{St}\{N \cdot \Box\Box(T^{\circ} \cdot W)\cdot N + (T^{\circ} \cdot W)(N \cdot \nabla N \cdot (\nabla N)^T \cdot N + H^2 - \nabla N{:}\nabla N) - 2N \cdot \Box(T^{\circ} \cdot W)H\}\delta X_n \delta X_n ds$$

When only the traction boundary changes, the Scale Hessian Matrix is

$$H_{nn} = N \cdot \Box F - FH + \Box \cdot (Nf \cdot W) - \Box \cdot (N\nabla W{:}D{:}\nabla U) - 2N \cdot \Box(T^{\circ} \cdot W)H + N \cdot \Box\Box(T^{\circ} \cdot W)\cdot N + (T^{\circ} \cdot W)(N \cdot \nabla N \cdot (\nabla N)^T \cdot N + H^2 - \nabla N{:}\nabla N)$$

This concept is developed in this paper corresponding to the Scale Gradient[1]. From above formulation, it is found that after the only one analysis of structure and adjoint structure is done, the first and second order sensitivity of shape optimization can be obtained. Therefore it overcomes the drawback of excessive computing amount in the second order sensitivity analysis. It is a great advantage to shape optimization.

CONCLUSION

By use of the variational method on a variable domain, the second order sensitivity of shape optimization has firstly been obtained in this paper. And the result has made clear that its computational amount is approaximately the same order magnitude as the first order sensitivity. The first order sensitivity is similar to that obtained by other methods. However for the general variational theory on a variable domain, the problems of objective functional defined on a variable boundary and the computing technique of results obtained in this paper can be seen in other papers that will be published.

REFERENCES

1. Botkin, M.E. Shape Optimization of Plate and Shell Structures. AIAA81–0553R
2. Wang.ShyYu, Sun. Yanbing and Gallagher, R.H. "Sensitivity Analysis in Shape Optimization of Continuum Structures" Computer Structures Vol.20, No.5, 1985.
3. Rousselet, B. "Shape Design Sensitivity Methods for Structural Mechanics" Optimization of Distributed Parameter Structure (1981, Ed. Haug, E.J. and Cea, J.)
4. Rousselet, B. and Haug, E.J. "Design Sensitivity Analysis of Shape Variation" Optimization of Distributed parameter Structure (1981, Ed. Haug, E.J. and Cea, J.)
5. Zolesio, J.P. " The Material Derivative (or speed) Method for Shape Optimization" Optimization of Distributed parameter structure (1981, Ed. Haug, E.J. and Cea, J.)
6. Braibant, V. "shape sensitivity by Finite Element" J. Struct. Mech. Vol. 14, 1986
7. Zolesio, J.P. "Domain Variational Formulation for Free Boundary Problem" "Optimization of Distributed Parameter Structure (1981, Ed. Haug, H.J. and Cea, J.)
8. Dems, K. and Mroz, Z. "Variational Approach by Means of Adjoint System to Structural Optimization and Sensitivity Analysis — II Structural Shape Variation" Int. J. Solid Structures" Vol.20 No.6, 1984
9. Cea, J. "Numerical Methods of Shape Optimal Design"Optimization of Distributed Parameter Structure (19181)
10. Dems, K. and Mroz, Z. "On a Class of Conservation Rules Associated with Sensitivity Analysis in Linear Elasticity" Int. J. Solids Structure Vol.22, No. 7, 1986

Approximation, Optimization and Computing:
Theory and Applications, A.G. Law and C.L. Wang (eds.)
Elsevier Science Publishers B.V. (North-Holland)

OPTIMAL DESIGN OF GETTING AND DELIVERING SYSTEM OF CRUDE OIL

Liu Yang* Cheng Geng Dong** Qian Ling Xi**

* Qinhuangdao Branch of Daqing Petroleum Institute, Qinhuangdao, P.R.C.
** Research Institute of Engineering Mechanics, DUT, Dalian, P.R.C.

ABSTRACT--This paper treats the optimal design problem of getting and delivering system of crude oil. The system is very important for crude oil production in oilfield and also it is a difficult design problem. Mathematical definitions of the system network are described in this paper and optimization problems which invole topology optimization and fuzzy optimization are formulated at different levels. Heuristic methods are presented for solving these problems and system optimization is achieved by dynamic programming. A real example illustrates the significant saving in practical costs of the system.

KEYWORDS: Getting and Delivering of Crude oil, Star Network, Optimum Design

1. INTRODUCTION

Getting and delivering of crude oil (GDCO) is a important production course in oil-field after reservoir exploration and oil recovery, aiming to the gathering and transportation of crude oil. The establishing of GDCO system involves a large capital expenditure and the costs of pipeline, intermediate stations and equipments are so great that it is necessary to optimize the GDCO system by determining a correct topology form of GDCO network and operating parameters, to obtain the maximum economic results at minimum costs.

GDCO network gathers crude oil and gas from hundreds of thousand of individual wells and transports it to a central point in the field for treating and storage. Intermediate stations are scattered in the network which make the transportation available by heating and pumping. The topology of the network is determined by the number of network levels, the setting number of stations, the linking relation of pipes and the position of stations. Due to limited space, only multilevel star style GDCO network is taken into consideration in this paper. In section 2, a general definitions of the network are given based on graph theory [1].

The process of GDCO is shown in Fig.1, which indicates a three levels GDCO network. Pipes bring crude oil from wells (nodes 1 to 51) to metering stations (nodes 52 to 58), then to dehytration stations (nodes 59 to 60), and finally to a combined station (node 61).

In section 3, the optimal design problem are formulated. The objective function is to minimize total costs of GDCO system. Design variables are temperature and quantity of hot water which makes crude oil to move, diameter of pipes and some discrete variables related to topology of the network.

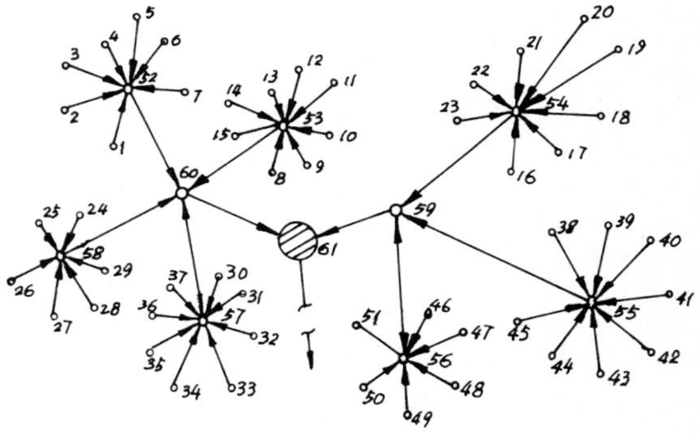

Fig.1

It is very hard to solve the design problem by exact solution method and its subproblem is shown to be a NP-hard problem for which developing good heuristic solution method is significant. Heuristic methods are presented for solving subproblems in section 4 and thereafter, optimization at system level is achieved by dynamic programming in section 5. A real design problem is solved which indicates the significant saving in capital expenditure.

2. DEFINITIONS OF GDCO NETWORK

Let G(V,E) to be a directed graph defined by a set of nodes V and a set of directed edges E. G(V,E) is a GDCO network if following conditions are satisfied:

(1) $G = \bigcup_{i=1}^{N} B_i (V_i, E_i)$

(2) $N >= 1$ Where N is the number of levels

(3) $B_i (V_i, E_i)$ is a bipartite graph with nodes set V_i and edges set E_i.

(4) $V_i = S_{i-1} \cup S_i$ and $S_{i-1} \cap S_i = \{\emptyset\}$ $\forall i$

(5) if $e_{kl} \in E_i$, then $k \in S_{i-1}$ and $l \in S_i$ $\forall i$

(6) $|S_{i-1}| > |S_i|$ $\forall i$

(7) $E_i \cap E_j = \{\emptyset\}$ $[i \neq j, j \in (1,2,\ldots,N)]$

(8) $$\bigcup_{i=1}^{N} E_i = E$$

(9) $$\bigcup_{i=1}^{N} V_i = V$$

(10) $|S_N| = 1$

(11) Indeg (i) = 0 $\forall_i \in S_0$

(12) Indeg (i) 1 $\forall_i \in S_1 \cup S_2 \ldots \cup S_N$

(13) Outdeg (i) = 1 $\forall_i \in S_0 \cup S_1 \ldots \cup S_{N-1}$

(14) Outdeg (i) = 0 $\forall_i \in S_N$

As a example shown in Fig.1, we find that N=3, $S_0 = (1,2,\ldots,51)$, $S_1 = (52,53,\ldots,58)$, $S_2 = (59, 60)$, $S_3 = (61)$. All equations are verified.

3. DESIGN PROBLEM OF GDCO NETWORK SYSTEM

The following decisions need to be made for GDCO network design (1) determining the topology of the network; (2) determining the geometric position of the network and (3) determining the operating parameters so as to minimize the total capital outlay and operating costs of the GDCO network system. This can be formulated as:

OP: Find X, P^D, δ, G, M

$$\text{Min} \quad F(X, P^D, \delta, G, M) = R(X, P^D, \delta, G, M) + \sum_{i=1}^{N-1} \sum_{j \in S_i} f_{ij} + \sum_{i=1}^{N} \sum_{j \in S_i} \sum_{k \in S_{i-1}} \xi_{ijk} W_{ijk} \delta_{ijk}$$

$$\text{s.t.} \quad C_i(X, P^D, \delta, G, M) \leq 0 \quad i=1,2,\ldots,H \quad (1)$$

$$\left.\begin{array}{ll} \sum_{j \in S_i} \delta_{ijk} = 1, & i=1,2,\ldots,N \quad \forall k \in S_{i-1} \\ |S_i| \geq m_i^L & i=1,2,\ldots,N \\ \sum_{j \in S_{i-1}} r_{ijk} \, \delta_{ijk} \leq H_j & i=1,2,\ldots,N \quad \forall j \in S_i \end{array}\right\} \quad (2)$$

$$\delta_{ijk} = \begin{cases} 1 & \text{if } S_{i-1}^{(k)} \text{ is connected to } S_i^{(j)} \\ 0 & \text{otherwise} \end{cases}$$

Where $\delta = \{\delta_{ijk} | \ i=1,2,\ldots,N, j \in S_i \ k \in S_{i-1}\}$

$M = \{m_i | \ i=1, 2,\ldots, N-1\}$

m_i^L stands for the lower bound of m

G for geometric design vector

X for continuous design vectors such as hot water temperature and its quantity.

P^D for diameter design vector

f_{ij} cost of establishing station $S_i^{(j)}$

H_j capacity of station $S_i^{(j)}$

r_{ijk} quantity through pipe

ξ_{ijk} length of pipe

W_{ijk} capital cost of a pipe with a unit length

The objective function is to minimize the sum of pipeline cost, operating cost and establishing cost of stations. Constraints (1) restric the operating parameters so as to make GDCO available and constraints (2) limit the linking route of pipes.

This problem is extremely large and complex. In a simple case (N=2), the problem reduces to the GAP which is a NP-hard problem and needs to be solved heuristically.

4. DESIGN PROBLEMS AT DIFFERENT LEVELS

The complex optimization problem of GDCO system can be decomposed into three subproblems:

4.1 Topology Optimization of GDCO Network

We assume that vector G and set M have been given and the topology optimization problem can be formulated as the following model.

P1: Find δ

$$\text{Min} \quad F = \sum_{i=1}^{N} \sum_{j \in S_i} \sum_{k \in S_{i-1}} V_{ijk} W_{ijk} \delta_{ijk}$$

$$\text{s.t.}(1) \quad \sum_{j \in S_i} \delta_{ijk} = 1; \quad i=1,2,\ldots,N; \ \forall k \in S_{i-1}$$

$$(2) \quad |S_i| \geq m_i^l; \quad i=1,2,\ldots,N-1$$

$$(3) \quad \delta_{ijk} = \begin{cases} 1 & \text{if } S_{i-1}^{(k)} \text{ is connected to } S_i^{(j)} \\ 0 & \text{otherwise} \end{cases}$$

By solving P1, a network with minimum pipeline cost based on given vector G and set M is generated. But it is still impractical to solve P1 by using integer programming technique because of the great computational efforts.

A heuristic method is presented here to solve P1 which takes the following steps : (1) giving a proper non-negative integer I and let $SS_i^{(j)}$ stands for a subset of S_i in which nodes are connected to $S_{i+1}^{(j)}$; (2) by ensuring the norm of subset equals to I, nodes in S_i are assigned to nodes in S_{i+1} greedily; (3) if $SS_i^{(j)} \cap SS_i^{(k)} = \{\phi\} (i \neq k)$ then goto step 4, otherwise I=I-1, goto step 2; (4) if I gets its maximum value, then goto step 5, otherwise I=I+1, goto step 2, (5) the nodes which are not in subset are assigned by solving a small size binary programming.

Numerical examples indicate that this heuristic method can efficiently solve relative large problem.

4.2 Geometric Optimization of the Network

After the determination of the topology of network, a further reduction of objective function can be made by the improvement of geometric vector G. This can be formulated as below.

P2:Find G

$$\text{Min } F(G)=\sum_{i=1}^{N-1}\sum_{j\in S_{i+1}}\sum_{k\in SS_i^{(j)}} V_{ijk}W_{ijk}$$

Based on optimum criterion, the optimal solution can be obtained by solving following equations.

$$\sum_{k\in SS_i^{(j)}}\frac{\partial V_{ijk}}{\partial x_{ij}}W_{ijk}\Bigg|_{j\in S_i}+\frac{\partial V_{i+1m,j}}{\partial x_{ij}}W_{i+1,m\cdot j}\Bigg|_{\substack{m\in S_{i+1}\\ j\in S_i}}=0$$

$$\sum_{k\in SS_i^{(j)}}\frac{\partial V_{ijk}}{\partial y_{ij}}W_{ijk}\Bigg|_{j\in S_i}+\frac{\partial V_{i+1m,j}}{\partial y_{ij}}W_{i+1,m\cdot j}\Bigg|_{\substack{m\in S_{i+1}\\ j\in S_i}}=0$$

$$i=1,\ 2,\ldots,\ N-1;\ j=1,\ 2,\ldots,\ m_i$$

4.3 Optimization of Operating Parameters

Another step to improve objective function is to reduce operating costs. Due to the transition stage from absolute permission to absolute impermission of operating conditions, it is more reasonable to formulate design problem as a fuzzy problem, that is

P3: Find $\{X\},\alpha$

Max α

s.t. $\mu_f[\{x\}]\geq\alpha$

$\mu_{P_i}[\{x\}]\geq\alpha$

$\mu_{T_j}[\{x\}]\geq\alpha \qquad \forall i,j$

Where $\{x\}$ denotes design vector of operating parameters and $\mu_f[\{x\}]$ stands for the menbership function of objective function, $\mu_{Pi}[\{x\}]$ for the one of pressure restric function and $\mu_{Tj}[\{x\}]$ for the one of temperature restric function. This model is the so called symmetric fuzzy optimization model and can be solved by SLP method because of the linearity of objective function.

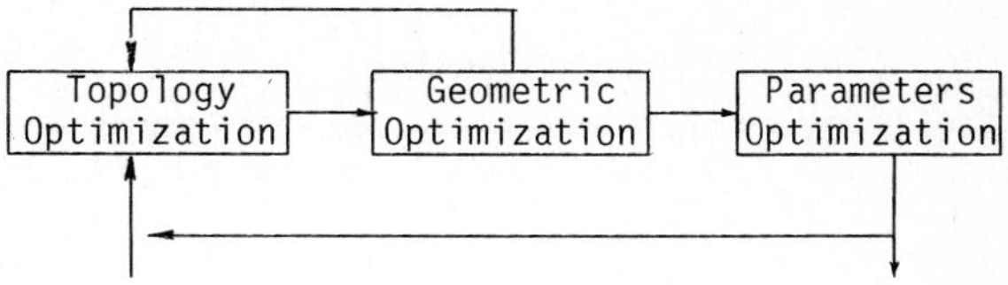

Fig.2

Now an iterative process can be carried out to reduce objective function by solving P1, P2, and P3 repeatedly until objective function value can not be improved.

5. SYSTEM OPTIMIZATION OF GDCO NETWORK

It is clear that the heuristic method presented above only generates suboptimal solution and the assumption that set M has been given which determines the setting number of stations is not tenable in the case that set M is design variable. To generate a "close to optimum" solution, it is necessary to optimize GDCO network at system level. This can be formulated as a dynamic programming model.

$$F(m_k)=\min_{\substack{u_k\in D_k\\ m_k\in A_k}}[L(m_k,u_k(m_k))+F(m_{k+1})];\quad F(m_N)=0$$

where u_k denotes decision variable in k stage and $F(m_k)$ is the cost function of k-level star network. $L[m_k^{(i)}, u_k(m_k)]$ stands for function value of k stage, D_k for feasible decision set of k stage and A_k for allowable state set of k stage.

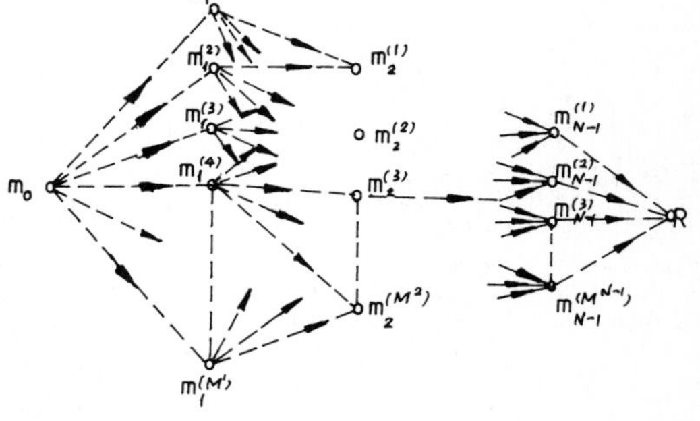

Fig.3

This problem can be equally turned to the shortest path problem in a digraph shown in Fig.3. The weight of edge, consists of partial pipeline cost and stations costs and operating cost, can be obtained by solving subproblem in section 4. The pipeline cost and the cost of establishing stations are often the most significant cost in GDCO network system.

The shortest path in the digraph gives the optimum design program which includes the setting number of intermediate stations, the linking relation of pipes and the operating parameters. For more detail, see reference [4].

6. COMPUTATIONAL EXAMPLE

The present method was used to a real design problem. In the problem, a three levels network needs to be designed to get and deliver

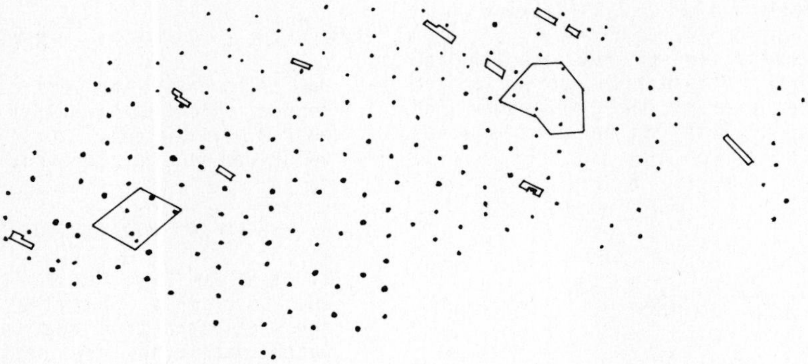

Fig.4

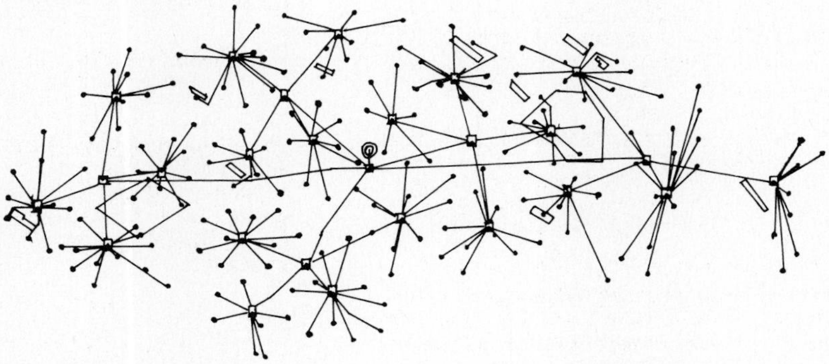

Fig.5

crude oil from 197 individual wells to a central point in field.

Well distribution is shown in Fig.4 in which polygons indicate the regions that usually can not be passed through. A program developed by authors was run on IBM PC/AT to find a final design which is shown in Fig.5. Detailed result is not listed here because of their size.

It is exciting that over 150 million Chinese yuan were saved in this example.

REFERENCES

[1] Wilson R J. Applications of Graph Theory. New York: Academic Press, 1979.

[2] Liu Yang and Cheng Geng Dong, Optimal Design of N-Level Star Network, Journal of Dalian University of Technology, 1989; Vol.29(2): pp 131-137.

[3] Cheng Geng Dong and Liu Yang, A Direct Solution Method for Fuzzy Optimization Problem under Symmetric Condition, Computational Structural Mechanics and Applications, 1988; Vol.5(3): pp 75-79.

[4] Liu Yang, Optimal Design of A Class of Multilevel Network, ph.D Dissertation, Research Institute of Engineering Mechanics, Dalian University of Technology 1988.

Approximation, Optimization and Computing:
Theory and Applications, A.G. Law and C.L. Wang (eds.)
Elsevier Science Publishers B.V. (North-Holland)

OPTIMAL WIND CHOICE IN RACES WITH FINISH

Vladimir MAZALOV and Sergey VINNICHENKO

Institute of Natural Resources, Siberian Department of Academy of Sciences, Butina 26, Chita, USSR

We consider the dynamic two-person game where a stopping time of the random sequence of the observations is Player's strategy. The procedure is looked upon as a random wind choice in races with finish.

1. INTRODUCTION

We consider the following dynamic two-person game. Suppose that two objects come from x_1, x_2 on the negative semi-axis begining to move to the origin with the random leaps. Value of the leap is defined as follows. Each player look at the observations from the finite sets $(x_i^{(1)})$, $(x_i^{(2)})$, $i=1,\dots,n$, of independent random variables with the same density $p(x)$ at the interval $(0,1)$. They have no possibility to return to the past observations and stop them at a moments t_1 and t_2. If not stopped for the n-period the latest observation should be chosen. This procedure may be considered as the choice of the wind. After that the objects are moved to the points $x_1+x_{t_1}^{(1)}$, $x_2+x_{t_2}^{(2)}$, each player is informed of the opponent position, and likewise. The winner is one whose object is firstly crossing the origin. Denote this game $G(x_1,x_2,p)$. Assume that $p(x) \le a$, $x\in(0,1)$. Playing such games we take interest in finding optimal behaviour strategies.

2. STRATEGIES

Stopping times [1] t_1 and t_2 of random sequences $(x_i^{(1)})$ and $(x_i^{(2)})$, $i=1,\dots,n$, are the strategies in this game with $t_1,t_2\le n$. Following one such game's rules its value at the point (x_1,x_2) satisfies the equalities of

$$H(x_1,x_2)=\begin{cases}1, & x_1\ge 0,\ x_2<0,\\ 0, & x_1\ge 0,\ x_2\ge 0,\\ -1, & x_1<0,\ x_2\ge 0,\end{cases}$$

and

$$H(x_1,x_2)=\sup_{t_1}\inf_{t_2} E\, H(x_1+x_{t_1}^{(1)},x_2+x_{t_2}^{(2)}),$$

if $x_1,x_2<0$.

The function $H(x_1,x_2)$ being monotonous with respect to both arguments and sequence of the observations consisting of independent identically distributed random variables, the optimal strategies are turned out to lie in the set of threshold's strategies [2] of the form

$$t_l(z)=\begin{cases}1, & x_1^{(l)}\ge z_1,\\ i, & x_1^{(l)}< z_1,\dots,x_{i-1}^{(l)}<z_{i-1},\ x_i^{(l)}\ge z_i,\\ n, & x_1^{(l)}< z_1,\dots,x_{n-1}^{(l)}<z_{n-1},\end{cases}$$

$l=1,2$, which prescribe to stop the observations at the i-step if the value of an observation overruns z_i, where $z=(z_1,\dots,z_{n-1})$ with $0\le z_i\le 1$, $i=1,\dots,n-1$. It is convinient to set $z_n=0$. The set of the threshold's strategies forms the unite $(n-1)$-dimensional cube denoted by Z.

$f(z,y)$, $0\le y\le 1$, denotes density of the distribution of leap $x_{t(z)}$ with the player using the strategy of $t(z)$ and the equation for $H(x_1,x_2)$ at the region $x_1,x_2<0$ being possibly presented as

$$H(x_1,x_2)=\sup_{z^1}\inf_{z^2}\int_0^1\int_0^1 f(z^1,y_1)f(z^2,y_2)\cdot H(x_1+y_1,x_2+y_2)\,dy_1dy_2.$$

Hence the dynamic game $G(x_1,x_2,p)$ dissimilates as a sequence
$G(x_1,x_2,p) \to G(x_1+x^{(1)}_{t_1}(z^1),x_2+x^{(2)}_{t_2}(z^2)) \to$
of games on compact set Z with continuous payoff function

$$F(z^1,z^2)=\int_0^1\int_0^1 f(z^1,y_1)f(z^2,y_2)H(x_1+y_1,x_2+y_2)dy_1dy_2$$

to be shown later on.
It is known [3] that the optimal strategies with such games lie in the set of the mixed strategies h(z) on the set Z. Thus the equation for $H(x_1,x_2)$ at the region $x_1,x_2<0$ can be transformed to

$$H(x_1,x_2)=\max_{h_1}\min_{h_2}\int_0^1\int_0^1 f_{h_1}(y_1)f_{h_2}(y_2)\cdot H(x_1+y_1,x_2+y_2)dy_1dy_2$$

where $f_h(y)=\int_Z f(z,y)dh(z)$ is the leap density for the mixed strategy h.
Next assertion can be easily proved by induction.
Theorem 1. The density f(z,y) for the threshold's strategy t(z) has the form

$$f(z,y)=\sum_{i=1}^{n}\prod_{j=1}^{i-1}u(z_j)I_{(y\geq z_i)}p(y),\ 0\leq y\leq 1,$$

where $u(z)=\int_0^z p(t)dt$ and I_A- indicator of the set A.
Now we have

$$F(z^1,z^2)=\sum_{i=1}^{n}\sum_{j=1}^{n}\prod_{l=1}^{i-1}u(z_l^1)\prod_{m=1}^{j-1}u(z_m^2)\cdot\int_{z_l^1}^1\int_{z_m^2}^1 H(x_1+y_1,x_2+y_2)p(y_1)p(y_2)dy_1dy_2$$

hence we get the continuity of the payoff function.

3. THE OPTIMALITY OF THE THRESHOLD'S STRATEGIES

We fix a strategy $t_2(z^2)$ of player II and find the best strategy $t_1(z^1)$ of player I which gives the greatest payoff $\max_z F(z,z^2)$. Such a strategy exists from the continuity of the function F. Thus there is dependence $z^1=g(z^2)$ (or in components $z_i^1=g_i(z^2)$, $i=1,\dots,n-1$.
To be somewhat diversifying we find partial derivatives of the function F.

$$\frac{\delta F}{\delta z_i^1}=p(z_i^1)\prod_{l=1}^{i-1}u(z_l^1)\left\{\sum_{s=i+1}^{n}\prod_{l=i+1}^{s-1}u(z_l^1)\cdot\sum_{j=1}^{n}\prod_{k=1}^{j-1}u(z_k^2)\int_{z_s^1}^1\int_{z_j^2}^1 H(x_1+y_1,x_2+y_2)\cdot p(y_1)p(y_2)dy_1dy_2-\sum_{j=1}^{n}\prod_{k=1}^{j-1}u(z_k^2)\cdot\int_{z_j^2}^1 H(x_1+z_i^1,x_2+y_2)p(y_2)dy_2\right\},\ i=1,\dots,n-1.$$

The expression in brackets denotes G_i. G_i is the function of $z_i^1,\dots,z_{n-1}^1$ and z^2 and we see that G_i is a decreasing function of z_i^1. Now the dependence $z_i^1=g_i(z^2)$ can be reformed as

$$z_i^1=\max\{z\in[0,d]:\ G_i(z,g_{i+1}(z^2),\dots,g_{n-1}(z^2);\ z^2)\geq 0\},\qquad (1)$$

where $d=\min(-x_1,1)$.
Using it we may prove the following assertions.

Theorem 2. Thresholds defined by (1) satisfy the inequalities

$$z_{n-1}^1\leq\dots\leq z_2^1\leq z_1^1 .$$

Theorem 3. Function $z^1=g(z^2)$ is continuous on set Z.

Next assertion gets out of Theorem 3.

Theorem 4. The equilibrium situation in the game on compact set Z with the payoff function F consists of the threshold's strategies.

Proof. The strategy $t_2(z^2)$ fixed, Theorems 3 leads to the best strategy $t_1(z^1)$ for player I giving the maximum

payoff $F(z^1,z^2)$ with the function $z^1=g(z^2)$ being continuous. By analogy with it the strategy $t_1(z^1)$ fixed, the best strategy $t_2(z^2)$ for player II giving the minimum payoff $F(z^1,z^2)$ with the function $z^2=q(z^1)$ being continuous too.
According to Brouwer Theorem the reflection $z^1=g(q(z^1))$ of convex closed set Z directed inside has a fixed point z_*^1. Denoting $z_*^2=q(z_*^1)$ we have

$$F(z_*^1,z_*^2)=\max_{z^1} F(z^1,z_*^2)=\min_{z^2} F(z_*^1,z^2).$$

This results in the optimality of strategies $t_1(z_*^1)$ and $t_2(z_*^2)$.

4. OPTIMAL CTRATEGIES IN THE VICINITY OF AN ORIGIN

Theorem 5. There exists such a vicinity of the origin $b\le x_1,x_2<0$ where optimal strategies of the both players in the game $G(x_1,x_2,p)$ are $t_1(-x_1,\dots,-x_1)$ and $t_2(-x_2,\dots,-x_2)$.

Proof. Let $x_1,x_2\ge -1$ and suppose that players I and II use the strategies $t_1(-x_1)$ and $t_2(-x_2)$. Using Theorem 1 we have here

$$f(-x_1,y)=\begin{cases}(u(-x_1))^{n-1}p(y), & 0\le y<-x_1,\\[2ex] \dfrac{1-(u(-x_1))^n}{1-u(-x_1)}\,p(y), & -x_1\le y\le 1,\end{cases}$$

$l=1,2.$
The inequalities

$$G_i(-x_1,\dots,-x_1;\ -x_2)\ge 0,\quad i=1,\dots,n-1,$$

are sufficient to be shown for the optimality of the strategy $t_1(-x_1)$ with small $|x_1|$ and $|x_2|$ present. The above inequalities are equivalent to

$$\int_0^{-x_1}\int_0^{-x_2} H(x_1+y_1,x_2+y_2)p(y_1)p(y_2)dy_1dy_2 \ge \frac{u(-x_1)-1}{(u(-x_2))^{n-1}} + \int_0^{-x_2}(u(-x_2-y))^n p(y)dy. \tag{2}$$

By analogy with it the inequality

$$\int_0^{-x_1}\int_0^{-x_2} H(x_1+y_1,x_2+y_2)p(y_1)p(y_2)dy_1dy_2 \le \frac{1-u(-x_2)}{(u(-x_1))^{n-1}} - \int_0^{-x_1}(u(-x_1-y))^n p(y)dy \tag{3}$$

is sufficient for the optimality of the strategy $t_2(-x_2)$.
If $x_1,x_2\to 0$ then $u(-x_1),u(-x_2)\to 0$. Hence the right sites of the inequalities (2), (3) converge to $-\infty$ and $+\infty$. On the other hand left sites of the (2), (3) are restricted by a unite. Hence we get the existence of the vicinity of an origin with the inequalities (2), (3) present. Theorem is proved.

The function F in the vicinity has the form

$$F(-x_1,-x_2)=(u(-x_2))^n-(u(-x_1))^n+(u(-x_1)u(-x_2))^{n-1}\int_0^{-x_1}\int_0^{-x_2} H(x_1+y_1,x_2+y_2)p(y_1)p(y_2)dy_1dy_2,$$

and then the conditions (2), (3) can be transformed to

$$(u(-x_2))^n-(u(-x_1))^{n-1}+(u(-x_1)\cdot u(-x_2))^{n-1}\cdot\int_0^{-x_2}(u(-x_2-y))^n p(y)dy \le F(-x_1,-x_2)\le (u(-x_2))^{n-1}-(u(-x_1))^n-(u(-x_1)u(-x_2))^{n-1}\cdot\int_0^{-x_1}(u(-x_1-y))^n p(y)dy \tag{4}$$

5. UNIFORM CASE

If $p(y)=1$, $0\le y\le 1$, the conditions (4)

have the form

$$(-x_2)^n-(-x_1)^{n-1}+(x_1x_2)^{n-1}(-x_2)^{n+1}/(n+1)$$

$$\leq F(-x_1,-x_2)\leq(-x_2)^{n-1}-(-x_1)^n-(x_1x_2)^{n-1}\cdot$$

$$(-x_1)^{n+1}/(n+1)$$

and the value of game $G(x_1,x_2,1)$ in this vicinity is equal to

$$H(x_1,x_2)=((-x_2)^n-(-x_1)^n)\cdot$$

$$\cdot\sum_{i=0}^{\infty}\frac{in+1}{\prod_{j=0}^{i}(jn+1)^2}(x_1x_2)^{in}.$$

REFERENCES

[1] Chow, Y.S. and Robbins, H. and Siegmund, D., Great Expectation: The Theory of Optimal Stopping (Houghton Mifflin, Boston, 1971).
[2] Mazalov, V.V., Game Moments of the Stopping (Nauka, Novosibirsk, 1987).
[3] McKinsey, J.C.C., Introduction to the theory of Games (McGray-Hill Company, Inc., New York, 1952).

Approximation, Optimization and Computing:
Theory and Applications, A.G. Law and C.L. Wang (eds.)
Elsevier Science Publishers B.V. (North-Holland)

EXTREMAL PROBLEMS AND A DUALITY PHENOMENON

Szilárd Gy. RÉVÉSZ

GAMF, Departement of Mathematics and Physics,
Kecskemét, Izsáki ut 10. 6000 HUNGARY

1. INTRODUCTION AND RESULTS

The extremal problem of calculating

$$\alpha(k):=\max\{a:\exists f\geq 0, f(x)=1+a\cos x+b\cos kx,\ a,b\in\mathbb{R}\}$$

arise in certain problems of analytic number theory. In another work [5] the calculation of

$$\gamma(k):=\sup\{a:\exists f\geq 0, f(x)=1+a\cos x+\sum_{n=k+1}^{N} a_n\cos nx, N\in\mathbb{N},\ a_n\in\mathbb{R}\ (n\in\mathbb{N})\}$$

has an important role. The latter problem reminds of the following extremal problem of Fejér: determine

$$\lambda(k):=\max\{a_1:\exists f\geq 0, f(x)=1+\sum_{n=1}^{k} a_n\cos nx,\ a_n\in\mathbb{R}\ (n\leq k)\}\ .$$

Fejér proved /see [2] or [3] I pp. 869--870/

$$(1)\quad \lambda(k)=2\cos\frac{\pi}{k+2}\ ,$$

and the values of the other quantities are /cf. Proposition 1 and [4]/

$$\alpha(k)=\frac{1}{\cos\frac{\pi}{2k}}\ ,\quad \gamma(k)=\frac{1}{\cos\frac{\pi}{k+2}}\ .$$

As a common generalization we consider for any $H\subset\mathbb{N}_2:=\{2,3,\dots\}$

$$(2)\quad \beta(H):=\sup\{a:\exists f\geq 0, f\in T, f(x)=1+a\cos x+\sum_{n\in H} a_n\cos nx, a_n\in\mathbb{R}(n\in H)\},$$

where T denotes the set of ordinary trigonometric polynomials. Another extremal quantity of particular importance is

$$(3)\quad \Delta(k):=\frac{1}{2}\sup\{\beta(H):H\subset\mathbb{N}_2, |H|=k\}\ .$$

We determine the order of magnitude of $\Delta(k)$ in Proposition 2. This result has an application in [5].
Let us denote $\overline{H}:=\mathbb{N}_2-H$. It is almost immediate that $\beta(H)\cdot\beta(\overline{H})\leq 2$. However, as the case of λ and γ suggests, even the following duality statement holds true.

THEOREM 1. Let $H\subset\mathbb{N}_2$ be arbitrary. We have

$$\beta(H)\cdot\beta(\overline{H})=2\ .$$

In these extremal problems, in particular in number theoretic applications, there are cases when we have to restrict ourselves to polynomials with nonnegative coefficients. More generally let us introduce for any $H,K\subset\mathbb{N}_2$

$$(4)\quad F(H,K):=\{f\in T, f\geq 0, f(x)=1+a\cos x+\sum_{n=2}^{\infty} a_n\cos nx, a_n\leq 0\ (n\notin H),\ a_n\geq 0\ (n\notin K)\}\ .$$

The corresponding extremal quantity is

$$(5)\quad \beta(H,K):=\sup\{a:\exists f\in F(H,K)\ \text{with}\ a=a_1=\frac{1}{\pi}\int_{-\pi}^{\pi} f(x)\cos x\,dx\}\ .$$

This setting generalizes the preceeding problem as $\beta(H,H)=\beta(H)$. For positive definite polynomials we write simply $\beta(H,\emptyset)$. Now let us denote for a pair $H,K\subset\mathbb{N}_2$

$$(6)\quad H^*:=\overline{(H\cap(2\mathbb{N}+1))\cup(K\cap 2\mathbb{N})}\ ,$$

$$K^*:=\overline{(H\cap 2\mathbb{N})\cup(K\cap(2\mathbb{N}+1))}\ .$$

Note that both sets H^* and K^* depend on both sets H and K . With these notations we formulate an even more general duality result.

THEOREM 2. Let $H,K\subset\mathbb{N}_2$ be arbitrary and H^*,K^* be as above. We have $\beta(H,K)\cdot\beta(H^*,K^*)=2$.

Clearly, Theorem 1 follows easily from Theorem 2 since if $K=H$ then $H^*=K^*=\overline{H}$.

2. PROOF OF THEOREM 2.
Let us denote $S=\mathbb{R}/2\Pi\mathbb{Z}$ and $\int=\int_S$. We put $G:=\{f(x+\Pi):f\in F(H,K)\}$.

Now if $f\in F(H,K)$ has coefficients a_n and $g\in F(H^*,K^*)$ has coefficients b_n then we have according to $0\le f,g$

$$0\le \frac{1}{\Pi}\int f(x+\Pi)g(x)dx=2-a_1b_1+\sum_{n=2}^{\infty}(-1)^n a_n b_n\le$$
$$\le 2-a_1b_1 \quad ,$$

since $(-1)^n a_n b_n\le 0$ $(n\in\mathbb{N}_2)$ as a quick reflection to (4) and (6) shows. Taking supremum according to (5), the above inequality entails

(7) $\beta(H,K)\cdot\beta(H^*,K^*)\le 2$.

Translation with Π leads to

(8) $\beta:=\beta(H,K)=\sup\{a_1:\exists f\in G,$
$\frac{1}{\Pi}\int f(x)\cos x\,dx=-a_1\}$.

Consider the Banach space $C(S)$ with the supremum norm. Denote

(9) $P:=\{f\in C(S):f>0\}$,
$A=\{\sum_{n=2}^{\infty}a_n\cos nx\in T:(-1)^n a_n\le 0$
$(n\notin H),(-1)^n a_n\le 0\ (n\notin K)\}$, $F=1-\beta\cos x+A$.

Plainly A and P are convex cones, further P is open and F is the translation of A by the "vector" $1-\beta\cos x$. Moreover, $F\cap P=\emptyset$. Indeed, if $f\in F\cap P$ exists, then with $\delta:=\min f>0$ we find a $g=(f-\delta)/(1-\delta)\ge 0$, $g\in G$, and for this g a_1 would be $-\beta/(1-\delta)<-\beta$, a contradiction in view of (8).
Hence $F\cap P=\emptyset$ and we can apply the separation theorem of convex sets in the Banach space $C(S)$ for the sets F and P , cf. [1] 2.2.2. Corollary, p. 118.
We get a nontrivial linear functional $L\ne 0$ and a $w\in\mathbb{R}$ satisfying

(10) $LP\ge w\ge LF$.

Since P is a cone, so is LP , ie. $LP=\{0\}$ or $[0,\infty)$ or $(0,\infty)$ and so we can suppose $w=0$. We get also from $0\in A$

(11) $L(1-\beta\cos x)\le 0$,

and from $h\in H$ and $k\in K$ similarly we get for any $A\ge 0$

$L(1-\beta\cos x+A(-1)^h\cos hx)\le 0$ $(h\in H)$,

$L(1-\beta\cos x+A(-1)^{k+1}\cos kx)\le 0$ $(k\in K)$

so with $A\to+\infty$ we obtain

(12) $(-1)^h L(\cos hx)\le 0$ $(h\in H)$,
$(-1)^k L(\cos kx)\ge 0$ $(k\in K)$.

Now apply the Riesz Representation Theorem, cf. [1] 4.10.1. Theorem, p. 203: $L=\int d\mu$ with some regular Borel measure μ on S . Here $\mu\ne 0$ since $L\ne 0$, μ is nonnegative since $LP\ge 0$, and considering $\mu(x)+\mu(-x)$ we can also suppose that μ is even. Since $L(1)\ge 0$, and $L(1)=0$ would imply $LP=0$ and $L=0$, we find $L(1)=\int 1\,d\mu>0$, hence we can normalize supposing $\mu(S)=2\Pi$. In all, the Fourier series of the nonnegative Borel measure μ is

(13) $\mu\sim 1+\sum_{n=1}^{\infty}b_n\cos nx$,

where in view of (11) and (12)

(14) $1-\frac{1}{2}\beta b_1\le 0$, $(-1)^h b_h\ge 0$ $(h\in H)$,
$(-1)^k b_k\ge 0$ $(k\in K)$.

Consider F_N , the Nth Fejér kernel, and $\sigma_N:=F_N*\mu$ the Nth Fejér polynomial of μ . Since both F_N and μ are nonnegative, so is σ_N , and we have

(15) $0\le\sigma_N(\mu,x)=1+\sum_{n=1}^{N}(1-\frac{n}{N})\,b_n\cos nx$.

It is easy to check that (14) and (15) entails $\sigma_N\in F(H^*,K^*)$, and so

$\beta(H^*,K^*)\ge(1-\frac{1}{N})\,b_1$.

According to the first part of (14) and letting $N\to\infty$ we obtain $\beta(H^*,K^*)\ge 2/\beta$, hence in view of (7) and (8) the proof is completed.

3. COMPUTATION OF SOME EXTREMAL QUANTITIES

Proposition 1. $\alpha(k)=1/\cos\frac{\Pi}{2k}$ and $\beta(\overline{\{k\}})=2\cos\frac{\Pi}{2k}$.

Proof. In view of Theorem 1 it suffices to show the first part.
Let $f\in F(\{k\})$ ie. $0\le f(x)=1+a\cos x+$ $+b\cos kx$. Considering $x=\Pi+\frac{\Pi}{2k}$ we get $0\le 1-a\cos\frac{\Pi}{2k}$ hence $\alpha(k)\le 1/\cos\frac{\Pi}{2k}$.

On the other hand the nonnegativity of the polynomial

$$f(x)=1+\frac{1}{\cos\frac{\pi}{2k}}\cos x+\frac{(-1)^k \operatorname{tg}\frac{\pi}{2k}}{k}\cos kx$$

proves the assertion.

Proposition 2. For any $k\in\mathbb{N}$ we have

$$1-\frac{5}{(k+1)^2}\le\Delta(k)\le 1-\frac{0.5}{(k+1)^2} .$$

Proof. Taking $H=\{2,3,\ldots,k+1\}$ Fejér's result (1) gives $\beta(H)=2\cos\frac{\pi}{k+3}$ and some calculation yields the lower estimate. On the other hand for any $H\subset\mathbb{N}_2$, $|H|=k$ we can take $n\notin H$ with $2\le n\le k+2$, and using Proposition 1 we get $\beta(H)\le\beta(\overline{\{n\}})=2\cos\frac{\pi}{2n}\le 2-(k+1)^{-2}$, whence the assertion.

REFERENCES

[1] Edwards, R.E., Functional Analysis, Holt-Rinehart-Winston, New York--Toronto-London, 1965.
[2] Fejér, L., Über trigonometrische Polynome, J. Angew. Math. 146 /1915/, 53-82.
[3] Fejér, L., Gesammelte Arbeiten I-II, Akadémiai Kiadó, Budapest, 1970.
[4] Révész, Sz.Gy., A Fejér-type extremal problem, Acta Math. Hung., to appear
[5] Révész, Sz.Gy., On Beurling's Prime Number Theorem, manuscript

Approximation, Optimization and Computing:
Theory and Applications, A.G. Law and C.L. Wang (eds.)
Elsevier Science Publishers B.V. (North-Holland)

THE EXPANSION OF FUNCTIONS UNDER TRANSFORMATION AND ITS APPLICATION TO OPTIMIZATION

Sui Yun-kang

Associate Professor
Research Institute of Engineering Mechanics, Dalian University of Technology, China

ABSTRACT. This paper proposes the expansion of functions under transformation, which represents infinite kinds of extensions of Taylor's expansion. Based on it, one develops from Duffin's condensation formula a condensation formula of high order. Furthermore, a theorem of efficiently solving mathematical programming by using the expansion of functions is proposed in the paper. According to the condensation formula of high order, one proposes two kinds of second order's primal algorithms for solving generalized geometric programming with rapid and stable convergence, which could be applied to a wide range of structural optimization problems. Finally, an efficient fully-stressed design method under the function transformation is proposed by using these expansions of functions.

1. HIGH ORDER EXPANSION OF FUNCTIONS UNDER TRANSFORMATION

Theorem 1. If 1) there exist groups of both single valued functions $H_i(t)$ and corresponding single valued inverse functions $H_i^{-1}(t)$ $(i=0, 1,\dots,n,\ t\in E^1)$, and the domain of function $y(x)$, $(x\in E^n)$ does not violate the domain of definition of function $H_0(t)$; 2) $H_i(t)$ $(i=1,\dots,n)$ have derivatives up to order m, then we have

$$\tilde{y}=H_0^{-1}\Big\{H_0[y(x^0)]+\sum_{l=1}^{m}\frac{1}{l!}\sum_{k_1=1}^{n}\cdots\sum_{k_l=1}^{n}\frac{\partial^l H_0[y(x^0)]}{\partial H_{k_1}(x_{k_1})\cdots\partial H_{k_l}(x_{k_l})}\prod_{i=1}^{l}[H_{k_i}(x_{k_i})-H_{k_i}(x_{k_i}^0)]\Big\} \qquad (1\text{--}1)$$

which gives the approximation of m-th degree at the current X^0 i.e.

$$\tilde{y}(x^0)=y(x^0),\quad \frac{\partial\tilde{y}(x^0)}{\partial x_i}=\frac{\partial y(x^0)}{\partial x_i},\ \dots,$$
$$\frac{\partial^m\tilde{y}(x^0)}{\partial x_{k_1}\cdots\partial x_{k_m}}=\frac{\partial^m y(x^0)}{\partial x_{k_1}\cdots\partial x_{k_m}}$$

The proof is neglected here[1]. If $H_0(t)=H_1(t)=\dots=H_n(t)=t$, then $\tilde{y}(x)$ is the Taylor's expansion. Thus the formula (1--1) is an extension of the Taylor's expansion. In other words, there is not only one Taylor's formula but an infinite number of ones. It is then possible to select better expansion when the Taylor's formula is used in practice; that is, using some of information of lower order derivatives, we may construct the best approximation enough approaching to primal function.

If $y(x)>0$, from the Theorem 1 taking $H_0(t)=H_1(t)=\dots=H_n(t)=\ln t$, we may get the following approximation of m-th degree about the function $y(x)$ at point x^0

$$\tilde{y}(x)=\exp\Big\{\ln y(x)+\sum_{l=1}^{m}\frac{1}{l!}\sum_{k_1=1}^{n}\cdots\sum_{k_l=1}^{n}\frac{\partial^l\{\ln[y(x^0)]\}}{\partial\ln x_{k_1}\cdots\partial\ln x_{k_l}}\prod_{i=1}^{l}\ln\frac{x_{k_i}}{x_{k_i}^0}\Big\} \qquad (1\text{--}2)$$

When m=1, formula (1--2) shows that $\tilde{y}(x)$ is an approximation of first degree of the function $y(x)$, i.e.

$$\tilde{y}=y(x^0)\prod_{i=1}^{n}\left[\frac{x_i}{x_i^0}\right]^{\partial\ln y(x_i^0)/\partial\ln x_i} \qquad (1\text{--}3)$$

Formula (1--3) is the Duffin's condensation formula which implies that formula (1--2) is an extension of Duffin's one from first to high order, named as high order condensation formula.

2. SOLUTION OF MATHEMATICAL PROGRAMMING USING FUNCTION TRANSFORMATION

According to the idea of function transformation, we may solve "mapping" mathematical programming instead of primal one. To this end, the Theorem 2 is proposed in the next paragragh.

Theorem 2. If 1) there exists group of monotone single valued functions $G_i(t)$, $(i=0, 1,\dots,m,\ t\in E^1)$ and the domain of functions $y_i(x)$ $(i=0, 1,\dots,m,\ x\in E^n)$ does not violate the domain of definition of functions $G_i(t)$; 2) there exist groups of both single valued function $H_i(t)$ and corresponding single valued function $H_i^{-1}(t)$ $(i=1, 2,\dots,n,\ t\in E^1)$, then there are following relations between optimum designs of programming PO and PM

$$z_i^* = H_i(x_i^*)$$

$$\text{PO}\begin{cases}\min & y_0(x_1,\ldots,x_n)\\ \text{s.t.} & y_i(x_1,\ldots,x_n)\le b_i \quad (i=1,\ldots,m)\end{cases}$$

$$\text{PM}\begin{cases}\text{opt} & G_0\{y_0[H_1^{-1}(z_1),\ldots,H_n^{-1}(z_n)]\}\\ \text{s.t.} & G_i\{y_i[H_1^{-1}(z_1),\ldots,H_n^{-1}(z_n)]\}\circledast G_i(b_i),\\ & (i=1,\ldots,m)\end{cases}$$

where b_i is in the domain of definition of function $G_i(t)$,

$$z_i = H_i(x_i)$$

$$\text{opt}=\begin{cases}\text{Minimize,} & \text{if } G_0(t) \text{ is monotone increasing function}\\ \text{Maximize,} & \text{if } G_0(t) \text{ is monotone decreasing function}\end{cases}$$

$$\circledast=\begin{cases}\ge & \text{if } G_i(t) \text{ is monotone increasing function}\\ \le & \text{if } G_i(t) \text{ is monotone decreasing function}\end{cases}$$

The proof is neglected [1].

Selecting suitable transformation function could make the nonlinear degree of PM lower than that of PO. Therefore, the solution of PM has higher efficiency compared with the solution of PO, when the approximete method is applied. The following section will be developed on the basis of the idea of the Theorem 2.

3. TWO PRIMAL ALGORITHMS OF SECOND ORDER CONDENSATION FOR SOLVING GGP

Many problems of structural optimization may be transformed into GGP (generalized geometric programming), which may be solved by dual algorithms through transforming them into approximate posynomial geometric programming. But the dual programming has many dual variables and a difficulty that some of the variables tend to zeros. Primal algorithms have not aforementioned drawbacks. However, if its features are not utilized in the solution procedure, they will not be different from general nonlinear programming. Reference [3] employs Duffin's condensation formula to solve primal GGP in logarithmic space, which shows higher efficiency than the dual algorithm. Due to the use of the Duffin's formula, the algorithms are classed as the algorithms of first order.

To search for more efficient ones, could the second order algorithm be employed? With the aid of the Theorem 1, this idea could be realized in the next paragragh.

Suppose there is a GGP problem:

$$P_1\begin{cases}\min & y_0(x)=P_0(x)-Q_0(x)\\ \text{s.t.} & y_m(x)=P_m(x)-Q_m(x)\circledast 1 \quad (m=1,\ldots,M)\\ & x_k \ge \underline{x}_k \quad (k=1,\ldots,n)\end{cases} \tag{3-1}$$

where $x\in E^n$, $P_m(x)$ and $Q_m(x)$ are posynomials, $\circledast$ denotes $\le$ or $\ge$. Programming P_1 can be transformed into P_2.

$$P_2\begin{cases}\min & \bar{y}(x)=P_0(x)-Q_0(x)+c \qquad (3\text{-}2)\\ \text{s.t.} & \dfrac{P_m(x)}{1+Q_m(x)}\le 1 \quad \text{or} \quad \dfrac{1+Q_m(x)}{P_m(x)}\le 1\\ & (m=1,\ldots,M)\\ & x_k\ge \underline{x}_k \quad (k=1,\ldots,n)\end{cases}$$

where c is a sufficiently large positive number to ensure $\bar{y}(x)>0$.

According to the formula (1-2), taking second order condensation of the objective and constraint functions of P_2 and then taking logarithmic transformation, we have an approximate problem P_3.

$$P_3\begin{cases}\min & \frac{1}{2}z^TAz+B^Tz\\ \text{s.t.} & \frac{1}{2}z^TH^{(m)}z + c^{(m)}z+d^{(m)}\le 0 \quad (m=1,\ldots,M)\\ & z_k\ge 0 \quad (k=1,\ldots,n)\end{cases} \tag{3-3}$$

where $z_k=\ln(x_k/\underline{x}_k)$. Taking first order condensation of the constraints and second order condensation of the objective function of P_2, we have another approximate problem P_4.

$$P_4\begin{cases}\min & \frac{1}{2}z^TAz+B^Tz\\ \text{s.t.} & c^{(m)}z + d^{(m)}\le 0 \quad (m=1,\ldots,M)\\ & z_k\ge 0 \quad (k=1,\ldots,n)\end{cases} \tag{3-4}$$

P_4 is a standard quadratic programming which may be solved by Lemke algorithm, but P_3 is a quadratic programming with the second order constraints.

To solve P_3, introducing Lagrange multipliers $\lambda\in E^M$ and $\mu\in E^n$, we have the Kuhn-Tucker necessary conditions as

$$\begin{cases}Az+B+\sum_{m=1}^{M}\lambda_m(c^{(m)}z+d^{(m)})-\mu=0\\ \frac{1}{2}z^TH^{(m)}z+c^{(m)}z+d^{(m)}\le 0\\ \lambda_m(\frac{1}{2}z^TH^{(m)}z + c^{(m)}z+d^{(m)}) = 0\\ \mu^Tz=0,\ z\ge 0,\ \lambda\ge 0,\ \mu\ge 0 \quad (m=1,\ldots,M)\end{cases} \tag{3-5}$$

Taking linear approximation of Taylor's expansion of formula (3-5) at an initial point z^0, λ^0 and μ^0 may set up a linear complementary problem

$$\begin{cases} w+\tilde{A}u=q \\ w^Tu=0,\ w\geq 0,\ u\geq 0 \end{cases} \tag{3-6}$$

where $w=[v^T\ \mu^T]^T$, $u=[\lambda^T\ z^T]^T$, $v\in E^M$ are slack variables of the second formula of (3-5),

$$\tilde{A}=\begin{bmatrix} 0 & \bar{c} \\ -\bar{c}^T & -\bar{A} \end{bmatrix}, \qquad \bar{A}=A+\sum_{m=1}^{M}\lambda_m^o H^{(m)},$$

$$\bar{c}=[c^{(1)}+P^{(1)},\ldots,c^{(M)}+P^{(M)}],$$

$$\bar{d}=\{-d^{(1)}+\tfrac{1}{2}z_0^T P^{(1)},\ldots,-d^{(M)}+\tfrac{1}{2}z_0^T P^{(M)}\}$$

and $\quad P^{(m)}=H^{(m)}z^o \qquad (m=1,\ldots,M)$.

Problem (3-6) may be solved by using Lemke algorithm. This procedure is a solution method of problem P_3, which is used iteratively to solve problem P_1.

Using P_4 to solve P_1 also requires sequential iteration. A number of numerical experiments show that the efficiency of solving problem P_1 by using P_3 or P_4 is satisfied. The number of iterations is 4 or 5. Owing to the fact that the constraints in P_3 are one order higher than those in P_4, the algorithm based on P_3 is more stable than that of P_4.

Example 1.[2]

$$\begin{aligned}
\min \quad & 10x_1(x_2x_3)^{-1}x_4+20(x_1x_4)^{-1}x_5x_6+30x_2x_3x_4 \\
& +100(x_1x_2x_3x_4x_5x_6x_7)^{-1}+5(x_1x_2x_7)^2x_3x_5x_6^{1.5} \\
& +50(x_3x_4x_5)^{-0.5}+25(x_3x_4)^2(x_5x_6x_7)^{-1} \\
& +10(x_3x_4)^{0.5}x_5x_6x_7
\end{aligned}$$

$$\begin{aligned}
\text{s.t.} \quad & 0.1(x_1x_2)^2x_3+0.05x_4x_5^{0.5}+0.15(x_6x_7)^{0.5}\leq 1 \\
& 0.1x_1x_4x_7+0.05x_1(x_2x_3)^{-1}x_5x_6x_7^{0.5}+0.05(x_2x_3)^2x_4^{-1} \\
& +0.15x_1^{-0.5}x_2^{-0.3}x_3x_5^{0.5}+0.1x_5x_6+0.1x_4^2+0.2x_1x_2x_3\leq 1 \\
& x_1+x_2+x_3+x_4+x_5+x_6+x_7\leq 10 \\
& x_1^2+x_1x_2+x_1x_3+x_1x_4+x_1x_5+x_1x_6+x_1x_7\leq 500 \\
& x_1x_2^{-1}+x_2x_3^{-1}+x_3x_4^{-1}+x_4x_5^{-1}+x_5x_6^{-1}+x_6x_7^{-1}\leq 100 \\
& x_1x_3^{-2}+x_2x_4^{-2}+x_3x_5^{-2}+x_4x_6^{-2}+x_5x_7^{-2}\leq 10 \\
& x_1^{-0.5}x_3+x_2^{-0.5}x_4+x_3^{-0.5}x_5+x_4^{-0.5}x_6+x_5^{-0.5}x_7\leq 50 \\
& x_1,\ldots,x_7\geq 0
\end{aligned}$$

Degree of difficulty of this problem is 40. Both dual algorithm and the algorithm based on P_4 fail to solve it, respectively. The calculation in terms of P_3 converges to an identical optimum solution from different initial points with 3 iterations. One of the iteration processes is given in table 1.

For structural optimization problems that require sensitivity analysis, sometimes we take $H_0(t)=H_1(t)=\ldots=H_n(t)=t$ in formula (1-1) for simplification, which amounts to using the Taylor's expansion, to construct second order approximation of objective and constraint functions for forming problem P_3. Based on this idea, a program of spatial truss with frequency range constraints is developed. This problem is stated mathematically as:

Table 1

Iter.	0	1	2	3
x_1	1.0	1.225	1.347	1.342
x_2	1.0	0.958	1.007	0.993
x_3	1.0	0.915	0.839	0.870
x_4	1.0	0.933	0.925	0.924
x_5	1.0	2.604	3.182	3.149
x_6	1.0	0.398	0.406	0.404
x_7	1.0	2.187	1.520	1.548
obj.		181.973	178.400	178.478

$$\begin{cases} \min \ w \\ \text{s.t.}\ \omega_u^2 \circledast \omega_u^{*2} \qquad (u=1,\ldots,U) \\ \qquad \underline{\sigma}_i\leq\sigma_i\leq\bar{\sigma}_i,\quad A_i\geq\underline{A}_i \quad (i=1,\ldots,n) \end{cases}$$

where w is structural weight, ω_u^2 are concerned frequencies, σ_i are bar's stresses, A_i are bar's sectional areas and $\circledast$ is $\leq$ or $\geq$. In the program, all of ω_u^2 are taken as second order approximation of their Taylor's expansions and the critical stress constraints are transformed into lower size limit. It is then solved in terms of problem P_3.

4. FULLY-STRESSED DESIGN METHOD UNDER THE FUNCTION TRANSFORMATION

In the past there is only one fully-stressed formula. Now we are in a position to develop infinite fully-stressed design formula, which can be selected to obtain the best result. In the following derivation, the truss structure is taken as an example.

Suppose the stress of i-th bar is

$$\sigma_i = N_i / A_i \tag{4-1}$$

According to the assumption of statically determinate structure, the internal force N_i is regarded as constant and σ_i is a function of A_i. From this, we have

$$\frac{\partial \sigma_i}{\partial A_j} = -\frac{\sigma_i\delta_{ij}}{A_i} \tag{4-2}$$

where δ_{ij} is Kronecker delta.

Supposing G(t), H(t) and $G^{-1}(t)$, $H^{-1}(t)$ are single valued functions and corresponding inverse ones, we introduce auxiliary variables

$$x_i=H(A_i) \qquad (i=1,\ldots,n) \tag{4-3}$$

From Theorem 1, the first order approximation of the stress under function transformation is taken as

$$\sigma_i^{(k+1)} = G^{-1}\left\{G(\sigma_i^{(k)}) + \sum_{j=1}^{n} \frac{\partial G(\sigma_i)}{\partial H(A_i)}\Bigg|_{A=A^{(k)}} [H(A_j^{(k+1)}) - H(A_j^{(k)})]\right\} \quad (4\text{-}4)$$

On the basis of formula (4-2), we have

$$\frac{\partial G(\sigma_i)}{\partial H(A_j)}\Bigg|_{A=A^{(k)}} = -\frac{G\ (\sigma_i^{(k)})\sigma_i^{(k)}\delta_{ij}}{H\ (A_i^{(k)})A_i^{(k)}} \quad (4\text{-}5)$$

With the aid of fully-stressed condition, the stress $\sigma_i^{(k+1)}$ in the next iteration should be equal to allowable stress σ_i^*. Thus from formula (4-4), the next area $A_i^{(k+1)}$ is

$$A_i^{(k+1)} = H^{-1}\left\{H(A_i^{(k)}) - \frac{[G(\sigma_i^{(k)}) - G(\sigma_i^*)]H'(A_i^{(k)})A_i^{(k)}}{G'(\sigma_i^{(k)})\sigma_i^{(k)}}\right\} \quad (4\text{-}6)$$

This is a fully-stressed formula under the function transformation. There are 16 cases under the combinations of linear, inverse, logarithmic and exponential functions. After numerical examination of few examples, 4 fully-stressed formulas with rapid and stable convergence are obtained which are listed in table 2.

Table 2

Case	1	2	3	4		
$G(\sigma_i)$	σ_i^{-1}	σ_i^{-1}	$\ln\sigma_i$	$\ln\sigma_i$	σ_i	σ_i^{-1}
$H(A_i)$	e^{A_i/A_i^0}	$\ln A_i$	A_i^{-1}	$\ln A_i$	A_i^{-1}	A_i

case

1 $A_i^{(k+1)} = A_i^{(k)} + A_i^{(0)} \ln\{[\sigma_i^{(k)}/\sigma_i^* - 1]A_i^{(k)}/A_i^{(0)} + 1\}$

2 $A_i^{(k+1)} = A_i^{(k)} e^{[\sigma_i^{(k)}/\sigma_i^* - 1]}$

3 $A_i^{(k+1)} = A_i^{(k)}[1 - \ln(\sigma_i^{(k)}/\sigma_i^*)]^{-1}$

4 $A_i^{(k+1)} = A_i^{(k)}\sigma_i^{(k)}/\sigma_i^*$

Using the same measure, the fully-stressed formula under function transformation of the planar frame is obtained. There are also similar 16 cases of which few good ones are selected after numerical experiments. We will not discuss this further.

5. CONCLUSIONS

The expansions of functions under transformation proposed in the paper have the advantages on constructing optimization models and efficiently solving them.

ACKNOWLEDGEMENTS

The author is grateful to graduate students Geng Shu-sen, You Zhong, Lin Yong-ming and Peng Ke-jian who were engaged in relevant research work.

REFERENCES

[1] Sui Yun-kang and Geng Shu-sen, Extension of Taylor's Expansion and its Application in Solving Mathematical Programming, Journal of Engineering Mathematics, Vol.2, No.1, 1985. (in Chinese).

[2] Rijckaert, M.J. and Martens, X.M., Comparison of Generalized Geometric Programming Algorithms, Advances in Geometric Programming, eds. by Avriel, M., Plenum Press, New York (1980).

[3] Avriel, M., Dembo, R.S. and Passy, U., Solution of Generalized Geometric Programming, Int. J. Num. Meth. Eng., 9 (1975).

Approximation, Optimization and Computing:
Theory and Applications, A.G. Law and C.L. Wang (eds.)
Elsevier Science Publishers B.V. (North-Holland)

A COMBINATORIAL ALGORITHM FOR 0-1 PROGRAMMING

Sun Huan-chun, Mao Sheng-gen

Dalian University of Technology, China.

ABSTRACT: In this paper a new algorithm for 0-1 programming, named the binary combinatorial algorithm, is presented and applied to solving some examples. Numerical results have shown that the proposed algorithm is more efficient than the existing ones. A computer code has been made with FORTRAN 77 and successfully used to solve a number of problems.

1. INTRODUCTION

Often design variables must be chosen from some given discrete values, such as rolled steel sections or some stipulated sections, so that the structural optimization is of discrete type. Usually this kind of problems is firstly solved by some method of continuous type, and then rounded to the nearest discrete ones. It is well known that this procedure does not assure us of obtaining the discrete optimum design. If the integer programming methods, such as the branch and bound technique, the cutting plane approach and the implicit enumeration method, is directly applied to the optimum design of engineering structures, the amont of computation work will be very large. Based on paper [1], [2], [3], a new algorithm for 0-1 programming is developed in this paper, which is named the binary combinatorial algorithm. A modified version of this algorithm based on the character of the structural design is also developed and used to solve some examples of the discrete optimum design of frames.

2. BINARY COMBINATORIAL ALGORITHM FOR 0-1 PROGRAMMING

2.1 The Algorithm For 0-1 Programming With Linear Objectives

The problem considered in this section is defined as:

$$\text{Min} \quad w=\sum_{i=1}^{n} c_i x_i \tag{2.1}$$

$$\text{s.t.} \quad f_j(x) \le 0 \quad (j=1,2,\dots,m) \tag{2.2}$$

$$x_i = 0 \text{ or } 1 \tag{2.3}$$

where $c_i \ge 0, c_1 \le c_2 \le \dots \le c_n$ (If $c_i < 0$, we can have the objective function expressed in y_i with a positive coefficient $(-c_i)$ only by letting $x_i = (1-y_i)$, and $f_j(x)$ may be any arbitrary nonlinear functions.

The fundamental idea of the algorithm can be described in terms of an example with three variables.

(x_3, x_2, x_1) is used to express any binary number with three bits, and it has a total of eight combinations, which are denoted by the binary number--(0 0 0), (0 0 1), (0 1 0),...(1 1 1), counting from (0 0 0) to (1 1 1). Because of $c_1 \le c_2 \le c_3$ it follows immediately that

$$w(000) \le w(001) \le w(010) \le w(011),$$

$$w(010) \le w(100) \le w(101) \le w(110) \le w(111).$$

But a comparison between W(1 0 0) and W(0 1 1) depends upon the magnitudes of C_1+C_2 and C_3. The basic procedure of this algorithm is given as follows:

At the beginning, the first feasible combination is sought by starting from (0 0 0) in the increasing order and taking the first combination that satisfies all constraints. If (0 1 0) is the first feasible combination, it is obviously impossible to find a better feasible one than (0 1 0), from (0 1 1), (1 0 0), (1 0 1), (1 1 0), (1 1 1). Similarly, if (1 0 0) is the first feasible combination, it is also impossible to find a better feasible one from (1 0 1), (1 1 0), (1 1 1). Thus we will not examine those combinations from which a better solution than the current optimum one (the optimum feasible solution obtained so far) can't be obviously obtained, so that the enumerative numbers are greatly reduced. But if (0 1 1) is the first feasible combination, (1 0 0) must be examined. The combination (1 0 0) is obtained by correcting the first nonzero element of (0 1 1) (counting from right to left) to zero and adding one to the second nonzero element of (0 1 1) (counting also from right to left). The procedure to find the solution is further explained clearly by an example with five variables:

1), Find the first feasible solution from (0 0 0 0 0) as above mentioned.

2), examine the combination obtained by

correcting the first (from right to left) nonzero element to zero of the binary number corresponding to the first feasible solution and adding one to the second nonzero element. This is named "correcting to zero and adding one" strategy. For example, a certain combination is (1 0 1 0 1), which may be the current feasible solution or whose objective value is larger than that of the current optimum feasible solution (no matter whether it is feasible or not). By the "correcting to zero and adding one" strategy, we first obtain (1 0 1 0 0) by correcting the first nonzero element of (1 0 1 0 1) to zero, and then have (1 1 0 0 0) by adding one to the second nonzero element of (1 0 1 0 1). (1 1 0 0 0) is the combination to be examined next.

3), If a certain combination is infeasible, but its objective value is smaller than that of the current optimum one, then the combination to be examined next is that obtained by adding one to the binary number corresponding to this combination. In particular, when there is only one nonzero element in the binary number corresponding to a feasible combination or to an infeasible combination for which the objective value is not less than the current optimum one, or the bit number of a combination to be examined is larger than n (the number of variables), the current optimum feasible solution (including this combination itself, if it is current optimum feasible) is just the optimum one.

Example 2-1,

$$\text{Min} \quad w = 42x_4 + 21x_3 + 14x_2 + 10x_1$$

$$\text{s.t.} \quad -18x_4 - 9x_3 - 11x_2 - 8x_1 \le -12$$

$$-14x_4 - 7x_3 - 2x_2 - 2x_1 \le -14$$

$$-6x_4 - 3x_3 - 6x_2 - 9x_1 \le -10$$

$$x_i = 0, \text{ or } 1 \quad (i=1,2,3,4)$$

Solving: We examine from (0 0 0 0) to (1 0 0 0), any one is infeasible among them, but the tenth combination (1 0 0 1) is feasible. Then the "correcting to zero and adding one" strategy is applied to it, and we find that the combination to be examined is (1 0 0 0 0). Because its bit number is five (greater than four), (1 0 0 1) is the optimum solution of this example. $x^*=(x_4, x_3, x_2, x_1)^*=(1\ 0\ 0\ 1)$, $w^*=52$, the number of combinations examined is ten compared with the total number of combinations of $2^4=16$.

Example 2-2,

$$\text{Min} \quad w = 10x_1 + 7x_2 + x_3 + 12x_4 + 2x_5 + 8x_6 + 3x_7 + x_8 + 5x_9 + 3x_{10}$$

$$\text{s.t.} \quad 3x_1 - 12x_2 - 8x_3 - x_4 - 7x_9 + 2x_{10} \le -2$$

$$x_2 - 10x_3 - 5x_5 + x_6 + 7x_7 + x_8 \le -1$$

$$5x_1 - 3x_2 - x_3 - 2x_8 + x_{10} \le -1$$

$$-5x_1 + 3x_2 + x_3 + 2x_8 - x_{10} \le 1$$

$$-4x_3 - 2x_4 - 5x_6 + x_7 - 9x_8 - 2x_9 \le -3$$

$$9x_2 - 12x_4 - 7x_5 + 6x_6 + 2x_8 - 15x_9 - 3x_{10} \le -7$$

$$-8x_1 + 5x_2 + 2x_3 - 7x_4 - x_5 - 5x_7 - 10x_9 \le -1$$

$$x_i = 0 \text{ or } 1 \quad (i=1,2,\ldots,10)$$

At first, the subscripts of design variables x_i are rearranged according to the magnititude of thier coefficients in the objective function in an increasing order and then the problem is solved by the algorithm proposed in this section. When the optimum solution is obtained, the original subscripts of the variables are resumed again. In this example, the solution is found as: $x^*=(x_{10}, x_9, \ldots, x_1)^*=(0\ 0\ 0\ 1\ 0\ 1\ 0\ 1\ 0\ 0)$, $w^*=6$. The number of combinations examined is 33 whereas the total number of combinations is $2^{10}=1024$.

Example 2-3,

$$\text{Min} \quad w = 6x_4 + 5x_3 + 3x_2 + x_1$$

$$\text{s.t.} \quad x_4 + 3x_4 \cdot x_3 + 2x_2 + x_1 \ge 4$$

$$2x_4 + 5x_3 + 3x_2 \cdot x_1 + 2x_1 \ge 7$$

$$x_i = 0 \text{ or } 1 \quad (i=1,2,3,4)$$

The first feasible solution is (1 0 1 1), and the corresponding objective value is 10. By applying the "correcting to zero and adding one" to (1 0 1 1), the combination (1 1 0 0) is to be examined next. Because the objective value corresponding to (1 1 0 0) is 11(>10), the next combination to be examined is (1 0 0 0 0). Due to the bit number of (1 0 0 0 0) being five (greater than four), we finally find the optimum solution as $x^*=(x_4, x_3, x_2, x_1)^*=(1\ 0\ 1\ 1)$, $w^*=10$. The number of combinations examined is 13 while the total number of combination is $2^4=16$.

There may be some nonlinear terms, such as $7x_3 \cdot x_2$, $6x_4 \cdot x_1$, in the above objective function. We can reduce them to linear terms only by letting $x_3 \cdot x_2 = x_6$, $x_4 \cdot x_1 = x_5$, and then adding two equality constraints $x_6 - x_3 \cdot x_2 = 0$ and $x_5 - x_4 x_1 = 0$.

2.2, The Algorithm for a Kind of Special Linear Zero-one Programming

Both objective and constraints are linear functions in the problems considered in this section, which can be defined as:

$$\text{Min} \quad w = \sum_{i=1}^{n} c_i x_i \tag{2.4}$$

$$\text{s.t.} \quad \sum_{i=1}^{n} a_{i,j} x_i \ge b_j \quad (j=1,2,\ldots,m) \tag{2.5}$$

$$x_i = 0 \text{ or } 1 \qquad (i=1,2,\ldots,n) \quad (2.6)$$

where

$$c_i \geq 0 \qquad c_1 \leq c_2 \leq \ldots \leq c_n, \quad a_{i,j} \geq 0.$$

An important character in this model is that if some combination is infeasible, the combination obtained by correcting any nonzero element in it to zero is certainly infeasible. Thus to find the first solution, we may proceed by the following steps. A 0-1 programming with six variables is taken as an example.

The procedure for solving the problem is:
$1 \Leftarrow i$

1). Starting from (1 1 1 1 1 1), examine the feasibility of each combination obtained by subtracting 1 bit by bit each time from the most left bit to the most right bit of (1 1 1 1 1 1) until an infeasible combination is met, then the 1 element which is lastly changed to zero, is reserved; And then do the same as above from the next bit of this reserved 1 element until a new infeasible combination is met, and then recover the last 1 element changed to zero. This procedure is repeated until the most right bit is subtracted and examined. As such the feasible combination corresponding to the smallest binary number is found. Thus the time to find the first feasible solution may be greatly saved. For example, if (0 1 0 1 1 1) is feasible, then (0 1 0 0 1 1) is examined next; If (0 1 0 0 1 1) is infeasible, then (0 1 0 1 0 1) is to be examined. If (0 1 0 1 0 1) is feasible, then examine (0 1 0 1 0 0); If (0 1 0 1 0 1) is infeasible, then (0 1 0 1 1 0) is to be examined. If (0 1 0 1 0 1) is feasible but (0 1 0 1 0 0) is infeasible, then (0 1 0 1 0 1) must be the feasible combination corresponding to the smallest binary number in this example; If (0 1 0 1 0 1) is infeasible then (0 1 0 1 0 0) can't be feasible. In a word, in the procedure of above examination, the last feasible combination is the first one required in section 2.1. The ith round for examination ends.

2). Use the "correcting to zero and adding one" strategy to the current optimum combination obtained in the ith round, then examine the combination thus obtained: If the objective value corresponding to it is not less than the current optimum one, the "correcting to zero and adding one" strategy is used again to it; If the objective value corresponding to it is less than the current optimum one, then execute the examination as in the first step beginning from the next bit of the element lastly corrected to one, until that after an element, which must be taken as one in order to satisfy the constraint conditions, is met, the objective value is no longer less than the current optimum one, or that the most right bit becomes zero. The (i+1)th round is over.

3). If the current optimum feasible combination is found in the (i+1)th round then let $i \Leftarrow i+1$ and return to step 2, otherwise the "correcting to zero and adding one" strategy is used for the combination on the end of (i+1)th round, corresponding to which the objective value is not less than the current optimum one. Then let $i \Leftarrow i+2$, return to step 1 and execute the examination in step 1 starting from the next bit of the element lastly changed to one.

The examination will be stoped until a combination, in which there is only one nonzero element, corresponding to an objective value not less than the current optimum one or being feasible is met. or that the bit number of a combination to be examined is larger than the number of design variables. Then the current optimum combination is the optimum solution to be found, the objective value of which corresponds to the optimum one.

In general, the maximum examination number m for solving a 0-1 programming can be approximately evaluated by the following formula based on the statistic and reduction methods assuming that the round number for examination is n-1; The examination number in the first round is certainly n+1: The examination number in each of other rounds is n, n-1,... and 3 respectivelly, but their order will not certainly coincide with the round order generally.

$$m = \frac{(n+4)(n-1)}{2}$$

where n is the number of design variables.

It must be pointed out that: 1), The above formula will not be true, when the coefficients of objective function and the coefficients in constraint condition are all the same respectively. Of course in this case we have more effective method to solve the problem; 2), If a 0-1 programming having multi-solution is solved by the method in section 2.1 or 2.2, generally, only a part of all solutions having same objective value can be obtained.

Example 2-4,

$$\text{Min } w = 5x_6 + 4x_5 + 3x_4 + 2x_3 + 2x_2 + x_1$$

$$\text{s.t.} \quad 6x_6 + 4x_5 + 3x_4 + 2x_3 + 3x_2 + 4x_1 \geq 14$$

$$3x_6 + 2x_5 + 4x_4 + 3x_3 + 2x_2 + 3x_1 \geq 10$$

This problem is solved by the method in this section as follows:

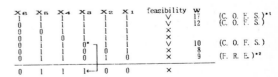

x_6	x_5	x_4	x_3	x_2	x_1	feasibility	w	
1	1	1	1	1	1	V	17	(C. O. F. S.)*1
0	1	1	1	1	1	V	12	(C. O. F. S.)
0	0	1	1	1	1	×		
0	1	0	1	1	1	×		
0	1	1	0*	1	1	V	10	(C. O. F. S.)
0	1	1	0	0	1	×	8	
0	1	1	0	1	0	×	9	(F. R. E.)*2
0	1	1	1←	0	0	×		

	x_6	x_5	x_4	x_3	x_2	x_1	feasibility	w	
	0*	1	1	1	0	1	×	10	(S. R. E.)*3
	1←	0	0	0	0	0	×	5	
	1	0	1	1	1	1	V	13>10	
	1	0	0*	1	1	1	V	10=10	
	1	0	0	0	1	1	×	8	
	1	0	0	1	0	1	×	8	
	1	0	0	1	1	0	×	9	(T. R. E.)
	1	0	1←	0	0	0	×	8	
	1	0	1	0	1	1	V	11>10	
	1	0	1	0	0	1	×	9	
	1	0*	1	0	1	0	×	10	(Fo. R. E.)
	1	1←	0	0	0	0	×	9	
	1	1	0	1	1	1	V	14	
	1	1	0	0	1	1	V	12	
0*	1	1	0	0	0	1	×	10	(Fi. R. E.)
1←	0	0	0	0	0	0			

*1 current optimum feasible solution.
*2 first round ends.
*3 second round ends,....

Therefore we obtained: w*=10 and x*=(0 1 1 0 1 1) or (1 0 0 1 1 1). The examination number is 23.

$$m = \frac{(n+4)\ (n-1)}{2} = 25$$

The total number of enumeration is $2^6=64$. Obviously, 23<25<64. 43 times examinations are needed by the method in section 2.1.

The above two combinatorial algorithms for 0-1 programming are originally developed for solving the problems of discrete structural optimization. The mathematical model of the discrete optimum design of steel frames or trusses can be reformulated as the form of (2.4)-(2.6), if the variable-linking is taken into account. Two examples of the discrete optimum design of a plane steel frame and a space steel frame are calculated by means of the algorithms presented in this paper, but omitted because of the page limit. The results of these two examples show that the algorithms described in section 2 are more effective than any other one known so far for solving the problems of discrete structural optimization.

3. CONCLUSIONS

The idea of the binary number combination algorithm can be easily understood and it is easy to be implemented in a computer code. The required memory is less than other methods for solving 0-1 programming. We store only the current optimum solution and the present combination for the algorithm in section 2.1. One more combination for which the objective value is not less than the current optimum one in the end of each round needs to be stored for the algorithm in section 2.2. Moreover the computing time is less than any other method known so for, especially for the problem with many constraints.

REFERENCES

[1] A.B. Templeman and D.F. Yates. "A Segmental Method for the Discrete Optimum Design of Structures," Eng. Opt., 5 (1983), No.6.

[2] Duan Ming-zhu, "An Improved Templeman's Algorithm for the Optimum Design of Trusses with Discrete Member Sizes", Eng. Opt., 1986, Vol.9 PP. 303-312.

[3] Sun Huan-chun and Chen Qin, "A Sequential Two-Level Algorithm for the Discrete Optimum Design of Structures", Proceedings of the 2nd International Conference on Computing in Civil Engineering, 1985, at Hang Zhou, China.

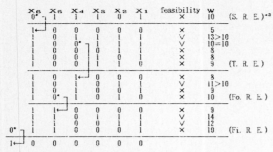

Approximation, Optimization and Computing:
Theory and Applications, A.G. Law and C.L. Wang (eds.)
Elsevier Science Publishers B.V. (North-Holland)

FIXED POINT ALGORITHMS AND ITS APPLICATIONS TO NONDIFFERENTIABLE PROGRAMMING

Tang Huanwen, Jiang Ye, Guo Jian and Hu Yunjiao

Dept. of Appl. Math., Dalian Univ. of Tcchnology, Dalian 116024, China

The main purpose of this paper is to apply the Merrill's algorithm and variable dimension restart algorithm to solve nondifferentiabli optimization problems. Some numerical results and extensions on the convergence of these algorithms used in nondifferentiable programming are also given.

1. INTRODUCTION

Let X be a nonempty set in R^n, and $f: X \to X$ a mapping, the following definitions are well known.

DEFINITION 1. A point $x^* \in X$ is called a (Brouwer) fixed point of f, if $f(x^*) = x^*$.

DEFINITION 2. Let $f: R^n_+ \to R^n_+$ be a (point to point) mapping, and $\epsilon > 0$, a point $x^* \in R^n_+$ is called an ϵ-fixed point of f, if $\|f(x^*) - x^*\| < \epsilon$.

It is clear that x may be taken as an approximate fixed point of f. And for the set-valued mapping, we have

DEFINITION 3. Let $F: R^n \to P(R^n)$ be a set-valued mapping in R^n, where $P(R^n)$ consists of all compact, convex and nonempty subsets of R^n. If there exists a point $x^* \in F(x^*)$, then x^* is called a Kakutani fixed point of F.

As early as in 1912, Brouwer proved the following theorem concerning the existence of fixed point for a point-to-point mapping.

THEOREM 1. (Brouwer) Let C be a compact, convex and nonempty subset of R^n. Then every continuous mapping $f: C \to C$ has at least one fixed point in C, i.e., $x^* \in C$, $f(x^*) = x^*$.

Then in 1942, Kakutani generalized this theorem to upper semicontinuous set-valued mapping.

THEOREM 2. (Kakutani) Let C be a compact, convex, nonempty subset of R^n, and $F: C \to P(C)$ an upper semicontinuous set-valued mapping, then F has at least one fixed point in C, i.e., $x^* \in C$, $x^* \in F(x^*)$.

We have recently employed the Merrill's algorithm and variable dimension fixed point algorithm in solving the system of nonlinear and nondifferentiable equations and some mathematical programmings, and obtained quite satisfactory results.

2. APPLICATION FOR FIXED POINT ALGORITHM IN NONSMOOTH OPTIMIZATION AND THE RELATED CONVERGENCE

Some results related to the generalized gradient are stated here without demonstration.

Let $f: R^n \to R$ be locally Lipschitzian.

DEFINITION 4. $f^0(x,v) = \lim_{\substack{y \to x \\ t \downarrow 0}} sup(f(y+tv) - f(y))/t$ is called the generalized directional derivative of f at point x with respect to direction v, in the sense of Clarke.

DEFINITION 5. $\partial f(x) = \{\xi \in R^n | f^0(x,v) \geq <\xi, v>, \forall v \in R^n\}$ is called the generalized gradient of f at point x.

THEOREM 3. $\partial f(x)$ is a compact, convex and nonempty set.

THEOREM 4. $\partial f(x)$ is upper semicontinuous in R.

THEOREM 5. If f is a finite convex function in a neighborhood of x, then $f(x)$ coincides with the subgradiential of f at $x \in U$, $Df(x)$, in the sense of convex analysis.

Proofs of these theorems can be found in [5].

I Unconstrained Minimization

(P_1) min $f(x)$ $x \in R^n$,

where $f: R^n \to R$ is locally Lipschitzian.

Define

$$G(x) = \{x\} - \partial f(x), \forall x \in R^n. \tag{1}$$

PROPOSITION 1. $G(x): R^n \to P(R^n)$ is an upper semicontinuous set-valued mapping.

PROPOSITION 2. Solving (P_1) is equivalent to finding the Kakutani fixed points of $G: R^n \to P(R^n)$.

II constrained Minimization

(P_2) min $f(x)$

s.t. $C_k(x) \le 0$, $k = 1, 2, \cdots, m$.

where f, $C_k: R^n \to R$ are locally Lipschitzian.

Define function C from R^n to R by $C(x) = \max_{k=1,2,\cdots,m.} C_k(x)$, $\forall x \in R^n$. We can write (P_2) as

$$(P_2') \quad \min \{f(x) | C(x) \le 0\}$$

Furthermore, let

$$G(x) = \begin{cases} \{x\} - \partial f(x), & for\ C(x) < 0, \\ \{x\} - conv\{\partial f(x) \cup \partial C(x)\}, & for\ C(x) = 0, \\ \{x\} - \partial C(x), & for\ C(x) > 0. \end{cases} \tag{2}$$

PROPOSITION 3. $G(x): R^n \to P(R^n)$ is an upper semicontinuous set-valued mapping.

PROPOSITION 4. Solving (P_2') is equivalent to finding the Kakutani fixed points of $G: R^n \to P(R^n)$.

In this way, we can solve an unconstrained or constrained minimization by finding the Kakutani fixed point of a set-valued mapping equivalently. In [7] a different way has been used to solve some nonsmooth minimizations by finding the zero-points of upper semicontinuous set-valued mappings, via homotopy algorithm.

The foregoing discussions suggest that it is necessary to find the generalized gradients of f and C in order to find the fixed points of G. It is well known that it is difficult to find the generalized gradients of $L.L.$ function. We have replaced the generalized gradients by ϵ-generalized gradients and a computational program in [1].

Now consider the following problem

$$(P_3) \quad \min f(x)$$

$$s.t. \quad C(x) \le 0$$

where $f: R^n \to R$ and $C: R^n \to R$ are all convex functions.

DEFINITION 6. Let $x^* \in S = \{x | C(x) \le 0\}$. For a given $\epsilon > 0$, if $f(x) \ge f(x^*) - \epsilon$, $\forall x \in S$, then we call x^* an ϵ–optimum of (P_3).

DEFINITION 7. Let $f: R^n \to R$ be continuous and $\epsilon \ge 0$. An ϵ– generalized gradient of f at point x is $\partial_\epsilon f(x) = \{\xi | f(y) - f(x) \ge <\xi, y - x> -\epsilon, \forall y \in R^n\}$.

When generalized gradients replaced by ϵ–generalized gradients, we can get the related mapping $G_\epsilon: R^n \to P(R^n)$ as in (1) and (2).

THEOREM 6. If $inf\{C(x) | x \in R^n\} < 0$, then for $\epsilon > 0$ small enough and $x^* \in B_\epsilon(x^*)$, x^* is a 2ϵ–optimum solution of (P_3).

Let

$$v_i = \lambda_i \frac{f(x + he_i) - f(x)}{h} + (1 - \lambda_i) \frac{f(x) - f(x - he_i)}{h},$$

where $\lambda_i \in [0, 1]$, $i = 1, 2, \cdots, n$. If f satisfies some kinds of convexity and h is sufficiently small, then $v = (v_1, v_2, \cdots, v_n) \in \partial_\epsilon f(x)$ [9], and $\overline{v} \in \partial_\epsilon f(x)$, where $\overline{v}$ is of the smallest norm of v_i's. Furthermore, if x^* is a minimum point of f and h is small enough, then $\overline{v} = 0$. Note that the above discussion may not be valid if f is an ordinary Lipschitz function, but in this case we can also handle the problem in an analogous way in practical computation, and some numerical experiments have shown that this method is feasible.

As for the global convergence of fixed point algorithms for nonsmooth optimization, we have the following theorems.

THEOREM 7. (Merrill's condition) Let F be an $U.S.C.$ set-valued mapping on R^n. Suppose there exist $x^0 \in R^n$, $\mu > 0$ and $\rho > 0$ such that for all x, $z \in R^n$, $f(x) \in F(x)$ with $x \notin B(x^0, \mu)$ and $z \in B(x, \rho)$, the following inequality holds

$$(f(x) - x)(x^0 - z) > 0, \tag{3}$$

then F has at least one fixed point in $B(x, \mu)$.

THEOREM 8. Let $f: R^n \to R$ be a convex function. If there exsits a bounded nonempty level set $Lev_\alpha f = \{x | f(x) \le \alpha\}$, then the condition (3) is satisfied, i.e., there exist $x^0 \in R^n$, $\mu > 0$ and $\rho > 0$ such that for all $x \notin B(x^0, \mu)$, $g \in G(x)$ and $z \in B(x, \rho)$, there holds

$(g - x)(x^0 - z) > 0.$

THEOREM 9. Let F be an $U.S.C.$ mapping on R^n, for some $\rho > 0$ exist $x^0 \in R^n$ and $\mu > 0$ such that for all x, $z \in R^n$ with $x \notin B(x^0, \mu)$ and $z \in B(x, \rho)$ the inequality $(f(x) - x)(x^0 - z) > 0$ holds for $f(x) \in F(x)$. Then the Merrill's algorithm and the variable dimension restart algorithm terminate in a completely labbled simplex if the mesh of the triangulation is less than ρ.

The rest of this section is to present some resułts obtained under weaker conditions than those given above.

DEFINITION 8. Let f be a real-valued function in R. If there exists a $\lambda \in [0,1]$ such that for any x, $y \in R^n$ satisfying $f(x) \leq f(y)$, one has

$$f(tx + (1-t)y) \leq \lambda t f(x) + (1 - \lambda t) f(y), \quad \forall t \in [0,1]$$

then $f(x)$ is called a λ-subconvex function in R.

DEFINITION 9. Let $f(x)$ be a Lipschitz function defined in R^n. Then $f(x)$ is called a generalized pseudo-convex function if for any $x_1, x_2 \in R^n$ such that $f(x_2) < f(x_1)$, the inequality $\xi^T(x_2 - x_1) < 0$, $\forall \xi \in \partial f(x_1)$ holds.

For unconstrained optimization (P_1), we have the following results.

THEOREM 10. Let $f : R^n \to R$ be Lipschitzian. If there exists a nonempty bounded level set $Lev_\alpha F$ such that

$$f_\alpha(x) = \begin{cases} f(x), & x \notin Lev_\alpha f, \\ \alpha, & x \in Lev_\alpha f. \end{cases}$$

is convex (λ-subconvex with $\lambda \in (0,1]$, or generalized pseudo-convex), then $G(x)$ (as defined in (1)) satisfies Merrill's condition.

THEOREM 11. If (P_1) satisfies the condition given in Theorem 10, then every cluster point of any infinite bounded sequence generalized by Merrill's algorithm or variable dimension restart algotithm is a stationary point of (P_1).

And for constrained optimization (P_2'), we have

THEOREM 12. Let f, $C : R^n \to R$ be Lipschitzian. If there exists a nonempty bounded level set $Lev_\alpha C$ such that

$$\phi(x) = \begin{cases} C(x) & x \notin Lev_\alpha C \\ \alpha & x \in Lev_\alpha C \end{cases}$$

is convex (λ-subconvex with $\lambda \in (0,1]$, or generalized pseudo-convex), then $G(x)$ (as defined in (2)) satisfies Merrill's condition.

THEOREM 13. Let f, $C : R^n \to R$ and $\phi(x)$ be defined as in Theorem 12. Assume that for some $\alpha < 0$, $LevC$ is bounded and nonempty, if $\phi(x)$ is a convex (λ-subconvex with $\lambda \in (0,1]$, or generalized pseudo-convex) function, then every cluster point of any infinite bounded sequence generated by Merrill's algorithm or variable dimension restart algorithm is a stationary point or a $K-T$ point of (P_2'). Furthermore, each minimizer of (P_2') must be a fixed point of $G(x)$.

In sum, these fixed point algorithms do not depend upon the choice of the initial points, in other words, they possess the properties of global cnvergences. On the other hand, the fixed point algorithm can be used for handling some more general functions including some nonsmooth functions. This implies that it is convenient to use fixed point algorithms for nonsmooth minimization.

3. NUMERICAL RESULTS

We have experienced with more than twenty standard optimization problems via Merrill's algorithm and variable dimension restart algorithm. These problems included differentiable and nondifferentiable, convex and nonconvex ones with or without constraints. For the purpose of comparision, we have computed the gradients with two different methods, one is the generalized gradients, the other is $\epsilon-$ generalized gradients, some of the numerical results are shown in Table 1, where (P) and $\epsilon - (P)$ represent the case in which the problem was solved by using exact generalized gradient and $\epsilon-$ generalized gradient, respectively. Note that some differentiable optimization algorithms failed to deal with these problems in that the convergence rates are slow or the precisions are not satisfactory. For example, for solving problem 1, if the descent direction were taken as the searching direction, the algorithm would converge to a non-minimum point.

The computational results show that the fixed point algorithms can be used to find a solution with desirable precision or even an exact solution and the convergence rate is fast, for problems with low dimensionalities. Thus it seems that the fixed point algorithms are promising for solving nondifferentiable optimization problems.

Table 1

No.	D.	(P) Optimum point	(P) t (s)	ϵ-(P) Optimum point	ϵ-(P) t (s)
1	2	(1.0003, 1)	3	(1, 1)	2
				(1, 1)	2
2	2	$(-1, -0.4998 \times 10^{-4})$	10	$(-1, 0)$	7
		$(-0.999, -0.3332\times10^{-4})$	8	$(-1, -0.3208\times10^{-4})$	6
3	10	$(1,1,\cdots,1)$	70	$(1,1,\cdots,1)$	160
				(1, 1, 1, 1, 1, 1, 0.9986 0.9986, 0.9991, 0.9957)	725
4	5	(1.013, 1.013, 1,1.019, 1.075)	235	(1.001, 1.001, 1, 1.011, 1.037)	432
		(0.9983, 1, 0.9983, 1, 1.171)	80	(1.007, 1.001, 0.9985, 1.007, 1.032)	176

Note: The problems in Table 1 are referred to [1], and D represents the dimensionality of a problem.

REFERENCES

[1] Guo Jian, Tang Huanwen and Jiang Ye, *J. of Dalian Univ. of Tech., Special Issue on math.*, in *Chinese*) 25(1986) 84.

[2] Tang Huanwen and Guo Jian, *J. Math. Res. Exposition*, (1987) 104.

[3] Jiang Ye, Guo Jian and Tang Huanwen, *Computational Math.* (in *Chinese*) 4(1988) 361.

[4] Hu Xinsheng and Tang Huanwen, *J. of Dalian Univ. of Tech.*, (in *Chinese*) 5(1989) 1.

[5] Clarke. F.H., Optimization and Nonsmooth Analysis (*Wiley-Interscience, New York, 1983)*

[6] Todd. M.J., The Computation of Fixed Points and Applications,*Lecture Notes in Economics and Mathematical Systems 124 (Springer,Berlin, West Germany*, 1976)

[7] Tang Huanwen and Hu Xinsheng, *J. of Dalian Univ. of Tech.*,(to appear)

[8] Wang Zeke, The Base of Simplicial Algotirhm for Computing Fixed Points *(Zhong Shan University Press, China*, 1986, in *Chinese*)

[9] J.J.Strodiot, *Math. Prog.*25(1983) 307.

Approximation, Optimization and Computing:
Theory and Applications, A.G. Law and C.L. Wang (eds.)
Elsevier Science Publishers B.V. (North-Holland)

OPTIMAL AND ADAPTIVE COMMUNICATION STRATEGIES IN DISTRIBUTED DECISION TREE ALGORITHMS

Alfred TAUDES

Department of Applied Computer Science, Vienna University of Economics and Business Administration, Austria, A-1090 Vienna, Augasse 2-6

We deal with communication problems which arise when solving combinatorial search problems in a distributed computing environment. In particular, we use Dynamic Programming in Markov Chains to determine optimal and adaptive strategies for the dissemination of intermediate results of decision tree algorithms. The method developed is applied to distributed sorting and searching.

1. INTRODUCTION

Decision Tree Algorithms solve *Selection Problems*. The solution of a selection problem is a set of elements of a *linearly ordered multi-set* the ranks of which have a specified property. Sort and search problems are the most important problems in this class. We deal with the distributed execution of decision tree algorithms on the *MRAM model of a computer*. A MRAM is a finite collection of RAMs that can communicate via message passing. It is a theoretical model for multiprocessor architectures without global shared memory. Thus a MRAM can be regarded as a simplified image of MIMD machines such as the increasingly popular networks of workstations or personal computers. The fundamental problem in programming combinatorial search algorithms on these machines is to find a "good" communication strategy. These programs typically use intermediate results of the search process as cut-off criteria to prune unpromising parts of the search space. On an MRAM machine no global shared memory is available for storing this information. It has to be communicated over the network. If too few message exchanges take place, superfluous searching can occur; if the implementation is too communication-intensive, time might be wasted for unnecessary memory coordination activities. In both cases, the speed-up of the distributed implementation can be reduced. Based on the representation of the execution of a decision tree algorithm as a *Markov Chain* we derive exact and simple asymptotic optimal policies for the communication of intermediate results of decision tree algorithms. Both the cases, the distribution of the elements of the linearly ordered multi-set is known and it has to be determined dynamically through the execution of the algorithm are analysed. The techniques developed are applied to the problem of distributed searching the maximum and to distributed sorting of the k largest elements. Furthermore, we discuss the application of this method to other combinatorial search problems.

2. BASIC DEFINITIONS

Selection Problems are problems on linearly ordered multi-sets. A linearly ordered multi-set is a finite multi-set $X = \{x_1, x_2, \cdots, x_n\}$ on which a compare operation $G(x_i, x_j)$ is defined which for each pair of elements of X either yields that $x_i > x_j$, that $x_i < x_j$ or that $x_i = x_j$. The *rank* $r(x_i)$ of $x_i \in X$ is defined as the number of elements $x_j \in X$ for which $G(x_i, x_j)$ yields $x_i > x_j$ plus the number of $x_k \in X$, $k < i$, for which $G(x_i, x_k)$ yields $x_i = x_k$. Let $I_{n,k} = \{(i_1, i_2, \cdots, i_k)\ i_j \neq i_l\ \forall j,\ l = 1, \cdots, k,\ 1 \leq i_j \leq n\ \forall j,\ k \leq n\}$, i.e. $I_{n,k}$ is the set of all k-tupels of integers less or equal n the elements of which are all distinct. A selection problem on X is specified by a set $J(X) \subseteq I_{n,k}$, which means that through the application of G one has to find k elements $\{x_{j_1}, x_{j_2}, \cdots, x_{j_k}\}$ of X such that $(r(x_{j_1}), r(x_{j_2}), \cdots, r(x_{j_k})) \in J(X)$. For example, if $J(X) = \{(n)\}$ one has to select the largest element of X; if $J(X) = \{(1, 2, \cdots, n)\}$ the ascending sort problem is specified; if $J(X)$ is the set of all permutations of $\{1, 2, \cdots, k\}, k \leq n$, the k smallest elements of X are requested.

The execution of an algorithm for a selection problem can be represented by a decision tree, a ternary tree whose nodes depict $G(x_i, x_j)$ and whose arcs resemble the three possible outcomes of this operation. Certain algorithms for selection problems such as heapsort or minimum search inspect each element of X in turn and store the x_i that are candidates for the solution according to the subset of X scrutinized so far in an appropriate data structure. The execution of these decision tree algorithms can be represented by a *state graph*, a simple, directed graph whose vertices resemble the possible intermediate results of the algorithm. Let $Y = \{y_1, \cdots, y_r\}$ be the set of values the x_i can take. [1] Then the vertex set $Z = \{z_1, \cdots, z_m\}$ of the state graph is the union of all combinations of r elements, taken $1, 2, \cdots, k$ times with repetitions, with or without reference to the ordering according to the specification of $J(X)$. There is an arc between two vertices ("states") $z_i = \{y_{i_1}, \cdots, y_{i_l}\}$ and $z_j = \{y_{j_1}, \cdots, y_{j_l}\}$ of the state graph iff ($j_l - i_l = 1$ and $i_l < k$) or ($j_l = k, i_l = k$ and $z_i \setminus z_j = \{y_{i_o}\}, z_j \setminus z_i = \{y_{j_o}\}$, with $y_{i_o} \leq y_{j_o}$ or $y_{i_o} \geq y_{j_o}$ according to the pruning criterium chosen).

[1] In the case of an uncountable set or for reasons of computational tractability, Y can also be specified as the set of intervalls the x_i can fall into.

Such an arc can be labelled with the probability of a state transition occuring, i.e. with the *transition probability* p_{ij}, which is the probability that an element of X takes a value y_{i_o} that transforms z_i to z_j, and with c_{ij}, the number of execution cycles it takes to move from state i to state j. Thus the state-graph can be represented by its *transition matrix* $P = \{p_{ij}\}$ and its *cost matrix* $C = \{c_{ij}\}$. [2] A particular execution of such a decision tree algorithm can be viewed as a random path leading from the initial state of the state-graph to a final state. Through proper sorting of Z, P can be brought into an upper triangular form: at first, sort the z_i according to their cardinality and then in lexikographic order according to the pruning criterium. Thus P is the transition matrix of an absorbing Markov Chain.

This representation leads us to the derivation of the expected running time of a decision tree algorithm on a RAM with uniform complexity measure (for the definition of a RAM program and of the uniform complexity measure see e.g. [Aho, Hopcroft, Ullman]):

$$e_i = r + P \times e_{i+1} \qquad i = 1, \cdots, n-1 \tag{1}$$

with

r - column-vector of length m denoting the expected running times for each state in one transition ($r_i = \sum_{j=1}^{m} c_{ij} \cdot p_{ij} \; \forall i$),

e_i - column-vector of length m denoting the expected running times for each state after i inspections ($e_n = 0$).

3. OPTIMAL AND ADAPTIVE COMMUNICATION STRATEGIES

In a distributed computing environment one has more than one RAM available to solve a selection problem. Therefore each RAM can scrutinize a subset of X. As can be seen from (1) the average time complexity of a decision tree algorithm depends on the state the algorithm is in. Thus it can "pay" to communicate an intermediate result between the RAMs, provided that the computational cost of such a communication in terms of execution cycles needed is offset by a gain in the local execution time. We set out to find situations when this is the case, i.e. we want to determine an *optimal decision function* that for every computational step and state yields whether a RAM should communicate in such a way that the average time complexity of the decision tree algorithm is minimized. In the sequel we assume that each RAM has a SEND instruction additional to its local instructions. SEND shall write the content of the local accumulator into the accumulator of the receiving RAM within $d, d \in N$ execution cycles. The communication protocol shall function in such a way that the receiving RAM sends its state to the sending RAM if the expected running time of the state received is greater than the one in the local state. This completes our definition of the MRAM-machine.

[2] Thus these decision tree algorithms are *random algorithms* in the terminology of [Kemp].

Using our representation of a decision tree algorithm as a Markov Chain the communication problem can be formalized as a choice problem between the continuation with normal transitions from the current local state and the transition with the state of another RAM at the price of the communication cost. The current state of another RAM in unknown due to the fact that our MRAM computer lacks global shared memory. However, as previous states are known, we can use the basic result of Markov Chain Theory which says that the state probabilities after w transitions are given by the product of the initial state probabilities with the w-step transition matrix $Q(w) = \{q_{ij}(u)\}$, $Q(w) = P^w$ (see e.g. [Howard1]).

If a RAM is in state l after i inspections and it has communicated with another RAM of the MRAM-machine w transitions ago, one can expect an effort due to communication of $\sum_{j=l+1}^{m} q_{uj}(w)(r_j + e_{i+1j}) + (\sum_{j=u}^{l} q_{uj}(w))(r_l + e_{i+1l}) + d$, where u denotes the state the receiving RAM has been in w transitions ago. The expected effort for the receiving RAM is $(\sum_{j=u}^{l} q_{uj}(w))(r_l + e_{i+1l}) + \sum_{j=l+1}^{m} q_{uj}(w)(r_j + e_{i+1j}) + d \cdot (\sum_{j=l+1}^{m} q_{uj}(w))$, and it is in state j with probability $\sum_{j=u}^{m} q_{uj}(w)$. Thus the unknown decision function must satisfy the following functional equation (see e.g. [Howard2]):

$$e_i^\star = \min \left\{ \begin{array}{l} r + P \times e_{i+1} \\ 2 \cdot (q(w) \times (r(l) + e_{i+1}(l))) + \\ d \cdot (1 + I(l) \times q(w)) - q(w) \times (r + e_{i+1}) \end{array} \right. \tag{2}$$

with

$i = 1, \cdots, n-1$,

$q(w) = (q_{u1}(w), \cdots, q_{um}(w))$,

$r(l) = (r_l, \cdots, r_l, r_{l+1}, r_{l+2} \cdots, r_m)$,

$e_{i+1}(l) = (e_{i+1l}, \cdots, e_{i+1l}, e_{i+1l+1}, e_{i+1l+2} \cdots, e_{i+1m})$,

$I(l)$ - vector of l zeroes and $m - l$ ones.

From this equation the decision function can be determined via backward induction using the fact that $e_n = 0$. However, for large n such a procedure is computationally intractable. Thus we resort to asymptotic results. In our case we can concentrate on "relative values" $s_i = (s_{i1}, \cdots, s_{im})^t$, $s_i = e_i - (c_{mm} \cdot (n - i))$, $i = 1, \cdots, n$ because the only result of a communication is a state-change and the state eventually reached is the trapping state z_m. Using e.g. the fact that $s_i \geq 0 \; \forall i$ and that $\lim_{i \to \infty} s_{ij} \leq \sum_{i=1}^{\infty} \max_{ij} c_{ij} \cdot (1 - p_{jm})^i$ $\forall j$ it follows that $\lim_{i \to \infty} s_i = v$ exists and we can obtain an upper bound for the expected running times per state by solving $v = (r - c_{mm}) + P \times v$ with $v_m = 0$.

Now assume a myoptic policy, i.e. that only the effects of one communication are considered. If the communication partner has been in state u w transitions before, a sending

RAM in state l can expect an effort due to communication that is lower than $\sum_{j=l+1}^{m-1} q_{uj}(w)v_j + (\sum_{j=u}^{l} q_{uj}(w))v_l + d$. For a RAM that has been in state u w transitions ago and that receives state l, one can expect an asymptotic effort of $(\sum_{j=u}^{l} q_{uj}(w))v_l + \sum_{j=l+1}^{m-1} q_{uj}(w)v_j + d \cdot (\sum_{j=l+1}^{m-1} q_{uj}(w))$. Thus a RAM of the MRAM-machine should communicate if it is in a state l where

$$2 \cdot (q(\acute{w}) \times v(l)) + d \cdot (1 + I(\acute{l}) \times q(\acute{w})) - q(\acute{w}) \times v - v_l < 0. \quad (3)$$

with

$$q(\acute{w}) = (q_{11}(w), \cdots, q_{1m-1}(w)),$$

$$v(l) = (v_l, \cdots, v_l, v_{l+1}, \cdots, v_{m-1}),$$

$I(\acute{l})$ - vector of l zeroes and $m - l - 1$ ones.

Further simplifications of this decision rule are possible. An easily computable upper bound can be obtained by neglecting eventual positive effects for the receiving RAM:

$$q(\acute{w}) \times v(l) + 2d - v_l \leq \quad (4)$$

$$\text{(Chauchy-Schwarz)}$$

$$[\sum_{j=1}^{m-1} q_{1j}(w)^2 \cdot \sum_{j=1}^{m-1} v_j(l)^2]^{\frac{1}{2}} + 2d - v_l \leq$$

$$[(m-1) \cdot \max_{j=1}^{m-1} q_{1j}(w)^2 \cdot (m-1) \max_{j=1}^{m-1} v_j(l)^2]^{\frac{1}{2}} + 2d - v_l \leq$$

$$[(m-1) \cdot \max_{i,j=1}^{m-1} q_{ij}(w)^2 \cdot (m-1) \max_{j=1}^{m-1} v_j(l)^2]^{\frac{1}{2}} + 2d - v_l \leq$$

$$\text{(properties of the } R_\infty\text{-Norm)}$$

$$(m-1) \cdot (\max_{i,j=1}^{m-1} p_{ij})^w \cdot \max_{j=1}^{m-1} v_j(l) + 2d - v_l.$$

If the distribution of the x_i's ist not (precisely) known in advance, one can start out with a prior distribution of P, and "learn" the transition probabilities through repeated executions using *Bayes Theorem*. Let $F = \{f_{ij}\}$ denote the matrix of the actual numbers of state changes during an execution. If the prior distribution of P is Matrix Beta with parameters $A = \{a_{ij}\}, a_{ij} > 0 \ \forall i, j$,

$$f(P|A) = \prod_{i,j=1}^{m} \frac{\Gamma(a_{i1} + a_{i2} + \cdots + a_{im-1})}{\prod_{j=1}^{m} \Gamma(a_{ij})} p_{ij}^{a_{ij}-1}, \quad (5)$$

it is shown in [Martin] that the posterior distribution $f(P|A, F)$ is again Matrix Beta with means $(a_{ij} + f_{ij})/(\sum_{j=1}^{m} a_{ij} + \sum_{j=1}^{m} f_{ij}) \ \forall i, j$. Thus one can, for instance, start out with an optimal policy based on a uniform distribution and redetermine the strategy after every execution based on F.

4. EXAMPLES

Let us start with a well-known problem: the determination of the maximum of an unsorted multi-set of integers. An obvious algorithm that solves this selection problem starts out with the first element as temporary solution, scans all elements of the multi-set and changes the temporary solution whenever a larger element is found. If one assumes that the elements of the multi-set are equally distributed in the interval $[1, \cdots, m]$, the state-graph of this decision tree algorithm can be represented by

$$p_{ij} = \begin{cases} \frac{1}{m} & \text{if } i < j \\ \frac{i}{m} & \text{if } i = j \\ 0 & \text{else,} \end{cases} \quad (6)$$

and

$$c_{ij} = \begin{cases} c_v & \text{if } i = j \\ c_v + c_c & \text{else,} \end{cases} \quad (7)$$

where c_v denotes the cost of a compare operation and c_c denotes the cost of a state-change (i.e. a variable assignment). Thus $v_i = \frac{c_c}{m-i} \sum_{j=i+1}^{m} v_j$ and it can be shown by induction that $v_i = c_c \cdot H_{m-i}$ with $H_i = \sum_{j=1}^{i} \frac{1}{j}$ - i-th harmonic number. This is nothing new (for the "standard solution" see. e.g. [Knuth]). However, our Markovian approach yields for instance, that another RAM should be contacted for its temporary solution if the local state l and the period of isolated activity w are such that that $v_l > 2d + v_l \cdot (m-1) \cdot (\frac{m-1}{m})^w$. This decision function is monotone, i.e. given d for every $l = 1, \cdots, m$ it either does not pay to communicate at all or there exists a critical period of isolated activity $w^\star$ such that $\forall w > w^\star$ it pays to communicate. This is a very important property from the implementation point of view. Once determined, just these cut-off points have to be stored and checked during an execution.

Now consider the problem of sorting the k largest elements of a multi-set. Assume that the x_i are uniformly distributed in the range $[1, \cdots, r]$. For reasons of simplicity we additionally postulate that k is small so that the "candidate elements" can be stored in a sorted linear list. Then it holds for $z_i = \{y_{i_1}, \cdots, y_{i_l}\}$ $i = 1, \cdots, m-1$ that

$$r_i = \begin{cases} c_{in} + c_c \cdot \sum_{j=1}^{i_l} j \cdot \prod_{o=1}^{j} (\frac{r - y_{io}}{r}) & \text{if } i_l < k \\ (c_{in} + c_{de}) \cdot (\frac{r - y_{i1}}{r}) + c_c \cdot \sum_{j=1}^{k} j \cdot \prod_{o=1}^{l} (\frac{r - y_{io}}{r}) & \text{else,} \end{cases} \quad (8)$$

and

$$v_i = \begin{cases} r_i + \frac{1}{r} \cdot \sum_{j=1}^{r} v_{o=Ind(z_i \cup j)} & \text{if } i_l < k \\ r \cdot (r_i + \frac{1}{r} \cdot \sum_{y_{i1}+1}^{r} v_{o=Ind(z_i \setminus y_{i1} \cup j)})/y_{i1} & \text{else,} \end{cases} \tag{9}$$

where $Ind(z_l)$ returns the index of a state that is defined by the set z_l and c_{in} denotes the cost of the insert operation and c_{de} the cost of the delete operation for the list. Again the decision function is monotone.

5. CONCLUSION

The method developed can be applied to combinatorial algorithms other than decision tree algorithms, too. For instance, if the x_i are vectors and G is defined as the comparison of the sum of the elements of two such vectors, we can use the techniques given above to determine optimal and adaptive communication strategies in the distributed search of the column of a matrix with minimal column sum. Similarly, if G is defined as the comparison of the maximum element of such a vector, we can analyze coordination patterns for a simple variant of alpha-beta search (see [Knuth, Moore]). Furthermore, if one approximates the search process on a search tree, e.g. generated through the application of a branch and bound method, by the process of searching the set of paths from the root of such a tree to a leaf, one can tackle communication issues of more complex combinatorial optimization problems, too.

REFERENCES

[Aho, Hopcroft, Ullman] Aho, A.V., J.E. Hopcroft und J.D. Ullman, *Data Structures and Algorithms*, Reading, Mass. u.a., Addison-Wesley, 1985

[Howard1] Howard R.A., *Dynamic Probabilistic Systems, Vol I: Markov Models*, Wiley, New York, 1971

[Howard2] Howard R.A., *Dynamic Probabilistic Systems, Vol II: Semi-Markov and Decision Processes*, New York, 1971

[Kemp] Kemp R., *Fundamentals of the Average Case Analysis of Particular Algorithms*, Stuttgart u.a., Teubner, Wiley, 1984

[Knuth] Knuth, D.E. *Fundamental Algorithms*, Reading, Mass. u.a., Addison-Wesley, 1. The Art of Computer Programming, 1973

[Knuth, Moore] Knuth D.E. und R.W. Moore, An Analysis of Alpha-Beta Pruning, *Artificial Intelligence*, Vol. 6, 293-326, 1975

[Martin] Martin J.J., *Bayesian Decision Problems and Markov Chains*, R.E. Krieger, New York, 1975

Approximation, Optimization and Computing:
Theory and Applications, A.G. Law and C.L. Wang (eds.)
Elsevier Science Publishers B.V. (North-Holland)

FUZZY STOCHASTIC MULTI-OBJECTIVE PROGRAMMING

Wang Caihua and Liu Pu

Department of Engineering Mechanics, Chongqing University
Chongqing, China

In this paper, we present a method of Fuzzy Stochastic Multi-objective Programming (FSMOP). Fuzziness is treated by establishing membership function of fuzzy function; Randomness is treated by using probability constrain method. Therfore, FSMOP is turned into comon multi-objective programming, three kinds of solution are given. Fuzzy Multi-objective Programming and Stochastic Multi-objective Programming can be solved as a special example by using this method.

1. INTRODUCTION

People are faced with more and more complex problems in modern engineering programming and design. Fuzziness, randomness and multi-objective are usually involved in these problems [1]—[5] . Especially, some great complex engineering projects are related to political situations, economic policies, environment conditions besides physical characteristic and geometry features. These factors are more of fuzziness and randomness, For the above-mentioned problems, we can solve the following fuzzy stochastic multi-objective programming:

$$
\begin{aligned}
\text{Fin}\quad & x=(x_1,x_2,\dots,x_n)^T\\
\min\quad & F(x,\tilde{\bar{A}}_j)=(f_1(x,\tilde{\bar{A}}_1),f_2(x,\tilde{\bar{A}}_2),\dots,f_m(x,\tilde{\bar{A}}_m))^T \qquad (1.1)\\
\text{s.t.}\quad & g_k(x,\tilde{\bar{A}}_k)\lessapprox \tilde{\bar{G}}_k,\quad k=1,2,\dots,t\\
& x_i\geqslant 0,\quad i=1,2,\dots,n
\end{aligned}
$$

where,

$$
f_j(x,\tilde{\bar{A}}_j)=\sum_{r=1}^{p}\tilde{\bar{A}}_{jr}x_1^{\beta_{jr1}}x_2^{\beta_{jr2}}\cdots x_n^{\beta_{jrn}}
$$

$$
g_k(x,\tilde{\bar{A}}_k)=\sum_{s=1}^{q}\tilde{\bar{A}}_{ks}x_1^{\beta_{ks1}}x_2^{\beta_{ks2}}\cdots x_n^{\beta_{ksn}}
$$

$$
j=1,2,\dots,m;\quad k=1,2,\dots,t
$$

$\tilde{\bar{A}}_{jr}$, $\tilde{\bar{A}}_{ks}$ and $\tilde{\bar{G}}_k$ are fuzzy stochastic numbers, the sign "~" means fuzziness and "–" means randomness; x_i is a variable with certainty; β_{jri} and β_{ksi} are arbitrary real numbers.
Now, studies on FSMOP are just beginning, This paper try to offer a basic method on the basis of reference[6]-[8].

2. TREATMENT OF FUZZINESS

Programming (1.1) can be equally expressed as the following:

$$
\begin{aligned}
\text{Find}\quad & x=(x_1,x_2,\dots x_n)^T\\
\text{goal}\quad & f_j(x,\tilde{\bar{A}}_j)=\sum_{r=1}^{p}\tilde{\bar{A}}_{jr}x_1^{\beta_{jr1}}x_2^{\beta_{jr2}}\cdots x_n^{\beta_{jrn}}\lessapprox \tilde{\bar{E}}_j \qquad (2.1)\\
\text{s.t.}\quad & g_k(x,\tilde{\bar{A}}_k)=\sum_{s=1}^{q}A_{ks}x_1^{\beta_{ks1}}x_2^{\beta_{ks2}}\cdots x_n^{\beta_{ksn}}\lessapprox \tilde{\bar{G}}_k\\
& i=1,2,\dots,n;\ j=1,2,\dots,m;\\
& k=1,2,\dots,t.
\end{aligned}
$$

where $\tilde{\bar{E}}_j$ is expectation level for each objective $f_j(x,\tilde{\bar{A}}_j)$ $(j=1,2,\dots,m)$. If it is difficult to give expectation level $\tilde{\bar{E}}_j$ we can let

$$
\tilde{\bar{E}}_j^{(k)}=f_j^{(k-1)}(x,\tilde{\bar{A}}_j)
$$

So, programming (2.1) can be written as the following standard form:

$$
\begin{aligned}
\text{Find}\quad & x=(x_1,x_2,\dots,x_n)^T\\
\text{s.t.}\quad & \tilde{\bar{Y}}_1=\tilde{\bar{A}}_{10}y_{10}+\tilde{\bar{A}}_{11}y_{11}+\dots+\tilde{\bar{A}}_{1p}y_{1p}\gtrsim 0\\
& \tilde{\bar{Y}}_2=\tilde{\bar{A}}_{20}y_{20}+\tilde{\bar{A}}_{21}y_{21}+\dots+\tilde{\bar{A}}_{2p}y_{2p}\gtrsim 0\\
& \cdots\cdots\cdots\cdots\\
& \tilde{\bar{Y}}_{m+t}=\tilde{\bar{A}}_{m+t,0}y_{m+t,0}+\dots+\tilde{\bar{A}}_{m+t,p}y_{m+t,p}\gtrsim 0 \qquad (2.2)
\end{aligned}
$$

$$
\text{where,}\quad y_{jr}=x_1^{\beta_{jr1}}x_2^{\beta_{jr2}}\cdots x_n^{\beta_{jrn}} \qquad (2.3)
$$

$$
\begin{aligned}
& y_{j0}=1,\quad x_i\geqslant 0\\
& i=1,2,\dots,n;\ j=1,2,\dots,m+t;\\
& r=0,1,\dots,p;\ p=\max(p,q).
\end{aligned}
$$

According to the extension principle membership function of $\tilde{\bar{Y}}_j$ is:

$$\mu_{\tilde{\bar{Y}}_j}(y_j)=\begin{cases}\max\limits_{(a_{jr}|y_j=f_j(y_{jr},a_{jr}))}(\min \mu_{\tilde{\bar{A}}_{jr}}(a_{jr})) \\ \quad ((a_{jr}|y_j=f_j(y_{jr},a_{jr}))=\emptyset) \\ 0 \quad (\text{others})\end{cases}$$
$$j=1,2,\ldots,m+t;\ r=0,1,\ldots,p \qquad (2.4)$$

where,

$$\mu_{\tilde{\bar{A}}_{jr}}(a_{jr})=\begin{cases}1-\dfrac{|a_{jr}-\bar{A}_{jr}|}{C_{jr}} \\ \quad (\bar{A}_{jr}-C_{jr}\leqslant a_{jr}\leqslant \bar{A}_{jr}+C_{jr}) \\ 0 \quad (\text{others})\end{cases}$$
$$j=1,2,\ldots,m+t;\ r=0,1,\ldots,p \qquad (2.5)$$

$\bar{A}_{jr}$ is the main value of $\tilde{\bar{A}}_{jr}$, C_{jr} are left and right spreads.

Substituting (2.5) into (2.4) we obtain

$$\mu_{\tilde{\bar{Y}}_j}(y_j)= 1-\frac{|y_j-\sum_{r=0}^{p}\bar{A}_{jr}y_{jr}|}{\sum_{r=0}^{p}C_{jr}y_{jr}} \qquad (2.6)$$
$$j=1,2,\ldots,m+t$$

$\tilde{\bar{Y}}_j \gtrsim 0$ stands for " $\tilde{\bar{Y}}_j$ is approximately positive", let $y_j=0$ from (2.6), it can equally expressed as the following[6]:

$$\mu_{\tilde{\bar{Y}}_j}(0)= 1-\frac{\sum_{r=0}^{p}\bar{A}_{jr}y_{jr}}{\sum_{r=0}^{p}C_{jr}y_{jr}}\leqslant 1 - h$$

$$\sum_{r=0}^{p}\bar{A}_{jr}y_{jr}\geqslant 0, \quad j=1,2,\ldots,m+t \qquad (2.7)$$

where $h\in[0,1]$. From (2.7), we obtain

$$\sum_{r=0}^{p}(\bar{A}_{jr}-hC_{jr})y_{jr}\geqslant 0 \qquad (2.8)$$
$$j=1,2,\ldots,m+t$$

So, programming (2.2) can be turned into the following stochastic programming:

Find $x = (x_1,x_2,\ldots,x_n)^T$

max $\bar{f}_s= \bar{A}_{s0}+ \bar{A}_{s1}y_{s1}+\ldots+ \bar{A}_{sp}y_{sp}$

s.t. $\bar{g}_j= (\bar{A}_{j0}- hC_{j0})+\sum_{r=1}^{p}(\bar{A}_{jr}-hC_{jr})y_{jr}\geqslant 0 \qquad (2.9)$

$0\leqslant h\leqslant 1$

$s\in(1,2,\ldots,m)$; $j=1,2,\ldots,m+t$

where $\bar{f}_s$ is the main value of s-th objective which is the most important among m objectives.

3. TREATMENT OF RANDOMNESS

The mean and the standard deviation of $\bar{f}_s$ are given by

$$\mu_f=\mu_{s0}+\mu_{s1}y_{s1}+\ldots+\mu_{sp}y_{sp} \qquad (3.1)$$

$$\sigma_f=(\sum_{r=0}^{p}(\frac{\partial f_s}{\partial A_{sr}}\Big|_{A_s=\mu_s})^2\sigma_{sr}^2)^{\frac{1}{2}} \qquad (3.2)$$

So, we have the new objective function:

$$F_s= k_1\,\mu_f + k_2\,\sigma_f \qquad (3.3)$$

where $k_1\geqslant 0$ and $k_2\geqslant 0$, respectively, represent the degree of importance of μ_f and σ_f for optimization.

Because of existence of stochastic number $\bar{A}_{jr}$, constraint condition of programming (2.9) can be turned into the following probabiblty constraint:

$$\int_0^\infty f_j(g_j)dg_j\geqslant p_j, \quad j=1,2,\ldots,m+t \qquad (3.4)$$

where $f_j(g_j)$ is probability density function of constraint $\bar{g}_j$. The mean value and the standard deviation of $\bar{g}_j$ are respectively:

$$\mu_{g_j}= \mu_{j0}-hC_{j0}+\sum_{r=1}^{p}(\mu_{jr}-hC_{jr})y_{jr} \qquad (3.5)$$

$$\sigma_{g_j}= (\sum_{r=0}^{p}(\frac{\partial\bar{g}_j}{\partial\bar{A}_{jr}}\Big|_{A_j=\mu_j})^2\,\sigma_{jr}^2)^{\frac{1}{2}} \qquad (3.6)$$

So, (3.4) can be rewritten as:

$$(2\pi)^{-\frac{1}{2}}\int_{-(\mu_{g_j}/\sigma_{g_j})}^{\infty}e^{-\frac{1}{2}\theta^2}d\theta\geqslant(2\pi)^{-\frac{1}{2}}\int_{-\Phi_j(p_j)}^{\infty}e^{-\frac{1}{2}z^2}dz \qquad (3.7)$$

where $\theta=(\bar{g}_j-\mu_{g_j})/\sigma_{g_j}$, $\Phi_j(p_j)$ is the value of standard normal state variabl of corresponding probability p_j.

From (3.7), we have:

$$\mu_{g_j}-\Phi_j(p_j)(\sum_{r=0}^{p}(\frac{\partial g_j}{\partial\bar{A}_{jr}}\Big|_{A_j=\mu_j})^2\sigma_{jr}^2)^{\frac{1}{2}}\quad 0$$
$$j=1,2,\ldots,m+t \qquad (3.8)$$

So, stochastic programming (2.9) can be changed to the following common programming:

Find $x = (x_1,x_2,\ldots,x_n)^T$

max $F_s= k_1(\mu_{s0}+\mu_{s1}y_{s1}+\ldots+\mu_{sp}y_{sp})+k_2(\sum_{r=0}^{p}(\frac{\partial\bar{f}_s}{\partial\bar{A}_{jr}}\Big|_{A_s=\mu_s})^2\,\sigma_{sr}^2)^{\frac{1}{2}}$

s.t. $\mu_{j0}-hC_{j0}+\sum_{r=1}^{p}(\mu_{jr}-hC_{jr})y_{jr}-$

$$\Phi_j(p_j)\left(\sum_{r=0}^{p}\left(\frac{\partial \bar{g}_j}{\partial \bar{A}_{jr}}\Big|_{A_j=\mu_j}\right)^2 \sigma_{jr}^2\right)^{\frac{1}{2}} \geqslant 0$$

$$0 \leqslant h \leqslant 1 \qquad (3.9)$$

$$s \in (1,2,\ldots,m); \; j=1,2,\ldots,m+t$$

Fuzzy stochastic number $\bar{A}_{jr}$ can be expressed by its main value $\bar{A}_{jr}$, spread C_{jr}, and standar deviation σ_{jr}

$$\underset{\sim}{\bar{A}}_{jr} = (A_{jr},\ C_{jr},\ \sigma_{jr}) \qquad (3.10)$$

For example, fuzzy stochastic number 5, if spread $C=2$, standard deviation $\sigma=0.5$, then $\underset{\sim}{\bar{5}} = (5,\ 2,\ 0.5)$.

4. SOME KINDS OF CONCRETE SOLUTION METHODS

4.1. Main Objective Method

It is a method of solving problem directly according to programming (3.9).

(1) Let $h^{(0)}=\delta$ (δ is a very small positive number), we get a corresponding optimal solution $x^{(0)}$.

(2) Let $h^{(k+1)}=h^{(k)}+\delta$, $k=0,1,\ldots$

(3) Put the optimal solution $x^{(k)}$ into the constraints of (k+1)th. If $x^{(k)}$ is feasible, let k+1=k and go to (2).

(4) Output the optimal solution $x^*=x^{(k)}$, stop.

4.2. Maxi-min Method

From programming (3.9) , under the condition of satisfying t constraints of the original problem, we single optimize each objective F_s $(s=1,2,\ldots,m)$, and then get the corresponding constraint optimial solution x_s^* and optimal value $F_s^*(s=1,2,\ldots,m)$. We make each optimal solution fuzzization and get fuzzization optimal solution $\underset{\sim}{N}_s$, its membership function is

$$\mu_{\underset{\sim}{N}_s}(x) = \left(\frac{F_s^h - F_s(x)}{F_s^h - F_s^*}\right)^q \qquad (4.1)$$

where F_s^h is the worst (largest) constraint value of F_s, $\mu_{\underset{\sim}{N}_s}(x)$ should be monotonous decreasing function on $[F_s^*, F_s^h]$, $q=1,2,\frac{1}{2},3,\ldots$

Let
$$\underset{\sim}{D} = \bigcap_{s=1}^{m} \underset{\sim}{N}_s \qquad (4.2)$$

$$\mu_{\underset{\sim}{D}}(x^*) = \max_{x\in R}\left(\bigwedge_{s=1}^{m} \mu_{\underset{\sim}{N}_s}(x)\right) \qquad (4.3)$$

where R is the feasible region.
In order to get optimale solution x^* from (4.3), the problem is turned into the following single objective programming:

Find $x = (x_1,x_2,\ldots,x_n)^T$

max λ

s.t. $\mu_{\underset{\sim}{N}_s}(x)$

$$\mu_{j0}-hC_{j0}+\sum_{r=1}^{p}(\mu_{jr}-hC_{jr})y_{jr} - \qquad (4.4)$$

$$\Phi_j(p_j)\left(\sum_{r=1}^{p}\left(\frac{\partial g_j}{\partial \bar{A}_{jr}}\Big|_{\bar{A}_j=\mu_j}\right)^2 \sigma_{jr}^2\right)^{\frac{1}{2}} \geqslant 0$$

$$0 \leqslant \lambda \leqslant 1, \quad 0 \leqslant h \leqslant 1$$

$$s=1,2,\ldots,m; \; j=m+1,m+2,\ldots,m+t$$

The solution of programming (4.4) is the same as that of main objective.

4.3. Synthetic Objective Method

Let
$$M = \sum_{s=1}^{m} w_s \cdot \mu_{\underset{\sim}{N}_s}(x) \qquad (4.5)$$

where w_s is the weight of objective F_s.
Therefore, the problem of solving optimal solution is the following programming problem:

Find $x = (x_1,x_2,\ldots,x_n)^T$

$$\max \quad M = \sum_{s=1}^{m} w_s \cdot \mu_{\underset{\sim}{N}_s}(x) \qquad (4.6)$$

$$\text{s.t.} \quad \mu_{j0}-hC_{j0}+\sum_{r=1}^{p}(\mu_{jr}-hC_{jr})y_{jr} -$$

$$\Phi_j(p_j)\left(\sum_{r=0}^{p}\left(\frac{\partial \bar{g}_j}{\partial \bar{A}_{jr}}\Big|_{A_j=\mu_j}\right)^2 \sigma_{jr}^2\right)^{\frac{1}{2}} \geqslant 0$$

$$0 \leqslant h \leqslant 1, \; j=m+1,m+2,\ldots,m+t$$

The solution is also the same as that of main objective.

In the above discussion , if tne problem is only of fuzziness, then just let $\sigma_{jr} = 0$; if only randomness then just let $C_{jr}=0$. The above solution is still suitable.

5. EXAMPLE

Solving FSMOP problem:

Find $x = (x_1,x_2,x_3)^T$

goal $\underset{\sim}{\bar{f}}_1 = \underset{\sim}{\bar{2}} x_1 x_2 x_3 \underset{\sim}{\geq} 10$

$$\underset{\sim}{\bar{f}}_2 = \underset{\sim}{\bar{1}}x_1^2 + \underset{\sim}{\bar{2}}(x_2-x_3)^2 \lesssim \underset{\sim}{\bar{16}}$$

$$\underset{\sim}{\bar{f}}_3 = \underset{\sim}{\bar{1}}(x_1-x_2)^2 + \underset{\sim}{\bar{1}}(x_2-x_3)^2 \lesssim \underset{\sim}{\bar{7}}$$

$$\text{s.t.}\quad \underset{\sim}{\bar{g}}_4 = \underset{\sim}{\bar{1}}x_1 + \underset{\sim}{\bar{1}}x_1x_2 + \underset{\sim}{\bar{2}}x_3 \lessapprox \underset{\approx}{\bar{20}}$$

$$\underset{\sim}{\bar{g}}_5 = \underset{\sim}{\bar{1}}x_1^2 + \underset{\sim}{\bar{1}}x_2^2 + \underset{\sim}{\bar{1}}x_3^2 \lesssim \underset{\approx}{\bar{35}}$$

$$x_i \geq 0, \quad i = 1,2,3.$$

where, fuzzy stochastic numbers are respectively:

$\underset{\sim}{\bar{2}}=(2,1,0.2)$ $\underset{\sim}{\bar{10}}=(10,4,2)$ $\underset{\sim}{\bar{1}}=(1,0.3,0.06)$

$\underset{\sim}{\bar{2}}=(2,0.8,0.4)$ $\underset{\sim}{\bar{16}}=(16,6,2)$ $\underset{\sim}{\bar{1}}=(1,0.6,0.2)$

$\underset{\sim}{\bar{1}}=(1,0.6,0.15)$ $\underset{\sim}{\bar{7}}=(7,3,2)$ $\underset{\sim}{\bar{1}}=(1,0.3,0.1)$

$\underset{\sim}{\bar{1}}=(1,0.2,0.05)$ $\underset{\sim}{\bar{2}}=(2,0.8,0.15)$ $\underset{\approx}{\bar{20}}=(20,8,4)$

$\underset{\sim}{\bar{1}}=(1,0.6,0.2)$ $\underset{\sim}{\bar{1}}=(1,0.4,0.2)$

$\underset{\sim}{\bar{1}}=(1,0.55,0.15)$ $\underset{\approx}{\bar{35}}=(35,14,6)$

Applying main objective method. Let $k_1=1$, $k_2=0.3$, $p_j=0.975$, While $h=0.45$, we have:

$$x_1^*=0.82 \quad x_2^*=2.85 \quad x_3^*=2.11$$

Applying maxi-min method. we get constraint optimal value and worst value: $F_1^*=-29.23$, $F_1^h=-5.32$; $F_2^*=0$, $F_2^h=7.51$; $F_3^*=0$, $F_3^h=8.0$. Let $w_1=0.3$, $W_2=0.4$, $w_3=0.3$, $p_j=0.975$, while $h=0.45$, we have:

$$x_1^*=1.32 \quad x_2^*=2.01 \quad x_3^*=1.73$$

Applying synthetical objective method. Let $w_1=0.3$, $w_2=0.4$, $w_3=0.3$, $p_j=0.975$, while $h=0.45$, we have:

$$x_1^*=0.92 \quad x_2^*=2.64 \quad x_3^*=2.04$$

The weight of each objective is not considered in the maxi-min method, so its solution has a little difference from the first and third method. How to choose the weight ,please refer to reference[10].

6. CONCLUDING REMARKS

The method can be used to solve diverse FSMOP problems; can automatically give a series of optimal results with different expectation levels in order to be selected by users; can give different value of h, and get the optimal results with different safty levels for user.

REFERENCES

[1] Wang Guangyuan The Initial Exploration of Structural Soft Design Theory, Herbin Architectural Engineering College, 1987.

[2] Wang Guangyuan and Ou Jinping, Fuzzy Stochastic Vibrations of Anti-seismic Sturcture, Herbin Architectural Engineering College, 1987.

[3] Qian Linxi, Computational Structural Mechanics, Now and Future, Computational Structural Mechanics and Applications, 3(1986),No.3, 1-6.

[4] Wang Caihua and Song Liantian, The Methodology of Fuzzy Theory, Publishing House of Chinese Architectural Industry, Beijing, 1988.

[5] Wang Caihua,Fuzzy Optimal Design of Engineering Structure, Chongqing University, 1986.

[6] Tanaka,H. and Asai,K., Fuzzy Linear Programming Problem with Fuzzy Numbers, Fuzzy Sets and Systems,13(1984) 1-10.

[7] Rao,S.S.,Optimization Theory and Applications (Second Edition),Wiley Eastern Limited, 1984.

[8] Feng Yingling, The Fuzzy Solutions of Multi-Object Optimization,Science Communication,1981, No.17.

[9] Pubois,D. and Prade, H., Fuzzy Sets and Systems Theory and Applications, Academic Press, New York, 1980.

[10] Wang Caihua and Zhu Yudong, A Processing Method of Objective Weights in Multi-objective Fuzzy Optimization, First Fuzzy Analysis and Design Conference, Chongqing, China, 1988.

Approximation, Optimization and Computing:
Theory and Applications, A.G. Law and C.L. Wang (eds.)
Elsevier Science Publishers B.V. (North-Holland)

Dynamic Programming and Basic Inequalities

Chung-lie Wang*
Department of Mathematics and Statistics
University of Regina
Regina, Saskatchewan S4S 0A2 Canada

1. INTRODUCTION

Dynamic programming (DP) has been designed to treat multistage processes possessing certain invariant aspects (e.g. see Bellman [4,XVI]). In accordance with the principle of optimality of DP (e.g. see Bellman [4.p,83]) suitable functional equations (e.g. see Bellman and Lee [7]) have been adopted as models to solve various optimization problems by many investigators, for example see Aris [1], Bellman [4,5], Bellman and Dreyfus [6], Beveridge and Schechter [8], Dreyfus and Law [9], Lee [16], Nemhauser [18], and White [33]. Since the eighties, Iwamoto [11-14], Iwamoto, Tomkins and Wang [15], and Wang [19-24], following the lead of Beckenbach and Bellman [2] and Bellman [3], have closed a link between the development of DP and that of the theory of inequalities. Very recently, Wang [28-33] has discovered that certain mathematical programming (MP) problems (such as fractional programming problems, etc.) can be reformed into the DP setting so that the DP approach can be used to solve them. Indeed, the main purpose for using the DP technique to solve an optimization problem with or without constraints is to reformulate the problem so that the complexities of multidimensional analysis are directly or indirectly avoided (e.g. see Bellman [4,p.7]).

For this reason, when we adopted functional equation (or DP) approach to establish certain inequalities (e.g. see Bellman [3], Iwamoto [11-14], Iwamoto, Tomkins and Wang [15], Wang [19-24]) we intentionally used parameters to replace some variables in order to lower the dimensionality of functions concerned. However, the flexibility, versatility and adaptability of the DP technique appear to be subject to no restriction. In other words, the DP technique is used not only to simplify the form of the problem but also to make the problem solvable by simpler or more basic means. This is the motivation of this paper.

In this paper, we state three basic inequalities - the arithmetic and geometric (AG) inequalities, the Hölder inequality and the Minkowski inequality in Section 2. In subsequent sections, we use the DP technique in a somewhat liberal manner to establish the three basic inequalities. As further examples we give a detailed analysis of the Cauchy inequality.

2. BASIC INEQUALITIES

We state the three basic inequalities (e.g. see [2,10,17]) as follows.

The AG inequality: For $x_j, y_j > 0$, $j = 1, \ldots, n$, we have

$$(x_1^{y_1} \cdots x_n^{y_n})^{1/\sum y_j} \leq \sum x_j y_j / \sum y_j \tag{1}$$

(Here and in what follows $\sum$ is used to designate $\sum_{j=1}^n$ whenever confusion is unlikely to occur.) The sign of inequality holds in (1) if and only if $x_1 = \cdots = x_n$.

The Hölder inequality: For $x_j, y_j > 0$, $j = 1, \cdots, n$, $q = p/(p-1)$, we have

$$\sum x_j y_j \leq (\sum x_j^p)^{1/p} (\sum y_j^q)^{1/q},\ p > 1$$
or
$$\sum x_j y_j \geq (\sum x_j^p)^{1/p} (\sum y_j^q)^{1/q},\ p < 1. \tag{2}$$

In either case, the sign of equality holds in (2) if and only if

$$\frac{x_1^p}{y_1^q} = \cdots = \frac{x_n^p}{y_n^q}.$$

The Minkowski inequality: For $x_j, y_j > 0$, $j = 1, \cdots, n$, we have

$$(\sum (x_j + y_j)^p)^{1/p} \leq (\sum x_j^p)^{1/p} + (\sum y_j^p)^{1/p},\ p > 1$$
or
$$(\sum (x_j + y_j)^p)^{1/p} \geq (\sum x_j^p)^{1/p} + (\sum y_j^p)^{1/p},\ p < 1. \tag{3}$$

In either case, the sign of equality holds in (3) if and only if

$$\frac{x_1}{y_1} = \cdots = \frac{x_n}{y_n}.$$

3. AG INEQUALITY

In this section and in what follows, we consider problems without any restriction on the dimensionality and the number of dynamic parameters (or state variable, see [4,p.81, 18,p.63]).

Consider the problem

$$\phi_n(a, b) = \min_{x,y} \sum x_j y_j \tag{4}$$

*The author was supported in part by the NSERC of Canada Grant A4091

subject to

$$\Pi_{j=1}^{n} x_j^{y_j} = a,\ a > 0,$$

and (5)

$$\sum y_j = b, \quad b > 0$$
$$x_j, y_j > 0 \quad 1 \le j \le n.$$

Since

$$\begin{aligned} \phi_1(a,b) &= \min_{(x_1^{y_1}, y_1)=(a,b)} x_1 y_1 = b a^{1/b} \\ \phi_2(a,b) &= \min_{x_2,y_2} [\phi_1(a x_2^{-y_2}, b - y_2) + x_2 y_2] \end{aligned} \qquad (6)$$

where

$$\begin{aligned} \phi_1(a x_2^{-y_2}, b - y_2) \;+\; x_2 y_2 &= (b - y_2)(a x_2^{-y_2})^{1/(b-y_2)} + x_2 y_2 \\ &= b[\frac{b - y_2}{b}(a x_2^{-y_2})^{1/(b-y_2)} + \frac{y_2}{b} x_2] \qquad (7) \\ &\ge b[(a x_2^{-y_2})^{1/b} x_2^{y_2/b}] = b a^{1/b}. \end{aligned}$$

The sign of equality holds in (7) if and only if

$$(a x_2^{-y_2})^{1/(b-y_2)} = x_2$$

or

$$x_2 = a^{1/b} \qquad (8)$$

From (6)-(8), it follows that the minimum

$$\phi_2(a,b) = b a^{1/b}$$

it attained at

$$x_1 = x_2 = a^{1/b}$$

In general, we can readily derive ϕ_{k+1} from ϕ_k for any $k \ge 1$ by exactly the same procedure as above ((6) and (7)) to derive ϕ_2 from ϕ_1. So, we obtain inductively that the minimum

$$\phi_n(a,b) = b a^{1/b} \qquad (9)$$

is attained at

$$x_1 = \cdots = x_n = a^{1/b}. \qquad (10)$$

It is now clear that the AG inequality (1) follows from (4), (5) and (9). Equality holds in (1) as expected (see (10)).

Remark. The elementary AG inequality used in (7) can be regarded as a variant of the Bernoulli inequality (e.g. see Mintronivić [17,p.34]).

4. HÖLDER INEQUALITY

Consider the problem

$$\text{opt} \sum x_j y_j \qquad (11)$$

subject to

$$\sum x_j^p = a,\ a > 0,$$

and

$$\sum y_j^q = b,\ b > 0, \qquad (12)$$
$$x_j, y_j > 0,\ 1 \le j \le n$$

where $q = p/(p-1)$.
There are two cases for the problem (11)-(12) to be considered: opt = max for $p > 1$; opt = min for $p < 1$.

First, consider the problem

$$\phi_n(a,b) = \max_{x,y} \sum x_j y_j \qquad (13)$$

subject to (12).

In this case, the concavity of the function $t^{1/p}$ for $p > 1$ is used. Since

$$\begin{aligned} \phi_1(a,b) &= \max_{(x_1^p, y_1^q)=(a,b)} x_1 y_1 = a^{1/p} b^{1/q} \\ \phi_2(a,b) &= \max_{x_2,y_2} [\phi_1(a - x_2^p, b - y_2^q) + x_2 y_2] \end{aligned} \qquad (14)$$

where

$$\phi_1(a - x_2^p, b - y_2^q) + x_2 y_2 = (a - x_2^p)^{1/p}(b - y_2^q)^{1/q} + x_2 y_2$$

$$= \left\{ \begin{array}{l} a[\frac{a - x_2^p}{a}(\frac{b - y_2^q}{a - x_2^p})^{1/q} + \frac{x_2^p}{a}(\frac{y_2^q}{x_2^p})^{1/q}] \\ \\ b[\frac{b - y_2^q}{b}(\frac{a - x_2^p}{b - y_2^q})^{1/p} + \frac{y_2^q}{b}(\frac{x_2^p}{y_2^q})^{1/p}] \end{array} \right\} \qquad (15)$$

$$\le \left\{ \begin{array}{l} a[\frac{a - x_2^p}{a}\frac{b - y_2^q}{a - x_2^p} + \frac{x_2^p}{a}\frac{y_2^q}{x_2^p}]^{1/q} \\ \\ b[\frac{b - y_2^q}{b}\frac{a - x_2^p}{b - y_2^q} + \frac{y_2^q}{b}\frac{x_2^p}{y_2^q}]^{1/p} \end{array} \right\} = a^{1/p} b^{1/q}.$$

The sign of equality holds in (15) if and only if

$$\frac{b - y_2^q}{a - x_2^p} = \frac{y_2^q}{x_2^p} \quad \text{or} \quad \frac{a - x_2^p}{b - y_2^q} = \frac{x_2^p}{y_2^q}$$

which is equivalent to

$$\frac{x_1^p}{y_1^q} = \frac{x_2^p}{y_2^q} = \frac{a}{b}. \qquad (16)$$

From (14)-(15), it follows that the maximum

$$\phi_2(a,b) = a^{1/p} b^{1/q}$$

is attained at x_1, x_2, y_1, y_2 satisfying (16).

Using the same argument as mentioned in the previous section, we obtain inductively that the maximum

$$\phi_n(a,b) = a^{1/p} b^{1/q} \qquad (17)$$

is attained at

$$\frac{x_1^p}{y_1^q} = \cdots = \frac{x_n^p}{y_n^q} = \frac{a}{b}. \qquad (18)$$

For the case $p < 1$, the above result can be duplicated by using the convexity of the function $t^{1/p}$ and opt = min.

It is now clear that the inequality (2) follows from (13), (12) and (17). Equality holds in (2) as expected (see (18)).

5. MINKOWSKI INEQUALITY

Consider the problem

$$\text{opt} \sum (x_j + y_j)^p \tag{19}$$

subject to

$$\sum x_j^p = a,\ a > 0,$$

and

$$\sum y_j^p = b,\ b > 0,$$

$$x_j, y_j > 0,\ 1 \le j \le n$$

There are also two cases for the problem (19)-(20) to be considered: opt = max for $p > 1$, opt = min for $p < 1$.

First, consider the problem

$$\phi_n(a,b) = \max_{x,y} \sum (x_j + y_j)^p \tag{20}$$

subject to (20).

In this case, the concavity of the function $(1 + t^{1/p})^p$, $t \ge 0$, for $p > 1$ is used.

Since

$$\begin{aligned} \phi_1(a,b) &= \max_{(x_1^p, y_1^q)=(a,b)} (x_1 + y_1)^p = (a^{1/p} + b^{1/p})^p \\ \phi_2(a,b) &= \max_{x_2,y_2} [\phi_1(a - x_2^p, b - y_2^p) + (x_2 + y_2)^p], \quad (21) \end{aligned}$$

where

$$\phi_1(a - x_2^p, b - y_2^p) + (x_2 + y_2)^p$$

$$= [(a - x_2^p)^{1/p} + (b - y_2^p)^{1/p}]^p + (x_2 + y_2)^p$$

$$= \left\{ \begin{array}{l} a[\frac{a-x_2^p}{a}(1 + (\frac{b-y_2^p}{a-x_2^p})^{1/p})^p + \frac{x_2^p}{a}(1 + (\frac{y_2^p}{x_2^p})^{1/p})^p] \\ \\ b[\frac{b-y_2^p}{b}(1 + (\frac{a-x_2^p}{b-y_2^p})^{1/p})^p + \frac{y_2^p}{b}(1 + (\frac{x_2^p}{y_2^p})^{1/p})^p] \end{array} \right\}$$

$$\le \left\{ \begin{array}{l} a[1 + (\frac{a-x_2^p}{a}\frac{b-y_2^p}{a-x_2^p} + \frac{x_2^p}{a}\frac{y_2^p}{x_2^p})^{1/p}]^p \\ \\ b[1 + (\frac{b-y_2^p}{b}\frac{a-x_2^p}{b-y_2^p} + \frac{y_2^p}{b}\frac{x_2^p}{y_2^p})^{1/p}]^p \end{array} \right\} = (a^{1/p} + b^{1/p})^p.$$

The sign of equality holds in (23) if and only if

$$\frac{b - y_2^p}{a - x_2^p} = \frac{y_2^p}{x_2^p} \quad \text{or} \quad \frac{a - x_2^p}{b - y_2^p} = \frac{x_2^p}{y_2^p}$$

which is equivalent to

$$\frac{x_1}{y_1} = \frac{x_2}{y_2} = \frac{a^{1/p}}{b^{1/p}} \tag{22}$$

From (22)-(23), it follows that the maximum

$$\phi_2(a,b) = (a^{1/p} + b^{1/p})^p$$

is attained at x_1, x_2, y_1, y_2 satisfying (24).

Using the same argument as above, we obtain inductively that the maximum

$$\phi_n(a,b) = (a^{1/p} + b^{1/p})^p \tag{23}$$

is attained at

$$\frac{x_1}{y_1} = \cdots = \frac{x_n}{y_n} = \frac{a^{1/p}}{b^{1/p}}. \tag{24}$$

For the case $p < 1$, the above result can be duplicated by using the convexity of the function $(1+t^{1/p})^p$ and opt = min. It is now clear that the inequality (3) follows from (21), (20) and (25). Equality holds in (3) as expected (see (26)).

6. ANALYSIS OF CAUCHY INEQUALITY

In view of all that we have presented in the previous sections, we have displayed more varieties to create inverse theorems in DP than those given in Iwamoto [11-14]. Moreover, reverse inequalities of the basis inequalities (1)-(3) and others (e.g. see [15,24]) can be likewise established. As examples, we present an analysis concerning the Cauchy inequality as follows in order to demonstrate the mentioned findings. In doing so, we give the Cauchy inequality

$$(\sum x_j y_j)^2 \le \sum x_j^2 \sum y_j^2 \tag{25}$$

and its reverse inequality

$$(x_1 y_1 - \sum_{j=2}^{n} x_j y_j)^2 \ge (x_1^2 - \sum_{j=2}^{n} x_j^2)(y_1^2 - \sum_{j=2}^{n} y_j^2) \tag{26}$$

with equality holding in each case if and only if

$$\frac{x_1}{y_1} = \cdots = \frac{x_n}{y_n}$$

6.1 Consider the problem

$$\phi_n(a,b) = \min_{x,y} \sum x_j^2 \tag{27}$$

subject to

$$\sum x_j y_j = a,\ a > 0,$$

and

$$\sum y_j^2 = b,\ b > 0 \tag{28}$$

$$x_j, y_j > 0, 1 \le j \le n.$$

Using the convexity of the function $t^k, t \ge 0$ for $k > 1$ or $k < 0$ and noting

$$\phi_1(a,b) = \min_{(x_1 y_1, y_1^2)=(a,b)} x_1^2 = \frac{a^2}{b},$$

we have

$$\phi_2(a,b) = \min_{x_2,y_2} [\phi_1(a - x_2 y_2, b - y_2^2) + x_2^2] \tag{29}$$

where

$$\phi_1(a - x_2y_2, b - y_2^2) + x_2^2 = \frac{(a - x_2y_2)^2}{b - y_2^2} + x_2^2$$

$$= \left\{ \begin{matrix} b[\frac{b-y_2^2}{b}(\frac{a-x_2y_2}{b-y_2^2})^2 + \frac{y_2^2}{b}(\frac{x_2y_2}{y_2^2})^2] \\ \\ a[\frac{a-x_2y_2}{a}(\frac{b-y_2^2}{a-x_2y_2})^{-1} + \frac{x_2y_2}{a}(\frac{y_2^2}{x_2y_2})^{-1}] \end{matrix} \right\}$$

$$\geq \left\{ \begin{matrix} b[\frac{b-y_2^2}{b}\frac{a-x_2y_2}{b-y_2^2} + \frac{y_2^2}{b}\frac{x_2y_2}{y_2^2}]^2 \\ \\ a[\frac{a-x_2y_2}{a}\frac{b-y_2^2}{a-x_2y_2} + \frac{x_2y_2}{a}\frac{y_2^2}{x_2y_2}]^{-1} \end{matrix} \right\} = \frac{a^2}{b} \tag{30}$$

The sign of equality holds in (32) if and only if

$$\frac{a - x_2y_2}{b - y_2^2} = \frac{x_2y_2}{y_2^2} \quad \text{or} \quad \frac{b - y_2^2}{y_2^2} = \frac{y_2^2}{x_2y_2}$$

which is equivalent to

$$\frac{x_1}{y_1} = \frac{x_2}{y_2} = \frac{a}{b} \tag{31}$$

From (31)-(32), it follows that the minimum

$$\phi_2(a, b) = \frac{a^2}{b}$$

is attained at x_1, y_1, x_2, y_2 satisfying (33).

Using the same argument as above, we obtain inductively that the minimum

$$\phi_n(a, b) = \frac{a^2}{b} \tag{32}$$

is attained at

$$\frac{x_1}{y_1} = \cdots = \frac{x_n}{y_n} = \frac{a}{b} \tag{33}$$

It is now clear that the inequality (27) follows from (29), (30), and (34). Equality holds in (27) as expected (see (35)).

6.2. Consider the problem

$$\phi_n(a, b) = \min_{x,y}[x_1y_1 - \sum_{j=2}^{n} x_jy_j] \tag{34}$$

subject to

$$x_1^2 - \sum_{j=2}^{n} x_j^2 = a, \ a > 0$$

and (35)

$$y_1^2 - \sum_{j=2}^{n} y_j^2 = b, \ b > 0$$

$$x_j, y_j > 0, \quad 1 \leq j \leq n.$$

Using the reverse relation of the concavity of the function t^p, $0 < p < 1$, $t \geq 0$ and noting

$$\phi_1(a, b) = \min_{(x_1^2, y_1^2)=(a,b)} x_1y_1 = a^{1/2}b^{1/2},$$

we have

$$\phi_2(a, b) = \min_{x_2,y_2}[\phi_1(a + x_2^2, b + y_2^2) - x_2y_2], \tag{36}$$

where

$$\phi_1(a + x_2^2, b + y_2^2) - x_2y_2 = (a + x_2^2)^{1/2}(b + y_2^2)^{1/2} - x_2y_2$$

$$= \left\{ \begin{matrix} a\left[\frac{a+x_2^2}{a}\left(\frac{b+y_2^2}{a+x_2^2}\right)^{1/2} - \frac{x_2^2}{a}\left(\frac{y_2^2}{x_2^2}\right)^{1/2}\right] \\ \\ b\left[\frac{b+y_2^2}{b}\left(\frac{a+x_2^2}{b+y_2^2}\right)^{1/2} - \frac{y_2^2}{b}\left(\frac{x_2^2}{y_2^2}\right)^{1/2}\right] \end{matrix} \right\}$$

$$\geq \left\{ \begin{matrix} a\left[\frac{a+x_2^2}{a}\frac{b+y_2^2}{a+x_2^2} - \frac{x_2^2}{a}\frac{y_2^2}{x_2^2}\right]^{1/2} \\ \\ b\left[\frac{b+y_2^2}{b}\frac{a+x_2^2}{b+y_2^2} - \frac{y_2^2}{b}\frac{x_2^2}{y_2^2}\right]^{1/2} \end{matrix} \right\} = a^{1/2}b^{1/2}. \tag{37}$$

The sign of equality holds in (39) if and only if

$$\frac{b + y_2^2}{a + x_2^2} = \frac{y_2^2}{x_2^2} \quad \text{or} \quad \frac{a + x_2^2}{b + y_2^2} = \frac{x_2^2}{y_2^2}$$

which is equivalent to

$$\frac{x_1}{y_1} = \frac{x_2}{y_2} = \frac{a^{1/2}}{b^{1/2}}. \tag{38}$$

From (38)-(39) it follows that the minimum

$$\phi_2(a, b) = a^{1/2}b^{1/2}$$

is attained at x_1, y_1, x_2, y_2 satisfying (40).

Using the same argument as above, we obtain inductively that the minimum

$$\phi_n(a, b) = a^{1/2}b^{1/2} \tag{39}$$

is attained at

$$\frac{x_1}{y_1} = \cdots = \frac{x_n}{y_n} = \frac{a^{1/2}}{b^{1/2}} \tag{40}$$

It is now clear that the inequality (38) follows from (36), (37) and (41). Equality holds in (28) as expected (see (42)).

7. CONCLUDING REMARK

The results given above have established the basic inequalities (1)-(3) and inverse and reverse counterparts of the Cauchy inequality (27) in an alternative manner by means of the DP technique. Consequently, we have revealed that the flexibility and versatility of the DP technique can not only reduce or remove the complexities of the demensionality of a problem but also simplify the overall analysis needed to solve the problem. As usual (e.g. see [28-32]) we only use simple inequalitis in the arguments of this paper, so the differentiability of functions is not required. Since almost all the basic inequalities are equivalent in the sense that they can be derived from one another (e.g. see [27]), the development of the DP technique in association with that of the theory of inequalities appears to be promising.

References

[1] R. Aris, "Discrete Dynamic Programming," Ginn (Blaisdell), Waltham, Mass., 1964.

[2] E. F. Beckenbach and R. Bellman, "Inequalities," 2nd rev. ed., Springer-Verlag, Berlin, 1965.

[3] R.Bellman, Some aspects of the theory of dynamic programming, Scripta Math. 21 (1955), 273-277.

[4] R. Bellman, "Dynamic Programming," Princeton Univ. Press, Princeton, N.J., 1957.

[5] R. Bellman, "Adaptive Control Process," Princeton Univ. Press, Princeton, N.J., 1961.

[6] R. Bellman and S. E. Dreyfus, "Applied Dynamic Programming," Princeton Univ. Press, Princeton, N.J., 1962.

[7] R. Bellman and E. S. Lee, Functional equations in dynamic programming, Aequationes Math. 17 (1978), 1-18.

[8] G. S. G. Beveridge and R. S. Schechter, "Optimization: Theory and Practice," McGraw Hill, New York, 1970.

[9] S. E. Dreyfus and A. M. Law, "The Art and Theory of Dynamic Programming," Academic Press, New York, 1977.

[10] G. H. Hardy, et al, "Inequalities," 2nd ed., Cambridge Univ. Press, Cambridge, 1952.

[11] S. Iwamoto, Inverse theorem in dynamic programming, I, J. Math. Anal. Appl. 58 (1977), 113-134.

[12] S. Iwamoto, Inverse theorem in dynamic programming, II, J. Math. Anal. Appl. 58 (1977), 247-279.

[13] S. Iwamoto, Inverse theorem in dynamic programming, III, J. Math. Anal. Appl. 58 (1977), 439-448.

[14] S. Iwamoto, Dynamic programming approach to inequalities, J. Math. Anal. Appl. 58 (1977), 687-704.

[15] S. Iwamoto, R.J. Tomkins and Chung-lie Wang, Some theorems on reverse inequalities, J. Math. Anal. Appl. 119 (1986), 282-299.

[16] E. S. Lee, "Quasilinearization and Invariant Imbedding," Academic Press, New York, 1968.

[17] D. S. Mitrinović, "Analytic Inequalities," Springer-Verlag, Berlin, 1970.

[18] G. L. Nemhauser, "Introduction to Dynamic Programming," Wiley, New York, 1967.

[19] Chung-lie Wang, Functional equation approach to inequalities, J. Math. Anal. Appl. 71 (1979), 423-430.

[20] Chung-lie Wang, Functional equation approach to inequalities, II, J. Math. Anal. Appl. 78 (1980), 522-530.

[21] Chung-lie Wang, Functional equation approach to inequalities, III, J. Math. Anal. Appl. 80 (1981), 31-35.

[22] Chung-lie Wang, Functional equation approach to inequalities, IV, J. Math. Anal. Appl. 86 (1982), 96-98.

[23] Chung-lie Wang, A generalization of the HGA inequalities, Soocho Journal of Mathematics 6 (1980), 149-152.

[24] Chung-lie Wang, Functional equation approach to inequalities, VI, J. Math. Anal. Appl. 104 (1984), 95-102.

[25] Chung-lie Wang, Inequalities and mathematics programming, in General Inequalities 3 (Proceedings, Third International Conference on General Inequalities, Oberwolfach), (E. F. Beckenbach, Ed.), pp.149-164, Birkhauser, Basel/Stuttgart, 1983.

[26] Chung-lie Wang, Inequalities and mathematical programming, II, in General Inequalities 4 (Proceedings , Fourth International Conference on General Inequalities, Oberwolfach), (W. Walter, Ed.) pp.381-393, Birkhauser, Basel/Stuttgart, 1984.

[27] Chung-lie Wang, A survey on basic inequalities, Notes Canad. Math. Soc. 12 (1980), 8-12.

[28] Chung-lie Wang, The principal and models of dynamic programming, J. Math. Anal. Appl. 118 (1986), 287-308.

[29] Chung-lie Wang, The principal and models of dynamic programming, II, J. Math. Anal. Appl., 135(1988), 268-283.

[30] Chung-lie Wang, The principle and models of dynamic programming, III, J. Math. Anal. Appl., 135(1988), 284-296.

[31] Chung-lie Wang, The principle and models of dynamic programming, IV, J. Math. Anal. Appl., 137(1989), 148-160.

[32] Chung-lie Wang, The principle and models of dynamic programming, V, J. Math. Anal. Appl., 137(1989), 148-160.

[33] D. J. White, "Dynamic Programming," Oliver & Boyd, London, 1969.

Approximation, Optimization and Computing:
Theory and Applications, A.G. Law and C.L. Wang (eds.)
Elsevier Science Publishers B.V. (North-Holland)

MULTIOBJECTIVE OPTIMIZATION IN SPILLWAY PROFILE *

Wang, Shu – yu   Zhou, Xiao – bao

Department of Civil Engineering, Zhejiang University,
Hangzhou, People's Republic of China

The paper discusses the shape optimization of overflow surface of hydraulic structures. A sychronous iterative method is used to solve the variational problem with variable domain and unknown discharge for 2 – D ideal potential flow. An ideal point approach is adopted to deal with this multiobjective optimization.

1. INTRODUCTION

The shape of overflow surface for hydraulic structures is generally designed via a series of hydraulic model tests. This trial – and – error procedure is time consuming and expensive. Since the computer technology and numerical analysis have been greatly developed, it is possible to calculate flow characteritics, such as the coefficient of discharge, the velocity and pressure distribution along the dam surface by numerical simulation, then, to seek an optimal shape of overflow surface using optimization technique. In this way, only a few model tests need to be done to check numerical results and to determine a final profile of spillway with less labour and cost.

2. DETERMING FLOW CHARACTERISTIC

2.1. MATHEMATICAL STATEMENT FOR OVERFLOW PROBLEM

The flow over a spillway is an accelerated fluid motion caused by the rapid contraction of boundary shape under gravitational influence. The fluid viscosity influence is quite small in comparison with inertia effects. It can be regarded as an ideal potential flow for cases where no serious separation of flow exists. The governing equation for ideal potential flow is

$$\psi_{xx}+\psi_{yy}=0 \quad \text{in } \Omega \qquad (1)$$

where ψ denotes the stream function taken as a basic unknown, Ω is the flow field considered. The boundary conditions can be stated as follows.

On the far – upstream a uniform velocity distribution along the depth of water is assumed

$$\psi|_{S_1}=(q/H_1)Y \qquad (2)$$

Where q is the unit discharge. H_1 is the total depth of water. Y is the height of considered point from the bottom. Similarly, at the nose sill

$$\frac{\partial\psi}{\partial n}|_{S_2}=0 \qquad (3)$$

The fixed boundaries (dam surface and reservoir bed) is a stream line where ψ is taken to be zero

$$\psi|_{S_3}=0 \qquad (4)$$

On the free surface two conditions, to be a streamline and the pressure on it to be zero (atmospheric pressure), have to be satisfied

$$\psi|_{S_4}=q$$
$$\frac{\partial\psi}{\partial n}|_{S_4}=\sqrt{2g(E_0-Y)} \qquad (5)$$

where n denotes the outward normal to the boundary of flow region, g is the acceleration of gravity. E_0 total energy head.

After solving the above boundary value problem, the velocities, u and v, can be calculated by derivatives of ψ with respect to x and y, and the pressure P can be found in terms of the Bernoulli's equation

$$p/\gamma+(u_2+v_2)/2g+y=E_0 \qquad (6)$$

2.2. VARIATIONAL PRINCIPLE AND FINITE ELEMENT FORMULATION

According to the variational principles, the boundary value problem is equivalent to a problem of minimizing the following functional [1]

*The project supported by National Natural Science Foundation of China

$$J=\frac{1}{2}\iint_{\Omega}(\psi_x^2+\psi_y^2)\,d\chi \tag{7}$$

in which ψ and the domain of integration Ω both vary. That is, the actual stream function ψ which satisfies Eq. (1) – (5) is such that the integral $J(\psi,\Omega)$ to be a minimum.

To avoid mathematical difficulties encountered in solving the nonlinear problem caused by the presence of unknown discharge and free surface location, we adopt a synchronous iterative method which consists of the following main steps: first, assuming an initial discharge and free surface location to convert the variational problem with variable domain and unknown discharge into that with a fixed domain under a specified discharge; second, by discretizing, minimizing Eq. (7) and imposing the stream line condition (or the zero pressure condition) on the free surface, to form a set of linear equations; third, checking whether the discharge is correct, if not, to calculate a higher approximate discharge for the next iteration, and modify the free surface profile to meet two boundary conditions (Eq. (5)). In brief, the solution to actual nonlinear problem is approached by successive corrections of a series of simple linear problems.

In terms of the element type, selecting the shape function Ni, the stream function ψ in a element can be expressed

$$\psi^e=N_i^e\psi_i^e \tag{8}$$

By discretizing and minimizing, we obtain derivatives of J with respect to nodal unknowns ψ_i^e

$$\frac{\partial J^e}{\partial\psi_i^e}=A_{ij}^e\psi_j^e-b_i^e \tag{9}$$

where

$$A_{ij}^e=\iint_{\Omega^e}(N_{i,x}N_{j,x}+N_{i,y}N_{j,y})\,d\Omega \tag{10}$$

$$b_i^e=\int_{S^e}N_i\sqrt{2g(E_0-Y)}\cdot ds \tag{11}$$

After assembling element by element, the following set of linear equations is formed

$$A_{mn}\psi_n=B_m \tag{12}$$

Notice that when the stream line condition is imposed prior on the free surface, the right side terms of Eq. (12) will vanish and the coefficient matrix A_{mn} must be modified by substituting the specified value q into the nodal unknowns ψ_i on the free surface.

2.3. ADJUSTING DISCHARGE AND FREE SURFACE LOCATION

The key to the synchronous iterative method is how to adjust the flow discharge and modify the location of free surface. Based on the displacement characters of the free surface which are analyzed by the pertubation method, the higher approximation of discharge will be estimated and the critical section dividing flow patterns can also be identified[2]. Then, the location of free surface can be updated by element to element for meeting two conditions (Eq. 5). With these data we can form and solve a modified variational problem which possesses a new fixed domain and new specified discharge. Repeat the above processes until the numerical results converge.

3. MODELLING FOR OPTIMIZATION

The profile of spillway usually consists of a crest curve, a straight stretch and a bucket arc. The parameters of crest curve have been investigated by hydraulic experiments for several decades. They are much more reasonable and precise than the others, and may not be easy to improve. Our emphases are put on the rest parts of spillway profile. The slope of straight stretch m, the radius of bucket R and the angle of nose sill θ are taken as the design variables X. The feasible region S is specified by the following constraints :

(1). Stability constraint

The coefficient of resistance to dam slide under the maximum water head must be greater than a value specified by the code and specification concerned

$$K=f\cdot\sum W/\sum P\geqslant[K] \tag{13}$$

(2). Strength constraints

$$\sigma_{i,min}\geqslant 0 \tag{14a}$$

$$\sigma_{i,max}\leqslant[R] \tag{14b}$$

Where $\sigma_{i,min}$ and $\sigma_{i,max}$ denote the minimum and maximum stress at a representative point i under the most disadvantageous loading condition. They are calculated via the method based on material mechanics. Eq. 14a means that tensile stresses are not allowed to occur in the dam.

(3). Dynamic water pressure constraints

To avoid cavitation damage, the minimum value of dynamic water pressure on the dam surface P_m must not be less than a specified limit [p], such as −4.0 M.

$$p_m/\gamma\geqslant[p/\gamma] \tag{15}$$

(4). Constraints for dissipation of energy and scour resistance

To ensure the stability of dam toe, the back slope i of scour hole must satisfy

$$i = t_k/L \leqslant [i_c] \qquad (16)$$

$$t_k = \alpha\sqrt{qH}$$

where tk is the depth of scour hole , L is the horizontal distance of trajectory from dam toe to the center of scour hole, H is the difference between the upstream and downstream water level. α is a coefficient depending on bed rock characters.

(5). Bounds of design variables.

The volume of spillway per width f1(X) and the negative water pressure on the overflow surface $-f2(X)$, obtained by solving Eq. (12) and (6) are taken as two objective functions. The former is to be minimized, the latter is to be maximized. Thus this multiobjective optimization problem can be stated as:

$$\min_{X \in S} . F(X) \qquad (17)$$

where $F: S \rightarrow R^2$ is a vector objective function given by

$$F(x) = [f1(X), f2(X)]^T.$$

4. MULTIOBJECTIVE TECHNIQUE

It can be easily noticed that the above two objective functions are contrary each other. There exists no unique solution which would give an optimum for both objective functions simultaneously. A new concept, Pareto optima, is introduced and defined as a design variable vector $\overline{X}$ if there exists no $X \in S$ such that $f_i(X) \leqslant f_i(\overline{X})$ for $i = 1,2,\ldots, n$, with $f_j(X) < f_j(\overline{X})$ for at least one j.

In the paper we use an ideal point approach[3] to solve the multiobjective problem. First, every objective function (f_i) is minimized separately in the feasible region by the complex mehtod. Let the solutions to be

$$X1: \min_{X \in S} f_1(X) = \overline{f}_1 \qquad (18)$$

and

$$X2: \min_{X \in S'} f_2(X) = \overline{f}_2 \qquad (19)$$

where

$$S' = \{X | X \in S, f_1(x) \leqslant \overline{f}_1 + B\}$$

B is a tolerance limit specified. The ideal point used as a reference solution in the image space is defined by $Z_{id} = [\overline{f}_1, \overline{f}_2]^T$. Generally, it is infeasible. Next, a weighted distance between $Z \in \Lambda$ and Z_{id} is formed and referred to a pseudo objective function

$$d(Z, Z_{id}) = (\sum_{j=1}^{2} \lambda_j((f_j(X) - \overline{f}_j)/\overline{f}_j)^2)^{1/2} \qquad (20)$$

in which Λ is the image of feasible region S in the objective function space. λ_j denotes the weighting coefficient for the jth objective function based on its importance or its sensitivity to design variables. Without loss of generality, λ_j can be normalized so that the summation of them, which are non-negative and not all zero, equals one. Then, we look for a point $Z^* \in \Lambda$ which is the nearest to the ideal point, that is

$$Z^*: \min_{Z \in \Lambda} . \ d(Z, Z_{id}) \qquad (21)$$

The corresponding version in the original space is

$$X^*: \min_{X \in S} (\sum_{j=1}^{2} \lambda_j((f_j(X) - \overline{f}_j)/\overline{f}_j)^2)^{1/2} \qquad (22)$$

5. NUMERICAL EXAMPLE

The profile of spillway of a hydroelectric project is illustrated in Fig. 1. There are 432 triangular finite elements and 259 nodes used for discretization of the flow region. The nodal coordinates can be automatically updated to avoid a seriously distorted mesh as the iteration proceeds. Results of the flow field calculation are also depicted in the figure. The velocity distributions on the cross sections and the pressure distributions on the dam surface are reasonable. The agreement is reached between the calculations and measurements from the hydraulic model test. Using the above numerical simulation of flow field for getting the pressure distribution to form constraints, a multiobjective optimization proceeds. The upper and lower bounds of each design variable are used as follows: $0.6 \leqslant m \leqslant 0.7$, $25 \leqslant R \leqslant 50$(m), $0.35 \leqslant \theta \leqslant 0.7$ (rad.). Because the objective functions and some constraints are implicit functions of design variables, the direct search method, such as the complex method, is used to seek Pareto optima. The initial value and Pareto optima for each objective function are shown in Tab. 1. The best compromise solution for multiobjective function problem with weighting coefficients $\lambda_1 = 0.4$ and $\lambda_2 = 0.6$ is shown in Tab. 2 and Fig. 2.

This paper discusses a numerical approach to design the profile of spillway. The primary application is desirable.

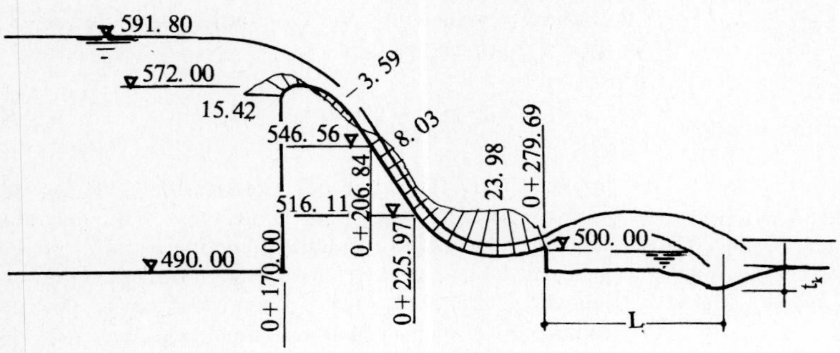

Figure 1 The Initial Profile of Spillway & Its Pressure Distribution On Dam Surface (p/γ , M)

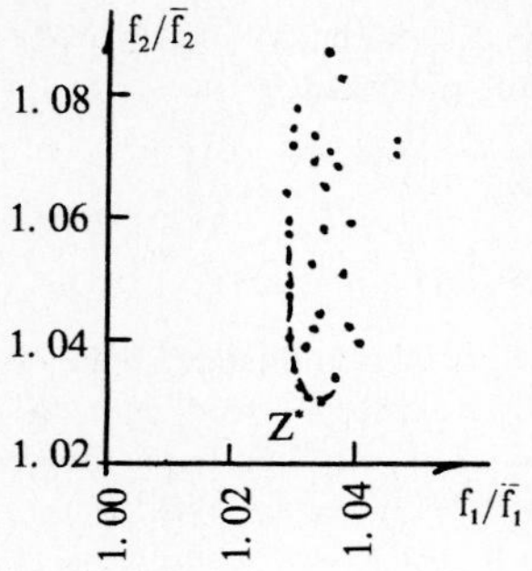

Figure 2 The Best Compromise Solution

Table 1. Results of Uniobjective Function Problems

		Design Varible			Volume of Spillway per Width(M^3/M)	Minimum Pressure p/γ (M)
		m	R	θ		
f1	initial value	0. 6500	40. 00	0. 5000	10884. 17	−3. 1279
	optimal value	0. 6107	25. 4097	0. 3502	9825. 0 ($=\bar{f}_1$)	−3. 6894
f2	initial value	0. 6107	25. 4097	0. 3502	9825. 0	−3. 6894
	optimal value	0. 6911	25. 9957	0. 3944	10131. 38	−2. 700 ($=-\bar{f}_2$)

Table 2. Results of Multiobjective Function Problem

	Design Variable			Pseudo Objective Function	Volume of Spillway per Width(M^3/M)	Minimum pressure p/γ (M)
	m	R	θ			
initial value	0. 6500	40. 0000	0. 5000	0. 1620	10884. 17	−3. 187
Optimum	0. 6839	25. 7531	0. 4378	0. 0304	10146. 30	−2. 779

REFERENCES

[1] Varoglu, E. and Finn, W. D. L, Variable Domain Finite Element Analysis of Free Surface Gravity Flow, Computers and Fluids, Vol. 6 (1978)

[2] Ding, D. Y., Application of displacement Characters of Free Surface to Overflow Problem in Hydraulic Structures, Chinese Science, No. 5, 1986

[3] Koski, J., Multicriterion Optimization in a Structural Design, in: Atrek, E. and others (eds.), New Directions in Optimum Structural Design, John Wiley and Sons, 1984.

Approximation, Optimization and Computing:
Theory and Applications, A.G. Law and C.L. Wang (eds.)
Elsevier Science Publishers B.V. (North-Holland)

A PARALLEL METHOD OF STRUCTURAL OPTIMIZATION

Wang Xicheng

(Dalian University of Technology)

Abstract-- A parallel method for structural optimization is described. Main operations of the algorithm are performed within elements, without assembling and solving the global system equations. The method has advantages such as simple formulation, easy programming, small storage space and calculation. Numerical results are presented to demonstrate the accuracy and effectiveness of this method.

INTRODUCTION

Computer systems have been undergoing progressive development. Notable peculiarity is their microminiaturization and parallelization. In the first, the use of microcomputer is becoming increasingly popular as the computer price over its capabilites decreases dramatically. Secondly, various parallel computers have emerged and developed. The basic motivation of the new advanced computer systems is to overcome the limitations of single instruction-single data computers (SISD), which execute instructions sequentially. Four broad classifications of parallel computers can be identified according to their machine organization: Single instruction-multiple data machines (SIMD); Multiple instruction-single data machines (MISD); Multiple instruction-multiple data machines (MIMD) and special purpose systolic machines[5-6].

New computer architectures have revolutionized the design of computer algorithms and promise to have significant influence on algorithms for structural optimization, which expend enormous computer time and main memory. The design of optimization algorithms for new computers has to pay attention to the following cases:

(1) There is a restriction on the main memory of most microcomputers available on the market.

(2) A parallel computer contains a number of interconnected processors, each of which can be programmable and execute its own instructions. The processors operate on shared memory or memories. The operations may be performed in parallel.

The traditional methods consist of alternative phases of analysis and optimization. In each step of iteration, the structural analysis is carried out completely by means of solving the F. E. M. global matrix equation. To improve the design, the optimization succeeds to minimize certain objective function.

This paper presents a parallel method for structural optimization. The method performs the analysis on the basis of minimizing potential energy $\Pi(U)$, which can be obtained by summing up the elements energy, so that the time-consumming work related to stiffness matrix can be avoided. The analysis task on the individual elements can be carried out in parallel. The minimization process in both phases may be treated similarly and could be done parallelly and interactionally. This is different from the traditional method in which the analysis phase should be done completely or nearly completely before switching to the optimization phase and vice versa.

PROBLEM FORMULATION

The structural optimization problem can be stated as problem A:

$$\text{Pb. A}\quad \begin{cases} \text{Minimize} \\ \quad W(A, U) = \sum_{j=1}^{m} W_j A_j \\ \text{Subject to} \\ \quad g_i(A, U) \leq 0 \quad i=1,2,\ldots, q. \\ \quad g_l(A, U) = 0 \quad l=1,2, \quad n. \end{cases} \tag{1}$$

where A is a vector of m design variables, W_j is the weight of element j when $A_j=1$, $U\in R^n$ is an n-dimensional nodal displacement vector. The constraints $g_i(A, U)$ include stress, displacement and minimum size constraints, and the equality constraints $g_l(A, U)$ represent the equilibrium equations:

$$K(A)\,U - P = 0 \tag{2}$$

The problem A may be considered as a bi-objective optimization problem B:

$$\text{Pb. B}\quad \begin{cases} \text{Minimize} & W(A, U) \text{ and } \Pi(U) \\ \text{Subject to} & g_i(A, U) \leq 0 \quad i= 1,\ldots,q \end{cases} \tag{3}$$

where $\Pi(U)$ is total potential energy of the structure. The displacement vector U has to satisfy kinematic boundary conditions which are taken care in the algorithm with the aid of node displacement qualitative numbers. It should be noted that the constraints $g_i(A, U)=0$ in problem A are replaced by minimizing $\Pi(U)$ in problem B.

If all the constraints are surrogated by a single constraint with appropriate weight coefficients $\lambda_i(i=1,\ldots,q)$, so that problem B converts to an equivalent problem C.

$$\text{Pb. C}\left[\begin{array}{ll} \text{Minimize} & W(A,U), \text{ and } \Pi(U) \\ \text{Subject to} & \sum_{i=1}^{q} \lambda_i \, g_i(A,U) \le 0 \\ & \sum_{i=1}^{q} \lambda_i = 1, \quad \lambda_i \ge 0 \\ & i=1,2,\ldots,q \end{array}\right. \tag{4}$$

in which the λ_i, $i=1,2,\ldots,q$ are non-negative multipliers, termed surrogate multipliers, forming a vector λ. Problem B and C are equivalent at the solution point.

The solution of problem C can be obtained by a series of iterations, each of which includes two parts: (1) analysis process; (2) optimization process. Estimates of the structural behavior and optimal design are evaluated in these processes alternatively.

ANALYSIS PROCESS

This process consists of minimizing the total potential energy

$$\Pi(U) = -U^T KU - U^T p \tag{5}$$

Some gradient-based optimization algorithms are used to minimize $\Pi(U)$ because its gradient can be calculated very easily as

$$D = \nabla\Pi(U) = KU - p \tag{6}$$

In this paper, only one (or a few) step is carried out in global iteration. The formulas in the simplest gradient method are given as

$$U = U + \alpha R \qquad R = -D \ , \quad \alpha = \frac{D^T D}{R^T K R} \tag{7}$$

If a conjugate gradient method is used, the following formulas are to be added

$$\overline{D} = K\overline{U} - p \qquad \overline{R} = -D + \frac{\overline{D}^T \overline{D}}{D^T D} R \tag{8}$$

All the calulations above can be performed on the element level, and summed up afterward. The operations on element can be carried out in parallel, independently of each other. The above equations can be expressed entirely in terms of the element matrix and element vectors as follows

$$P_e = A\,P, \quad d_e = k_e \quad u_e - p_e \tag{9}$$

$$\alpha = \frac{\delta}{r^T k \, r^T} \ , \qquad R = -A^T d_e \tag{10}$$

$$\overline{R} = -\overline{s}_e + \frac{\overline{\delta}}{d_e^T s_e} r_e \tag{11}$$

where

$$s_e = A^T A\, d_e, \quad \delta = d_e^T s_e \ , \quad r_e = A\,R,$$
$$\overline{\delta} = \overline{d}_e \overline{s}_e \tag{12}$$

The lower case letlers represent element quantities, such as

$$k_j = \{k_1, k_2, \ldots, k_j, \ldots, k_m\}^T,$$

in which k_j is the jth element stiffness matrix. The matrix A is the transformation matrix. It should be notice that only one (or a few) above processes is performed in each interation, instead of a complete minimization of $\Pi(U)$.

OPTIMIZATION PROCESS

The weight minimization of structures is can be written as the following optimization problem [1]

$$\text{Minimize} \qquad W(X) = \sum_{j=1}^{m} W_j X_j^{-1}$$

$$\text{Subject to} \qquad \sum_{i=1}^{q} \lambda_i \left(\sum_{j=1}^{m} \tau_{ji} X_j - \overline{\Delta}_i\right) = 0 \tag{13}$$

$$\sum_{i=1}^{q} \lambda_i = 1, \quad \lambda_i \ge 0$$

Where X_j is inverse design variable $1/A_j$, τ_{ji} is contribution of design variable J to structral behavior or size bounds, when $X_j = 1$ [1]. Formulas of the optimization process based on Larage multiplier method are given as

$$X_j = \left(\frac{W_j}{\eta \sum_{i=1}^{q} \lambda_i \tau_{ji}}\right)^{1/2} \tag{14}$$

$$\eta = \left[\frac{\sum_{i=1}^{q} \lambda_i \sum_{j=1}^{m} \tau_{ji} \left(\frac{W_j}{\sum_{i=1}^{q} \lambda_i \tau_{ji}}\right)}{\sum_{i=1}^{q} \lambda_i \overline{\Delta}_i}\right]^2 \tag{15}$$

Where η is Lagrange multiplier, $\overline{\Delta}_i$ is a bound of the jth constraint function. The λ can be expressed .

$$\lambda = \frac{\exp \rho g_i}{\sum_{i=1}^{q} \exp(\rho g_i)} \tag{16}$$

Where ρ is a positive constant with an initial value equal to 1.0. For subsequent iteration $\Delta\rho$(0.1-1.0) will be added to ρ. As done in the analysis process, the optimization process above is carried out only once in each iteration.

COMPUTER PROGRAM DESCRIPTION

Each iteration cycle consists of the following steps:

1 Give initial values A, U=0, ρ =1.0 and tolerance ε_1 , ε_2; Calculate structural weight W.

2 Perform analysis process (one or a few step).

3 Perform optimization process (one step).

4 Calculate modified structural weight $\bar{W}$ and $\varepsilon_w = |(W-\bar{W})/W|$. Calculate $\varepsilon_p = d_e^T s_e$.

5 Chek convergence: if $\varepsilon_p \leq \varepsilon_1$ and $\varepsilon_w \leq \varepsilon_2$, then optimal design is obtained, otherwise go to 3.

The flow chart is shown in Fig.1, and the parallelism of the algorithm is shown in Fig.2.

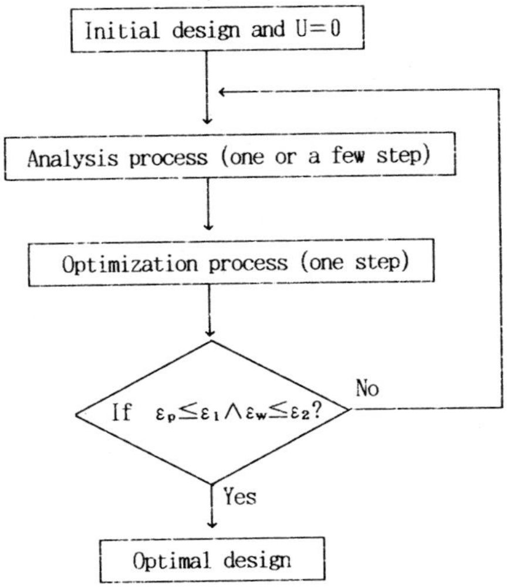

Fig.1

EXAMPLE PROBLEMS

Example 1 Four Bar Space Truss

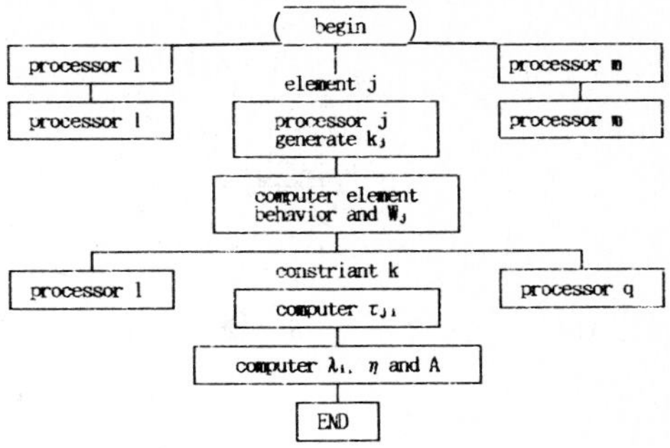

Fig.2

The structure is a four bar pyramid truss shown in Fig.3. The material is aluminum with γ =0.1 lb/in^3 and E=10*10^6 psi. Stress limits of $\bar{\sigma}$ =±25000 psi are imposed on all members. The problem involve a sigle load case, consisting of P_x =40 K, P_y=100 K, P_z=-30 K. The displacement limits at the top joint are ±0.3 inch in the

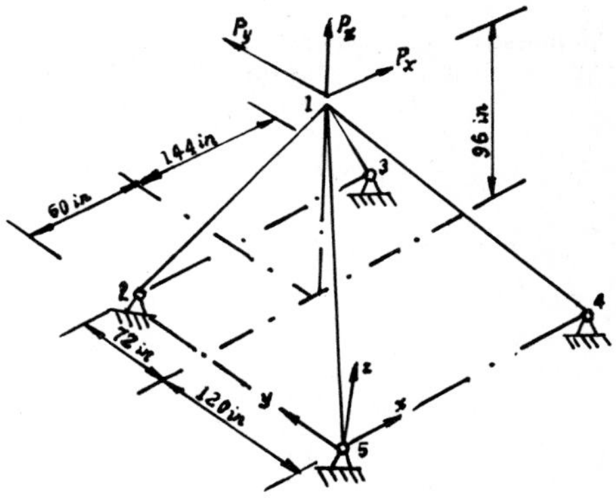

Fig.3

X-direction, ±0.5 inch in the Y-direction and ±0.4 inch in the Z-direction. Results are given in Table 1-2. Table 1 shows modified values of displacement vector and improved values of the design variables in each iterations. Table 2 gives comparisons with others.

Example 2 Seventy-two Member Space Truss [5]

This structure, shown in Fig.4, has been studied previously in many references. All members are aluminum with E=10*10^6 psi and γ=0.1 lb/in. Stress limits of $\bar{\sigma}$=±25000 psi are imposed on all members. Displacement limits of $\bar{u}$=±0.25 inch in the X and Y directions are imposed on the 4 top nodes. A lower limit of

Table 1

iterations	a_1	a_2	a_3	a_4	u	v	w	weight
1	10.0	10.0	10.0	10.0	0.05360	0.13400	-0.0402	697.182
2	5.001	5.001	5.001	5.001	0.11497	0.26769	-0.07499	348.661
3	3.164	3.343	2.500	3.271	0.16961	0.41316	-0.11132	211.730
4	2.491	2.774	1.340	2.680	0.19438	0.51753	-0.13156	157.838
5	2.611	2.570	0.802	3.107	0.18621	0.49512	-0.14868	151.349
6	2.636	2.359	0.466	3.090	0.20895	0.50810	-0.15550	140.327
7	2.700	2.250	0.270	3.294	0.22941	0.48237	-0.17678	138.426
8	2.616	2.198	0.145	3.141	0.23266	0.50425	-0.17804	131.171
9	2.663	2.183	0.080	3.164	0.23655	0.49940	-0.17717	130.531
10	2.653	2.170	0.044	3.163	0.23800	0.50133	-0.17856	129.365
11	2.672	2.164	0.024	3.183	0.24028	0.49783	-0.17970	129.439
12	2.657	2.162	0.013	3.172	0.23953	0.50014	-0.17980	128.775
Theroetical solution					0.23958	0.49995	-0.17988	

Table 2

	a_1	a_2	a_3	a_4	weight	reanalysis
[2]	2.511	2.159	0.000	3.419	130.625	7
[4]	2.614	2.159	0.000	3.210	128.53	14
This paper	2.657	2.162	0.013	3.172	128.775	

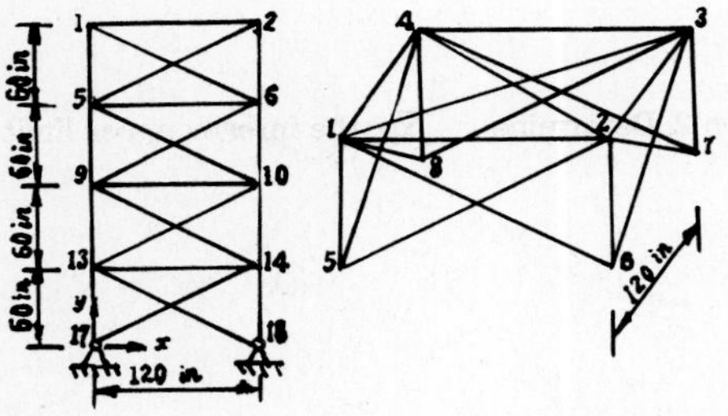

Fig.4

Table 3

Iteretions	initial	1	2	3	4	5
	6	7	8	9	10	11
	12	13	14	15	16	17
Weights	853.09	426.54	213.27	136.11	90.30	99.21
	153.20	209.50	280.68	356.60	390.35	485.66
	376.00	353.94	375.51	383.91	383.56	380.84
No. of. D. V	1	2	3	4	5	6
	7	8	9	10	11	12
	13	14	15	16		
Final design	0.157	0.537	0.411	0.571	0.509	0.522
	0.100	0.100	1.286	0.516	0.100	0.100
	1.905	0.518	0.100	0.100		

0.1 in^2 is imposed on all members. Two load conditions are considered. Case 1 has a loading of $P_x = P_y = 5000$ 1b, $P_z = -5000$ 1b at the node 1. Case 2 has a loading of $P_z = -5000$ 1b at node 1-4 respectively. Table 3 shows the structural weights in each iterations and final design.

CONCLUSION

In previous sections, we described a parallel method of structural optimization, which carried out minimization of potential energy for the structural analysis and the minimization of structural weight for optimization in terms of a bi-objective model. The main memory capacity was greatly reduced. The alternation between analysis and optimization could be done in each iteration to improve the computational efficiency. This method is suitable for microcomputers, especially for parallel computer systems. Numerical results have demonstrated the effectiveness of this method.

REFERENCES

[1] Qian Lingxi, Optimal Design of Engineering Structures, Hydro-Electric Power Press, 1983 (In Chinese).

[2] Khan, M.R., Willmert, K.D., and Thornton, W.A., A New Optimality Criterion Method for Large Scale Structures, AIAA/ASME 19th Structures, Structural Dynamics and Materials Conference, pp. 47-58, 1978.

[3] Venkayya, V.B., Design of Optimum Structures, Computer & Structures Vol.1, No.1-2, 265-279, 1971.

[4] Schmit, L.A., Farshi, B., Some Approximation Concepts for Structural Synthesis, AIAA, 12(2) 231-233, 1974.

[5] Miranker, W.L., A Survey of Parallelism in Numerical Analysis, SIAM Rew. 13, 524-547, 1971.

[6] Kincho, H.LAW., A Parallel Finite Element Solution Method, Computer & Structures Vol.23, No.6, pp.845-858, 1985.

Approximation, Optimization and Computing:
Theory and Applications, A.G. Law and C.L. Wang (eds.)
Elsevier Science Publishers B.V. (North-Holland)

INTEGER SEARCH METHOD FOR INTEGER LINEAR PROGRAMMING

WU XINGBAO

Wuhan College of Metallurgic Managerial Cadre
Renjia Road, Wuhan, Hubei, P.O.Box 430081, P.R.China

A new approach to solving Integer Linear Programming (ILP) is considered in the paper, and its computational efficiency has been tested by a vast amount of numerical examples.

1. METHOD

Without lossing the generality, let us consider the following ILP:

$$\max z = c_1x_1 + c_2x_2 + \cdots + c_nx_n$$
$$\text{s.t. } g_i(x_1, x_2, \cdots\cdots, x_n) \leq 0, \quad i=1, \cdots, m$$
$$x_j \in Z^+, \quad j=1, 2, \cdots, n$$

where Z^+ denotes the set of non-negative integer numbers and all $g_i(x_1, x_2, \cdots\cdots, x_n)$ are linear functions. Furthermore, there is no harm in the presumption:

$$|c_1| \geq |c_2| \geq \cdots\cdots \geq |c_n|.$$

The Integer Search Method (ISM) is firstly to seek a feasible integer solution of the original ILP by solving simple LPs with progressive decreasing dimension. Once the feasible integer solution is found, the corresponding constraint can be imposed in which the objective value is required better than the current such that the searching range is contracted by a big margin. Then the improving feasible integer solution will be found in the modified space. The optimum(s) has been reached right away until no further improvement can be achieved. By the way,what has no solution and what is unbounded can be judged in the searching process.

The following additional remarks should be made. Why $|c_1| \geq |c_2| \geq \cdots\cdots \geq |c_n|$ is assumed in the above mention is that the better feasible integer solution can be found sooner in the general case. This is similar to the idea selecting the entering variable in the simplex algorithm for LP.

For simplicity, the continuous feasible solution space can be assumed in the bounded case. The steps of ISM can be detailed as follows:

Step 1. Solve ULP1:

$$\max z = x_1$$
$$\text{s.t. } g_i(x_1, x_2, \cdots\cdots, x_n) \leq 0, \quad i=1, \cdots, m$$
$$x_j \in R^+, \quad j=1, 2, \cdots, n$$

where R^+ denotes the set of non-negative real numbers. If none of the feasible solution exists, terminate; neither does the original ILP. Otherwise, the optimal value $x_1^{(1)}$, i.e. the upper limit of x_1 in the continuous feasible solution space, is obtained. Go to step 2.

Step 2. Determine $\bar{\bar{x}}_1$ – the integer upper limit of x_1

$$\bar{\bar{x}}_1 = [x_1^{(1)}]$$

where [a] denotes the maximal integer number that is not greater than a. Go to step 3.

Step 3. Solve LLP1:

$$\min z = x_1$$
$$\text{s.t. } g_i(x_1, x_2, \cdots\cdots, x_n) \leq 0 \quad i=1, \cdots, m$$
$$x_j \in R^+, \quad j=1, 2, \cdots, n$$

The optimal value $x_1^{(2)}$, i.e. the lower limit of x_1 in the continuous feasible solution space, is obtained. Furthermore, $\underline{x_1}$ – the integer lower limit of x_1 is determined:

$$\underline{x_1} = \begin{cases} [x_1^{(2)}], & \text{if } x_1^{(2)} = [x_1^{(2)}] \\ [x_1^{(2)}]+1, & \text{if } x_1^{(2)} \neq [x_1^{(2)}] \end{cases}$$

Go to step 4.

Step 4.If $\bar{\bar{x}}_1 < \underline{x_1}$, terminate, the original ILP has no solutions for x_1 cannot be integer. Otherwise, go to step 5.

Step 5. If $c_1>0$ then $x_1 := \bar{\bar{x}}_1$, i.e. searching for x_1 is from large to small; otherwise, $x_1 := \underline{x_1}$, i.e. searching for x_1 is from small to large. Go to step 6.

Step 6. Solve ULP2 with n−1 decision variables $x_2, \cdots\cdots, x_n$ (x_1 is already settled):

$$\max z = x_2$$
$$\text{s.t. } g_i(x_1, x_2, \cdots\cdots, x_n) \leq 0, \quad i=1, \cdots, m$$
$$x_j \in R^+, \quad j=2, \cdots, n$$

If none of the feasible solution exists then $x_1 := x_1 - 1$ or $x_1 := x_1 + 1$, respectively, according to $c_1 > 0$ or $c_1 < 0$, and return to solve ULP2 until the optimal value $x_2^{(1)}$ is found. $\bar{\bar{x}}_2$ is determined and go to step 7: $\bar{\bar{x}}_2 = [x_2^{(1)}]$
Otherwise, if x_1 cannot be modified yet, i.e. $x_1 = \bar{x}_1$ or $x_1 = \bar{\bar{x}}_1$, respectively, according to $c_1 > 0$ or $c_1 < 0$, then the original ILP has no solution for x_2 cannot be integer, terminate.

Step 7. Solve LLP2 with n−1 decision variables (x_1 is already settled):

$$\min z = x_2$$
$$\text{s.t. } g_i(x_1, x_2, \cdots\cdots, x_n) \leqslant 0, \quad i=1, \cdots, m$$
$$x_j \in R^+, \quad j=2, \cdots, n$$

The optimal value $x_2^{(2)}$ is obtained and $\bar{x}_2$ is determined:

$$\bar{x}_2 = \begin{cases} [x_2^{(2)}], & \text{if } x_2^{(2)} = [x_2^{(2)}] \\ [x_2^{(2)}]+1, & \text{if } x_2^{(2)} \neq [x_2^{(2)}] \end{cases}$$

Go to step 8.

Step 8. If $\bar{\bar{x}}_2 < \bar{x}_2$,then $x_1 := x_1 - 1$ or $x_1 := x_1 + 1$, respectively, according to $c_1 > 0$ or $c_1 < 0$, and go to step 6. Otherwise, go to step 9.

Step 9.If $c_2 > 0$ then $x_2 := \bar{\bar{x}}_2$, otherwise, $x_2 := \bar{x}_2$. Go to step 10.

Step 10.As above, an additional variable is settled as a feasible integer value and the range of the next variable is searched progressively by solving LPs with decreasing dimension. Until a feasible integer solution $(x_1^{(0)}, x_2^{(0)}, \cdots, x_n^{(0)})$ is found. Go to step 11. Otherwise, the original ILP has no solution; terminate.

Step 11. Modify the searching limit by imposing the constraint where the original objective value is required better than the current. Go to step 12.

Step 12. If $x_1 > \bar{x}_1$ or $x_1 < \bar{\bar{x}}_1$, respectively, according to $c_1 > 0$ or $c_1 < 0$ then the further improving solution will be searched, i. e. $x_1 := x_1 - 1$ or $x_1 := x_1 + 1$, respectively, according to $c_1 > 0$ or $c_1 < 0$ and go to step 6 with the imposed constraint throughout the whole process searching the improving solution. Otherwise, if the searching limit is achieved yet or the improving solution does not exist then the current integer solution $(x_1^{(0)}, x_2^{(0)}, \cdots, x_n^{(0)})$ (as well as the reserving solution(s) with the same objective value as the current one) is optimum(s); terminate.

2. APPLICATIONS

Example 1.
$$\min z = 4x_1 + x_2$$
$$\text{s.t. } x_1 + 2x_2 \leqslant 8$$
$$2x_1 + x_2 \geqslant 6$$
$$x_j \in Z^+, \quad j=1, 2$$

Instead of the original ILP, let us consider the following equivalent ILP:

$$\max z = -4x_1 - x_2$$
s.t. ibid.
$$x_j \in Z^+, \quad j=1, 2$$

At first, solving ULP1:

$$\max z = x_1$$
s.t. ibid.
$$x_j \in R^+, \quad j=1, 2$$

obtain the optimal value $x_1^{(1)} = 8$, therefore, $\bar{\bar{x}}_1 = 8$.
Secondly, solving LLP1:

$$\min z = x_1$$
s.t. ibid.
$$x_j \in R^+ \quad j=1, 2$$

the optimal value $x_1^{(2)} = 4/3$ is obtained, therefore, $\bar{x}_1 = 2$. Owing to $c_1 = -4 < 0$, let $x_1 := 2$. Solving ULP2:

$$\max z = x_2$$
$$\text{s.t. } x_2 \leqslant 3 \quad (2 + 2x_2 \leqslant 8)$$
$$x_2 \geqslant 2 \quad (4 + x_2 \geqslant 6)$$
$$x_2 \in R^+$$

and LLP2:

$$\min z = x_2$$
$$\text{s.t. } x_2 \leqslant 3$$
$$x_2 \geqslant 2$$
$$x_2 \in R^+,$$

obtain $x_2^{(1)} = 3$ and $x_2^{(2)} = 2$, therefore, $\bar{\bar{x}}_2 = 3$ and $\bar{x}_2 = 2$. Owing to $c_2 = -1 < 0$; let $x_2 := 2$. In such a case, a feasible integer solution (2,2) is found. Modifying the searching upper limit, by solving MULP1:

$$\max z = x_1$$
$$\text{s.t. } x_1 + 2x_2 \leqslant 8$$
$$2x_1 + x_2 \geqslant 6$$
$$4x_1 + x_2 \leqslant 10 \quad (-4x_1 - x_2 \geqslant -10)$$
$$x_j \in R^+ \quad j=1, 2$$

obtain $x_1^{(1)} = 2$, therefore, $\bar{\bar{x}}_1 = 2$. Now, the searching upper limit has been achieved yet. So the current solution (2,2) is the optimum of ILP and the optimal value of original ILP is $z^* = 10$.

It is worth noting that the range of the last decision variable can be determined by the constraints with the previous n−1 variables settled.

Example 2.
$$\max z = 2x_1 + x_2 + x_3$$
$$\text{s.t. } -4x_1 + 5x_2 + 2x_3 \leqslant 4$$
$$4x_1 - x_2 + x_3 \leqslant 3$$
$$2x_1 - 3x_2 + x_3 \leqslant 1$$
$$x_j \in Z^+, j=1, 2, 3$$

At first, solving ULP1:

$$\max z = x_1$$
s.t. ibid.
$$x_j \in R^+ \quad j=1, 2, 3$$

obtain the optimal value $x_1^{(1)} = 19/16$, therefore, $\bar{\bar{x}}_1 = 1$.
Secondly, solving ILP1:

min $z=x_1$
s.t. ibid.
$x_j \in R^+$ $j=1, 2, 3$

obtain the optimal value $x_1^{(2)}=0$, therefore, $\underline{x}_1=0$. Owing to $c_1=2>0$, let $x_1:=1$. Solving ULP2:

max $z=x_2$
s.t. $5x_2+2x_3\leqslant 8$ $(-4+5x_2+2x_3\leqslant 4)$
$x_2-x_3\geqslant 1$ $(4-x_2+x_3\leqslant 3)$
$3x_2-x_3\geqslant 1$ $(2-3x_2+x_3\leqslant 1)$
$x_j \in R^+$ $j=2, 3$

and LLP2:

min $z=x_2$
s.t. $5x_2+2x_3\leqslant 8$
$x_2-x_3\geqslant 1$
$3x_2-x_3\geqslant 1$
$x_j \in R^+$, $j=2, 3$

obtain $x_2^{(1)}=10/7$ and $x_2^{(2)}=1$, therefore, $\bar{x}_2=\underline{x}_2=1$. Let $x_2:=1$ (as well as, $x_1=1$), the constraints are turned into
$2x_3\leqslant 3$ $(-4+5+2x_3\leqslant 4)$
$x_3\leqslant 0$ $(4-1+x_3\leqslant 3)$
$x_3\leqslant 2$ $(2-3+x_3\leqslant 1)$
$x_3 \in Z^+$

obtain $\bar{x}_3=\underline{x}_3=0$. In such a case,(1,1,0) is a feasible integer solution of ILP. Then, modifying the searching lower limit by solving MLLP1:

min $z=x_1$
s.t. $-4x_1+5x_2+2x_3\leqslant 4$
$4x_1-x_2+x_3\leqslant 3$
$2x_1-3x_2+x_3\leqslant 1$
$2x_1+x_2+x_3\geqslant 3$
$x_j \in R^+$, $j=1, 2, 3$

obtain $x_1^{(2)}=7/16$, therefore, $\underline{x}_1=1$. Now, the searching lower limit has been achieved yet, so the current solution (1,1,0) is the optimum of ILP and its optimal value is $z*=3$.

Example 3. max $z=x_1+x_2$
s.t. $x_1+2x_2\leqslant 6$
$5x_1+4x_2\leqslant 20$
$x_j \in Z^+$, $j=1, 2$

Solving ULP1 and LLP1, obtain $x_1^{(1)}=4$ and $x_1^{(2)}=0$, therefore, $\bar{x}_1=4$ and $\underline{x}_1=0$. Owing to $c_1>0$, let $x_1:=4$. Then, the constraints are turned into
$x_2\leqslant 1$ $(4+2x_2\leqslant 6)$
$x_2\leqslant 0$ $(20+4x_2\leqslant 20)$
$x_2 \in Z^+$
In this way, obtain $\bar{x}_2=\underline{x}_2=0$. Hence,(4,0) is a feasible integer solution and its objective value is 4. Modifying the searching lower limit, by solving MLLP1:
min $z=x_1$
s.t. $x_1+2x_2\leqslant 6$
$5x_1+4x_2\leqslant 20$
$x_1+x_2\geqslant 4$
$x_j \in R^+$ $j=1, 2$

obtain $x_1^{(2)}=2$, therefore, $\underline{x}_1=2$. Because the searching lower limit has not been achieved yet, let $x_1:=4-1=3$. In such a case, the extended constraints with imposed restriction are turned into
$2x_2\leqslant 3$ $(3+2x_2\leqslant 6)$
$4x_2\leqslant 5$ $(15+4x_2\leqslant 20)$
$x_2\geqslant 1$ $(3+x_2\geqslant 4)$
$x_2 \in Z^+$

obtain $\bar{x}_2=\underline{x}_2=1$. So a new feasible integer solution (3, 1) is found and its objective value is also 4. Therefore, the imposed restriction is not changed, and the searching lower limit is not changed, too. Once more, let $x_1:=3-1=2$, the extended constraints are turned into
$x_2\leqslant 2$ $(2+2x_2\leqslant 6)$
$2x_2\leqslant 5$ $(10+4x_2\leqslant 20)$
$x_2\geqslant 2$ $(2+x_2\geqslant 4)$
$x_2 \in Z^+$

obtain $\bar{x}_2=\underline{x}_2=2$. In this instance, another new feasible integer solution (2, 2) is found and its objective value is still 4. Now, the searching lower limit has been achieved yet. Hence, all the obtained solutions (4,0),(3,1) and (2,2) are the optimums of ILP and its optimal value is $z*=4$.

Example 4. max $z=-5x_1-2x_2-2x_3$
s.t. $2x_1+x_2-x_3\geqslant 4$
$x_1+3x_2+2x_3\geqslant 5$
$4x_1-x_2+x_3\geqslant 7$
$x_j \in Z^+$, $j=1, 2, 3$

The feasible solution space is unbounded for ULP1 is unbounded. Solving LLP1, obtain $x_1^{(2)}=11/6$, therefore, $\underline{x}_1=2$. Owing to $c_1=-5<0$, let $x_1:=2$. ULP2 is unbounded, too. Solving LLP2, obtain $x_2^{(2)}=3/5$, therefore, $\underline{x}_2=1$. Owing to $c_2=-2<0$, let $x_2:=1$. Under $x_1=2$ and $x_2=1$, the constraints are turned into
$x_3\leqslant 1$ $(4+1-x_3\geqslant 4)$
$x_3\geqslant 0$ $(2+3+2x_3\geqslant 5)$
$x_3\geqslant 0$ $(8-1+x_3\geqslant 7)$
$x_3 \in Z^+$,

obtain $\bar{x}_3=1$ and $\underline{x}_3=0$. Owing to $c_3=-2<0$, let $x_3:=0$. Then, a feasible integer solution (2,1,0) is obtained and its objective value is -12. Modifying the searching upper limit, by solving MULP1:

max $z=x_1$
s.t. $2x_1+x_2-x_3\geqslant 4$
$x_1+3x_2+2x_3\geqslant 5$
$4x_1-x_2+x_3\geqslant 7$
$5x_1+2x_2+2x_3\leqslant 12$ $(-5x_1-2x_2-2x_3\geqslant -12)$
$x_j \in R^+$ $j=1, 2, 3$

obtain $x_1^{(1)}=2$, therefore, $\bar{x}_1=2$. Now, the searching upper limit has been achieved yet, so the current solution (2,1,0) is the optimum of ILP and its optimal value is $z*=-12$.

Example 5. max $z=16x_1+12x_2+10x_3$

$$\text{s.t. } 13x_1+2x_2+4x_3\leqslant 22$$
$$6x_1+10x_2+9x_3\leqslant 28$$
$$x_1\in R^+,\quad x_j\in Z^+,\qquad j=2,3$$

At first, solving ULP2:

$$\max z=x_2$$
s.t. ibid.
$$x_j\in R^+ \qquad j=1,2,3$$

obtain $x_2^{(1)}=14/5$, therefore, $\overline{\overline{x}}_2=2$. Secondly, solving LLP2:

$$\min z=x_2$$
s.t. ibid.
$$x_j\in R^+ \qquad j=1,2,3$$

obtain $x_2^{(2)}=0$, therefore, $\overline{x_2}=0$. Owing to $c_2=12>0$, let $x_2:=2$. Solving ULP3:

$$\max z=x_3$$
$$\text{s.t. } 13x_1+4x_3\leqslant 18 \quad (13x_1+4+4x_3\leqslant 22)$$
$$6x_1+9x_3\leqslant 8 \quad (6x_1+20+9x_3\leqslant 28)$$
$$x_j\in R^+ \qquad j=1,3$$

and LLP3:

$$\min z=x_3$$
$$\text{s.t. } 13x_1+4x_3\leqslant 18$$
$$6x_1+9x_3\leqslant 8$$
$$x_j\in R^+, \qquad j=1,3$$

obtain $x_3^{(1)}=8/9$ and $x_3^{(2)}=0$, therefore, $\overline{\overline{x}}_3=\overline{x_3}=0$. Let $x_3:=0$ (as well as, $x_2:=2$) then the constraints are turned into

$$13x_1\leqslant 18 \ (13x_1+4+0\leqslant 22)$$
$$3x_1\leqslant 4 \ (6x_1+20+0\leqslant 28)$$
$$x_1\in R^+$$

and $x_1^{(1)}=4/3$ and $x_1^{(2)}=0$ are obtained. Owing to $c_1=16>0$, let $x_1:=4/3$. In such a case, a feasible solution (4/3, 2, 0) of the original mixed ILP is obtained. Modifying the searching lower limit, by solving MLLP2:

$$\min z=x_2$$
$$\text{s.t. } 13x_1+2x_2+4x_3\leqslant 22$$
$$6x_1+10x_2+9x_3\leqslant 28$$
$$16x_1+12x_2+10x_3\geqslant 45\frac{1}{3}$$
$$x_j\in R^+ \quad j=1,2,3$$

obtain $x_2^{(2)}=65/36$, therefore, $\overline{x_2}=2$. Now, the searching lower limit has been achieved yet. Hence, the current solution (4/3,2,0) is the desired optimum of mixed problem and its optimal value is $z^*=136/3$.

3. NOTES

(1) The Integer Search Method (ISM) is closely combining the cutting method with the search method in the integer case. It greatly explores the affect of the own character of ILP on the solving process and breaks free from conventions of available methods for ILP. These are the obvious characters of ISM.

(2) All the examples in [1]-[7] and others have been tested by ISM one by one. The results have shown that computation is more efficient by ISM than by the previous methods. In the general case, ISM is better than the available methods such as Gomory's cutting method and all the branch-and -bound methods which have been presented by Land and Doig, Dakin, Driebeek, Benders and so on. Furthermore. It appears that the potential advantage of ISM is more powerful as the dimension (number of the decision variables) is more large.

(3) All ILPs, including the pure ILP and the mixed ILP, can be solved by ISM. For the mixed problem, the integer decision variables are required to search one by one but the continuous decision variables are not necessary to search one by one and can be solved by the simplex method as the continuous problem.

(4) All the optimums can be obtained by ISM if the optimums of ILP are more than one.

(5) ISM can be applied to the case where the feasible solution space is unbounded. In practical computation, the upper limit, for example ,can be set an enough large integer number in the beginning if it is necessary.

(6) For some special case the computational technique can be applied to searching such that a feasible integer solution can easily be obtained by the constraints where solving LPs can be avoided even.

ACKNOWLEDGE

I am gratefully indebted to Prof. George B. Dantzig for going over my manuscript.

REFERENCES

[1] Wu Xingbao, Practical Algorithms in Modern Management (Hubei Publishing House of Science and Technique, 1987)

[2] C. H. Papadimitriou etc., COMBINATORIAL OPTIMIZATION: Algorithms and Complexity (Prentice-Hall Inc. 1982).

[3] G.B.Dantzig,Linear Programming and Extensions (Princeton Univ. Press, 1963)

[4] H.A. Taha, INTEGER PROGRAMMING: Theory, Applications, and Computations (Academic Press, 1975).

[5] H. A. Eiselt etc., OPERATIONS RESEARCH HANDBOOK: Standard Algorithms and Methods (Walter de Gruyter & Co. 1977)

[6] Xia Delin, A New Method of Integer Linear Programming-Branch Direction Search Method, Applied Mathematics and Mechanics, 6:3 (1985), 277-284.

[7] Wu Xingbao,Some Notes on 《A New Method of Integer Linear Programming-Branch Direction Search Method 》, Journal of Wuhan College of Metallurgic Managerial Cadre, 3(1987), 23-27.

Approximation, Optimization and Computing:
Theory and Applications, A.G. Law and C.L. Wang (eds.)
Elsevier Science Publishers B.V. (North-Holland)

NONLINEAR ANALYSIS BY PARAMETRIC QUADRATIC PROGRAMMING METHOD

W. X. ZHONG

Res. Ins. of Eng. Mechanics
Dalian University of Technology
Dalian 116024, PRC

R. L. ZHANG

Dept. of Engineering Mechanics
Shanghai Jiaotong University
Shanghai 200030, PRC

ABSTRACT

A unified numerical method is introduced by the PQP (Parametric Quadratic Programming) method to different tension-compression stiffness, nonlinear elasticity, plasticity and friction contact problems. Emphases are laid on the constructions of constitutive state-equations appearing to be linear complementary forms and the selection of control variables (i.e. parameters). No iteration is necessary by the PQP method to arrive at the exact solution in the FEM sense. Some sample examples are illustrated.

1. INTRODUTION

A series of papers [1-19] about the theories and applications of PVP (parametric variational principle) in contact problems, elastoplasticity and geomechanics ,have been published. A PVP features in that the energy functional constructed contains two kinds of variables, the state variables which are subject to variation as the quantities in classical variational principles, and the control variables which control the variational process to make the nonlinear constitutive state equations satisefied [1,13]. A PQP problem is formulated from a PVP counterpart. Refs. 12 and 17 have contrived a two-step algorithm to solve such quadratic programming problem with large discretized FEM system, demanding less computer costs suitable for calculation on microcomputers.

Ref. 13 has illustrated the fundamentals of PVP and its applications with many analytical and numerical examples, demostrated the usefulness of the new variational method. Firstly, it can solve some nonlinear problems such as non-associated flow, tangential slip with friction, and elasto-plastic coupling irrecoverable flow problems, generally called non-normality flow problems, to which classical principles of variation are no longer valid. Secondly, nonlinear problems are used to be solved with various kinds of iterative schemes. The iterative scheme and the convergence criterion selected may greatly affect the accuracy, convergent speed and stability of a solution, which is still a hard problem. The essence of FE solution of PVP is to solve linear equations by two steps without iterative process . Thus, the accuracy and the convergence are fairly great. For the incremental problem, bigger load step will not result in unstable oscillation. Compared with the existing iterative methods, its computational effort is also much less. It is attractive to use PVP with PQP method to study nonlinear problems and its methodology is still developing. The following of this paper is to give a summerized introduction to PQP method through some simple nonlinear examples.

2. DIFFERENT STIFFNESS PROBLEM

Let us consider a structure with three bars and a rigid plate shown in Fig. 1a. Each bar has a tensile stiffness K^+ and a compressive stiffness K^- which may be unequal, Fig. 1b. The

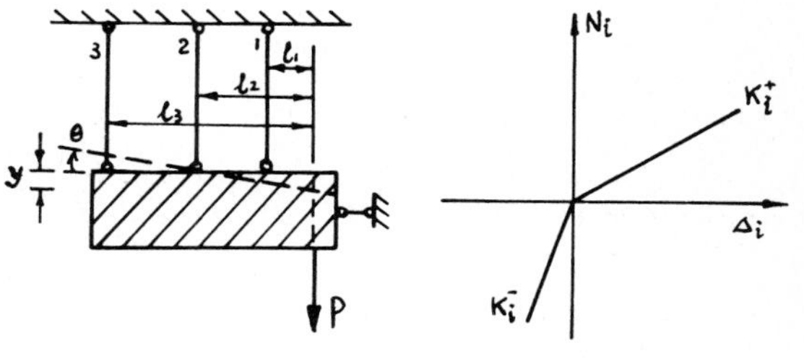

(a) (b)

Fig.1. Three bar structure

analysis of the structure has to satisfy three requirements as follows.

Equilibrium

$$\left.\begin{aligned} &N_1+N_2+N_3=P \quad \text{or} \quad \sum_i N_i-P=0 \\ &N_1\ell_1+N_2\ell_2+N_3\ell_3=0 \quad \text{or} \quad \sum_i N_i\ell_i=0 \end{aligned}\right\} \tag{1}$$

Compatibility

$$\Delta_i=y-\theta\ell_i \quad (i=1,2,3) \tag{2}$$

Constitutive

$$\Delta_i=\begin{cases} N_i/K_i^+ & \text{for } N_i>0 \\ N_i/K_i^- & \text{for } N_i\le 0 \end{cases} \quad (i=1,2,3) \tag{3}$$

where N_i are the axial force and Δ_i are the

elongation. If the displacements $\{y,\theta\}$ are taken to be the unkowns, then the problem can be solved via the following way.

(i) Find the N_i-$\{y,\theta\}$ relations by substituting constitutive relations (3) into compatibility conditions (2) and eleminating Δ_i;

(ii) Substitution of N_i-$\{y,\theta\}$ into equlibrium equations (1) results in two equations by which y and θ are found. Or, alternatively they are found through the potential energy functional $\Pi(y,\theta)$ and the variational principle, $\delta\Pi=0$;

(iii) Finally replacement of y and θ from N_i-$\{y,\theta\}$ relations leads to N_i.

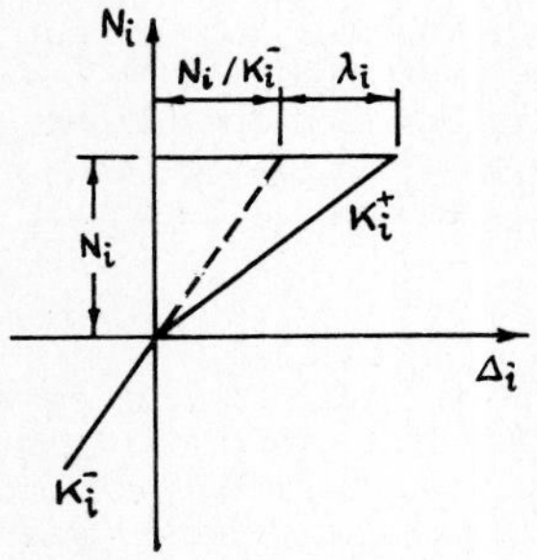

Fig. 2

No difficulty arises for the linear case when $K_i^+=K_i^-$. But if the constitutive relations are nonlinear when $K_i^+\neq K_i^-$, it is difficult to predict which bar is in tension or compression *a priori*, i.e. we do not know whether K_i^+ or K_i^- should be used at the first step. If this problem is solved by conventional method, generally the "Test and iteration" loop is inevitable. Now let us see how the PQP method works.

Suppose the ith bar may be subject to tension, then (from Fig. 2)

$$\Delta_i=N_i/K_i^-+\lambda_i \tag{4}$$

where λ_i are the supplementary elongation

$$\lambda_i=N_i/K_i^+-N_i/K_i^- \tag{5}$$

Since

$$N_i=K_i^-(\Delta_i-\lambda_i)=K_i^-(y-\theta\ell_i-\lambda_i) \tag{6}$$

substituting (6) into (5), we find

$$f_i(y,\theta,\lambda_i)=(y-\theta\ell_i)(K_i^--K_i^+)-\lambda_iK_i^- \tag{7}$$

when $f_i=0$, the ith bar is subject to tension and $\lambda_i\geq 0$, otherwise if $f_i<0$ then the ith bar is in compression and $\lambda_i=0$. Therefore λ_i are non-negtive scalors. It is seen that f_i and λ_i resemble the yield functions and flow multipliers respectively. Introducing slack variables υ_i into (7), we find a unified constitutive relations

$$\left.\begin{aligned} &N_i=K_i^-(\Delta_i-\lambda_i)\\ &f_i(y,\theta,\lambda_i)+\upsilon_i=0\\ &\upsilon_i\lambda_i=0,\ \upsilon_i,\lambda_i\geq 0,\ i=1,2,3, \end{aligned}\right\} \tag{8}$$

for both tension and compression, which is called the constitutive state-equation. According to parametric variational theory [1], y and θ are the state variables subject to variation, λ_i are the control variables without being subject to variation, thus the functional of the total potential energy of the system is

$$\Pi[\lambda_i(\cdot)]=\sum_i\Big[\frac{1}{2}K_i^-(y-\theta\ell_i)^2-\lambda_iK_i^-(y-\theta\ell_i)\Big]-Py \tag{9}$$

where notation $\lambda_i(\cdot)$ emphasizes that λ_i are the control variables without subject to variation. Consequently the solution can be obtained from the following PQP problem

$$\left.\begin{aligned} &\textit{minimize}\quad \Pi[\lambda_i(\cdot)]\\ &\textit{subject to}\quad f_i(y,\theta,\lambda_i)+\upsilon_i=0\\ &\qquad\qquad \upsilon_i\lambda_i=0,\ \upsilon_i,\lambda_i\geq 0,\ i=1,2,3 \end{aligned}\right\} \tag{10}$$

We have illustrated the concept and methodology of PQP method through a simple nonlinear bar system. PQP problem (10) is a typical formula by which more complicated problems can be solved. If we use the classical variational method, letting y, θ and λ_i be the quantities subject to variation simultaneously, neither the equilibrium nor the constitutive relations can be satisfied. For instance, setting $\partial\Pi/\partial\lambda_i=0$ and $\partial\Pi/\partial y=0$ leads to

$$\lambda_i=y-\theta\ell_i \tag{11}$$

and

$$\sum_i[K_i^-(y-\theta\ell_i)-\lambda_iK_i^-]-P=0 \tag{12}$$

Eqn (12) would be the equilibrium equations, but substitution (11) for λ_i gives $P\equiv 0$ which is in conflict with real situation. Therefore variation with respect to control variables may yield meaningless results.

Example 1. Consider the three bar structure in Fig.1a with P=-10(kg), K_i^-=10(kg·cm), K_i^+=6(kg·cm) (i=1,2, 3), and $\ell_1=0$, ℓ_2=2(cm), ℓ_3=4(cm).

Sustituting these data into eqn (9) then taking the first variation with respect to state variables to vanish, the PQP problem becomes a linear complementary problem given in Table 1.

Table 1

basis	υ_1	υ_2	υ_3	λ_1	λ_2	λ_3	P/K_i^-
υ_1	1	0	1	-40/6	8/6	-4/6	-20/6
υ_2	0	1	0	8/6	-52/6	8/6	-8/6
υ_3	0	0	1	-4/6	8/6	-40/6	4/6

Using LEMKE's complementary pivot scheme [27,29] we obtain the solution

$$\lambda_1=\lambda_2=0,\ \lambda_3=0.1(\text{cm})$$

$$y=-51/60=-0.85(\text{cm}),\ \theta=-11/40=-0.275(\text{rad})$$

which indicates that bar 1 and 2 are in compression while bar 3 in tension.

If we let K_i^-=0 then the PQP scheme results in

a ray solution which implies that no unique solution exists. The rigid plate may rotates freely since the bars can not be subject to tension.

3. MULTI-BRANCH PIECEWISE LINEARIZED PROBLEMS

The different stiffness problem discussed in last section has one turning point at origin and belongs to one-branch piecewise linear constitutive relations. In this section we will discuss the multi-branch piecewise linearized constitutive relations which can also be generalized to other nonlinear elasticity problems with multiaxial stress state as well as deformation theory of plasticity [19].

For simplicity, let us consider again a simple bar system with a constitutive relation shown in Fig 3. which consist of n turning points, i.e., the ith bar has n+1 different stiffnesses $K_i^{(1)}$, $K_i^{(2)}, \cdots, K_i^{(n)}$. Let $P_i^{(0)}=0$ and $|P_i^{(n+1)}|=\infty$, then

$$\left.\begin{aligned} &N_i=K_i^{(1)}(\Delta_i-\lambda_i),\\ &\lambda_i=\sum_{\alpha=1}^{j}\lambda_i^{(\alpha)},\quad j=0,1,2,\cdots,n\\ &\lambda_i^{(\alpha)}=(N_i-P_i^{(\alpha)})(1/K_i^{(\alpha+1)}-1/K_i^{(\alpha)}) \end{aligned}\right\} \tag{13}$$

for $P_i^{(j)}\le N_i\le P_i^{(j+1)}$. Consequently the yield functions can be written by

$$f_i^{(j)}(\Delta_i,\lambda_i)=\frac{K_i^{(j)}K_i^{(j+1)}}{K_i^{(j+1)}-K_i^{(j)}}\lambda_i^{(j)}+K_i^{(1)}\Big(\Delta_i-\sum_{\alpha=1}^{j}\lambda_i^{(\alpha)}\Big)-P_i^{(j)}$$
$$\text{for } P_i^{(j)}\le N_i\le P_i^{(j+1)},\ j=1,2,\cdots,n. \tag{14}$$

which satisfy the nonnegative conditions

$$\lambda_i^{(\alpha)}\begin{cases}\ge 0 & \text{for } f_i^{(\alpha)}=0\\ =0 & \text{for } f_i^{(\alpha)}<0\end{cases}\qquad \alpha=1,2,\cdots,n \tag{15}$$

Finally the constitutive state equations are found

$$\left.\begin{aligned} &f_i^{(j)}(\Delta_i,\lambda_i^{(j)})+\upsilon_i^{(j)}=0,\\ &\upsilon_i^{(j)}\lambda_i^{(j)}=0,\quad \upsilon_i^{(j)},\lambda_i^{(j)}\ge 0\\ &j=1,2,\cdots,n \end{aligned}\right\} \tag{16}$$

The PQP formulation can be obtained by the way similar to last section.

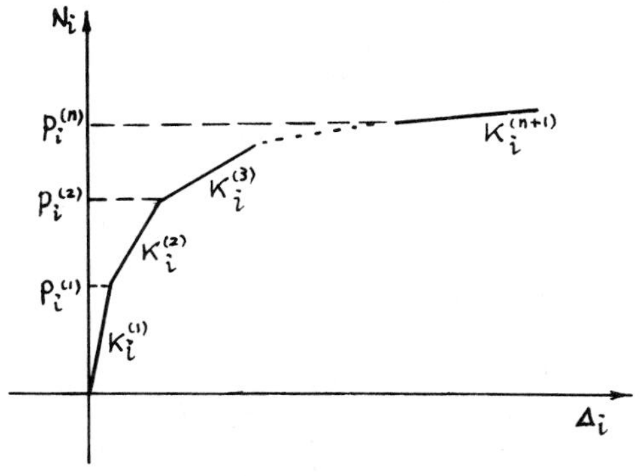

Fig. 3. Multi-branch constitutive eqn.

4. ELASTO-PLASTICITY PROBLEM

The flow theory is based on three fundamental assumptions, the definition of yield criterion, the flow rule and the elastic response, i.e.,

$$\left.\begin{aligned} &f(\sigma_{ij},\varepsilon^P_{ij},k)\le 0\\ &d\varepsilon^P_{ij}=\lambda\frac{\partial g}{\partial\sigma_{ij}}\\ &d\sigma_{ij}=D_{ijkl}(d\varepsilon_{ij}-d\varepsilon^P_{ij}) \end{aligned}\right\} \tag{17}$$

From which the PQP formula can be found [1]

$$\left.\begin{aligned} &\mathit{min.}\quad \Pi_P[\lambda(\cdot)]\\ &\mathit{s.t.}\quad f(du_i,\lambda)+\upsilon=0\\ &\qquad\ \upsilon\lambda=0,\ \upsilon,\lambda\ge 0, \end{aligned}\right\} \tag{18}$$

where

$$f(du,\lambda)=f(\sigma^0,\varepsilon^0,k^0)+\frac{\partial f}{\partial\sigma_{ij}}D_{ijkl}du_{k,l}-\left[\frac{\partial f}{\partial\sigma_{ij}}D_{ijkl}\frac{\partial g}{\partial\sigma_{ij}}-\frac{\partial g}{\partial\sigma_{ij}}\cdot\frac{\partial k}{\partial\varepsilon_{ij}}\right]\lambda \tag{19}$$

$$\Pi_P[\lambda(\cdot)]=\int_\Omega\left\{\frac{1}{2}D_{ijkl}du_{i,j}du_{k,l}-\lambda\frac{\partial g}{\partial\sigma_{ij}}D_{ijkl}du_{k,l}\right\}d\Omega-\left[\int_\Omega db_i du_i d\Omega+\int d\bar{p}_i du_i dS\right] \tag{20}$$

5. CONTACT PROBLEM WITH FRICTION

Coulomb's friction contact problem is in fact a special kind of material of plasticity and can be solved by the similar way as that for flow plasticity. Refs 13 and 18 gives the formula for this problem by

$$\left.\begin{aligned} &\mathit{min.}\quad \Pi_C[\lambda(\cdot),\Lambda(\cdot)]\\ &\mathit{s.t.}\quad f(du_i,\lambda)+\upsilon=0\\ &\qquad\ \mathcal{F}(du^A_\alpha,du^B_\alpha,\Lambda)=0\\ &\qquad\ \upsilon\lambda=0,\quad \nu\Lambda=0\\ &\qquad\ \upsilon,\nu,\lambda,\Lambda\ge 0, \end{aligned}\right\} \tag{21}$$

where λ and Λ are the control variables of plasticity and contact slip respectively, υ and ν are the complementary variables, and

$$\Pi_C[\lambda(\cdot),\Lambda(\cdot)]=\Pi_P+\int_{S_C}\sum_{\beta=\rho,\vartheta,\eta}\left\{\sum_{\alpha=\rho,\vartheta,\eta}\frac{1}{2}E_{\alpha\beta}d\epsilon_\alpha d\epsilon_\beta-\Lambda R_\beta d\epsilon_\beta\right\}d\Omega \tag{22}$$

6. CONCLUSION

PQP method based on parametric variational Principle is a simple and efficient numerical scheme for nonlinear studies. It at last is reduced to be a linear complementary problem which can be solved readily. To apply the PQP method it is necessary to turn the constitutive relations of the nonlinear problem under consideration to the form of constitutive state-equations with complementary conditions, to correctly select the control and state variables, and to form a performance index, i.e. the energy functional containing both kinds of variables.

ACKNOWLEDGEMENTS

The work presented in this paper was supported by the National Natural Science Fundation of China.

REFERENCES

[1] Zhong, W. X. and Zhang, R. L., Parametric Variational Principles and Their Quadratic Programming Solutions, Comp. Struct., 30, 887-896 (1988).
[2] Zhong, W. X., On Variational Principles of Elastic contact Problems and Parametric Quad-ratic Programming (in Chinese), Comp. Struct. Mech. Appl., **2**,1-10 (1985).
[3] Zhong, W. X. and Zhang, R. L., On a New Variational Method and Mathematical Programming solution, Report on 2nd National Conference of Computational Mechanics, Shanghai, Aug. (1986).
[4] Zhong, W. X. and Sun, S.M., A Finite Method for Elasto-plastic Structures and Contact Problems by Parametric Quadratic Programming, Int. J. Numer. Methods Eng., **26**, 2723-2738 (1988).
[5] Zhong, W. X. and Zhang, R.L., The Parametric Variational Principles for Elastoplasticity, Acta Mechanica Sinica, **4**, 98-104 (1988).
[6] Zhong, W. X. and Zhang, R. L., Quadratic Programming with Parametric Vector in Plasticity and Geomechanics, Prc. NUMETA, GK, **2**, C11 (1987).
[7] Zhong, W. X. and Zhang, R. L., On the Second Stage Minimum Principles in Plasticity, Comp. Struct. Mech. Appl., **4**(4),1-12 (1987).
[8] Zhong, W. X., On the Potential Energy Principle for Elastic Contact Problems in Chinese), Comp. Struct. Mech. Appl., **26**(3) (1983).
[9] Zhong, W. X. and Zhang, R. L. and Sun, S.M., Parametric Quadratic Programming Applications in Computational Mechanics (in Chinese), part 1, Comp. Struct. Mech. Appl., **5**(4), 80-100 (1988), part 2, **6**(1), 98-110 (1989), **6**(2), 113-123 (1989)
[10] Qian, L. X., Zhang, R. L., Zeng, P., Numerical Analyses of Retaining Structures by PQP Method, to be published recently.
[11] Zhang, R. L. and Zhong,W.X., The Parametric Minimum Potential Energy Principle of Flow Theory (in Chinese), Acta Mechanica Solida Sinica, **10**(1), 90-96 (1989).
[12] Zhang, R. L. and Zhong, W.X., The Numerical Solution for PMPEP by Parametric Quadratic Programming (in Chinese), Comp. Struct. Mech. Appl., **2**(1), 1-12 (1987).
[13] Zhang,R. L., Parametric Variational Principles: Theory and Applications (in Chinese), Ph.D. Thesis, RIEM, Dalian University of Technology, Dalian, (1987).
[14] Zhang, R. L. and Zhong, W.X., Guide to the Computer Program for Analyses of Elastoplastic and Contact Structures DDJ-W(S), (in Chinese), Liberary of RIEM, Dalian University of Technology, (1986).
[15] Zhang, R. L. and Ling, X. J., Excavation Analysis by PVP Method for the People's Square Underground Car Parking in Shanghai, Proc. of 3rd International Conference on Underground Space and Earth Sheltered Buildings, Shanghai, **2**, 714-719 (1988).
[16] Zhang, R. L., Limit Analysis for Non-associated Flow Rule Materials by PVP Method (in Chinese), Report on 5th East China Conference on Solid Mechanics, Wuxi, Sep. (1988), to be published on J. Shanghai Jiaotong Uni. immediately.
[17] Zhang, R. L., A New Scheme for Quadratic Programming with Free Design Variables(in Chinese), Comp. Struct. Mech. Apll., **4**(1), (1987).
[18] Zhang, R. L., Analyses of Elastic and Elasto-plastic Contact Problems with Friction by Parametric Minimum Potential Energy Principles, to be published.
[19] Zhang, R. L., Constitutive State-Equation and Parametric Variational Principles for Deformation Theory, to be published.
[20] Y. C. Fung, A First Course In Continuum Mechanics, Prentice-Hall, Englewood Cliffs, NJ, 1977.
[21] Bellman, R. and Dreyfus, S. E., Applied Dynamic Programming, Princeton Uni. Press, Princeton, NJ.(1962).
[22] Pontryagin, L. S. *et al*, The Mathematical Theory of Optimal Processes, Interscience, Wiley, NY.(1962).
[23] Washizu, K., Variational Methods in Elasticity and Plasticity, Pergamon Press, Oxford. (1982).
[24] Bao, S. S., Optimization: Theory and Applications, 2nd ed., John Wiley & Sons, NY. (1984).
[25] William, L. B., Modern Control Theory, Prentice-Hall, Englewood Cliffs, NJ. (1985).
[26] Xie,X.S., Optimal Control Theory and Application, Qinghua Uni. Press, Beijing (in Chinese 1986).
[27] Lemke, C. E., "Bimatrix Equilibrium Points and Mathematical Programming", Manage. Sci., 11, 681. (1965).
[28] Cottle, R. W. and Dantzig, G. B., "Complementary Pivot Theory and Mathematical Programming", J.Linear Algebra Appl., 1,105. (1968).
[29] Reklaitis, G.V., Ravindran,A. and Ragsdell, K. A., Engineering Optimization, Methods and Applications,Jhon Wiley & Sons,NY (1983).
[30] Hill, R., The Mathematical Theory of Plasticity, Clarendon Press, Oxford (1950).
[31] Maier, G., Complementary Plastic Work Theorems in Piecewise-linear Elasto-plasticity, Int. J. Solids & Structs (1969).

Approximation, Optimization and Computing:
Theory and Applications, A.G. Law and C.L. Wang (eds.)
Elsevier Science Publishers B.V. (North-Holland)

GENERALIZED FRITZ-JOHN NECESSARY CONDITIONS FOR QUASIDIFFERENTIABLE MINIMIZATION

Zun-Quan Xia (Tzun-Chyuarn Shiah)

Dept. of Appl. Math., Dalian Univ. of Technology, Dalian 116024, China

This paper is to explore necessary conditions of Fritz-John form for constrained minimization of quasidifferentiable functions. The generalized Fritz-John conditions associated with Lagrangian multipliers depending on direction are derived only under ordinary assumptions.

1. INTRODUCTION

In the past several years necessary conditions found for constrained minimization of quasidifferentiable functions in the sense of [4] were given in cone structures [3], [5], [8], and some forms of generalized Fritz-John necessary conditions for quasidifferentiable functions were also described [6], [7] and [9], but Lagrangian multipliers depend on the particular elements in superdifferentials of the objective function and constraint functions of a problem. A new form of generalized Fritz-John necessary conditions for a quasidifferentiable function will be proposed. This kind of form is very close to the normal form of generalized Fritz-John necessary conditions. A normal form of generalized Fritz-John necessary conditions for quasidifferentiable functions should be of the form $-\overline{\partial}_x L(x^*,\ \lambda^*) \subset \underline{\partial}_x L(x^*,\ \lambda^*)$ where $L(x,\ \lambda)$ is the Lagrangian function for constrained minimization of a quasidifferentiable function, x^* is a local minimizer, and λ^* is a corresponding Lagrangian multiplier vector. Unfortunately, generally speaking,there is no Lagrangian multiplier vector λ^* satisfying this inclusion relation and corresponding slackness conditions at a local minimizer, see Sec. 2 and [6]. What is an appropriate form for generalized Fritz-John necessary conditions of constrained minimization of quasidifferentiable funotions?

The generalized Fritz-John necessary conditions for constrained minimization of Lipschitz functions, given in [2], can be replaced equivalently by the following form

$$\delta^*(d|\partial_x L(x^*,\ \lambda^*)) \geq 0,\ \forall d \in R^n,$$

Slackness conditions.

Naturally, it can be conjectured that an appropriate form of an extension to quasidifferentiable minimization is of the form

$$\delta^*(d| -\overline{\partial}_x L(x^*,\ \lambda^*)) \leq \delta^*(d|\underline{\partial}_x L(x^*,\ \lambda^*)), \forall d \in R^n. \qquad (1-1)$$

A simple example, [6], used again in next section shows that the form (1-1) does not hold for quasidifferentiable functions in the sense of [4]. However, the from (1-1) is valid in direction, i.e., Lagrangian multipliers depend on direction. In other words, for each $d \neq 0, d \in R^n$, there exists at least one directionally Lagrangian multiplier vector λ^* associated with d such that $\delta^*(d| -\overline{\partial}_x L(x^*,\ \lambda^*)) \leq \delta^*(d|\underline{\partial}_x L(x^*,\ \lambda^*))$ holds.

2. FRITZ-JOHN NECESSARY CONDITIONS

Consider the following problem

$$\begin{aligned} &min f(x),\quad x \in R^n \\ &g_i(x) \leq 0,\quad i = 1, \cdots, m \qquad (2-1) \\ &g_i(x) = 0,\qquad i = m+1, \cdots, m+p, \end{aligned}$$

where $f,\ g_i,\ i = 1, \cdots,$ m+p, are continuous and quasidifferentiable in R^n. Let $\wedge$ be the set of all $\lambda = (\lambda_0,\ \lambda_1, \cdots, \lambda_m \cdots, \lambda_{m+p})^T$ such that $\lambda_i \geq 0,\quad i = 0,\ 1, \cdots, m,$ and $\sum_{i=1}^{m+p} \lambda_i^2 = 1$. For the convenience of writing, let $g_0(x) := f(x) - f(x^*),\ x^* \in R^n$, where $x \in R^n$ is a local minimum point. Define the Lagrangian function

$$L(x, \lambda) := \lambda_0 f(x) + \sum_{i=1}^{m+p} \lambda_i g_i(x).$$

Furthermore, we define $G(x) := max_{\lambda \in \wedge}\ [\lambda_0(f(x) - f(x^*)) + \sum_{i=1}^{m+p} \lambda_i g_i(x)]$, [1]. Therefore, we have $G(x) = max_{\lambda \in \wedge_0} \langle\ \lambda,\ g(x)\rangle = \delta^*(g(x)|\wedge_0)$, where $\wedge_0 := co\wedge$ and $g(x) := (g_0(x), \cdots, g_{m+p}(x))^T$. Note that $G(x^*) = 0$ and $G(x) \geq 0$ for all x near x^*. From convex analysis it follows that a finite convex function in R^n is locally Lipschitzian.

THEOREM 2.1. Suppose x^* is a local minimum point to the problem (2-1). Then for each $d \in R^n$ there exisits a vector $\lambda^* = (\lambda_0^*, \lambda_1^*, \cdots, \lambda_{m+p}^*)^T$ such that

$$\delta^*(d| -\overline{\partial}_x L(x^*, \lambda^*)) \leq \delta^*(d|\underline{\partial}_x L(x^*, \lambda^*)) \qquad (2-2)$$

$$\lambda_i^* g_i(x^*) = 0, \quad i = 1, \cdots, m, \qquad (2-3)$$

$$\lambda_i^* \geq 0, \quad i = 0, \cdots, m, \qquad (2-4)$$

$$\sum \lambda_i^{*2} = 1 \ \ or \ \ \lambda^* \neq 0, \qquad (2-5)$$

where λ^* is a Lagrangian multiplier vector depending on direction d.

Proof. Let $F(\cdot) := \delta^*(\cdot|\wedge_0)$ be a support function defined in R^{m+p+1}. Evidently, F is convex and $G(\cdot) = F(g(\cdot))$. F is uniformly quasidifferentiable in terms of [10, Lem. 3.2], g is quasidifferentiable, and it follows from [5, Th. 12.2] that $G(x) = F(g(x))$ is quasidifferentiable at $x \in N(x^*)$, where $N(x^*)$ means a neighborhood of x^*. From [1, Lem.1] one has that there exists an $N(x^*)$ such that $G(x) \geq G(x^*)$, $\forall x \in N(x^*)$. In view of [3] we obtain $-\overline{\partial}G(x^*) \subset \underline{\partial}G(x^*)$. According to [5,Th.12.2], one has

$$\underline{\partial}G(x^*) = \{u | u = \sum_{i=0}^{m+p} [\lambda_i(\alpha_i + \beta_i) - \lambda_i'\alpha_i - \lambda_i''\beta_i,$$

$$\lambda \in \underline{\partial}F(\cdot)g(x^*), \alpha_i \in \underline{\partial}g_i(x^*), \beta_i \in \overline{\partial}g_i(x^*)\},$$

$$\overline{\partial}G(x^*) = \{u | u = \sum_{i=0}^{m+p} [\overline{\lambda}_i(\alpha_i + \beta_i) + \lambda_i'\alpha_i + \lambda_i''\beta_i,$$

$$\overline{\lambda} \in \overline{\partial}F(\cdot)g(x^*), \alpha_i \in \underline{\partial}g_i(x^*), \beta_i \in \overline{\partial}g_i(x^*)\},$$

where $\lambda = (\lambda_0, \lambda_1, \cdots, \lambda_{m+p})^T$, $\overline{\lambda} = (\overline{\lambda}_0, \overline{\lambda}_1 \cdots, \overline{\lambda}_{m+p})^T$, and λ' is a lower bound of λ and λ'' is an upper bound of λ. Since $F(\cdot)$ is convex, $\overline{\lambda}$ can be taken as $\theta_{m+p+1} = (0, \cdots, 0)^T \in R^{m+p+1}$. Taking $\lambda' = (-1, \cdots, -1)^T \in R^{m+p+1}$ and $\lambda'' = (1, \cdots, 1)^T \in R^{m+p+1}$, we have

$$\sum_{i=0}^{m+p} \underline{\partial}g_i(x^*) + \sum_{i=0}^{m+p} -\overline{\partial}g_i(x^*) \subset$$

$$W := \{u | u \in \sum_{i=0}^{m+p} (1 + \lambda_i)\underline{\partial}g_i(x^*)$$

$$+ \sum_{i=0}^{m+p} (1 - \lambda_i)(-\overline{\partial}g_i(x^*), \ \ \lambda \in \partial F(\cdot)g(x^*)\}.$$

It follows from this that

$$\delta^*(\cdot | \sum_{i=0}^{m+p} \underline{\partial}g_i(x^*) + \sum_{i=0}^{m+p} -\overline{\partial}g_i(x^*))$$

$$\leq \delta^*(\cdot|W) = \sum_{i=0}^{m+p} \delta^*(\cdot|\underline{\partial}g_i(x^*)) - \overline{\partial}g_i(x^*))$$

$$+ \sup_{\lambda \in \partial F(\cdot)g(x^*)} [\sum_{i=0}^{m+p} \lambda_i \delta^*(\cdot|\underline{\partial}g_i(x^*))$$

$$+ \sum_{i=0}^{m+p} -\lambda_i \delta^*(\cdot| -\overline{\partial}g_i(x^*))] \qquad (2-6)$$

Since $\underline{\partial}F(\cdot)g(x^*) \subset \wedge_0$ and for each direction the supremum of the second item on the right-hand side of (2-6) can be attained at an extreme point of $\wedge_0$, for each $d \in R^n$ there exists at least one $\lambda^* \in \wedge_0$ such that $||\lambda^*|| = 1$ and

$$\sup_{\lambda \in \wedge_0} [\sum_{i=0}^{m+p} \lambda_i \delta^*(d|\underline{\partial}g_i(x^*))$$

$$+ \sum_{i=0}^{m+p} -\lambda_i \delta^*(d| -\overline{\partial}g_i(x^*))] = \sum_{i=0}^{m+p} \lambda_i^* \delta^*(d|\underline{\partial}g_i(x^*))$$

$$+ \sum_{i=0}^{m+p} -\lambda_i^* \delta^*(d| -\overline{\partial}g_i(x^*)) \geq 0. \qquad (2-7)$$

Let $I := \{m+1, \cdots, m+p\}$, $I^+ := \{i \in I | \lambda_i^* \geq 0\}$ and $I^- := \{i \in I | \lambda_i^* < 0\}$. The left-hand side of (2-7) can be rewritten as

$$\sum_{i=0}^{m} \lambda_i^* \delta^*(d|\underline{\partial}g_i(x^*)) + \sum_{i \in I^+} \lambda_i^* \delta^*(d|\underline{\partial}g_i(x^*))$$

$$+ \sum_{i \in I^-} \lambda_i^* \delta^*(d|\underline{\partial}g_i(x^*)) + \sum_{i=0}^{m} -\lambda_i^* \delta^*(d| -\overline{\partial}g_i(x^*))$$

$$+ \sum_{i \in I^+} -\lambda_i^* \delta^*(d| -\overline{\partial}g_i(x^*))$$

$$+ \sum_{i \in I^-} -\lambda_i^* \delta^*(d| -\overline{\partial}g_i(x^*)).$$

Finally for each $d \in R^n$, one has

$$\delta^*(d| -[\sum_{i=0}^{m} \lambda_i^* \overline{\partial}g_i(x^*) + \sum_{i \in I^+} \lambda_i^* \overline{\partial}g_i(x^*) + \sum_{i \in I^-} \lambda_i^* \underline{\partial}g_i(x^*))]$$

$$\leq \delta^*(d| \sum_{i=0}^{m} \lambda_i^* \underline{\partial}g_i(x^*) + \sum_{i \in I^+} \lambda_i^* \underline{\partial}g_i(x^*) + \sum_{i \in I^-} \lambda_i^* \overline{\partial}g_i(x^*)).$$

In consequence of the last inequality, one has $\delta^*(d| -\overline{\partial}_x L(x^*, \lambda^*)) \leq \delta^*(d|\underline{\partial}_x L(x^*, \lambda^*))$. Since x^* is a local solution to the problem (2-1), the conditions in (2-3) are satisfied. The condititions in (2-4) and (2-5) can be obtained easily from the demonstration given above. The proof is completed.

Given a $d \in bdB_1(0)$. Let $M(x, d)$ be the set consisting of all Lagrangian multipliers in the direction d at a point x, satisfying the conditions (2-2)–(2-5). This set is called directionally Lagrangian multiplier set in direction d at x. It is possible for $M(x, d)$ to be empty.

COROLLARY 1. If $M^*(x^*) := \cap_{d \in bdB_1(0)} M(x^*, d) \neq \phi$, then

$$-\overline{\partial}L(x^*, \lambda) \subset \underline{\partial}L(x^*, \lambda), \forall \lambda \in M^*(x^*). \qquad (2-8)$$

EXAMPLE [6].$min\{f(x) = x | g(x) \leq 0\}$, where $g(x) := min\{-x, -2x\}$. The point $x^* = 0$ is a local solution, and $\underline{\partial}f(0) = 1$, $\overline{\partial}f(0) = 0$, $\underline{\partial}g(0) = 0$, $\overline{\partial}g(0) = [-2, -1]$. Let $L(x, \lambda) = \lambda_0 f(x) + \lambda g(x)$. We have then $\delta^*(1| - \overline{\partial}L(0, \lambda)) = 2\lambda$, $\delta^*(1|\underline{\partial}L(0, \lambda)) = \lambda_0$, and $\delta^*(-1| - \overline{\partial}L(0, \lambda)) = -\lambda$, $\delta^*(-1| - \underline{\partial}L(0, \lambda)) = -\lambda_0$. In addition, for $d = 1$ or -1, one has $M(0, 1) = \{(\lambda_0, \lambda) \geq 0 \; |2\lambda \leq \lambda_0, \lambda_0^2 + \lambda^2 = 1\}$, $M(0, -1) = \{(\lambda_0, \lambda) \geq 0 | \lambda \geq \lambda_0, \lambda_0^2 + \lambda^2 = 1\}$. Therefore there is no Lagrangian multiplier vector satisfying(2-8) in the normal sense, but there are directionally Lagrangian multipliers such that $M(x^*, 1) = M(0, 1) \neq \phi$ and $M(x^*, -1) = M(0, -1) \neq \phi$.

COROLLARY 2. For the problem (2-1) without equality constraints or without inequality constraints, the Th. 2.1 is still valid.

COROLLARY 3[6, Th.1] Suppose the problem (2-1) does not contain equality constraints and $f, g_i, i = 1, \cdots, m$ are subdifferentiable. Then there is a $\lambda^* = (\lambda_0^*, \lambda_1^*, \cdots, \lambda_m^*)^T$ such that

$$0 \in \partial L(x^*, \lambda^*), \quad (\partial := \underline{\partial})$$

$$\lambda_i^* g_i(x^*) = 0, \;\; i = 1, \cdots, m,$$

$$\lambda^* \geq 0, \; \sum (\lambda_i^*)^2 = 1 \;\; or \;\; \lambda^* \neq 0.$$

REMARK. The way used above can be also used to derive Fritz-John necessary conditions for constrained optimization of Lipschitz functions. It can be done in terms of [4, Th. 2.3.9], [1] and the demonstration in Th. 2.1.

COROLLARY 4 [5, Prop. 16.2]. Consider the problem $min\{f(x) \quad |g(x) \leq 0, \; x \in R^n\}$, where f, g are quasidifferentiable. Let x^* be a local minimal point and $g(x^*) = 0$. Then $-[\overline{\partial}f(x^*) + \overline{\partial}\, g(x^*)] \subset co\{\underline{\partial}f(x^*) - \overline{\partial}g(x^*), \underline{\partial}g(x^*) - \overline{\partial}f(x^*)\}$.

Note that the condition (2-2) can be replaced by

$$L_x'(x^*, \lambda^*; d) \geq 0,$$

λ^* depending on $d \in R^n$, since

$$\delta^*(d|\underline{\partial}_x L(x^*, \lambda^*)) - \delta^*(d| - \overline{\partial}_x L(x^*, \lambda^*)) = L_x'(x^*, \lambda^*; d).$$

For a Lipschitz function the only difference from a quasidifferentiable function on Fritz-John necessary conditions is that there exists at least one Lagrangian multiplier vector not depending on direction. For a quasidifferentiable function, Lagrangian multipliers do depend on direction. This is mainly because the basic definition of quasidifferentiable functions is based upon the concept of directional derivative and the form of differential is defined by a pair of sets. It is clear from Corol. 1 that the problem (2-1) has the normed form of Fritz-John necessary conditions at a local solution point, if there exists a common Lagrangian multiplier vector for each direction at the point.

REFERENCES

[1] R.W.Chaney, *SIAM J. Control and Optimization* 25 (1987) 1072.
[2] F.H.Clarke, *Math . of Operations research* 1(1976) 165.
[3] V.F.Demyanov and L.N.Polyakova, *USSR Comput. Maths. Math. Phys.* 20 (1981) 34.
[4] V.F.Demyanov and A.M.Rubinov, *Doklady Akademii Nauk USSR* 250 (1981) 34.
[5] V.F.Demyanov and A.M.Rubinov, *Quasi-differential Calculus(Optimization Software, Inc., Publication Division, New York*, 1986).
[6] K.Eppler and B.Luderer, *Wiss.d. D. TU Karl-Marx-Stadt* 29 (1987) 187.
[7] V.V.Gorokhorsik, *Vests, Akad. Navuk BSSR Ser-Fiz-Mat Hayka* (1986) 3.
[8] L.N.Polyakova, *Mathematical Programming Study* 29 (1986) 44.
[9] A.Sharpiro, *Mathematical Programming Study* 29 (1986) 56.
[10] V.F.Demyanov and A.M.Rubinov, *Math. Operationsforsch. u. Statist., Ser. ptimization*, Vol.14 (1983) 3.

III
COMPUTING
Theory and Techniques

Approximation, Optimization and Computing:
Theory and Applications, A.G. Law and C.L. Wang (eds.)
Elsevier Science Publishers B.V. (North-Holland)

ORTHOGONAL POLYNOMIALS IN A-PRIORI IMAGE RECONSTRUCTION: THEORY AND COMPUTATION

A.G. Law and A.D. Strilaeff *

Inner product data $\{< f, g_k >\}$ about an element f can be used in different ways to construct an approximation to f. If p is an element which embodies known features of f, such as its support, then $\sum < g_k, pg_j > c_j =< g_k, f >$, $1 \le k \le n$. The monomial base $g_k(x) = x^{k-1}$ is considered, with an integral inner product. By means of an associated orthogonal polynomial family ($\{\phi_j\}$, with $< \phi_j, p\phi_k >= 0$ when $j \ne k$), this approximation may be obtained, and evaluated, in $0(n^2)$ time, from the recurrence coefficients. These coefficients themselves can also be calculated in $0(n^2)$ time - or much less in certain cases - thus avoiding an $0(n^3)$ process for obtaining the solution, $\{c_j\}$, of the linear system directly. Two examples of polynomial, a-priori reconstruction are presented, one with fast identification of the recurrence coefficients and a second that uses moments in an $0(n^2)$ process. A third example compares accuracy with that from a Fourier basis.

1. INTRODUCTION

An image reconstruction problem is one of estimating an unknown element, f, from a set of data which involve f. In emission tomography, for example, the measured data are values of certain line integrals and these may be expressed as inner products $< g_k, f >$, relative to some known base functions g_k [1]. Or, in signal processing, observed samples can be viewed as inner products which are evaluated in Fourier transform space instead [1]. If f is a member of a Hilbert space $\{H, < , >\}$ that contains (known) linearly independent elements $g_1, g_2, g_3, \cdots, g_n$, a common estimator for f is

$$g = \sum_{j=1}^{n} c_j g_j,$$

where

$$\sum_{j=1}^{n} < g_k, g_j > c_j =< g_k, f >, \ 1 \le k \le n. \qquad (1)$$

That is, $\| f - \sum c_j g_j \|$ is minimized. Since the Gram matrix $(< g_k, g_j >)$ is non-singular, linear system (1) has a unique solution that could be produced by, say, a Gauss elimination procedure. This is a costly $O(n^3)$ process. As well, it can encounter ill-conditioning: the classic example has monomials $g_k(x) = x^{k-1}$ and inner product

$$< f, g >= \int_0^1 f(x)g(x)dx,$$

to produce the notorious Hilbert matrix $(1/(j+k-1))$. The least squares approximation, g, is consistent with the given data in the sense that $< g_k, g >=< g_k, f > \forall k$, and globally optimal. It is possible, however, that some desirable *local* detail about f might be suppressed in the procedure. Thus, for example, if the support of f were known to be a proper subset of that for each g_k, could this knowledge be exploited within the reconstruction process to provide additional detail about f's behavior? Suppose some feature of $f \in H$ is known a-priori, and let p be an element which embodies this knowledge. (In the example, p could be the characteristic function for the support set of f.) Let $g_1, g_2, g_3, \cdots, g_n$ be selected, and assume the data $\{< g_k, f >\}$ are known. Then, provided multiplication and division are defined, alternative base elements $\{pg_k\}$ may be used with a new inner product

$$< f, g >_p \stackrel{d}{=} < f/p, g >, \qquad (2)$$

and the corresponding expression $\| f - \sum_{j=1}^{n} c_j(pg_j) \|_p$ is minimized iff

$$\sum_{j=1}^{n} < pg_k, pg_j >_p c_j =< pg_k, f >_p .$$

In other words,

$$R_p(f) \stackrel{d}{=} \sum_{j=1}^{n} c_j p g_j$$

is the a-priori reconstruction for f, where

$$\sum_{j=1}^{n} < g_k, pg_j > c_j =< g_k, f >, \ 1 \le k \le n. \qquad (3)$$

It is easy to check that p constant embodies minimal prior knowledge about f (i.e. $R_p(f) = g$) while $p = f$ embodies maximal prior knowledge ($R_p(f) = f$) [1].

For any prescribed approximation base $\{g_k\}$, any inner product $< , >$, and a selected a-priori element p, once the linear system (3) has been constructed it can be solved in $O(n^3)$

*University of Regina, Saskatchewan, Canada. S4S 0A2.

Base Elements	Inner Product	Matrix ($< g_k, g_j >$)	Cost of Solution
$g_k(j) = e^{i2\pi kj/n}$	$\sum_{j=0}^{n-1} f(j)\overline{g(j)}$	circulant	$O(n \log_2 n)$
$g_k(x) = e^{ikx}$	$\int_{-\pi}^{\pi} f(x)\overline{g(x)}dx$	Toeplitz	$O(n^2)$
$g_k(x_j) = x_j^k$	$\sum_{j=0}^{n-1} f(x_j)g(x_j)$	Hankel	$O(n^2)$
$g_k(x) = x^k$	$\int_{-1}^{1} f(x)g(x)dx$	Hankel	$O(n^2)$

Table 1. Operational costs for solving linear system (1).

arithmetic operations. In certain cases, with $p \equiv 1$, tailored procedures can improve efficiency of solution. Table 1 lists some common situations, and algorithms for the first three cases are well known [2], [5], [3]. In this paper, we consider the problem of image reconstruction with the monomial base $g_k(x) = x^{k-1}$, an a-priori function $p(x)$ and some (real) given inner product

$$< f, g >= \int_a^b f(x)g(x)w(x)dx. \quad (4)$$

Polynomials orthogonal with respect to a modified inner product whose weight is $p(x)w(x)$ are employed to provide algorithms for indirect solution of linear system (3) in $O(n^2)$ time. In test cases, use of a prior function can enhance local detail and polynomial a-priori reconstruction can be more accurate than that using a trigonometric counterpart.

2. ORTHOGONAL POLYNOMIALS AND A-PRIORI RECONSTRUCTION COMPUTATIONS

Given an inner product $< \, , \, >$ and projection data

$$\{< f, x^k >\}, \; 0 \leq k \leq n-1,$$

if $p(x) \geq 0$ is a selected, piecewise-continuous prior knowledge function, then the goal is to obtain a reconstruction of polynomial type

$$\sum_{j=0}^{n-1} c_j p(x) x^j, \quad (5)$$

without having to solve directly a linear system of the form

$$\sum_{j=0}^{n-1} < x^k, p(x)x^j > c_j =< f, x^k >, \; 0 \leq k \leq n-1. \quad (6)$$

Indirect, but more efficient ($O(n^2)$) techniques for approximating f will hinge on the monic polynomials $\{\phi_k(x)\}$ which are orthogonal on $a \leq x \leq b$ with respect to the associated inner product $< \, , \, >_{1/p}$. That is, $\deg \phi_k = k \; \forall k$ and $< p\phi_k, \phi_j >= 0$ if $k \neq j$. Thus, there exist scalars α_k and β_k, with $\beta_k > 0$, $k \geq 1$, such that

$$\phi_{-1}(x) \equiv 0,$$
$$\phi_0(x) \equiv 1, \quad \text{and}$$

$$\phi_k(x) = (x - \alpha_k)\phi_{k-1}(x) - \beta_k \phi_{k-2}(x), \; k \geq 1. \quad (7)$$

The approximation sought for f is written

$$R_p(f) = \sum_{j=0}^{n-1} d_j p \phi_j$$

where the least squares coefficients, d_j, are determined by

$$\sum_{j=0}^{n-1} < p\phi_j, p\phi_k >_p d_j =< f, p\phi_k >_p, \; 0 \leq k \leq n-1.$$

Hence,

$$\sum_{j=0}^{n-1} < \phi_j, p\phi_k > d_j =< f, \phi_k >,$$

so that

$$d_k = \frac{< f, \phi_k >}{< p\phi_k, x^k >}, \; k = 0, 1, 2, \ldots, (n-1). \quad (8)$$

As indicated below, once the coefficients in the recurrence (7) have been determined, the d_k can be computed, recursively, by an $O(n^2)$ algorithm (Algorithm 1). Thereafter, $R_p(f)$ may be evaluated at a prescribed x_0 by an $O(n)$ algorithm (Algorithm 2).

ALGORITHM 1. Given the data $\{< f, x^j >\}$, $0 \leq j \leq n-1$, and given the recurrence coefficients β_1, α_k and β_{k+1} for $k = 1, 2, 3, \ldots, n-1$, $\{d_j\}_{j=0}^{n-1}$ may be generated in n^2 multiplications, as follows:

Initialization: $c_{0,1} = < f, x^0 > / \beta_1$

Continuation: For $i = 1, 2, 3, \ldots, n-1$,

$$c_{i,1} = < f, x^1 > / \beta_1$$
$$c_{i-2,2} = (c_{i,1} - \alpha_1 c_{i-1,1}) / \beta_2.$$

Continuation: For $j = 2, 3, 4, \ldots, n-1$,

$$c_{i,j+1} = (c_{i+1,j} - \alpha_j c_{i,j} - c_{i,j-1}) / \beta_{j+1}, \; i = 0, 1, 2, \ldots, 2n-j-4.$$

Continuation: For $j = 0, 1, 2, \ldots, n-1$,

$$d_j = c_{0,j+1}.$$

ALGORITHM 2. Given the recurrence coefficients α_1, α_k and β_k for $k = 2, 3, 4, \ldots, n-1$ and given the coefficients $\{d_j\}_{j=0}^{n-1}, \sum_{j=0}^{n-1} d_j p(x) \phi_j(x)$ may be evaluated at a prescribed x_0 in $2(n-1)$ multiplications, as follows:

Initialization: $c_{n-1} = d_{n-1}$

$$c_{n-2} = d_{n-2} + (x_0 - \alpha_{n-1})c_{n-1}.$$

Continuation: For $j = n-3, n-2, n-1, \ldots, 0$,

$$c_j = d_j + (x_0 - \alpha_{j+1})c_{j+1} - \beta_{j+2} c_{j+2}.$$

The final multiplication $c_0 p(x_0)$ gives the desired value.

The recurrence coefficients α_k and β_k play a fundamental role in efficient reconstruction of f, and the ease with which they are obtained depends on the form of $p(x)$. Methods for generating these coefficients are discussed in the next section.

3. GENERATING THE RECURRENCE COEFFICIENTS α_k AND β_k

The recurrence coefficients α_k and β_k, in (7), may be computed directly using $<\ ,\ >_{1/p}$, in $O(n^2)$ time. Two general algorithms that have been used for this purpose are the discrete Stieltjes procedure (which involves numerical integration) and the modified Tchebicheff scheme (it employs moments), but these can lead to numerical instability [4]. The approach taken here involves, instead, linear translation of the interval $[a,b]$ in $<\ ,\ >_{1/p}$ into $[-1,1]$. Subsequently, if the translation of the weight $p(x)w(x)$ is "recognizable" (e.g., if $p(cx+d)w(cx+d) = \sqrt{1-x^2}$ so that Tchebicheff polynomials of the second kind emerge) then the translated-recurrence coefficients are known by identification and an $O(n)$ calculation will yield α_k and β_k. If, on the other hand, the translate of $p(x)w(x)$ is not a tabulated one on $-1 \le x \le 1$, then an $O(n^2)$ computation is required. *

Let $c = (b-a)/2$ and $d = (b+a)/2$, and define polynomials P_k by:

$$P_k(x) = c^{-k}\phi_k(cx+d),\ k \ge -1.$$

Then, from (7),

$$\begin{aligned} P_{-1}(x) &\equiv 0, \\ P_0(x) &\equiv 1, \quad \text{and} \\ P_k(x) &= (x-a_k)P_{k-1}(x) - b_k P_{k-2}(x),\ k \ge 1, \end{aligned} \tag{9}$$

where

$$\begin{aligned} \beta_1 &= cb_1, \\ \alpha_k &= ca_k + d \\ \text{and}\quad \beta_{k+1} &= c^2 b_{k+1},\ k \ge 1. \end{aligned} \tag{10}$$

Thus, the α_k and β_k can be obtained from a_k and b_k in $O(n)$ arithmetic operations.

If a_k and b_k are not known by identification with a recognizable polynomial recurrence, they can be computed directly in several ways. One scheme is given below - it is a variant of the (possibly unstable) Tchebicheff algorithm, and employs the first $2n-1$ moments of the relevant weight that has been shifted to $[-1,1]$:

$$\mu_i \stackrel{d}{=} \int_{-1}^{1} x^i [p(cx+d)w(cx+d)]dx,\ 0 \le i \le 2n-2. \tag{11}$$

ALGORITHM 3. Given the moments $\{\mu_i\}_{i=0}^{2n-2}$, then the corresponding recurrence coefficients $\{a_i\}_{i=1}^{n-1}$ and $\{b_i\}_{i=1}^{n}$ may be computed in $(n-1)(2n-1)$ multiplications as follows:

Initialization: $b_1 = \mu_0,\ c_{0,-1} = 0$

Continuation: For $i = 1,2,3,\ldots,2n-2$

$$\begin{aligned} c_{i,-1} &= 0 \\ c_{i,0} &= \mu_i/b_1 \end{aligned}$$

Continuation: For $i = 1,2,3,\ldots,n-1$

$$\begin{aligned} a_i &= c_{i,i-1} - c_{i-1,i-2} \\ b_{i+1} &= c_{i+1,i-1} - a_i c_{i,i-1} - c_{i,i-2} \\ c_{j,i} &= (c_{j+1,i-1} - a_i c_{j,i-1} - c_{j,i-2})/b_{i+1}, \\ & \quad j = i+1, i+2, \ldots, 2n-2-i. \end{aligned}$$

4. EXAMPLES AND DISCUSSION

To illustrate how a-priori information can enhance detail about an image f, two examples (with $w(x) \equiv 1$ here, for simplicity) are presented below.

EXAMPLE 1. Suppose the inner product prescribed is

$$< f,g > = \int_0^2 f(x)g(x)\cdot 1 \cdot dx,$$

the data sequence $\{< f, x^k >\}_{k=0}^4$ is $\left\{\frac{3}{4}, \frac{3}{4}, \frac{53}{64}, \frac{63}{64}, \frac{6303}{5120}\right\}$, and it is assumed that f's support is $[0,2]$. Selecting $p(x) = 1$ on $[0,2]$ leads to the requirement for the recurrence coefficients of the monic orthogonal polynomials corresponding to the weight $p(x)w(x) = 1$ on $[-1,1]$. As these are Legendre polynomials, standard tables provide $b_1 = 2, a_i = 0$ and $b_{i+1} = i^2/(4i^2-1)$ for $i = 1,2,3,4$. Since $c = d = 1$ here, (10) yields

$$\{\alpha_1,\alpha_2,\alpha_3,\alpha_4,\beta_1,\beta_2,\beta_3,\beta_4,\beta_5\} = \left\{1,1,1,1,2,\frac{1}{3},\frac{4}{15},\frac{9}{35},\frac{16}{63}\right\},$$

and then Algorithm 1 produces

$$\{d_0,d_1,d_2,d_3,d_4\} = \left\{(3)2^{-3}, 0, (-495)2^{-9}, 0, (108,675)2^{-17}\right\}.$$

Subsequently, Algorithm 2 generated the data to plot Figure 1, showing little detail about f other than it appears symmetric and that much of it is positive and lies between, say, $\frac{1}{4}$ and $\frac{7}{4}$. Using this newer information, we re-select $p(x) = 1$ on $[\frac{1}{4}, \frac{7}{4}]$ and zero elsewhere. Now, $c = \frac{3}{4}$ and $d = 1$ and, by (10),

$$\{\alpha_1,\alpha_2,\alpha_3,\alpha_4,\beta_1,\beta_2,\beta_3,\beta_4,\beta_5\} = \left\{1,1,1,1,\frac{3}{2},\frac{3}{16},\frac{3}{20},\frac{81}{560},\frac{1}{7}\right\}.$$

With these revised recurrence coefficients and the original data about f, Algorithm 1 now yields

$$\{d_0,d_1,d_2,d_3,d_4\} = \left\{\frac{1}{2}, 0, (-40)3^{-3}, 0, (-1,400)3^{-5}\right\}.$$

The corresponding plot, Figure 2, reveals that f appears globally symmetric, with a local minimum near $x = 1$. The

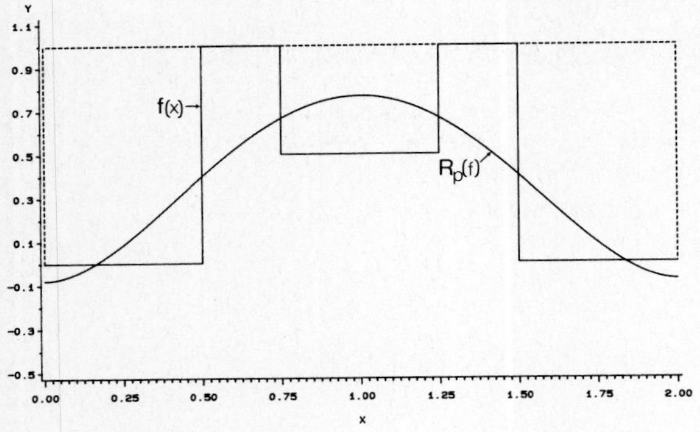

Figure 1. Reconstruction from the data in Example 1, using $p(x) = 1$ on $[0, 2]$.

test function which was used to generate the data for this example is indicated in the figure. The prior function, $p(x)$, is indicated by the dashed curve in Figures 1 and 2, and in Figures 3 and 4.

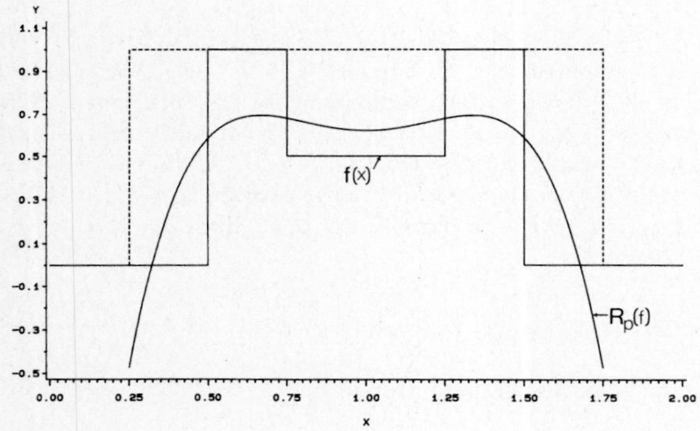

Figure 2. Reconstruction from the data in Example 1, using $p(x) = 1$ on $[\frac{1}{4}, \frac{7}{4}]$.

EXAMPLE 2. Consider the same situation, and data, as in Example 1. In the first step of that analysis, it was deduced that the unknown function, $f(x)$, should be nonnegative over, say, $[\frac{1}{4}, \frac{7}{4}]$, and this information was used to generate a second choice for $p(x)$. Other forms for $p(x)$ on $\frac{1}{4} \le x \le \frac{7}{4}$ are possible, of course. For example:

(a) $p_2(x) = (1 - [\frac{4}{3}(x-1)]^2)^{\frac{1}{2}}$,

or

(b) $p_3(x) = 1 - \frac{4}{3}|x - 1|$.

The shift of $p_2(x)w(x)$ is recognized as the weight for the Tchebicheff polynomials of the second kind, so the corresponding α_k and β_k can be found just as in Example 1. The polynomials associated with $p_3(x)w(x)$ require the more general process, however. The moments of $(1 - |x|)$ on $-1 \le x \le 1$ are

$$\mu_i = \begin{cases} \frac{2}{(i+1)(i+2)}, & \text{if } i \text{ is even} \\ 0, & \text{if } i \text{ is odd.} \end{cases}$$

With these moments, Algorithm 3 produces $a_i = 0$ for $1 \le i \le 4, b_1 = 1, b_2 = \frac{1}{6}, b_3 = \frac{7}{30}, b_4 = \frac{57}{245}$ and $b_5 = \frac{683}{2793}$. Relations (10) may then be employed to generate the coefficients α_k and β_k for (7), and the corresponding reconstruction $R_p(f) = \sum_{j=0}^{4} d_j p\phi_j$ has

$$\{d_0, d_1, d_2, d_3, d_4\} = \{1, 0, \frac{1120}{1323}, 0, -37.89816\}.$$

Algorithm 2 may now be invoked to generate plot data. Figures 3 and 4 display these two additional reconstructions.

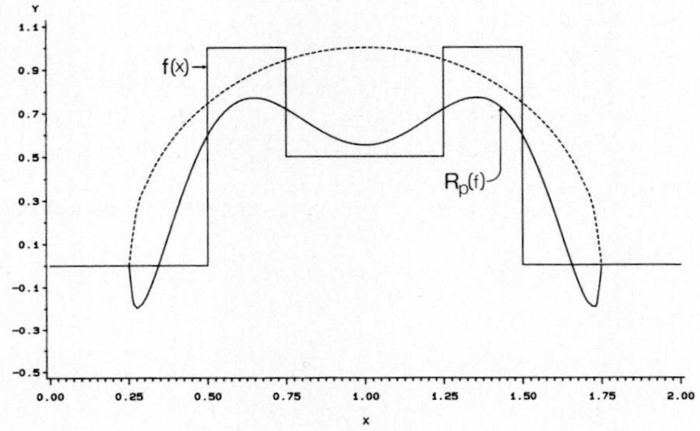

Figure 3. Reconstruction from the data in Example 1, using $p(x) = (1 - [\frac{4}{3}(x-1)]^2)^{\frac{1}{2}}$ on $[\frac{1}{4}, \frac{7}{4}]$.

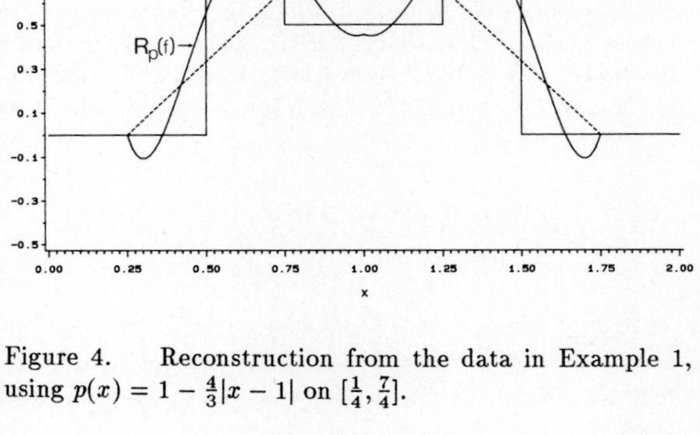

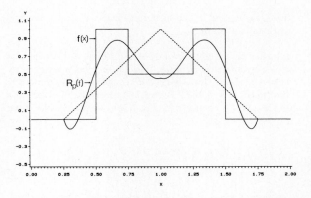

Figure 4. Reconstruction from the data in Example 1, using $p(x) = 1 - \frac{4}{3}|x - 1|$ on $[\frac{1}{4}, \frac{7}{4}]$.

It is of interest that the illustrations indicate (correctly) that the support of f is a strict subset of $[0,2]$ and that there is some minimum near $x = 1$. These conclusions are reached using a base of polynomials with only very low degrees.

The objective above has been to use polynomials for demonstrating how certain a-priori information about an image can be exploited for uncovering some local detail. The polynomial machinery illustrates the potential of the general method (3), but it is the recurrence property of orthogonal polynomials which permits an efficient implementation here. Algorithms 1, 2, and 3 also provide computational efficiency of the polynomial techniques - arithmetic cost, as well as accuracy of approximation, is a basic question that accompanies whichever choice for the base functions g_k. With the standard Fourier elements

$$g_1(x) = 1, g_2(x) = \cos x, g_3(x) = \sin x, \ldots,$$

computational efficiences of various algorithms have been well studied. It is natural to compare *accuracy* of reconstruction in test cases without regard to computational costs, between the polynomial and Fourier approaches to reconstruction. Figure 5 illustrates that the least squares approximation $P(x) = (3)(2^{-12})(569 + 2310x^2 - 1575x^4)$ competes favourably with the approximation $S(x) = \frac{3}{4} - \frac{1}{\pi}\cos \pi x$ produced by the Fourier base $\{1, \sin \pi x, \cos \pi x, \sin 2\pi x, \cos 2\pi x\}$, for the indicated function $f(x)$ (again, $p(x) \equiv 1$ in this illustration).

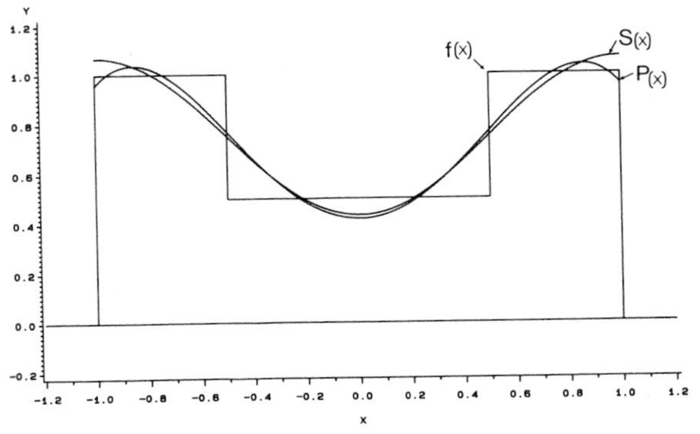

Figure 5. A-priori polynomial and Fourier reconstruction, using $p(x) = 1$ on $[-1, 1]$.

ACKNOWLEDGEMENTS

This research was supported in part by the Natural Sciences and Engineering Research Council of Canada: Grants 0GP0007186 and Scholarships PGS 1 and PGS 2.

FOOTNOTES

* An alternative under development is an $O(n^2)$ iterative improvement method for generating recurrence coefficients using the simpler ordinary Tchebicheff algorithm, in conjunction with an extended precision version of Algorithm 1 for improved stability.

REFERENCES

[1] Andrews, R.A., Law, A.G., Strilaeff, A.D. and Sloboda, R.S., "Application of Orthogonal Elements in A-priori Reconstruction: Fourier and Polynomial Techniques," Proceedings of the IEEE Pacific Rim Conference on Communications, Computers and Signal Processing, Victoria, B.C., Canada, June 1-2, 1989, pp. 83-86.

[2] Cooley, J.W. and Tukey, J.W., "An Algorithm for the Machine Calculation of Complex Fourier Series," Math. Comp., Vol. 19, No. 90, 1965, pp. 297-301.

[3] Forsythe, G.E., "Generational and Use of Orthogonal Polynomials for Data Fitting with a Digital Computer," SIAM Journal, Vol. 5, 1957, pp. 74-88.

[4] Gautschi, W., "On Generating Orthogonal Polynomials," SIAM J. Sci. Stat. Comput., Vol. 3, No. 3, 1982, pp. 289-317.

[5] Justice, J.H., "An Algorithm for Inverting Positive Definite Toeplitz Matrices," SIAM J. Appl. Math., Vol. 23, No. 3, 1972, pp. 289-291.

Approximation, Optimization and Computing:
Theory and Applications, A.G. Law and C.L. Wang (eds.)
Elsevier Science Publishers B.V. (North-Holland)

OPL: A Notation and Solution Methodology for Hierarchically Structured Optimization Problems

Bruce MacLeod * Robert Moll +

* Computer Science Department, University of Southern Maine, Portland, ME 04103
+ Computer Science Department, University of Massachussets, Amherst, MA 01003

1 Introduction

This paper reports on a domain independent method for solving complex discrete optimization problems. The method is well-suited for solving problems with a hierarchical structure, such as those which arise in facility layout, scheduling, and routing. Other researchers have proposed domain independent methods which conduct an intelligent, but nonetheless complete search of a solution space for such problems [1,7]. Unfortunately, most discrete optimization problems have solution spaces which are too large to be searched completely. Our approach acknowledges this limitation and attempts instead to locate a solution of high quality using an approximation/local search paradigm.

A two-part notation underlies our approach. It consists of a static graph-theoretic structure called a ***hierarchical constraint graph*** or *HCG*, and a control structure called a *policy*. An HCG represents the underlying hierarchical structure of the problem. A policy describes the ways in which the assignment relationships identified in the HCG are to be satisfied and optimized. OPL (Optimization Programming Language) is the programming environment in which our notation is realized. In OPL, a user must 1) construct an appropriate HCG; and 2) write a suitable policy program to solve the problem. Once a problem is understood, the OPL notation often allows an experienced user to construct a reasonably well-behaved solution in a day or so.

2 OPL Specification Machinery

An HCG is a directed acyclic graph that identifies the hierarchical relationships among the objects that make up a discrete optimization problem. Nodes in the graph represent classes of objects. Arcs identify the nature and form of generalized assignment relationships between classes.

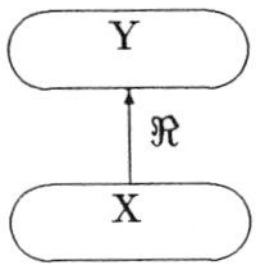

Figure 1: A Generalized Assignment Relationship

The notation in Figure 1 indicates that objects of type "X" are to be assigned to objects of type "Y" according to assignment relation "ℜ". Intuitively, such a relationship can be thought of as asserting: "if there is space, place object A of type X into a container B of type Y". Indeed, we will frequently use "object-container" terminology when we refer to the elements of our generalized assignment relationships.

Much of OPL's power is achieved by limiting the class of assignment relation primitives. In the current version of OPL, only four are available. These primitives have the property that many well known approximation and local search algorithms have a meaningful interpretation for each primitive type. The primitives are: **partition** – cluster a set of objects into a collection of unordered subsets; **bounded order** – assign a set of objects to positions in one or several fixed length lists; **unbounded order** – assign a set of objects to relative positions in one or several lists of unbounded length; and **unbounded ring** – assign a set of objects to relative positions in one or several rings of unbounded size. These four primitive relations, when composed together, are sufficient to describe a broad class of discrete optimization problems. Thus, OPL specifications are expressed in terms of interrelated kernel optimization problems (rather than, say, in terms of equations), and the problem decompostion induced by an HCG specification allows an OPL user to exploit a problem's inherent structure in a natural way.

A feasible solution to an optimization problem is obtained when all assignment relationships in a problem's HCG have been satisfied for every object. Relationships are satisfied incrementally by the application of approximation algorithms to the relations of an HCG. Local search routines are also part of the solution process; they are responsible for improving the quality of a solution or partial solution incrementally.

We have modeled some of our primitive approximation procedures on the family of approximation algorithms available for bin packing problems. Thus for each primitive we formulate algorithms which we identify as *first-fit*, *best-fit*, and *next-fit*. Each is supplied with an order in which containers (and positions in a container) are considered and – especially in the case of best-fit – a metric that evaluates a particular assignment. Thus *first-fit* applied to a bounded order relation considers each of the containers and each slot in each container in some order, and makes the indicated assignment to the first slot that can legally "hold" the object being assigned.

Local search procedures such as *2-opt* [2] can also be associated with our primitive relationships. For example, 2-opt for bounded order means: reverse the subsequence between two elements in a slotted container, and check for improvement. Similarly, 2-opt for ring relations means: reverse the order of a subsequence of entries between two ring elements and check for improvement.

OPL programming primitives are called *improvement policies*. They consist of approximation and/or local search routines that are associated with the primitive relations of the notation. An improvement policy seeks to place or rearrange a collection of objects in such a way that the new solution extends or improves the old solution. More concretely, an improvement policy consists of:

- a collection of objects, called *primary objects*, which the improvement policy acts upon.
- a sequence $(t_1,...,t_k)$ of transformations called an *improvement sequence*, where each t_j is an approximation or local search routine which is associated with one of the relations in the HCG.
- a collection of HCG arcs called *bound* arcs, which identify those relationships in an existing partial solution that may not be altered by the improvement sequence.
- a *dispatch function* which, each time it is called, chooses the next primary object to be considered by the improvement sequence.

A *policy program* is an optimizing algorithm for a problem instance that has improvement policies as primitives, and includes in addition elementary programming constructs (if, do-while, for, ..). Data structuring functions (sorting, extracting, merging, ...) are also available.

3 An Example

We demonstrate the workings of OPL by illustrating its application to instances of vehicle routing, a classical problem in operations research.

In the simplest version of the Vehicle Routing Problem (VRP), vehicles deliver packages to a collection of geographically dispersed customers. Each vehicle may have a maximum carrying capacity and a maximum travel time. The goal of a VRP instance is to find a solution that respects all constraints and at the same time minimizes the total travel time of all the vehicles. In the Multiple Depot Vehicle Routing Problem (DVRP), customers are serviced from one of a collection of depots.

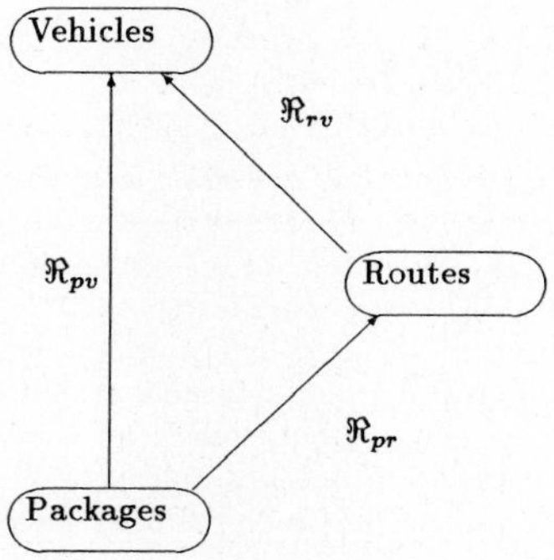

Figure 2: HCG for the VRP Problem

The HCG in Figure 2 represents the relationships that exist between the packages, the routes, and the vehicles that make up the VRP problem. Let us assume that vehicles are identical and that vehicle capacity for packages is limited in the sense that no single vehicle can carry a significant fraction of the packages. (This capacity constraint is considered to be a part of the relation $\Re_{pv}$ and is identified when $\Re_{pv}$ is first specified.) We assume that there are no limits on vehicle tour distance. Solution cost is the total distance traveled by all vehicles.

Informally here is our solution. First we associate an angle with each package delivery site, using the dispatch depot of the vehicles as origin. We sort packages by angle and then assign them to vehicles –a partition relationship– on a "next-fit" basis. That is, we attempt to place a package in the current vehicle. If it won't fit, we place it instead in the next unoccupied vehicle. This assignment crudely clusters packages with nearby destinations into the same vehicle. Next we route each vehicle: we examine each package in a vehicle, in turn, and insert it in that vehicle's route in a position closest to a package that has already been placed in the route.

Once these routes have been constructed we apply 2-opt local search to optimize each route.

Our algorithm does a reasonable job of solving this simple version of VRP. Below we describe how to build a policy program that realizes this algorithm. We construct this policy program in four steps.

- **Step One**:

  Step one establishes the packages to vehicles assignment. First we sort by polar angle, as described above. Next we apply our first improvement policy, which we call IP_1. Its improvement sequence consists of a single approximation routine, which realizes the partition relation $\Re_{pv}$ using a "next-fit" algorithm. In abbreviated form we write this policy as:

  next-fit($\Re_{pv}$)

  Thus, IP_1 is a bona-fide policy consisting of a transformational sequence with one entry (the call to next-fit). The collection of primary objects to which IP_1 is applied consists of all packages. The set of bound arcs associated with IP_1 is empty.

- **Step Two:**

  Improvement policy IP_2 routes the packages associated with each vehicle. The algorithm considers each package in a vehicle and places that package in the best possible position in the route–the position which incurs the least additional travel distance. The improvement sequence is again one approximation procedure, which we abbreviate as follows:

  best-fit($\Re_{pr}$, least-travel-distance)

  The underlying OPL machinery will maintain consistency automatically in the following sense: packages in different vehicles cannot be assigned to the same route.

  IP_2 will be called as many times as there are vehicles. Each time it is called, the packages belonging to the vehicle under consideration are designated as the primary objects. This process is captured in the FOR loop of the policy program presented in the next column.

- **Step Three:**

  IP_3's singleton improvement sequence improves upon the existing partial configuration by performing local search on each of the routes using 2-opt local search. The packages to vehicles and packages to routes arcs are designated as bound. We write

  2-opt($\Re_{pr}$, decreased-travel-distance)

  to describe IP_3. A 2-opt interchange is accepted if the total travel distance over all vehicles decreases.

- **Step Four:**

  IP_4 attempts to satisfy the remaining relation, $\Re_{rv}$. Since the packages in a route have already been assigned to a vehicle, each route is bound to the vehicle that its packages are assigned to. In addition, it has been assumed that no travel time constraints are associated with this relation. Thus, the required assignment has been made implicitly, and we make the assignment explicit using the following policy:

  first-fit($\Re_{rv}$)

  For each route, the underlying OPL machinery allows only the single feasible vehicle to be considered, and so each route is trivially assigned to the proper vehicle.

The preceding steps yield the following program.

```
packages = sort(packages ,polar-coordinates)
IP1(packages)
FOR each vehicle in vehicles
    begin
        packages = packages in vehicle
        IP2(packages)
        IP3(packages)
    end
IP4(routes)
```

The program can now be applied to the problem instance. It could be augmented further to obtain additional improvement. For example, after the FOR loop has finished, packages could be interchanged between routes (and vehicles) according to a suitable metric and then 2-opt local search could be applied again to individual routes. This strategy would involve two local search procedures operating together in a single improvement policy. The ability to compose multiple local search and approximation routines in a single improvement policy adds considerable power to OPL.

OPL can handle more complex constraint information. For instance, suppose certain packages require refrigeration both when transported by vehicles and when installed in depots, and suppose further that only certain vehicles and depots have refrigeration equipment. Such object characteristics can be included in the specification, and appropriate constraint information can be added to the relevant HCG arcs. This kind of constraint information will distort but frequently not destroy the nice performance characteristics of the approximation and local search algorithms in use in a policy program.

OPL is flexible, and functions well as a prototyping tool. Indeed, it's easy to create policy programs which are sensitive to the characteristics of a particular problem instance. For instance, suppose that a particular VRP problem instance is known to have tight restrictions on the travel time of vehicles but unlimited capacity in the vehicles. A solution method which constructs routes first and then assigns packages to vehicles and routes works well here. Two improvement policies realize this solution method. The first policy specifies a method for the $\Re_{pr}$ assignments; the second specifies a method for the $\Re_{pv}$

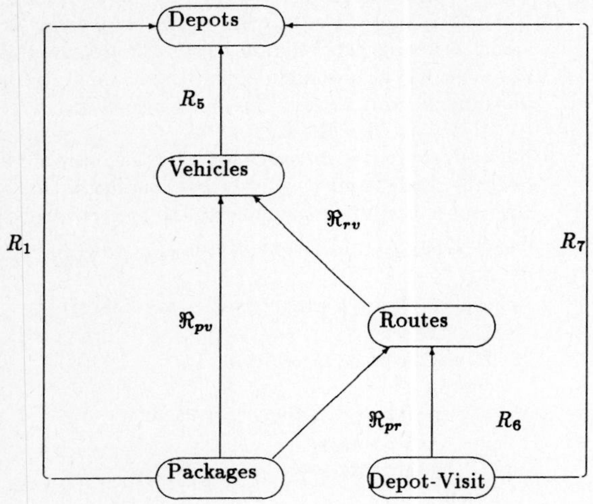

Figure 3: HCG for the DVRP Problem

assignments.

Now suppose that both travel time and vehicle capacity significantly constrain a problem instance. A solution method is needed that both places a package in a vehicle and incrementally routes the vehicle. A single improvement policy with four elements in its transformational sequence will realize this strategy. The policy will: 1) assign a package to a vehicle; 2) add that package to the vehicle's route; 3) optimize the route using 2-opt; and 4) complete the placement by making sure the route is assigned to the vehicle.

The multiple depot vehicle routing problem, or DVRP, has a more complicated structure (see Figure 3). Its HCG includes the HCG for the simpler VRP in a natural way. Additional nodes and arcs capture the necessity of making a depot assignment. In Figure 3, R_1 partitions packages among depots and R_5 partitions vehicles among depots. The node *Depot-Visit* accounts for the fact that a vehicle must begin and end its route at a depot. R_6 is therefore a ring assignment since it places a depot-visit on a route, and R_7 is a (trivial) partition relation which associates each depot-visit with the corresponding depot.

We have represented and solved DVRP using OPL. A comparison of our solutions for three problem instances shows a 7.4%, 2.4%, 2.8% improvement over published results [3,5,6]. We stress that solutions were constructed very quickly using OPL.

4 Conclusion

Our OPL notation is a viable domain independent method for solving a broad class of discrete optimization problems. In [3] we report on three very different OPL applications: DVRP, a warehousing problem, and a check processing problem in the financial industry. For each problem, the hierarchical structure is made explicit using the OPL specification machinery and good solutions are developed with OPL policy programs. Based on these results, we believe that OPL is a promising notation and programming environment for the rapid development of effective optimization algorithms.

References

[1] Lauriere, J.L., "Alice: A Language for Intelligent Combinatorial Exploration" *A.I. Journal*, 1978.

[2] Lin, S. "Computer Solutions of the TSP", BSTJ, No. 10, December, 1965, pp. 2245-2269

[3] MacLeod, B. "OPL: A Notation and Solution Methodology for Hierarchically Structured Optimization Problems", Ph.D. Thesis, University of Massachusetts, 1989.

[4] Moll R.N. and B. MacLeod "Optimization Problems in a Hierarchical Setting", COINS Technical Report 88-87, October, 1988, University of Massachussets, Amherst, MA 01003

[5] Perl, J. and M. Daskin "A Warehouse Location Routing Problem", *Transporation Research-B* Vol 19B, No 5. pp. 381-396, 1985.

[6] Perl, J. "The Multi-Depot Routing Allocation Problem" *American Journal of Mathematical and Management Sciences* Vol 7, pp. 7-34, 1987.

[7] Smith, D. "Structure and Design of Global Search Algorithms" *Kestrel Institute Technical Report* July 1988. Palo Alto, California 94304, 1988.

Approximation, Optimization and Computing:
Theory and Applications, A.G. Law and C.L. Wang (eds.)
Elsevier Science Publishers B.V. (North-Holland)

APPROXIMATION TO THE SOLUTION OF THE THOMAS-FERMI EQUATION

Ichizo NINOMIYA
Chubu University,Kasugai,Aichi,Japan 487

A high precision numerical solution of Thomas-Fermi equation is obtained. Function tables, rational minimax approximations and FORTRAN programs are worked out.

1. INTRODUCTION

Thomas-Fermi equation ,a quasi-classic approximation for electron density of complex atoms is the parameter-free non-linear ordinary differential equation of the second order

$$\phi'' = \frac{\phi^{3/2}}{x^{1/2}} \tag{1}$$

satisfying the boundary conditions

$$\phi(0) = 1, \tag{2}$$

$$\phi(\infty) = 0. \tag{3}$$

The solution has been of interest for half a century, and there exist several contributions for its approximation [1][2][3]. In this paper, we follow largely Krutter's [2] description and notation. First,ϕ is expanded in a series, each containing a slope parameter,both at $x = 0$ and $x = \infty$. Numerical solutions are then generated by integrating the equation forward form $x = 0$ and backward from $x = \infty$.Fitting them at an intermediate point, the parameters are determined and the solution ϕ from zero to infinity is obtained. A series of rational minimax approximations for ϕ and ϕ' for regions $(0,1)$ and $(1,\infty)$ are constructed by Remes' algorithm and FORTRAN subprograms for these functions are written. The computation is carried out on a FUJITSU M-382 computer using quadruple precision.

2. SOLUTION IN THE VICINITY OF THE ORIGIN

The equation (1) can be rewritten, by use of the variable transformation

$$t = x^{1/2}, \tag{4}$$

as

$$\frac{d^2\phi}{dt^2} - \frac{1}{t}\frac{d\phi}{dt} = 4t\phi^{3/2} \tag{5}$$

Assuming power series expansions of the form

$$\phi = \sum_{k=0}^{\infty} a_k t^k, \tag{6}$$

$$\phi^{1/2} = \sum_{k=0}^{\infty} b_k t^k, \tag{7}$$

we obtain, from the boundary condition (2) and the identity $\phi = \phi^{1/2}\phi^{1/2}$, the result

$$a_0 = 1, a_1 = 0, a_2 = -B, b_0 = 1, b_1 = 0, b_2 = -\frac{B}{2}, \tag{8}$$

and the relation

$$b_n = \frac{1}{2}(a_n - \sum_{k=1}^{n-1} b_k b_{n-k}), n = 3, 4, \ldots \tag{9}$$

where B is an undetermined constant. When the series are inserted into (5) and the identity $\phi^{3/2} = \phi\phi^{1/2}$ is taken into account,we obtain the relation

$$a_n = \frac{4}{n(n-2)} \sum_{k=0}^{n-3} a_k b_{n-k-3}, n = 3, 4, \ldots \tag{10}$$

If B is given numerically, then using (10) and (9) alternately, the coefficients $a_n, n = 3, 4, ..$ can be obtained numerically, and thus the computation of ϕ and ϕ' in the vicinity of the origin becomes feasible. The constant B has the meaning of slope $\phi'(0) = -B$ and it is our problem to determine B so that the continuation of the solution ϕ may take on the value $\phi(\infty) = 0$ at infinity.

3. SOLUTION IN THE VICINITY OF THE INFINITY

Changing variables from (x, ϕ) to (v, y) by the transformation

$$v = Ax^{-\nu}, \phi = \frac{144}{x^3} y, \tag{11}$$

where A and ν are constants to be determined appropriately afterward, the equation (1) is transformed into

$$\nu^2 v^2 \frac{d^2y}{dv^2} + \nu(\nu + 7)v\frac{dy}{dv} + 12y - 12y^{3/2} = 0. \tag{12}$$

It is observed that (12) is independent of the constant A and the boundary condition (3) is satisfied automatically.Assume now that y and $y^{1/2}$ are expanded in power series of v arround $v = 0$, or in the vicinity of $x = \infty$, as follows:

$$y = \sum_{k=0}^{\infty} c_k v^k, \tag{13}$$

$$y^{1/2} = \sum_{k=0}^{\infty} d_k v^k. \tag{14}$$

Substituting them into the identity $y = y^{1/2}y^{1/2}$ and (12), and equating like powers of v, we obtain the relations:

$$c_n = \sum_{k=0}^{n} d_k d_{n-k}, \quad (15)$$

$$(\nu^2 n(n-1) + \nu(\nu+7)n + 12)c_n = 12 \sum_{k=0}^{n} c_k d_{n-k} \quad (16)$$

for every non-negative integer n. Especially, from the case $n = 0$ and $n = 1$, we obtain

$$c_0 = d_0 = 1, \quad (17)$$

$$c_1 = 2d_1, c_1(\nu(\nu+7) - 6) = 0. \quad (18)$$

Solving the above quadratic equation, ν is given by the positive, physically meaningful root

$$\nu = \frac{\sqrt{73} - 7}{2} = 0.77200\ 18726\ 58765\ 58394. \quad (19)$$

Although c_1 is arbitray, we put $c_1 = -1$ for convenience, since we have another free constant A. Now (15) and (16) can be rewritten as simultaneous linear equations concerning c_n and d_n for $n > 1$ as follows:

$$c_n - 2d_n = \sum_{k=1}^{n-1} d_k d_{n-k}, \quad (20)$$

$$n((n-1)\nu^2 + 6)c_n - 12d_n = 12 \sum_{k=1}^{n-1} c_k d_{n-k}. \quad (21)$$

These equations can be readily solved for c_n and d_n in terms of c_k and d_k for $k < n$ and, thus a power series solution around $x = \infty$, which is independent of A, is determined numerically. The remaining task is to determine A so that the continuation of the solution may take on the boundary value $\phi(0) = 1$ at the origin.

4. DETERMINATION OF CONSTANTS A AND B

Numerical determination of A and B is worked out on a FUJITSU M-382, an IBM compatible machine, using quadruple precision (33D). Employing the Broyden's method [5] for simultaneous nonlinear equations, we proceed as follows. First, select starting approximation

$$A = 13.2710, \quad (22)$$

$$B = 1.58807, \quad (23)$$

from the contributions of forerunners [1][2]. Repeat the following three steps until sufficient convergence is attained.

4.1. Forward Solution

Compute coefficients $a_k, k = 0, 1, 2, \ldots 50$ from (9) and (10). Evalute ϕ and $\frac{d\phi}{dt}$ at $x = 0.1$ by the series (6) and its term-by-term differentiation. Integrate the equation(5) numerically by the Bulirsch-Stoer[4] rational extrapolation method from $x = 0.1$ forward up to a prescribed intermediate point x_0.

4.2. Backward Solution

Compute coefficients $c_k, k = 0, 1, 2 \ldots 50$ as described in 2. Note that they need not be recomputed any more, since they are independent of A. Evaluate y and $\frac{dy}{dv}$ at $x = 20$ by the series (13) and its term-by-term differentiation. Integrate the equation (12) numerically by the Bulirsch-Stoer method from $x = 20$ backward (forward in v) down to the point x_0.

4.3. Fitting

Compare ϕ and ϕ' of the forward and backward solutions at x_0 and examine the discrepancies. Update A and B in accordance with Broyden's scheme.

4.4. Result

It turns out that, by virtue of the close initial guess, six trials for the cases $x_0 = 4.0, 4.2, 4.4, 4.6, 4.8, 5.0$ are all successful, converging within a few iterations and yielding excellent results which coincide with each other at leading 24 decimal digits. The result is given by

$$A = 13.27097\ 38480\ 26935\ 15409\ 03, \quad (24)$$

$$B = \ 1.58807\ 10226\ 11375\ 31255\ 606. \quad (25)$$

With A and B in hand, we are now able to compute ϕ and ϕ' at any given point by the methods described in the present and the preceding sections. We show 20D values of them in Table 1 and Table 2 and their graphs in Figure 1 below.

X	PHI(X)	-PHI'(X)
0.0	1.00000000000000000000	1.58807102261137531260
0.1	0.88169707675049963557	0.99535464609157190042
0.2	0.79305943203421563308	0.79422700919556131658
0.3	0.72063947608907787549	0.66179978009014688119
0.4	0.65954116083401080874	0.56464244406998148562
0.5	0.60698638335597990950	0.48941161257453808865
0.6	0.56116202361482947044	0.42917187169701265686
0.7	0.52079145650926045564	0.37979474528748045724
0.8	0.48493098798369233500	0.33860715613257085122
0.9	0.45285871536120349444	0.30377575605662584253
1.0	0.42400805208070560023	0.27398905159330625199
2.0	0.24300850716111955527	0.11824319162548762055
3.0	0.15663267321649584132	0.06245713085412097622
4.0	0.10840425691890771108	0.03694375782412348635
5.0	0.07880777925136990425	0.02356007495470051288
6.0	0.05942294925042258079	0.01586754953340707981
7.0	0.04609781860449858987	0.01114253181486708840
8.0	0.03658725526467680239	0.00808860296964547432
9.0	0.02959093527054687372	0.00603307471445739244
10.0	0.02431429298868086419	0.00460288187126925450
11.0	0.02025036497993102055	0.00357981515693432478
12.0	0.01706392230000933536	0.00283053641912945253
13.0	0.01452651761197413921	0.00227052464662819215
14.0	0.01247840598705962019	0.00184450137582626667
15.0	0.01080535875582389202	0.00151532308202360630
16.0	0.00942407890577514863	0.00125743533815926715
17.0	0.00827276394280789993	0.00105288677682984217
18.0	0.00730484590171563776	0.00088883110995551761
19.0	0.00648474643871812591	0.00075592141518874300
20.0	0.00578494119156694044	0.00064725433277769203
21.0	0.00518389337037112101	0.00055766158386766336
22.0	0.00466457575334431480	0.00048322574079376467
23.0	0.00421339805457804712	0.00042094370056865410
24.0	0.00381941807113599723	0.00036848922006930106
25.0	0.00347375441676563247	0.00032404299776975115
26.0	0.00316914439208047115	0.00028616951697050513
27.0	0.00289960765048920568	0.00025372671673447631
28.0	0.00266018787093669836	0.00022579900812887555
29.0	0.00244675256198795368	0.00020164709231848144
30.0	0.00225583661620285588	0.00018067000647699264

Table 1

1/X	PHI(X)	-PHI'(X)
0.01	0.0001002425681394073З	0.0000027393510686783З
0.02	0.0006322547829849047З	0.0000324989020482588l
0.03	0.0017477759034049304З	0.0001279821311915013О
0.04	0.0034737544167656324 7	0.0003240429977697511 5
0.05	0.0057849411915669404 4	0.0006472543327776920 3
0.06	0.0086341186813063374 4	0.0011160205810318260 3
0.07	0.0119662467436519396 8	0.0017418614574988552 8
0.08	0.0157253594534612394 3	0.0025308532471787585 2
0.09	0.0198579100546650900 6	0.0034849172678019424 2
0.10	0.0243142929886808641 9	0.0046028818712692545 0
0.11	0.0290494092659584169 4	0.0058813251719862058 5
0.12	0.0340227294723587172 0	0.0073152301025521229 5
0.13	0.0391980998510676931 6	0.0088984871171461506 8
0.14	0.0445434263339945315 7	0.0106242764424342064 1
0.15	0.0500303109744253621 6	0.0124853564507349713 6
0.16	0.0556336814581888939 6	0.0144742794632128855 7
0.17	0.0613314352137999114 9	0.0165835517307793038 0
0.18	0.0671041087040136889 3	0.0188057506151960552 2
0.19	0.0729345762511755927 9	0.0211336090377002820 4
0.20	0.0788077792513699042 5	0.0235600749547005128 8

Table 2

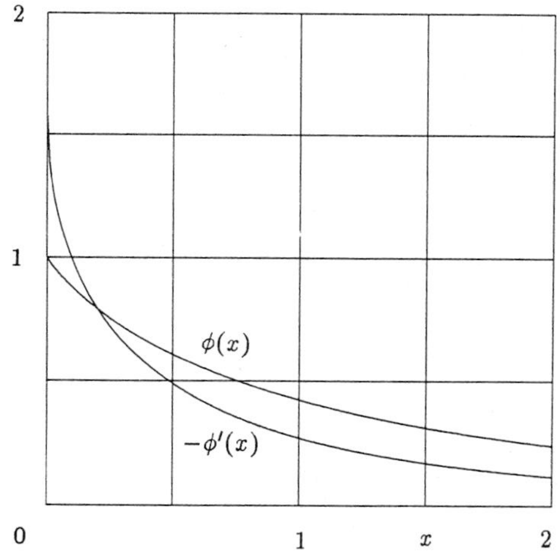

Figure 1 $\phi(x)$ and $-\phi'(x)$ versus x

5. RATIONAL MINIMAX APPROXIMATIONS

Although we can manage to compute ϕ and ϕ' at an arbitray point,necessary numercal work is tedious and time-consuming .Thus it is worth while to construct explicit and simple approximations for them once for all. We divide the whole region $(0, \infty)$ into two parts $(0,1)$ and $(1, \infty)$,in each of which we construct a series of rational minimax approximations with maximum relative errors ranging from 10^{-3} to 10^{-18} by means of the Remes' algorithm [6]. We adopt the form $\phi = P(x^{1/2})$,$\phi' = Q(x^{1/2})$ in the region $(0,1)$ and the form $\phi = \frac{144}{x^3}R(x^{-\nu})^4$, $\phi' = \frac{1}{x}S(x^{-\nu})\phi$ in the region $(1, \infty)$ corresponding to the behavior of the approximands in the respective regions,where P, Q, R and S are all rational functions.Table 3 below summarizes the results obtained. Integers separated by a slash under the heading ORDR are the orders of numerator and denominator respectively of rational approximations. Each entry under the title PREC,being the minus of base 10 logarithm of maximum relative error,indicates precision in number of decimal places.

P		Q		R		S	
ORDR	PREC	ORDR	PREC	ORDR	PREC	ORDR	PREC
2/ 2	3.48	0/ 4	3.33	1/ 1	3.65	1/ 1	3.25
2/ 3	4.39	0/ 5	4.50	1/ 2	5.15	1/ 2	4.81
3/ 3	5.76	1/ 5	6.10	2/ 2	6.38	2/ 2	6.00
3/ 4	7.17	1/ 6	7.53	2/ 3	7.71	2/ 3	7.37
4/ 4	8.50	1/ 7	8.74	3/ 3	8.89	3/ 3	8.51
4/ 5	10.19	2/ 7	9.30	3/ 4	10.14	3/ 4	9.79
5/ 5	10.49	2/ 8	10.07	4/ 4	11.28	4/ 4	10.91
5/ 6	10.92	2/ 9	11.88	4/ 5	12.49	4/ 5	12.14
6/ 6	12.37	3/ 9	12.98	5/ 5	13.61	5/ 5	13.24
6/ 7	15.24	3/10	14.50	5/ 6	14.78	5/ 6	14.43
7/ 7	16.33	3/11	15.78	6/ 6	15.89	6/ 6	15.52
7/ 8	16.59	4/11	16.17	6/ 7	17.04	6/ 7	16.69
8/ 9	17.96	4/12	18.61	7/ 7	18.14	7/ 7	17.77
9/ 9	19.26	-	-	7/ 8	19.27	7/ 8	18.91

Table 3 Rational Minimax Approximations

6. FORTRAN PROGRAMS

Utilizing rational minimax approximations obtained above, it is a simple matter to write FORTRAN programs.Function subprograms TMFRM and TMFMP shown in Figure 2 and Figure 3 compute ϕ and ϕ' respectively with 6D precision. TMFRM uses the approximations P_4^3 and R_3^2,while TMFMP uses Q_6^1 and S_3^2,where ,for example,P_4^3 stands for a ratio of polynomials of degree 3 and 4. Here some remarks concerning technical details seem in order: Every constant appearing in the DATA statement is represented in double precision to minimize decimal-binary conversion error.The leading coefficient of the denominator of every rational approximation is set to 1 for the purpose of saving one multiplication. Constants BIG and XMAX control flow of algorithm to avoid unnecessary computations and difficulties such as overflow or underlow for large arguments.They are inplementation-dependent and should be adjusted appropriately in accordance with scheme of floating number representation. Subroutine MGRR prints error message when a negative argument is entered.We also have double precision version DTMFRM and DTMFMP with 16D accuracy.The programs named above are all members of NUMPAC(Nagoya University Mathematical Package) [7], a comprehensive and excellent numerical library constructed by Nagoya University Numerical Group.It is installed and available at 50 principal computing institutes in Japan.

```
      FUNCTION TMFRM(X)
      DATA BIG/ 6.5E+10/
      DATA XMAX/6.43E+26/
      DATA FN/ -0.772001872658765 6D+00/
      DATA A0/  5.902873657995040 7D+00/
      DATA A1/  6.038446163407781 5D+00/
      DATA A2/ -3.624155407654569 9D+00/
      DATA A3/  5.956452142601739 9D-01/
      DATA B0/  5.902873259428629 3D+00/
      DATA B1/  6.038509440072823 4D+00/
      DATA B2/  5.748404524890351 5D+00/
      DATA B3/  2.330589587962611 3D+00/
      DATA C0/  1.842342147113523 2D+01/
      DATA C1/  1.242719504372033 4D+01/
      DATA C2/  1.113332938957306 1D+00/
      DATA D0/  5.318383698684896 0D+00/
      DATA D1/  2.123245168205828 6D+01/
      DATA D2/  1.206026538361426 0D+01/
      IF(X.LT.0.0) GO TO 40
      IF(X.GT.1.0) GO TO 10
      T=SQRT(X)
      TMFRM=(((A3*T+A2)*T+A1)*T+A0)/
     *((((T+B3)*T+B2)*T+B1)*T+B0)
      RETURN
   10 IF(X.GT.BIG) GO TO 20
      T=X**FN
      TMFRM=(((C2*T+C1)*T+C0)/
     *(((T+D2)*T+D1)*T+D0))**4/X**3
      RETURN
   20 IF(X.GT.XMAX) GO TO 30
      TMFRM=144.0/(X*X)/X
      RETURN
   30 TMFRM=0.0
      RETURN
   40 TMFRM=0.0
      CALL MGRR(5HTMFRM,X,TMFRM,8HARG LT 0,2)
      RETURN
      END
```

Figure 2 Function Subprogram TMFRM

```
      FUNCTION TMFMP(X)
      DATA BIG/ 6.5E+10
      DATA XMAX/1.68E+20/
      DATA FN/ -0.7720018726587656D+00/
      DATA P0/ -3.2002805581554181D+01/
      DATA P1/  5.1593106562011941D+00/
      DATA Q0/  2.0151999539135485D+01/
      DATA Q1/  2.2130333755962752D+01/
      DATA Q2/  2.7873117568464952D+01/
      DATA Q3/  1.4926502284947164D+01/
      DATA Q4/  9.4877278925191828D+00/
      DATA Q5/  2.4031961701822779D+00/
      DATA R0/ -1.1880908624300817D+01/
      DATA R1/ -7.5178426317355123D+00/
      DATA R2/ -4.8944742810794874D-01/
      DATA S0/  3.9603027049752886D+00/
      DATA S1/  1.6030688822835828D+01/
      DATA S2/  9.7867201102222410D+00/
      IF(X.LT.0.0) GO TO 40
      IF(X.GT.1.0) GO TO 10
      T=SQRT(X)
      TMFMP=(P1*T+P0)/
     *((((((T+Q5)*T+Q4)*T+Q3)*T+Q2)*T+Q1)*T+Q0)
      RETURN
   10 IF(X.GT.BIG) GO TO 20
      T=X**FN
      TMFMP=((R2*T+R1)*T+R0)*TMFRM(X)/
     *((((T+S2)*T+S1)*T+S0)*X)
      RETURN
   20 IF(X.GT.XMAX) GO TO 30
      XX=X*X
      TMFMP=-432.0/XX/XX
      RETURN
   30 TMFMP=0.0
      RETURN
   40 TMFMP=0.0
      CALL MGRR(5HTMFMP,X,TMFMP,8HARG LT 0,2)
      RETURN
      END
```

Figure 3 Function Subprogram TMFMP

REFERENCES

[1] Sommerfeld,A.,Z.Phys.(1932) 78.

[2] Krutter,H.,J.Comp.Phys.,47(1982) 308.

[3] Civan,F. et al.,J.Comp.Phys.,56(1984) 343.

[4] Bulirsch,R.and Stoer,J.,,Numer.Math.,8 (1966) 1.

[5] Broyden,C.G.,Math.Comp.,19(1965) 577.

[6] Ralston,A.and Rabinowitz,P.,A First Course in Numerical Analysis,2nd Ed.(McGraw-Hill, New York,1978).

[7] Ninomiya,I and Hatano,Y,Johoshori,26(1985) 1033(In Japanese).

Approximation, Optimization and Computing:
Theory and Applications, A.G. Law and C.L. Wang (eds.)
Elsevier Science Publishers B.V. (North-Holland)

NUMERICAL SOLUTION OF PLATE BENDING PROBLEMS IN TWO DIMENSIONS

Terenzio SCAPOLLA

Institute for Physical Science and Technology, University of Maryland, College Park, MD 20742, USA

Summary. We consider the numerical solution of two–dimensional plate bending problems with finite element methods. We solve several problems with different types of finite elements in order to assess their performances. We mainly focus on the relation between accuracy of the approximate solution and number of degrees of freedom. We also consider the computational cost needed to achieve a given accuracy. The outline of the paper is the following. First we briefly recall the formulation of the plate bending problem in two dimensions according to the Kirchhoff theory. Then we present the numerical results we obtained solving different problems: clamped square plate, simply supported square plate and simply supported rhombic plate. Finally some conclusion are drawn.

Notations. We use $f_{/x}$ to denote the derivative of a function f respect to the variable x, $f_{/xy}$ to denote the derivative repsect to the variable x and y, $f_{/n}$ to denote the outward normal derivative.

1. TWO–DIMENSIONAL PLATE PROBLEM

We refer to Washizu [1] for a complete derivation of the two–dimensional problem under the Kirchhoff hypotheses. We just recall that, under suitable assumptions that take into account the smallness of the thickness respect the other two dimensions, all the components of the strain and stress tensors can be expressed in terms of only one function $u(x, y)$, the normal displacement. Let Ω be the domain occupied by the plate, $\partial\Omega$ its boundary and $f(x, y)$ the transversal load applied to the plate. The Kirchhoff model of plate bending problem lead to the following equation:

$$(1)\qquad \begin{cases} \Delta^2 u = f & \text{in } \Omega \\ u = g_1 & \text{on } \partial\Omega \\ \dfrac{\partial u}{\partial n} = g_2 & \text{on } \partial\Omega \end{cases}$$

where Δ^2 is the biharmonic operator.

An application of the principle of minimum potential energy shows that the displacement $u(x, y)$ minimizes the total energy of the plate

$$(2)\qquad \mathcal{E}(v) = F(v) - \int_\Omega fv\,dxdy$$

where

$$(3)\qquad \begin{aligned} F(v) =& \frac{1}{2}\int_\Omega \Big[\big(v_{/xx} + v_{/yy}\big)^2 \\ &+ 2(1-\nu)\big(v_{/xy}^2 - v_{xx}v_{yy}\big)\Big]\,dxdy \end{aligned}$$

and

$$(4)\qquad \begin{aligned} D &= \frac{Et^3}{12(1-\nu^2)} \\ E &= \text{Young's modulus} \\ t &= \text{thickness of the plate} \\ \nu &= \text{Poisson's ratio} \end{aligned}$$

The function v belong to a space of type $H_0^2(\Omega)$, since second derivatives are involved in relation (3). Moreover, the function v will take into account the boundary conditions to which the plate is subject. The boundary of the plate can be clamped (displacement and its normal derivative vanish on the boundary), simply supported (only the displacement vanishes) or free (no constrain on the displacement). The minimum problem for the functional (2) can be written in the following equivalent variational form:

$$(5)\qquad \begin{cases} \text{Find } u \in V \text{ such that} \\ B(u, v) = \displaystyle\int_\Omega fv\,dxdy \qquad \forall v \in V_0 \end{cases}$$

where, for each pair u, v of functions belonging to $H^2(\Omega)$, $B(\cdot,\cdot)$ is the following bilinear form:

$$(6)\qquad \begin{aligned} &B(u, v) = D\times \\ &\int_\Omega \Big[u_{/xx}v_{/xx} + u_{/yy}v_{/yy} + 2(1-\nu)u_{/xy}v_{/xy} \\ &+ \nu\big(u_{/xx}v_{/yy} + u_{/yy}v_{/xx}\big)\Big]\,dxdy \end{aligned}$$

and V_0 is a space of virtual displacements verifying the corresponding homogeneus boundary conditions.

2. CLAMPED SQUARE PLATE

Let us consider a square plate subject to a uniform load. We have solved this model problem with different types of finite elements. We give a short description of the finite elements used in the computations of the two–dimensional plate.

The first element is theit reduced Hsieh–Clough–Tocher element (RHCT). It is a composite element derived from the complete element. The complete triangle is divided into three subtriangles. On each subtriangle a cubic polynomial function is given. The initial 30 degrees of freedom are reduced to 12 imposing suitable conditions to get a C^1 function. The normal derivatives at the midpoint of the sides are eliminated by requiring it to be linear along each side. The degrees of freedom are: $u, u_{/x}, u_{/y}$ at the vertices. We refer to Ciarlet [2] and Chinosi [3] for further details. The second element is a *hybrid element* (HIBR). This element belongs to a large family of finite elements proposed by Pian [4] and analyzed by Brezzi and Marini [5]. The approximate space consists of C^1 functions that are cubic polynomials on each side with linear normal derivative. Moreover the functions satisfy a biharmonic equation on the triangle. The degrees of freedom are: $u, u_{/x}, u_{/y}$ at the vertices. The third element is the so–called *Argyris element* (ARGY). The element has been introduced by Argyris [6]. The approximation space consists of complete space of polynomials of degree 5. The high degree allows the conformity of the element. The degrees of freedom are: $u, u_{/x}, u_{/y}, u_{/xx}, u_{/xy}, u_{/yy}$ at the vertices, $u_{/n}$ at the midpoint of the sides.

We have solved different problems. First we consider a *clamped* plate, that is we impose the boundary conditions

$$(7) \qquad u = \frac{\partial u}{\partial n} = 0 \qquad \text{on } \partial\Omega$$

The clamped square plate is a widely used test for plate bending problems. It is known that the exact solution has the regularity

$$(8) \qquad u \in H^3(\Omega) \cap H_0^2(\Omega)$$

Let us conside a uniform mesh of the type shown in Fig.1.

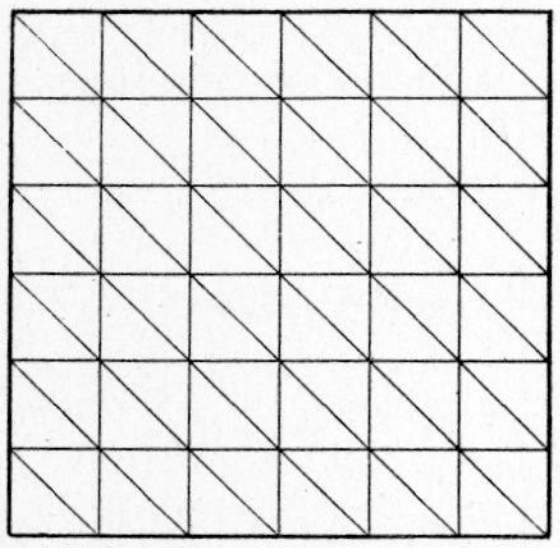

Fig.1: Square plate with uniform mesh

For each problem the displacement at the center C of the plate and the strain energy have been computed. Whenever exact solutions were available, the following errors have been computed. Let $u_{ex}(C)$ denote the exact displacement at the center of the plate and $u_{fe}(C)$ the finite element solution. The relative displacement error is defined as

$$(9) \qquad d = \frac{u_{ex}(C) - u_{fe}(C)}{u_{ex}(C)} \times 100$$

Let $\mathcal{E}_{ex}$ denote the exact energy of the plate and $\mathcal{E}_{fe}$ the finite element energy. The relative energy norm $||e||$ of the error $e = \mathcal{E}_{ex} - \mathcal{E}_{fe}$ can be expressed in the following way:

$$(10) \qquad ||e|| = \left(\frac{\mathcal{E}_{ex} - \mathcal{E}_{fe}}{\mathcal{E}_{ex}}\right)^{\frac{1}{2}} \times 100$$

In all the pictures the symbol '◇' denotes ARGY element, 'o' RHCT and '□' denotes HYBR. In Fig.2 the value of the displacement error against the number of degrees of freedom is shown. We used the value proposed by Timoshenko [7] as exact value in relation (9).

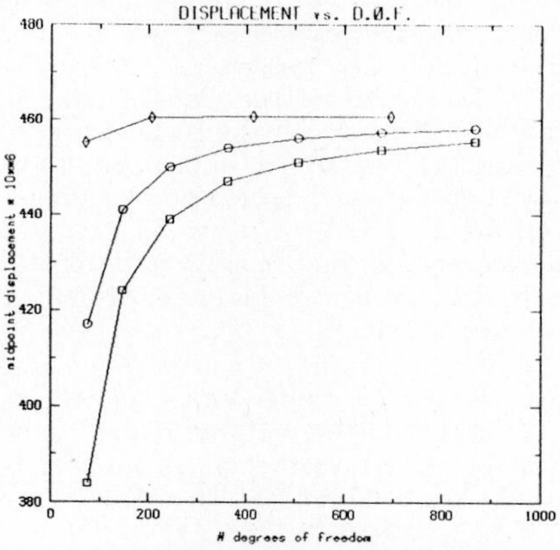

Fig.2: Clamped square plate
Displacement *vs.* degrees of freedom

Fig.3 gives the energy norm error against number of degrees of freedom. Both the scales are logarithmic. This allows to evaluate the rate of convergence of the finite elements.

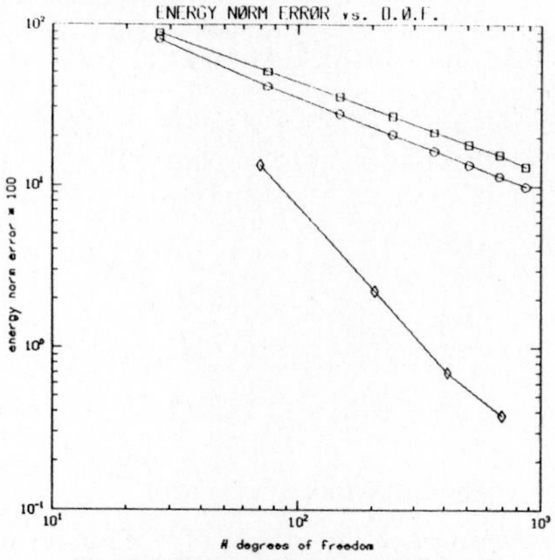

Fig.3: Clamped square plate
Energy norm error *vs.* degrees of freedom

The pictures show that for this problem all the three elements give good performances with a reasonable number of degrees of freedom. The ARGY element performs better, as expected due to the higher de-

gree. Good accuracy is achieved by all the elements. As we noted previously, the exact solution of the problem is very smooth and this is the reason for a good convergence for all the elements.
A fair comparison between the elements has to consider the computational time needed in order to obtain a prescribed accuracy. In Fig.4 the energy norm error is shown against the computational time. We considerered as computational time the time to evaluate the elemental stiffness matrices, the time to assembly the matrices and the time required by the solver (we used a direct method).

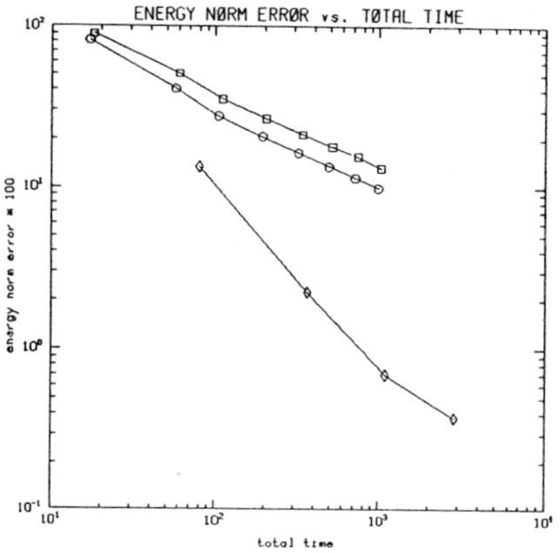

Fig.4: Clamped square plate
Energy norm error *vs.* computational time

It is well known that higher order finite elements require more time to compute the elemental stiffness matrix and to solve the final linear system. However, since few elements are required, it clearly appears from Fig.4 that the use of high order degree finite elements, such as ARGY, is more convenient respect to the lower degree ones.

3. THE SIMPLY SUPPORTED SQUARE PLATE

Now we consider a *simply supported* square plate, that is we impose on the boundary the constrain

$$u = 0 \quad \text{on } \partial\Omega \tag{11}$$

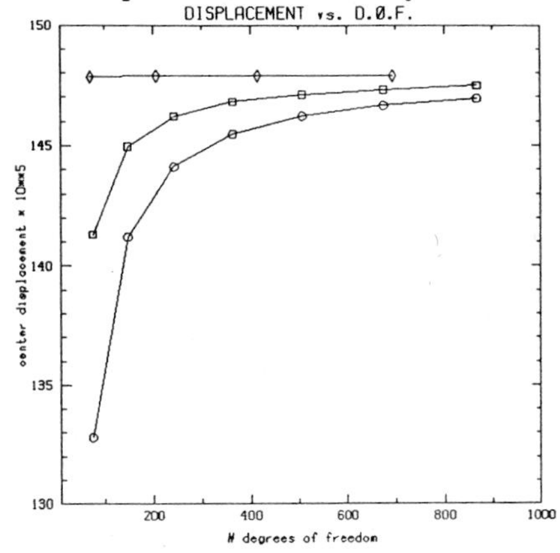

Fig.5: Simply supported square plate
Displacement *vs.* degrees of freedom

In Fig.5 the value of the displacement error against number of degrees of freedom is shown.
Fig.6 gives the energy norm error against number of degrees of freedom.

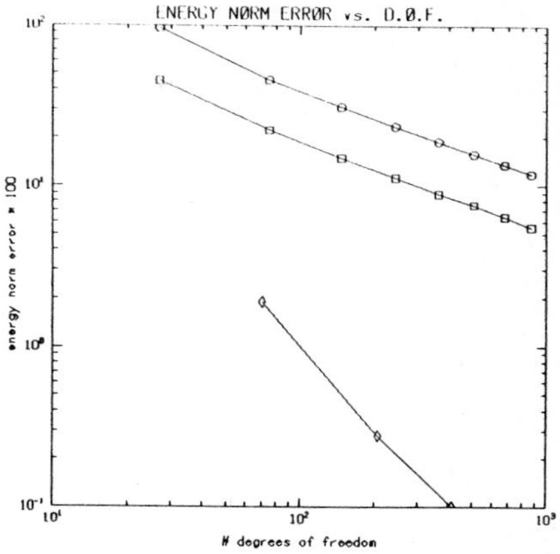

Fig.6: Simply supported square plate
Energy norm error *vs.* degrees of freedom

A good convergence is obtained by all the elements but still the ARGY element performs better. When computational time is considered we have the same type of behaviour we found for the clamped square plate.

4. SIMPLY SUPPORTED RHOMBIC PLATE

We now consider the bending of a simply supported rhombic plate (see Fig.7).

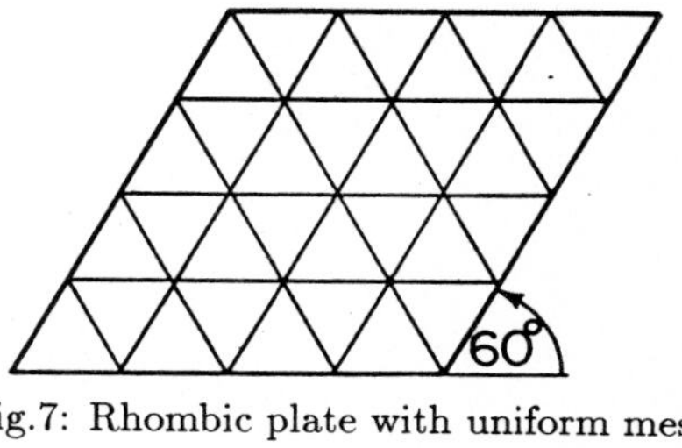

Fig.7: Rhombic plate with uniform mesh

The problem was proposed and analyzed by Morley [8]. The exact solution presents a singularity at the obtuse corners that affect severely the performance of the finite element solution. The singularity becomes stronger as the skew angle becomes smaller. We refer to Babuška and Scapolla [9] for a more detailed study of the mathematical aspects of the problem together with an analysis of the regularity of the exact solution. Let us consider $\alpha = 60°$.

In Fig.8 we show the energy norm error against number of degrees of freedom.
The results show that the ARGY element give better results, despite the widespread opinion that high order finite elements should not be used for the ap-

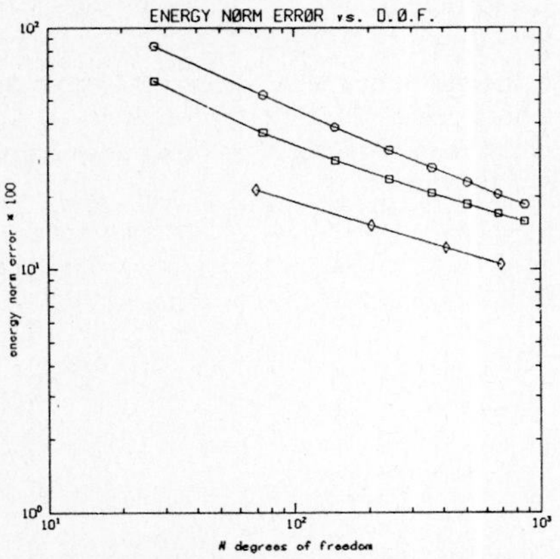

Fig.8: Simply supported rhombic plate
Energy norm error *vs.* degrees of freedom

proximation of problems where the solution is unsmooth. Of course the rate of convergence is limited by the smoothness of the solution (see [9]) but the value of the constant appearing in the abstract estimate depends on the degree of the element and this explains the different behaviors of the elements. The rate of convergence is the same (the lines in Fig.9 are parallel) but the value of the constant plays an important role in giving high order element better approximation results.

As regards the relation between accuracy of the solution and computational cost (see Fig.9) we find that ARGY element is more convenient to use, compared with RHCT and HYBR element.

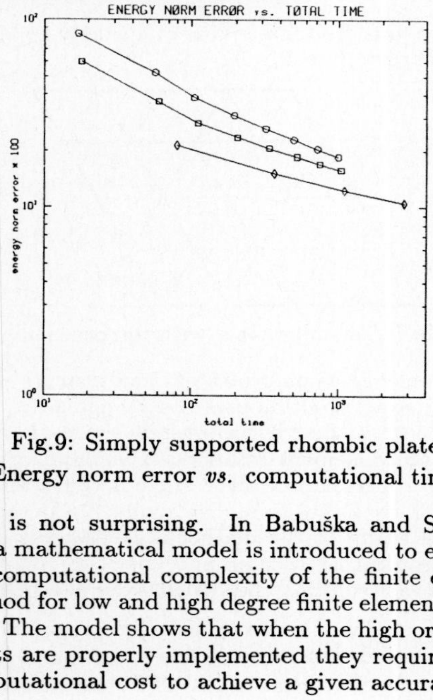

Fig.9: Simply supported rhombic plate
Energy norm error *vs.* computational time

This is not surprising. In Babuška and Scapolla [10] a mathematical model is introduced to evaluate the computational complexity of the finite element method for low and high degree finite element methods. The model shows that when the high order elements are properly implemented they require lower computational cost to achieve a given accuracy.

5. CONCLUSIONS

Several plate bending problems have been solved with three types of finite elements. The numerical results show that the higher degree finite element performs better compared with the lower degree finite elements. The behaviour is expected, due to well known abstract error estimates, when a smooth problem is solved. This is the case, i.e., of a clamped square plate bending problem. It is shown that even when problems with unsmooth solution are solved (e.g., simply spported rhombic plate) the high degree element gives better performances. The important relation between accuracy of the solution and computational cost has been also analyzed. The results show that high order elements can achieve a desired accuracy with lower cost than low degree element.

The result indicate that high order finite elements give better results and are more convenient. The introduction of the p–finite element method (see Babuška [11]) for second order elliptic equations has proved to be very effective. In this direction new high degree elements with special features for plate bending are being developed (we refer to Chinosi *et al.* [12] for details).

REFERENCES

[1] WASHIZU K.: *Variational Methods in Elasticity and Plasticity*, Pergamon Press, Oxford (1968)

[2] CIARLET P.G.: *The Finite Element Method for Elliptic Problems*, North–Holland, Amsterdam (1979)

[3] CHINOSI C.: The reduced Hsieh–Clough–Tocher element, Pubblicazione 401, Istituto di Analisi Numerica del Consiglio Nazionale delle Ricerche, Pavia (1984)

[4] PIAN T.H.H.: Finite element formulation by variational principles with relaxed continuity requirements, in *The Mathematical Foundations of The Finite Element Method*, I. Babuška and K. Aziz (eds.), Academic Press, New York (1972), 689–710

[5] BREZZI F., MARINI L.D.: On the numerical solution of plate bending problems by hybrid methods, *RAIRO*, 9–R3 (1975), 5–50

[6] ARGYRIS J.H., FRIED I., SCHARPF D.W.: The TUBA family of plate elements for the matrix displacement method, *The aeronautical J. of the Royal Aeronautical Society*, 72 (1968), 701–709

[7] TIMOSHENKO S., WOINOWSKI–KRIEGER S.: *Theory of Plates and Shells*, McGraw–Hill, New York (1959)

[8] MORLEY L.S.: Bending of a simply supported rhombic plate under uniform loading, *Quart. J. Mech. and Appl. Math.*, 15 (1962), 413–426

[9] BABUŠKA I., SCAPOLLA T.: Benchmark computation and performance evaluation for a rhombic plate bending problem, *Int. J. Numer. Met. Engrg.*, 28 (1989), 155–179

[10] BABUŠKA I., SCAPOLLA T.: Computational aspects of the h, p and h–p versions of the finite element method, in *Advances in Computer Methods in Partial Differential Equations–VI*, R. Vicnevetsky and R.S. Steplemen (eds.), IMACS (1987), 234–240

[11] BABUŠKA I.: The p and h–p versions of the finite element method. The state of the art, in *Finite Elements - Theory and Application* , Springer-Verlag, New York (1988), 199-239

[12] CHINOSI C., SACCHI G., SCAPOLLA T.: Hierarchic conforming finite elements for plate bending problems, in preparation

Approximation, Optimization and Computing:
Theory and Applications, A.G. Law and C.L. Wang (eds.)
Elsevier Science Publishers B.V. (North-Holland)

MINIMUM-WEIGHT DESIGN OF BEAMS FOR DEFLECTION BY ODE TECHNIQUES AND SOFTWARE*

Si YUAN

Dept. of Civil Engineering, Tsinghua University, Beijing, 100084, P. R. China

Abstract

The non-standard ordinary differential equations(ODEs),resulting from minimum-weight design problems of beams, are converted to an equivalent standard nonlinear ODE system by using trivial ODE and interval mapping techniques. The ODE system is then directly and efficiently solved by robust ODE code COLSYS. The sought longitudinal shapes and switching interface points are explicitly given as part of the solutions with the accuracy satisfying the user pre-specified error tolerances.

1. INTRODUCTION

The present paper makes an attempt to apply Ordinary Differential Equation (ODE) techniques and the available ODE codes to solving problems of optimal design. To focus on the basic idea of the ODE approach, a comparatively simple example, the minimum-weight design of a simply supported beam, has been chosen as the working problem. The problem of minimum-weight design of beams is formulated into ODEs, in a natural way, by means of variational methods. Some of the ODEs require interface boundary conditions at unknown switching points which delimit the constant cross-sections near both ends from the variable ones of the central portion of beams. These original ODEs are then converted to an equivalent standard two-point boundary-value nonlinear ODE system by using trivial ODE and interval mapping techniques, so that it can be solved directly by means of any general-purpose ODE codes. In the present work, ODE code COLSYS [1][2] is adopted and found very efficient. As a by product, the unknown switching interface points are explicitly given as a part of the solution. The advantages of the present approach are the ease in application and implementation, the rich information from the results and the high accuracy of the solutions. The present research shows that the ODE techniques and software have the potential to play a more important role in the practice of optimal design.

2. A SIMPLE BEAM PROBLEM

In this section, we consider the problem of minimum weight design of a simply supported beam with a rectangular cross-section subject to a concentrated load [4],as shown in Fig.1(a).

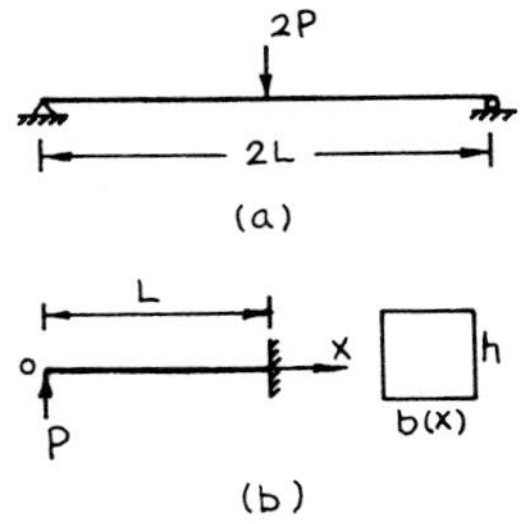

Fig.1 Simply Supported Beam

Taking account of the symmetry, we shall analyze only the left part, as shown in Fig.1(b). The differential equilibrium equation for the deflection w and the associated boundary conditions (BCs) are

$$(EIw'')'' = 0 \qquad x \in [0,L] \qquad (1a)$$

$$-EIw'' = 0 \qquad \text{at } x = 0 \qquad (1b)$$

$$-(EIw'')' = -P \qquad \text{at } x = 0 \qquad (1c)$$

$$w = 0 \qquad \text{at } x = L \qquad (1d)$$

$$w' = 0 \qquad \text{at } x = L \qquad (1e)$$

where E is the Young's modulus of elasticity and I is the moment of inertia. For a rectangular cross-section with given height h, I is equal to $b(x)h^3/12$. The variable width b(x) in Eq.(1) should be so determined that the weight of the beam

*The project supported by National Natural Science Foundation of China

$$w=\rho h\int_0^L b(x)\,dx = \text{Minimum} \qquad (2)$$

where ρ is the density of weight. An additional BC is needed to determine b(x) from Eq. (2). This BC is provided by specifying the deflection at the free end

$$w = w_o \qquad \text{at } x = 0 \qquad (3)$$

Therefore the shape optimization problem of the beam is well defined by Eqs. (1)-(3).

Eqs. (1), however, can be equivalently derived by minimizing the total potential energy of the beam defined by

$$\Pi = \frac{1}{2}\int_0^L \frac{Eh^3}{12}\, b(x)\, (w'')^2 dx \qquad (4)$$

To derive the ODE for the shape function b(x), we combine the two minimization procedures defined in Eqs. (2) and (4) by means of Lagrange multiplier λ to write a new functional

$$* = \frac{1}{2}\int_0^L \frac{Eh^3}{12}\, b(x)\, (w'')^2 dx$$

$$-\lambda\int_0^L \rho h\, b(x)dx+Pw(0) \qquad (5)$$

Taking variations of both the deflection w and the variable width b(x) yields

$$\delta\Pi^*= \int_0^L \delta b(x)\, [\, \frac{Eh^3}{24} (w'')^2 - \lambda\rho h]\, dx$$

$$+\int_0^L (\delta w'')\, \frac{Eh^3}{12} b(x)(w'')\, dx + P\,\delta w(0) \qquad (6)$$

and from the stationary condition, $\delta\Pi^*=0$, we can obtain both Eqs. (1) and the following

$$w'' = k \qquad x \in [0,L] \qquad (7a)$$

where

$$k^2 = (\, \frac{24\lambda P}{Eh^2}\,) \qquad (7b)$$

Substituting the above equation back into Eqs. (1), we obtain

$$b'' = 0 \qquad x \in [0,L] \qquad (8a)$$

$$b = 0 \qquad \text{at } x = 0 \qquad (8b)$$

$$b' = 12P/(Eh^3 k) \qquad \text{at } x = 0 \qquad (8c)$$

$$w = 0 \qquad \text{at } x = L \qquad (8d)$$

$$w' = 0 \qquad \text{at } x = L \qquad (8e)$$

As a result, we have an ODE problem consisting of two 2nd-order ODEs, (7a) and (8a), and five BCs defined in (3) and (8b)-(8e), together with an unknown constant k to be determined, which means at least an additional first order ODE is needed to make the ODE problem well defined. On account of the fact that k is constant, we set up the required additional ODE by using trivial ODE technique [3]

$$k' = 0 \qquad x \in [0,L] \qquad (9)$$

The addition of Eq.(9), which guarantees k to be a constant, to the ODE problem forms a 'standard' nonlinear ODE system [3], which is summarized as follows

ODEs: $x \in [0,L]$

$$k' = 0 \qquad b'' = 0 \qquad w'' = k \qquad (10)$$

BCs:

$$w = w_o \qquad \text{at } x = 0 \qquad (11a)$$

$$b = 0 \qquad \text{at } x = 0 \qquad (11b)$$

$$b' = 12P/(kEh^3) \qquad \text{at } x = 0 \qquad (11c)$$

$$w = 0 \qquad \text{at } x = L \qquad (11d)$$

$$w' = 0 \qquad \text{at } x = L \qquad (11e)$$

The ODE system defined by Eqs. (10) and (11)can directly be solved by using the ODE code COLSYS. The COLSYS solution to this trivial problem is exactly the same as the following exact solution

$$k = 2w_o/L^2 \qquad (12a)$$

$$b(x) = 12Px/(kEh^3) \qquad (12b)$$

$$w = w_o[1-2(x/L)+(x/L)^2] \qquad (12c)$$

The maximum width b_c is

$$b_c = 12PL/(kEh^3) \qquad (12d)$$

3. MINIMUM CROSS-SECTIONAL AREA

While the above solution is quite simple, it is far from practical since the width b is zero at the free end. In what follows, we shall consider the problem with a constraint

$$b(x) \geqslant b_o \qquad x \in [0,L] \qquad (13)$$

where b_o is the given minimum width. Let the deflection in the section [0,c] be $\hat{w}$ and the one in [c,L] be w. The functional Π^* (see Eq. (5)) takes the following form

$$\Pi^*=\frac{1}{2}\int_0^c \frac{Eh^3}{12}\, b_o\, (\hat{w}'')^2 dx + \frac{1}{2}\int_c^L \frac{Eh^3}{12} b(x)(w'')^2 dx$$

$$+P\hat{w}(0) - \lambda[\int_0^c \rho h b_o dx + \int_c^L \rho h b(x)dx] \qquad (14)$$

Following the similar variation procedure and the trivial ODE technique described above, the following ODE system is obtained

ODEs:

$$k' = 0 \tag{15a}$$

$$\hat{w}'''' = 0 \qquad x \in [0,c] \tag{15b}$$

$$b'' = 0 \qquad x \in [c,L] \tag{15c}$$

$$w'' = k \qquad x \in [c, L] \tag{15d}$$

BCs:

$$\hat{w} = w_o \quad \text{at } x = 0 \tag{16a}$$

$$\hat{w}'' = 0 \quad \text{at } x = 0 \tag{16b}$$

$$\hat{w}''' = 12P/(Eh^3 b_o) \quad \text{at } x = 0 \tag{16c}$$

$$b = b_o \quad \text{at } x = c \tag{16d}$$

$$b' = b_o \hat{w}'''/k \quad \text{at } x = c \tag{16e}$$

$$\hat{w}'' = k \quad \text{at } x = c \tag{16f}$$

$$\hat{w} = w \quad \text{at } x = c \tag{16g}$$

$$\hat{w}' = w' \quad \text{at } x = c \tag{16h}$$

$$w = 0 \quad \text{at } x = L \tag{16i}$$

$$w' = 0 \quad \text{at } x = L \tag{16j}$$

where k is the same as defined in (7b) and the interface point c remains to be determined. Making the following interval mapping to map both sections [0,c] and [c,L] to a unit interval [0,1]

$$t = \begin{cases} x/c & x \in [0,c] \\ (L-x)/(L-c) & x \in [c,L] \end{cases} \tag{17}$$

and adding another trivial ODE c'=0, we can have the following standard nonlinear ODE system

ODEs: $t \in [0,1]$

$$k' = 0$$

$$c' = 0$$

$$b'' = 0$$

$$w'' = (L-c)^2 k$$

$$\hat{w}'''' = 0 \tag{18}$$

BCs:

$$\hat{w} = w_o \quad \text{at } t = 0$$

$$\hat{w}'' = 0 \quad \text{at } t = 0$$

$$\hat{w}''' = c^3 12P/(Eh^3 b_o) \quad \text{at } t = 0$$

$$w = 0 \quad \text{at } t = 0$$

$$w' = 0 \quad \text{at } t = 0$$

$$b = b_o \quad \text{at } t = 1$$

$$b'c^3 = b_o (c-L)\hat{w}'''/k \quad \text{at } t = 1$$

$$\hat{w} = w \quad \text{at } t = 1$$

$$\hat{w}'' = c^2 k \quad \text{at } t = 1$$

$$cw' = (c-L)\hat{w}' \quad \text{at } t = 1$$

(19)

The Eqs. (18) (19) are now ready to be input to the ODE code COLSYS. The analytical solutions to (15) (16) are readily derived as

$$\hat{w}(\xi) = w_o + (\hat{w}_c/2)\,[3\eta^2 - 6\eta)\xi + \xi^3] \qquad \xi \in [0,\eta]$$

$$b(\xi) = b_o$$

(20a)

$$w(\xi) = (\hat{w}_c/2)\, 3\eta(1-\xi)^2$$

$$b(\xi) = b_o\, \xi/\eta \qquad \xi \in [\eta, 1]$$

(20b)

where

$$\xi = x/L \qquad \eta = c/L$$

$$\hat{w}_c = PL^3/(3EI_o) \qquad I_o = b_o h^3/12 \tag{21}$$

and the interface point η is determined by solving the following cubic equation

$$\eta^3 - 3\eta + 2w_o/\hat{w}_c = 0 \tag{22}$$

This problem has been solved by the code COLSYS on an IBM PC with single precision. The data for COLSYS are: No. of collocation points = 4; No. of subintervals = 1; Error tolerance=0.0001.

Table 1 Numerical Results

b_o/b_c	η	b_{max}/b_c
0.0	0.0	1.0
0.2	0.2027794	0.9862936
0.4	0.4257186	0.9395878
0.6	0.7292991	0.8227074

The input data for the beam are simply: $P=L=h=w_o=1$, $E=6$. The numerical results are shown in Table 1 and Fig. 2, where b_c is defined by (12d). Substituting the η values into Eq. (22), it can be verified that the results are accurate up to the sixth digit.

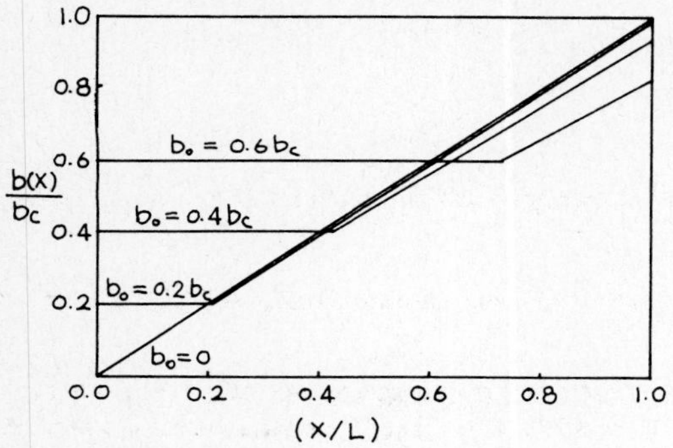

Fig.2 Shapes of the width

4. CONCLUDING REMARKS

Taking advantage of the availability and reliability of general-purpose ODE codes, a new approach, i.e. the ODE approach, to structural optimization problems is presented through a simple example. It is straightforward to apply this method to more complicated problems of beams and columns, and it is extendable to 2-dimensional problems by employing certain semi-discretization procedures, which is at the present time undergoing extensive investigation.

REFERENCES

[1] U.Ascher, J. Christiansen and R. D. Russell; "Collocation Software for Boundary-Value ODE", ACM Trans. Math. Software, Vol.7, No.2, (June 1981), pp. 209-222.

[2] U. Ascher, J. Christiansen and R. D. Russell; "Algorithm 569, COLSYS: Collocation Software for Boundary-Value ODEs [D2]", ACM Trans. Math. Software, Vol. 7, No.2, (June 1981), pp. 223-229.

[3] U. Ascher and R. D. Russell; "Reformulation of Boundary Value Problems into'Standard' Form", SIAM Rev. 23 (April 1981), pp. 238-254.

[4] H. C. Hu; "Variational Theorems and its Applications in Elasticity", Science Press, 1981, pp. 50-54, (in Chinese).

Approximation, Optimization and Computing:
Theory and Applications, A.G. Law and C.L. Wang (eds.)
Elsevier Science Publishers B.V. (North-Holland)

A CONTINUOUS/DISCRETE NONLINEAR OPTIMIZATION DESIGN METHOD

Xia De Lin and Zhou Jingfang

Huazhong University of Science and Technology, Wuhan, P. R. China

Abstract A new method for continuous/discrete nonlinear optimization is presented in this paper, which is called "Variable Subspace Intersecting Iteration Algorithm". The principle of this method is firstly to divide the set of design variables into two subsets. After finding out a initial feasible point, we can search for local optimal point in the two subsets consequently. At last, we can search for the final optimal point.

1. INTRODUCTION

In engineering optimization design, the optimization method with continuous-discrete variable is a difficult topic and prevalent study has not come to be premature, whereas many problems contain two kinds of design variables. This paper presents a general method toward the end of solving this kind of problems efficietly.

2. STATEMENT OF THE PROBLEM AND BASIC DEFINITIONS

A nonlinear optimization problem with continuous/discrete variables can be stated as:

$$\min f(x^c, x^d), \quad X = [x^c, x^d] \in R^n$$

$$\text{S.to: } g_i(x^c, x^d) \leq 0, \quad i = 1, 2, \cdots, m$$

$$x^c \in R^{n-k} \tag{1}$$

$$x^d \in S, \quad S = [S_1, S_2, \cdots, S_K]^T$$

$$X \geq 0$$

where x^c, x^d, refer to continuous and discrete variable vectors, S is the set of k's discrete variables' elements, each subset $S_j (j = 1, 2, \cdots, k)$ is a vector, their numbers of dimension may be different and be arranged from little to large, $f(x^c, x^d)$, $g_i(x^c, x^d)$ are nonlinear real functions.

For this kind of optimal design problems, the goal is to find out a vector $X = [x^c, x^d]$ within n-dimensional vector space which can not only satisfy constraints $g_i(x^c, x^d) \leq 0$, but also fit discrete variable set S and additionally make objective function least. Obviously this problem is different from general optimal problems on feasible solution set, basic definitions and convergance criteria. For example, though feasible solution set of this problem is infinite, which is the same with continuous variable optimal problem (whereas the set of discrete variable optimal problem is finite), but it isn't continuous, either. On convergance criteria, it is when objective function is the least both in continuous variable subspace and discrete one at the same time that it is optimal. For these reasons, we present some basic definition and convergance criteria for this problem.

2.1. Basic definitions and convergance criteria

2.1.1 Feasible point and feasible range

If a point X in n-dimensional space satisfies following conditions:

$$X \in R^{n-k} \cap S$$

$$g_i(x) \leq 0 \quad i = 1, 2, \cdots, m$$

then point X is a feasible point, subset $R^{n-k} \cap S | g_i(x) \leq 0$ is feasible area referring to X. For 2-dimensional space, feasible range will consist of finite grid lines (see Fig. 1); whereas for 3-dimensional space, if there are noly two continuous variables, then feasible range will consist of two grid planes (see Fig. 2)

For n-dimensional problem, feasible range consist of infinite number of super grid planes. Obviously searching of optimization is on these super-planes (or planes and straight lines).

2.1.2. Neighbourhood of a feasible point

The neighbouring area of a feasible point X_i is

defined as intersecting set of discrete subset S and continuous subset R^{n-k}. The neighbourhood of point P_i in Fig.1 consists of finite points on segment P_iA and P_iB of unit length and segment $A_{i-1}B_{i-1}$ and $A_{i+1}B_{i+1}$; in n-dimensional space, neighbourhood of point P_i consists of ε which is the neighbouthood of point P in continuous-variable subspace and neighbourhood grid discrete subspace. The neighbourhood of point P_i is $\varepsilon + \Sigma_{i-1} + \Sigma_{i+1}$ with

condition of $\{\begin{matrix} n=3 \\ K=1 \end{matrix}$ (see Fig. 2).

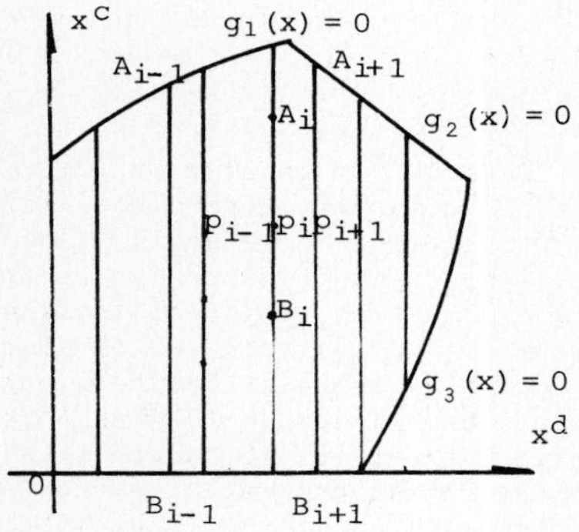

Fig. 1

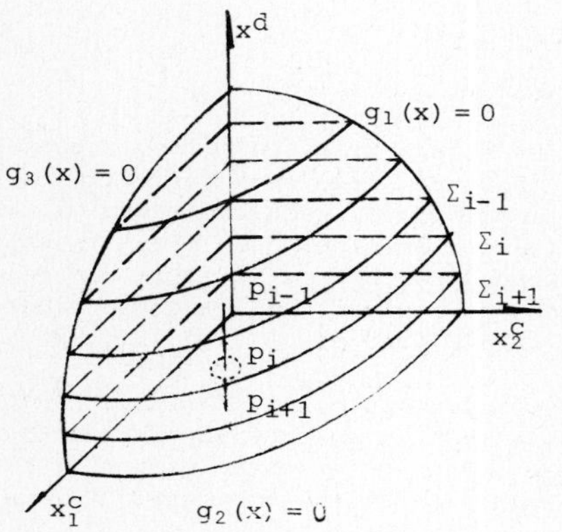

Fig. 2

2.1.3. Optimzation convergance criteria

The method in this paper will search respectively in continuous variable subset and discrete variable subset. It starts from an initial feasible point to find out a best improved feasible point in continuous variable subspace with discrete variables constant, and then take it as a beginning point, go on with first and second search in discrete variable subspace with continuous variable constant. If there is no new improved point, the third search will be carried out in neighbourhood of the point. If a new improved point hasn't been found out yet, search should be ended. The first and second search principle in discrete variable subspace is referring to "A Discrete Nonlinear Stepping Optimization Method And Its Application" [5] by authors.

2.2. Basic principle

Firstly an initial feasible point should be found out timely, which is not easy for there are discrete variables. For this reason, an efficent operating is presented here to get this initial feasible point quickly and simply, which is stated below.

2.2.1. Discussion about obtaining an initial feasible point

At first, a point x_0' is given, it is a starting point to find out a feasible discrete point. The method is shown below:

Assunimg that, for problem (1), point x_0' makes the constraints $g_j(x_0') \le 0$ $(j=1,2,\cdots,m_1)$, $g_k(x_0') > 0$ $(k=m_1+1,\cdots,m)$ then solve:

$$\begin{aligned} &\text{Min } G(x) = \sum_k (g_k(x)+\varepsilon)^2, \quad 0<\varepsilon<1 \\ &\text{S.to: } g_j(x) \le 0, \quad j=1,2,\cdots,m_1 \\ &\qquad g_l(x) = -x_l \le 0, \quad l=1,2,\cdots,n \end{aligned} \tag{2}$$

to obtain a point x_c which satisfies all the constraints, and discrete it in set S so as to obtain an initial feasible discrete point. Iterating search of problem (2) is based on formula $x_n = x_0' + \lambda \cdot d$, where search direction d is:

$$d = \sum_k - \frac{g_k(x_0)}{\nabla g_k^T(x_0)\cdot\nabla g_k(x_0)}\,\overline{\nabla g_k(x_0)}, \quad k=m_1,\cdots,m$$

where $\overline{\nabla g_k(x_0)} = \nabla g_k(x_0)/\|\nabla g_k(x_0)\|$.

Step length increment $\Delta\lambda$ is used to determine a new point x_n when the procedure is unidimensional search in direction d, then check the constraints which is not yet satisfied, so that G(x) will be reformed, afterward a new programming problem (2) should be solved until the point x_c is obtained which can satisfy all the constraints of original problem. Finally point x_c is discreted in set S and thus a discrete feasible beginning point is obtained.

2.2.2. Search in continuous variable subspace

For initial feasible point x_0, x_0^d will keep constant when searching in continuous variable subspace.

Projective gradient method [7] is used to find out an optimal point in continuous variable subspace according to nonlinear objective function and constraints.

2.2.3. Searching in discrete variable subspace

For a given feasible initial point x_0, the optimization method only for discrete variable problems is used. It is detailed below:

I. First step search: initial point is $[x_0^c, x_0^d]$, decreasing searching direction may be:

$$\begin{aligned} D_1 &= [\mathrm{sign}(-\nabla f_1(x_0)),\ 0,0,\cdots,0]^T, \\ D_2 &= [0,\ \mathrm{sign}\ (-\nabla f_2(x_0)),\ 0,\cdots,0]^T, \\ &\cdots \qquad \cdots \qquad \cdots \qquad \cdots \\ D_k &= [0,0,\cdots,0,\ \mathrm{sign}(-\nabla f_k(x_0))]^T. \end{aligned} \tag{3}$$

where $\nabla f_i(x_0)$ $(i = 1,2,\cdots,k)$ is ith component of gradient of objective function at point x_0.

The quickest decreasing direction D in the search procedure is used, that is:

$$D = D_i \,|\, \min\{D_i^T\}\nabla f(x_0) \leq 0,\ i = 1,2,\cdots,k \tag{4}$$

then unidimensional search and iteration equation $x_1 = x_0 + \lambda \cdot D$ are used to gain the feasible optimal point in this direction and to take it as a new initial point to carry out the whole procedure; if there isn't feasible point, then go into second decreasing direction to search.

II. second step search: initial point is $[x_0^c, x_0^d]$. Search is going on in neighbouring grid set, total number of these grids is $3^k - (2 \cdot k + 1)$ and grids only in decreasing direction are considered. This paper presents a method to determine the position of these grids, which is to order the components of Sj with each specified design variable and specify them according to their natural ordinal number, whereas the neighbouring grids' ordinal numbers are different from the feasible discrete initial point with a difference + 1 or - 1, so these neighbouring grids' ordinal number vector and position can be obtained from the initial point. If ith grid's ordinal number vector defined as $\{D_p\}_i$, then the inequality:

$$\{D_p\}^T \nabla f(x_0) < 0,\ i = 1,2,\cdots,3^k - (2 \cdot k + 1)$$

can determine whether the ith grid is a better point, and select a point which can make the objective function value minimum and satisfy all the constraints. Therefore the point is taken as one of the second step and define it as a new initial point and process the iteration again. If this optimal point does not exist, go into the third step.

III. The third step is to take respectively neighbouring grids x_i of discrete feasible initial point x_0 as starting one and determine search direction according to the character of design variables: for discrete variables, search direction is chosen from:

$$\begin{aligned} \{D_{p_1}\}_i &= [d_{p_1},\ 0,0,\cdots,0]^T \\ \{D_{p_2}\}_i &= [0,\ d_{p_2},\ 0,\cdots,0]^T \\ &\cdots \qquad \cdots \qquad \cdots \\ \{D_{p_k}\}_i &= [0,0,\cdots,0,d_{p_k}]^T \\ i &= 1,2,\cdots,3^k - (2 \times k + 1) \end{aligned}$$

(but for continuous variables, search direction should consider all of the continuous variables and go into optimization with projective gradient method). For each neighbouring point x_i, the total number of this type of direction is k, but they are not all feasible and it is necessary to determine feasible directions to find out the quickest decreasing direction. For feasible neighbouring grids, descreasing direction is gained from:

$$\{D_{p_j}\}_i^T \nabla f(x_0) < 0 \qquad i = 1,2,\cdots,3^k - (2 \times k + 1)$$
$$j = 1,2,\cdots,k$$

if x_0 is unfeasible, the direction will be determined by

$$\nabla g_{k_1}^T(x_i) \cdot \{D_{p_j}\}_i > 0, \quad k_1 = 1,2,\cdots,k_2$$
$$\nabla f^T\ (x_0) \times \{x_i + \lambda\{D_{p_j}\}_i - x_0\} < 0$$
$$i = 1,2,\cdots,3^k - (2 \times k + 1), \quad j = 1,2,\cdots,k$$

where $g_{k_1}(x_i)$ is the constraint which neighbouring point x_j doesn't satisfy. The third step will optimize along these directions. If an optimal one can be found out, it will be taken as a new initial point and run the whole process again; or iteration ends, when the initial point is the local optimal one of the problem.

By the way, unidimensional search method of discrete variable is used in the first and third steps. Based on the character of montone function and discrete variables, Fibonacci method is used to get local optimal point.

IV. The whole algorism procedure:

(1). Input the number of variables and constraints n, m, and specify a initial point x_0'.

(2). Input a matrix to indicate the character of each variables (continuous or discrete).

(3) To find out an initial feasible point x_0 from x_0'.

(4). To optimize with projective gradient method in continuous variable subspace with x_0^d constant and result is $x_1' = [x_0^c, x_1^d]$.

(5). To take x_1' as an initial point and go on with first step search in discrete variable set with x_1^c constant. If succeed, an improved point $x_1 = [x_1^c, x_1^d]$, ND = 1; or ND = 0.

(6). ND = 1? If so, goto (3).

(7). To take x_1' as a starting point and to go on with second step in discrete variable subspace with x_1^c constant. If succeed, $X = [x_1^c, x_1^d]$, $ND_1 = 1$, or $ND_1 = 0$.

(8). $ND_1 = 1$? If so, goto (3).

(9). To take x_1' as a starting point and to go on with third step in whole variable space. If succeed, $x_1 = [x_1^c, x_1^d]$, $ND_2 = 1$, or $ND_2 = 0$.

(10). $ND_2 = 1$? If so, goto (3).

(11). To output the optimal solution x_{op}, and $f(x_{op})$.

(12). Stop

3. COMPUTATIONAL ILLUSTRATION

A lot of engineering practical problems have been calculated to prove the method in this paper. Here, one of them were presented.

There is a spring used in internal-combustion engine's valve. Its maxmum deformation $\lambda = 16.59$ mm. working loading F = 680N, working frequency fr = 25Hz, the highest working temperature is 150°C, the material is 50CrVA steel wire, the required life time is $N = 10^6$ (number of iterating), the spring's structure must satisfy that: the wire's diameter d, the spring's out-diameter D, working ring number $n(\geq 3)$ must be in the definition set S to make the spring's weight lowest.

Assuming that vector [d, D, n] refers to $[x_1, x_2, x_3]$ and according to these demands, this problem can be stated as:

$$\text{Min } f(x) = 1.925\ 10^{-5}(x_3+1.8)x_2 \times x_1^2$$
$$\text{S.to } g_1(x) = -404.9 + 3503.96\ x_2^{0.86}/x_1^{2.86} \leq 0$$
$$g_2(x) = -x_2/x_1 + 6 \leq 0$$
$$g_3(x) = [(x_3+1.3)\times x_1 + 18.25]/x_2 - 5.3 \leq 0$$
$$g_4(x) = 2.5 - x_1 \leq 0$$
$$g_5(x) = x_1 - 9.5 \leq 0$$
$$g_6(x) = 30 - x_2 \leq 0$$
$$g_7(x) = x_2 - 60 \leq 0$$
$$g_8(x) = 375 - 3.56\times 10^5 \times x_1/(x_2^2 \times x_3) \leq 0$$
$$g_9(x) = 3 - x_3 \leq 0$$
$$g_{10}(x) = 680 - 10000\times 16.59 \times x_1^4/(x_3 \times x_2^2) \leq 0$$
$$x_1 \in S, X_3 = \text{positive integer}$$
$$S = [2.5, 3.0, 3.5, 4.0, 4.5, 5.0, 5.5, 6.0, 6.5, 7.0, 7.5, 8.0, 8.5, 9.0, 9.5, 10.0]^T$$
$$x_0 = [6.0, 48.0, 6.0]^T$$

number of iteration	x	f(x)
0	[8, 48, 3]	0.2839
1	[7.5, 48, 3]	0.2495
2	[7.5, 45.037, 3]	0.2341
3	[6.5, 39.001, 3]	0.1523

4. CONCLUSION

Continuous/discrete variable optimization is different from pure discrete variable or continuous variable optimization in mathematics model, definition area, neighbouring area, convergance criteria and solving method. So this kind of problems must be solved by the method defferent from general ones. A method by some designers is to take the problem as a continuous variable one at first, after optimization, to descrete the variables to given values so as to get socalled optimal solution, Which is not correct. For instance, see example, from initial point $x_0 = [6.3, 50, 5.5]^T$, we get optimal point $x^* = [6, 38, 7.25]^T$, after discreting it, we obtain that $x_{op} = [6, 38, 7.]^T$ objective function value is 0.2317, but if using the method in this paper, a better solution will appear; again, some solutions do not satisfy certain constraints in some other examples. These show that using the method in this paper can not only find out optimal point which satifies all of the constraints but also gain better solution quickly and simplely for this kind of probems.

REFERENCE

[1] R. S. Garfinhel and G. L. Nemhauser, "Integer Programming", John Wiley & Sons, 1972.

[2] Xia De Lin and C. L. Wang,"On an Approximation Method of Beometric Programming". Congressus Numerantium, Vol, 34, (1982), canada.

[3] Xia De Lin, "A New Method of Integer Linear Programming-Branch direction search method". Applied mathematics and Mechanics, Vol, No. 3, 1985.

[4] Xia De Lin,"Discrete Variable Method in Ship structure Optimization" Conference on Ship Building Engeering of China, 1984.

[5] Xia De Lin and Zhou Jingfang,"A Discrete Nonlinear stepping optimization method and its Application", Applied mathematics and Mechanics, Vol. 9, No. 9, 1988.

[6] Zhang Yu Kai, Basics of Mechanics Optimization Design, Mechanical Industry, China.

[7] Edward J. Hang; Jasbir S. Arora Applied Optimal Design-Mechanical and Structure Systems, John Wiley and Sons, 1979.

Approximation, Optimization and Computing:
Theory and Applications, A.G. Law and C.L. Wang (eds.)
Elsevier Science Publishers B.V. (North-Holland)

EXPLICIT SENSITIVITY ANALYSIS AND ITS APPLICATIONS IN 2-D STRUCTURAL OPTIMIZATION *

Zhang Kefeng   Gao Shibao   Wang Shuyu

Department of Civil Engineering , Zhejiang University, Hangzhou, P. R. C.

This paper presents a trapezoid element with variable thickness for 2-D continuum structures. The explicit expressions of the element stiffness matrix and their derivatives with respect to nodal coordinates are derived precisely. Because of the explicit formulations for design sensitivity analysis, the computational efforts are reduced, the accuracy and stability of sensitivities are enhanced.

1. INTRODUCTION

Sensitivity analysis is one of the main subjects in shape optimization. The finite difference method and semi-analysis method are extensively used in sensitivity analysis, but those methods need a lot of CPU time and cannot ensure the accuracy of results. Therefore, how to reduce the computational efforts and how to improve the accuracy have become a key problem in sensitivity analysis. Many references have reported the developments in this aspect. This paper presents a part of our work of doing this in hydraulic structural optimization.

A variable thickness trapezoid element with explicit expressions presented in this paper is suitable for 2-D continuum structures. The explicit stiffness matrix is derived. Especially, the derivatives of it with respect to nodal coordinates are also derived analytically. Then in virtue of mapping relations between design variables and nodal coordinates, the derivatives of behaviour constraints with respect to design variables can be obtained without using numerical itegration and finite difference method. To verify the validity of formulations and computational efficiency, a large number of numerical tests for behaviour sensitivities are made by the analytical method and the finite difference method. Results show that the analytical method is superior to the finite difference method either in accuracy or in computational efforts.

Finally, the optimal shape of a concrete gravity dam with broad joints is conducted by using the method of SLP and analytical sensitivities.

2. EXPLICIT EXPRESSIONS OF A VARIABLE THICKNESS TRAPEZOID ELEMENT

It is very attractive to obtain an explicit element stiffness matrix of an arbitrary quadrilateral because of its extensive application, but there is not any regulation among four nodes of an arbitrary quadrilateral, the explicit expressions of its element stiffness matrix which has been obtained is too complicated to be used efficiently. This paper gives explicit expressions of a variable thickness trapezoid element.

For a 4-node isoparametric element, a shape function is

$$N_i=(1+\xi_i\xi)(1+\eta_i\eta)/4 \quad (i=1,2,3,4) \qquad (1)$$

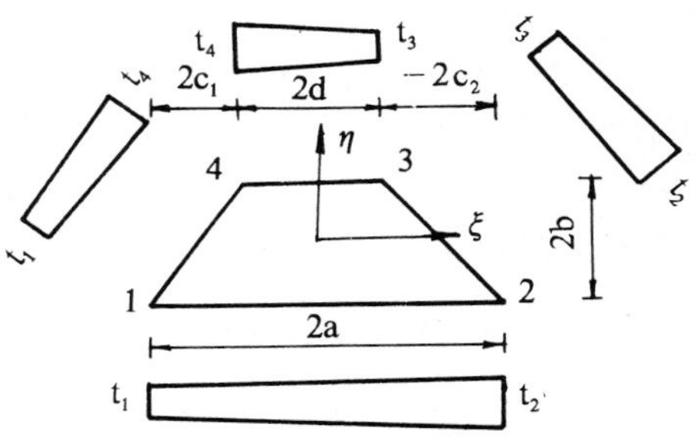

Fig 1 A variable thickness trapezoid element

For a trapezoid element shown in Fig 1, because its two sides are parallel to each other, the derivatives of global coordinates x, y with respect to local coordinates ξ, η can be written as

$$\begin{aligned} x_{,\xi} &= (a+d)/2+(d-a)\eta/2 \\ y_{,\xi} &= 0 \\ x_{,\eta} &= (c_1+c_2)/2+(c_2-c_1)\xi/2 \\ y_{,\eta} &= b \end{aligned} \qquad (2)$$

*The project supported by National Natural Science Foundation of China

in which

$$a=(x_2-x_1)/2 \qquad b=(y_3-y_2)/2$$
$$c_1=(x_4-x_1)/2 \qquad c_2=(x_3-x_2)/2 \qquad (3)$$
$$d=(x_3-x_4)/2$$

Hence, the Jacobian matrix becomes

$$J=\begin{bmatrix} x,_\xi & y,_\xi \\ x,_\eta & y,_\eta \end{bmatrix} = \begin{bmatrix} A_1+A_2\eta & 0 \\ A_3+A_2\xi & b \end{bmatrix} \qquad (4)$$

where

$$A_1=(a+d)/2 \qquad A_2=(d-a)/2 \qquad (5)$$
$$A_3=(c_1+c_2)/2$$

The determinant of the Jacobian matrix is

$$|J|=b\ (A_1+A_2\eta) \qquad (6)$$

and derivatives of shape function with respect to global coordinates are

$$N_{i,x}=N_{i,\xi}/(A_1+A_2\eta)$$
$$N_{i,y}=(A_1N_{i,\eta}+A_2N_{i,\eta}\cdot\eta-A_3N_{i,\xi}-A_2N_{i,\xi}\cdot\xi)/(bA_1+bA_2\eta) \qquad (7)$$

Now, Consider the element thickness to be a linear function of ξ, η. That is

$$t=\sum_{i=1}^{4}N_it_i=m+e\xi+f\eta+r\xi\eta \qquad (8)$$

In which, m, e, f, r can be expressed by t_1, t_2, t_3, t_4 (see Fig 1)

The element stiffness sub-matrix $[k_{ij}]^e$ is as follows

$$[k_{ij}]^e=\int\int_{-1}^{1}[B_i]^T[D][B_j]t|J|d\xi d\eta$$
$$=[k_{ij}^e]+[\Delta k_{ij}^\xi]+[\Delta k_{ij}^\eta]+[\Delta k_{ij}^{\xi\eta}] \qquad (9)$$
$$(i=1,2,3,4;\ j=1,2,3,4)$$

Substitution of Eq. (6) into Eq. (9) yields

$$[k_{ij}^0]=\frac{Em}{4(1-\mu^2)}\begin{bmatrix} KM_{11} & KM_{12} \\ KM_{21} & KM_{22} \end{bmatrix}$$

$$[\Delta k_{ij}^\xi]=\frac{Ee}{4(1-\mu^2)}\begin{bmatrix} KE_{11} & KE_{12} \\ KE_{21} & KE_{22} \end{bmatrix} \qquad (10)$$

in which

$$KM_{11}=\frac{b}{2}\xi_i\xi_jF+\frac{1-\mu}{2b}G_1$$
$$KM_{12}=\mu\xi_i\eta_j+\frac{1-\mu}{2}\xi_j\eta_i-G_2$$
$$KM_{21}=\mu\xi_j\eta_i+\frac{1-\mu}{2}\xi_i\eta_j-G_2$$
$$KM_{22}=\frac{1-\mu}{4}b\xi_i\xi_jF+\frac{1}{b}G_1$$
$$KE_{11}=\frac{1-\mu}{2}G_3$$
$$KE_{12}=\frac{\mu}{3}\xi_i\xi_j\eta_j+\frac{1-\mu}{6}\xi_i\xi_j\eta_i-G_4$$
$$KE_{21}=\frac{\mu}{3}\xi_i\xi_j\eta_i+\frac{1-\mu}{6}\xi_i\xi_j\eta_j-G_4$$
$$KE_{22}=G_3$$

and

$$G_1=\eta_i\eta_j\left(1+\frac{\xi_i\xi_j}{3}\right)A_1-\left[(\xi_i\eta_j+\xi_j\eta_i)A_3+\frac{\eta_i+\eta_j}{3}\xi_i\xi_jA_2\right]$$
$$+\frac{\xi_i\xi_j}{2}\left(A_3^2+\frac{1}{3}A_2^2\right)F$$

$$G_2=\frac{1+\mu}{4}\xi_i\xi_jA_3F$$

$$G_3=\frac{A_1}{3b}\eta_i\eta_j(\xi_i+\xi_j)-\frac{1}{3b}[(\xi_i\eta_j+\xi_j\eta_i)A_2+\xi_j\xi_i(\eta_i+\eta_j)A_3]$$
$$+\frac{\xi_i\xi_j}{3}A_3A_2F$$

$$G_4=\frac{1+\mu}{12}\xi_i\xi_jA_2F$$

$$F=\frac{1}{A_2}\left[1+\left(\frac{A_1}{A_2}\right)^2\eta_i\eta_j-\frac{A_1}{A_2}(\eta_i+\eta_j)\right]\ln\frac{d}{a}+\frac{2}{A_2}\left[(\eta_i+\eta_j)-\frac{A_1}{A_2}\eta_i\eta_j\right] \qquad (11)$$

Similar expressions can be obtained for $[\Delta k_{ij}^\eta]$ and $[\Delta k_{ij}^{\xi\eta}]$.

3. SENSITIVITY ANALYSIS

In structure optimization the behavioural constraints, such as displacement and stress constaints, can be calculated by solving nodal displacemens from the well-known stiffness equation, That is

$$\{u\}=[K]^{-1}\{P\} \qquad (12)$$

$$\{\sigma\}^e=[D][B]\{u\}^e \qquad (13)$$

Takin the derivatives of Eq. (12) and Eq. (13) with respect to any design variable a_i, we will get displacement and stress sensitivities as follows

$$\frac{\partial\{u\}}{\partial a_i}=[K]^{-1}\left[\frac{\partial\{p\}}{\partial a_i}-\frac{\partial[K]}{\partial a_i}\{u\}\right] \qquad (14)$$

$$\frac{\partial\{\sigma\}^e}{\partial a_i}=[D][B]\frac{\partial\{u\}^e}{\partial a_i}+[D]\frac{\partial[B]}{\partial a_i}\{u\}^e \qquad (15)$$

$$(i=1,2,\cdots n)$$

The key to $\frac{\partial\{u\}}{\partial a_i}$, $\frac{\partial\{\sigma\}^e}{\partial a_i}$ is the calculation of derivatives of the stiffness matrix and load vectors with respect to design variables. Because the stiffness matrix [K] and load vectors {P} are explicit functions of the nodal coordinates when the trapezoid elements are used, it's easy to obtain their derivatives with respect to the nodal coordinates. For an example, the derivatives of $[k_{ij}^0]$ with respect to the nodal coordinates x_k, y_k will be

$$\frac{\partial[k_{ij}^0]}{\partial y_k} = \frac{Em}{4(1-\mu^2)}\begin{bmatrix} \frac{\partial KM_{11}}{\partial y_k} & 0 \\ 0 & \frac{\partial KM_{22}}{\partial y_k} \end{bmatrix}$$

$$\frac{\partial[k_{ij}^0]}{\partial x_k} = \frac{Em}{4(1-\mu^2)}\begin{bmatrix} \frac{\partial KM_{11}}{\partial x_k} & -\frac{\partial G_2}{\partial x_k} \\ -\frac{\partial G_2}{\partial x_k} & \frac{\partial KM_{22}}{\partial x_k} \end{bmatrix} \qquad (16)$$

where

$$\frac{\partial KM_{11}}{\partial y_k} = \frac{1}{2}\xi_i\xi_j F\frac{\partial b}{\partial y_k} - \frac{1-\mu}{2b^2}G_1\frac{\partial b}{\partial y_k}$$

$$\frac{\partial KM_{22}}{\partial y_k} = \frac{1-\eta}{4}\xi_i\xi_j F\frac{\partial b}{\partial y_k} - \frac{1}{b^2}G_1\frac{\partial b}{\partial y_k}$$

$$\frac{\partial KM_{11}}{\partial x_k} = \frac{b}{2}\xi_i\xi_j\frac{\partial F}{\partial x_k} + \frac{1-\mu}{2b}\frac{\partial G_1}{\partial x_k}$$

$$\frac{\partial KM_{22}}{\partial x_k} = \frac{1-\mu}{4}b\xi_i\xi_j\frac{\partial F}{\partial x_k} + \frac{1}{b}\frac{\partial G_1}{\partial x_k}$$

It is easy to convert the derivatives of [K] and {P} with respect to the nodal coordinates into those with respect to the design variables by the mapping relations between the nodal coordinates and the design variables. Thus, the derivatives of the displacemant vector and the stress vector with respect to the design variables can also be obtained. Because of the explicit sensitivity analysis, the computational efficiency is high.

4. NUMERICAL EXAMPLES

4. 1. SENSITIVITY ANALYSIS

A concrete dam with broad joints is shown in Fig 2, nine design variables are adopted. The initial values of them are a=[0. 2,0. 2,0. 7,0. 4, 0. 7,0. 5,0. 6,0. 1,0. 6]^T The principal stress sensitivities are calculated with the analytical method and the finite difference method. Results are listed in Table 1. They are in close agreement. But it should be pointed out that the difference method uses the different step size for each design variable to obtain good results, and it is difficult to choice the rational step size for each design variable.

Stress sensitivities calculated with a series of different step sizes are listed in Table 2. The step size of the 8th design variable varies from 3‰ to 9 %. The results show that the signs of sensitivities corresponding to the step

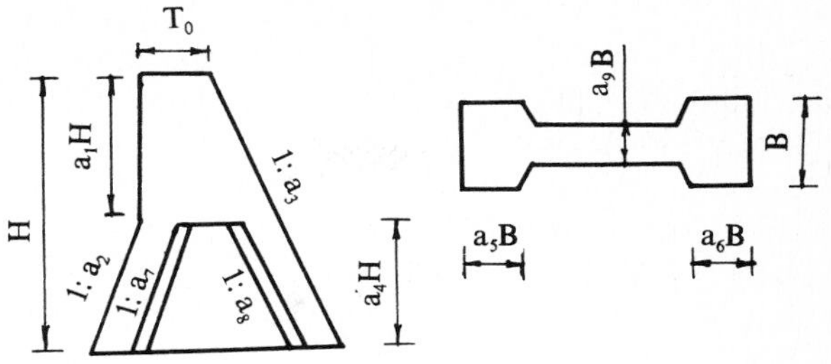

Fig 2 A concrete dam with broad joints

Table 1 comparsion between analytical method and difference method

	analytical method		difference method		
	heel	toe	heel	toe	step size
$\frac{\partial\sigma}{\partial a_1}$	37. 69	−330. 55	37. 54	−332. 57	2%
$\frac{\partial\sigma}{\partial a_2}$	−56. 97	95. 33	−57. 24	95. 22	7. 5‰
$\frac{\partial\sigma}{\partial a_3}$	−121. 73	494. 55	−122. 07	493. 03	1‰
$\frac{\partial\sigma}{\partial a_4}$	28. 18	111.84	27. 52	110. 92	7‰
$\frac{\partial\sigma}{\partial a_5}$	−38. 85	5. 78	−38. 31	5. 91	8. 6‰
$\frac{\partial\sigma}{\partial a_6}$	−9. 04	47. 08	−9. 04	46. 80	1. 9%
$\frac{\partial\sigma}{\partial a_7}$	−49. 75	48. 72	−50. 05	48. 41	8. 3‰
$\frac{\partial\sigma}{\partial a_8}$	−14. 59	2. 55	−14. 19	2. 60	22%
$\frac{\partial\sigma}{\partial a_9}$	−7. 79	−27. 76	−7. 80	−27. 51	1. 4%

size 3‰ and 9% are not the same. Therefore, it's necessary to use analytical sensitivities.

Table 2 sensitivities influenced by different step sizes

Step Size	9. 0%	6. 0%	3. 0%	9. 0‰	6. 0‰	3. 0‰
heel	−16. 1	−11. 8	−12. 7	−24. 4	8.1	2. 8
toe	2. 7	2. 6	2. 4	3. 2	0. 8	−1. 2

4. 2. OPTIMUN DESIGN

For the optimum design of the concrete dam with broad joints shown in Fig 2, six design variables are adopted, a=[a_1,a_2,. . . , a_6]^T, the volume of concrete is considered as the objective function. The constraints involve geometric constraints, stress constraints and stability constraints. The trapezoid elements combined with the triangle elements are used to form a finite element mesh in structural analysis, and the method of SLP is also adopted to optimize the shape of the dam. The initial design point is a=(0. 2,0. 16,0. 65,0. 63,0. 62, 0. 35)^T, after 13 iterations , the objective value decreases from 74903 m^3 to 67500 m^3. The principal tension stress achieves its limit. The optimal shape and its corresponding data are shown in Fig 3 and Table 3. Fig 4 illustrates the history of iterations.

Table 3 optimun design of the dam

No. of iteration		initial	1	5	10	13
design variables	a_1	0. 200	0. 240	0. 350	0. 358	0. 362
	a_2	0. 160	0. 130	0. 128	0. 100	0. 100
	a_3	0. 650	0. 630	0. 600	0. 600	0. 600
	a_4	0. 630	0. 670	0. 650	0. 642	0. 638
	a_5	0. 620	0. 590	0. 574	0. 534	0. 529
	a_6	0. 350	0. 325	0. 250	0. 250	0. 250
stress (MPa)	heel	0. 660	0. 755	0. 758	0. 798	0. 800
	toe	−2. 86	−3. 03	−3. 16	−3. 21	−3. 21
Objective Value (m^3)		74903	71679	68204	67546	67500

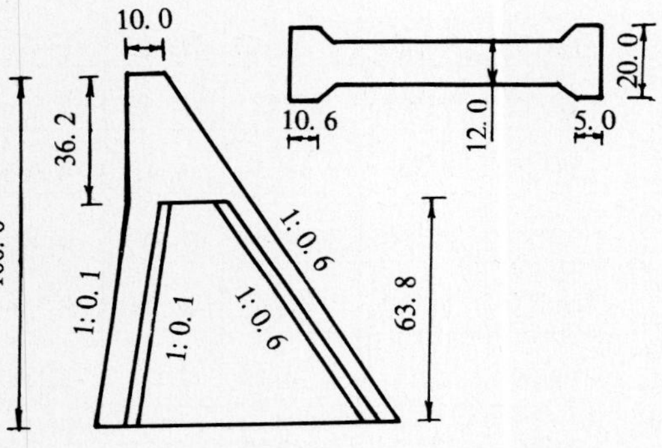

Fig 3 Optimal shape

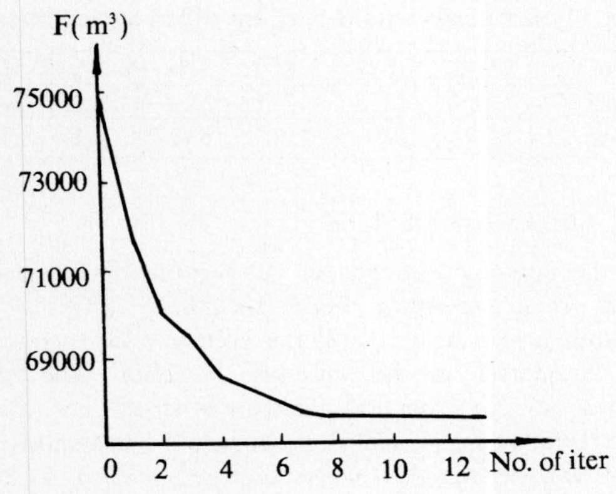

Fig 4 History of iterations

5. CONCLUSIONS

The trapezoid element with variable thickness is suitable to solve the structures with arbitrary linear boundaries and provides more accurate solution with a fairly coarse idealization. The analytical formulations for design sensitivities derived here are correct. In comparison with the difference method, the analytical method has high accuracy and efficiency.

REFERENCES

[1] Wang, S. Y. , Sun, Y. B. , and Gallagher, R. H. , Sensitivity analysis in shape optimization of continuum structures, Computers & Structures, 1985. 5, pp855 − 867

[2] Wang, S. Y. , and Zhang, K. F. , Explicit Analysis of Trapezoid Element, Journal of zhejiang university, 1989. 8 . (in chinese)

Approximation, Optimization and Computing:
Theory and Applications, A.G. Law and C.L. Wang (eds.)
Elsevier Science Publishers B.V. (North-Holland)

A DIRECT ALGORITHM FOR LINEAR COMPLEMENTARY PROBLEM OF QUADRATIC PROGRAMMING — Application to the Parametric Variational Principle, PVFEM

Zeng Pan*

Research Institute of Engineering Mechanics
Dalian University of Technology, Dalian, China

ABSTRACT

The parametric variational principle, PVP, is powerful to solve the non-linear unspecified boundary problems in continuum mechanics. Lemke method is usually used to solve the resulting linear complementary problem of Quadratic Programming. A direct algorithm is proposed here in the hope of higher efficiency in computational work.

Key words: parametric variational principle,
finite element method,
quadratic programming,
Lemke method.

1. PARAMETRIC VARIATIONAL PRINCIPLE——PVP

The PVP is a modern variational principle which is a development of optimum control theory to solve the non-linear unspecified boundary problems in continuum mechanics, such as problems of elasto-plasticity and problems of contact[1-7]. All these problems have to be solved step by step for incremental loading. The finite element formulation for each step of PVP for elastoplasticity problem is a quadratic programming as follows:

$$\text{Find } \delta, \quad \underset{(\delta)}{\text{Min}}[\frac{1}{2}\delta^T K\delta - \delta^T(\phi\lambda + q)] \tag{1a}$$

Subject to:

$$c\delta - M\lambda - d + \nu = 0 \tag{1b}$$

$$\nu^T\lambda = 0, \quad \nu \geq 0, \qquad \lambda \geq 0 \tag{1c}$$

in which:

$\delta \in R^n$: state variable incremental vector;
$\lambda \in R^m$: parametric or control incremental vector, which does not participate as arguments in the functional variation, but controls the process of variation in satisfying the constraints (1b) (1c);
$\nu \in R^m$: complementary vector with λ ;
$q \in R^n$: load incremental vector;
$M \in R^{1\times 1}$: hardening matrix;
$K \in R^{n\times n}$: elastic stiffness matrix;
$c \in R^{1\times n}$: yielding constraint matrix;
$\phi \in R^{n\times 1}$: plastic potential matrix; ($\phi = c^T$ for associated flow rule, $\phi \neq c^T$ for non-associated flow rule)
d: yielding tolerance vector.

* Post-Doctoral Fellow

The parametric quadratic programming (1) have been applied to solve a lot of problems[1-8] in using the well-known Lemke algorithms. This paper proposes a direct algorithm PVFEM, which is more economical and efficient in computing work as expected.

2. DIRECT ALGORITHM FOR PARAMETRIC QUADRATIC PROGRAMMING—PVFEM

The Kuhn— Tucker conditions convert the quadratic programming (1) to:

$$(M - cK^{-1}\phi)\lambda = (-d + cK^{-1}q) - \nu \tag{2}$$

$$\lambda^T\nu = 0, \quad \lambda \geq 0, \quad \nu \geq 0 \tag{3}$$

$$\delta = K^{-1}(\phi\lambda + q) \tag{4}$$

Let $A = M - cK^{-1}\phi$, and $B = cK^{-1} - d$, so that (2) and (3) become:

$$\begin{cases} \nu + A\lambda = B & (5) \\ \lambda^T\nu = 0, \quad \lambda \geq 0, \quad \nu \geq 0 & (6) \end{cases}$$

in which A is matrix (1 x 1); B, ν, λ are vectors (1 x 1).

The linear compementary problem (5)—(6) gives the solution of λ and ν which define the state of each element i:

elastic state when $\lambda_i = 0, \ \nu_i \geq 0$
plastic state when $\lambda_i > 0, \ \nu_i = 0$

The direct algorithm consists in separating the vector $\underset{(1\times 1)}{\lambda}$ into two sub-vectors: $\underset{(k\times 1)}{\lambda} > 0$ and $\underset{[(1-k)\times 1]}{\lambda} = 0$. This aim will be achieved if the matrix A could be transformed into:

$$TA\overline{T} = \begin{bmatrix} I & \vdots & D \\ \cdots & \vdots & \cdots \\ 0 & \vdots & 0 \end{bmatrix} \begin{matrix} (k) \\ \\ (l-k) \end{matrix} \qquad (7)$$

with columns headed (k) (l-k).

(k is the maximum rank of A)

in which T is the transformation matrix of elimination process:

$$T = \begin{bmatrix} T_1 & \vdots & T_2 \\ \cdots & \vdots & \cdots \\ T_3 & \vdots & T_4 \end{bmatrix} \begin{matrix} (k) \\ \\ (l-k) \end{matrix} \qquad (8)$$

with columns headed (k) (l-k).

and $\overline{T}$ is the transformation matrix to rearrange the order of columns i and j, in order to make the upper left submatrix of $TA\overline{T}$ being unit matrix $I_{(k\times k)}$.

$$\overline{T} = \begin{bmatrix} 1 & & i & & j & & \\ & \ddots 1 & \vdots & & \vdots & & \\ & & \ddots \vdots & & \vdots & & \\ \cdots & \cdots & 0 & \cdots & 1 & \cdots & \\ & & \vdots & 1 \ddots & \vdots & & \\ & & \vdots & 1 & \vdots & & \\ \cdots & \cdots & 1 & \cdots & 0 & \cdots & \\ & & \vdots & & \vdots 1 & \ddots & 1 \end{bmatrix} \begin{matrix} \\ \\ \\ i \\ \\ \\ j \\ \\ \end{matrix} \qquad (9)$$

D — matrix resulting from Jordan-Gauss elimination.

According to the transformation of A, Eq.(5) has to be transformed to

$$T\nu + TA\overline{T}(\overline{T}^{-1}\lambda) = TB \qquad (10)$$

Denoting $\lambda' = \overline{T}^{-1}\lambda$, (10) becomes

$$T\nu + TA\overline{T}\lambda' = TB \qquad (11)$$

or

$$\begin{bmatrix} T_1 & \vdots & T_2 \\ \cdots & \vdots & \cdots \\ T_3 & \vdots & T_4 \end{bmatrix} \begin{Bmatrix} \nu_1 \\ \cdots \\ \nu_2 \end{Bmatrix} + \begin{bmatrix} I & \vdots & D \\ \cdots & \vdots & \cdots \\ 0 & \vdots & 0 \end{bmatrix} \begin{Bmatrix} \lambda'_1 \\ \cdots \\ \lambda'_2 \end{Bmatrix} = \begin{Bmatrix} W_1 \\ \cdots \\ W_2 \end{Bmatrix} \qquad (12)$$

where

$$\begin{Bmatrix} \lambda_1 \\ \cdots \\ \lambda_2 \end{Bmatrix} = \begin{Bmatrix} \overline{T}^{-1}\lambda \\ \cdots\cdots \\ 0 \end{Bmatrix} \quad \text{with} \quad \lambda > 0, \qquad (13)$$

$$\begin{Bmatrix} W_1 \\ \cdots \\ W_2 \end{Bmatrix} = TB \qquad (14)$$

$$\begin{Bmatrix} \nu_1 \\ \cdots \\ \nu_2 \end{Bmatrix} = \nu, \quad \text{with } \nu_1 = 0, \ \nu_2 \geq 0, \qquad (15)$$

From (12), we have

$$\lambda'_1 = \overline{T}^{-1}\lambda_1 = W_1 - T_2 T_4^{-1} W_2 \qquad (16)$$

Finally, Eq.(4) gives δ, the solution of problem:

$$\delta = K^{-1}\left[\phi\overline{T}\begin{Bmatrix} W_1 - T_2 T_4^{-1} W_2 \\ \cdots\cdots\cdots \\ 0 \end{Bmatrix} + q\right] \qquad (17)$$

3. REALIZATION OF THE TRANSFORMATION OF A

Jordan-Gauss elimination is here used to get (16) from (5). The linear independent rows of matrix A is gradually determinated in the elimination process. $T_1 - T_4$ in matrix T are also included in process of elimination. In following step a–c, the choice of main element for elimination actually considers non-negativity and complementary condition (6). The main element, a_{ij}, for elimination is selected as follows:

a. choose the row i by

$$\max_i \sqrt{\sum_j a_{ij}^2}$$

b. choose the column j for elimination by the rule

$$\min_{1 \leq j \leq l} \left\{ \frac{b_{ij}}{a_{ij}}, \quad a_{ij} > 0 \right\}$$

c. The element a_{ij} is chosen as the main element to carry out the elimination of Jordan-Gauss.

Repeating step a—c, after treatment for k times, we can directly get the solution of λ_1.

EX.1. A linear complementary problem of quadratic programming is

$$\begin{cases} \nu + A\lambda = B \\ \lambda^T\nu \geq 0, \quad \lambda \geq 0, \quad \nu \geq 0 \end{cases}$$

in which

$$A = \begin{bmatrix} -2 & -1 & 0 & -1 \\ 0 & -2 & -1 & 0 \\ -2 & -3 & -1 & -1 \\ -2 & -5 & -2 & -1 \end{bmatrix}, \quad B = \begin{Bmatrix} -3 \\ -2 \\ -4 \\ -5 \end{Bmatrix}$$

If Lemke method is used, artificial variable Z_o has to be added. The complementary variables, ν, are firstly considered as basis variables. The exchange of basis variables between ν, Z_o and λ gradually make $Z_o = 0$, and force $\lambda_i (>0)$ entering in the basis. The initial tableau for exchanges of Lemke method is as follow:

	$\nu_1\nu_2 \cdots\cdots \nu_l$	$\lambda_1\ \lambda_2 \cdots\cdots \lambda_l$	Z_0	Right item
ν_1	1	$a_{11}\ a_{12} \cdots\cdots a_{1l}$	-1	b_1
ν_2	1	$a_{21}\ a_{22} \cdots\cdots a_{2l}$	-1	b_2
$\vdots$	$\ddots$ 0	$\cdots\cdots\cdots$	$\vdots$	$\vdots$
	0 $\ddots$	$\cdots\cdots\cdots$		
ν_l	1	$a_{l1}\ a_{l2} \cdots\cdots a_{ll}$	-1	b_l
	↑ complementary variables	↑ parametric variables	↑ artificial variables	

In the direct algorithm proposed here, no artificial-variable, nor exchanges of the basis are needed. The algorithm consists simply in Jordan-Gauss elimination process to transform the matrix A. The initial bableau of elimination is as follow:

	$\lambda_1\ \lambda_2 \cdots\cdots\cdots\cdots \lambda_l$	Right item
λ_1	$a_{11}\ a_{12} \cdots\cdots\cdots\cdots a_{1l}$	b_1
λ_2	$a_{21}\ a_{22} \cdots\cdots\cdots\cdots a_{2l}$	b_2
$\vdots$	$\cdots\cdots\cdots\cdots\cdots\cdots$	$\vdots$
	$\cdots\cdots\cdots\cdots\cdots\cdots$	
λ_l	$a_{l1}\ a_{l2} \cdots\cdots\cdots\cdots a_{ll}$	b_l

For the example (1), the elimination process is as follow

λ_1	λ_2	λ_3	λ_4	b
[-2]	-1	0	-1	-3
0	-2	-1	0	-2
-2	-3	-1	-1	-4
-2	-5	-2	-1	-5
1	$\frac{1}{2}$	0	$\frac{1}{2}$	$\frac{3}{2}$
0	[2]	1	0	2
0	2	1	0	1
0	4	2	0	2
1	0	$\frac{1}{4}$	$\frac{1}{2}$	1
0	1	$\frac{1}{2}$	0	1
0	0	0	0	1
0	0	0	Q	2

The result is

$$\lambda_{1-4} = 1,\ 1,\ 0,\ 0,$$

$$\nu_{1-4} = 0,\ 0,\ 1,\ 2,$$

Ex.2. $\min_{x} f(x) = \frac{1}{2} X^T A X + BX$

s.t. $X \geq 0$

where, the number of variables is n=50,

$$A_{(50,50)} = \begin{bmatrix} 1100\ldots0 \\ 0110\ldots0 \\ 0011 \\ \vdots\vdots\vdots\vdots\ \ddots \\ \vdots\vdots\vdots\ \ \ddots\ 1 \\ 000\ldots\vdots 1 \end{bmatrix}, \qquad B_{(50\times1)} = \begin{Bmatrix} -2 \\ -2 \\ \vdots \\ \vdots \\ -2 \\ -1 \end{Bmatrix}$$

In this example, Lemke method requires more than 4 min. on the micro-computer IBM PC while the direct algorithm requires a little less than 3 min.

REFERENCES

[1] Zhong, W.X., On Variational Principle of Elastic Contact Problems and Parametric Quadratic Programming Solution, Comp. Struct. Mech. Appl.2. 1985 (in Chinese)

[2] Zhong, W.X., On Parametric Complementary Energy Variational Principle in Soil Mechanics, Acta Mechanica Sinica, 18, 1986, (in Chinese)

[3] Zhong, W.X., Zhang, R.L., Parametric Variational Principle and Their Quadratic Programming Solution in Plasticity, Proceedings ICCEM'87, Beijing, 1987, Computer & Structures, Vol.50, 1988

[4] Zhang, R.L., Zhong, W.X., The Numerical Solution of PMPEP by Parametric Quadratic Programming, Comp. Strut. Mech. Appl.,4, 1987 (in Chinese)

[5] Zhang, R.L., A New Quadratic Programming Scheme for Computation, Comput. Struct. Mech. Appl.4, 1987 (in Chinese)

[6] Zhong, W.X., Zhong, R.L., Quadratic Programming with Parametric Vector in Plasticity and Geomechanics, NUMETA, UK, 1987

[7] Zhong, W.X., Sun, S.M., A Finite Element Method for Elasto-plastic Structures and Contact Problems by Parametric Quadratic Programming, Int.J. Numerical Method in Engineering, Vol.26, 1988

[8] Boot, J.C., Quadratic Programming, 1964

IV
APPLICATIONS

Approximation, Optimization and Computing:
Theory and Applications, A.G. Law and C.L. Wang (eds.)
Elsevier Science Publishers B.V. (North-Holland)

INTEGRATIVE OPTIMIZATION FOR ARCHITECTURAL STRUCTURE DESIGN AND ITS PRACTICAL EXAMPLES

Hou Chang

Nanjing Architectural and Civil Engineering Institute, Nanjing, P.R.China

As far as structure integrity option design is concerned, it still depends upon experience nowdays, so one – sided view is unavidable. In order to make design options more reasonable, this paper applies structure optimum analysis and present value analysis method and tries to search optimum overall results so as to implement combination of technology, economic and decision. It has certain significances to avoid or reduce malfuntions of design , or to redo the design.

1. Introduction

One of the features in the development of modern science and technology is to highlight the methodology, from the viewpoint of designing a building (or structure). It is necessary to study and analyze it based on " Integral Engineering", so as to get optimum overall results. What so – called " Integral Engineering" from the viewpoint of structure selection, it means a building is an overall system, from the top roof down to each part of foundation there is an organic connection, so the designers must take into account the relationship between structure selection and internal forces, to coordinate it and to utilize the material property resonably.

With respect to structure type, it not only should consider the initial investment of structure, but also the maintenance cost of it, that is to say, to establish the relationship between initial investment and periodical maintenance cost, in order to make out the economical evalution of the said option.

2. Method of Integral Optimum Design

To decompose the integral optimum problems into several smaller subproblems, such as roof truss, beam, column and foundation etc. The calculating process can be divided into two iterative procedure, the first procedure is to optimize each subproblem solely, it implies the designer can adopt different optimum method based on the features of each subproblem for various kind of structure; the second procedure is to adjust the solution of first procedure for the subproblems. The iterative procedure can be ended through several cycles. Such method is called structure optimum decomposed method.

In case there are different design options to be selected, the designer has to pay attention to the existence of maintenance fee and the difference of expected value of loss caused by earthquake. If there is, it is necessary to carry out present value analysis based on time difference of capital input. The simplest way is to transform the constant investment present value, and to take the minimum present value of overall expenses at the standard of selection.

3. A Practical Question

Large-scale product warehouse (span 60M, length 135M) is needed more and more recently for fertilizer plant, a new structure-" Λ " type steel truss with R. C. "Γ " type columns has been proposed, this new structure is shown in the following sketch, and it is intended to be a typical design of this kind of building.

An optimum design on the cross-section of member of this large span steel struss has been studied[1]. Practically, it is very difficult to " typify " it, because it can't be sure that this option is the better one from overall viewpoints.

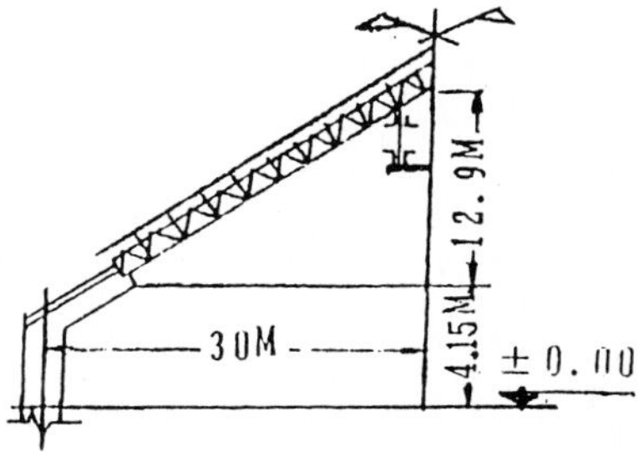

Figure 1

1). From the viewpoint of structure selection, this building consists of roof- struss, columns and foundations etc, they are connected and restricted each other and forming an integrity. For example, the above mentioned option has resulted in oversize foundation and column stability owing to relative big horizontal thrust an the baseplate of roof-struss, and there is also some argument referring to the span, elevation even truss figure (straight line or arch form).

2). From the viewpoint of structure type, if the designer defines the minimum investment as target function. In general, the structure optimization will only consider the initial investment. This warehouse is for storing urea, and the urea has certain corrosive characteristics, in case the designer selected steel structure, then it is necessary to pay additional annual maintenance fee; on the other hand, if the designer selected Reinforced Concrete structure mixed with compactive additives, then it is not necessary to maintain it.

Therefore, with respect to the design option of a structure, in order to get a satisfactory result, the designer should combine structural mechanics, engineering and economics etc various factors altogether. It will not be convincible if the design option is decided by experience.

Therefore, we proposed the following two options by means of integral optimization method for this urea warehouse, and it has greatly changed the original design option.

4. Taking Roof-Truss, Column and Foundation As a Whole

Following the above structure optimum decomposed method, to carry out optimization on three interrelated subproblems separately, i.e. steel roof plate beam, reinforced concrete column and foundation. For example, to use sequence linear programming (SLP) method for steel roof plate beam, and then combine all the three subproblems together, you will find out that "Γ" type column is not so reasonable and try to change the column type. By means of two stage repeated iterative calculating treatment, you will get a new outline of column with variable cross-section and slope inward, and at the same time, to raise the elevation of arch footing in order to reduce the span. According to the principle of minimizing the vertical force as well as moment acting on the foundation, to calculate the antiskidding self- supporting foundation, the final solution is shown on Diagram 2.

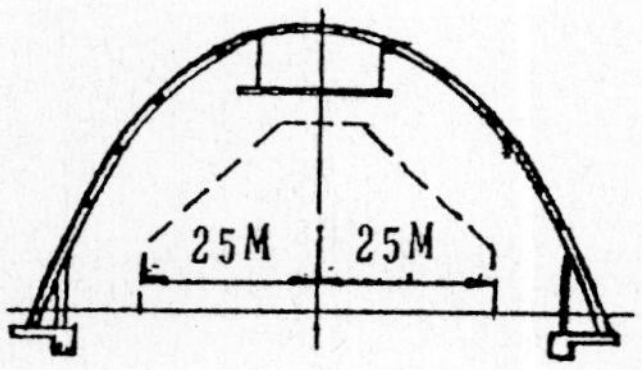

Figure 2

1). To change the "Λ" frame into section degree parabolic arch: it reflects the relation between structure type and internal forces. The main load of thia building is the self weight of structure and overhead conveyer platform acted as concentrated load. Through calculation and comparison, it is known, the relation between max. Moment occured in three hinges arch frame (or "Λ" type structure) and the relevant distributed load q is as follows:

$$\text{For arch frame}: \quad Mmax = 12.0q \tag{1}$$

$$\text{For "}\Lambda\text{" frame}: \quad Mmax = 86.0q \tag{2}$$

2). To raise the arch footing elevation so as to reduce the span.

It reflects the relation between span and internal force. The most disadvantageous combination of moment M_D has certain relation with span l. It can be expressed by a curve line getting through the origin point. While the span is within $50 - 60M$, it can be expressed by linear equation as follows:

$$M_D \approx 2.6l - 104 \tag{3}$$

To substitute different span l, it can show that the reduction of span can greatly reduce the moment of the structure and correspondent horizontal thrust so as to reduce the steel consumption. Yet the arch footing elevation has to be raised in order to meet the requirement of process (storage volume, the outline of portal scrapper frame), to remain the original storage volume. Special attention should be paid to the height of the column for not causing any construction difficulties.

3). To change "Γ" type column into smaller top and bigger bottom shape with slope inward. From Energy Criterion Method, it is easy to derive the optimum energy criterion of axial compressive (tensile) member, i.e. the following term must be satisfied in order to minimize the weight of axial force members.

$$D'^2 = C\ (constant) \tag{4}$$

In the above formula, D is the real displacement of axial force member , D' is the derivative of D with respect to X (length in axial force direction).

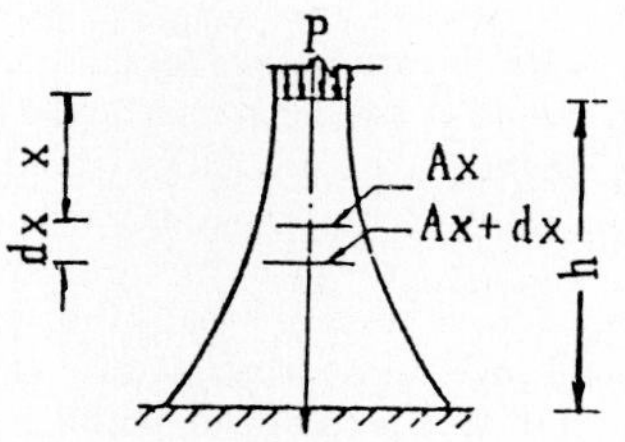

Figure 3

Relating to the column under the action of self weight and external load, to use this criterion, we can get an equal strength column shape (figure 3), to divide the column into numerous defferential sections, we can get the relationship between the displacement and the force on differential section dx according to Hook's Law:

$$dD/dx = Ax \cdot rdx/(EdAx) = C\ (constant) \tag{5}$$

In the formula:

r – density of material

others referring to figure 3.

Changing formula (5) into following expression:

$$dAx/Ax = r/EC \cdot dx \tag{6}$$

To integrate both side of formula (6)

$$\int 1/Ax \cdot dAx = \int r/EC \cdot dx \tag{7}$$

We can get the rule of change on the cross-section Ax with respect to the height of column.

$$lnAx = r/EC \cdot X + C_1$$

or

$$Ax = exp(r/EC \cdot X + C_1) = exp(r/EC \cdot X) \cdot e^{C_1}$$

$$= C_0 exp(r/EC \cdot X) \quad (8)$$

The integrate constant can be fixed by following term, i.e. when $X = 0$

$$A_0 = P/EC \quad (9)$$

then, we can get:

$$C_0 = A_0 = P/EC \quad (10)$$

and finally:

$$Ax = A_0 \cdot exp(r/EC \cdot X)$$

$$= P/EC \cdot exp(r/EC \cdot X) \quad (11)$$

Following formula (11), we can get the equal strength column shape shown in figure 3.

This conclusion can be extended to this design option easily, for such smaller top and bigger bottom shape, it is obviously we can take advantagers of the lower portion gravity for antiskid. (to resist the horizontal displacement of arch footing).

In order to offset major portion of Mv (caused by Vertical force) by M_H (caused by horizontal thrust), we can change the column into slope inward shape , and use 0.618 method (one dimensional search) to optimize the column slope. Owing to major portion of Mv and M_H have been offset each other so the antiskidding gravity foundation will be reduced correspondingly. Through calculation and comparison, such kind of slope column, its moment is only 343 $KN-M$, about 1/5 moment of Vertical column.

5. Taking Roor Covering, Roof-Truss, Column and Foundation As a Whole and Considering The Relation Between Constant Maintenance and Initial Investment

From the review of more than 10 existing similar warehouse, there are two problems: The first problem is there is no ideal roof covering material. Furthermore, the steel structure is disadvantageous for meeting the requirement of anti-corrosion, therefore it has to spend money for anti-corrosion maintenance annually.

In order to solve the above two problems, we are intended to change the combined structure – steel roof truss plus roof slab, into reinforced concrete net-plate structure still by using structure optimum analysis method based on the previous structure option. The sketch of R. C. net-plate structure is shown in figure 4.

Obviously, R. C. net-plate structure integrates the bearing structure with roof covering structure as single unity and serves dual functions, and also easy to optimization. The reduction of roof system weight will further reduce the infrastructure size so as to obtain overall maximum economic results.

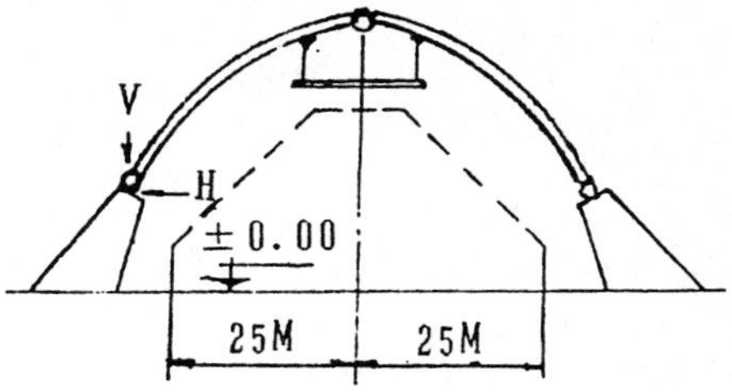

Figure 4.

In case the steel structure is adopted, then it is necessary to take into account constant investment, that is the maintenance cost per year, or we name it as annual cost expressed by Py. Py will be happened in the course of n years (the service period of the building), suppose the lifetime of this building is n years, then, after 1, 2,..., n years, Py will be turned into present value in the following sequence:

$$Py(1+i)^{-1}, Py(1+i)^{-2}, \ldots, Py(1+i)^{-n}$$

Thus, within n years with annual payment Py, its summation of present value is:

$$P_1 = Py(1+i)^{-1} + Py(1+i)^{-2} + \cdots + Py(1+i)^{-n} \quad (12)$$

According to its regularity, we can use the summation formula of equal proportion numeric sequence to obtain the brief relation between P_1 and Py by means of equal proportion series formula:

$$1 + a + a^2 + \cdots + a^n = (1 - a^{n+1})/(1-a)$$

then, formula (14) can be changed into

$$P_1 = Py(1+i)^{-1}[1 + (1+i)^{-1} + (1+i)^{-2} + \cdots + (1+i)^{1-n}]$$

$$= Py(1+i)^{-1}[1 - (1+i)^{-n}]/[1 - (1+i)^{-1}]$$

$$= Py[(1+i)^n - 1]/i(1+i)^n \quad (13)$$

But the constant investment does not keep unchanged, usually it will be increased year by year. Suppose the increasing difference per year (or once time) is G, and named as gradient, then the first year cost is Py, and the second year will be $(Py + G)$ and the third year is $(Py + 2G)$ etc. If the lifetime of building is set as n years (e.g. $n = 50$), then formula (13) is the present value formula for the cost after ending of n years without considering the gradient condition.

Taking gradient G into consideration, it is easy to derive the present value formula for equal differential gradient in terms of considering gradient only:

$$P_2 = G/i \cdot [(1+i)^n - 1]/i - n][1/(1+i)^n] \quad (14)$$

Obviously, to turn constant investment into initial investment only, then the cost should be paid at present is:

$$P = P_1 + P_2 = Py[((1+i)^n - 1)/(i(1+i)^n)]$$

$$+ G[((1+i)^n - 1)/i - n] \cdot [1/(i(1+i)^n)] \quad (15)$$

Based on formula (15), we can make quantitative comparison between these two investment options, and incorporate it into overall option evaluation and make the final economic calculation more reasonable.

Reference Literature

[1]. Hou Chang, Optimum design of large span steel truss, ≪ Optimization of Capital Construction ≫, No. 1, 1983.
[2]. Hou Chang, The selection of optimum investment option,≪ Chinese Journal of Operations Research ≫, Vol. 6, No. 2, Dec., 1987.
[3]. The summary of Urea Warehouse Structure Design by Hubei Chemical Design Institute, Dec., 1976.
[4]. A. D. Russell, K. T. Choudhary: Cost optimization of buildings, J. Struct. Div ASCE, V. 106, N. ST1, Jan., 1980.
[5]. K. A. Andam, J. Knapton: Optimum cost design of precast concrete framed structures, Engineering optimization V.5, N.1, 1980.

Approximation, Optimization and Computing:
Theory and Applications, A.G. Law and C.L. Wang (eds.)
Elsevier Science Publishers B.V. (North-Holland)

GRONWALL-BIHARI TYPE INEQUALITIES FOR DISTRIBUTIONS

James Conlan and Chung-lie Wang
University of Regina, Regina, Canada

1. INTRODUCTION

The extension of Gronwall-Bellman and Bihari ($G-B, B$) type inequalities to distributions is motivated by the following considerations. First, it unifies the theory of continuous and discrete $G-B, B$ inequalities; and second, it plays the same role in studying stability and boundedness of impulsive differential equations as their classical analogues play in the theory of classical differential equations [4], [8].

In sections 2 and 3 we will give a detailed treatment of the one dimensional case; and in section 4 we will briefly indicate how the results extend to $\mathbf{R^n}$. Finally in section 5 we consider more complicated cases, whose classical analogues have been the subject of much recent study [1-3], [5-7], [9,10].

2. SOME PRELIMINARY REMARKS

We will consider an interval $I(x) = (0,x), x > 0$. For $0 = x_0 < x_1 < \cdots < x_k < x$ we let $I_j = (x_j < s \leq x_{j+1})$ for $j < k, I_k = (x_k < s < x)$. Thus $I(x) = \cup_{0 \leq j \leq k} I_j$.

Definitions:

i $f(x) : R^1 \to R^1$ is of class A if f is continuous and non-negative for $x \geq 0$.

ii $w(t) : R^1 \to R^1$ is of class B if w is continuous and non-decreasing for $t \geq 0$, and positive for $t > 0$.

iii For $h : R^1 \to R^1$ and $x \geq 0$, we let $h^*(x) = \sup\{h(t) : 0 \leq t \leq x\}$.

iv For $f(x) : R^1 \to R^1$ and letting P be the partition $0 = x_0 < \cdots < x_k < x$, we say f is of class $J(P)$ if f is non-negative and continuous on each interval (x_j, x_{j+1}) or (x_j, x), and if $\lim_{h\to 0^+} f(x_{j-h}) = f(x_j)$ for $j = 1, \cdots, k$.

The following well known theorems are stated for future reference.

Theorem 1 Assume u is of class $J(P)$ and f is of class A, and a, b are two non-negative constants. If u satisfies

$$u(x) \leq a + b\int_0^x f(s)u(s)ds \quad \text{for} \quad x \geq 0$$

then

$$u(x) \leq a exp\left\{b\int_0^x f(s)ds\right\}$$

Theorem 2: Assume u is of class $J(P)$, and f, a, b are of class A. If for $x \geq 0$, u satisfies

$$u(x) \leq a(x) + b(x)\int_0^x f(s)u(s)ds,$$

then

$$u(x) \leq a(x) + b(x)\int_0^x a(s)f(s)exp\left\{\int_s^x b(t)f(t)dt\right\}ds.$$

Theorem 3: Assume u is of class $J(P)$ and f, a, b are of class A, and w is of class B. If, for $x \geq 0$, u satisfies

$$u(x) \leq a(x) + b(x)\int_0^x f(s)w(u(s))ds,$$

then

$$u(x) \leq G^{-1}\left\{G(a^*(x)) + b^*(x)\int_0^x f(s)ds\right\}$$

where

$$G(t) = \int_{t_0}^t \frac{ds}{w(s)} \quad ,0 < t_0 \leq t.$$

Note: The above theorems are usually stated under the additional assumption that u is continuous. However the usual proofs go through at any point of continuity. If x_j is a point of discontinuity, the conclusions follow from $u(x_j) = \lim_{h\to 0^+} u(x_j - h)$.

We let δ be the Dirac delta function, defined by

$$\int_{\mathbf{R^1}} \delta(x)f(x)dx = f(0) \quad , \int_{(a,b]} \delta(x-b)f(x)dx = f(b)$$

for any continuous f.

We let $\{g_n(x)\}$ be a "δ approximating sequence" defined by: g_n is continuous, and $g_n(x) = 0$ for $|x| > 1/n$,

$$\int_{(-\frac{1}{n},0)} g_n(x)dx = 1 - \frac{1}{n}, \int_{(0,\frac{1}{n})} g_n(x)dx = \frac{1}{n}.$$

Lemma 1: If $f \varepsilon J(P)$, then for any $\varepsilon > 0$, there is an $N > 0$ s.t. for any $n \geq N$,

$$\left| \int_{\mathbf{R}^1} g_n(x-a)f(x)dx - \int_{\mathbf{R}^1} \delta(x-a)f(x)dx \right| \leq \varepsilon \quad (1)$$

Proof: Consider the case where $a = x_j$ (if $a \neq x_j, j = 1, \cdots, k$, the result is trivial). By the mean value theorem

$$\int_{\mathbf{R}^1} g_n(x-x_j)f(x)dx = \left\{ \int_{(a-\frac{1}{n},a)} + \int_{(a,a+\frac{1}{n})} \right\}$$
$$g_n(x-x_j)f(x)dx = (1-\frac{1}{n})f(\xi_1) + \frac{1}{n}f(\xi_2),$$

where

$$\xi_1 \varepsilon (a - \frac{1}{n}, a), \quad \xi_2 \varepsilon (a, a+\frac{1}{n}).$$

For any $\varepsilon > 0$, there is an $N > 0$ such that for $n \geq N$, $|(1-\frac{1}{n})f(\xi_1) + \frac{1}{n}f(\xi_2) - f(a)| < \varepsilon$ (by left continuity of f). This is equivalent to (1) □.

3. G-B, B INEQUALITIES IN $\mathbf{R}^1$ Theorem 4:

Assume that a, b are non-negative constants, that f and ϕ are of class A, and u is of class $J(P)$. Then if

$$u(x) \leq a + b\int_0^x \{f(s) + \sum_{0<x_j\leq x} \delta(s-x_j)\phi(s)\}u(s)ds, \quad (2)$$

we have

$$u(x) \leq a\exp\{\int_0^x b[f(s) + \sum_{0<x_j\leq x} \delta(s-x_j)\phi(s)]ds\}. \quad (3)$$

Proof: By (2), lemma 1, and theorem 1, for any $\varepsilon \geq 0$, and all n greater than some N_0,

$$u(x) \leq (a+\varepsilon) + b\int_0^x \{f(s) + \sum_{0<x_j\leq x} g_n(s-x_j)\phi(s)\}ds$$
$$\leq (a+\varepsilon)\exp\{b\int_0^x [f(s) + \sum_{0<x_j\leq x} g_n(s-x_j)\phi(s)\}ds\}$$
$$\leq (a+2\varepsilon)\exp\{b\int_0^x [f(s) + \sum_{0<x_j\leq x} \delta(s-x_j)\phi(s)\}ds\}$$

from which (3) follows. □

Corollary 1: If a, b, f, ϕ and u are as in theorem 4, and if

$$u(x) \leq a + b\int_0^x f(s)u(s)ds + b\sum_{0<x_j\leq x} \phi(x_j)u(x_j),$$

then

$$u(x) \leq a\exp\{b\int_0^x f(s)ds\} \prod_{0<x_j\leq x} \exp\{b\phi(x_j)\}.$$

Theorem 5: Assume u is of class $J(P)$, and a, b, ϕ, ψ are of class A. If u satisfies

$$u(x) \leq a(x) + b(x)\int_0^x \{\psi(s) + \phi(s)\sum_{0<x_j\leq x} \delta(s-x_j)\}u(s)ds,$$

then

$$\begin{aligned} u(x) \quad &\leq \quad a(x) + b(x)\int_0^x a(s)[\psi(s) \\ &+ \quad \phi(s)\sum_{0<x_j\leq x} \delta(s-x_j)] \\ &\exp\{\int_s^x b(t)[\psi(t) \\ &+ \quad \phi(t)\sum_{s<x_m\leq x} \delta(t-x_m)]dt\}ds \end{aligned}$$

Proof: Using theorem 2, and proceeding as in the proof of theorem 4,

$$\begin{aligned} u(x) \quad &\leq \quad (a+\varepsilon) + b\int_0^x \{\psi + \phi\sum_{0<x_j\leq x} g_n(s-x_j)\}u(s)ds \\ &\leq \quad (a+\varepsilon) + b\int_0^x (a+\varepsilon)\{\psi + \phi\sum_{0<x_j\leq x} g_n(s-x_j)\} \\ &\exp \quad \{\int_s^x b[\psi + \phi\sum_{s<x_m\leq x} g_n(t-x_m)]dt\}ds \\ &\leq \quad (a+2\varepsilon) + b\int_0^x (a+\varepsilon)[\psi + \phi\sum_{0<x_j\leq x} \delta(s-x_j)] \\ &\exp \quad \{\int_s^x b[\psi + \phi\sum_{s<x_m\leq x} g_n(t-x_m)]dt\}ds \end{aligned}$$

Consider

$$I_1 = \int_0^x (a+\varepsilon)\psi\exp\{\int_s^x b[\phi\sum_{s<x_m\leq x} g_n(t-x_m)]dt\}ds$$

For $s < x_m \leq x$, and some $\xi \varepsilon (x_m - \frac{1}{n}, x_m + \frac{1}{n})$

$$\begin{aligned} \int_s^x b\phi g_n(t-x_m)dt \quad &= \quad b(\xi)\phi(\xi)\int_s^x g_n(t-x_m)dt \\ &\leq \quad b(\xi)\phi(\xi) \leq b(x_m)\phi(x_m) + \varepsilon_m \\ &= \quad \int_s^x b(t)\phi(t)\delta(t-x_m)dt + \varepsilon_m. \end{aligned}$$

Hence

$$I_1 \le \int_0^x (a+\varepsilon)\psi \exp\{\int_s^x b[\phi \sum_{s<x_j\le x} \delta(t-x_j)]dt + \sum_m \varepsilon_m\}ds$$

Since ε and $\sum \varepsilon_m \to 0$ as $n \to \infty$,

$$I_1 \le \int_0^x a\psi \exp\{\int_s^x b[\phi \sum_{s<x_j\le x} \delta(t-x_j)]dt\}ds.$$

Moreover,

$$\int_0^x (a+\varepsilon)\phi \sum_{0<x_j\le x}$$

$$\delta(s-x_j)\exp\{\int_s^x b[\psi + \phi \sum_{s<x_m\le x} g_n(t-x_m)]dt\}ds$$

$$\le \int_0^x (a+\varepsilon)\phi \sum_{0<x_j\le x}$$

$$\delta(s-x_j)\exp\{\int_s^x b[\psi + \phi \sum_{s<x_m\le x} \delta(t-x_m)]dt + \sum \varepsilon_m\}ds.$$

and, since $\varepsilon, \sum \varepsilon_m \to 0$ as $n \to \infty$, the conclusion follows. □

An easy consequence of theorem 5 is

Corollary 2: Under the hypotheses of theorem 5,

$$\begin{aligned} u(x) \;\le\; & a(x) + b(x)\int_0^x a(s)\psi(s)\exp[\int_s^x b(t)\psi(t)dt] \\ & \exp\{\sum_{s<x_r<x} b(x_r)\phi(x_r)\}ds \\ +\; & b(x)\sum_{0<x_j<x} a(x_j)\phi(x_j)\exp[\int_{x_j}^x b(t)\psi(t)dt] \\ & \exp[\sum_{x_j<x_r\le x} b(x_r)\phi(x_r)]. \end{aligned}$$

Theorem 6: Assume u is of class $J(P)$ and a, b, ψ, f, ϕ are of class A, and w, h are of class B. If u satisfies

$$\begin{aligned} u(x) \le a(x) \;+\; & b(x)\int_0^x \psi(s)w(u(s))ds \qquad (4) \\ +\; & f(x)\sum_{0<j\le k} \phi(x_j)h(u(x_j)), \end{aligned}$$

then

$$\begin{aligned} u(x) \;\le\; & G^{-1}\{G[a^*(x)] + b^*(x)\int_0^x \psi(s)ds \qquad (5) \\ +\; & f^*(x)\sum_{0<j\le k} \phi(x_j)h(u(x_j))\} \end{aligned}$$

where

$$G(t) = \int_{t_0}^t [w(s)+h(s)]^{-1}ds, \quad 0 < t_0 \le t.$$

Proof: Proceeding as in the proof of theorem 4,

$$\begin{aligned} u(x) \;\le\; & a(x) + \int_0^x [b(x)\psi(s)w(u(s)) \\ +\; & f(x)\sum_{0<j\le k} \phi(s)\delta(s-x_j)h(u(s))]ds \end{aligned}$$

$$\begin{aligned} \le\; & (a+\varepsilon) + \int_0^x \{b^*(x)\psi(s) \\ +\; & f^*(x)\sum_{0<j\le k} g_n(s-x_j)\phi(s)\} \\ \{\; & w(u(s)) + h(u(s))\}ds, \end{aligned}$$

where $\varepsilon \to 0$ as $n \to \infty$. Since $w+h$ is of class B, theorem 3 gives

$$\begin{aligned} u(x) \;\le\; & G^{-1}\{G[a^*(x)+\varepsilon] + \int_0^x [b^*(x)\psi(s) \\ +\; & f^*(x)\sum_{0<r\le x} \psi(s)g_n(s-x_r)]ds\} \end{aligned}$$

Letting $n \to \infty$, we get (4).

4. EXTENSION TO $\mathbf{R^n}$

In $\mathbf{R^2}$ let $S(X) = S(X_1, X_2) = \{(x_1, x_2) : 0 < x_i < X_i; i = 1, 2\}$. Let $S(X)$ be sub-divided into rectangles by the lines $x_1 = t_{1,j}, x_2 = t_{2,k} (j = 1, \cdots, m; k = 1, \cdots, n)$ and let $S_{j,k}$ be the rectangle

$$S_{j,k} = \{(x_1, x_2); t_{1,j} < x_1 \le$$

$$t_{1,j+1}, t_{2,k} < x_2 \le t_{2,k+1}\},$$

$$S_{m,k} = \{(x_1, x_2) : t_{1,m} < x_1 < X_1,$$

$$t_{2,k} < x_2 \le t_{2,k+1}\}, \quad \text{etc.}$$

(we assume $t_{1,0} = t_{2,0} = 0$). We let $J(m,n)$ be partition of S into the above-mentioned sub-rectangles.

We modify definitions $i - iv$ as follows: In (i) and (iii) we let f and h be $\mathbf{R^n} \to \mathbf{R^1}$. Def. (iv) becomes Def $(iv*)$: For $f : \mathbf{R^n} \to \mathbf{R^1}$ we say f is of class $J(m,n)$ if f is non-negative and continuous at each interior point of each $S_{j,k}$, and is left continuous at boundry points of $S_{j,k}$ (i.e., at $x_1 = t_{1,j+1}$ or $x_2 = t_{2,k+1}$).

This extends in an obvious way to $\mathbf{R^n}$.

We will use the following notation:

$$x = (x_1, \cdots, x_n) \quad ; x \ge y \quad means \quad x_i \ge y_i;$$

$$\int_0^x f(s)ds = \int_0^{x_n} \cdots \int_0^{x_1} f(s_1, \cdots, s_n)ds_1 \cdots ds_n;$$

$$\sum_{0<y<x} f(y) = \sum_{u<y_1<x_1} \cdots \sum_{0<y_n<x_n} f(y_1, \cdots, y_n).$$

With the above notation, theorems (1 - 3) continue to hold.

We let $\bar{\delta}$ be the n-dimensional Dirac delta function defined by $\bar{\delta}(x) = \delta(x_1)\cdots\delta(x_n)$ where δ is as in section 2, so that

$$\int_{(a,b]} f(x)\bar{\delta}(x)dx =$$

$$\left\{\int_{(a_1,b_1]}\cdots\int_{(a_n,b_n]}\right\} f(x)\bar{\delta}(x)dx = f(b).$$

Similarly, we define $\bar{g_r}$ by $\bar{g_r}(x) = g_r(x_1)\cdots g_r(x_n)$.

Routine checking then shows that the results of section 3 also hold in $\mathbf{R^n}$.

5. FURTHER GENERALIZATIONS

Motivated by applications to functional and integro-differentia equations, a number of generalizations of the Gronwall-Bellman and Bihari inequalities and their discrete analogues have been developed in recent years. As a simple example, investigating boundedness and stability of the integro-differential equation

$$\frac{du}{dx} = f(x, u(x), \int_0^x u(s)ds),$$

with f satisfying a Lipschitz cond. in the last 2 variables, leads, in the usual way, to an inequality of the form

$$v(x) \leq a(x) + \int_0^x b(s)\left\{v(s) + \int_0^s g(s)v(t)dt\right\} ds.$$

From this it can be shown that $v(x)$ has a bound depending only on a, b, g. See [2], [3], [5], [6], [7], [10], [11] for more details, and generalizations.

We give two examples of the extension of such considerations to inequalities involving distributions.

Notation: If $f_1, \ldots, f_k$ are of class A we let

$$f_{1,\ldots,k}(x) = \max\{fi(x); 1 \leq i \leq k\}.$$

Theorem 7: Assume: (i) u is of class $J(P)$, and

$$a(x), f_i(x), g_i(x), h_i(x)$$

are of class A,

$$\begin{aligned}(ii)\quad u(x) \;&\leq\; a(x) + f_1(x)\int_0^x \{g_1(s)\\ &+ \sum_{0\leq x_j<x} h_1(s)\delta(s-x_j)\}u(s)ds\\ &+ f_2(x)\int_0^x g_2(s)\{\int_0^s [g_3(t)\\ &+ \sum_{0\leq x_j\leq s} h_2(t)\delta(t-x_j)]u(t)dt\}ds. \qquad (6)\end{aligned}$$

Then

$$\begin{aligned}u(x) \;&\leq\; a(x) + [f_{1,2}(x)+1]\int_0^x \{g_{1,2,3}(s)\\ &+ \sum_{0\leq x_j<x} h_{1,2}(s)\delta(s-x_j)\}a(s)\\ &\exp\ \{\int_s^x [f_{1,2}(t)+1][g_{1,2,3}(t)\\ &+ \sum_{s\leq x_j<x} h_{1,2}(t)\delta(t-x_j)]dt\}ds. \qquad (7)\end{aligned}$$

Proof: From (6),

$$\begin{aligned}u(x) \;&\leq\; a(x) + f_{1,2}(x)\int_0^x \{g_{1,2}(s)\\ &+ \sum_{0\leq x_j<x} h_1(s)\delta(s-x_j)\}\{u(s)\\ &+ \int_0^s [g_3(t) + \sum_{0\leq x_j\leq s} h_2(t)\delta(t-x_j)]u(t)dt\}ds.\end{aligned}$$

Set

$$z(s) = u(s) + \int_0^s [g_3(t) + \sum_{0\leq x_j<s} h_2(t)\delta(t-x_j)]u(t)dt.$$

Then

$$\begin{aligned}u \leq z \;&\leq\; a + f_{1,2}\int_0^x \{g_{1,2}(s)\\ &+ \sum_{x_j} h_1(s)\delta(s-x_j)\}z(s)ds\\ &+ \int_0^x [g_3(t) + \sum_{x_j} h_2(t)\delta(t-x_j)]u(t)dt\\ &\leq\; a + [f_{1,2}+1]\int_0^x \{g_{1,2,3}\\ &+ \sum_{x_j} h_{1,2}(s)\delta(s-x_j)\}z(s)ds,\end{aligned}$$

and (7) follows from theorem 5.

Theorem 8: Assume:

(i) u is of class $J(P)$, and a, f_i, g_i, h_i are of class A, and m, r are of class B;

$$\begin{aligned}(ii)\quad u(x) \;&\leq\; a(x) + f_1(x)\int_0^x [g_1(s)\\ &+ \sum_{0\leq x_j<x} h_1(s)\delta(s-x_j)]u(s)ds\\ &+ f_2(x)\int_0^x [g_2(s)\\ &+ \sum_{0<x_j<x} h_2(s)\delta(s-x_j)]m\{u(s)\\ &+ f_3(s)\int_0^s [g_3(t)\\ &+ \sum_{0<x_j<s} h_3(t)\delta(t-x_j)]r(u(t))dt\}ds \quad (8)\end{aligned}$$

Then

$$u(x) \le P^{-1}\{P(a^*(x)) + f^*_{1,2,3}(x)\int_0^x [g_{1,2,3}(t) + \sum_{x_j} h_{1,2,3}(t)\delta(t-x_j)]dt\} \quad (9)$$

where

$$P(t) = \int_{t_0}^t \frac{ds}{s+m(s)+r(s)} \quad ,0 < t_0 \le t$$

and $a^*(x) = \max a(t), 0 \le t \le x$, and similarly for $f^*_{1,2,3}$.

Proof: Set $z(x) = u(x) + f_3(x)\int_0^x [g_3(s) + \sum_{x_j} h_3(s)\delta(s - x_j)]r(u(s))ds$.

Then

$$\begin{aligned} u &\le z \le a + f_1 \int_0^x [g_1 + \sum_{x_j} h_1(s)\delta(s-x_j)]zds \\ &+ f_2(x)\int_0^x [g_2 + \sum_{x_j} h_2(s)\delta(s-x_j)]m(z)ds \\ &+ f_3 \int_0^x [g_3 + \sum_{x_j} h_3\delta(t-x_j)]r(z)ds \\ &\le a + f_{1,2,3}(x)\int_0^x [g_{1,2,3}(s) \\ &+ \sum h_{1,2,3}(s)\delta(s-x_j)][z(s) \\ &+ m(z(s)) + r(z(s))]ds. \end{aligned}$$

and (8) follows, from theorem 6.

Our final example will be an extension of a theorem of Greene [9].

Theorem 9: Assume: (i) u, v, k_i, f_i, g_i, h_i, are of class A;

$$\begin{aligned} u &\le k_1 + f_1 \int_0^x [g_1 + \sum_{x_j} h_1\delta(s-x_j)]uds \\ &+ f_z \int_0^x e^{\mu s}[g_2 + \sum_{x_j} h_2\delta(s-x_j)]vds \\ v &\le k_2 + f_3 \int_0^x e^{-\mu s}[g_3 + \sum_{x_j} h_3\delta(s-x_j)]uds \\ &+ f_4 \int_0^x [g_4 + \sum_{x_j} h_4\delta(s-x_j)]vds. \end{aligned} \quad (10)$$

Then there exist non-neg. constants C_1, C_2, M_1, M_2 s.t. $u(x) \le M_1 \exp(C_1 x), v(x) \le M_2 \exp(C_2 x)$.

Proof: Proceeding as in Wang [9], multiply the first of (9) by $e^{-\mu x}$ and add to second, getting

$$\begin{aligned} e^{-\mu x}u + v &\le (k_1 e^{-\mu x} + k_2) + f_{1,2,3,4}(x)\int_0^x \{[g_1 \\ &+ \sum_{x_j} h_1\delta(s-x_j)]e^{-mux}u(s) \\ &+ e^{\mu(s-x)}[g_2 + \sum_{x_j} h_2\delta(s-x_j)]v(s) \\ &+ e^{-\mu s}[g_1 + \sum_{x_j} h_3\delta(s-x_j)]u(s) \\ &+ [g_4 + \sum h_4\delta(s-x_j)]v(s)\}ds \\ &\le (k_1 e^{-\mu x} + k_2) + f_{1,2,3,4}\int_0^x [g_{1,2,3,4}(s) \\ &+ \sum h_{1,2,3,4}(s)\delta(s-x_j)] \\ &\quad \{(e^{-\mu x} + e^{-\mu s})u(s) + (e^{\mu(s-x)}+1)v(s)\}ds \\ &\le (k_1 e^{-\mu x} + k_2) + f_{1,2,3,4}(x)\int_0^x [g_{1,2,3,4}(s) \\ &+ \sum h_{1,2,3,4}(s)\delta(s-x_j)](e^{\mu(s-x)}+1)(e^{-\mu s}+v)ds. \end{aligned}$$

and the conclusion follows from this and theorem 5.

References

[1] Agarwal, R.P., and Thandapani, E.: On discrete generalizations of Gronwall's inequality. Bull-Inst. Math. Acad. Sin. 9, No. 2, (1981)

[2] Conlan-Wang: A Unified Approach to Continuous and Discrete Gronwall-Bellman Inequalities. Submitted to Applicable Analysis.

[3] Conlan-Wang: A Unified Approach to Gronwall- Bihari Inequalities. Submitted to Mathematica Applicata

[4] Lakshmikanthan: Unpublished notes on Impulsive Differential Equations.

[5] Pachpatte, B.G. On some integral inequalities similar to Bellman-Bihari inequalities. Jour. Math. Anal. Appl. 49 (1975).

[6] Pachpatte, B.G. A note on Gronwall type integral and integro-differential inequalities. Tamkang Journ. Math 8 (1977).

[7] Pachpatte, B.G. On some new integral and integro-differential inequalities in two indep. variables and their applications. Jour. Diff. Eqns. 33, 1979.

[8] Rao, V.S.H.: Integral inequalities of Gronwall type for distributions. Jour. Math. Anal. Appl. 72 (1979)

[9] Wang, C.L.: A short proof of a Greene theorem. Proc. Am. Math. Soc. 69, (1978)

[10] Yeh, C.C. and Shih, M.H.: The G-B inequalities in several variables. Jour. Math. Anal. Appl. 86 (1982)

[11] Yeh, C.C.: On some integral inequalities in n-indep. variables and their applications. Jour. Math. Anal. App. 86 (1982)

Approximation, Optimization and Computing:
Theory and Applications, A.G. Law and C.L. Wang (eds.)
Elsevier Science Publishers B.V. (North-Holland)

EQUILIBRUM DISTRIBUTION OF DISKS ON THE ROTATING TABLE — A sort of two dimensional geometric layout optimization with behavioural constraints*

S.E. Deng H.F. Teng W.X. Zhong

Department of Mechanical Engineering,
Dalian University of Technology, China

ABSTRACT: This paper proposes an algorithm to solve the problem of equilibrum distribution of disks on a rotating table — a sort of two dimensional geometric layout optimization with behavioural constraints. The algorithm is divided into three phases, i.e., packing phase, phase to find a feasible initial layout and optimal layout phase. Based on the principles of quasi-physical method and the generalized force directed relaxation method and searching directed tree and CAD technology, a method to contract geometric distributive area successively on the area of the feasible geometric layout and at the same time to carry on the process of iteratively improving layout is proposed. Finally, we have made corresponding software and calculated several examples. The results are satisfactory.

Keywords: Geometry, Layout, Behavioural constraints, Optimization, Quasi-physical method, Directed tree, CAD. The generalized force directed relaxation method.

1 INTRODUCTION

The geometric layout optimization is a problem of NP (Nondeterministic polynomial). It's study is of great theoretical significance. In recent years, the problem of geometric layout optimization has been studied by some authors [1-4]. Huang Wen-qi and Zhan ShuHao [1] have presented a quasi-physical method to solve the problem of geometric packing. Zvi Drezner [2] also presented a two-phase algorithm called dispersion-concentration algorithm to solve facitilies optimal layout problem. However, at present, there see only a few papers studying the problem of geometric layout optimization with behavioural constraints. This paper studies the problem of geometric layout optimization with behavioural constraints and proposes a valid algorithm.

2 PROBLEM FORMULATION AND OPTIMIZATION MODEL

2.1 Problem Formulation:

In this paper, the problem of two dimensional geometric layout optimization with behavioural constraints is introduced. We call it "Equilibrum Distribution of Disks on a Rotating Table", i.e., there is a large rotating circular table (called graph plane) with radius R and uniform thickness. Now, there are n disks (called graph units), each of which has a radius r_i and mass m_i $(i=1,\ldots,n)$, to be distributed on the rotating table on the following conditions:

(1) There is not any overlap among all the disks.

(2) Each disk does not overtake the edge of the table.

(3) The remnants of static equilibrum is smaller than a permissible value after geometric layout.

(4) Each disk should be as closed as possible to the central of the table.

2.2 Optimization Model

Let $X=[X_1, X_2]$, consider the following problem:

Minimize

$$F(X)=\mathrm{Max}\,[(\sqrt{x_{11}^2+x_{12}^2}+r_1),\ldots,(\sqrt{x_{n1}^2+x_{n2}^2}+r_n)]$$

Subject to:

1). $$\sqrt{(x_{i1}-x_{j1})^2+(x_{i2}-x_{j2})^2}\geq r_i+r_j \quad (i\neq j,\ i,j=1,\ldots,n)$$

2). $$\sqrt{x_{i1}^2+x_{i2}^2}+r_i\leq R \qquad (i=1,\ldots,n)$$

3). $$\sqrt{(\sum_{i=1}^{n} m_i x_{i1})^2+(\sum_{i=1}^{n} m_i x_{i2})^2}\leq[\delta]$$

Where $X_1=[x_{11}, x_{21},\ldots,x_{n1}]$ $X_2=[x_{12}, x_{22},\ldots,x_{n2}]$ $[\delta]$ is a permissible value of the static equilibrum; x_{i1} and x_{i2} are the central coordinates

---

* The projects supported by the National Natural Science Fundation of China.

of disk i (see Fig.1).

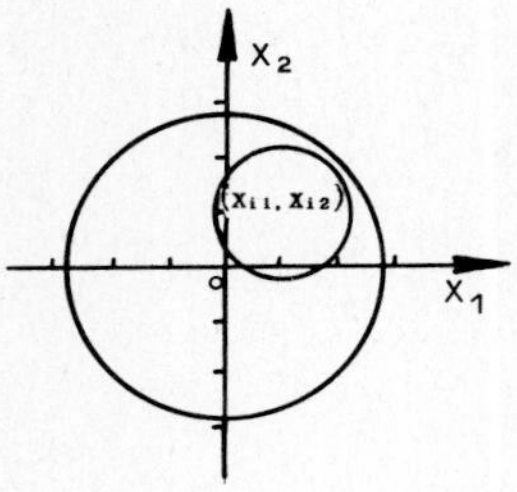

Fig.1

3 MAIN DESIGN THOUGHT

The problem of this geometric layout optimization will be divided into three phases to be solved, i.e., the packing phase, the phase to find a feasible initial layout and the optimal layout phase. Each phase will be stated as follows.

Phase 1: The Packing Phase

Functional formulation: The topological relation of two arbitrary graph units on the graph plane has four types: 1. Intersection; 2. consistence; 3. contact (externally tangent); 4. disjoint. According to the quasi-physical principle[1], if we assume the two graph units to be elastic bodies, for the topological relations of intersection and consistence, elastic deformation will be occured, for the other topological relations, the elastic deformation is not occured. Thus, for any two graph units on the graph plane, the elastic potential energy will be expressed as follows

$$U_{ij} = \begin{cases} L_{ij}^2 \sqrt{r_i^2 + r_j^2} & (i \neq j, i, j=1, \ldots, n) \ (L_{ij} > 0) \\ 0 & (L_{ij} \leq 0) \end{cases}$$

$$L_{ij} = (r_i + r_j) - \sqrt{(x_{i1} - x_{j1})^2 + (x_{i2} + x_{j2})^2} \qquad (i \neq j)$$

Similarly, the elastic potential energy between a graph unit and the graph plane is given by

$$U_{io} = \begin{cases} L_{io}^2 \sqrt{r_i^2 + R^2} & (i=1, \ldots, n) \ (L_{ij} > 0) \\ 0 & (L_{ij} \leq 0) \end{cases}$$

$$L_{io} = \sqrt{x_{i1}^2 + x_{i2}^2} + r_i - R$$

Thus, the total elastic potential energy of the whole elastic system is expressed as follows

$$U = \sum_{i=1}^{n} U_{io} + w \sum_{i=1}^{n-1} \sum_{j=i+1}^{n} U_{ij} \qquad (1)$$

Where w is a weighting factor

Solving method: For arbitrary given initial location of all graph units, we minimize the total elastic potential energy U of equation (1). If $U_{min} = 0$, then the problem of the graph units packing is solved. If $U_{min} \neq 0$, the process of iteratively improving layout is carried on. If once, for a geometric layout, $U_{min} = 0$, the problem of graph units packing has been solved. Finally, If the geometric layout with a zero U_{min} has not been found out, the relative informations will be given.

Phase 2: The Phase to Find a Feasible Initial Layout

Functional formulation: Considering static equilibrous demands and adding gravitational potential energy in equation (1), we can find that equation (1) will be changed as follows

$$\overline{U} = \sum_{i=1}^{n} U_{io} + w \sum_{i=1}^{n-1} \sum_{j=i+1}^{n} U_{ij} + \lambda M \qquad (2)$$

$$M = [(\sum_{i=1}^{n} m_i x_{i1}) / (\sum_{i=1}^{n} m_i)]^2 + [(\sum_{i=1}^{n} m_i x_{i2}) / (\sum_{i=1}^{n} m_i)]^2$$

Where λ is a weighting factor and M is the gravitational potential energy.

Solving method: Based on geometric layout given in phase 1, the total potential energy $\overline{U}$ of equation (2) is minimized. Thus a new geometric layout is formed. We check this new layout in order to know whether it satisfies the static equilibrous requirement. If this layout satisfies the static equilibrous requirement, the feasible initial geometric layout has been found. If this layout does not satify the static equilibrous requirement, the process of the iteratively improving layout is carried on. If once a feasible layout is found, the phase to find a feasible initial layout has been finished. Finally, if the feasible layout has not been found, the relative informations will be given.

Phase 3: Optimal Layout phase

Within the area of the feasible geometric layout, we first contract the area of the graph plane and at the same time minimize the total potential energy $\overline{U}$ of equation (2) until present layout reaches the bounds of feasible layout field. Then, on the basis of the present layout, we carry on the process of the iteratively improving layout, If the layout is improved, this improved layout will become present layout. Above method is used again until the terminative criterion is satisfied.

4 TARGET AND QUASI-TARGET POSITION OF A GRAPH UNIT

According to the principle of the force directed relaxation method [3], The target point position of present graph unit i is defined as a position where the total force applied on graph unit i by all the other graph units equates zero. We denote this target point by T. Then position $T(x_{iT1}, x_{iT2})$ is given by

$$x_{iT1} = (\sum_{\substack{j=1 \\ j \neq i}}^{n} x_{j1})/(n-1)$$

$$x_{iT2} = (\sum_{\substack{j=1 \\ j \neq i}}^{n} x_{j2})/(n-1)$$

For the target point T, when it is located at a position where another graph unit is located, we choose the central position of this graph unit to be target position of graph unit i. When point T is located at a position where no any graph unit is located, we choose position T to be target position of graph unit i. The graph unit at target position is regarded to be a empty graph unit.

The quasi-target position is regarded as some positions where some other graph units, which are some graph units in the neighbourhood of the target position, are located. We choose the number of the quasi-target position to be ε^*.

5 THE ITERATIVELY IMPROVING LAYOUT ALGORITHM

5.1 The Convergence Algorithm

This paper adopts the generalized force directed relaxation method [3] . By constructing directed tree successively and at the same time searching every directed tree to find out an optimal geometric layout. At first, the directed tree (shown in Fig.2) is constructed, each node of it represents a graph unit and each edge represents a trial transformation of interchange chain. On this directed tree,

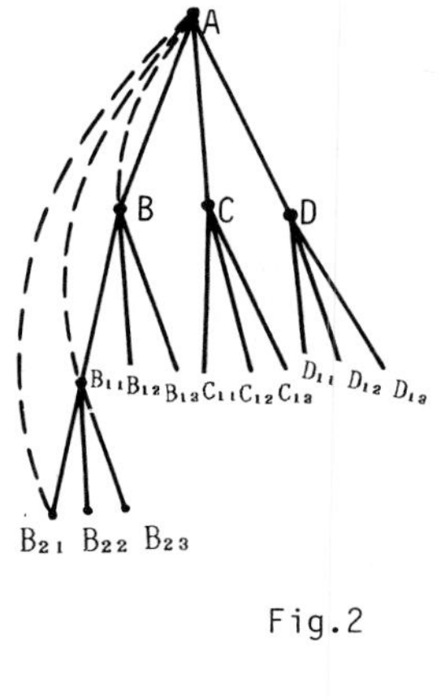

Fig.2

graph unit on the root node is a graph unit chosen to initiate the trial interchange and is refered to be an iterative object (The graph unit, which is the farthest from its target point among all the graph unit and is not marked with asterisk (*), is chosen to be the iterative object). The graph unit on the other node of it is a graph unit located on the target position or quasi-target position of its above level graph unit on same path. In order to get a good quasi-optimal solution quickly, the height of the directed tree is limited and denoted by λ^*. On the directed tree, the process of the trial transformation is carried on. On the process of searching the directed tree, if there is a trial interchange which reduces the value of the correspondingly evaluating objective function, this trial interchange is accepted and replaces the old layout as present layout. Then we again construct a new directed tree to carry on searching it. If no trial interchange contributes to reducing the value of this evaluating objective function on the whole directed tree, the graph unit on the root node is marked with asterisk (*). In the way to construct and search the directed tree continually, finally, an optimal geometric layout will be found. (Notice: If once layout is improved, all the asterisk marks (*) are cleared away.)

5.2 Terminative Criterion

Here, we introduce two parameters. We define N_ε^* as a fixed positive integer in advance and N_ε as the total number of graph units which are marked with asterisk(*).

Phase 1: Packing Phase

When one of two terminative criterions below is satisfied, this phase will be terminated.

(1) $\bar{R} = \mathrm{Max}[(\sqrt{x_{11}^2 + x_{12}^2} + r_1), \ldots, (\sqrt{x_{n1}^2 + x_{n2}^2} + r_n)] \leq R$

(2) $N_\varepsilon \geq N_\varepsilon^*$

Phase 2: Phase to Find a Feasible Initial Layout

When one of two terminative criterions below is satisfied, this phase will be terminated.

(1) $\bar{W} = \sqrt{(\sum_{i=1}^{n} m_i x_{i1})^2 + (\sum_{i=1}^{n} m_i x_{i2})^2} \leq [\delta]$

and $\bar{R} = \mathrm{Max}[(\sqrt{x_{11}^2 + x_{12}^2} + r_1), \ldots, (\sqrt{x_{n1}^2 + x_{n2}^2} + r_n)] \leq R$

(2) $N_\varepsilon \geq N_\varepsilon^*$

Phase 3: The Optimal Layout Phase

If $N_\varepsilon \geq N_\varepsilon^*$, this phase will be terminated.

5.3 The Evaluating Objective Function

Phase 1: Packing Phase

$$f = \sum_{i=1}^{n-1} \sum_{j=i+1}^{n} \sqrt{(x_{i1} - x_{j1})^2 + (x_{i2} - x_{j2})^2}$$

Phase 2: Phase to Find a Feasible Initial Layout

$$f = \sqrt{(\sum_{i=1}^{n} m_i\, x_{i1})^2 + (\sum_{i=1}^{n} m_i\, x_{i2})^2}$$

Phase 3: The Optimal Layout Phase

$$f = \mathrm{Max}[(\sqrt{x_{11}^2 + x_{12}^2} + r_1), \ldots, (\sqrt{x_{n1}^2 + x_{n2}^2} + r_n)]$$

7 Examples

TABLE I

n=5	R=35(mm)	$\varepsilon^*=3$	$\lambda^*=4$	[δ]=1.0(g.mm)	$N\varepsilon^*=4$									
No.	r_i (mm)	Phase 1				Phase 2				Phase 3				
		x_{i1} (mm)	x_{i2} (mm)	$\overline{R}$ (mm)	$\overline{W}$ (g.mm)	x_{i1} (mm)	x_{i2} (mm)	$\overline{R}$ (mm)	$\overline{W}$ (g.mm)	x_{i1} (mm)	x_{i2} (mm)	$\overline{R}$ (mm)	$\overline{W}$ (g.mm)	
1	6.0	8.10	-7.49			-5.99	-13.51			-3.14	2.18			
2	7.0	-11.09	13.00			9.54	-7.67			-12.72	10.98			
3	8.0	-23.89	1.95	34.63	314.06	-18.53	2.04	34.96	0.77	-13.53	-7.35	24.32	0.69	
4	9.0	-9.00	-24.00			3.76	-25.69			6.88	13.35			
5	11.0	18.31	11.09			5.20	23.24			8.63	-10.15			
Run on IBM AT, Total number of trial transformation=109, Computation time=3.5 hours														

TABLE II

n=7	R=50(mm)	$\varepsilon^*=3$	$\lambda^*=4$	[δ]=3.4(g.mm)	$N\varepsilon^*=5$									
No.	r_i (mm)	Phase 1				Phase 2				Phase 3				
		x_{i1} (mm)	x_{i2} (mm)	$\overline{R}$ (mm)	$\overline{W}$ (g.mm)	x_{i1} (mm)	x_{i2} (mm)	$\overline{R}$ (mm)	$\overline{W}$ (g.mm)	x_{i1} (mm)	x_{i2} (mm)	$\overline{R}$ (mm)	$\overline{W}$ (g.mm)	
1	10.0	27.78	27.85			20.23	34.37			16.80	-13.51			
2	11.0	32.80	0.00			38.33	-2.14			-3.33	-20.81			
3	12.0	20.00	-25.20			8.08	-36.19			-19.70	-4.65			
4	11.5	-4.99	-30.20	49.34	4640.1	-15.02	-21.20	49.95	0.06	3.05	20.72	32.44	3.12	
5	9.5	-25.00	-4.99			-31.85	1.23			1.63	-0.79			
6	8.5	-29.01	26.91			3.93	25.34			-17.34	15.73			
7	10.5	-5.10	10.00			-29.45	26.25			20.52	6.65			
Run on IBM AT, Total number of trial transformation=194, Computation time=7.0 hours														

Table I, II show two examples of the geometric layout optimization with behavioural constraints and the result of each phase. Here, $m_i = r_i$ (g) $(i=1,\ldots,n)$. Fig.3 is results of second example (its parameters and results are shown in Table II).

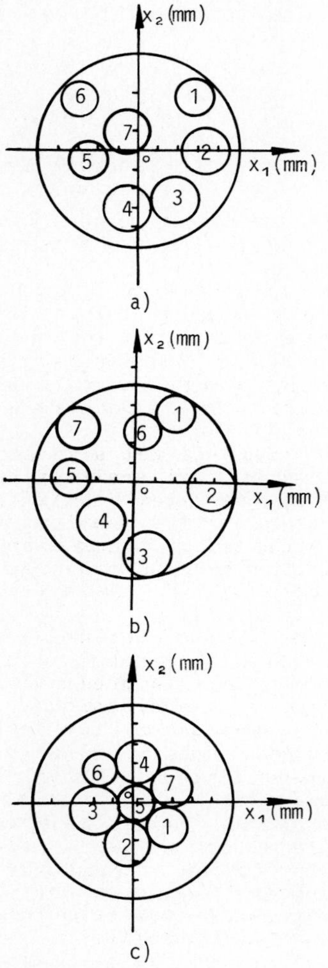

a) result of phase 1
b) result of phase 2
c) result of phase 3

Fig.3

REFERENCES

[1] Huang Wei-qi and Zhan Shu-hao, A Quasi-physical method for sloving the packing problem, ACTA Mathematical Applicatae Sinica, Vol.2, No.2, May, 1979, P176-180.

[2] Zvi Drezner, DISCON: A new method for the layout problem, Operations Research, Vol.28, No.6, 1980, P1375-1384.

[3] Zhuang Wen-jun and Li Yu-xing, Layout Design Automation of IC, Shanghai JiaoTong University Publishing, China, Aug., 1986, P111-156.

[4] Kalman Peleg and Elizabeth Peleg, Container dimensions for optimal utilization of storage and transportation space, Computer Aided Design, July, 1976, Vol.8, No.3, P175-180.

Approximation, Optimization and Computing:
Theory and Applications, A.G. Law and C.L. Wang (eds.)
Elsevier Science Publishers B.V. (North-Holland)

YIDOYU--AN OPTIMIZATION PROGRAM SYSTEM FOR WING STRUCTURES

Ding Huiliang, Sun Xianxue, Long Qihao, Chen Wenpu, Pan Bingchen

Aircraft Strength Research Institute(ASRI), CAE

SUMMARY

A structural optimization program based on FEM and mathematical programming technique (SUMT) has been developed and well verified by some aeronautical applications of wing-type structures.A brief introduction and sample applications are presented to show the potential benefits of structural optimization and the efficiency of this system.

1. INTRODUCTION

It is a lengthy process to design a wing structure of aircraft to meet many often contrary requirements simultaneously. To speed up this process and get a satisfactory design, a structural analysis and optimization program system YIDOYU (Wing Structural Design Optimization under Multiple Constraints) has been developed in CAE (The Chinese Aeronautical and Astronautical Establishment) since 1979 [1,2].
The finite element formula and mathematical programming technique (SUMT) are used to predict the structural behaviour and search for the optimum. This program system has been successfully used for several wing-type structures in China and show its ability to solve different kind of problems, its reliability and high computational efficiency in preliminary design and detailed conceptual design stages of the aircraft. This system has now been integrated into a CAD/CAM software, the Computer Integrated Engineering and Manufacturing (CIEM) developed by CAE.
In this paper the methods and main features of YIDOYU are briefly introduced and some representative sample applications are given to show its ability and efficiency in structural applications.

2. MAIN FEATURES

YIDOYU is a program system for analysis and design optimization for structural size and balance weight of structures.
To consider all the design constrainst, including stress,displacement, natural frequency, aeroelasticity (e.g.control efficiencies, divergence speed and flutter speed) and minimum size simultaneously, the mathematical programming technique is used in stead of the engineering optimality criterion.
As a self-contained system, besides those for FE analysis and optimization, other modules for aerodynamics (both steady and unsteady),load calculation, preprocessor for structural model preparation and postprocessor for report and figure output are included to facilitate the user. An excecutive and data management subsystem with problem-oriented user interface is also provided.

APPROXIMATE CONCEPTS AND ANALYTICAL DERIVATIVES

In mathematical programming it is possible to consider all design requirements simultaneously

$$\text{Min} \quad W(D)$$
$$\text{s.t.} \quad G_j(D) > 0 , \; j = 1,2,\ldots,J$$

to search for optimum design D.
In structural application constraint, $G_j(D)$ are normally complex implicit functions of D and a FE structural analysis is necessary for constraint evaluation. Besides, the dimension of the problem (N) may be of the order of thousands. It is very difficult (if not possible) to solve this primary optimization problem because number of the constraint and objective evaluation is about N^2 to N^3.
The approximate concepts due to L. Schmit [3] are therefore adopted. By variable linking, thousands of the structural parameters X is logically controlled by a few design variables

$$\{ D \} = [T] \{ X \}$$

In addition, the active set of variables is adopted for each constraint to lower the variables further. In this way the dimension N could be reduced to a reasonable level (say less than 100) even for a complex large structure.
An ε-effective set strategy of constraints is used during optimization and 'key elements' are assigned for stress constraints in each region ('block').
An (accurate and effect) analytical scheme for constraint derivatives is devised for stress, displacement, frequency, control efficiency, divergence and flutter speed and explicit expressions for behaviour constraints,

$$G_j(D) = G_j(D_0) + \sum \partial G/\partial D_i * (D_i - D_{i0})$$

are used in the optimization model to reduce hundreds of structural analysis to a reasonable level of about ten.
It is the use of the approximate concepts and the analytical scheme of derivative calculation make it possible to conjugate the FEM and the numerical optimization to form the basis of YIDOYU.

STRUCTURAL ANALYSIS

A FE package performs the static, dynamic, static aeroelasticity and flutter analysis. A finite element library with conventional elements, scale elements, generalized elements and mass elements, a material library, an airfoil library and typical structural design data are also provided to ease the modeling and strength evaluation.
The variable bandwidth Cholesky decomposition is used for static analysis. In dynamic analysis the QL method is used. The modified LR decomposition is chosen for complex eigenvalues and a modified Lagueree iteration for the flutter speed.

AERODYNAMICS AND LOADS

Aerodynamic distribution could be converted into grid point load by the program. The inertia load is calculated according to the mass and overload factor at each grid point. The steady aerodynamic forces for static aeroelasticity are obtained by the kernel function method while unsteady aerodynamic forces for flutter analysis are calculated by the subsonic spatial doublet lattice method.

OPTIMIZATION METHODS

The Sequential Unconstrained Minimization Technique (SUMT) is used with different options on unconstrained algorithms such as the steepest descent direction, the modified Newton's method, the variable merit methods of DFP and BFGS etc. In one dimension search, the golden-section and quadratic fitting could be chosen.
An extended second-order interior-penalty functionis used and therefore an infeasible initial design is admissible and the violation of constraints in intermediate designs due to the use of approximate concepts could be treated effectively.

PRE- AND POST-PROCESSING

FE model with optimized numbering of grid points could be generated by specific problem-oriented statements to ease data preparation and to minimize the error.
Several options for output format including table and figure (chart) are also available.
Interfaces with some commercial available interactive graphics software such as CADAM and SUPERTAB has been developed.

PROGRAM CONSTRUCTION

This system has a hierarchical program structure with four levels namely the executive, the main programs for specified job flow, the independent functional modules and the subroutines for common usage. Date management and transformation are realized by the data files and/or the dynamic arrays in common blocks. A simplified flow chart is shown in fig.1.

3. SAMPLE APPLICATIONS

Sample applications by YIDOYU are shown in fig.2. Three typical sample problems are presented in this paper.

TARGET WING STRUCTURE

YIDOYU is used to determine the best location of tip equipment cabin from three alternatives and then to optimize the size of a target wing shown in fig.3.
The wing is idealized with a FE model of 220 grid nodes and designed under displacement, frequency, control efficiency, divergence and flutter speed constraints and the lower bounds on design variables. Ten linked design variables are chosen for the thickness of the skins, webs and spar flanges.
The baseline wing is chosen as starting point for fully stressed design

(FSD) and then a numerical optimization is performed.
The design history are shown in fig.4 and table 1. It can be seen that performences have been improved. For example, the flutter speed is increasing 5% while the objective (weight) gets an improvement of 3.6%.
This optimum design provided a guideline for a successful new version of target,which has passed all ground and flight tests and behaviour well in service.

FLUTTER DESIGN OF A VERTICAL TAIL

Because of the deficiency in flutter speed, a vertical tail with control surface is optimized under the flutter speed and minimum size constraints. The FE model consists of 257 nodes, 1013 elements (fig.5). Total 13 structural size variables and 5 mass variables are chosen during optimization search.
In this design study 17 situation and more than 60 combination are carefully evaluated. A better design is obtained by increasing the skin thick at top with 49 m/sec increasing in flutter speed and only 3.1 kg penalty in structural weight.
This result has been verified and accepted in a production aircraft.

PRELIMINARY DESIGN OF A WING

A preliminary design of a wing structure with three rear control surfaces is studied. Only upper half of which is modeled because of the symmetry. Total 317 grid nodes and 1267 elements are used and 7 load conditions are considered.
Different structural schemes are analyzed and compared and a better scheme was chosen. The FE model and the displacement of the wing is shown in fig.6. A set of typical analysis results is listed in Table 2.

Table 1 Design Iteration Results of Target Wing

		Total Weight W(kg)	Control Effic. η(%)	Diverg. Speed V_D(%)	Frequency ω(Hz)	Flutter Speed V_f(%)	Displacement U(%)
Baseline		121.3				100.5	106.4
F S D		111.4	105.1	625	6.80	83.5	92.1
Design Requirement			>100	>100	>6.50	>100	>100
M.P.	1	112.5	104.7	675	6.81	88.7	94.1
	2	113.6	104.4	725	6.81	93.3	95.3
iter.	3	114.7	104.1	775	6.82	97.6	97.3
	4	115.8	103.9	825	6.82	101.6	98.7
	5	116.9	103.7	875	6.83	105.3	100.0

4. CONCLUSION

It has been proven that the conjunction of mathematical optimization technique with the FE analysis and engineering approximate concepts is an effective way to promote the practical application of the structural design optimization. Based on the experience gained from YIDOYU, some improvement/enhancement is now underway including the optimization of composite structures (aeroelasticity tailoring) to explore this more hopeful field.

REFERENCES

[1] Ye Kejia,Lin Menghe and Lian Mengjie, Structural Design Optimization System for Aircraft Under Multiple Constraints - Introduction to YIDOYU, International Aviation, No.7,July 1983; (in chinese) also: N84-10768

[2] Ding,Huiliang, Principles and Design Specification of YIDOYU, ASRI Internal Report, 1980 (in Chinese)

[3] Schmit Jr. L.A., and Farshi, Some Approximation Concepts for Structural Syatheis AIAA.J.12, NO.5 692-699, 1974

[4] Li Guansen and Ding Huiliang, Theoretical Manual of YIDOYU. Chinese Aeronaultical Science-Technology Literature(CASTL)HJL86. (in chinese)

[5] Long Qihao, User's Manual of YIDYOU, (CASTL) HJL850024, 1985 (in chinese)

[6] Pan Binson, Long Qihao,Chen Wenpu and Li Zhoumin, User's Guide to YIDOYU, Chinese CASTL JHL860046, 1986 (in chinese)

[7] Sun Xianxue, Manual for the Demonstration Examples of YIDOYU, CASTL HJL860047, 1986(in chinese)

Table 2 Analysis Results of Wing

	Frequency (Hz)	Flutter Speed (m/sec)	Control Efficiency	Weight (kg)
Metal Structure	12.18, 31.5, 38.16 48.26, 56.01, 69.33	655.56	0.16277 0.35546	1201.96
Metal + Composite	12.38, 32.24, 39.76 56.03, 59.83, 73.53	682.56	0.16580 0.35546	1187.42

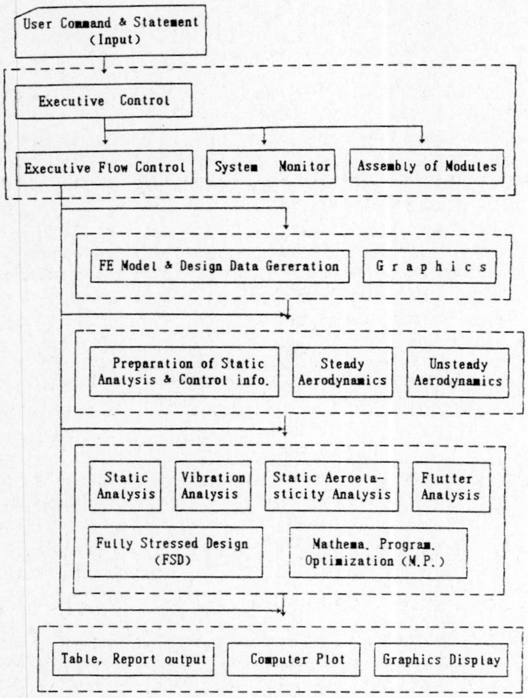

Fig. 1 System Construction

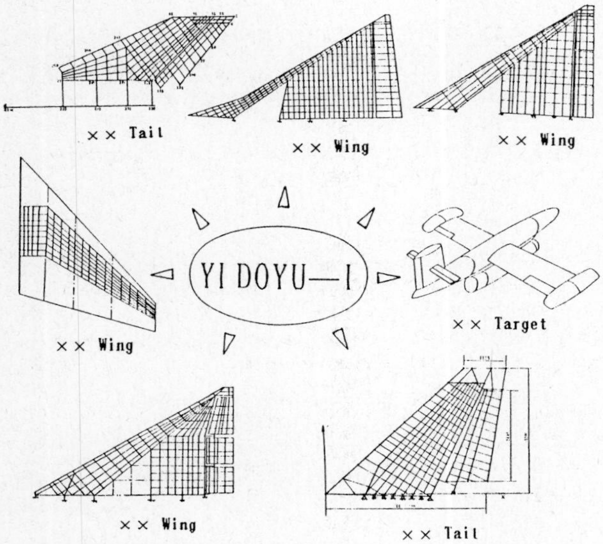

Fig. 2 Sample Applications

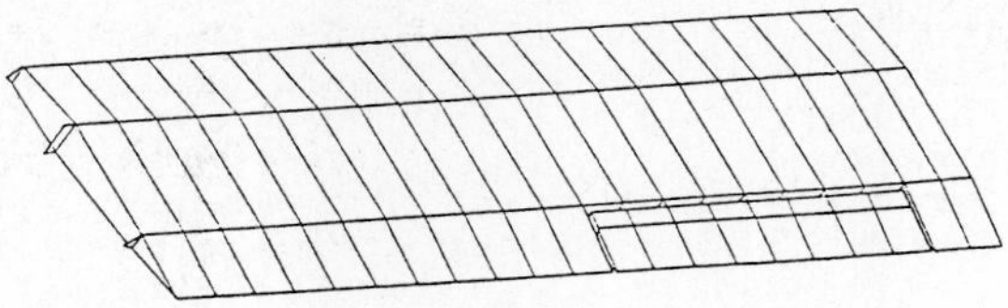

Fig. 3 FE Model of the Target wing

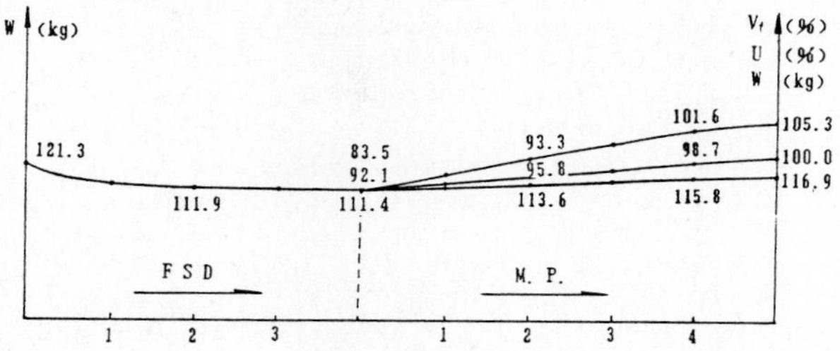

Fig. 4 Design History of Target Wing

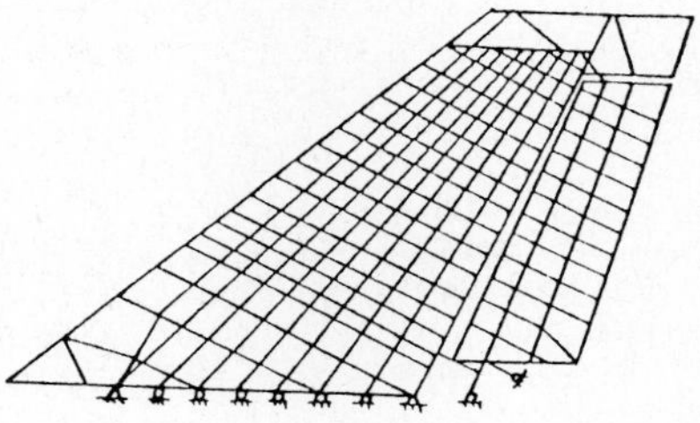

Fig. 5 FE Model of A Virical Tail

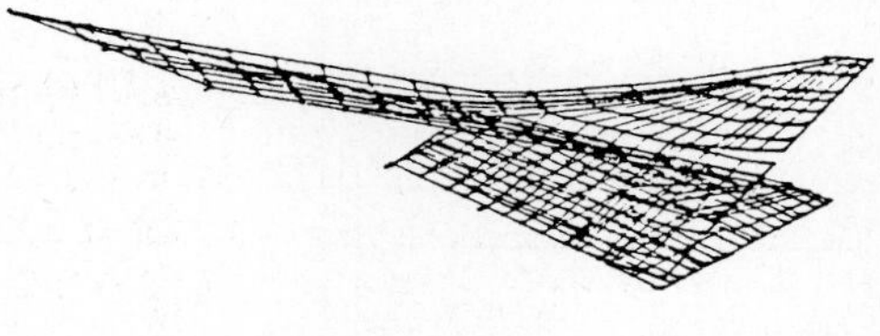

Fig. 6 FE Model and Displacement of Wing

Approximation, Optimization and Computing:
Theory and Applications, A.G. Law and C.L. Wang (eds.)
Elsevier Science Publishers B.V. (North-Holland)

UTILIZATION OF THE MULTI-LEVEL APPROACH IN OPTIMAL DESIGN OF R.C. STRUCTURES

PROF. DR. K.S. DINNO

Saddam University for Engineering and Science, Baghdad, Iraq.+

The optimal design of most practical structures involves a large number of variables and constraints which make wholemeal optimization intractable. The paper describes the utilization of the multi-level approach to achieve optimal designs. Examples are given of two types of structures of wide application.

1. INTRODUCTION

An optimal structural design is one which satisfies the structural requirements of stability, strength and serviceability with optimum cost, weight or any other chosen optimality criterion. The solution must clearly also satisfy prescribed functional requirements. In reinforced concrete work however, practical problems involve many design variables and constraints. The objective function, which is usually taken as cost of structure, and the constraints are inherently non-linear and hence non linear programming methods are involved.

In most early work on optimal design of R.C. [1][2][3] concrete dimensions were considered fixed and the design variables were steel areas at critical sections with volume of the steel being the objective function to be minimised. Later work [4] saw expansion in the number of variables to include member dimensions which were however varied in groups, again in order to limit the number of variables.

In order to widen the scope of optimal design, optimization formulations to be used must cope with multiplicity of variables and constraints yet at the same time retain feasibility of convergence and computational efficiency. These objectives are fulfilled in this paper by utilising the recent advances in the development and utilisation of the multi-level approach to optimization [5][6][7][8].

Following this approach the optimal designs of two types of R.C. structures of wide usage but essentially differing structural forms are formulated here, namely multi-storey, multi-bay frames and rectangular ground storage tanks.

2. OPTIMIZATION AND DECOMPOSITION BY THE MULTI-LEVEL APPROACH

An optimum design solution involves determining a vector X of the design variables

$$X = X(x_1, x_2, x_3, \dots, x_n) \qquad \dots(1)$$

which minimises an objective function Z considered here to be cost

$$Z = F(x_1, x_2, x_3, \dots, x_n) \qquad \dots(2)$$

subject to the design constraints

$$g_i(X) \geq 0 \qquad \dots(3)$$

$$i = 1, 2, \dots, m,$$

where n and m are total number of variables and constraints respectively.

Description of the design variables, design constraints and objective function for each of the two types of problems considered here is given later.

In principle it may be possible to solve the optimization problem involving all variables and constraints in one integrated process. For large problems, such as the ones considered here, this procedure would be excessively time consuming and would most likely encounter convergence difficulties. On the other hand decomposition of the overall problem into suboptimization problems by the multi-level approach offers an effective means of solution. The multi-level approach embodies hierarchial relations between the subsystems and involves operating at various levels with appropriate interaction links. Its success is based on the premise that in a design problem, the objective function

+ On secondment from College of Engineering, University of Baghdad.

represented by cost, is more sensitive to some design variables and constraints than to others. Hence, design variables and their often implicitly related constraints are classified into groups with each group being assigned to a particular level. The effectiveness of this approach relies on the ability to subdivide the overall problem into multi-levels with optimization being carried out at each level for the variables assigned thereto. Engineering judgment and knack are involved in performing the grouping. In a structural design problem the outer level is concerned with optimization of the topological variables related to shape and/or number of members, the next level with optimization of overall geometrical variables, followed by the level in which element or member dimensions are optimized. Finally optimization at the sub-element (steel reinforcement) is performed at the innermost level.

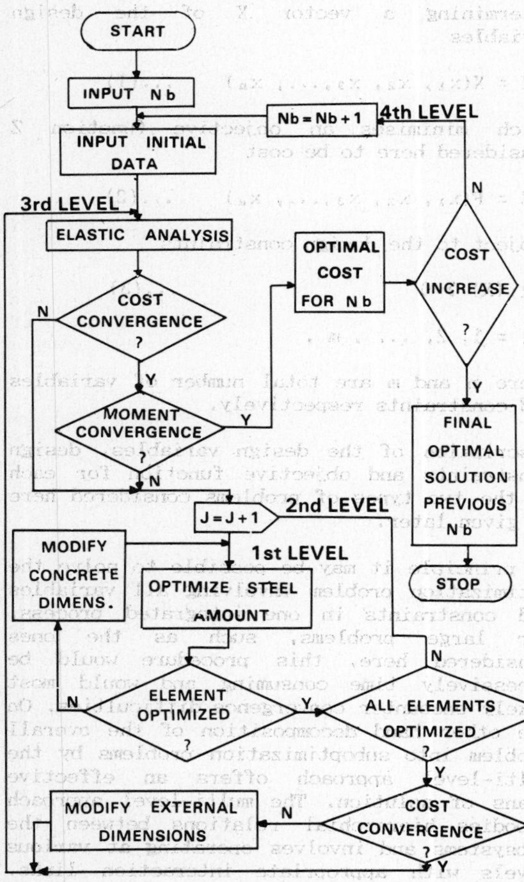

FIG. 1 : FLOWCHART_ OPTIMIZATION BY MULTI - LEVEL APPROACH - TANK PROBLEM

Model coordination method [6] is followed for creating the proper interaction between the various levels. Fig.1 shows a flow chart for the multi-level approach for application problem No.1 with further description given later.

3. PROBLEM No.1 - OPTIMAL DESIGN OF RECTANGULAR GROUND STORAGE TANKS

3.1 Problem Description, Loading and Analysis

A layout of a typical ground storage tank for purified water is shown in Fig.2. The structural loading includes the following cases: Empty tank before embankment of sides and roof with soil i.e. under dead load condition, tank under internal hydrostatic pressure with various compartments being full i.e. representing hydraulic test condition before embankment with soil, and tank empty under service conditions with full external earth and ground water pressures acting.

The tank is analysed elastically and is idealised as a one way frame in the plane of the cross section (AA) in Fig.2. The stiffness matrix method is employed in this analysis. Longitudinal walls are analysed as plates under the various load cases.

3.2 Design Variables, Objective Function and Constraints

For a given storage volume, the problem variables are:

i) Number of internal baffle walls N_b
ii) Clear internal dimensions B and H
iii) Effective depth d_i for each member (wall or slab units)
iv) Area of longitudinal steel reinforcement at six critical locations for each member, namely at two ends and at maximum span moment location for each of two member faces.
v) Area of transverse steel reinforcement for each member.

The objective function comprises cost of concrete, reinforcement, formwork, as well as costs of earthwork, waterproofing, movement and construction joints. The constraints are those derived from design in accordance with BS 5337 (Now replaced by BS 8007), as well as practical side constraints.

3.3 Multi-Level Formulation and Optimization Algorithm

Fig.1 shows a flow chart from which it may be observed that the number of internal (baffle) walls is optimized in the fourth level, tank overall dimensions in the third level, member thicknesses in the second and steel reinforcement in the first level. The analysis is carried out in the third level, with re-analysis being performed only when all members have been through the optimization cycles in the second and first levels. Convergence limits are placed in the third level on both design forces and on total cost. In the second level, the concrete dimensions of each member are optimized whereby after each modification in dimensions entry is made to the first level to determine the amount of steel reinforcement in the critical sections of that member. When the optimum cost of that member is achieved in accordance with a pre-assigned tolerance, the next member is optimized.

The optimization algorithm used here is due to Rosenbrock for constrained optimization [9]. Being a direct search technique, this method does not involve the often very difficult calculation of derivatives of the objective function and constraints. This method has been widely used for solving different types of non-linear constrained and non-constrained optimization problems with satisfactory results [10].

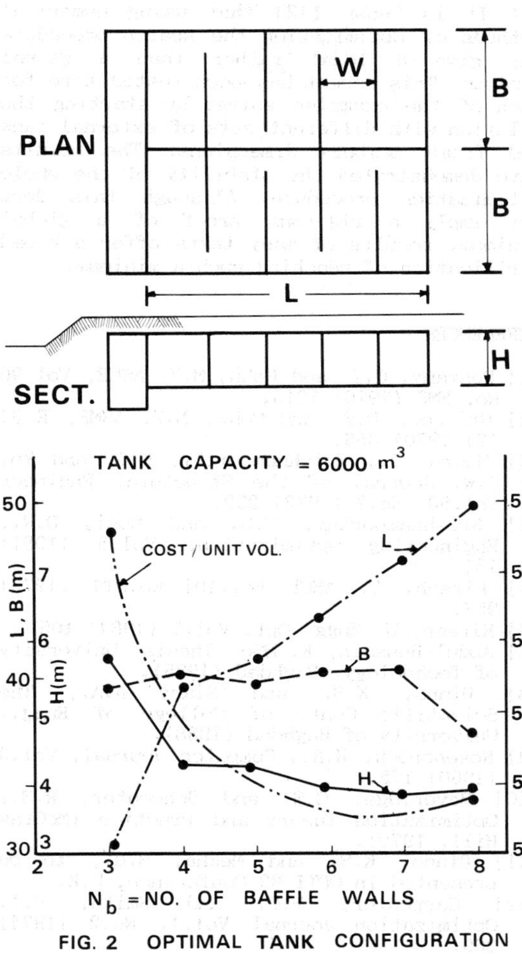

FIG. 2 OPTIMAL TANK CONFIGURATION

3.4 Numerical Results

The optimal design was determined for tanks of capacities 500, 1000, 1500, 2000, 2500, 3000, 4000, 5000, 6000 cu.m. for assumed typical technical data and unit prices. The number of members and hence number of variables and constraints depends on the number of internal baffle walls. For a minimum of 2 baffle walls 135 variables and 360 constraints are involved increasing to 261 variables and 1670 constraints for 8 baffle walls. A typical optimal configuration is shown in Fig.2 for a 6000 cu.m tank.

4. PROBLEM No.2 OPTIMAL DESIGN OF R.C. FRAMES

4.1 Problem Description, Loading and Analysis

A typical frame is shown in Fig.3. Factored dead, live and wind loads appropriate for the ultimate as well as the serviceability limit states of design are considered. The frame is analysed elastically by the stiffness matrix method, but may be analysed inelastically [11]. Design force and moment envelopes are developed in order to ensure that each member is designed for the most critical loading.

4.2 Design Variables, Objective Function and Constraints

Spans and storey heights are considered pre-assigned. Design variables for beams comprise width, depth and areas of reinforcing steel at three critical sections taken to be at the two ends and at location of maximum span moment envelope. Following the ACI 318-83 design rules, steel areas are calculated directly from internal stress resultants and cross sectional dimensions. This allows for treating the nine unknown steel areas as dependent variables. For columns the independent variables are column dimensions while dependent variables comprise longitudinal and transverse reinforcement at the two end critical sections.

The design is made to satisfy various ACI code requirements as constraints in addition to side constraints which are included to reflect construction and functional requirements.

4.3 Multi-Level Formulation and Optimization Algorithm

The flow chart for this problem is similar to that shown in Fig.1 except that only three levels are involved. In the third or outer level, design forces are evaluated from structural analysis (elastic or inelastic as the case may be). When an elastic analysis is followed for the purpose of ultimate strength design moment redistribution is permitted according to design codes. This is performed in the third level. The analysis is repeated only when all members are optimized in the second and first levels. Convergence criteria and optimization algorithm are as for problem No.1.

4.4 Numerical Results

For the frame shown in Fig.2, 56 independent variables and 588 constraints are involved.

Fig.3 shows the moment and cost convergence behaviours of the frame.

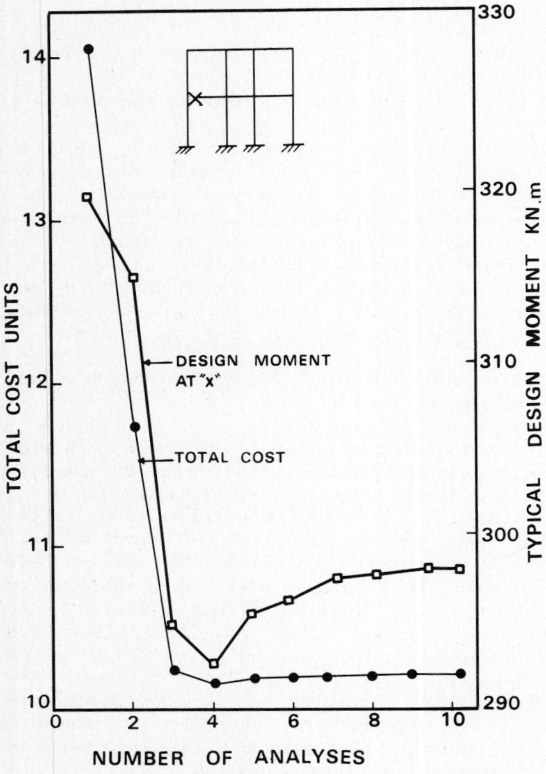

FIG. 3 FRAME CONVERGENCE BEHAVIOUR

5. DISCUSSION AND CONCLUSIONS

5.1 Using the multi-level approach it has been possible to handle the optimal design of problems with as many as 261 variables and 1670 constraints. In spite of the large number of variables, decomposition of the problem has ensured that each sub-optimization problem involves only few variables with enormous ensuing computational advantages.

5.2 Considering the overall convergence behaviour of the solution, two convergence controls were placed at the third level namely on overall cost and on internal moments at salient points in the structure. The convergence control on the internal moments has been found to be important to ensure that an optimum cost solution is also statically admissible and in most cases this determined the need for further re-analysis cycles. Controls on moments were found to be more critical than on shear forces and were therefore adopted.

5.3 The Rosenbrock method of constrained optimization requires starting with feasible values of the variables. However, from several trial solutions it has been found that the efficiency of solution is greatly enhanced by adopting near minimum values of the variables as initial feasible values rather than arbitrarily chosen values. In carrying out a re-analysis it is important to start again with feasible values (again preferably near min. values) and not with the values reached in the previous cycle; for such values may not be feasible in the new cycle.

5.4 It is known [12] that using numerical methods of optimization the search procedure may give a local rather than a global minimum. This aspect has been tested here for each of the examples solved by starting the solution with different sets of external tank and frame members dimensions. The results have demonstrated the stability of the whole optimization procedure. Although this does not imply a rigorous proof of a global minimum, results of many tests offer a broad verification of reaching such a minimum.

REFERENCES

[1] Rosvany, G.I. and Cohn, M.Z. ASCE, Vol 96 No. EM6 (1970) 1013.
[2] Grieson, D.E. and Cohn, M.Z. ASME, E 37 (2)(1970) 356.
[3] Munro, J., Krishnamoorthy, C.S. and Yu, C.W. Journal of the Structural Engineer Vol.50, No.7 (1972) 259.
[4] Krishnamoorthy, C.S. and Mosi, D.R., Engineering Optimization, Vol.5 (1981) 151.
[5] Kirsch, U. ASCE Vol.101 No.ST4 (1975) 957.
[6] Kirsch, U. Engg. Opt. Vol.7 (1984) 105.
[7] Abdul-Hussain, K. MSc. Thesis, University of Technology, Baghdad (1985).
[8] Dinno, K.S. and Slaby A.A., The Scientific Conf. of College of Engg., University of Baghdad (1988).
[9] Rosenbrock, H.H., Computer Journal, Vol.3 (1960) 175.
[10] Beveridge, G.S. and Schechter, R.S., Optimization Theory and Practice (McGraw Hill, 1970).
[11] Dinno, K.S. and Mekha, B.B., to be presented in OPTI 89 Conference, U.K.
[12] Carpenter, W.C. and Smith, E.A. Optimization Journal Vol.1, No.2 (1974) 269.

Approximation, Optimization and Computing:
Theory and Applications, A.G. Law and C.L. Wang (eds.)
Elsevier Science Publishers B.V. (North-Holland)

RELIABILITY ASSESSMENT IN FRACTURE MECHANICS*

Gong, Yaonan and Qian, Chun

Solid Mechanics Research Center, Beijing University of Aeronautics and Astronautics, P.O.Box 85, Beijing 100083 CHINA

In the present paper, a modified second–moment reliability index β criterion for evaluating the reliability of predicted fatigue life (or, the da / dN) is suggested. Instead of the common used Gauss normal distribution, the Weibull distribution is used in this paper. Correspondingly, a new definition of the reliability index β, different from that proposed earlier, is needed and proposed in this paper. Together with the modified second–moment reliability index criterion, the direct Monte Carlo simulation is also used to verify the accuracy of the suggested method. Results obtained by two methods agreed very well with each other.

1. INTRODUCTION

The reliability analysis of the predicted crack growth life is an important task, since it is closely related to the safety design of engineering structures such as flight vehicle, naval vessel, submarine, offshore plateform, pressurized container, nuclear reactor, etc.. The reliability of a cracked structural component is mainly influenced by three parameters, namely, the magnitude and distribution of load (or stress), material property, and size and distribution of defect. For a veriety of causes related to the metallurgy and machining process, more or less defects (say, cracks or flaws) exist in a structural component. In fact, as well as the load and material property, the size, number and distribution of the crack are uncertain. However, its mechanical effects can be simulated with a single equivalent crack of random size.

The conventional analysis and design procedures are based on the deterministic parameter assumption, though almost all of the design parameters are uncertain. In the past, in order to ensure a safe structure, a safety factor has been used in the design and analysis process. Nevertheless, it has been verified in the practice and will be shown below that this technique cannot ensure the structure to be safe. Recently, one has recognized that it is reasonable and feasible to ensure the structure to be safe through an acceptable failure probability P_f (say, 0.01% or less). In this case the reliability of structure system will be

$$P_R = 1 - P_f \tag{1}$$

The most intuitive method can be used in reliability assessment is one of the statistical approaches, namely, the direct Monte Carlo simulation, a kind of numerical test on the computer. However, due to its immense expenditure on computational efforts, the direct Monte Carlo simulation is not applicable to a large scale engineering problem, especially as nonlinearity is involved. In 1969, Cornell proposed the concept of reliability index β. In 1974, Hasofer and Lind [2] further proposed the second–moment reliability index criterion. Later on, Ove Ditlevsen [3] proposed a generalized second –moment reliability index and its narrow bounds. In these methods, the definition of reliability index β is based on the assumption that the probability density function follows the Gauss normal distribution. However, in probabilistic fracture mechanics, Gauss normal distribution is not a favorite candidate, thus the Weibull distribution is adopted. Correspondingly, a modified definition for the reliability index β is needed and suggested in this paper.

The present paper focuses attention on suggesting a feasible approach for the reliability assessment of the predicted crack growth rate da / dN under cyclic loading. Again, to verify its validity, the direct Monte Carlo simulation is used as well.

2. APPROACHES

In what follows, for simplicity but no loss of generality, it is assumed that both the load and size of defect are random parameters, but the material property is deterministic. Thus the basic (input) variables of the problem to be considered are

$$X = [x_1, x_2, x_3]^T = [S_{max}, R, a]^T$$

* Research project supported by the Aeronautical Science Foundation under grant no. 87 B51403

where S_{max} is the maximum nominal stress under cyclic loading, R the stress ratio ($=S_{min}/S_{max}$), and "a" the size of a single equivalent crack. Let $(da/dN)_{sp}=const$ be a design value of the crack growth rate. Then the failure criterion can be written as

$$G^*(X)=(da/dN)-const \geqslant 0 \quad (2)$$

or alternatively, the safe criterion will be

$$G(X)=(da/dN)-const<0 \quad (3)$$

The crack growth rate is a function of S_{max}, R, and crack size 2a, and can be simulated with Walker's formula

$$da/dN=C\,[K_{max}(1-R)^q]^p, \; with \; R\geqslant 0 \quad (4)$$

for linear case, where the material constants C, q, p in Eq. (4) should be determined experimentally. Seeing its great versatility and that it can be adjusted to fit many different distribution fashions, Weibull model is adopted in this paper for the interpretation of the scatter in load and crack size. Location Parameter x_o, scale parameter B and shape parameter m in Weibull distribution

$$f(x)=[m\frac{(x-x_0)^{m-1}}{B^m}]\,exp[-(\frac{x-x_0}{B})^m] \quad (5)$$

are assigned as

Table 1. Parameter values

variable	X_0	B	m
S_{max}(MPa)	0.0	50.0	15
R	0.0	0.40	15
a(m)	0.0	0.00015	1

Two approaches for evaluating the probability of failure–free operation of a cracked plate are presented as below:

2.1. Modified Reliability Index β Criterion (MRI)

Let

$$Y=[y_1,y_2,y_3]^T$$

be normalized orthogonal basic variables where

$$y_i=(x_i-x_{0i})/B_i \quad (6)$$

Assume that x_i and x_j are unrelated variables in the present case, namely

$$cov[y_i,y_j]=A$$

where A is a diagonal matrix. Define

$$Z=[z_1,z_2,z_3]^T$$

$$with \; z_i=y_i^{mi} \quad (7)$$

$$and \; \beta=(Z^TZ)^{\frac{1}{2}} \quad (8)$$

where β is the shortest distance from the origin of the normalized space to the boundary of failure region $G^*(X)=O$. Now the problem may be stated as

$$find \; Z$$

$$min \; \beta=(Z^TZ)^{\frac{1}{2}} \quad (9)$$

$$s.t. \; G^*(Z)=0,$$

By use of penalty function method, the above constrained minimization problem can be reduced to a unconstrained one,

$$min \; F(Z,r)=\beta(Z)-G^*(Z)/r \quad (10)$$

where r is a penalty factor. Searching the minimal point Z^* of β by, say, the Powell method [4], the probability to failure–free operation of the cracked plate, in the sense of G(Z) less than 0, can be obtained.

2.2. Direct Monte Carlo Simulation (DMCS)

Direct Monte Carlo simulation is essentially a kind of numerical test. At first, a set of evenly distributed random number r are, according to

$$i=mod\,[i\times j+k,\; m]$$

and

$$r=(real(i)+0.5)/real(m),$$

generated on the computer [5,6]. Giving an initial value of i, a new random value i (integer) and r(real) are generated. Conducting this process successively, a set of evenly distributed random number r can be obtained. Then, in light of the probability density function of maximum nominal stress S, stress ratio R, and crack size a, this set of evenly distributed random number r are transformed into corresponding set of Weibull distribution S, R and a by [1], for example,

$$S = B_s(-\ln(1 - R_s))^{(1/m_s)}, \ etc.$$

where subscript s denotes that the parameter concerned is relative to the nominal stress S. The computed results are compared with that from Eq.(6) in Figure 1.

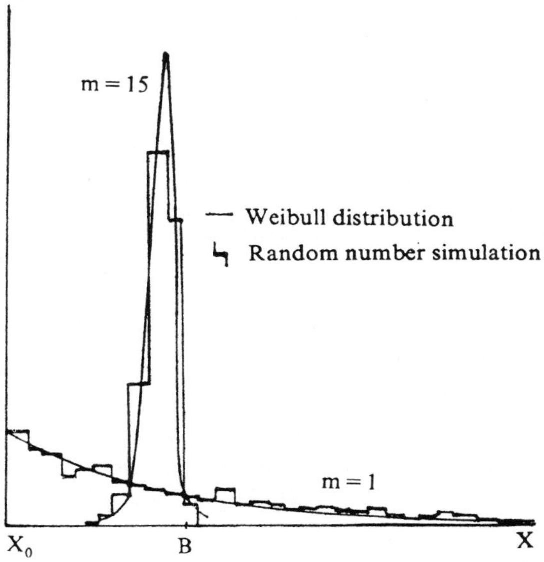

Fig.1 comparison of computed results with that from Eq.(6)

Each da / dN obtained from a particular set of S, R and a is compared with the design value const; if da / dN is less than const, this parameter set would not lead the cracked plate to failure, thus counted as a " miss" , otherwise a "hit" is counted. Consequently, the reliability can be evaluated as

$$P_R = \frac{total\ number\ of\ \text{"miss"}}{total\ number\ of\ simulation}$$

Results of reliability assessment from two different approaches agreed well with each other, as shown in Table 2. The relative error lies within 0.2% – 1.0%, depending on the design value const of da / dN. The authors considered that, by figure 1 and Table 3, the most part of relative error comes from the direct Monte Carlo simulation.

Table 2. comparison of Monte Carlo simulation and penalty function method for total number of simulation n = 27000, B_a = 0.00015

const	Reliability		
(10^{-7}mm / cycle)	DMCS	MRI	relative error
6.0	0.9990371	0.9895387	0.95%
7.0	0.9994815	0.9927956	0.67%
8.0	0.9996667	0.9949113	0.48%
9.0	0.9998149	0.9963305	0.35%
10.0	0.9998519	0.9978075	0.25%
11.0	0.9999260	0.9979948	0.19%
12.0	0.9999630	0.9984874	0.15%
13.0	1.0000000	0.9988460	0.12%

Table 3. influence of simulation number n and initial random number i on the predicted reliability

n	i_s	i_R	i_a	P_R
64000	10	10	10	0.9698281
64000	10	1	100	0.9805938
64000	10000	1	300	0.9798282
27000	10	10	10	0.9710371
8000	10	10	10	0.9731250
1000	1	1	1	0.9740000
1000	10000	1	300	0.9810000

REFERENCES

1. Probabilistic Fracture Mechanics and Reliability, edited by J.W.Provan, Martinus Nijhoff Publishers, Dordrecht (1987)

2. Hasofer, A.M. and Lind, N.C.: An exact and invariant second–moment code format, J. of the Engineering Mechanics Div., vol.100, EM1:111–121 (1974)

3. Ditlevsen, Ove: Generalized second–moment reliability index, J Struct. Mech. 7(4) 435–451 (1979)

4. Masano Aoki, Introduction to optimization techniques, The MacMillan Company, New York (1971)

5. Merchant, M.J.: FORTRAN 77 – Language and Style, Wadsworth Publishing Co., (1981) 252–254

6. Kruth, D.E.: Art of computer programming, vol.2, 2nd ed., Addison–Wesley (1973)

Approximation, Optimization and Computing:
Theory and Applications, A.G. Law and C.L. Wang (eds.)
Elsevier Science Publishers B.V. (North-Holland)

ON ROUGH MODELS OF HISTORY RECORDS

Adam KOWALCZYK

Telecom Australia Research Laboratories, P.O. Box 249, Clayton, Vic. 3168, Australia

Applicability of many machine learning techniques in a number of areas of practical interest (like telecommunication network, econometric models) are severely limited by the complexity of their mathematical models which involve thousands of attributes. The aim of this note is to show that information theory and global differential geometry provide techniques for dealing with complementary aspects of simplification of such models.

1. INTRODUCTION

States of many systems of practical interest (like telecommunication networks, econometrics models, etc.) are naturally described in terms of considerable numbers of parameters. This leads to a combinatorial explosion in the total number of options to be analyzed, with the result that only rather small size problems can be directly handled by present day computers, and future improvements in computing power will change the situation only marginally. These limitations are extremely severe in the case of machine learning, when we want to analyze and extract dominant global features of the system. In this particular case prospects are seriously dependent on whether practical techniques can be found for automatic introduction of far reaching simplifications and abstractions of the system modelling.

To make the discussion more concrete we concentrate on the example of history records (HR) for a telecommunication network. (Ultimately HR is to be used for discovery of new heuristics like a strategic change corresponding to 1-3 jump in Fig. 1 for network management). A complete HR can be in practice quite large. For example, let us consider a 40-node network. To describe all possible demand states in this case we need $40 \cdot 30 = 1560$ variables. Even if we restrict the number of calls between any pair of nodes to be either 0 or 1 we end up with the huge number ($2^{1560} \approx 10^{468}$) of possible demands. (A present day supercomputer can perform during the Age of Universe only $\approx 3.8 \cdot 10^{27}$ simple operations.) Although in practice we encounter only a minute number of the total number of Network states, each state will be practically unique and before we gather sufficient experience the size of the record will be practically unmanageable. A cure for this problem may involve the following:

(*i*) introduce clustering of data,

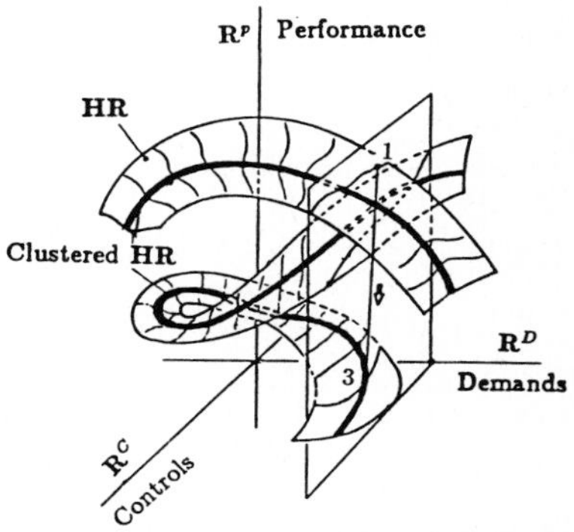

Figure 1: Example of simplified history record.

(*ii*) record only selected information,

(*iii*) for learning of dominant global features use only maximally simplified (although adequate) models.

Obviously implementation of the above recommendations will depend heavily on a particular HR under consideration. In this paper we try to estimate how much could be achieved along these lines.

2. DATA CLUSTERING

Clustering is obviously one of the main methods of creating simplified versions of HR. Different aspects of data clustering are currently intensively researched in different areas of artificial intelligence (e.g. in pattern recognition, in the rough sets theory). But some specific adaptation of this research to the case of HR with emphasis on dynamics, in the sense of continuous evolution in time, will yet need some attention. We do not intend to pursue this difficult subject any further in this paper for the following reason. Although the data clustering can cause a significant drop in the number of individual records, it does not have much potential for reduction of dimensionalities of the demands and control spaces which is the main theme of this paper.

3. SELECTION OF SIGNIFICANT ATTRIBUTES

We define a *formal history record* (FHR) as a triple

$$\mathbf{H} \stackrel{\text{def}}{=} (\mathbf{B}, \mathbf{R}^D \times \mathbf{R}^C \times \mathbf{R}^p, \mu), \quad (1)$$

where $\mathbf{R}^D$, $\mathbf{R}^C$ and $\mathbf{R}^p$ are the D, C and p dimensional Cartesian spaces interpreted as demand, control and performance spaces (Fig. 1), respectively. The *body of the record*, $\mathbf{B}$, is a finite subset of $\mathbf{R}^D \times \mathbf{R}^C \times \mathbf{R}^p$ and μ is a probability measure (density) on $\mathbf{B}$. Ideally $\mathbf{H}$ should be *deterministic*, which means that each point of $\mathbf{B}$ is uniquely determined by its projection on the demand-control space $\mathbf{R}^D \times \mathbf{R}^C$.

Let $I \stackrel{\text{def}}{=} (i_1, ..., i_c)$ and $J \stackrel{\text{def}}{=} (j_1, ..., j_d)$ be two multi-indices such that $1 \le i_1 < ... < i_c \le C$ and $1 \le j_1 < ... < j_d \le D$. Let π_{IJ} denote the projection

$$\pi_{IJ} : \mathbf{R}^D \times \mathbf{R}^C \times \mathbf{R}^p \to \mathbf{R}^d \times \mathbf{R}^c \times \mathbf{R}^p, (x, y, z) \mapsto (x_I, y_J, z),$$

$\mathbf{B}_{IJ} \stackrel{\text{def}}{=} \pi_{IJ}(\mathbf{B})$ and μ_{IJ} be the probability measure induced on $\mathbf{B}_{IJ}$. The FHR $\mathbf{H}_{IJ} \stackrel{\text{def}}{=} (\mathbf{R}^d \times \mathbf{R}^c \times \mathbf{R}^p, \mathbf{B}_{IJ}, \mu_{IJ})$ will be called *IJ-reduct* of $\mathbf{H}$ (or, more generically, *reduct* of $\mathbf{H}$).

EXAMPLE 3.1. Let us consider a FHR $\mathbf{H} \stackrel{\text{def}}{=} (\mathbf{B}, \mathbf{R}^2 \times \mathbf{R}^0 \times \mathbf{R} \approx \mathbf{R}^2 \times \mathbf{R}, \mu)$ with the trivial control space ($C = 0$), the body $\mathbf{B}$ having six points $(0,0,0)$, $(0,1,0.1)$, $(1,0,1)$, $(1,2,1.2)$, $(2,1,2)$ and $(2,2,2.3)$ and the uniform probability density.

There are only two proper reducts in this case, $\mathbf{H}_1 \stackrel{\text{def}}{=} \mathbf{H}_{(1)()}$ with the body

$$\mathbf{B}_{(1)()} \stackrel{\text{def}}{=} \{(0,0), (0,0.1), (1,1), (1,1.2), (2,2), (2,2.3)\}$$

corresponding to the projection $\pi_{(1)()} : (x_1, x_2, z) \mapsto (x_1, z)$ and $\mathbf{H}_2 \stackrel{\text{def}}{=} \mathbf{H}_{(2)()}$ having the body

$$\mathbf{B}_{(2)()} \stackrel{\text{def}}{=} \{(0,0), (0,1), (1,1.1), (1,2), (2,1.2), (2,2.3)\}$$

and corresponding to the projection $\pi_{(2)()} : (x_1, x_2, z) \mapsto (x_2, z)$. By inspection we find that $\mathbf{H}$, but none of its reducts, is deterministic. Further, $\mathbf{H}_1$ is much 'closer to being deterministic' than $\mathbf{H}_2$. For instance, $\mathbf{H}_1$ allows one to determine the performance as a function of demand up to 0.15 (by approximation by averages) while for $\mathbf{H}_2$ such an error reaches .85. In this sense $\mathbf{H}_1$ contains *'more information'* about $\mathbf{H}$ than $\mathbf{H}_2$ does. □

3.1 MODIFIED SHANNON'S ENTROPY APPROACH

Heuristically, a reduct is better, i.e. closer to being deterministic, if on average it:

(*i*) clusters less fibres of $\mathbf{B}$ over the demand-control space $\mathbf{R}^C \times \mathbf{R}^D$,

(*ii*) the clustered fibres have more similar values of the performance parameters.

It is known that Shannon's entropy can be used to quantify (*i*) (c.f. Ben-Bassat in [7, pp. 773–791]), but it does not take into account (*ii*) at all. In order to quantify both (*i*) and (*ii*) we introduce the following generalisation of the information entropy.

For the FHR defined by (1) let us denote by Π the projection of $\mathbf{B}$ into $\mathbf{R}^p$. A function $\psi : \Pi \times \Pi \to [0,1]$ such that $\psi(z,z) = 1$ and $\psi(z,z') = \psi(z',z)$ for every $z, z' \in \Pi$ will be called *a similarity relation on* $\mathbf{H}$ (c.f. [12] [13]). We interpret $\psi(z,z')$ as degree of similarity between z and z' with the following convention: the bigger $\psi(z,z')$ the more similar z and z' are. An example of similarity relation is *the trivial similarity relation* τ defined as 1 on the diagonal $\{(z,z)\ ;\ z \in \Pi)\}$ of $\Pi \times \Pi$ and as 0 elsewhere.

Let $\mathcal{C}_{\mathbf{H}}$ and $\mathcal{P}_{\mathbf{H}}$ be partitions of $\mathbf{B}$ according to values of projections in $\mathbf{R}^C \times \mathbf{R}^D$ and $\mathbf{R}^p$. The number

$$-\sum_{\substack{(A,B)\\ A\cap B\neq\emptyset}} \mu(A\cap B)\log\left(\sum_{C}\frac{\mu(A\cap C)}{\mu(A)}\psi(\pi(B),\pi(C))\right),$$

where $A \in \mathcal{C}_{\mathbf{H}}$ and $B, C \in \mathcal{P}_{\mathbf{H}}$,

will be called ψ*-entropy* of $\mathbf{H}$ and denoted by $H_\psi(\mathbf{H})$. This definition is analogous to the notion of rough entropy considered in [6] and is a straightforward modification of the classical conditional entropy of partitions [1,9,11].

PROPOSITION 3.1

(i) The entropy $H_\tau(\mathbf{H})$ *is equal to the the classical conditional entropy* $H(\mathcal{P}_{\mathbf{H}}|\mathcal{C}_{\mathbf{H}})$ *of the partition* $\mathcal{P}_{\mathbf{H}}$ *assuming the partition* $\mathcal{C}_{\mathbf{H}}$ [9].

(ii) $0 \le H_\psi(\mathbf{H}) \le H_\tau(\mathbf{H})$. □

In analogy to classical conditional entropy we can interpret $H_\psi(\mathbf{H})$ as a measure of uncertainty about values of performance parameters if values of demands and controls are known. Theorem 3.2 below provides another heuristic interpretation of $H_\psi(\mathbf{H})$.

We say that a FHR $\mathbf{H}$ is ψ*-deterministic* if $\psi \equiv 1$ on all demand-control fibres, i.e if $\psi(z,z') = 1$ for all pairs of $\mathbf{B}$ of the form (x,y,z) and (x,y,z'). The following results are straightforward to prove.

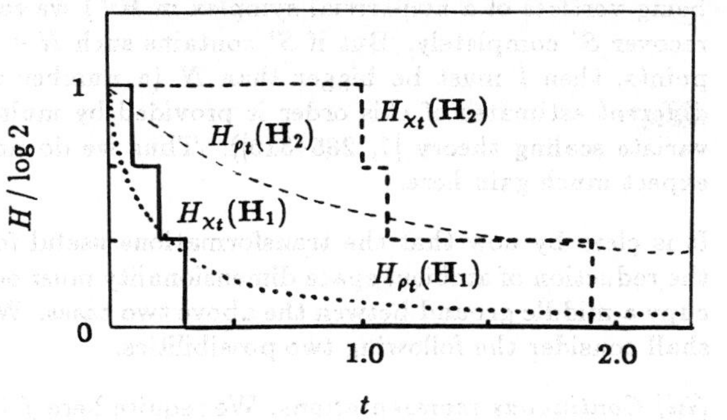

Figure 2: Plots of entropies for Example 3.2.

THEOREM 3.2 *$H_\psi(\mathbf{H}) = 0$ iff $\mathbf{H}$ is ψ-deterministic.* □

COROLLARY 3.3 *$H_\tau(\mathbf{H}) = 0$ iff $\mathbf{H}$ is deterministic.* □

By virtue of the above results $H_\psi(\mathbf{H})$ can be viewed as a measure of how far $\mathbf{H}$ is from being ψ-deterministic. We postulate:

HYPOTHESIS *ψ-entropy of reducts, $H_\psi(\mathbf{H}_{IJ})$, is a plausible measure of quality of reducts: the smaller it is, the better (i.e. 'closer' to being deterministic), is the reduct.* □

Usefulness of this hypothesis is a matter for future verification. The following example is a simple illustration how all this can work.

EXAMPLE 3.2 Let $\mathbf{H}$, $\mathbf{H_1}$ and $\mathbf{H_2}$ be as in Example 3.1 and $\chi_t(z, z')$, $z, z' \in \Pi \subset \mathbf{R}$, be a similarity relation on $\mathbf{H}$ parametrised by $t > 0$ and defined as 1 for $|z - z'| \le t$ and 0, otherwise.

Entropies $H_{\chi_t}(\mathbf{H_1})$ and $H_{\chi_t}(\mathbf{H_2})$ are plotted in Fig. 2. Inspection of the Figure shows that $H_{\chi_t}(\mathbf{H_1}) < H_{\chi_t}(\mathbf{H_2})$ for most values of t. Thus, according to the above hypothesis the reduct $\mathbf{H_1}$ is better than $\mathbf{H_2}$. This agrees fully with our previous heuristic conclusion reached in Example 3.1.

Note that χ_0 is the trivial similarity relation τ. Since $H_{\chi_0}(\mathbf{H_1}) = H_{\chi_0}(\mathbf{H_2})$ it follows that classical conditional entropy (and classical information theory approach) is unable to resolve properly the question which of these two reducts contains 'more information'. We remark that Shannon's entropy could perform even worse by giving a counter-intuitive answer [6].

In Fig. 2 we also plot ρ_t-entropies of $\mathbf{H_1}$ and of $\mathbf{H_2}$ for the following similarity relation

$$\rho_t(z, z') \stackrel{\text{def}}{=} \frac{1}{1 + |z - z'|/t} \quad \text{for } z, z' \in \Pi \text{ and } t > 0.$$

Although the shapes of the ρ_t-entropy curves are different fromχ_t-entropies, the final conclusion that $\mathbf{H_1}$ contains 'more information' about $\mathbf{H}$ than $\mathbf{H_2}$ is the same. □

3.2. DECISION THEORY AND SECOND MOMENT APPROACHES

Alternative approaches for selection of significant attributes can be considered. The decision theory and the second moment approaches will be discussed here briefly, mainly for the sake of comparison with the entropy approach and with emphases on their shortcomings. These shortcomings should be overcome if these techniques are to be applied to simplification of history records.

(i) Decision theory approach. This approach uses the frame of statistical decision theory [8] and includes approximation theory based methods in particular. It requires preselection of both the cost function and the class of decision mappings. Thus it is more restrictive than the entropy approach which requires only preselection of a similarity relation (being, in a sense, analogous to the choice of cost function).

(ii) Second moment methods. Here we refer to the group of methods like Karhunen-Loeve transformation in pattern recognition [7], the principal component method in classical statistics [8] and intrinsic dimensionality extraction in multidimensional scaling [7, pp. 347–360].

These methods are very sensitive to the choice of the numerical representation for discrete (non-numerical in their nature) attributes. Additionally, the final results will be almost completely determined by the requirement of 'good resolution' of the bulk of the attributes, i.e. control and demand attributes. The most important to us - good resolution of the performance attributes which are typically in a relatively small number - will play a negligible role in this approach and so the final result will not very useful.

4. GEOMETRICAL VIEW

The methods for simplification of FHR by recording only a small number of essential attributes discussed in the previous Section are obviously more efficient if applied to sufficiently small chunks of the record rather than globally. Such an approach will lead to a specific, locally 'low dimensional' FHR. However, this FHR is still imbedded in the ambient space of the original high dimension. One can now turn to more general transformations (than projections) of such records into lower-dimensional spaces. An estimate of the potential gain from such approaches derived from some general imbedding results of topology and differential geometry is presented in this section. At first we must suitably extend the notions of continuous and smooth transformations to finite subsets of Euclidean spaces.

4.1. Transformations of Finite subsets

For a multi-index $I = (i_1, ..., i_n)$, $1 \leq i_1 < ..., i_n \leq N$, we denote by $\mathbf{R}_I^N \subset \mathbf{R}^N$ the n-dimensional coordinate subspace

$$x_i = 0 \quad \text{for every } i \in \{1, ..., N\} - \{i_1, ..., i_n\}.$$

The subset $D_I(\epsilon, x_o) \stackrel{\text{def}}{=} \mathbf{R}_I^N \cap B(\epsilon, x_o)$, where $B(\epsilon, x_o)$ is an ϵ-ball centered at x_o, will be called an *n-ϵ-disc*. A subset $S \subset \mathbf{R}^N$ is called *n-ϵ-local* if for every $x \in S$ there exists an n'-ϵ-local disk $D_{I(x)}(\epsilon, x)$ such that $1 \leq n' \leq n$ and

$$S \cap B(\epsilon, x_o) \subset D_{I(x)}(\epsilon, x).$$

A subset $S' \subset \mathbf{R}^N$ will be called an *n-ϵ-dlocal extension of S* if (i) it is n-ϵ-local, (ii) $S \subset S'$ and (iii) for every $x \in S$ there exist $x' \in S$ and two multi-indices I, I' such that

$$D_{I'}(\epsilon, x') \subset S' \cap B(2\epsilon, x) \subset D_I(2\epsilon, x).$$

Now we focus on a finite n-ϵ-local subsets and their (always infinite!) n-ϵ-dlocal extensions.

THEOREM 4.1 [5] *Any n-ϵ-local subset of $\mathbf{R}^N$ has an n-δ-dlocal extension for every $\delta, \epsilon/2$.* □

Now let us consider transformations $f : S' \to \mathbf{R}^l$, $l \in \{1, 2, ...\}$. (We say also that $f(S')$ is a representation of S' in $\mathbf{R}^l$.) Such a transformation is equivalent to the choice of l abstract attributes (functions) on S'. We are interested in minimizing l subject to certain restrictions on f. Some possibilities are:

(i) Enumeration. We require f to be 1-1 (one-to-one) only. In this case $l = 1$ is always sufficient. However, such an approach has serious drawbacks. Typically we encounter here discontinuity and complete loss of information on the structure of S' (close points are randomly separated, distant points become close, etc.).

(ii) Isometries. We require that f is an isometry, i.e.

$$|f(x) - f(x')|_l = |x - x'|_N \quad \text{for every } x, x' \in D.$$

In such a case $f(S')$ determines S' up to a rigid motion of $\mathbf{R}^N$. It means, in particular, that knowing $f(S')$ and additionally $N+1$ points of S' in general position (i.e. being vertices of a non-trivial symplex in $\mathbf{R}^N$) we can recover S' completely. But if S' contains such $N+1$ points, then l must be bigger than N (a number of different estimates of this order is provided by multivariate scaling theory [7, 285–316]). Thus we do not expect much gain here.

It is clear by now that the transformations useful for the reduction of ambient space dimensionality must occupy a middle ground betwen the above two cases. We shall consider the following two possibilities.

(iii) Continuous representations. We require here f to be 1-1 and 'continuous' or even 'smooth', in the sense of having a C^∞-extension onto an open neighbourhood of S', in order to preserve applicability of some numerical algorithms. In this case points which are close will remain close and so it will be possible to model evolution and dynamical changes on $f(S')$ to a certain extent. However, working on the level of the representation $f(S')$ we cannot determine how big changes, in terms of S', are required to modify the system from one state to another (speed of change, in particular).

(iv) Local determination of geometry (LGD-representations). We require f to be 1-1, smooth and to preserve full information about local geometry, i.e. about original distances between different states in sufficiently small chunks of S'. More precisely, in this case we require that for every (linear) hyperplane $\Lambda \subset \mathbf{R}^N$ and every connected and open in Λ component V of $\Lambda \cap S'$, the set $f(V)$ determines V up to a rigid motion of R^N. This eliminates the main drawbacks of continuous transformations (iii). Note, however, that in this case we lose information about global geometry in S', but this is inevitable if $l < N$ and is of little interest, in general.

From the imbedding theorems for differentiable spaces [2] [4] combined with elements of Riemannian geometry the following results have been derived [5].

THEOREM 4.2 *Every n-ϵ-dlocal subset of $\mathbf{R}^N$ has a $(2n+1)$-dimensional smooth representation and an $(n+2)(2n+1)$-dimensional smooth LGD-representation.* □

REMARK There are examples of of n-dimensional manifolds which cannot be imbedded into $\mathbf{R}^{2n-1}$. On this ground it is reasonable to expect that examples of n-ϵ-local data bases which cannot be smoothly imbedded into $\mathbf{R}^{2n-1}$ also exist. Thus the above estimate for smooth representations is probably close to optimal. □

EXAMPLE 4.1 Every 10-ϵ-dlocal subset $S' \subset \mathbf{R}^N$ has a smooth representation in $\mathbf{R}^{21}$ and an LGD representation in $\mathbf{R}^{12 \cdot 21} = \mathbf{R}^{252}$ (independently of N). □

4.2. THE CASE OF HISTORY RECORDS

Let $\mathbf{H} = (\mathbf{B}, \mathbf{R}^D \times \mathbf{R}^D \times \mathbf{R}^p, \mu_{\mathbf{H}})$ be a FHR and Δ and Γ be the projections of $\mathbf{B}$ into $\mathbf{R}^D$ and $\mathbf{R}^C$, respectively. We consider below imbeddings of $\mathbf{H}$ in $\mathbf{R}^{D'} \times \mathbf{R}^{C'} \times \mathbf{R}^p$ of the form $(x, y, z) \mapsto (\phi(x), \psi(y), z)$.

THEOREM 4.3 *Let the FHR be such that the projections $\Delta \subset \mathbf{R}^D$ and $\Gamma \subset \mathbf{R}^C$ are d-ϵ-local and c-ϵ-local subsets, respectively. Then $\mathbf{B}$ has*

(i) an imbedding into $\mathbf{R}^{D'} \times \mathbf{R}^{C'} \times \mathbf{R}^p$, $D' = 2d + 1$ and $C' = 2c + 1$ with ϕ and ψ smooth on some n-$\epsilon/2$-extensions of Δ and Γ, respectively,

(ii) an imbedding into $\mathbf{R}^{D'} \times \mathbf{R}^{C'} \times \mathbf{R}^p$, $D' = (d+2)\times \times(2d + 1)$, $C' = 2(c + d)(c + d + 1) + 5c + 3$ with ϕ and ψ smooth and LGD on some n-$\epsilon/2$-extensions of Δ and Γ, respectively. □

PROOF.

(i) Such smooth representations are obtained by applying Theorem 4.2 twice, to $\Delta \subset \mathbf{R}^D$ and to $\Gamma \subset \mathbf{R}^C$.

(ii). Such a representation can be derived from *(i)* by adding $(d + c + 1)(2d + 2c + 3)$ functions representing coefficients of a respective Riemannian metric on $\Delta \times \Gamma$ in terms of the natural coordinates on $\mathbf{R}^{2d+1} \times \mathbf{R}^{2c+1} \times \mathbf{R}^p$ (c.f. the proof of Theorem 4.5 in [5]). □

5. CONCLUSIONS

Information theory based methods discussed in Section 3 offer promising algorithms for selection of the set of significant attributes. Topological and geometrical approaches to the subject in Section 4 emphasised the importance and relevance of some fundamental topological and geometrical results to the task of the construction of simplified models of performance. The estimates for the minimal numbers of attributes required for abstract global simplifications obtained in that Section demonstrate a possibility of further significant enhancement of the results of the information theory approach.

All these results contribute to the development of final, robust simplified models of performance, by offering some ideas for practical solutions and by showing the extents of possible gains and fundamental limitations.

ACKNOWLEDGEMENTS

I would like to thank Dr. Andrew Jennings for his continuous interest and numerous discussions. The permission of the Executive General Manager, Telecom Research, to publish this material is also gratefully acknowledged.

BIBLIOGRAPHY

[1] Aczél J. and Daróczy, Z. (1975) *On Measures of Information and their Characterization*, Academic Press, New York.

[2] Breur, M. and Marshall, C. R. (1978) " Banachian Differentiable Spaces", Math. Ann., vol. **237**, pp. 105–120.

[3] Hurewicz, W. (1941) *Dimension Theory*, Princeton.

[4] Kowalczyk, A. (1980) " Whitney's and Nash's Embedding Theorems for Differential Spaces", Bull. Acad. Polon. Sci., Sér. Math. Astronom. Phys., vol. **28**, pp. 385–390.

[5] Kowalczyk, A. (1988) "Geometrical Perspective on Representation of a Data Base by a Reduced Number of Attributes", Telecom Tech. Branch Paper **88/1**, Australian Telecommunication Commission.

[6] Kowalczyk, A. and Szymański J. (1988) "Rough Simplifications of Decision Tables", Telecom Tech. Branch Paper **88/17**, Australian Telecommunication Commission.

[7] Krishnaiah, P. R. and Kanal, L. N., eds. (1987) *Handbook of Statistics, Vol. 2*, North Holland.

[8] Lloyd, E. (1984) *Handbook of Applicable Mathematics*, J. Wiley & Sons Ltd., Chichester.

[9] Papoulis, A. (1984) *Probability, Random Variables and Stochastic Processes*, Mc Graw-Hill, Inc., Auckland.

[10] Pawlak, Z., Wong, S. K. M. and Ziarko, W. (1987) "Rough Sets: Probabilistic Versus Deterministic Approach", technical report: Department of Computer Science, University of Regina.

[11] Shannon, C. E. (1948) "A mathematical theory of communication", Bell System Technical Journal, Vol. 4.

[12] Trillas, E. and Valverde, L. (1984) "An inquiry into indistinguishability operators", *in* Skala, H. J., Termini, S. and Trillas, E. ed., *Aspects of Vagueness*, D. Reidel Publishing Company, Dordrecht, Holland.

[13] Zadeh, L. A. (1971) "Similarity relations and fuzzy orderings", Inf. Sci., Vol. **3**, pp. 177–200.

Approximation, Optimization and Computing:
Theory and Applications, A.G. Law and C.L. Wang (eds.)
Elsevier Science Publishers B.V. (North-Holland)

WATER CONING IN OIL WELLS VIA A NEW VARIATIONAL TECHNIQUE FOR FREE BOUNDARY PROBLEMS

S. M. Makky and A. J. Hamad

Math. Dept., College of Science, University of Baghdad, Baghdad, Iraq
Mech. Enging. Dept., Military Technical College, Baghdad, Iraq

Water coning in oil wells is formulate as a free boundary problem. An equivalent variational problem is found, then an approximate solution for the variational problem is sought by representing both the unknown velocity potential and the unknown boundary as linear combinations of functions belonging to certain complete sequences of functions. Thus, the variational problem is recast to the realm of mathematical programming. A computer programme is developed and used successfully to solve the water coning problem. The same procedure is used to solve gas coning, gas and water coning in oil wells. The results seem to be very realistic.

1. INTRODUCTION

Oil reservoirs consist mainly of three zones: gas, oil, and water bounded above by impermeable stratum. Due to differences of these materials there are horizontal gas-oil and water-oil interfaces. When an oil well is drilled so that it partially penetrates the oil zone and oil is removed at a constant rate u, the oil-water intereface and the gas-oil interface will assume steady state shapes as a result of pressure gradients in the oil zone, [1].

In oil production it is advisable to obtain oil unmixed with either gas or water. Consequently, determining the shapes of the interface surfaces as functions of oil production rate and well penetration is of vital importance in oil production industry.

Since the interface surfaces are unknown and their determination is a major objective of the study these problems represent examples of free boundary problems, [2]. The procedure used here to solve such problems comprises four steps.

The first is to find an equivalent extremum variational formulation for the free boundary problem.

The second step is to express the function representing the unknown boundary as a linear combination of the elements of a preassigned complete sequence of functions.

The third step is to use Ritz method to solve the variational problem directly. This is accomplished by prescribing for the unknown variable in the boundary value problem (the velocity potential) the same treatment as for the unknown boundary. Thus, the whole problem is transferred to the realm of mathematical programming.

The last step is to solve the resulting nonlinear programming problem. Two algorithms, steepest descent and Hooke & Jeeve are used for this purpose.

The shape of each interface is found as a function of u and of well penetration.

2. STATEMENT OF THE PROBLEM

We consider first water coning. In this case it is assumed that the reservoir has a water zone beneath the oil zone with no upper gas zone. Due to the axisymmetry of the problem it is advisable to use the usualy cylindrical coordinate system (r,θ,z), whose z-axis coincides with the central axis of the well, and whose origin is the intersection of the z-axis with the horizontal water oil interface (before oil production), see Fig (1). The symmetry of the problem also indicates that all variables must be independent of θ, and hence the problem can be considered two dimensional. It follows that the velocity potential $\emptyset$ in the oil zone satisfies the following form of Laplace equation [3]:

$$r\emptyset_{rr} + \emptyset_{r} + r\emptyset_{zz} = 0 \qquad (1)$$

For a uniform production rate and when a steady state prevails the boundary conditions associated with D.E. (1)

read as follows

$V_z(r,1)=0$, for $r_w < r < r_e$ (2)

$V(r_e,z)=L$, for $0 < z < L$ (3)

$V_z(r,0)=0$, for $r_1 < r < r_e$ (4)

$V[r,h_1(r)]=H_1(r)$, for $0<r < r_1$ (5)

$V_n[r,h_1(r)] = 0$, for $0 < r < r_1$ (6)

$V_r(0,z) = 0$, for $h_{o1}<z<L-d$ (7)

$V_z(r,L-d) = u$, for $0 < r < r_w$ (8)

$V_r(r_w,z) = 0$, for $L-d<z < L$ (9)

In the above relations [4]:

$$V= - \mu \varnothing / k\, o_o \ , \quad u= Uk o_o / \pi\mu\, r_w^2 \qquad (10)$$

$$H_1(r)=L+[1-(\rho_w/\rho_o)]\, h_1(r) \qquad (11)$$

d: depth of oil well penetration

$h_1(r)$: a function representing the height of water-oil interface (an unknown)

h_{o1}: height of water-oil interface at r=0

k: coefficient of permeability

L: thickness of the oil zone

r_e: very large radius (assumed)

r_w: radius of well

U: oil production rate

V_n: directional derivative along normal

V_r, V_z: partial derivatives with respect to r and z respectively

ρ_o , ρ_w: density of oil and water

μ: dynamical viscosity of oil

The problem is to find h_1 and V as functions of oil production rate u, and well penetration d, when the other specifications of the oil reservoir are given, and such that relations (1)-(11) are satisfied.

3. SOLUTION OF THE PROBLEM

3.1 Variational Formulation:

It can be shown that [5] the variational problem which is equivalent to the free boundary problem described above has the following form: Minimize

$$Q[V(r,z),h_1(r)]=\iint_R r(V_r^2+V_r^2)\,drdz \qquad (12)$$

$$= \int^{r_1}\int_{h_1}^{d} r(V_r^2)\, dzdr + \int_{r_w}^{r_1}\int_{d}^{L} r(V_r^2 + V_z^2)\,dzdr + \int_{r_1}^{r_e}\int_{o}^{L} r\,(V_r^2 + V_z^2)\, dzdr$$

where r_1 is the point of intersection of the water-oil interface with the line z=0.

3.2 Equation of the Free Surface

As mentioned earlier, the equation satisfied by points on the interface is

$z= h_1(r)$, for $0 < r < r_1$

The function h_1 must satisfy certain conditions, these are

$$h_1(0)=h_{o1}, h_1'(0)=0, h_1(r_1)=h_1'(r_1)=0 \qquad (13)$$

The first condition of (13) gives the height of the water-oil interface at r=0 (under neath the center of the well). The second condition is due to the axisymmetry of the problem. The last two conditions of (13) indicates that the interface meets the line z=0 smoothly.

Approximating the function h_1 by a cubic polynomial in r that satisfies conditions (13) it follows the equation of the interface is

$$z=h_1^*(r)=h_{o_1}[1-3(r/r_1)^2+2(r/r_1)^3] \qquad (14)$$

In the above relation h_{o1} and r_1 are two parameters to be found.

REMARK

A better approximation for h_1 can be achievedif the interface surface is approximated by

$$z= h_1^*(r) + \sum_{\substack{n=2\\m=2}}^{M,N} \alpha_{mn}\, r^n\, (r-r_1)^m \qquad (15)$$

where h_1^* is given by (14), and α_{mn} are parameters to be found.

3.3 Form of the Velocity Potential:

To satisfy the boundary conditions(2)-(9) it is assumed that V is the sum of two differentiable functions

$$V(r,z) = G(r,z) + F(r,z) \quad (16)$$

One of these functions, say F satisfies boundary conditions (2)-(9), while the other satisfies corresponding homogeneous boundary conditions. To fullfil this requirement the region R is subdevided to two regions R_1 and R_2, where

$$R_1 = \left[(r,z): \begin{array}{l} 0 < r < r_1,\ h_1(r) < z < L-d \\ r_w < r < r_1,\ L-d < z < L \end{array} \right]$$

$$R_2 = [(r,z): r_1 < r < r_e,\ 0 < z < L]$$

Define F in R_1 and R_2 by

$$F = H_1(r) + (ul/d)[z-h_1)] - (u/2d)[z^2 - h_1^2(r)], \text{ in } R_1 \quad (17a)$$

$$F = (L-F_1)[(r-r_1)/(r_e-r_1)]^2 + F_1, \text{in } R_2 \quad (17b)$$

The function G can be expressed by

$$G = r^2(r-r_e)(r-r_w)^2 (z-L)^2 [z-(L-d)]^2 \cdot [z-h_1(r)]^2 \left(\sum_{i=0,j=0}^{i=1,j=J} B_{ij}\, r^i z^j \right) \quad (18)$$

where B_{ij} are unknown parameters to be found.

3.4 Equivalent Nonlinear Programming Problem:

When relations (14)-(18) are used in connection with (12) and the indicated integrals are computed it follows that Q becomes a function of ho_1, r_1, α_{mn}, B_{ij} for $m=2,\dots,M$, $n=2,\dots,N$, $i=0,\dots,1$, $j=0,\dots,J$

Thus the free boundary problem of finding the unknown boundary represented by $z=h_1(r)$, and the velocity potential $\emptyset$ is reduced to determining the minimum of Q as a function of the above mentioned parameters.

Two methods, Hooke & Jeeve's, and the steepest descent method are used to solve the unconstrained nonlinear programming problem, [6], [7].

REMARK

The convergence of the above described procedure is ensured by the convergene of the Ritz method for the solution of variational problems and the convergence of the algorithm used to solve the resulting nonlinear programming problem. The convergence of the Ritz method is proved in [8] when the variational problem has fixed boundaries. The proof is generalized [9] to cover the case when the variational problem is equivalent to a free boundary problem.

4. NUMERICAL EXAMPLES

A computer programme based upon the 4 step procedure presented in this paper is developed and used successfully to determine an approximate solution for the coning problem.

Since the controlable specifications of a well are the well penetration and the rate of production, the interfaces and velocity potential are determined for various combinations of oil production and well penetration while keeping other specifications fixed. The results are given here in graphical forms (see Figs. 2&3). A comparison is made between our results and the results of [4] for the water coning problem. As Fig.4 shows there is an agreement between both results.

Furthermore, a glance at the graphs indicates that as the rate of production increases interface comes closer to the well. In other works, when u increases ho_1 becomes larger.

Also, increasing the depth of penetration d causes a rise in the water-oil, i.e. ho_1 becomes larger. These conclusions are in conformity with one's expectations.

The same treatment is used for the study of gas coning, and water & gas coning.

5. CONCLUSION

Even though there are some attempts to solve the coning problem in oil wells [11]-[15] the novel variational techique described in this paper for solving free boundary problems seem to be an effective tool towards this goal.

WELL SPECIFICATIONS

$u = 1.25$ CP, $\rho_o = .7$ gm/cc, $L = 38$ m, $r_w = 0.076$ m, $r_e = 152$ m

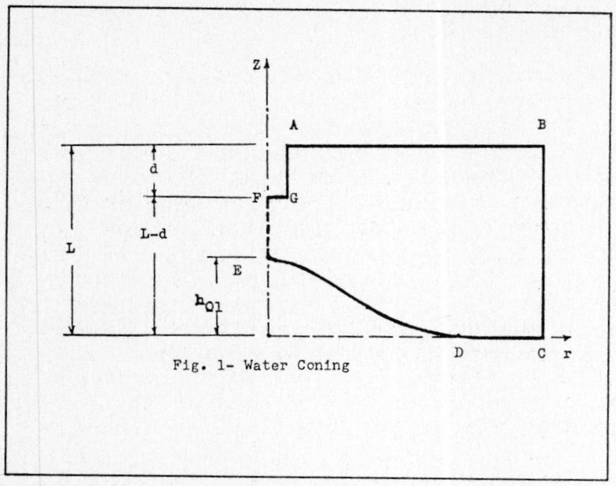

Fig.1

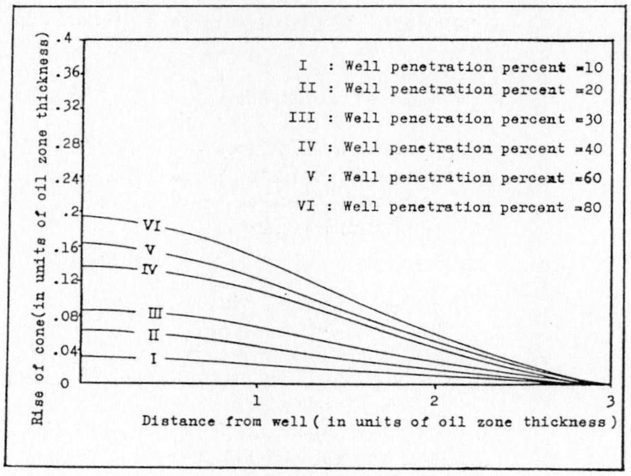

Fig. 2

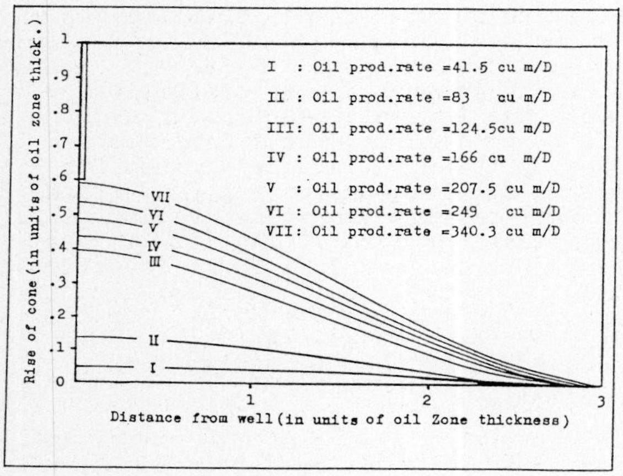

Fig. 3

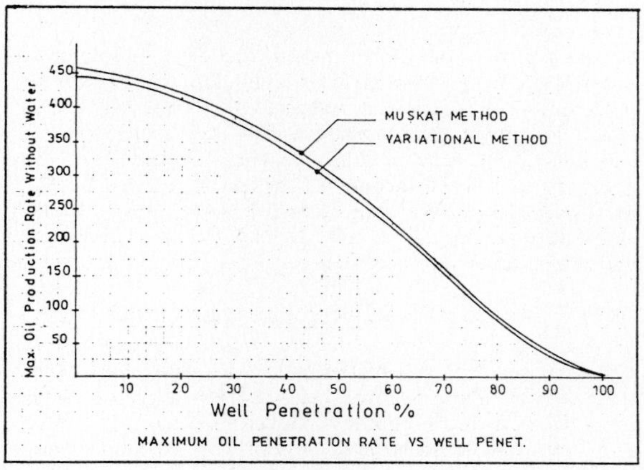

Fig. 4

REFERENCES

[1] Karplus, W.J., J. AIME, Pet. Branch Vol. 207, 240, 1956.

[2] Crank.J., Free and Moving Boundary Problems (Clarendon Press., 1984).

[3] Forsythe, G.E., Wasow, W.R., Finite-Difference Methods for PDEs (John Wiley & Sons, 1960)

[4] Muscat, M., Wyckoff, R.D., J.AIME, Pet. Branch, Vol.114, 144, 1935.

[5] Gelfand, I.M., Fomin, S.V., Calculus of Variations (Prentice-Hall, 1963).

[6] Bazaraa, M.S., Shetty, C.M., Non-linear Programming: Theory and Algorithms (John Wiley & Sons, 1979)

[7] Luenberger, D., Introduction to Linear and Nonlinear Programming, (Addison-Wesley, 1973).

[8] Kantorovich, L.V., Krylov, V.1., Approximate Methods of Higher Analysis (Noordhoff Ltd., 1958).

[9] Kitan, A.1., On the Seepage Problem, M.Sc. Thesis, University of Baghdad, 1987, (in Arabic).

[10] Hamad, A.J., Direct Variational Methods for Solution of Engineering Problems, M.Sc. Thesis, Military Engineering College, 1988, (in Arabic).

[11] Arthur, M.G., J.AIME, Pet. Branch, Vol. 155, 184, 1944.

[12] Chappelear, J.E., Hirasaki, G.J., J.Soc. of Pet. Engng., Vol. 65, 1976.

[13] Chierici, G.L., Ciucci, G.M., J. Pet. Tech., Vol. , 923, 1964.

[14] Coasts, K.H., Dempsy, J.R., J.AIME, Pet. Branch, Vol. 251, 63, 1971.

[15] Welge, H.J., Weber, A.G., J. Soc. of Pet. Engineers, Vol. 4, 345, 1964.

Approximation, Optimization and Computing:
Theory and Applications, A.G. Law and C.L. Wang (eds.)
Elsevier Science Publishers B.V. (North-Holland)

Economic Strategies in Outport Fishing Communities in Newfoundland§

S. J. May‡, H. S. Gaskill*‡, and J. D. May†

Department of Mathematics and Statistics‡, Memorial University of Newfoundland, St. John's, Newfoundland, Canada; Department of Economics†, Memorial University of Newfoundland, St. John's, Newfoundland, Canada; Newfoundland Oceans Research and Development Corporation, St. John's, Newfoundland, Canada*

This paper considers the implications of the current income support programs in Canada on the economic well-being of rural households which are engaged in productive activities in the home. Specifically, a study of economic behaviour in outport fishing communities in Newfoundland is presented. The study is based on two microeconomic models of behaviour. Simulations of the models are performed and used to develop a series of hypotheses about economic behaviour in relation to the rules governing unemployment insurance, the income support program of primary interest. The study provides an illustration of the use of simulation models as tools which can be used by governments in the formulation of policies affecting the fishing industry.

1. Introduction

In this paper a study of economic strategies employed in outport fishing communities in Newfoundland is presented. Newfoundland outports are small communities which have traditionally based their economies on the harvest of fish. In many of these communities the main source of cash income is through an inshore fishery, or through the processing of fish at small plants within the communities which are dependent on the inshore fishery for their source of fish.

Work at outport fishing plants, and in the inshore fishery generally, tends to be of a seasonal nature. Thus, cash-generating work related to the inshore fishery tends to be of a seasonal nature, and a variety of responses by individuals who are coping with seasonal unemployment and low cash incomes have developed. Such strategies, which include the production of goods at home, embrace the seasonal aspects of work in the fishery, and government income support programs. The principal such support program, and the one of prime interest in this paper, is the unemployment insurance program.

In this paper we present two models describing the economic behaviour of an individual householder in a limited work situation over a period of one year. These models were created to examine the interaction between unemployment insurance (UI) benefits and behavior strategies involving home production which have been observed in the outport communities in Newfoundland. The main focus of this portion of the study was in determining the economic impact on the general wellbeing of households which would result from possible policy changes governing the availability of UI.

The significance of the reported models is not in the deep mathematics contained therein, since the mathematics involved is fairly simple. Rather, it lies in the fact that there has been a shift in the use applied to these types of models. Our methodology has been to use the models, in the form of computer simulations, to make predictions about the general behavior of the householder at a microeconomic level and to identify the main factors which influence this behavior. The results of these studies lead to a series of qualitative conclusions which have consequences relating to the impact of possible changes in policy on the outport communities.

2. Two Models of Household Economic Behavior

Model 1

The first model, hereafter referred to as Model 1, initially described in May [1], permits a householder to make optimal consumption choices between two commodities, a and b. The first commodity, a, is thought of as comprising food and clothing. The second, b, comprises shelter. The model runs for a period of 52 weeks in one week cycles. At the start of a cycle, the householder has monetary resources available to him/her in the form of savings and transfer payments. As well, if work is available in a market place over which the householder has no control, the householder can perform work, thereby acquiring additional money which can be used during the cycle. The behavior of the householder is determined by maximizing the value of a wellbeing function which depends on four variables: consumption of goods a and b, leisure, and savings.

The main variables employed in the model are described in the following table:

Table 1

Variable	Description
avalwk	available work (hrs)
savs	available money including transfer payments
a	units of commodity 1
b	units of commodity 2
l	units of leisure (hrs)
wk	units of work (hrs)
savso	units of savings at end of a cycle ($)
wtot	accumulated wellbeing

§ This work was supported by two contracts from the Royal Commission on Employment and Unemployment in Newfoundland and by NSERC grants #A9098 and #A3464.

In addition to the variables listed, the model employs the constants shown in Table 2. The values adopted for

Table 2

Description	Name of Constant	Value
p_a	price of a	\$20.26
p_b	price of b	\$93.75
wg	wage rate	\$7.45
c_a	utility coefficient of a	8
c_b	utility coefficient of b	6
c_l	utility coefficient of leisure, l	1
c_s	utility coefficient of $savso$	2

the constants are based on economic realities in Newfoundland. A full discussion of rationale underlying the choice of these values can be found in NORDCO [2]. The coefficients used in the utility function reflect the authors' *a priori* beliefs about the relative importance of food/clothing, shelter, savings and leisure.

Within any time cycle, the problem is to maximize $U(a,b,l,savso)$ where:

$$U(a,b,l,savso) = c_a \ln(a) + c_b \ln(b) + c_l \ln(l) + c_s \ln(savso). \quad (1)$$

The function U is a utility, or wellbeing, function.

The problem is constrained by the following equations and inequalities:

$$0 \leq wk \leq avalwk \leq 40. \quad (2)$$

$$0 \leq l = 105. - wk \quad (3)$$

$$savso = savs + wg \times wk - p_a \times a - p_b \times b \quad (4)$$

$$0 \leq a, b, savso \quad (5)$$

The constraint on the amount of work, wk, in Equation 2 is intended to model limited employment opportunities through shift work in seasonally operated fish plants. Equation 3 contains the obvious relation between work and leisure and sets a reasonable upper bound on the total available time. Equation 4 effectively limits consumption of either good to be determined by the available money supply; borrowing is not permitted.

If the quantity of work is fixed, then the problem reduces to finding an appropriate division between a and b. Our method for generating the optimum is simply to calculate the partial derivatives of U with respect to a and b and set them equal to 0. This leads to a system of two linear equations in the two unknowns, a and b. The absolute optimum, which involves the three variables, a, b and wk is found by searching along the wk axis within the feasible region. Needless to say, since the solution scheme reduces to a linear system in two unknowns, the solution scheme is computationally fast, even though a search along the wk axis may be required. Since in practice, $wk = avalwk$, the solution scheme is almost instantaneous. Indeed, a closed form solution can be written down.

Model 2

The second model, Model 2, employs the same description of wellbeing as that in Model 1, based on the same four variables a, b, l, $savso$. The weekly cyclic structure is identical. However, in the second model, the householder can engage in home production, that is, perform work within the home which will produce a quantity of good b which can either be retained or sold for cash. To do this, the householder can invest cash from savings, transfer payments or earnings in commodity d. This commodity is used as a direct material in the production of commodity b. The produced good can either be retained, or it can be sold and the money generated can be passed forward as savings available in the next cycle. It is to be emphasized that money generated by the sale of b cannot be used to acquire commodities in the same cycle. In the present model, we have associated home production with the shelter commodity, b. This ensures a consistency between Models 1 and 2, and also minimizes the complexity of the home production model by not requiring additional variables. As in Model 1, the first commodity is thought of as representing food/clothing. The principal question which this model is intended to address is that of assessing the degree of protection offered by a home production strategy to the uncertainty of work shortages in the fishery.

The variables employed in the model which are either new or different from those in Model 1 are described in Table 3.

Table 3

Variable	Description
d	units of direct materials
wk	units of work place work (hrs)
hwk	units of home work (hrs)

In addition to the variables listed, the model employs various constants which are listed in Table 4.

Table 4

Description	Name of Constant	
p_d	price of d	\$62.50
cp	coefficient of production	1
r	labour exponent	.24

As with Model 1, within each time cycle the problem is to maximize

$$U(a,b,l,savso) = c_a \ln(a) + c_b \ln(h_b + b) + c_l \ln(l) + c_s \ln(savso), \quad (6)$$

where the amount of good b produced at home, h_b, is given by:

$$h_b = cp\ d^{1-r}\ hwk^r. \quad (7)$$

The only difference between the wellbeing functions for

the two models occurs in the term related to good b. Now there are two sources of this commodity, purchase in the market place and production at home. Production at home is governed by Equation 7, which is a Cobb-Douglas production function. These functions describe the conversion of input materials and labor into output. The functional form of the Cobb-Douglas technology involves both labor and materials, which seems most appropriate in the home production situation. The constant in the function was chosen so that the marginal value of additional labour rapidly decreased. This had the effect of inhibiting substitution of labor for materials, since the amount of labour was not as limited as cash to purchase materials.

The problem is constrained by the following equations and inequalities:

$$0 \leq wk \leq avalwk \leq 40. \quad (8)$$

$$0 \leq l = 105. - wk - hwk \quad (9)$$

$$0 \leq a, d, savso \quad (10)$$

$$savso = savs + wg \times wk - p_a \times a - p_b \times b - p_d \times d \quad (11)$$

The constraints listed in 8-11 play similar roles to the analogous constraints in Model 1.

In this model b is permitted to be either positive or negative. If $b \geq 0$, then the interpretation is that b is bought in the market place. If $b < 0$, then the interpretation is that b is sold. This leads to the addition of two constraints which prevent money generated from the sale of b during a particular cycle being used to acquire more of any commodity in the same cycle.

Several methods for finding the optimum were explored. A methodology similar to that applied to the model without home production yielded a system of non-linear equations which proved computationally difficult to handle. For this reason, two methods based on versions of a gradient algorithm were employed. (A discussion of these methods can be found in Wismer and Chattergy [3].) While alternative algorithms were considered, they were hampered by the fact that the surface associated with the objective function in the region of interest was very flat. Thus, small changes in the value of the objective function required large changes in the feasible solution. This in turn has the effect of increasing the number of iterations required to find a solution.

One algorithm was found to produce a solution in less than 15 iterations. The stopping criterion used was to compare $| U^p - U^{p+1} |$ with a small test value. Since the value of interest was the optimal value of U and not the values comprising the optimal bundle of consumption goods, this was a suitable approach. However, if the interest was in accurately determining these values, this criterion would be inadequate due to the flatness of the objective surface around the optimum.

The solution found by this method proved to be dependent on the initial feasible solution. Specifically, the initial value of the number of hours of homework, hwk, which was the ultimate determinant of the value of leisure, was critical. The reason for this was the contribution to the gradient in the direction of leisure, l, was small in comparison with the other components. Thus, very little motion along the leisure axis took place prior to encountering the boundary. Since the boundary encountered did not involve the leisure constraint, the direction of motion was not influenced by l. This problem was easily solved since sensitivity studies on the initial feasible solution showed that the optimal solution could be found by assigning to hwk an initial value of about 65% of the potential hours of leisure.

The alternative method for finding the maximum value of wellbeing was a direct application of the hemstitch algorithm. While this method generally found slightly larger values for the objective function, it required on the order of 100 times as many iterations to achieve an optimum. In general, the differences between the maximums for the two methods were less than 1%. Since good approximate maximums could be found in a more computationally effective way using the first algorithm, the wellbeing curves were generated by using the first algorithm.

3. Model Results

The full models simulated the behavior of an outport householder over a complete year. The main tests conducted with the two models were aimed at looking at the relationship between available work, on an annual basis, and total annual wellbeing. These calculations were repeated for various scenarios related to the rules governing the qualification for unemployment insurance.

Two distinct types of unemployment insurance schemes were examined. The first type involved a fixed number of weeks of entitlement to benefits. If the householder had accumulated the required number of weeks work, then he/she received unemployment insurance payments for the remaining portion of the year in which he/she was not working (minus a two week waiting period). The qualifying periods ranged from ten weeks to twenty weeks. The second type gave 1 week of benefits for each week worked. No waiting period was involved.

For all the scenarios mentioned, a total annual wellbeing was calculated as a function of weeks of available work. In Figure 1, curves of total annual wellbeing versus weeks of available work are presented for all unemployment insurance (UI) scenarios. Model 1, in which no home production was available, was used to construct the curves. Each of the curves for the scenarios involving a qualifying period has a jump discontinuity. (The discontinuity is indicated by the vertical dashed lines.) The jump occurs at the point at which the number of weeks of available work exactly matches the number required to qualify for UI. In all cases, the jump is from a negative, or marginally positive, annual wellbeing to an annual wellbeing of more than 1300. The existence of this jump discontinuity in the wellbeing curve makes obvious the fact that the single most important factor in the annual life cycle of an outport householder is whether there will be enough available work to qualify for UI.

Also presented in Figure 1 is the curve for the second type of UI scenario. This curve is continuous; however, it shows a sharp change in its slope at 26 weeks of avail-

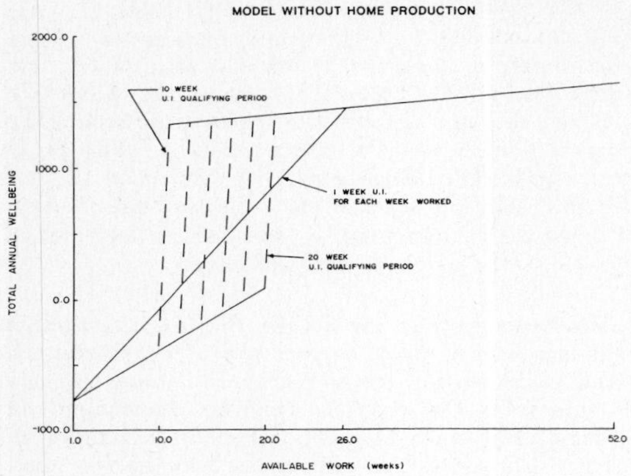

Figure 1

able work. This is due to the fact that at 26 weeks of available work, the individual has accumulated enough work to qualify for UI for the remainder of the year.

In Figure 2 the graphs presented in Figure 1 are repeated, except the computations were made with Model 2 which permits home production. As can be seen, the availability of home production offers some protection from the lack of available work. That is, the curves are shifted upwards on the wellbeing scale so that they now range from 200 to almost 2200. However, the continued existence of the jump discontinuities in the wellbeing curves mean that in spite of the protection offered by home production, the critical factor in the outport householder's life is still whether there is enough available work to qualify for UI.

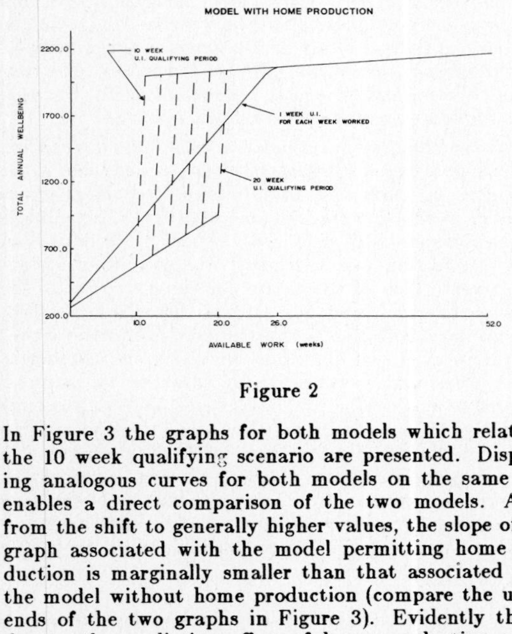

Figure 2

In Figure 3 the graphs for both models which relate to the 10 week qualifying scenario are presented. Displaying analogous curves for both models on the same axis enables a direct comparison of the two models. Aside from the shift to generally higher values, the slope of the graph associated with the model permitting home production is marginally smaller than that associated with the model without home production (compare the upper ends of the two graphs in Figure 3). Evidently this is due to the mediating effect of home production which partially makes up for lack of work in the market place. However, as can be seen from the large discontinuity in the upper curve, home production can in no way replace work in a fish plant or the fishery. Moreover, UI is still an essential component of the householders life.

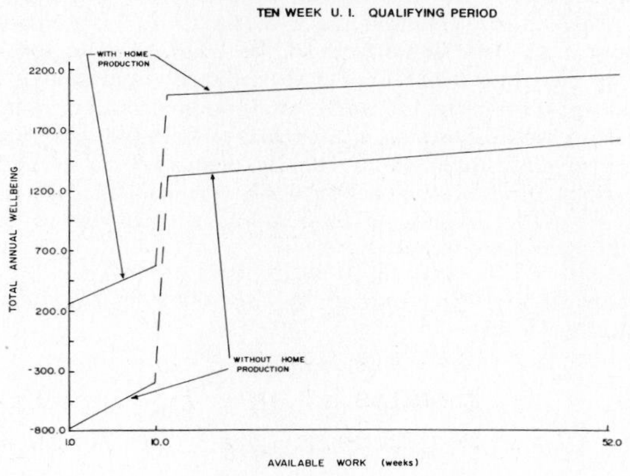

Figure 3

The conclusions we draw from this analysis are:

- in small outports, the most important determinant of annual wellbeing is qualifying for UI;
- longer UI qualifying periods can have a serious negative effect on annual wellbeing;
- while home production ameliorates well-being it does not replace employment.

4. Summary

The purpose of this paper has been to identify the main factors influencing economic behaviour in outport communities having economies based on the inshore fishery, and to illustrate the use of simulation models as tools to evaluate economic policies affecting these communities. Of specific interest was the effect of income support programs. The results demonstrate that changes in unemployment insurance regulations can have a devastating effect on these individuals. The use of simulation models such as those described in this paper, can lead to a better understanding of such effects. Studies like these performed prior to the implementation of new policies could play an important role in the formulation of sensible public policies.

References

[1] May, S., 1986. *An economic model of household production.* Report to the Royal Commission on Employment/Unemployment in Newfoundland, 11pp.

[2] NORDCO, 1986. *A sensitivity analysis of two models of household behavior.* Report to the Royal Commission on Employment/Unemployment in Nfld., 30pp.

[3] Wismer, D. A., Chattergy, R., 1978. *Introduction to nonliner optimization: a problem solving approach.* Elsevier North-Holland, Inc., New York, 395pp.

Approximation, Optimization and Computing:
Theory and Applications, A.G. Law and C.L. Wang (eds.)
Elsevier Science Publishers B.V. (North-Holland)

HALF FLEXIBLE SIMPLEX METHOD AND APPLICATIONS IN ENGINEERING

Niu Tiejun* Li Shoufan+ Wang Zaishen* Liang Yanchun*

In this paper, a new method (half flexible simplex method) of optimum structural design is presented. The advantages of feasible direction method and flexible simplex method in mathematical programming problems are absorbed and the shortcomings in both of them are counteracted in the method of this paper. The three optimal methods are compared by means of concrete structural computations. The calculations show that the optimal results of half flexible simplex method can be used to make evaluations and demonstrations in advance for structural designs.

1. INTRODUCTION

There are many practical methods in mathematical programming problems[1],[2]. One of the most general methods in structural optimization is feasible direction method and its transformed forms. The advantages of feasible direction method are that the concept is concise and the computational speed is fast. The shortcoming of feasible direction method is that the treatment of non-technical constraints is difficult. The advantage of flexible simplex method[3] is that the treatment of various constraints is all suitable for the method. Hence flexible simplex method possesses prospects for spreading applications. However, flexible simplex method requires more times in computing objective functions than those of feasible direction method. Sometimes, the objective functions of optimum structural problems cannot be written as analytical expressions of design variables. Thus the reanalysis is required for computing an objective function. It is an important problem to reduce the number of times of reanalysis in optimal designs.

In this paper, a slightly changed flexible simplex method which is called half flexible simplex method is proposed. A computing program is worked out with standard FORTRAN IV. The optimal design of package cushioning in packaging dynamics is taken as a design example. The comparisons of the computing results among the three methods are made. The results show that good effectiveness is achieved by means of the method in this paper.

2. STATEMENT OF THE OPTIMUM PROBLEM

In general, an optimum structural problem can be summed up a mathematical programming problem:

$$\begin{cases} \min & \varphi(y(t)) \\ \text{s.t.} & h_i(y(t))=0 \quad (i=1,2,\ldots,r) \\ & g_j(y(t))\leq 0 \quad (j=1,2,\ldots,s) \end{cases} \qquad (1)$$

Where $y(t)$ is a n-dimension vector, $\varphi(y(t))$ is an objective function, $h_i(y(t))$ is a technical constraint and $g_j(y(t))$ is a non-technical constraint.

The optimum design problem of a package structure is taken as an example to

* Jilin University, Changchun, China. + Unit 86003, Changchun, China.

illustrate Eq.1. The equations of motion for a package structure containing critical elements can be written as[4]

$$[M]\{\ddot{x}(t)\} + [C]\{\dot{x}(t)\} + [K]\{x(t)\} = \{f(t)\} \quad (2)$$

Where [M], [C] and [K] are the mass (or inertia), damping and stiffness matrices of the system, respectively; $\{x(t)\}$ and $\{f(t)\}$ are the displacement and force vectors, respectively.

Taking Fourier transform both sides of Eq.2, we obtain

$$(-\omega^2[M] + j\omega[C] + [K])\{X(\omega)\} = \{F(\omega)\} \quad (3)$$

Thus

$$\{X(\omega)\} = [H(j\omega)]\ \{F(\omega)\} \quad (4)$$

Where

$$[H(j\omega)] = (-\omega^2[M] + j\omega[C] + [K])^{-1} \quad (5)$$

is the transfer function matrix of the system over the frequency domain.

The arbitrary component of the frequency response $\{X(\omega)\}$ can be expressed as

$$X_i(\omega) = \sum_{k=1}^{n} H_{ik}(j\omega)\ F_k(\omega) \quad (6)$$

Where $H_{ik}(j\omega)$ is the element of the transfer function matrix $[H(j\omega)]$ which can be found by means of Lagrange's interpolation[5]. The time domain response $\{X(t)\}$ can be obtained by means of taking inverse Fourier transform.

The impact response of the total system is considered in packaging dynamics. Without lossing generality, let the coordinate of the element which is subjected to a shock load directly is x_1 and the impulse is f_1. Then the force vector is

$$\{f(t)\} = \{f_1\,\delta(t),\ 0, \ldots, 0\}^T$$

Thus

$$X_i(\omega) = H_{i1}(j\omega) f_1 \quad (i=1,2,\ldots,n) \quad (7)$$

This is the frequency displacement response of the packaged container.

The acceleration is a target measuring the damage of a packaged container. To avoid the damage of the container resulting from the overburdansome internal stress owing to a large inertia force, the appropriate cushioning damping and stiffness should be selected. For this reason, the cushioning stiffness k_1 and damping c_1 are taken as design variables. The objective function is taken as

$$\varphi(c_1, k_1) = \max_{\omega} \sum_{i=1}^{n} \alpha_i \omega^2 |X_i(\omega)| \quad (8)$$

where $\alpha_i > 0$ is a weighted coefficient, which satisfies $\sum_{i=1}^{n}\alpha_i = 1$. The value of α_i may adjust the 'proportion' which each component of the container possesses in the objective function. Therefore, the optimal problem of the package cushioning can be written as a mathematical programming problem as mentioned in Eq.1

$$\begin{cases} \min \max_{\omega} \sum_{i=1}^{n} \alpha_i\ \omega^2\ |X_i(\omega)| \\ \text{s.t.}\quad \underline{c} \le c_1 \le \bar{c}, \quad \underline{k} \le k_1 \le \bar{k} \end{cases} \quad (9)$$

or

$$\begin{cases} \min \max_{\omega} \sum_{i=1}^{n} \alpha_i\ \omega^2\ |X_i(\omega)| \\ \text{s.t.}\quad \underline{c} \le c_1 \le \bar{c}, \quad \underline{k} \le k_1 \le \bar{k} \\ \qquad \omega^2\ |X_j\ (\omega)| \le b_j \end{cases} \quad (10)$$

There are only technical constraints in optimal problem (9), which has been solved by means of feasible direction method[4]. Besides technical constraints, there are acceleration constrants for some critical elements in optimal problem (10). The non-technical constraints in Eq. 10 are the same property as stress constraints in minimizing weight design[6].

Thus the mathematical model is set up, in which the container containing critical elements is a multi-degree system and the optimum package cushioning design is considered. It should be pointed out that the similar mathematical models not only appear in packaging dynamics but also appear in optimal design of vibration isolators[7] and in vehicle ride comfort[8]. Therefore, the mathematical programming problems such as Eq.9 and Eq.10 possess certain general significance.

3. SOLVING PROCESS AND EXAMPLE

Although the optimal problem (10) is a standard mathematical programming problem, reanalyses are required in flexible simple method and the calculation of every reanalysis is complicated. Therefore, in order to reduce the number of times of reanalyses, the negative gradient method is adopted when finding the optimal objective function and the flexible simplex method is used when treating constraints. A FORTRAN program is worked out based on this idea. The program can be operated in microcomputers. The computing results show that good effectiveness is achieved.

Let us take a five-degree-of-freedom package structure as an example.

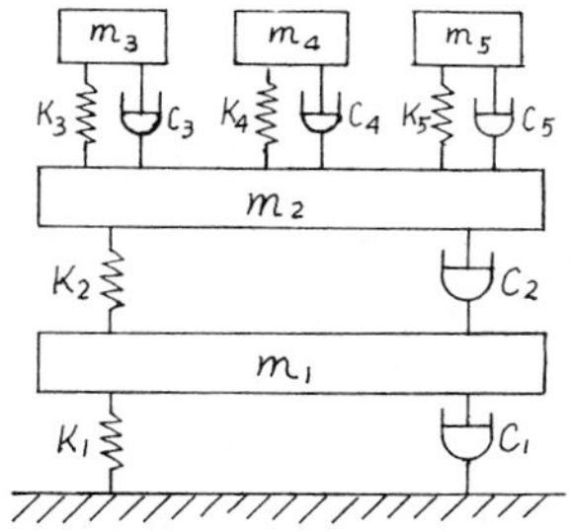

Fig.1 simplified model of packaged item with critical elements

Consider the structure shown in Fig.1. The matrices $[M]$, $[C]$ and $[K]$ in Eq.2 can be written as

$$[M] = \begin{bmatrix} m_1 & 0 & 0 & 0 & 0 \\ 0 & m_2 & 0 & 0 & 0 \\ 0 & 0 & m_3 & 0 & 0 \\ 0 & 0 & 0 & m_4 & 0 \\ 0 & 0 & 0 & 0 & m_5 \end{bmatrix}$$

$$[C] = \begin{bmatrix} c_1 + c_2 & -c_2 & 0 & 0 & 0 \\ -c_2 & \sum_{i=2}^{5} c_i & -c_3 & -c_4 & -c_5 \\ 0 & -c_3 & c_3 & 0 & 0 \\ 0 & -c_4 & 0 & c_4 & 0 \\ 0 & -c_5 & 0 & 0 & c_5 \end{bmatrix}$$

$$[K] = \begin{bmatrix} k_1 + k_2 & -k_2 & 0 & 0 & 0 \\ -k_2 & \sum_{i=2}^{5} k_i & -k_3 & -k_4 & -k_5 \\ 0 & -k_3 & k_3 & 0 & 0 \\ 0 & -k_4 & 0 & k_4 & 0 \\ 0 & -k_5 & 0 & 0 & k_5 \end{bmatrix}$$

Where the data are taken as

m_1=50kg, m_2=200kg, m_3=10kg, m_4=15kg, m_5=10kg, c_2=150Ns/m, c_3=50Ns/m, c_4=20Ns/m, c_5=20Ns/m, k_2=1000N/m, k_3=2000N/m, k_4=1000N/m, k_5=300N/m.

The technical constraints are as follows

$\underline{c}$=100Ns/m, $\bar{c}$=1000Ns/m,

$\underline{k}$=500N/m, $\bar{k}$=10000N/m or $\bar{k}$=20000N/m.

The non-technical constraint in Eq.10 is

$\omega^2 |X_3(\omega)| \leq 2\text{m/s}^2$

The weighted coefficients are

α_1=0, α_2=0.1, α_3=0.2, α_4=0.3, α_5=0.4

The impulsive is taken as

f_1=3000Ns

The initial values of the design variables are

c_i^0=200Ns/m, k_i^0=8000N/m.

The initial value of the objective function is

$\varphi(c_i^0, k_i^0) = 9.42\text{m/s}^2$.

The optimum designs of the damping and stiffness in the package cushioning are made by means of half flexible simplex method, feasible direction method and flexible simplex method, respectively. In order to compare the computing results, the trend diagrams of the design variables for the three optimal methods are drawn in Figs. 2-4, where half flexible simplex method is expressed with —▸—▸—, feasible direction method with – – – and flexible simplex method with -.-.- . The optimized objective functions for half flexible simplex method, feasible direction method and flexible simplex method are noted with φ_1, φ_2 and φ_3, respectively. The non-technical constraint is considered in the comparison between half flexible simplex method and flexible simplex method.

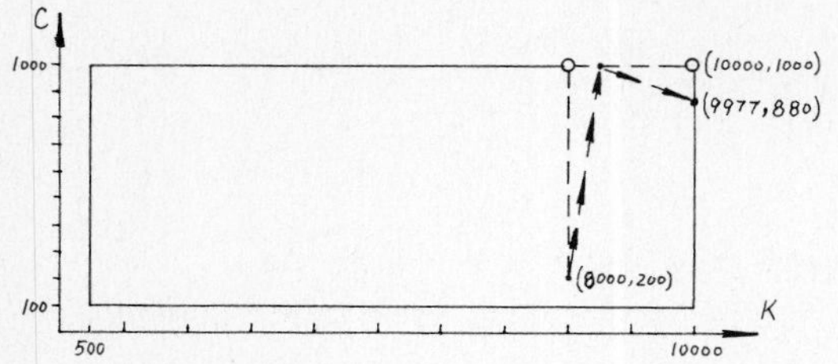

Fig.2 Comparison of computing results between half flexible simplex method and feasible direction method.
φ_1 =5.11m/s^2, φ_2 =5.05m/s^2.

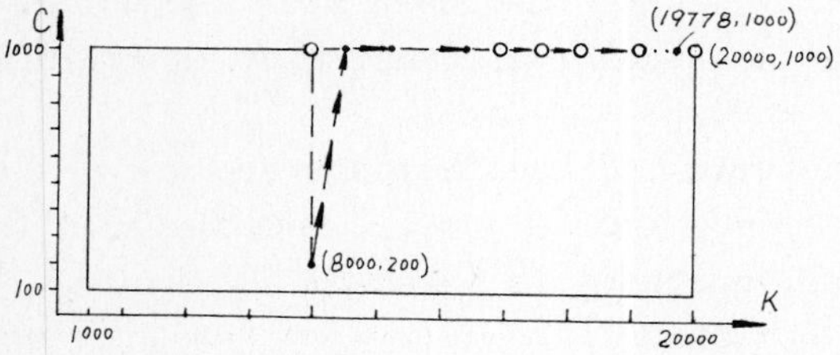

Fig.3 Comparison of computing results between half flexible simplex method and feasible direction method.
φ_1 =2.89m/s^2, φ_2 =2.87m/s^2.

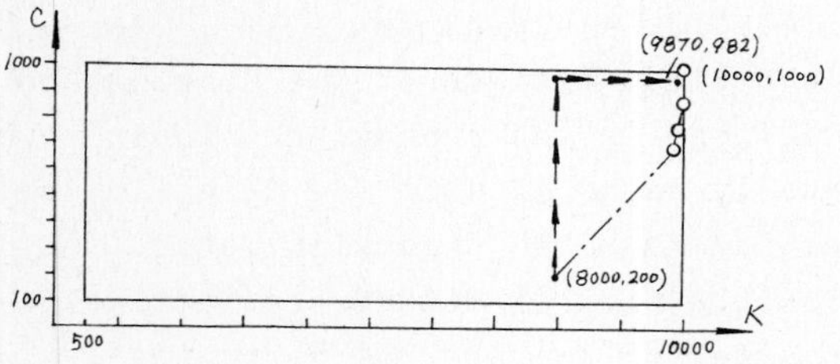

Fig.4 Comparison of computing results between half flexible simplex method and flexible simplex method.
φ_1 =5.11m/s^2, φ_3 =5.05m/s^2.

4. CONCLUSIONS AND RECOMMENDATIONS

a. The computing results show that the optimal trend and optimized results obtained by means of half flexible simplex method and feasible direction method are identical basically. Therefore, half flexible simplex method is reliable.

b. The method in this paper can be realized in microcomputer. It can be used for comparisons in a good many design programmes.

c. In the optimal problem of this paper, the relations between objective functions and design variables are complicated; the feasible regions are complicated too. Therefore, the optimized design is only one which is better than the original design, but not the most optimum design. This problem is a pending one that needs studying further.

REFERENCES

[1] R.H., Gallagher, O.C., Zienkiewicz, Optimum Structural Design, John Wiley & Sons, 1973.

[2] Wang Deren, Solving Processes of Non-linear Equtions and Optimum Methods, 1979.

[3] D.M., Himmelblau, Applied Nonlinear Programming, McGraw-Hill Book Company, 1972.

[4] Liang Yanchun, Wang Zaishen and Li Hongli, Optimal Design of Physical Parameters in Package Cushioning for Products Containing Critical Elements, Second International Packaging Conference Proceedings, Beijing, China. 1988.

[5] Liang Yanchun, Wang Zaishen, A Method for Calculating Transfer Functions of Mechanical System—Lagrange's Interpolation, Computational Structural Mechanics and Applications, Dalian, China,(4) 1986.

[6] Sui Runkang, Zhong Wanxie and Qian Lingxi, Optimum Design of Composed of Bars-Membranes-Beams Program System DDDU-2, Journal of Dalian Institute of Technology, (1) 1983.

[7] Liang Yanchun, Wang Zaishen, Optimum Design of Dynamic Absorbers for the Vibration of Multi-Degree-of-Freedom Systems, Chinese Journal of Mechanical Engineering, (3) 1988.

[8] Wang Zaishen, Liang Yanchun, Damped Least Square Feasible Direction Method for Optimal Design of Physical Parameters of Mechanical Systems, Acta Scientiarum Naturalium Universitatis Jilinensis,(4) 1987.

Approximation, Optimization and Computing:
Theory and Applications, A.G. Law and C.L. Wang (eds.)
Elsevier Science Publishers B.V. (North-Holland)

An Absorbing Boundary Algorithm for the FD-TD Method with Arbitrary Incidence Angle

Jasmin E. Roy and Dennis H. Choi
Faculty of Engineering, University of Regina
Regina, Saskatchewan, Canada, S4S 0A2

This paper presents the development of a simple, yet effective, plane absorbing boundary for the FD-TD method in three spatial dimensions. The algorithm is based on understanding of the wave propagation on a discrete mesh, with the concepts of numerical, phase and group velocities. The efficacy of the absorbing boundary is demonstrated by measuring the reflection level in the frequency domain for homogeneous metallic rectangular waveguides operated in their dominant mode.

1. Introduction

The Finite Difference Time Domain method developed by Yee [1] some years ago has been widely applied in electromagnetic engineering area. In order for the FD-TD method to simulate wave propagation in open space, an absorbing boundary algorithm must be used. The purpose of an absorbing boundary consists in simulating a boundary that permits the electromagnetic waves to propagate through it with a minimum of reflection so as to limit the size of the computational domain required for characterizing open structures.

There is a handful of schemes for reducing the level of reflection off the termination plane of a computational domain in which propagation is simulated. An obvious method consists in introducing gradually artificial losses in the medium near the boundary. Taflove and Brodwin [2] recommended that such losses be introduced over a thickness of one wavelength. This scheme is usually considered prohibitive for it increases unduly the size of the computational domain.

Another scheme that comes readily to mind is to derive the field values at the boundary by extrapolation from the field values in the interior computational domain. Taylor et al [3] used a two-point linear extrapolation scheme for the axial E field component. The level of accuracy afforded by this scheme was not mentioned.

Recently, a simple algorithm [4] similar in nature to the one proposed in this paper was used for simulating the propagation of the quasi-TEM wave in a microstrip structure with the wave impinging at normal incidence onto the absorbing boundary. The amount of reflection was reported to be of the order of 3% to 5%. The use of a super-absorbing boundary [5] was said to further reduce the amount of reflection, although no specific figure was given.

In fact, such simple algorithm had already been proposed by Taflove and Brodwin in a paper [2] written many years ago. The algorithm is a departure from the previous schemes in that it is not a direct discretization of the advective differential operator. This paper presents this ad hoc development and generalizes the algorithm to encompass the case of an arbitrary incidence angle. Yet, the algorithm remains very simple and very local, requiring the knowledge of field values at points located only one space increment away from the absorbing boundary.

2. Development of a simple absorbing boundary algorithm

The development of this absorbing boundary stems from the realization that the velocity at which a signal propagates on a cartesian grid, in terms of phase and group velocities, is different from the velocity $v_n = \Delta l/\Delta t$ at which a numerical disturbance propagates on the same grid. Δl is the spatial increment and Δt is the temporal increment.

For low frequency components propagating in a homogeneous medium, both phase and numerical velocities are related simply by a constant factor $s = c/v_n$ where c is the intrinsic propagation velocity for the given medium. The value of s must be chosen small enough to keep the algorithm stable.

Let us condider the case of a plane wave impinging at normal incidence onto a plane absorbing boundary lying at the face $x = 0$ of a rectangular computational domain in three dimension. On each of these consecutive two planes($x = 0$ and Δl) three of the six electromagnetic field components, namely here the two electric field components E_y and E_z parallel to these planes and the magnetic field component H_x perpendicular to these planes. This absorbing boundary algorithm consists simply in transferring after a delay of $(1/s)\Delta t$ the values of the two field components parallel to the absorbing boundary, from the grid points lying at plane $x = \Delta l$ to the grid points lying at plane $x = 0$.

Since updating a field value takes up $1\Delta t$, the algorithm must then introduce a delay of $(1/s) - 1$ additional time steps. Naturally, this algorithm works best only for integer values of $(1/s)$ since delays can be readily implemented only in integer number as a result of the time discretization. But this restriction is not serious since the choice of s can be adjusted, so long as the stability criterion $S \leq (1/\sqrt{3})$ is satisfied.

The case of a plane wave impinging onto a plane absorbing boundary at an arbitrary incidence angle θ and with propagation velocity c can be treated simply by considering in the plane next to the absorbing boundary the field pattern resulting from the off-normal incidence angle. This field pattern can now be considered in its own right as a wave impinging onto the aborbing boundary at normal incidence and with propagation velocity v_p where v_p is the phase velocity of the original plane wave in the direction normal to the absorbing boundary.

The number of delays required for implementing the absorbing boundary becomes v_n/v_p instead of v_n/c, thus becomes $(1/s)/(v_p/c)$ instead of $(1/s)$. In other words (see Fig. 1), the field value at point R of the absorbing boundary can be obtained either as the field value at point P delayed by $(v_n/c) = (1/s)$ time steps or as the field value at point Q delayed by

$$(v_n/v_p) = (1/s)/(v_p/c)$$

time steps. This procedure holds true providing that the wavefront amplitude is nearly constant for both points P and Q. This latter condition is satisfied readily if the spatial increment Δl is small enough when the wavefront is nonuniform, and is satisfied exactly regardless of the value of Δl when the wavefront is uniform.

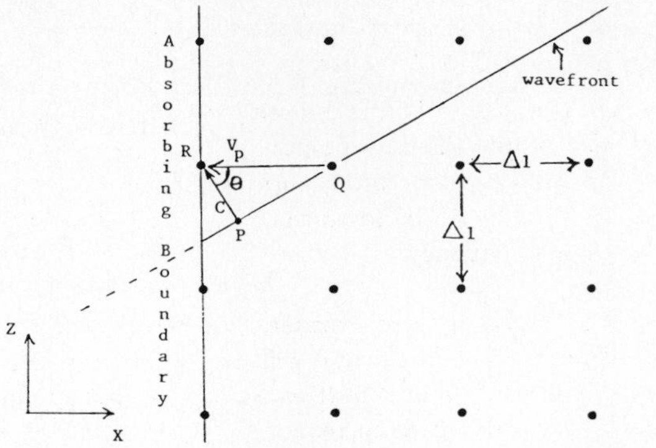

Fig. 1 Treatment of an arbitrary incident angle

For higher frequencies, the dispersion of the FD-TD method causes the simulation's phase velocity v_p to decrease with increasing frequency. The number of delays that must be introduced artificially must then be modified accordingly.

Defining the parameter W as $W = v_n/v_p$, the absorbing boundary algorithm for an x plane wave impinging from the right $(x > i)$ onto the absorbing plane $x = i$ at an arbitrary incidence angle θ becomes simply:

$$U^{n+1}(i,j,k) = U^{n+1-[W]}(i+1,j,k) \tag{1}$$

where U represents the field value, and [W] means Nearest Integer value of W and $v_p = \zeta c/cos\theta$. The factor ζ accounts for the dispersive behavior of the FD-TD method and is function of the normalized frequency $\Delta l/\lambda$. If θ is not known a priori, it can be estimated by monitoring the field values in the vicinity of the absorbing boundary.

3. Application to a homogeneous rectangular waveguide

In order to verify the concept of the present simple absorbing boundary algorithm, we present the case of wave propagation in a rectangular waveguide. A homogeneous rectangular waveguide is discretized on a uniform Yee lattice of spacing Δl. The transversal dimensions of the waveguide are $10\Delta l$ x $1\Delta l$, yielding a TE_{10} normalized cutoff frequency $\Delta l/\lambda_c = 0.050$.

The excitation was generated at the end plane $x = 0$ of the computational domain by driving the two transversal E field components, E_y and E_z, with a spatial distribution corresponding to that for the TE_{10} mode. The absorbing boundary was implemented at the end plane $x = x_l$ of the computational domain.

The excitation waveform consisted of a cosine function modulated by a gaussian envelope. The carrier frequency represented the desired tuning frequency and was chosen at different points within the frequency range exhibiting non-zero group velocity. Most figures were obtained with bandwidth of 0.010 centered about the tuning frequency.

The Discrete Time Fourier Transform was computed for both incident and reflected time waveforms and the percentage of reflection for any normalized frequency $\Delta l/\lambda$ to which corresponded a significant spectrum magnitude, was measured by forming the ratio of the spectrum magnitude for the reflected pulse over the spectrum magnitude for the incident pulse.

The dispersive and the anisotropic behaviors of the FD-TD method as well as the dispersive effect of the waveguide structure were accounted for at once by substituting into the dispersive equation of the FD-TD method the two transversal wave numbers $\kappa_y = 0$ and $\kappa_z = \pi/10$ corresponding to the TE_{10} mode for the present waveguide. The general dispersive equation for the FDTD method with

$$\Delta x = \Delta y = \Delta z = \Delta l$$

reads as follows:

$$\frac{1}{s^2}\sin^2(s\pi\Delta l/\lambda) = \sin^2(\kappa_x\Delta l/2) + \sin^2(\kappa_y\Delta l/2) + \sin^2(\kappa_z\Delta l/2) \quad (2)$$

The phase velocity in the longitudinal direction x was then computed as $v_p = \omega/\kappa_x$ where $\omega = 2\pi c/\lambda$ is the radian frequency, λ and $c = 1/\sqrt{\mu\varepsilon}$ are respectively the intrinsic wavelength and the intrinsic propagation velocity for the propagation medium with constitutive parameters μ and ε, and κ_x is the longitudinal wave number. The expression for the phase velocity becomes:

$$\frac{v_p}{c} = \frac{\omega}{\kappa_x} = \frac{(\pi\Delta l/\lambda)}{arcsin(\sqrt{A})} \quad (3)$$

and the expression for the group velocity becomes:

$$\frac{v_g}{c} = \frac{\partial\omega}{\partial\kappa_x} = \frac{s\sin(\kappa_x\Delta l)}{sin(2\pi s\Delta l/\lambda)} = \frac{2s\sqrt{A}\sqrt{1-A}}{sin(2\pi s\Delta l/\lambda)} \quad (4)$$

where:

$$A = \frac{1}{s^2}\sin^2(\pi s\Delta l/\lambda) - \sin^2(\pi\Delta l/\lambda_c)$$

Finally, the amount of delay W required by the absorbing boundary algorithm was computed as:

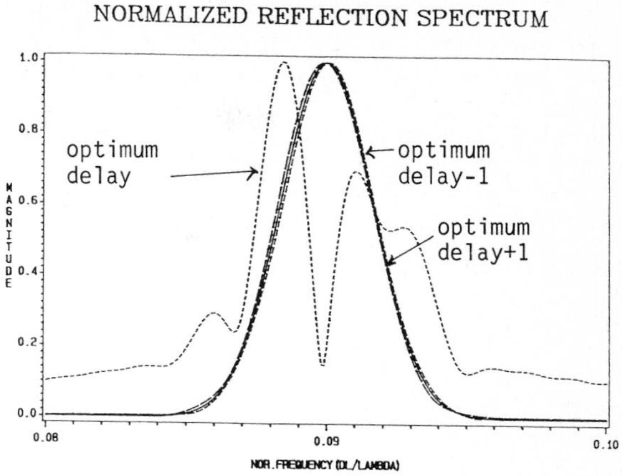

Fig. 2 Normalized Reflection Spectrum

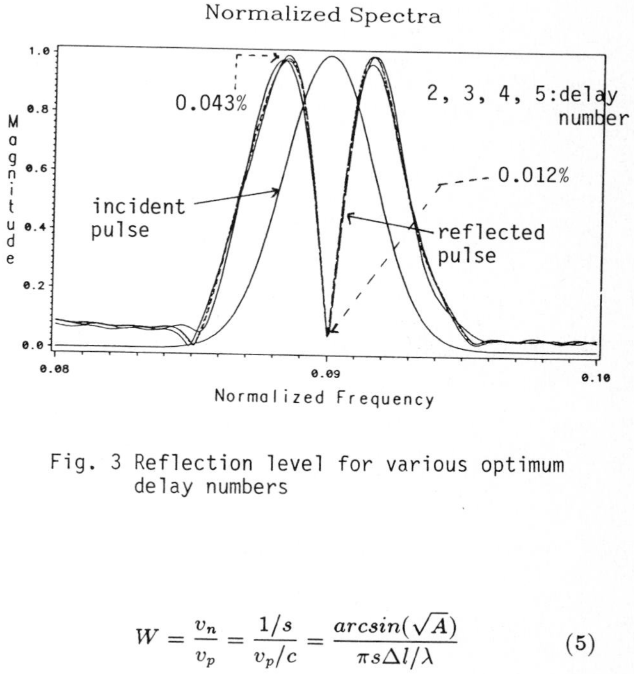

Fig. 3 Reflection level for various optimum delay numbers

$$W = \frac{v_n}{v_p} = \frac{1/s}{v_p/c} = \frac{arcsin(\sqrt{A})}{\pi s\Delta l/\lambda} \quad (5)$$

Fig. 2, 3, and 4 display the spectrum magnitude of the pulse reflected off the absorbing boundary which was tuned to various normalized frequencies for various numbers of delays by proper selection of the s value. All spectra are normalized with respect to their maximum in order to display adequately the effect of the absorbing boundary. The figures show the percentage of reflection for all cases considered herein. Fig. 2 also displays the incidence angle $\theta = \arcsin((\Delta l/\Lambda)/(\Delta l/\lambda))$

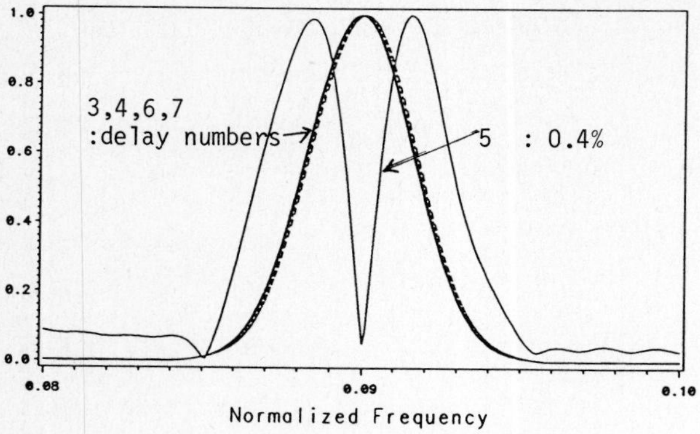

ig. 4 Reflection level for different delay numbers which are not optimized
(The delay number is optimum for the number 5)

at which the tuning frequency component of a simulated wavefront impinges onto the absorbing boundary.

A glance at the results reveals that the amount of reflection at the tuning frequency is less than 0.02% for most cases considered herein. The absorbing boundary is seen to behave very well for a wide range of tuning frequencies, hence for a correspondingly wide range of incidence angle. The results also indicate that the algorithm performs equally well for different values of optimum delay number. Therefore the number of iterations can be kept minimum without degradation of the absorbing boundary's performance by selecting the largest S value that satisfies both the stability criterion and the requirement of having an integer delay number W.

The effect of truncating the excitation waveform was also investigated in terms of truncation magnitude and waveform symmetry. The minimum and maximum levels of reflection were seen to decrease with smaller truncation magnitude. The asymmetry of the excitation waveform was found to distort the spectrum magnitude of the reflection by the presence of sidelobes, without affecting significantly the maximum and minimum levels of reflection. The proximity effect of the absorbing boundary was also investigated by comparing the field value seen at the observation point when the absorbing boundary was positioned at different distance values from the observation point, with the field value seen at the same observation point when the computational domain was so long as to simulate effectively a perfectly absorbing boundary. The proximity effect was found to be unnoticeable at the tuning frequency, thus confirming that the absorbing boundary generates no evanescent wave of its own at that frequency. For other frequencies, the absorbing boundary presents a complex impedance, mostly resistive, giving rise to small reflection levels.

4. Conclusion

This paper has presented the development of a very simple yet very effective plane absorbing boundary for the FDTD method with arbitrary incidence angle in three dimensions. The efficacy of this absorbing boundary algorithm was then demonstrated for the case of a homogeneous rectangular wavequide.

References

1. K. S. Yee, "Numerical solution of initial boundary value problems involving Maxwell's equations in isotropic media", IEEE Trans. Vol. AP-14, pp.302-307, May 1966.

2. A. Taflove and M. E. Brodwin, "Numerical Solution of Steady-State Electromagnetic Scattering Problems Using the Time-Dependent Maxwell's Equations", IEEE Trans. Microwave Theory and Techniques, Vol. MTT-23, No. 8, pp. 623-630, August 1975.

3. C. D. Taylor, D-H. Lam and T.H. Shumpert, "Electromagnetic Pulse Scattering in Time-Varying Inhomogeneous Media", IEEE Trans. Antennas Prop. Vol. AP-17, No. 5, pp. 585-589, Sept. 1969.

4. X. Zhang and K. K. Mei, "Time-Domain Finite Difference Approach for the Calculation of the Frequency-Dependent Characteristics of the Microstrip Discontinuities", IEEE Trans. MTT, Vol. 36, No. 12, pp. 1775-1787, Dec. 1988.

5. J. Fang. and K. K. Mei, "A Super-Absorbing Boundary Algorithm For Solving Electromagnetic Problems By FD-TD Method", AP-S Symposium, Syracuse, pp. 472-475, June 1988.

Approximation, Optimization and Computing:
Theory and Applications, A.G. Law and C.L. Wang (eds.)
Elsevier Science Publishers B.V. (North-Holland)

APPROXIMATE SIMULATION MODEL FOR ANALYSIS AND OPTIMIZATION IN ENGINEERING SYSTEM DESIGN

Jaroslaw Sobieszczanski-Sobieski
NASA Langley Research Center, Hampton, Virginia, U.S.A.

ABSTRACT

Computational support of the engineering design process routinely requires mathematical models of behavior to inform designers of the system response to external stimuli. However, designers also need to know the effect of the changes in design variable values on the system behavior. For large engineering systems, the conventional way of evaluating these effects by repetitive simulation of behavior for perturbed variables is impractical because of excessive cost and inadequate accuracy. This paper describes an alternative based on recently developed system sensitivity analysis that is combined with extrapolation to form a model of design. This design model is complementary to the model of behavior and capable of direct simulation of the effects of design variable changes.

1. INTRODUCTION

Mathematical models are a well established means for simulation of the behavior of engineering systems to support design decisions. These models usually employ a network of disciplinary behavioral models such as structural analysis, aerodynamic analysis, propulsion analysis, etc. and have one common characteristic: they answer the question "what will be the behavior (response) of the system to a given external stimulus ?" The answers these models provide are typically cast in a numerical form of some behavior variables, for instance, stress and displacement fields induced by a given load acting on a structure.

In design, however, engineers must decide how to change design variables in order to effect a desired change in behavior. To do so they need answers to "what if" questions, e.g. "what will be the change in the behavior if a particular design variable is altered ?" Indeed, one may assert that the design process is not completed until all such questions are answered, at least for the major design variables. The "what if" questions may be answered by repeated use of behavior models combined with design variable perturbations to obtain finite-difference approximations to design derivatives. In large applications, the computational cost of behavior models interconnected in a network and the accuracy problems intrinsic in the finite-difference methods render that approach impractical. As a result, practical applications of systematic, mathematically based optimization in the design of complete engineering systems have been lagging [1], relative to the progress of engineering optimization theory noted in recent times.

In contrast, optimization applications in structural engineering have been growing steadily in number, size, and complexity of successfully solved cases. This growth may be attributed, at least in part, to the concept of decoupling the design space search from the full analysis (the behavior model), and coupling it instead with an approximate analysis based on derivatives of behavior with respect to design variables. That concept, introduced in [2], spurred the development of sensitivity analysis to the point where it became routine for structures [3]. The approximate analysis concept was also a basis for the development of a body of efficient structural optimization procedures, e.g., [4], that have also diversified to applications in other disciplines, e.g., [5] and [6].

The purpose of this paper is to show that the recent development of algorithms for the sensitivity analysis of complex, internally coupled systems that comprise several subsystems and that involve many disciplines made it possible to use the approach initiated in [2] to develop a combination of extrapolation and sensitivity analysis for such systems. That capability will be referred to as a model of design. The purpose of the model of design, complementary to the commonly used model of behavior, is to simulate efficiently the effect of design variable changes on behavior, so it may be used to answer the "what if" questions quickly and inexpensively in order to support formal optimization as well as judgmental decisions in design.

2. PHYSICAL SYSTEM VERSUS SIMULATION SYSTEM

Engineering design of contemporary aircraft, spacecraft, and other complex systems is a prolonged and complicated process that involves human creativity, ingenuity, and judgment, all supported by massive computations. The computations involve a collection of computer programs, each representing a physical subsystem of the engineering system at hand or a particular aspect of that system behavior. By virtue of passing the data to each other the computer programs in that collection become modules in a coupled system that simulates the physical one and is its mathematical model of behavior, just as each module by itself is a mathematical model of behavior for a discipline or a part of the physical system.

For instance in aircraft design (figure 1) we may distinguish the wing and the fuselage as separate subsystems in the aircraft system. We may also consider the structural and aerodynamic analyses as separate modules in a system of programs that supports the design. This example points out that, in general, no one-to-one correspondence exists between the modules and the subsystems. In fact, the structural finite element model processed in the structural analysis may reflect the wing and the fuselage as separate substructures in the airframe, but the aerodynamic analysis may operate on a single digitized model of the aerodynamic surface that extends over the entire wing-fuselage assembly.

In reality, the wing and the fuselage interact aerodynamically and structurally. As shown in figure 2, they modify each other flow fields through the conduit of aerodynamics, and through the conduit of structures they exert forces on each other at the fuselage-wing junction. The corresponding mathematical model may represent

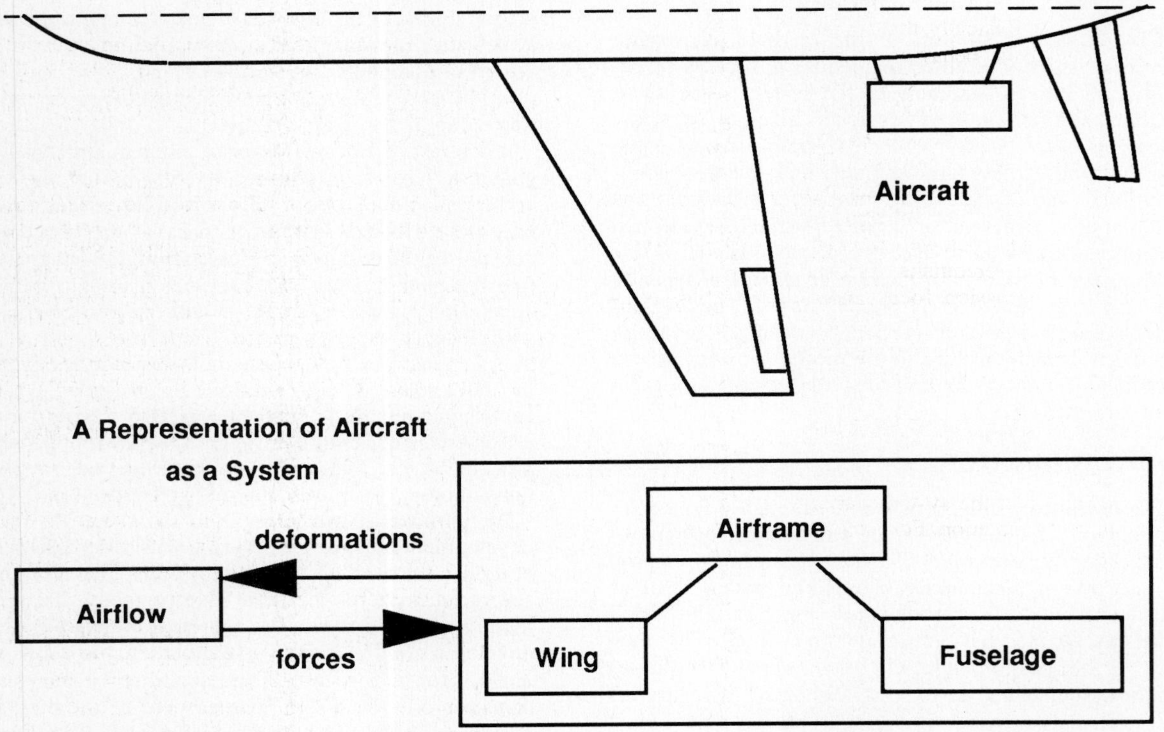

Figure 1 Aircraft system and its aspect and object decompositions.

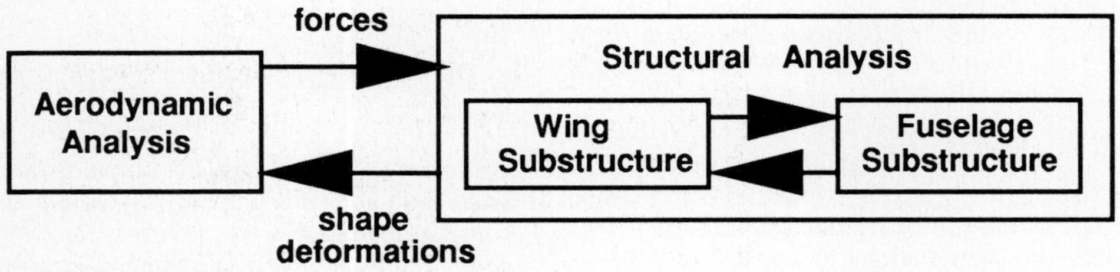

Figure 2 Simulation system.

the interactions by the force-deformation compatibility conditions at the wing-fuselage junction, and by the aerodynamic forces and airframe deformations that couple the aerodynamic and structures analysis modules. The example shows also that for purposes of mathematical simulation, the physical system may be decomposed into smaller parts (the object decomposition) or each aspect of behavior may be assigned a module in the simulation (the aspect decomposition). Both types of decomposition may be used simultaneously, as they are in this example.

Once the real, physical system has been conceptually decomposed and the corresponding mathematical simulation system has been assembled, it is the simulation system that provides computational support for the design process. Basically, that support has two functions: first, to reveal the behavior of the physical system in response to external stimuli; and second, to find out how that behavior may be modified by changing physical attributes (design variables) of the system. The first function calls for a system analysis. The latter answers the "what if" types of questions which is what design is all about, and is performed by sensitivity analysis. In the remainder of this paper, we will examine system analysis only briefly as a prerequisite to sensitivity analysis which will be the discussion focus and will show how the sensitivity analysis can be formalized as a basic ingredient in the mathematical model of design that complements the model of behavior.

3. SYSTEM ANALYSIS AS A MODEL OF BEHAVIOR

Each module in the system may be represented by a function vector notation. For the i-th module in a system of NM modules, we have

$$F^1(Z,Y^1) = 0 \tag{1}$$

where F is a vector of functions, NF long, Y is a vector of dependent variables, and Z is a vector of independent variables. The set of NF simultaneous equations in eq. 1 yield NF elements of Y for a given Z and are the governing equations for the physical phenomena simulated by the module. No assumptions are made as to the mathematical nature of eq. 1, they may be nonlinear, transcendental, etc., so that an iterative algorithm may be required to solve for Y. In practice, a module is an entity comprising eq. 1 together with its solution algorithm coded as a computer program, usually including embellishments such as graphics. Then, Z is the input and Y is the output of the program which may be treated as a black box.

Structural finite element analysis and aerodynamic analysis of an aircraft wing are examples of the above. In structural finite element analysis, eq. 1 is

$$KY - P = 0 \tag{2}$$

and represents an equilibrium of internal forces, the KY term, and the external loads P. Both P and the stiffness matrix K may be functions of input Z that describes the overall wing shape, cross-sectional dimensions, and the loading conditions. These equations and their solution for Y (the displacements and the resulting stress) are typically implemented in large computer programs, yielding tens of thousands of elements of Y and comprising hundreds of thousands of lines of code (e.g., Program NASTRAN, [7]).

Similarly, an example of governing equations for the aerodynamic analysis is

$$(J^{-1}U)_t + F_x + G_y + H_z - (G_v)_y = 0 \tag{3}$$

where the subscripts indicate differentiation with respect to time t and the coordinates x, y, and z correspond to the streamwise, normal, and chordwise directions. In these equations, the terms are defined in [8], the vector U corresponds to Y in the generic notation used in this paper and the other terms contain input corresponding to Z. The equations yield pressure data for hundred of thousands points over the wing surface (270,000 points were used in [8]). Due to viscosity and compressibility effects, eq. 3 are distinctly nonlinear. Again, implementation of these equations and their solution took form of a large computer code that in an elastic wing system analysis appears as an aerodynamic module.

To simulate an elastic wing behavior, the structural and aerodynamic modules are assembled to make them to interact with each other, simulating the aerodynamics-structure coupling illustrated in figure 2. In reality, the coupling occurs because the aerodynamic forces deform the elastic wing. In turn, the deformation modifies the aerodynamic forces. In the simulating system, the coupling is realized by entering the aerodynamic pressure output Y from the aerodynamic module into the input Z of the structural module as the load data, and by using the deformation output Y from the structural module as the new shape data in the input Z of the aerodynamic module. In the presence of nonlinearities, the simulating system operates by iterating between the modules until the governing equations in each are satisfied to a desired tolerance.

At this point, it is necessary to distinguish three parts in the input Z of any module. The first part of Z consists of the physical quantities X that the designers change to influence the system behavior, the second part includes the constants Q, and the third part comprises the outputs Y from the other modules in the simulating system. Of course, both X and Q remain constant for the duration of analysis and the X elements are changed between the consecutive system analyses. The division between X and Q is not rigid, it is up to the designer to move physical quantities from Q to X and vice versa.

With the above definition of Z, it is now possible to generalize eq. 1 to a set of governing equations for a simulating system consisting of NM modules identified by subscripts:

$$\begin{aligned}
&\ldots\ldots\ldots\ldots\ldots\ldots \\
&F^{i-1}((X^{i-1},Q^{i-1},Y^j),Y^{i-1}) = 0; \quad j \neq i-1 \\
&F^i((X^i,Q^i,Y^j),Y^i) = 0; \quad j \neq i \qquad j = 1 \ldots NM \\
&F^{i+1},((X^{i+1},Q^{i+1},Y^j),Y^{i+1}) = 0; \quad j \neq i+1 \\
&\ldots\ldots\ldots\ldots\ldots\ldots
\end{aligned} \tag{4}$$

Since each module represents a set of NF ^{i} equations solvable for NF ^{i} elements of Y ^{i}, it follows that the number of equations in eq. 4 is equal to the number of elements in Y concatenated of the Y ^{i} vectors.

In addition to aerodynamic and structural modules discussed in the wing example, support of a complete aircraft design would require modules for propulsion, control, electromagnetics, interior environment control, fuel management, avionics, weaponry, aircraft, performance, and more, all forming a system represented by eq. 4, and coupled internally by the Y cross-feed. To simulate the real system behavior the simulating system has to be iterated to convergence, assuming that nonlinearities exist that preclude solution by a linear algebraic elimination algorithm. The iterations may be nested because some of the modules may require internal iterations for their own solutions. Given the computational size of each module, to converge the solution for one setting of X is a formidable undertaking, even when using present-day supercomputers.

In design, the computational expense of producing the behavior data for one setting of X has to be incurred repeatedly as the designers change that setting in search of one that makes the system behave in an acceptable manner and then again in pursuit of a behavior that is better than merely acceptable.

Thus, the expense of the behavior model operation motivates a proposition that another mathematical model, capable of revealing directly and inexpensively how the behavior will change if a design variable is altered, should be added to the designer's tool box to complement the behavior model already there. That additional model will be referred to as the model of design.

4. MODEL OF DESIGN

One way to create a model of design is to use the model of behavior to obtain the data at several settings of X, each setting interpreted as a point in a hyperspace defined by X. The number of points is limited by the budget available for computational expenses, and their locations are strategically chosen throughout the intervals of interest using methods known as experiment design methods, e.g., [9]. Once the behavior data at these points have been generated, the behavior model is replaced by the hypersurface fitted to the points. The model is invoked as an explicit interpolating functions to obtain data on various aspects of the behavior between the points at essentially no cost, as needed in the course of design. The literature notes a number of applications of that type of design model in support of formal optimization, e.g., [10].

4.1. Extrapolation

One alternative is to obtain the behavior data from a behavior model at a single judiciously chosen setting of X, including among the data their derivatives with respect to the design variables X. The data is used an extrapolation formula, for instance, a Taylor series

$$Y(X) = Y_o + \nabla Y^T(X-X_o) + \tfrac{1}{2}(X-X_o)^T [\nabla^2 Y](X-X_o) + (\textit{higher-order terms}); \quad (5)$$

where Y quantifies a particular aspect of the behavior of interest, e.g., stress in the wing structure, propulsion thrust, or maximum flight range. Once the expense of using the behavior model to generate the data needed in the series above has been paid, the information about the effect of X on Y is available essentially at no cost, albeit its accuracy deteriorates as one moves away from the reference point where the analysis took place. The advantage of this design model over the one described previously is that one does not need to saturate the entire potentially interesting design space with analysis points. Instead, one starts at a single point and lets the extrapolation formula guide the search for the next point where the behavior data and the derivatives need to be refreshed by new analysis. Structural optimization applications with nearly 100 design variables have been reported, e.g, [11], where good results required only 10 to 20 repetitions of analysis that included calculation of derivatives.

The usefulness of this extrapolation-based design model critically depends on the computational cost of derivatives. Finite differencing as a means to obtain the derivatives may be prohibitively expensive since it adds another outer loop around the iterative analysis loops required for the solution of eq. 4, some of which may be nested already. Moreover, the finite differencing of iteratively obtained solutions may be meaningless for a small difference due to computational noise, while for a larger difference errors due to nonlinearities set in. Therefore, one may assert that the extrapolation-based, design model is incomplete for the purposes of large-scale applications if it does not include a means for efficient and accurate computation of derivatives.

To this end, [12] introduced a direct system sensitivity analysis that bypasses the finite differencing of the system analysis.

4.2. System Sensitivity Analysis

As shown in [12], the derivatives of behavior with respect to a design variable may be obtained from a set of simultaneous linear algebraic equations generated by application of the implicit function theorem to eq. 4. These equations are rewritten to show the Y's as implicit functions of X

....

$$\begin{aligned} Y^{i-1} &= f^{i-1}(X^{i-1}, Q^{i-1}, Y^j); \quad j \neq i-1 \\ Y^i &= f^i(X^i, Q^i, Y^j); \quad j \neq i \qquad j = 1 \ldots NM \\ Y^{i+1} &= f^{i+1}(X^{i+1}, Q^{i+1}, Y^j); \quad j \neq i+1 \end{aligned} \quad (6)$$

....

The derivatives of Y with respect to a particular design variable X k appear as unknowns in a set of equations

$$[A]\{\partial Y/\partial X\} = \{RHS\}_k; \quad k = 1 \ldots NX \quad (7)$$

where:

$$[A] = \begin{bmatrix} I & \cdot & \cdot & \cdot & \cdot \\ \cdot & I & \cdot & \cdot A^{ij} & \\ \cdot & \cdot & I & \cdot & \cdot \\ \cdot & \cdot & \cdot & \cdot & \cdot \\ \cdot & \cdot & \cdot & \cdot & I \end{bmatrix};$$

$$Y^T = [Y^{1T}, Y^{2T}, \ldots Y^{iT}, \ldots Y^{NM\,T}];$$

$$[A^{ij}] = [\partial f^i / \partial Y^j]$$

$$RHS^T{}_k = [\{\partial f^1/\partial X_k\}^T, \{\partial f^2/\partial X_k\}^T, \ldots \{\partial f^i/\partial X_k\}^T, \ldots \{\partial f^{NM}/\partial X_k\}^T]$$

The vectors and matrices in the above have the following dimensions

$$\{Y^i\},\ NF^i \times 1;\quad [A^{ij}],\ NF^i \times NF^j;\quad [A],\ NA \times NA;$$

$$NA = \sum_{i=1}^{NM} NF^i;\quad \{RHS\},\ NA \times 1;\quad \{X\},\ NX \times 1; \tag{8}$$

By virtue of the implicit function theorem, eq. 7 are always linear, therefore they may be efficiently solved for many different vectors Y by factoring the A matrix once and storing it. The stored matrix may then be back-substituted over by the RHS vectors, each corresponding to one particular X_k.

The A^{ij} is a Jacobian matrix of the partial derivatives of the output from the i-th module with respect to the output from the j-th module that is received as input in the i-th module. The submatrices on the diagonal of A are identity matrices, and the element of the RHS_k are partial derivatives of the output from the modules with respect to a particular X_k. By definition the partial derivatives in A^{ij} and in RHS_k may be computed for each i-th module independently of each other. This enables one to use specialized methods for sensitivity analysis that have been developed for many engineering disciplines, e.g., [3], and [13], or even use finite differencing but on one module at a time, thus avoiding the cost of repetitive solution of the system equations, eq. 4. On the other hand, the computational cost of generating and solving eq. 7 grows superlinearly with the volume of the coupling data passed from one module to another as measured by NF^i in eq. 8. Indeed, in the limiting case of all A_{ij} being null matrices there is no coupling and the trivial solution of eq. 7 is Y = RHS because A = I. A later will show how one may keep NF^j from growing inordinately by using physical insight in selecting a minimal number of the pieces of data to be transmitted among the modules. Numerical conditioning of A was examined in [12] which concluded that singularity is not a danger if eq. 4 represent a well-posed problem.

Once the system sensitivity analysis has been reduced to solving the linear equations (eq. 7) it is possible to calculate the higher-order derivatives of Y with respect to X as derivatives of the first derivatives obtained from these equations. This approach was implemented in [14] by applying the same implicit function theorem to eq. 7 that was used to derive eq. 7 from eq. 4. Repetitive use of that theorem generates a recursive chain of formulas for the higher-order derivatives shown below in a compact notation which is defined first

$$(\ )^r{}_{klm\ldots} = \partial^r(\ )/\partial X_k \partial X_l \partial X_m \ldots$$

$$\begin{aligned} Z^o &= Y^1{}_k \\ Z^1{}_l &= Y^2{}_{kl} \\ Z^2{}_{lm} &= Y^3{}_{klm} \\ &\vdots \\ Z^N{}_{lm\ldots} &= Y^{N+1}{}_{klm\ldots} \end{aligned} \tag{9}$$

where any subscript may be repeated as required to form a mixed derivative with respect to any combination of variables X.

In the above notation, the second- and higher-order derivatives are

$$\begin{aligned} A Z^1{}_l &= R^1{}_l - A^1{}_l Z^o; \\ A Z^2{}_{lm} &= R^2{}_{lm} - A^1{}_m Z^1{}_l - D^1{}_m(A^1{}_l Z^o); \\ A Z^3{}_{lmn} &= R^3{}_{lmn} - A^1{}_n Z^2{}_{lm} - D^1{}_n(A^1{}_m Z^1{}_l) \\ &\quad - D^2{}_{mn}(A^1{}_l Z^o); \\ A Z^4{}_{lmnp} &= R^4{}_{lmnp} - A^1{}_p Z^3{}_{lmn} - D^1{}_p(A^l{}_n Z^2{}_{lm}) \\ &\quad - D^2{}_{np}(A^1{}_m Z^1{}_l) - D^3{}_{mnp}(A^1{}_l Z^o); \end{aligned} \tag{10}$$

etc

where $D^q{}_{lmn}(\ )$ is a shorthand for the q-th mixed derivative of the product of the pair of functions in the parentheses, obtained by the usual rules of product differentiation. Once the derivatives of Z are obtained, the derivatives of Y are available from eq. 9. The above regular pattern can extended easily beyond the first four derivatives shown above.

It is apparent from eq. 7 and 10, that the computational cost may be reduced by factoring A only once, since A is the matrix of coefficients in equations for derivatives of every order. On the other hand, that cost escalates super-linearly with the derivative order because of the increase of the number of the derivatives to be computed and the accumulation of the prerequisite data required by the recursivity of eq. 10. By weighing that computational cost against the accuracy improvement attained by the use of the higher-order derivatives in the extrapolation (eq. 5) one decides to which order the sensitivity analysis should be extended. The current practice tends to include only the first, and occasionally, the second derivatives in large scale applications, but these practical limits are likely to go up as the progress in computer technology continues to lower the computational cost.

The extrapolation in eq. 5 together with the sensitivity analysis defined by eq. 7 through 10 define a model of design complementary to the model of behavior represented by eq. 4. Once the derivatives have been calculated and substituted in eq. 5, one may compute the effect of any X_k on the behavior practically instantaneously and at relatively negligible cost, provided that the increment of X_k is kept within extrapolation bounds (move limits) consistent with the problem nonlinearity and the order of extrapolation.

4.3. Enhancing the Design Model by Replacement Variables

The extrapolation bounds (move limits) may be widened by introducing artificially a degree of nonlinearity into the design model by a judicious replacement of the design variables. One such replacement is described in [15]. The Y^i_j behavior variable is tested for the sign of its first derivative with respect to X_k. If that derivative is positive, the extrapolation of Y^i_j continues to be done with respect to X_k, but for a negative derivative the extrapolation is done with respect to a replacement variable $R_k = 1/X_k$.

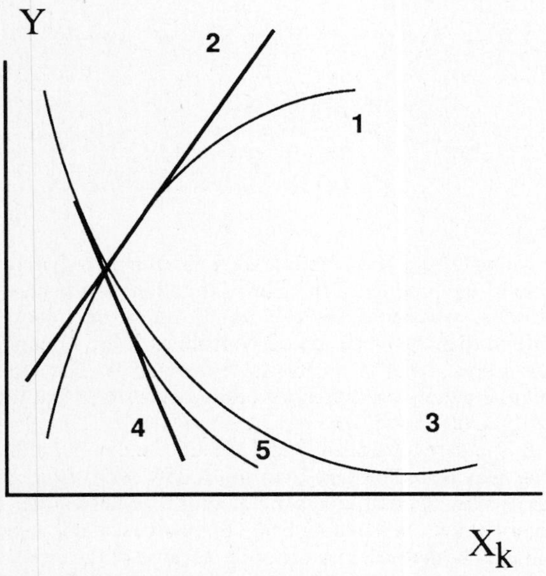

Figure 3 Replacement variables in extrapolation.

Justification for such selective replacement may be explained graphically as shown in figure 3. A positive derivative characterizes the function as ascending. An ascending function in a physical system is likely to obey the law of diminishing returns, hence it should look like curve 1. Extrapolation with respect to X_k by means of tangent 2 is likely to overestimate, and thus be a conservative approximation - a desirable feature in engineering design. Conversely, a negative derivative identifies a descending function portrayed by curve 3. Again, the law of diminishing returns is likely to render that curve asymptotic to the X-axis, so that tangent 4 would be an undesirable nonconservative approximation that underpredicts the value of the function. To reduce the error, one may extrapolate with respect to the reciprocal X_k, in effect following curve 5, and thus preserving the asymptotic nature of the true function 3 and at least some of its nonlinearity.

5. EXAMPLES

The first system optimization using a design model of the type described above was reported in [16]. It was a simple test case of a cantilever beam (structural analysis module) whose dynamic response to a ramp-shaped load impulse was controlled by exerting forces on the beam with actuators commanded by a control system (control module). Successful optimization for the minimum weight of the entire system, including the weights of the beam and of the actuators, employed a linear design model.

Four examples described in this section were selected to represent information accumulated since the above case was completed. Each of these examples illustrates different aspects of the foregoing discussion. The first example shows the role of physical insight in setting up the system sensitivity analysis. The second example demonstrates how greatly the system derivatives, with respect to design variables, may differ from the partial derivatives. The third example addresses the issue of the

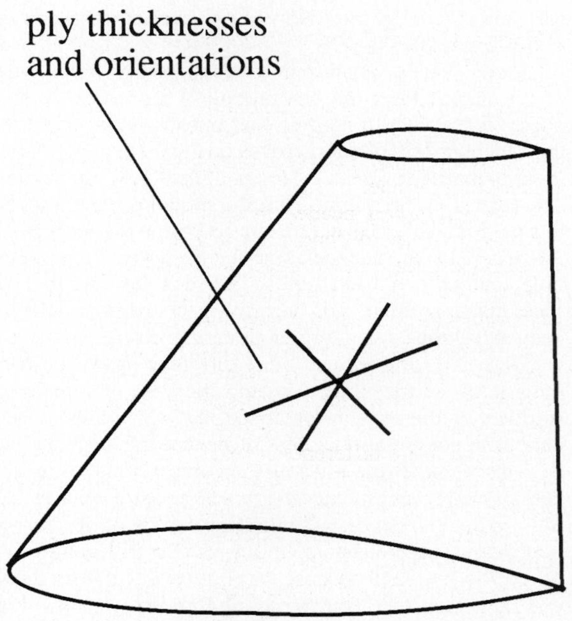

Figure 4 A transport aircraft wing.

accuracy of the extrapolation. Finally, the last example shows the effect of including the second derivatives in the model of design used in a formal optimization.

5.1. Elastic, Slender, Composite Wing

As the first example, consider a slender wing structure

of a transport aircraft (figure 4) with skin made of a composite material. Suppose that optimization of the composite skin involves 30 design variables comprising the ply thicknesses and orientation angles. The behavior model of the wing comprises the aerodynamic and structural modules that exchange the aerodynamic force and structural deformation data. The aerodynamic module uses a nonlinear method of analysis for a transonic flow and outputs an aerodynamic pressure value for each of 1000 discrete points over the wing surface. The structural analysis employs a finite-element method and outputs 1500 discrete displacement values for the finite-element model grid. A gradient-guided optimization program uses derivatives of aerodynamic pressure and structural displacement to guide the search for a constrained optimum in the design space.

If the derivatives needed in optimization were to be obtained by finite differencing performed on the model of behavior, the optimization would be organized as shown by the flowchart in figure 5. First, for a trial setting of X, one would iterate the aerodynamics and structures, loop 1, until convergence of the aerodynamic loads and structural deformations is obtained to a satisfactory tolerance. This iteration constitutes the system analysis for this case.

Next, the finite differencing would proceed in loop 2. Under the simplest finite-difference, one-step-forward scheme, each pass through loop 2 involves reanalysis of the system for the design variables perturbed one at a time. In that reanalysis, the aerodynamics-structure iteration in loop 1 must be reconverged. Assuming 5 passes to converge in loop 1, the total number of executions of each of the modules to carry out the system analysis once to obtain the reference solution and once for each design variable to obtain the solution first derivatives would be, assuming 30 design variables: 5 x (1 + 30) = 155. The solution and its derivatives are used in the extrapolation employing eq. 5 curtailed to the linear part of the series. The extrapolation is coupled to the optimization program in loop 3. The optimization yields a new, presumably improved, design for which the system solution and its derivatives have to be refreshed so loop 4 has to be traversed several times until overall convergence. Estimating that loop 4 would have to be repeated 15 times, one gets 4 x 155 = 620 executions of each module, if optimization is coupled directly to the model of behavior. The aggregate cost of these executions accounts for nearly the total cost of the entire optimization because the cost of executing loop 3 is trivial.

Introduction of a model of design that comprises the system sensitivity analysis by eq. 7 and the extrapolation by eq. 5 enables one to change the optimization organization from the one illustrated in figure 5 to the one shown

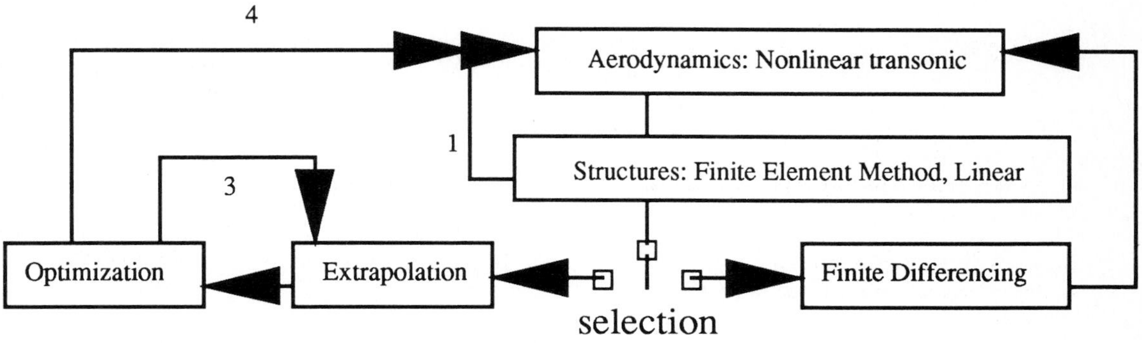

Figure 5 Optimization with dervatives obtained by finite differencing of the behavior model.

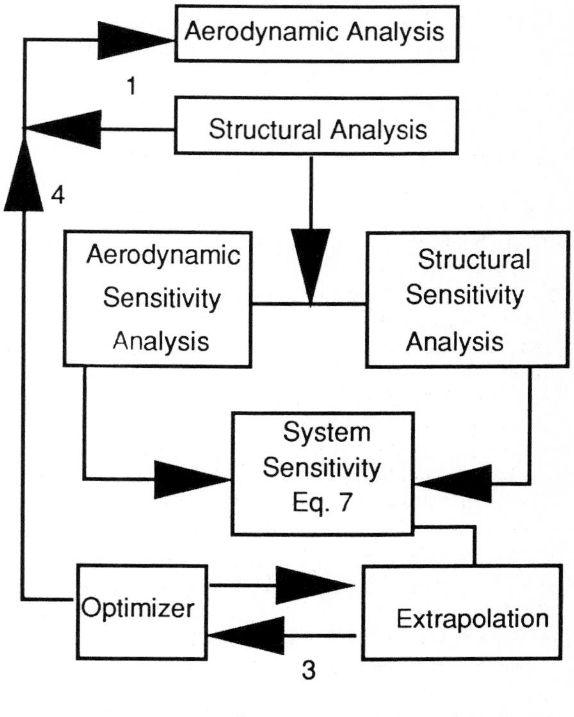

Figure 6 Optimization based on derivatives obtained by system sensitivity analysis.

in figure 6. Loop 1 is the same as in figure 5, but the finite differencing that engages the system analysis, loop 2 in figure 5, is replaced by two disciplinary sensitivity analyses, one for aerodynamics and one for structures, that are executed independently of each other (an opportunity for parallel processing). As mentioned before, algorithms for disciplinary sensitivity analyses become routine in structures [3], begin to be available in aerodynamic analysis [17], and are generally much less computationally costly than finite differencing.

The volume of data to be exchanged between the modules governs the computational cost of the disciplinary sensitivity analyses. As assumed in this example, there are 1000 pressure data output from the aerodynamic module and 1500 displacement data output from the structures module. Labeling the aerodynamic and structures modules as 1 and 2 respectively, the Jacobian matrices in eq. 7 would have the dimensions 1000 x 1500 for A^{12} and 1500 x 1000 for A^{21}. In other words, one would have to take the partial derivatives of each pressure datum with respect to each displacement datum and vice versa. Despite the efficiency of disciplinary sensitivity analysis, computation of that many partial derivatives would still be economically prohibitive. However, the number of derivatives needed may be radically reduced by physical insight.

Since the wing is slender (high aspect ratio), its chordwise, plate-like bending is negligible relative to the spanwise, beam-like bending. Furthermore, the wing aerodynamic forces are affected by the changes of the streamwise airfoil angle of attack caused by the structural twist (and also bending in the case of a swept wing. Since the wing twist angle and bending deflections are known to be distributed quite smoothly spanwise, it is a reasonable assumption to transmit only, say, 5 values of the angle-of-attack changes at 5 spanwise wing locations as a description of structural deformations. Conversely, the 1000 aerodynamic pressure data may be collected into a vector of, say, 10 concentrated forces at 10 spanwise wing locations.

This condensation may easily be implemented as postprocessing so it becomes a part of each module. Then, the dimensions of the Jacobians in eq. 7 reduce to 10 x 5 and 5 x 10 for A^{12} and A^{21}, respectively. By the same token, RHS for this example would be reduced to only 15 elements. In this RHS, all elements are null except the bottom partition of 5 elements that contains the partial derivatives of displacement with respect to the a structural design variable.

Once the partial derivatives from the two disciplinary sensitivity analyses are obtained, the system derivatives are calculated from eq. 7 and optimization using eq. 5 executes in loop 3 followed by loop 4, the same as in figure 5.

In summary, the use of a model of design that embeds a system sensitivity analysis has reduced the number of executions of each modules in this example from 620 to merely 5 x 15 = 75 (loop 1 nested in loop 4), at the price of adding the cost of the disciplinary sensitivity analyses and the cost of solving eq. 7 (nested in loop 4 hence incurred 15 times under the assumptions used in the example). Even with that added cost, the overall optimization cost reduction is likely to be more than an order of magnitude. An additional potential benefit is time saved due to the parallel processing of the independently executed disciplinary sensitivity analyses.

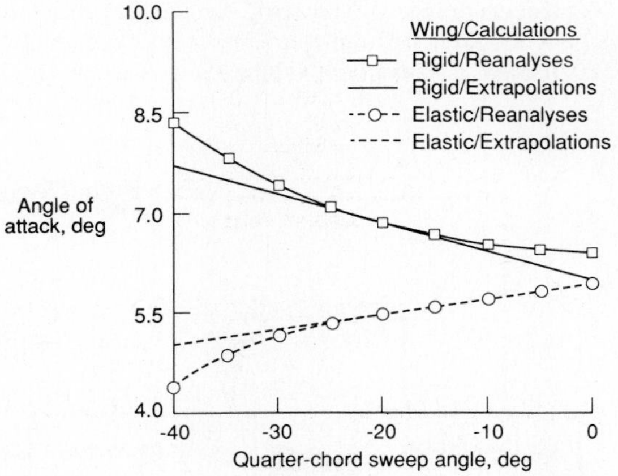

Figure 7 Trimmed angle of attack as a function of sweep angle.

5.2. Forward-Swept, Elastic Wing

A result reported for a forward-swept, elastic wing in (18], is an example of a drastic difference between the values of the system derivative and the partial derivative of a behavior variable with respect to a design variable. The system again comprises the aerodynamics and structures modules, the behavior variable is the trimmed angle of attack (the angle of attack at the reference chord required to maintain a prescribed lift), and the design variable is the sweep angle.

The sweep angle is the horizontal coordinate in figure 7 (negative degrees indicate forward sweep) and the trimmed angle of attack is the vertical coordinate. The trimmed angle of attack as a function of the sweep angle is shown for the rigid and flexible wing. The slopes of the tangents represent the derivatives at an arbitrarily chosen sweep angle of 20 degrees. The partial derivative corresponds to the tangent slope of the rigid wing curve because it was obtained from the aerodynamics alone. The system derivative is represented by the tangent slope of the flexible wing curve since it reflects the interaction of two modules in the wing system. It is apparent that the interaction was so strong in this case that the derivative and the partial derivative have opposite signs.

The lesson from this example is that a trend predicted on the basis of only one module in a coupled system may completely misguide the design decisions.

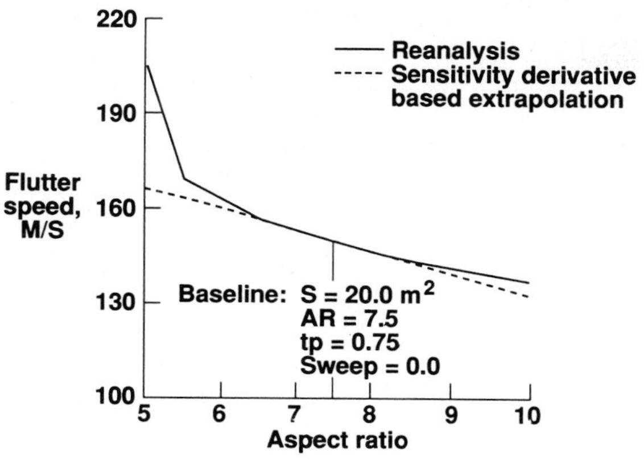

Figure 8 Flutter speed as a function of wing aspect ratio.

5.3. Wing Flutter

Usefulness of the design model as a predictor of the effects of the design variable changes depends on the degree of nonlinearity of these effects. Results that shed light on that issue were reported in [18] and [19]. The previous example showed good accuracy of the linear extrapolation over a broad numerical range of the design variable for a static type of the behavior. Reference [19] included results for a dynamic type of aeroelastic behavior. An example of such a result is given in figure 8 that shows the flutter speed as a function of the wing aspect ratio (slenderness).

The function exhibits a low degree of nonlinearity for the aspect ratio greater than 6.5 which makes the linear extrapolation in that range a very good predictor of the aspect ratio influence on the flutter speed, as illustrated by the dotted line tangent. On the other hand, the function is strongly nonlinear for the aspect ratio smaller than 6.5, and the linear extrapolation in that interval would have to be subject to fairly narrow move limits to be reliable. Slope discontinuities in the function caused by the flutter mode switching are also detrimental to the extrapolation accuracy. This example indicates that care is needed in setting the extrapolation move limits and that the use of higher derivatives in the model of design should be considered as means to avoid excessively narrow move limits.

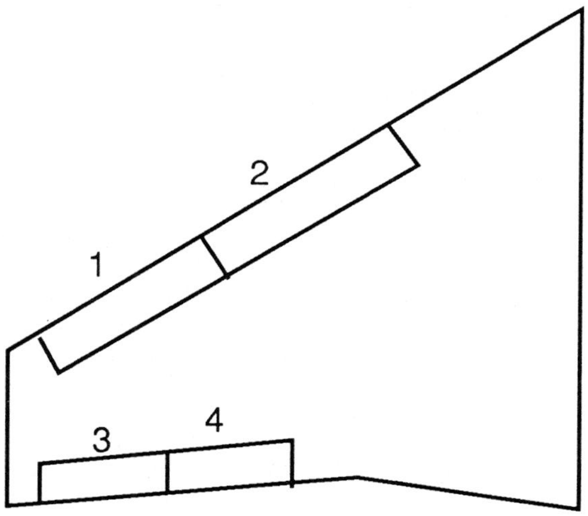

Figure 9 A fight wing configuration with control surfaces.

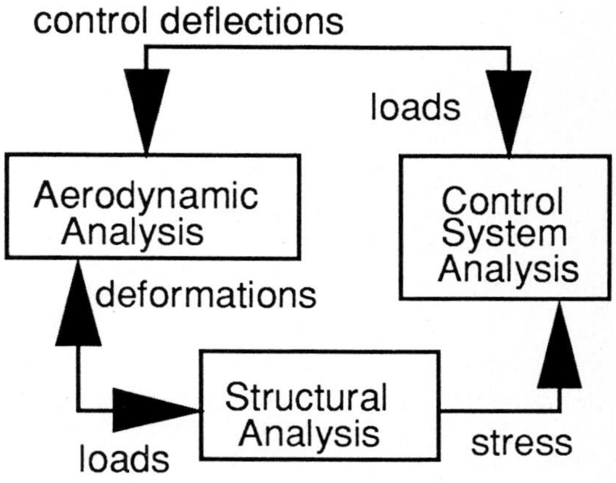

Figure 10 A simulation system for the fighter wing.

5.4. Optimization Using a Quadratic Design Model

An example of the use of the second-order derivatives in the extrapolation coupled with an optimization program was given in [20]. The object of the study was a wing shown in figure 9, that was equipped with two control surfaces on the leading edge and two on the trailing edge. A control system was programmed to deflect these surfaces to reduce the wing-root bending moment while maintaining a constant lift. The model of behavior included the modules representing aerodynamic, structural, and control analyses coupled as depicted in figure 10. Optimization used the wing-root bending moment as the objective function to be minimized by manipulating the control surface deflection as design variables under the constraint of the constant lift.

The volume of data communicated between the aerodynamics and structures modules was judiciously limited by physical insight in a manner similar to that discussed in the first example. For instance, the airfoil lift coefficients were computed at only four spanwise locations.

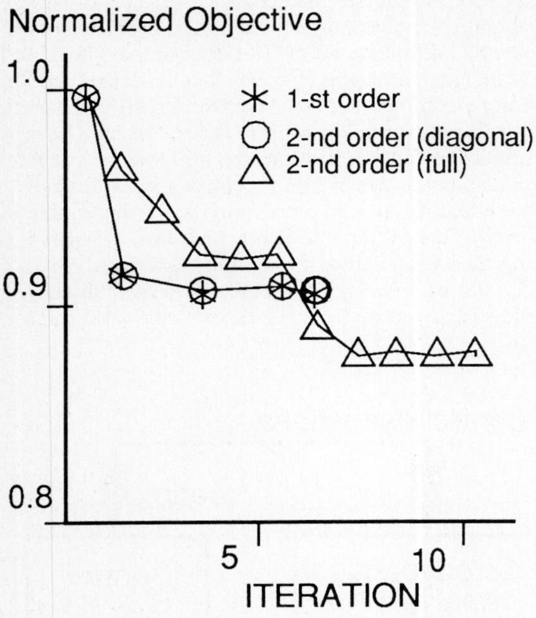

Figure 11 Histogram of optimization of the fighter wing.

Optimization results obtained with the use of a model of design based on the first derivatives, the diagonal terms of the second derivative matrix, and the full second derivative matrix are illustrated in figure 11. They show that in this particular case a meaningful reduction of the minimum objective function was brought about by including a full set of the second derivatives.

6. CONCLUDING REMARKS

Mathematical simulation of complex engineering systems is commonly used in the design of these systems to obtain information about the system response to external stimuli, in effect it is used as a mathematical model of behavior. This paper shows that means exist for such simulation to be extended to answering the "what if" questions concerning effects of the design variables on behavior - the questions that must be answered in the quest for an optimal design. System sensitivity analysis quantifies answers to such questions by computing derivatives of behavior with respect to design variables without the costly finite differencing of system analysis. The algorithm for system analysis offers accuracy and an opportunity for parallel processing. The algorithm also allows the use of specialty methods for partial-sensitivity analysis in the disciplines involved in the system at hand.

Sensitivity analysis yields derivatives of the first- and higher-orders that may be coupled with an extrapolation based on these derivatives to form a model of design. That model is capable of answering the designer's queries about the effect of design variables practically instantaneously, and at a negligible cost comparing to the use of finite differencing on the model of behavior.

Complementing the model of behavior with the model of design extends the array of tools that assist an engineer in the design of a physical system. As shown in figure 12, that array affords the designer the option of getting answers to three basic questions that occur in the design process. The "what now" question about the system response will be answered by the model of behavior. The "what if" question about the effects of a design variable will be answered by the model of design. Finally, the question "what is the best" in search for an optimal setting of many design variables, under complex and possibly competing considerations, may be answered by a formal optimization that calls on both models.

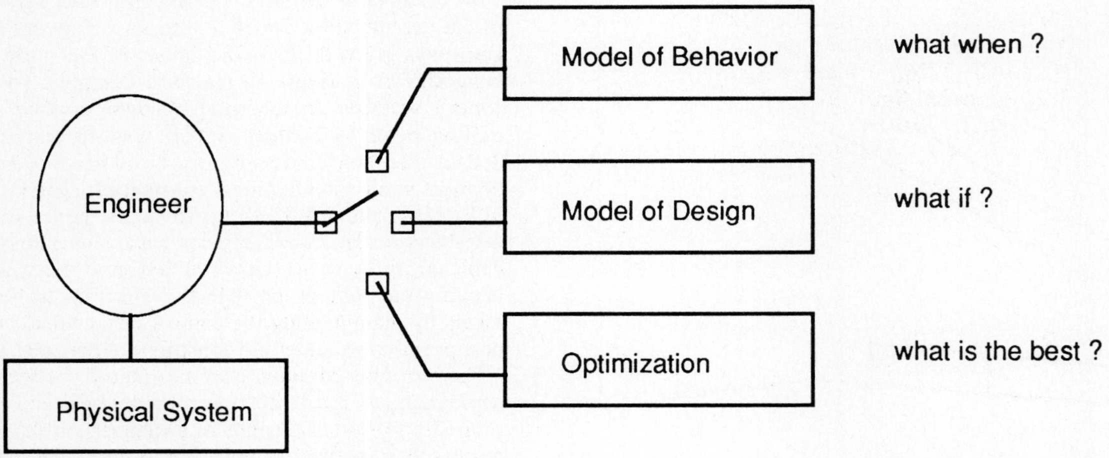

Figure 12 Array of tools for system design.

7. REFERENCES

[1] Ashley, H.: On Making Things the Best - Aeronautical Uses of Optimization; J. of Aircraft, vol.19, No.1, Jan. 1982, pp.5-28.

[2] Schmit, L. A.; and Miura, H.: Approximation Concepts for Efficient Structural Synthesis. NASA CR-2552, 1976.

[3] Adelman, H. A; and Haftka, R. T.: Sensitivity Analysis of Discrete Structural Systems, AIAA J., Vol.24, No.5, May 1986, pp.823-832.

[4] Foundations of Structural Optimization - A Unified Approach; editor: Morris, A. J.; John Wiley and Sons, 1982.

[5] Recent Experiences in Multidisciplinary Analysis and Optimization; Proceedings of a Symposium held at NASA Langley Research Center, April 1984; editor: Sobieski, J.; NASA CP-2327,

[6] Second NASA/Air Force Symposium on Recent Advances in Multidisciplinary Analysis and Optimization; Hampton, Virginia, September 28-30 1988; Proceedings in NASA CP - No. 3031; editor: Barthelemy, J. F..

[7] The NASTRAN User's Manual; MSC-NASTRAN Version 64. MSR-39; MacNeal-Schwendler Corp., 815 Colorado Blvd, Los Angeles, CA, 90041, July 1984.

[8] Vatsa, V. N.; Wedan, B. W.; and Turkel, E.: 3-D Euler and Navier-Stokes Calculations for Aircraft Components. NASA CP-3020, Vol. 1, Part 2. NASA Langley Research Center, Hampton, Va, April 1988.

[9] Ryan, T. P.: Statistical Methods for Quality Improvement (ch.13); John Wiley and Sons, 1989.

[10] Free, J.W.; Parkinson, A. R.; Bryce, G. R.; and Balling, R. J.: Approximation of Computationally Expensive and Noisy Functions for Constrained Nonlinear Optimization; ASME J. of Mechanisms, Transmission, and Automation in Design, Vol. 109, No. 4, pp. 528-532, Dec. 1987.

[11] Fleury, C.; and Schmit, L. A.: Dual Methods and Approximation Concepts in Structural Synthesis; NASA CR-3226, Dec. 1980.

[12] Sobieszczanski-Sobieski, J.; On the Sensitivity of Complex, Internally Coupled Systems; AIAA/ASME/ ASCE/AHS 29th Structures, Structural Dynamics and Materials Conference, Williamsburg, Va, April 1988; AIAA Paper No CP-88-2378, also published as NASA TM 100537, January 1988.

[13] Proceedings of the Symposium on Sensitivity Analysis in Engineering, NASA Langley Research Center, Hampton Va, Sept. 1986; Adelman, H. M.; and Haftka, R.T. - editors. NASA CP-2457, 1987.

[14] Sobieszczanski-Sobieski, J.: Sensitivity Analysis of Complex Coupled Systems Extended to Second and Higher-Order Deritivatives, to appear in AIAA J., also published as NASA TM 101587, April 1989.

[15] Fleury, C.; and Braibant, V.: Structural optimization, A New Dual Method Using Mixed Variables; Aerospace Laboratory, Univ. of Liege, Belgium, Report SA-115, March 1984.

[16] Sobieszczanski-Sobieski, J.; Bloebaum, C. L.; and Hajela, P.: Sensitivity of Control-Augmented Structure Obtained by a System Decomposition Method; AIAA Paper No. 88-2205, AIAA 29th Structures, Structural Dynamics, and Materials Conference, Williamsburg, Va., April 1988, to appear in AIAA J.

[17] Yates, E.C.: Aerodynamic Sensitivities from Subsonic, Sonic, and Supersonic Unsteady, Nonplanar Lifting-Surface Theory; NASA TM 100502, September 1987.

[18] Barthelemy, J. F.; and Sobieszczanski-Sobieski, J.: Optimum Sensitivity Derivatives of Objective Functions in Nonlinear Programing; AIAA J, Vol.22, No.6, June 1983, pp.913-915.

[19] Kapania, R; and Bergen, F.: Shape Sensitivity Analysis of Flutter Response of a Laminated Wing; AIAA Paper No. 89-1267, 30th AIAA/ASME/ASCE/AHS/ASC Structures, Structural Dynamics, and Materials Conference, Mobile, Al., April 1989.

[20] Ide, H.; and Levine, M.: Use of Second-Order CFD Generated Global Sensitivity Derivatives for Coupled Problems; AIAA Paper No. 89-1178, 30th AIAA/ASME/ASCE/AHS/ASC Structures, Structural Dynamics, and Materials Conference, Mobile, Al., April 1989.

Approximation, Optimization and Computing:
Theory and Applications, A.G. Law and C.L. Wang (eds.)
Elsevier Science Publishers B.V. (North-Holland)

OPTIMAL DESIGN OF MINIMAX DEFLECTION OF A CYLINDRICAL SHELL *

Yu huan−ran Liang bin

Department of mechanics, Lanzhou university, Gansu, China.

ABSTRACT. In this paper, we study the optimal design of minimax deflection of a thin cylindrical shell with axisymmetrical arbitrary boundary and distributed load, under the condition of the volume being constant by the stepped reduction method. Some examples are given.

1. INTRODUCTION

In the recent twenty years, the optimal design of elastic structure is noticed by many designers. In the feild, the studies of the problem of optimal design of shells are far less than other[1] which is due to the fact that the governing equations of shell are very complex. Here we are interested to solve the design problem of the thin cylindrical shell to find out the thickness function which causes minimax deflection of the shell, under the condition of volume being constant and middle surface shape being given. In this problem, the explicit formulation of objective function can't be given out by traditional method, so it leads to meet many difficulties for optimal computation. By using of the stepped reduction method [3−4], we can give out the solution of deflection of cylindrical shell with variable thickness, further can simplify optimal computation. Now we consider a thin elastic cylindrical shell with axisymmetrical varying thickness under the assumption that it's boundary conditions and distributed load are axisymmetrical. According to the above mentioned stepped reduction method, discreting the thickness function into step function, the explicit formulation of objective function will be written out. So that the problem will be reduced to a standard nonlinear programing with an equlity constraint.

2. SOLUTION

A axisymmetrical varied thickness cylindrical shell as shown in Fig. 1, with uniform material, elastic constant E, μ, length L, thickness $h(x)$, is subject to radial axisymmetrical distributed load $q(x)$. Divide the shell into N shell ele ments shown in Fig. 2. The order of

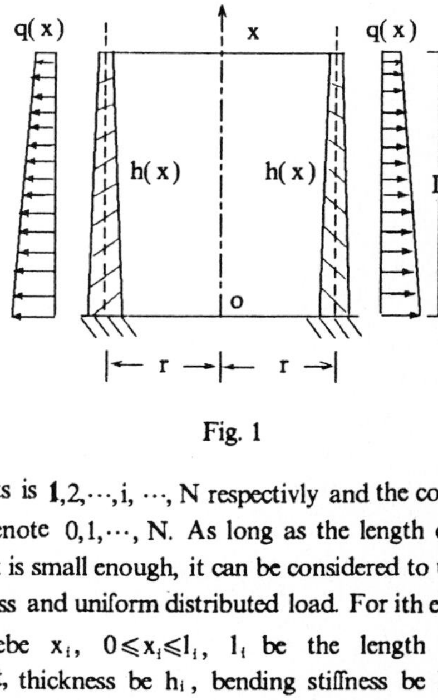

Fig. 1

elements is $1,2,\cdots,i,\cdots,N$ respectivly and the connected lines denote $0,1,\cdots,N$. As long as the length of each element is small enough, it can be considered to uniform thickness and uniform distributed load. For ith element, variablebe x_i, $0 \leqslant x_i \leqslant l_i$, l_i be the length of the element, thickness be h_i, bending stiffness be D_i and the load be q_i. The differential equation of the middle surface radial deflection $w_i(x_i)$ of the element is

---

*Project supported by the Science Fund of the National Education Committee.

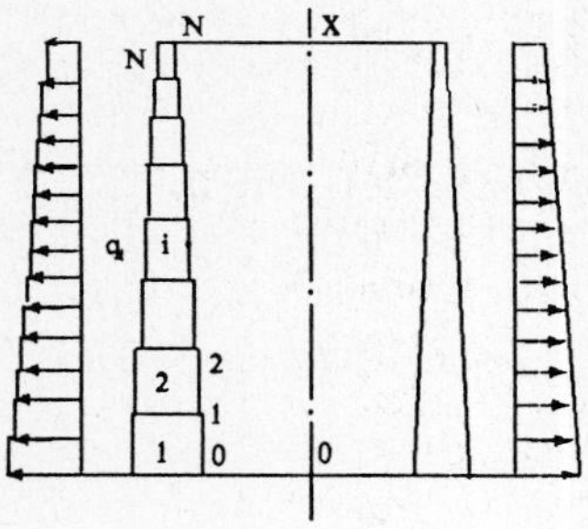

Fig . 2

$$d^4w_i(x_i)/dx_i^4 + 4k_i^4 w_i(x_i) = q_i/D_i \qquad (1)$$

where $k_i^4 = Eh_i/4r^2D_i$, $D_i = Eh_i^3/(1-\mu^2)$. For convenience of deducing and numerical computation, let

$$y_i = x_i/L, \quad Z_i = h_i/h_o,$$

where h_o is the thickness of uniform shell in which the given volume is V_0. Furthermore, make corresponding dimensionless change of the deflection, slope, bending moment and shear force. Then, the solution of the equation (1) can be written as:

$$W_i(y_i) = W_i(0)F_{1i}(y_i) + W_i'(0)F_{2i}(y_i) + M_i(0)F_{3i}(y_i)$$

$$+ Q_i(0)F_{4i}(y_i) + F_{5i}(y_i)$$

where $W_i(0)$, $W_i'(0)$, $M_i(0)$ and $Q_i(0)$ respectivly is deflection, slope, bending moment and shear force at $x_i=0$. For brevity the complicated description of the $F_{ki}(y_i)$ ($k=1,2,3,4,5$) will not be presented here. Readers may refer to Ref. [3]. According to the relations among W_i, W_i', M_i and Q_i, we have

$$S_i(y_i) = P_i(y_i)S_i(0) + T_i(y_i) \qquad (2)$$

where $S_i(y_i) = (W_i(y_i),\ W_i'(y_i),\ M_i(y_i),\ Q_i(y_i))^T$ $P_i(y_i)$ is a 4×4 matrix that consist of $F_{ki}(y_i)$ and thier differential, $T_i(y_i)$ is a 4×1 matrix. Continuity conditions at the step conjunction of two neighboring elements must be satisfied, so we have

$$S_i(y_i) = PP_i(y_i)S_1(0) + TT_i(y_i) \qquad (3)$$

where

$$a_i = l_i/L, \quad i=1,2,\cdots,N,$$

$$PP_1(y_1) = P_1(y_1), \quad PP_i(y_i) = P_i(y_i)\prod_{j=1}^{i-1}P_j(a_j),$$

$$TT_1(y_1) = T_1(y_1), \qquad TT_i(y_i) = T_i(y_i) +$$

$$P_i(y_i)[T_{i-1}(a_{i-1}) + \sum_{j=1}^{i-2}T_j(a_j)\prod_{l=j+1}^{i-1}P_l(a_l)],$$

$$i=2,3,\cdots N.$$

When $i=N$ and $y_N=a_N$, then, (3) will give W_N, W_N', M_N and Q_N. Under the condition of thickness function being given, if the boundary conditions are given out, the $S_1(0)$ will can be obtained. Further, W, W′, M and Q at all points will can also be obtained.

3. OPTIMAL DESIGN

Our optimal design problem can be stated as follows: Determine the thickness function $h(x)$ which minimizes

$$\underset{0\leqslant x\leqslant L}{\text{MAX}}\ W(h(x),\ x)$$

subject to (4)

$$\int_0^L h(x)\,dx = V_0/2\pi r \qquad h(x) \leqslant h_{min}$$

where h_{min} is the given minimum thickness.

As previously stated, after diving the shell into N shell elements, our optimal problem (4) becomes:

Determine N+1 variables $Z_1, Z_2, \cdots, Z_N, y$, to minimize

$$\text{MAX}\ W(Z_1, Z_2, \cdots, Z_N, y)$$

subject to (5)

$$\sum Z_i a_i = 1, \quad Z_i \leqslant Z_{min}, \qquad i=1,2,\cdots,N$$

where $Z_{min} = h_{min}/h_0$.

Under the condition of $Z_1, Z_2, \cdots Z_N$ being given, if the boundary conditions are giving out, the solution W will can be obtained. For finding the point of the maximum deflection, let the length of each element be small enough to guarantee that exist no more than one stationary point of deflection at each element. First, find all of element $\{J\}$ which $W_j \cdot W_{j-1} \leqslant 0$. By means of the average section method, the stationary points of deflection at these elements will be obtained. Denote these points as $\{y_J\}$. Second, compare the deflection corresponding to points $\{y_J\}$, and find the point of the maximum deflection at whole shell. Denote the point as y_m, and denote the deflection of the point as W_m, $W_m = W_m(Z_1, Z_2, \cdots, Z_N, y_m)$, where m indicate point y_m in the mth element. So that the optimal problem (5) can be reduced as:

Determine the $Z_1, Z_2, \cdots Z_N$, minizing

$$W_m(Z_1, Z_2, \cdots, Z_N, y_m)$$

subject to (6)

$$\sum Z_i a_i = 1, \quad Z_i \leqslant Z_{min} \qquad i = 1, 2, \cdots, N.$$

The problem (6) is a nonlinear programming with an equality constraint, and the explicit formulation of the objective function was given out. Since the deritives of the objectiv function with design variables can be obtained easily, we can solve the problem (6) using various multidimensional gradient method. In order to show the method developed in this paper, an iterative is conducted as follows:

For convenience, let diving be equi-section $a_i = 1/N$.

Step 1: Let $k=0$, $\varepsilon > 0$, select $Z_i(0)$, $i = 1, 2, \cdots, N$.

Step 2: For $y_i = 1/N$, compute $P_i(1/N)$, $T_i(1/N)$, $PP_i(1/N)$ and $TT_i(1/N)$.

Step 3: According to boundary conditions, compute the $S_1(0)$.

Step 4: Find point y_m and write expression $W_m = W_m(Z_1, Z_2, \cdots, Z_N, y_m)$

Step 5: Find optimal solution of the problem (6), Z_1, $Z_2, \cdots, Z_N$

Step 6: If $\| Z_i(k+1) - Z_i(k) \| < \varepsilon$ stop.

Step 7: Let $k = k+1$, goto step 2

4. ILLUSTRATIVE EXAMPLES

In the follow examples, we choose the initial thickness $Z_i(0) = 1$ $(i = 1, 2, \cdots N)$ and let $\mu = 0.3$, $r/L = 0.5$ $h_0/L = 0.1$

The computation show that the results for $N = 80$ are hardly any different from the results for $N = 160$. Therefore, we take $N = 80$ in our examples. The ratio is denoted kk that between maximum deflection of the optimal shell and maximum deflection of the uniform shell with same volume.

In the Fig. 3a, 4a and 5a, the solid lines represent the thickness of the optimal shell. In the Fig. 3b, 4b and 5b, the solid lines represent the deflection curves of the optomal shell and the dotted lines represent the deflection curves of the uniform shell with same volume.

Example 1. built and built under the uniform load. kk=.760.

Example 2. Simply supported and simply supported under the uniform load. kk=.798.

Example 3. built and free under the linear distributed load. kk=.619.

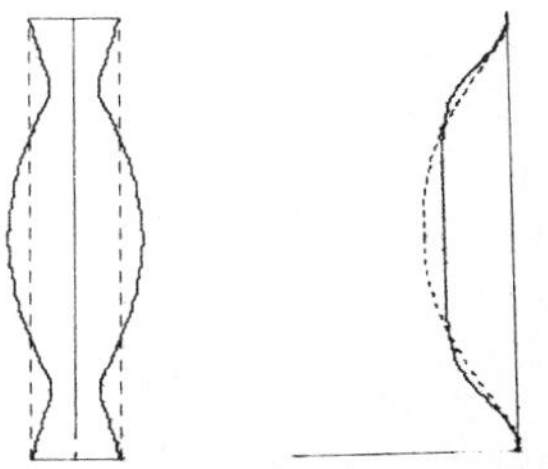

(a) (b)

Fig. 3. example 1

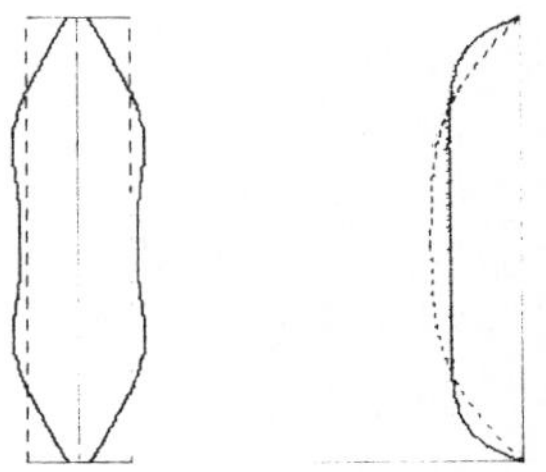

(a) (b)

Fig. 4. example 2

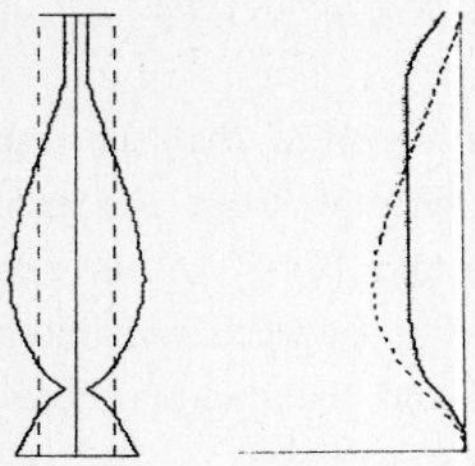

(a) (b)

Fig. 5. example 3

REFERENCES

[1] Edward J. H., A Review of Distributed Parameter Structural Optomization Literature, Proc. Advanced Study Institute Optimization of Distributed parameter Structures, Iowa, 1980 .

[2] Yu H. R. and Yeh K. Y., Optimal Design of Minimax Deflection of An Annular plate, Applied Mathematics and Mechanics, Vol. 9, No. 1 Jan. 1988.

[3] Huang Y. Y. and Liang G. J., Axisymmetric Deformation of Variable thickness Cylindrical Shell. Shanghai Mechanics, No. 1, 1983.

[4] Yeh K. Y., and Ji Z. J., The general soultion on nonliear axisymmetrical deflection of nonhomogeneous cylindrical shell, Applied mathematics and mechanics, 3, 10 (1989).

Approximation, Optimization and Computing:
Theory and Applications, A.G. Law and C.L. Wang (eds.)
Elsevier Science Publishers B.V. (North-Holland)

APPLICATION OF OPTIMIZATION TO AIRCRAFT ENGINE DISKS

Yu Xiao–ming and Ma Mei

Group 405, Department of Jet Propulsion,
Beijing University of Aero. & Astro., Beijing 100083, P.R.China

An automated optimum disk synthesis technique for aircraft gas turbine engines is developed using a mathematical programming method. The shape of a disk is defined by the combined function. The stress analysis of the disk is carried out by finite element method using isoparametric elements. The optimization algorithm used is the constrained variable metric method. A semi–analytical method of sensitivity analysis is discussed. A application is presented.

1. INTRODUCTION

The design of an advanced gas turbine engine begins in a preliminary design phase when the engine designer is challenged with selecting the best engine configuration for the defined requirements. The disk is a key component in aircraft engines. The engine designer faces a very difficult problem in preliminary design phase of disks. It is that the designer should reduce the weight of disk as could as possible while meeting burst speed, stress and geometric constrains. Traditionally, this disk with ″minimum″ weight is the result of an extensive study of numerous candidate disks of varying sizes. The designer will expend lots of time to review a large number of these candidate disk against the criteria of aircraft engine disks design.

During the last two decades, significant efforts have been made in the application of optimization to the aircraft engine disk synthesis. Mathematical programming approach to the optimum design of rotating disks may be found in Res.[1–4]. Ali Seireg and Surana [1] have optimized the cross section of disk. The disk is approximated by a number of rings, the optimum thicknesses of which are determined by mathematical programming. However, in these investigations, the methods of stress analysis are not very accurate, particularly in the region of stress concentration. S.S. Bhavikatti and C.V. Ramakrishnan [2] have defined the shape of disk with polynomial curves, which depend on thicknesses at four predefined points. The stress analysis is carried out by finite element method and optimization by improved move limit method of sequential linear programming.

The aim of this paper is to apply the optimization technique in synthesizing gas turbine engine disks with minimum weight while meeting burst margin, stress and geometric constrains. In this paper , the boundary of disk is described by the combined function of arc and straight line. The optimization is carried out using a powerful constrained variable metric method. A suitable subroutine is developed to generate automatically a isoparmetric finite elements mesh for prescribed design variables. The derivatives of the displacement U or of constraints based on U is calculated by the semi–analytic method of sensitivity.

2. OPTIMUM TECHNIQUE

The general nonlinear programming problem can conveniently be expressed as finding a local solution X^* of

$$\left.\begin{array}{lll} \text{minimize} & f(X), & X\in R^n \\ \text{subject to} & C_i(X)=0, & i\in E \\ & C_i(X)\leqslant 0, & i\in I \end{array}\right\} \qquad (1)$$

where E and I are finite index sets of equality and inequality constraints respectively. Constrained variable metric method is used to solve (1).

From (1), a Lagrangian function is introduced

$$L(X,\lambda)=f(X)-\lambda^T C(X) \qquad (2)$$

At a iterative point X^k, $d=X-X^k$, f(X) is approximated by the quadratic function and C(X) is approximated by the linear function, then (1) is replaced by a sequence of quadratic programming (QP) subproblems of the type

$$\left.\begin{array}{lll} \text{minimize} & [\nabla f(X^k)]^T d + \frac{1}{2} d^T H d & d \in R^n \\ \text{s.t.} & C_i(X^k) + [\nabla C_i(X^k)]^T d = 0 & i \in E \\ & C_i(X^k) + [\nabla C_i(X^k)]\ d \leqslant 0 & i \in I \end{array}\right\} \quad (3)$$

where H is a (positive definite) approximation to $\nabla^2 L$, the Hessian matrix of the Lagrangian function, by BFGS formula in variable metric method.

Given initial approximations the basic method is:
For k = 1,2,3, • • •
i) Solve (3) to get d and set d^k = d as a search direction,
ii) Set $X^{k+1} = X^k + \alpha d^k$, the step size α is determined by the "watchdog" technique which has been suggested by Chamberlain, Lemarechal, pedersen and Powell [3].

3. SENSITIVITY COMPUTATION

A critical step to producing an efficient shape optimization programming is based on the sensitivity analysis of the structural response. Finite difference schemes for calculating the sensitivities of the optimization constrains limit the problem sizes to 4–5 design variables and the same number of constrains because the required computer time quickly becomes prohibitive.

This problem can be conceptually resolved quite easily by using the well know analytic method for sensitivity calculation. For simplicity, let us consider the static structural response R related to the nodal displacement vector $U = (u_1, u_2, \cdots, u_n)^T$ by

$$R = Q^T U \quad (4)$$

where Q is virtual load vector and U satisfies the static equilibrium equations

$$KU = P \quad (5)$$

in which K is the global stiffness matrix of the structure, P is the external force vector. The formulas for the analytic method of sensitivity calculation are

$$\frac{\partial R}{\partial x_i} = \frac{\partial Q^T}{\partial x_i} U + Q^T \frac{\partial U}{\partial x_i} \quad (6)$$

$$\frac{\partial U}{\partial x_i} = K^{-1}\left(\frac{\partial P}{\partial x_i} - \frac{\partial K}{\partial x_i} U\right) \quad (7)$$

Equation (7) indicates that the computation of the gradients $\frac{\partial U}{\partial x_i}$ reduces to the evaluation of the gradients of the stiffness matrix and of the load vectors P with respect to shape. In reference [4] , a analytic method for the calculus of the shape derivatives $\frac{\partial P}{\partial x_i}$ and $\frac{\partial K}{\partial x_i}$ has been devised, that is based on the isoparametric formulation of finite element. However, this kind of analytical expressions for sensitivity are not only lengthy but also strongly dependent on the type of design variable, response and the element formulation, which needs knowledge of the details of the existing finite element program.

G.D.Cheng and Y.W.Liu [5] have proposed a new computation scheme for sensitivity analysis, that is semi–analytic method.

In this method, the structural response R is considered as a function of the design variables X and the nodal displacement vector U, we prefer to write R = R(X,U). Finite difference scheme is used to calculate the $\frac{\partial P}{\partial x_i}$ and $\frac{\partial K}{\partial x_i}$, then, solve the Eq.(7) to get $\frac{\partial U}{\partial x_i}$, determine the first–order approximation of U, $U(X+\Delta X_i) = U(X) + \frac{\partial U}{\partial x_i} X$, so, we can obtain the sensitivity of R to design variables

$$\frac{\partial R}{\partial x_i} = \frac{R(X + \Delta X_i, U(X + \Delta X_i)) - R(X,U)}{\Delta x_i} \quad (8)$$

The detail formulation may be found in Ref.[5] or [6].

4. MATHEMATICAL MODEL

Fig.1 shows a typical rotating disk with hub and rim. Generally, engineering structure design consideration specify the inner radius of hub and the dimensions of rim. The width of hub and the outer radius of hub can be variables, also the disk can have varying thickness between hub and rim. The present investigation is concerned with the design of optimum shape of this portion of rotating disk. Authors [6] have proposed the idea of control node which can identify the basic shape of disk, such as A. B. C. D in Fig.2. We take the coordinates of control nodes $B(x_1,y_1)$, $C(x_2,y_2)$and $D(x_3,y_3)$ as design variables. According to some given engineering design considerations, the shape of part E (in Fig.2) is modified and does not vary in optimization process. Moreover, the web cross section is symmetric about the axis Q–Q. The curved profile of the web is expressed in the combined function of arc and straight line. The straight line BC is tangent to arcs AB and CD. Ref.[6] has given the detail formulations of the combined function.

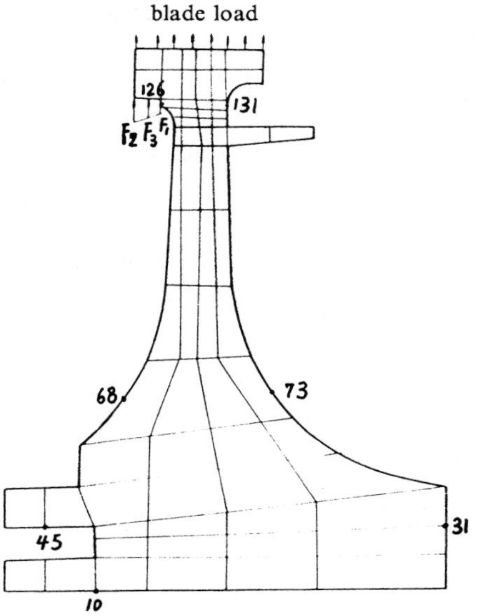

Fig.1 The model of a gas turbine disk

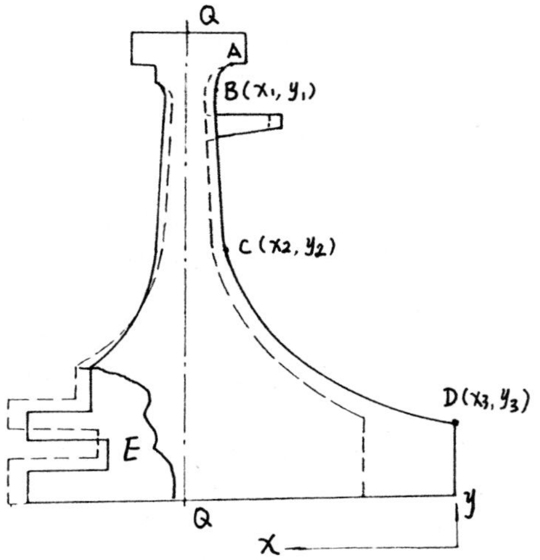

FIg.2 The result of optimization
– – – – final; ——— initial

The stress analysis of gas turbine disk is carried out by finite element analysis using linear elastic analysis with isoparametric element. Not only are centrifugal and thermal loads considered, but also the loads on the rim are considered. To obtain an effective finite element mesh in optimization process, authors [6] have developed an adaptive mesh refinement technique, Fig.1 shows an automated mesh generated for finite element analysis. Two finer meshes are used near the rim and hub since the stress concentration in these regions is high.

In the design of gas turbine disk, the best objective function is not well defined, in this paper , weight minimization is considered. The constrains are MCRS (maximum centrifugal radial stress), MTS (mean tangential stress), MTSI (maximum tangential stress at the inner radius of hub) and MTSW (maximum tangential stress at web).

5. RESULTS AND CONCLUSIONS

The parameters considered for detailed investigations is shown following. The disk spins at 13,520 rpm, blade load is 171.2N / mm^2,external force F_1= 111.5 N / mm^2, F_2= 126.6 N / mm^2, F_3= 119.1 N / mm^2, the material properties are given as Young's modulus E = 1.764 × 10^5N / mm^2, Poisson's ration is 0.30. Fig2. shows the optimum shape of the gas turbine disk. The number of iteration is 14. The detailed results are clearly listed in Table1–3. From Table 3, The stress distribution of optimum gas turbine disk is better than initial disk.

Table 1: Design variables of disk

	x_1 mm	y_1 mm	x_2 mm	y_2 mm	x_3 N/mm^2	y_3 N/mm^2	Weight kg
initial	55.00	213.00	54.00	166.67	0.00	113.50	43.87
final	56.50	213.42	55.29	155.04	15.02	115.30	38.22
lower	50.00	200.00	50.00	120.00	15.00	114.00	
upper	60.00	220.00	60.00	190.00	20.00	118.00	

Table 2: Constraints of disk

	MTS N/mm^2	MTSI N/mm^2	MTSW N/mm^2	Maximum N/mm^2
initial	669.22	1030.86	890.69	1663.22
final	721.53	896.79	853.69	1181.81
upper	775.80	983.70	879.30	1200.00

Table 3: Stress distribution (units: N/mm^2)

number of node	10	45	31	68	73	126	131
initial	1663.32	1089.87	151.32	783.08	477.95	980.19	981.95
final	1181.81	738.30	602.00	766.72	890.31	949.92	1072.12

REFERENCES

[1] Ali Seireg and Surana, K.S., Optimum Design of Rotating Disks, J. Eng. Industry 1–8 (Feb. 1970)

[2] Bhavikatti, S.S. and Ramakrishnan, C.V., Optimum Shape Design of Rotating Disk, J. Computer & Structure, vol. 11, pp397–401 (1980)

[3] Chamberlain, R.M., Leemarechal,C., Pedersen,H.C. and Powell, M.J.D., The Watchdog Technique For Forcing Convergence In Algorithms For Constrained Optimization, Report DAMTP 80 / NA9, University of Cambridge

[4] Wang, S.Y., Sun, Y. and Gallagher, R.H., Sensitivity Analysis In Shape Optimization of Continuum Structures, J. Computer & Structure, vol. 20, No:3, pp855–865

[5] Cheng,G.D. and Liu Y.W., A New Computation Scheme for Sensitivity Analysis, J. of Engineering Optimization, vol.12 pp.219–234 (1987)

[6] Yu, X.M., Ma,M. and Cheng, G., Optimum Shape Design of Rotating Disks And Experimental Research, 89–GT–40, Presented at the 34th International Gas Turbine and Aeroengine Congress and Exposition June 4–8,1989, Toronto, Ontario, Canada.

Approximation, Optimization and Computing:
Theory and Applications, A.G. Law and C.L. Wang (eds.)
Elsevier Science Publishers B.V. (North-Holland)

Principle of Maximum Information, Holography and Application

Zhang Haoran

(Shandong Academy of Sciences, PRC)

From the viewpoint of holography, the principle of maximum information has been interpreted and a depth distribution of characteristic X-ray production in microprobe analysis has accordingly been deduced. Some experimental works show that it nearly coincides with Philibert absorption method which is widely used for absorption corrections. Therefore it reveals some mysteries of entropy in this aspect.

1. Introduction

There exist some random events whose probabilities cannot be directly evaluated and there exist random events which do not allow us to compute the corresponding ralative frequencies. often, we only have mean values of some random variables which depend on the random events. When the mean value of a random variable is given, there exist many random distributions compatible to this mean value. The problem is how to select the best one. R. S. Ingarden(1963), E.T.Jaynes(1957), and S. Kullback and R. A.Leiber(1951) established a selection criterion, i.e. the principle of maximum information[1].

According to the above principle, we construct the random distribution which maximizes the entropy or the conditional entropy, subject to some set of restraints (usually some mean values of some random variables). In other words, the principle of maximum information enables us to reproduce the original distribution only from a piece of information. It is something like mathematical holography.

In 1980, Zhang Yingqing, a chinese biologist, found that there exist holographic phenomenon on organism. A piece of branch of tree, for example, is similar to the tree. Thus, the holographic biology was established.

As a matter of fact, there also exist some holographic phenomenon in the natural world and in the human society. Further, Wang Tsunzen and Yan Chunyou, both are chinese scholar, have suggested "universe holography". According to the theory, each part of thing is contained in other parts while contains the other parts in it, and part contains entire thing in it.

Therefore, the principle of maximum information can be interpreted in such a way that in random environment a piece of information (such as the mean values), i.e, the part, contains the distribution pattern, i.e, the entirety. The principle of maximum information is a important mathematic tool, which is used to reproduce the entirety from the part.

2. Application

An example, which shows the principle can be utilized to obtain an interesting result, is the determination of the depth distribution of characteristic X-ray production in microprobe analysis using the principle of maximum information.

In electron microprobe analysis, electron bombardment is used to generate X-ray in the sample to be analysed. From the wavelength and intensity of the lines in the X-ray spectrum the elements present can be identified and their concentrations estimated.

In quantitative analysis the intensities of the X-ray lines from the specimen are compared with those from standard samples of known composition. Because of specimen absorbing X-ray the measured intensities require absorption correction. Therefore the depth distribution of the generated X-ray intensities must be determined.

Suppose that the specimen is the alloy containing element A, the mass concentration of element A is C_A, then, the total X-ray intensities generated in the specimen is given by

$$I_A=C_A\int_0^\infty \phi_A(\rho z)d(\rho z) \qquad (1)$$

where $\phi_A(\rho z)$ is the depth distribution of characteristic X-ray production of element A.

It is often difficult to deduce the depth distribution theoretically. The commonly adopted method of determing it depends on experimental data. But, because of randomness of electron traveling and scattering in solids, with the principle of maximum information the depth distribution can be easily reproduced from a piece of information. From (1), we obtain

$$\int_0^\infty C_A/I_A\phi_A(\rho z)d(\rho z)=1 \qquad (2)$$

let us denote

$$C_A/I_A\phi_A(\rho z)=p(\rho z) \qquad (3)$$

We have

$$\int_0^\infty p(\rho z)d(\rho z)=1 \qquad (4)$$

of course

$$p(\rho z)\geqslant 0 \quad ; \qquad p(\rho z)\in(0, +\infty]$$

Obviously, $p(\rho z)$ is just the probability density function, the corresponding entropy is

$$H=H(p)=-\int_0^\infty p(\rho z)\ln p(\rho z)d(\rho z) \qquad (5)$$

Suppose X_d is the diffusion depth of electron in specimen. We have known that $\phi(\rho z)$, therefore $p(\rho z)$, increases when $\rho z \leqslant X_d$, and falls off when $\rho z > X_d$ [2]. Thus the mean value of the random distribution is X_d

According to the principle of maximum information, the probability density function $p(\rho z)$, which maximizes the corresponding entropy (5) compatible to the mean value X_d, is the exponential distribution

$$p(\rho z)=(1/x_d)\exp(-\rho z/x_d) \qquad (6)$$

Taking into account that $p(\rho z)$ reaches maximum at X_d, the $p(\rho z)$ has the form as

$$p(\rho z)=k\ \exp(-\mathrm{sign}(\rho z-x_d)(\rho z-x_d)/x_d) \qquad (7)$$

Where k is constant, sign() is sign function. considering with (3), we have

$$\phi_A(\rho z)=m\ \exp(-\mathrm{sign}(\rho z-x_d)(\rho z-x_d)/x_d)$$

where m is constant, in the absorption correction calculation, the constant is canceled , the final expression for $\phi(\rho z)$ can he writen as

$$\phi_A(\rho z)=\exp(-\mathrm{sign}(\rho z-x_d)(\rho z-x_d)/x_d) \qquad (8)$$

Thus, from the diffusion depth X_d, a piece of information, we can easily obtain the depth distribution.

Therefore, using the equality (1) and (8), we obtain the total X-ray intensities generated in the specimen

$$I_A=C_A\int_0^\infty \exp(-\mathrm{sign}(\rho z-x_d)(\rho z-x_d)/x_d)d(\rho z)$$
$$=C_A x_d(2-e^{-1}) \qquad (9)$$

Taking into account the specimen absorbing X-ray, the measured X-ray intensities is obtained

$$I_A'=C_A\int_0^\infty \phi_A(\rho z)\exp(-\chi\rho z)d(\rho z)$$

$$=C_A(1/ex_d/(1-x_d\chi)(\exp(1-x_d\chi)-1)$$
$$+ex_d/(1+x_d\chi)\exp(-(1+x_d\chi))) \qquad (10)$$

Let us denote

$$I_A'/I_A=f(\chi)$$

and for simplifying the expression, let

$$x_d\chi=y$$

we have

$$f(\chi)=1/(2e-1)(1/(1-y)(\exp(1-y)-1)$$
$$+1/(1+y)\exp(1-y)) \qquad (11)$$

It's the ratio of the measured X-ray intensities and the generated X-ray intensities.

In order to adjust (11) we introduce a modifying constant c in the following way

$$f(\chi)=1/(2e-1)(1/(1-cy)(\exp(1-cy)-1)$$
$$+1/(1+cy)\exp(1-cy)) \qquad (12)$$

Calculating $f(\chi)$ for standard sample and unknown specimen respectively, the ratio

$$Fa=f^{std}(\chi)/f^{unk}(\chi)$$

is called absorption factor.

3. Example

In order to illustrate the efficacy of (12), we now consider an alloy containing iron and nickel, analysed at 25kv , and a compound TiNi, analysed at 20kv.

The results of absorption correction with expression (12) (C=0.47) and the Philibert absorption method widely used for absorption corrections are given in following Table.

		philibert		expression(12)	
		f(χ)	Fa	f(χ)	Fa
Fe	Std.	0.952	1.003	0.952	1.003
(25kv)	unk.	0.949		0.949	
Ni	std.	0.962	1.162	0.962	1.162
(25KV)	unk.	0.828		0.828	
TI	std.	0.960	1.024	0.955	1.029
(20kv)	unk.	0.938		0.928	
N i	std.	0.079	1.032	0.979	1.031
(20kv)	unk.	0.949		0.50	

The most interesting thing here is that we only have the diffusion depth then the depth distribution is easily obtained which is usually obtainied by some tedious experiments and deduction.

Imagining that if we can express the information of a system with something like entropy, maybe we can reproduce the system only from "a piece of information" and hence find new mysteries of entropy.

Reference

[1] Silviu Guiasu. Information Theoty with Application PP294—301.
[2] S.J.B Reed, Electron Microprobe Analysis PP240—260.

Approximation, Optimization and Computing:
Theory and Applications, A.G. Law and C.L. Wang (eds.)
Elsevier Science Publishers B.V. (North-Holland)

DYNAMIC PROGRAMMING METHOD FOR THE ANALYSIS OF THE MOST UNFAVOURABLE LIVE LOAD

Wanxie Zhong

Institute of Engineering Mechanics
Dalian University of Technology
Dalian, PRC

Weiping Shen

Department of Engineering Mechanics
Shanghai Jiao Tong University
Shanghai, PRC

SUMMARY

This paper provides a dynamic programming model to analyze the maximum stress of the highway bridge structures under truck train loading. Bridge structures are idealized with finite element model and the influence surface is calculated using the generalized Müller-Breslau's principle [1]. The locations of trucks in train can be determined automatically to produce the maximum stress by the dynamic programming method [2].

1. INTRODUCTION

According to the standard for highway bridges of China, two kinds of live loads are recommended [3]. One of them is the truck train loading. The truck train consists of one heavy truck and a number of standard trucks (Fig. 1) The minimal spacing between one truck and another adjacent truck after it is a prescript value. The trucks should be so located along the traffic lane, asto produce the maximum stress in the member of the bridge structure under consideration. In the Russian and the American standards there are similar description about the truck train loadings (Fig. 2 and Fig. 3)[4,5].

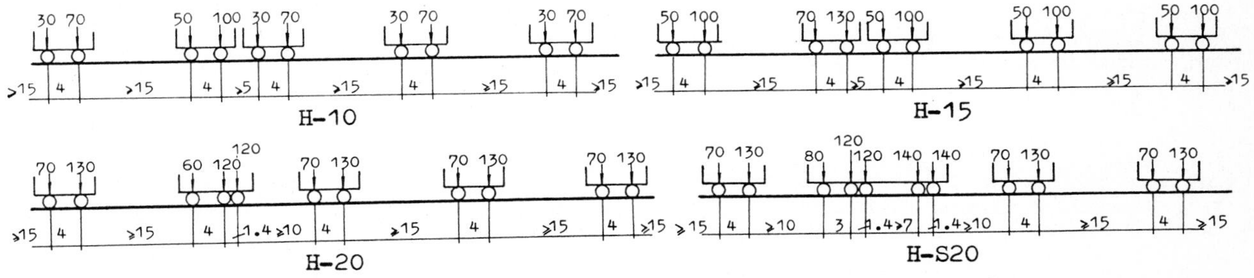

Fig. 1 Truck Train Loadings Defined in the Chinese Standard

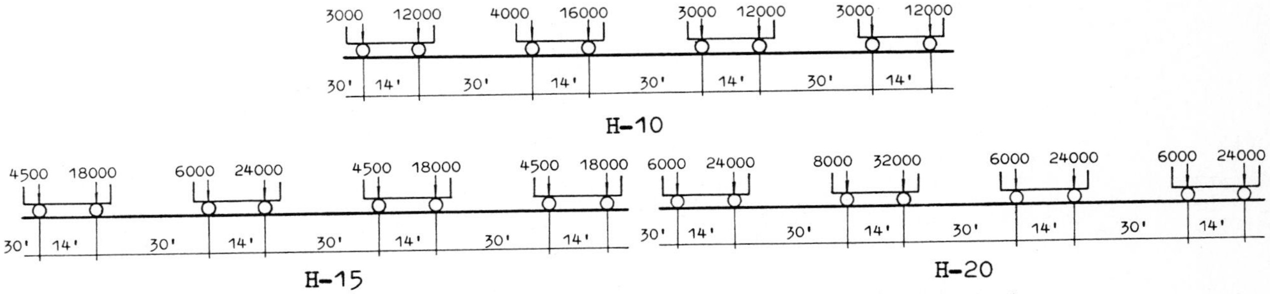

Fig. 2 Truck Train Loadings Defined in the American Standard

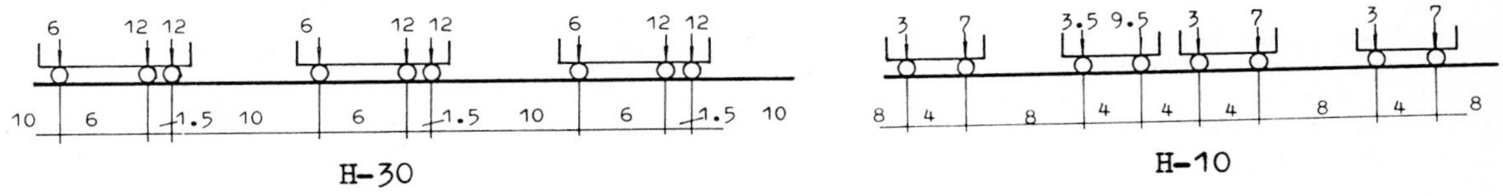

Fig. 3 Truck Train Loadings Defined in the Russian Standard

It is obviously a programming problem to find the most unfavourable truck train loading. And the dynamic programming is most suitable to solve such a problem[6].A series of objective functions $\mathcal{F}_n(x)$ is defined as the maximum stresses subject to the following constraints: the number of trucks in the train is not greater than n and the location of the first truck does not exceed x. The properties of these functions are discussed. It leads to a recurrence formulation.

2. THE FORMULATION FOR THE SIMPLEST TRAIN

The simplest train consists of only standard trucks. The influence function along the traffic lane is $f(x)$. Where x is the natural coordinate, i.e. the distance to the origin along the lane and $f(x)$ is the value of stress component produced by a unit force acting at x. $f(x)$ can be calculated by use of the generalized Müller-Breslau's principle [1].

From $f(x)$ and the weight distribution illustrated in Fig. 1-3, the influence function of single standard truck $F^1(x)$ can be obtained. $F^1(x)$ is defined as the value of stress component produced by a single standard truck, whose front wheel is located at x. The domain of variable x is [0,L], L=L'+l, where L' is the length of the lane and l is the distance between the front and the rear wheel. The domain can be prolongated to x < 0 or x > L, where $F^1(x)$ is equal to 0.

The objective function $f_n(x)$ is defined as following: $f_n(x)$ is the maximum value of stress component produced by a train of standard trucks subject to the following constraints: the truck spacings are not larger than the prescript value dx_a; the number of trucks in the train is not greater than n; and the location of the first truck in the train does not exceed x. Obviously $f_n(x)$ has following properties:

$$f_0(x) = 0 \ ; \quad f_n(x) = 0 \ , \quad \text{if } x < 0 \ ;$$
$$f_n(x_1) \le f_n(x) \ , \quad \text{if } x_1 < x \ ; \qquad (1)$$
$$f_n(x) \le f_m(x) \ , \quad \text{if } n < m \ .$$

To perform the numerical analysis, the domain of variable x is discretized to d_m-1 sections with the same length Δx. The coordinates of d_m nodes are $x_i=(i-1).\Delta x$, i=1,2,......,d_m. dx_a=da.Δx+ remainder, take Δx as a length unit, then $dx_a \cong d_a$.$F^1(x)$ and $f_n(x)$ can be rewritten into the discretized form $F^1[I]$ and $f_n[I]$, I=1,2,....,d_m.

If $f_n[I]$ has been solved, then $f_{n+1}[I]$ can be derived from $f_n[I]$. There are altogether two possibilities of locating the first truck in the train, which determine $f_{n+1}[I]$:

(1). The first truck is located at I, then the second truck cannot exceed I-da, and the train except the first truck involves only n trucks. Therefore one obtains

$$f_{n+1}[I]=F^1[I]+f_n[I-da] \ ; \qquad (2.1)$$

(2) The first truck is not located at I, then one obtains

$$f_{n+1}[I]=f_{n+1}[I-1] \ . \qquad (2.2)$$

According to the definition of $f_{n+1}[I]$, the value $f_{n+1}[I]$ should be equal to the greatest one of expressions (2.1) and (2.2). At last one obtains

$$f_{n+1}[I]=\max\{F^1[I]+f_n[I-da] \ ; \ f_{n+1}[I-1]. \} \qquad (3)$$

3. THE FORMULATION FOR THE GENERAL CASE OF A TRAIN

In this case there are different types of trucks with individual weight distributions and minimal truck spacings. All the types are divided into two kinds. If the number of a type truck in the train is unrestricted, it belongs to the first kind of trucks. Otherwise if the number is restricted to a limited amount, it belongs to the second kind. In this section the most general train composition will be analyzed. It consists of s types of the first kind and t types of the second kind. The number of trucks of each type of the second kind is restricted to m_j, j=1,2,......,t. The influence functions of single trucks and the minimal truck spacings of the first kind are $F^1{}_i[I]$ and da_i, i=1,2,......,s, I=1,2,......,d_m. The influence functions of single trucks and the minimal spacings of the second kind are $F^2{}_j[I]$ and db_j, j=1,2,......,t, I=1,2,......,d_m.

The objective function $\mathcal{F}_n(l_1l_2....l_t)[I]$ is defined as following: $\mathcal{F}_n(l_1l_2....l_t)[I]$ is the maximum value of stress component produced by a train of s truck types of the first kind and t truck types of the second kind. l_j, j=1,2,...,t, is the restricted amount for the jth truck type of the second kind in the train. The function is subject to the same constraints as $f_n[I]$ with the minimal spacings da_i, i=1,2,....,s, and db_j, j=1,2,....,t, corresponding to s types of the first kind and t types of the second kind. Similarly one has

$$\mathcal{F}_0(l_1l_2....l_t)[I]=0 \ ;$$
$$\mathcal{F}_n(l_1l_2....l_t)[I]=0, \quad \text{if } I\le 0 \ ;$$
$$\mathcal{F}_n(l_1l_2..l_t)[I_1]\le\mathcal{F}_n(l_1l_2..l_t)[I], \text{ if } I_1<I; \qquad (4)$$
$$\mathcal{F}_n(l_1l_2..l_t)[I]\le\mathcal{F}_m(l_1l_2..l_t)[I], \text{ if } n<m;$$
$$\mathcal{F}_n(l_1'l_2'..l_t')[I]\le\mathcal{F}_n(l_1l_2..l_t)[I],$$
$$\text{if } l_j'\le l_j, \quad j=1,2,......,t.$$

There are three groups of possibilities of locating the first truck:

(1) The first truck belongs to the first kind and is located at I,

$$\mathcal{F}_{n+1}(l_1l_2..l_t)[I]=F^1_i[I]+\mathcal{F}_n(l_1l_2..l_t)[I-da_i],$$
$$i=1,2,......,s; \qquad (5.1)$$

(2) The first truck belongs to the second kind and is located at I,

$$\mathcal{F}_{n+1}(l_1l_2..l_t)[I]=F^2_j[I]+$$
$$\mathcal{F}_n(l_1l_2..l_j-1..l_t)[I-db_j],\ 1\le j\le t,\ l_j>0; \qquad (5.2)$$

(3) The first truck is not located at I,

$$\mathcal{F}_{n+1}(l_1l_2..l_t)[I]=\mathcal{F}_{n+1}(l_1l_2..l_t)[I-1] \qquad (5.3)$$

The recurrence formula is

$$\mathcal{F}_{n+1}(l_1l_2....l_t)[I]=$$
$$\max\begin{Bmatrix} F^1_i[I]+\mathcal{F}_n(l_1l_2..l_t)[I-da_i],i=1,2,..,s; \\ F^2_j[I]+\mathcal{F}_n(l_1l_2..l_j-1..l_t)[i-db_j], \\ 1\le j\le t,\ l_j>0 \qquad ; \\ \mathcal{F}_{n+1}(l_1l_2....l_t)[I-1] \qquad . \end{Bmatrix} \qquad (6)$$

where $0\le l_j\le m_j$. If all the $l_j=m_j$, $j=1,2,....t$, the final result is solved.

4. THE FORMULATIONS FOR 2-DIMENSIONAL AND MULTI-LANES

In this section the formulation for the 1-dimensional lane will be extended to parallel 2-dimensional multi-lanes. We take the direction vertical to x-axis in the road plane as the y-axis. The locations of the single truck in the discretized form are $F^1_i[I,J]$ and $F^2_j[I,J]$, $1\le I\le dm$, $1\le J\le dk$.

Suppose the locations of trucks are arbitrary in the y direction and the minimal truck spacings are independent of y coordinate. The objective function can be expressed in the same form as that in section 3. The corresponding discretized form is $\mathcal{F}_n(l_1l_2....l_t)[I]$. It must be equal to the maximum value of all the possibilities. And one obtains

$$\mathcal{F}_{n+1}(l_1l_2....l_t)[I]=$$
$$\max\begin{Bmatrix} \max\limits_J F^1_i[I,J]+\mathcal{F}_n(l_1l_2..l_t)[I-da_i], \\ 1\le J\le dk,\ i=1,2,....,s \qquad ; \\ \max\limits_J F^2_j[I,J]+\mathcal{F}_n(l_1l_2..l_j-1..l_t)[I-db_j], \\ 1\le J\le dk,\ 1\le j\le t,\ l_j>0 \qquad ; \\ \mathcal{F}_{n+1}(l_1l_2....l_t)[I-1] \quad . \end{Bmatrix} \qquad (7)$$

It is assumed that the traffic cases in different lanes are independent of each other, hence the maximum stress in the case of parallel multi-lanes is simply the sum of all the maximum stresses of the individual lanes. Suppose the total number of lanes is N, the maximum stress of kth lane is $\mathcal{F}_{n+1}^{(k)}(l_1l_2....l_t)[I]$, k=1,2,.., N, then one obtains

$$\mathcal{F}_{n+1}(l_1l_2..l_t)[I]=\sum_{k=1}^{N}\mathcal{F}_{n+1}^{(k)}(l_1l_2..l_t)[I] \qquad (8)$$

5. ALGORITHM

Using some computational techniques, $\mathcal{F}_n(l_1l_2....l_t)[I]$ in formula (7) can be replaced by a new function $\mathcal{F}_n(l'_1l'_2...l'_{t'})[I]$, where $l'_j=1$ or 0, $j=1,2,......,t'$, $t'=\sum_{j=1}^{t} m_j$. The computational process for the latter is far simpler than that for the former.

The flow chart is illustrated in Fig. 4.

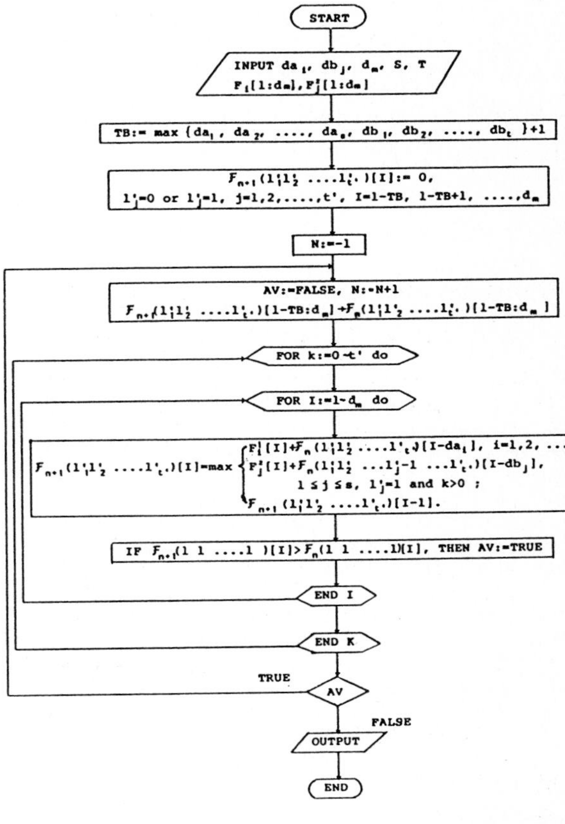

Fig. 4 Flow Chart of Computational Process

This process will be briefly explained:

(1) It is a recurrence process. If all the $\mathcal{F}_n(l'_1l'_2....l'_{t'})[I]$, I=1,2,......,dm and $\mathcal{F}_{n+1}(l'_1l'_2....l'_{t'})[I_1]$, $I_1=0,1,....,I-1$ have been solved, $\mathcal{F}_{n+1}(l'_1l'_2....l'_{t'})[I]$ can be calculated from them. Expression (4) shows that $\mathcal{F}_0(l'_1l'_2...l'_{t'})[I]=0$ and $\mathcal{F}_n(l'_1l'_2...l'_{t'})[0]=0$. They can be used as the starting value.

(2) In function $\mathcal{F}_n(l'_1l'_2...l'_{t'})[I]$, $l'=1$ or 0, the total number of different objective functions is $2^{t'}$.

(3) Let $k=\sum l'_j$, it may equal to 0,1,....,t', According to the different value of k, the $2^{t'}$ functions can be divided into t'+1 different levels. The total number of functions in kth level is $C^k_{t'}$, and $\sum C^k_{t'} = 2^{t'}$. To calculate the functions in kth level, only the functions in the kth and (k-1)th level are needed.

(4) There are three loops in the whole computational process. The iterative variable in the outside loop is n, the number of trucks in the train, the starting value is n=0. In the middle loop the calculation is repeated from the 0th to t'th level of the objective functions. In the inside loop the value of each function is computed from I=1 to I=dm. When $\mathcal{F}_{n+1}(1\ 1 \ldots 1)[I] = \mathcal{F}_n(1\ 1\ \ldots.1)[I]$, I=1,2,...., dm, the whole computation is completed, $n^* = n$ is the total number of trucks in the train and the solved maximum stress is $\mathcal{F}_n(1\ 1\ \ldots.1)[dm]$.

6. NUMERICAL EXAMPLE

Fig. 5 illustrates a numerical example. The bridge structure is a plate of four spans, which is idealized with 686 triangular plate bending elements (Fig. 5.1). The influence surface was calculated by use of the generalized Müller-Breslau's principle. The live load is truck H-10 of Fig. 1 moving on two lanes (Fig. 5.2). Fig. 5.3 and Fig. 5.4 show the influence lines, the influence lines of single standard truck and the influence lines of single heavy truck of lane No. 1 and lane No. 2 respectively. The maximum positive bending moment of the element under consideration $|M_x+|$ equals to 13.91 (T-M/M), which is produced by the truck train of 6 trucks, three trucks on each lane. The locations of all the trucks are demonstrated by the upward arrows in Fig. 5.3 and Fig. 5.4. The maximum negative bending moment $|M_x-|$ equals to 3.866 (T-M/M). The locations of 6 trucks, which produce the maximum negative moment are illustrated by the downward arrows.

ACKNOWLEDGEMENT

The financial support was provided by the Fund for Young University Teachers of the State Education Commission of PRC.

REFERENCES

[1] Shen, W., Liu, G. and Zhong, W., A New Principle for Analysing the Influence Function of Complicated Structure, in: K.-Nukulchai, W., Ueda, T., Wieland, M. and Chauhan, R.L. (eds.), Structural Engineering and Construction (Vp Press, Bangkok,1989) Vol. 1, pp. 3-8.

[2] Zhong, W., Shen, W. and Shi, P., Analysis of the Maximum Stress of Highway Bridges by Dynamic Programming, in: K.-Nukulchai, W., Ueda, T., Wieland, M. and Chauhan, R.L. (eds.), Structural Engineering and Construction (Vp Press, Bangkok, 1989) Vol. 1, pp. 186-191.

[3] Standard Specifications for Highway Bridges (in Chinese), Adopted. by the Traffic Ministry of China, (People's Press, Beijing, 1985).

[4] Standard Specifications for Highway Bridges, Published by American Association of State Highway Officials (12th Edn., 1977).

[5] Standard Specifications for Railway and Highway Bridges of the Soviet Union (CH200-62)(in Chinese), (People's Press, Beijing, 1963).

[6] Bellman, R.E. and Dreyfas, S.E., Applied Dynamic Programming (Princeton University Press, 1962).

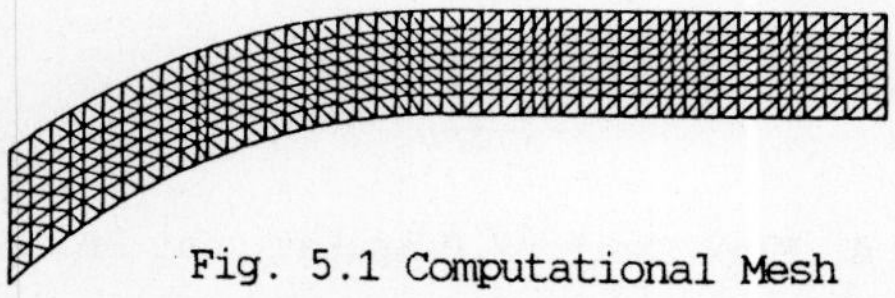

Fig. 5.1 Computational Mesh

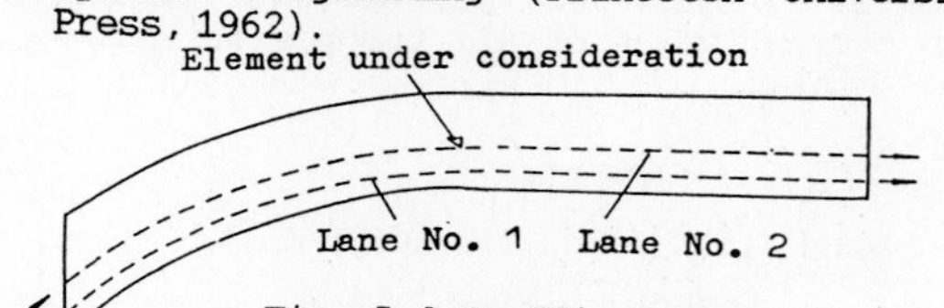

Fig. 5.2 Traffic Lanes

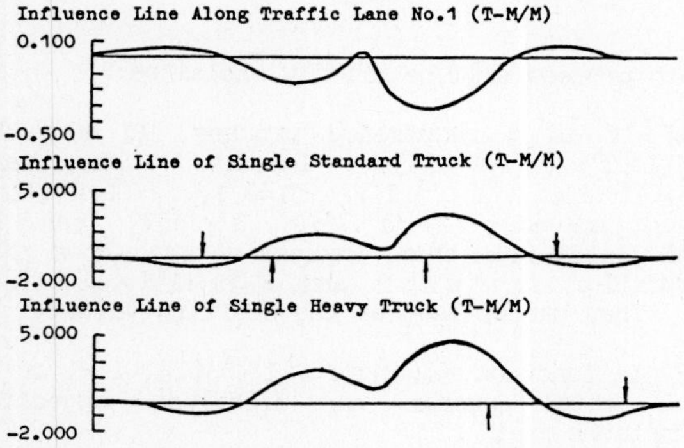

Fig. 5.3 Influence Lines of Lane No. 1

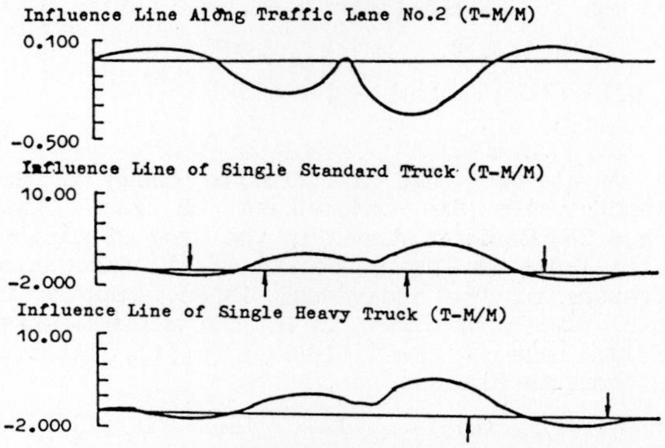

Fig. 5.4 Influence Lines of Lane No. 2

AUTHOR INDEX